Lecture Notes in Artificial Intelligence   5572

Edited by R. Goebel, J. Siekmann, and W. Wahlster

Subseries of Lecture Notes in Computer Science

T0190025

Emilio Corchado   Xindong Wu   Erkki Oja
Álvaro Herrero   Bruno Baruque (Eds.)

Hybrid Artificial Intelligence Systems

4th International Conference, HAIS 2009
Salamanca, Spain, June 10-12, 2009
Proceedings

 Springer

Series Editors

Randy Goebel, University of Alberta, Edmonton, Canada
Jörg Siekmann, University of Saarland, Saarbrücken, Germany
Wolfgang Wahlster, DFKI and University of Saarland, Saarbrücken, Germany

Volume Editors

Emilio Corchado
Álvaro Herrero
Bruno Baruque
Universidad de Burgos, Grupo de Investigación GICAP
Área de Lenguajes y Sistemas Informáticos, Departamento de Ingeniería Civil
Escuela Politécnica Superior - Campus Vena, Francisco de Vitoria
09006 Burgos, Spain
E-mail: {escorchado; ahcosio; bbaruque@ubu.es}

Xindong Wu
University of Vermont, Department of Computer Science
33 Colchester Avenue, Burlington VT, USA
E-mail: xwu@cs.uvm.edu

Erkki Oja
Helsinki University of Technology, Computer and Information Science
P.O. Box 5400, 02015 HUT, Finland
E-mail: Erkki.Oja@hut.fi

Library of Congress Control Number: Applied for

CR Subject Classification (1998): I.2.6, I.2, H.3, H.4, H.2.8, F.2.2, I.4-6

LNCS Sublibrary: SL 7 – Artificial Intelligence

ISSN   0302-9743
ISBN-10   3-642-02318-5 Springer Berlin Heidelberg New York
ISBN-13   978-3-642-02318-7 Springer Berlin Heidelberg New York

springer.com

© Springer-Verlag Berlin Heidelberg 2009
Printed in Germany

Typesetting: Camera-ready by author, data conversion by Scientific Publishing Services, Chennai, India
Printed on acid-free paper   SPIN: 12689453   06/3180   5 4 3 2 1 0

Preface

The 4th International Conference on Hybrid Artificial Intelligence Systems (HAIS 2009), as the name suggests, attracted researchers who are involved in developing and applying symbolic and sub-symbolic techniques aimed at the construction of highly robust and reliable problem-solving techniques, and bringing the most relevant achievements in this field. Hybrid intelligent systems have become increasingly popular given their capabilities to handle a broad spectrum of real-world complex problems which come with inherent imprecision, uncertainty and vagueness, high-dimensionality, and nonstationarity. These systems provide us with the opportunity to exploit existing domain knowledge as well as raw data to come up with promising solutions in an effective manner. Being truly multidisciplinary, the series of HAIS conferences offers an interesting research forum to present and discuss the latest theoretical advances and real-world applications in this exciting research field.

This volume of *Lecture Notes in Artificial Intelligence* (LNAI) includes accepted papers presented at HAIS 2009 held at the University of Salamanca, Salamanca, Spain, June 2009.

Since its inception, the main aim of the HAIS conferences has been to establish a broad and interdisciplinary forum for hybrid artificial intelligence systems and associated learning paradigms, which are playing increasingly important roles in a large number of application areas.

Since its first edition in Brazil in 2006, HAIS has become an important forum for researchers working on fundamental and theoretical aspects of hybrid artificial intelligence systems based on the use of agents and multiagent systems, bioinformatics and bio-inspired models, fuzzy systems, artificial vision, artificial neural networks, and optimization models.

HAIS 2009 received 206 technical submissions. After a thorough peer-review process, the International Program Committee selected 85 papers, which are published in these conference proceedings. In this edition a special emphasis was put on the organization of special sessions. Eight special sessions, containing accepted 44 papers, were organized related to actual topics: Real-World HAIS Applications and Data Uncertainty, Applications of Hybrid Artificial Intelligence in Bioinformatics, Evolutionary Multiobjective Machine Learning, Hybrid Reasoning and Coordination Methods on Multi-agent Systems, Methods of Classifiers Fusion, Knowledge Extraction Based on Evolutionary Learning, Hybrid Systems Based on Bioinspired Algorithms and Argumentation models and Hybrid Evolutionary Intelligence in Financial Engineering. In addition, a tutorial with the title "Evolutionary Algorithms for Clustering" was included. The selection of papers was extremely rigorous in order to maintain the high quality of the conference and we would like to thank the Program Committee for their hard work in the reviewing process. This process is very important to the creation of a conference of high standard and the HAIS conference would not exist without their help.

The large number of submissions is certainly not only a testimony to the vitality and attractiveness of the field but an indicator of the interest in the HAIS conferences themselves.

As a follow-up of the conference, we anticipate further publication of selected papers in special issues of the *Journal of Information Science*. Our thanks go to the keynote speaker, Witold Pedrycz from the University of Alberta (Canada). We would like to fully acknowledge support from the BISITE Group at the University of Salamanca. We especially thank Juan Manuel Corchado for his support in the organization of this conference and also Ajith Abraham for his guidance and continuing support of the HAIS series of conferences.

We wish to thank Alfred Hoffman, Anna Kramer and Ursula Barth from Springer for their help and collaboration during this demanding publication project.

June 2009

Emilio Corchado
Xindong Wu
Erkki Oja
Álvaro Herrero
Bruno Baruque

Organization

Honorary Chairs

Carolina Blasco   Director of Telecommunication,
                                   Regional Government of Castilla y León (Spain)
Erkki Oja   Helsinki University of Technology (Finland)

General Chair

Emilio Corchado   University of Burgos (Spain)

International Advisory Committee

Ajith Abraham	Norwegian University of Science and Technology (Norway)
Carolina Blasco	Director of TelecommunicationRegional Government of Castilla y León (Spain)
Pedro M. Caballero	CARTIF (Spain)
Andre de Carvalho	University of Sao Paulo (Brazil)
Juan M. Corchado	University of Salamanca (Spain)
José R. Dorronsoro	Autonomous University of Madrid (Spain)
Petro Gopych	Universal Power Systems USA - Ukraine LLC (Ukraine)
Francisco Herrera	University of Granada (Spain)
Lakhmi Jain	University of South Australia (Australia)
Samuel Kaski	Helsinki University of Technology (Finland)
Daniel A. Keim	University of Konstanz (Germany)
Isidro Laso	D.G. Information Society and Media (European Commission)
Xin Yao	University of Birmingham (UK)
Hujun Yin	University of Manchester (UK)

Program Committee

Xindong Wu	University of Vermont (USA) (PC Chair)
Agnar Aamodt	Norwegian University of Science and Technology (Norway)
Ajith Abraham	Norwegian University of Science and Technology (Norway)
Rafael Alcalá	University of Granada (Spain)
Ricardo Aler	University Carlos III of Madrid (Spain)

Davide Anguita	University of Genoa (Italy)
Angel Arroyo	University of Burgos (Spain)
Fidel Aznar	University of Alicante (Spain)
Antonio Bahamonde	Universidad de Oviedo (Spain)
Javier Bajo	Pontifical University of Salamanca (Spain)
Bruno Baruque	University of Burgos (Spain)
Ester Bernadó	Ramon Llull University (Spain)
Josh Bongard	University of Vermont (USA)
José Manuel Benitez	University of Granada (Spain)
Juan Botía	University of Murcia (Spain)
Vicente Boti	Polytechnic University of Valencia (Spain)
Andrés Bustillo	University of Burgos (Spain)
Oscar Castillo	Tijuana Institute of Technology (Mexico)
Jonathan Chan-King	Mongkut's University of Technology Thonburi (Thailand)
Richard Chbeir	Bourgogne University (France)
Enhong Chen	University of Science and Technology of China (China)
Huajun Chen	Zhejiang University (China)
Sung Bae Cho	Yonsei University (Korea)
Juan Manuel Corchado	University of Salamanca (Spain)
Emilio Corchado	University of Burgos (Spain)
Rafael Corchuelo	University of Seville (Spain)
Jose Alfredo F. Costa	Federal University UFRN (Brazil)
Leticia Curiel	University of Burgos (Spain)
Bernard De Baets	Ghent University (Belgium)
Rónán Daly	University of Glasgow (UK)
Theodoros Damoulas	University of Glasgow (UK)
Andre de Carvalho	University of São Paulo (Brazil)
Marcilio de Souto	UFRN (Brazil)
María J. del Jesús	University of Jaén (Spain)
Ricardo del Olmo	University of Burgos (Spain)
Nicola Di Mauro	University of Bari (Italy)
José Dorronsoro	Autonomous University of Madrid (Spain)
George Dounias	University of the Aegean (Greece)
António Dourado	University of Coimbra (Portugal)
Enrique de la Cal	University of Oviedo (Spain)
Aboul Ella	University of Cairo (Egypt)
Juan José Flores	University of Michoacana (Mexico)
Richard Freeman	Capgemini (UK)
Kunihiko Fukushima	Kansai University (Japan)
Bogdan Gabrys	University of Bournemouth (UK)
Inés M. Galván	University Carlos II of Madrid (Spain)
Matjaz Gams	Jozef Stefan Institute Ljubljana (Slovenia)
Jun Gao	Hefei University of Technology (China)
Salvador García	University of Jaén (Spain)
Mark Girolami	University of Glasgow (UK)

Albert Orriols	Ramon Llull University (Spain)
José Otero	University of Oviedo (Spain)
Joaquín Pacheco	University of Burgos (Spain)
Vasile Palade	Oxford University (UK)
Juan Pavón	Complutense University of Madrid (Spain)
Witold Pedrycz	University of Alberta (Canada)
Carlos Pereira	University of Coimbra (Portugal)
Lina Petrakieva	Glasgow Caledonian University (UK)
Gloria Phillips-Wren	Loyola College in Maryland (USA)
Julio Ponce	Autonomous University of Aguascalientes (Mexico)
Khaled Ragab	King Faisal University (Saudi Arabia)
B Ribeiro	University of Coimbra (Portugal)
Ramón Rizo	University of Alicante (Spain)
Fabrice Rossi	TELECOM ParisTech (France)
Ozgur Koray Sahingoz	Turkish Air Force Academy (Turkey)
Wei Chiang Samuelson	Oriental Institute of Technology (Taiwan)
José Santamaría	University of Jaén (Spain)
Pedro Santos	University of Burgos (Spain)
Robert Schaefer	AGH University of Science and Technology (Poland)
Javier Sedano	University of Burgos (Spain)
Dragan Simic	Novi Sad Fair (Serbia)
Dominik Slezak	University of Regina (Canada)
Ying Tan	Peking University (China)
Dacheng Tao	Wuhan University (China)
Ke Tang	University of Science and Technology of China (China)
Nikos Thomaidis	University of the Aegean (Greece)
Dacheng Tao	Nanyang Technological University (Singapore)
Eiji Uchino	Yamaguchi University (Japan)
José M. Valls	University Carlos III of Madrid (Spain)
Vassilios Vassiliadis	University of the Aegean (Greece)
Sebastian Ventura	University of Córdoba (Spain)
José Ramón Villar	University of Oviedo (Spain)
Guoyin Wang	Chongqing University of Posts and Telecommunications (China)
Jie Wang	Minnesota State University, Mankato (USA)
Michal Wozniak	Wroclaw University of Technology (Poland)
Zhuoming Xu	Hohai University (China)
Ronald Yager	Iona College (USA)
Zheng Ron Yang	University of Exeter (UK)
Hujun Yin	The University of Manchester (UK)
Huiyu Zhou	Brunel University (UK)
Rodolfo Zunino	University of Genoa (Italy)

Special Sessions Program Committees

Real-World HAIS Applications and Data Uncertainty

Camelia Chira	University of Babes, Bolyai (Romania)
Enrique de la Cal	University of Oviedo (Spain)
Richard T. Freeman	Capgemini (UK)
Isaías García	University of León (Spain)
Lars Graening	Honda Research Institute Europe GmbH
Luis Junco	University of Oviedo (Spain)
Gerardo M. Méndez	Technological Institute of Nuevo León (Mexico)
José Otero	University of Oviedo (Spain)
Ana Palacios	University of Oviedo (Spain)
Camelia Pintea	University of Babes, Bolyai (Romania)
Adolfo Rodríguez	University of León (Spain)
Luciano Sánchez	University of Oviedo (Spain)
Javier Sedano	University of Burgos (Spain)
Mª del Rosario Suárez	University of Oviedo (Spain)
José Ramón Villar	University of Oviedo (Spain)

Applications of Hybrid Artificial Intelligence in Bioinformatics

Bruno Baruque	University of Burgos (Spain)
Andrés Bustillo	University of Burgos (Spain)
Emilio Corchado	University of Burgos (Spain)
Álvaro Herrero	University of Burgos (Spain)
Zheng Ron Yang	University of Exeter (UK)
Hujun Yin	University of Manchester (UK)

Evolutionary Multiobjective Machine Learning

Henrik Bostrom	University of Skövde (Sweden)
Juan C. Fernández	University of Córdoba (Spain)
César Hervás	University of Córdoba (Spain)
Andrew Hunter	University of Lincoln (UK)
Pedro Isasi	University Carlos III of Madrid (Spain)
Yaochu Jin	Honda Research Institute Europe / Bielefeld University (Germany)
David Quintana	University Carlos III of Madrid (Spain)
Peter Rockett	University of Sheffield (UK)
Katya Rodríguez	UNAM (Mexico)
El-Ghazali Talbi	INRIA / Futurs University of Lille (France)
Yago Sáez	University Carlos III of Madrid (Spain)

Hybrid Reasoning and Coordination Methods on Multi-agent Systems

Estefanía Argente	Polytechnic University of Valencia (Spain)
Javier Bajo	Pontifical University of Salamanca (Spain)

Juan Botía	University of Murcia (Spain)
Vicente Botti	Polytechnic University of Valencia (Spain)
Juan Manuel Corchado	University of Salamanca (Spain)
Marc Esteva	IIIA-CSIC (Spain)
Alberto Fernández	Rey Juan Carlos University (Spain)
Rubén Fuentes	Complutense University of Madrid (Spain)
Adriana Giret	Polytechnic University of Valencia (Spain)
Jorge Gómez	Complutense University of Madrid (Spain)
Jose Manuel Molina	University Carlos III of Madrid (Spain)
Rubén Ortiz	Rey Juan Carlos University (Spain)

Methods of Classifiers Fusion

Robert Burduk	Wroclaw University of Technology (Poland)
Emilio Corchado	University of Burgos (Spain)
Giorgio Fumera	University of Cagliari (Italy)
Bogdan Gabrys	Bournemouth University (UK)
Konrad Jackowski	Wroclaw University of Technology (Poland)
Marek Kurzynski	Wroclaw University of Technology (Poland)
Elzbieta Pekalska	University of Manchester (UK)
Konstantinos Sirlantzis	University of Kent (UK)
Krzysztof Walkowiak	Wroclaw University of Technology (Poland)
Michal Wozniak	Wroclaw University of Technology (Poland)

Knowledge Extraction Based on Evolutionary Learning

Jesús Alcalá-Fdez	University of Granada (Spain)
Salvador García	University of Jaen (Spain)
Joaquín Derrac Rus	University of Granada (Spain)
Sebastián Ventura	University of Córdoba (Spain)
Alberto Fernández	University of Granada (Spain)
Antonio Aráuzo-Azofra	University of Córdoba (Spain)
Leila Shafti	Autonomous University of Madrid (Spain)
Julián Luengo	University of Granada (Spain)
Antonio Peregrín	University of Huelva (Spain)
Guiomar Corral	Ramon Llull University (Spain)
Pedro González	University of Jaén (Spain)
José Santamaría	University of Jaén (Spain)
Núria Macià	Ramon Llull University (Spain)
Andreea Vescan	Babes-Bolyai University (Romania)
José Ramón Villar	University of Oviedo (Spain)
José Otero	University of Oviedo (Spain)
Romain Raveaux	University of La Rochelle (France)
María del Rosario Suárez	University of Oviedo (Spain)
Jaume Bacardit	University of Nottingham (UK)
Pietro Ducange	University of Pisa (Italy)

Albert Orriols	Ramon Llull University (Spain)
Yannis Marinakis	Technical University of Crete (Greece)
Cristóbal José Carmona	University of Jaén (Spain)
Rafael Alcalá	University of Granada (Spain)

Hybrid Systems Based on Bioinspired Algorithms and Argumentation Models

Lucíana Buriol	UFRGS (Brazil)
Simoneé Suaren	Technical University (Mauritius)
Stella Heras	Polytechnic University of Valencia (Spain)
Samer Hassan	Surrey University (UK)
Arturo Hernández	CIMAT (Mexico)

Hybrid Evolutionary Intelligence in Financial Engineering

George Dounias	University of the Aegean (Greece)
Nikos Thomaidis	University of the Aegean (Greece)
Michael Doumpos	Technical University of Crete (Greece)
Constantin Zopounidis	Technical University of Crete (Greece)
John Beasley E.	Brunel University (UK)
Jovita Nenortaite	Vilnius University (Lithuania)
Vassilios Vassiliadis	University of the Aegean (Greece)

Organizing Committee

Emilio Corchado	University of Burgos (Chair)
Bruno Baruque	University of Burgos (Co-chair)
Álvaro Herrero	University of Burgos (Co-chair)
Angel Arroyo	University of Burgos
Pedro Burgos	University of Burgos
Andrés Bustillo	University of Burgos
Jacinto Canales	CPIICyL
Juan Manuel Corchado	University of Salamanca
Leticia Curiel	University of Burgos
Carlos López	University of Burgos
Miguel Ángel Manzanedo	University of Burgos
Raúl Marticorena	University of Burgos
David Martín	University of Burgos
Juan Vicente Martín	University of Burgos
Juan Carlos Pérez	University of Burgos
Jose Manuel Sáiz	University of Burgos
Lourdes Sáiz	University of Burgos
Pedro Santos	University of Burgos (Spain)
Javier Sedano	University of Burgos
Belén Vaquerizo	University of Burgos

Table of Contents

Agents and Multi Agents Systems

HAIS Applications

Cluster Analysis

Data Mining and Knowledge Discovery

Evolutionary Computation

Learning Algorithms

Special Session

Real World HAIS Applications and Data Uncertainty

Applications of Hybrid Artificial Intelligence in Bioinformatics

Evolutionary Multiobjective Machine Learning

Hybrid Reasoning and Coordination Methods on Multi-Agent Systems

Methods of Classifiers Fusion

Knowledge Extraction Based on Evolutionary Learning

Hybrid Systems Based on Bioinspired Algorithms and Argumentation Models

Hybrid Evolutionary Intelligence in Financial Engineering

Agents in Home Care: A Case Study

Juan A. Fraile[1], Dante I. Tapia[2], Sara Rodríguez[2], and Juan M. Corchado[2]

[1] Pontifical University of Salamanca, c/ Compañía 5, 37002 Salamanca, Spain
jafraileni@upsa.es
[2] Departamento de Informática y Automática, University of Salamanca,
Plaza de la Merced s/n, 37008 Salamanca, Spain
{dantetapia,srg,corchado}@usal.es

Abstract. Home Care is the term used to refer to any kind of care to a person at his own home. This article presents a case study of the HoCa hybrid multiagent architecture aimed at improving of dependent people in their homes. Hoca architecture uses a set of distributed components to provide a solution to the needs of the assisted people and its main components are software agents that interact with the environment through a distributed communications system. This paper describes the hybrid multiagent system in a home care environment and presents the results obtained.

Keywords: Dependent environments, Hybrid Multiagent Systems, Home Care.

1 Introduction

"Divide and conquer" is a technique that has been widely used to resolve large and complex problems, where each part of the problem is easier to handle and to find its solution separately. To that end, it is necessary to act cooperatively, and follow some order of execution in the way of communication between tasks. In order to resolve a task it will be necessary to know the type of inputs and outputs expected, as well as distributed computing methods. One of the alternatives is the use of hybrid systems. Multiagent hybrid systems [3] try to combine the runtime reactive agents with the rationality of the deliberative agents. These hybrid multi-agent systems are specially used when the requirements of the problem can not be satisfied by neither reactive multiagents systems nor deliberative multiagent systems independently. The integration of reactive and deliberative agents requires three key concepts: communication, cooperation and coordination. The HoCa architecture focuses in these concepts to facilitate the development of home care environments.

Home Care requires the improvement of the services offered to the users as well as the way they can be accessed [3]. Moreover, it is necessary to adopt the trends already tested and proven in technological environments [1]. Intelligent environments are focused on the user, since the user is the centre of the new technological facilities and demands access to unified services [2]. The importance acquired by the dependency people sector has dramatically increased the need for new home care solutions [5]. Besides, the commitments that have been acquired to meet the needs of this sector, suggest that it is necessary to modernize the current systems. Multiagent systems [14], and intelligent devices-based architectures have been recently explored as supervisor

E. Corchado et al. (Eds.): HAIS 2009, LNAI 5572, pp. 1–8, 2009.

systems for health care scenarios [1] [7] for elderly people and for Alzheimer patients [5] [13]. These systems allow providing constant care in the daily life of dependent patients [4], predicting potentially dangerous situations and facilitating a cognitive and physical support for the dependent patient [2]. Taken into account these solutions, it is possible to think that multi-agent systems facilitate the design and development of home care environments [6] and improve the services currently available, incorporating new functionalities. Multi-agent systems add a high level of abstraction regarding to the traditional distributed computing solutions.

The aim of this paper is to present a case study where the HoCa hybrid multiagent architecture is used to develop a multiagent system to monitor the routine tasks of daily life of the patients and detect dangerous situations at home. It is necessary to integrate new technologies into the patient's home to achieve this objective, using techniques of artificial intelligence, intelligent agents and wireless technologies. The purpose is to optimize the effectiveness and management of the home care to facilitate the working conditions of the medical staff and improve the patient's quality of life.

The rest of the paper is structured as follows: Section 2 describes the proposed architecture and agents types to resolve the problem. Section 3 describes a case study to test the architecture and finally, Section 4 presents the results and conclusions obtained.

2 Agent Types in HoCa

The HoCa multi-agent architecture uses a series of components to offer a solution that includes all levels of service for various systems [9]. It accomplishes this by incorporating intelligent agents, identification and localization technology, wireless networks and mobile devices [10].

The Agents platform is the core of the architecture and integrates two types of agents as show the Figure 1, each of which behaves differently for specific tasks. The first group of agents is made up of deliberative BDI agents [8], who are in charge of the management and coordination of all system applications and services [5]. However, there are pre-defined agents which provide the basic functionalities of the architecture:

- CoAp Agent: This agent is responsible for all communications between applications and the platform. Manages the incoming requests from the applications to be processed by services. It also manages responses from services to applications. CoAp Agent is always on "listening mode". Applications send XML messages to the agent requesting for a service, then the agent creates a new thread to start communication using sockets.
- CoSe Agent: It is responsible for all communications between services and the platform. The functionalities are similar to CoAp Agent but backwards. This agent is always on "listening mode" waiting for responses of services. Manager Agent indicates CoSe Agent the service that must be invoked.
- Directory Agent. Manages the list of services that can be used by the system. For security reasons, the list of services is static and can only be modified manually, however services can be added, erased or modified dynamically. The list contains the information of all trusted available services.

- Supervisor Agent. This agent supervises the correct functioning of the agents in the system. Supervisor Agent verifies periodically the status of all agents registered in the architecture by means of sending ping messages.
- Security Agent. This agent analyzes the structure and syntax of all incoming and outgoing XML messages. If a message is not correct, the Security Agent informs the corresponding agent (CoAp or CoSe) that the message cannot be delivered.
- Manager Agent. Decides which agent must be called taking into account the users preferences. Users can explicitly invoke a service, or can let the Manager Agent decide which service is better to accomplish the requested task. Manager Agent has a routing list to manage messages from all applications and services.
- Interface Agent. This kind of agent has been designed to be embedded in users' applications. The requests are sent directly to the Security Agent, which analyzes the requests and sends them to the Manager Agent. These agents must be simple enough to allow them execute on mobile devices, such as cell phones or PDA's.

Figure 1 along with a simple example helps to understand the communication between different types of agents in the architecture. A patient is visited by the medical service due to a feverish that suffers by an infection. The medical service went to the house before the patient's explicit request made through the alerts system. The patient through an application has inserted the alert in the system. The CoAp agent is responsible for registering this information into the system and notifies the supervisor agent. The security agent confirmed the credentials of the user who enters the information and validates the information entered. The supervisor agent at the same time performs two tasks, requests the directory agent you select the service to run to launch the alert and through the interface agent informs the manager agent of operations performed. The CoSe agent runs the service that launches the alert and finally the alert is sent through the reactive agents of the architecture. At all times the manager agent is informed of the steps being taken in the system and is responsible for validating the alert sending through the interface agent to the corresponding reactive agent. Once the patient enters the information into the system, this process seems very laborious and slow is running in a few thousandths of a second.

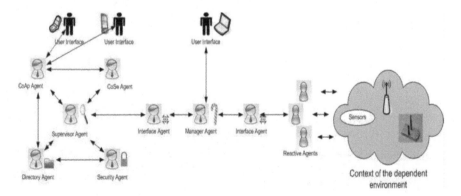

Fig. 1. Agents Workflow in the HoCa architecture

The second group is made up of reactive agents [4]. Most of the research con-ducted within the field of multi-agent systems focuses on designing architectures that incorporate complicated negotiation schemes as well as high level task resolution, but don't focus on temporal restrictions. In general, the multi-agent architectures assume a reliable channel of communication and, while some establish deadlines for the inter-action processes, they don't provide solutions for limiting the time the system may take to react to events.

3   Using HoCa to Development a Hybrid Multi-Agent System for Dependent Environment

HoCa has been employed to develop a hybrid multi-agent system aimed to enhance assistance and care for low dependence patients at their homes. The house has 89 m² and it live two dependents persons. As shown in Figure 2 are installed 33 passive infrared motion detectors for roof of the SX-360 series and 11 mechanisms for auto-matic door opening. The detectors movements and mechanisms for opening doors, in-teract with the microchip Java Card & RFID [11] users to offer services in run time. Each dependent user is identified by a Sokymat ID bracelet Band Unique Q5 which has an antenna and a chip RFID-Java-Crypto-Card with 32K Module and Crypto-CoProzessor (1024 bit RSA) compatibel to SUNs JavaCard 2.1.1 [15]. The sensors or actuators are placed in strategic positions from home as shows Figure 2 plane. All these devices are controlled by platform agents. This sensors network through a sys-tem of alerts is responsible for generating alarms comparing the user current state with the parameters of the user daily routine who has stored the system. The system can generate alarms if it is determined the parameters for example if the user in a non-working day stands before a certain hour, or if the user spends more time than speci-fied on the door of your home without entering, or the user is a long time motionless in the hallway, etc.

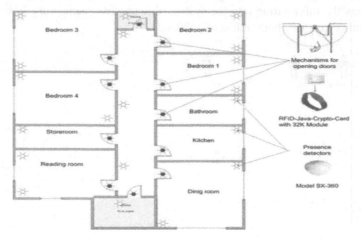

Fig. 2. Home plane

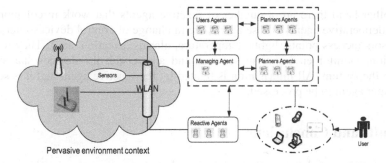

Fig. 3. HoCa structure in a dependent environment

Main functionalities in the system include reasoning, planning mechanisms, management alerts and responses in execution time offered to certain stimuli, as shown in Figure 3. Also in Figure 3 shows as they relate the different types of agents and the connection between the user devices for the execution of services and applications. The pervasive environment there is configured wireless devices and RFID readers also managed by the agents' platform designed. These functionalities allow the system the use of several context-aware technologies to acquire information from users and their environment.

Each agent in the system has its own functionalities. If an agent needs to develop a task in collaboration with other agent a request form is send. There are priority tasks that a set of agents can perform. This ensures that the priority tasks are always available. There are four types of deliberative BDI agents, as shown in Figure 3:

- User Agents. This agent manages the users' personal data and behaviour. The beliefs and goals used for every user depend on the plan or plans defined by the super-users. User Agent maintains continuous communication with the rest of the system agents, especially with the ScheduleUser Agent and with the SuperUser Agent. There is one agent for each patient registered in the system.
- SuperUser Agent. It also runs on mobile devices (PDA) and inserts new tasks into the Manager Agent to be processed by a Case-Based Reasoning mechanism. It also needs to interact with the User Agents to impose new tasks and receive periodic reports, and with the ScheduleUser Agents to ascertain plans' evolution. There is one agent for each supervisor connected to the system.
- SheduleUser Agent. It is a BDI agent with a Case-Based Planning (CBP) mechanism embedded in its structure. It schedules the users' daily activities obtaining dynamic plans depending on the tasks needed for each user. Every agent generates personalized plans depending on the scheduled-user profile. There are as many ScheduleUser Agents as nurses connected to the system.
- Manager Agent. It runs on a Workstation and plays two roles: the security role that monitors the users' location and physical building status trough a continuous communication with the Devices Agent; and the manager role that handle the databases and the tasks assignment. It must provide security for the users and ensure the tasks assignments are efficient. There is just one Manager Agent running in the system.

On the other hand there are a number of reactive agents that work in collaboration with the deliberative agents. These agents are in change of control devices interacting with sensors (access points, lights, temperature, alarms detection, etc.). They receive information, monitor environment services and also check the devices status connected to the system. All information is treated by the reactive agent and it is sent to the manager agent to be processed.

4 Results and Conclusions

The high number of dependent population makes it necessary to develop new approaches to provide specialized care. Technologies such as multi-agent systems, systems for information management, secure communications, multimedia interfaces, domotics or collaborative virtual environments, represent an important factor in the design and development of new home care facilities. In particular, a platform based on agents in combination with mechanisms for information management provides a flexible infrastructure for special care [12]. The collective vision of the HoCa architecture allows the system to create patterns that serve to further responds to the patient's problems.

HoCa architecture has been used in this study to develop a real environment for home care. The multiagent system obtained has been compared to the previous human-based system taking into account the amount of necessary medical staff before and after the implementation of the prototype. Some of the experiments focused on determining the time spent in indirect tasks for patient's care and the number of medical staff working simultaneously monitoring and attending possible emergencies. The results have been satisfactory, showing that the HoCa-based system reduces the response time and optimizes the use of resources. It has been also analyzed the impact of such a system in a real environment, checking the way traditional services can be optimized, enhancing their quality.

Figure 4 shows the comparison of the average medical personnel required in the two studies made with ALZ-MAS and HoCa. The ALZ-MAS architecture [6] allows the monitoring of patients in geriatric residences, but home care is carried out through

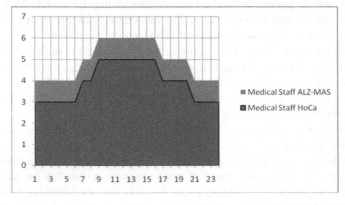

Fig. 4. Comparison of medical staff between the HoCa and the ALZ-MAS architectures

traditional methods. The case study presented in this work consisted of analysing the functioning of both architectures in a test environment. The HoCa architecture was implemented in the home of 5 patients and was tested for 30 days. Figure 4 represents the number of people who are in need each hour of the day. The graph seen as the medical personnel necessary for both systems is lower during the hours of lesser activity of the patients. On the contrary during the hours of the morning and afternoon will need more health staff. It also looks like the HoCa system does not need as personal as ALZ-MAS.

As the result HoCa architecture creates an environment that facilitates intelligent and distributed and provides services to dependents at home. Automating tasks and patient monitoring improve the system security and efficiency of care to dependents. The use of RFID technology, JavaCard and mobiles with people provides a high level of interaction between users and patients through the system and is essential in building an intelligent environment. Moreover the good use of mobile devices can facilitate social interactions and knowledge transfer.

References

1. Angulo, C., Tellez, R.: Distributed Intelligence for smart home appliances. Tendencias de la minería de datos en España. Red Española de Minería de Datos. Barcelona, España (2004)
2. Augusto, J.C., McCullagh, P.: Ambient Intelligence: Concepts and Applications. Invited Paper by the International Journal on Computer Science and Information Systems 4(1), 1–28 (2007)
3. Bajo, J., et al.: Hybrid Architecture for a Reasoning Planner Agent. In: Apolloni, B., Howlett, R.J., Jain, L. (eds.) KES 2007, Part II. LNCS, vol. 4693, pp. 461–468. Springer, Heidelberg (2007)
4. Carrascosa, C., Bajo, J., Julian, V., Corchado, J.M., Botti, V.: Hybrid multi-agent architecture as a real-time problem-solving model. Expert Systems With Applications 34(1), 2–17 (2008)
5. Corchado, J.M., Laza, R.: Constructing Deliberative Agents with Case-based Reasoning Technology. International Journal of Intelligent Systems 18, 1227–1241 (2003)
6. Corchado, J.M., Bajo, J., de Paz, Y., Tapia, D.: Intelligent Environment for Monitoring Alzheimer Patients, Agent Technology for Health Care. Decision Support Systems 34(2), 382–396 (2008)
7. Corchado, J.M., Bajo, J., Abraham, A.: GERAmI: Improving the delivery of health care. IEEE Intelligent Systems. Special Issue on Ambient Intelligence (March/April 2008)
8. Corchado, J.M., Gonzalez-Bedia, M., De Paz, Y., Bajo, J., De Paz, J.F.: Replanning mechanism for deliberative agents in dynamic changing environments. Computational Intelligence 24(2), 77–107 (2008)
9. Fraile, J.A., Tapia, D.I., Sánchez, M.A.: Hybrid Multi-Agent Architecture (HoCa) applied to the control and supervision of patients in their homes. In: Proceedings of HAIS 2008. Hybrid Artificial Intelligence Systems. Third International Workshop, Burgos, Spain, pp. 54–61 (September 2008) ISBN: 978-3-540-87655-7
10. Fraile, J.A., Bajo, J., Pérez, B., Sanz, E.: HoCa Home Care Multi-Agent Architecture. In: International Symposium on Distributed Computing and Artificial Intelligence 2008 (DCAI 2008), pp. 52–61 (October 2008) ISBN: 978-3-540-85862-1

8   J.A. Fraile et al.

11. ITAA, Radio Frequency Identification. RFID coming of age. Information Technology Association of America (2004), http://www.itaa.org/rfid/docs/rfid.pdf
12. Segarra, M.T., Thepaut, A., Keryell, R., Poichet, J., Plazaola, B., Peccatte, B.: Ametsa: Generic Home Control System Based on UPnP. In: Independent living for persons with disabilities and elderly people, pp. 73–80. IOS Press, Amsterdam (2003)
13. Tapia, D.I., Bajo, J., De Paz, F., Corchado, J.M.: Hybrid Multiagent System for Alzheimer Health Care. In: Proceedings of HAIS 2006. Solange Oliveira Rezende, Antonio Carlos Roque da Silva Filho Eds, Ribeirao Preto, Brasil (2006)
14. Weiser, M.: The Computer for the Twenty-First Century. Scientific American 265, 94–104 (1991)
15. ZhiqunChen (Sun Microsystems). Java Card Technology for Smart Cards. Addison Wesley Longman, ISBN 0201703297

EP-MAS.Lib: A MAS-Based Evolutionary Program Approach

Mauricio Paletta[1] and Pilar Herrero[2]

[1] Centro de Investigación en Informática y Tecnología de la Computación (CITEC),
Universidad Nacional de Guayana (UNEG), Av. Atlántico. Ciudad Guayana, Venezuela
mpaletta@uneg.edu.ve
[2] Facultad de Informática, Universidad Politécnica de Madrid, Campus de Montegancedo
S/N. 28.660 Boadilla del Monte, Madrid, Spain
pherrero@fi.upm.es

Abstract. Evolutionary/Genetic Programs (EPs) are powerful search techniques used to solve combinatorial optimization problems in many disciplines. Unfortunately, depending on the complexity of the problem, they can be very demanding in terms of computational resources. However, advances in Distributed Artificial Intelligence (DAI), Multi-Agent Systems (MAS) to be more specific, could help users to deal with this matter. In this paper we present an approach in which both technologies, EP and MAS, are combined together aiming to reduce the computational requirements, allowing a response within a reasonable period of time. This approach, called EP-MAS.Lib, is focusing on the interaction among agents in the MAS, and emphasizing on the optimization obtained by means of the evolutionary algorithm/technique. For evaluating the EP-MAS.Lib approach, the paper also presents a case study based on a problem related with the configuration of a neural network for a specific purpose.

Keywords: Evolutionary Program, Multi-Agent System, Combinatorial Optimization Problem, JADE.

1 Motivation and Related Work

Evolutionary/Genetic Programs (EPs) [5, 8] are powerful searching techniques used to solve Combinatorial Optimization Problems (COP) in many disciplines. Its objective is to achieve the convergence to a good solution (the optimal or close to it) to those problems. Depending on the complexity of the problem, EPs could require high computational capabilities such as CPU time, memory, etc. This problem increases considerably if the calculation of the fitness function also requires high computational capabilities for each of the potential solution. The combination of EP and MAS technologies [6] is an alternative to deal with this situation, aiming to distribute the computational recourses in many sources.

This paper presents EP-MAS.Lib, a framework designed to cover the necessity previously mentioned by integrating the EP and MAS technologies. This integration is possible by designing some agents in a MAS scenario as well as any specific COP based on the EPs specifications. We also present some JADE [2], [4] based implementation details for this proposal and results obtained from experiments done on a

E. Corchado et al. (Eds.): HAIS 2009, LNAI 5572, pp. 9–17, 2009.

particular COP where the calculation of the fitness of a potential solution is very demanding on computational capabilities. In fact, this problem is related with the configuration of a neural network for learning cooperation in a collaborative grid environment.

Some examples of making evolutionary algorithms in distributed environments can be consulted on [1], [11], [13]. In [1] authors describe the DREAM (Distributed Resource Evolutionary Algorithm Machine) framework for the automatic distribution of evolutionary algorithms. This research is more focused on an evolution algorithm approaches more than in the MAS inter-agent communication aspects. G2DGA [3] is another example of a framework concerning development of Peer-to-Peer distributed computing systems.

In the same order of ideas, authors in [11] present an approach to an addressing automatic test generating system in distributed computing context by the integration of genetic algorithms and MAS. Authors do not use a communication protocol for taking care of scheduling the genetic operations. Instead, they use a control agent.

Finally, in [13] authors use a MAS guided by a multi-objective genetic algorithm to find a balance point in the means of a solution of the Pareto front. Agents interaction in this research is not properly done with a communication protocol, instead authors use functions of communication.

Using JADE as a middleware to propose a multi-agent synchronous evolutionary system can be reviewed on [9], [10], [11]. However, as we know, no work has focused on the main purpose of this proposal, which embrace the design of a multi-agent system for cooperation to implement an EP to resolve any COP properly configured.

The rest of the paper is organized as follows. Section 2 describes the EP-MAS.Lib approach. Details of the implementation and evaluation of this proposal is presented in section 3. Finally, section 4 reaches some conclusions and presents some future works.

2   The EP-MAS.Lib Approach

2.1   General Description

Generally EP-MAS.Lib is a second-layer framework used to define and resolve any particular COP by using the combination of EP and MAS technologies. JADE is the first-layer framework. Fig. 1-a shows this general layer-architecture. As one of the differences from previous works in this same context, EP-MAS.Lib allows the definition of any problem through the incorporation of basic EP operations: the calculation of fitness and genetic operation of crossing and mutation (see Section 2.3 for details). This corresponds to the highest layer showed in the Fig. 1-a.

The proposed EP-MAS framework basically consists of two different types of agents (see Fig. 1-b): 1) EPEnvAgent (Env): It informs to the rest of the agent about the EP parameters and problem specifications. It also takes control of the population growth. On the other hand, it allows the evolution among multiple system nodes; 2) EPAgent (Ag): it represents a potential solution to the problem as well as carry out the main steps of evolutionary computation: selection and variation (crossover and mutation) - to generate descendants (new agents and therefore potential solutions).

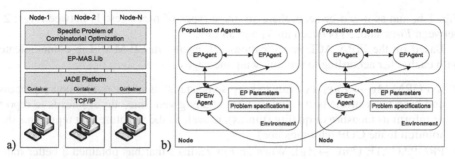

Fig. 1. a) EP-MAS.Lib general layer-architecture; b) EP-MAS.Lib agent interactions

Each computational node may have multiple Ag (a population) but only one Env is needed. The number of Ag in the system (population size) may vary, as the EP evolves. Ag can be created as well as eliminated from the system. The creation of a new Ag is the result of applying genetic operators, while the disposal is due to self-destruction or suicide. Some parameters are used to control these functions (see details below).

As it is happening with the agent containers in JADE, there are two execution modes for the Env: 1) main, and 2) auxiliary. There is only one main Env in the entire system and corresponds to the first instantiated agent. For each node in which the main Env is missing, an auxiliary Env is necessary. Therefore, it is possible to distribute the EP execution between multiple computational nodes.

It is important to highlight that, unlike traditional evolutionary algorithm, there are aspects in EP-MAS.Lib to consider and that were defined aiming to improve the efficiency of the algorithm: 1) Population size is not fixed; it can grow/decrease during the execution of the algorithm. Controlling population growth is done by adjusting the EP parameters. 2) EP parameters are not fixed during the execution; it can be changed according to the current situation observed. 3) There aren´t a specific selection operation or mechanism; instead, each Ag by itself looks for another Ag for crossing and/or decides to mutate by itself. This is done through the communication protocol used by agents and defined in this framework (see Section 2.2 for details).

On the other hand, the EP parameters permit to control the evolutionary algorithm and can be changed dynamically according to the current results and population growth. Parameters are the following: 1) Probability in which each Ag must decide self-destruction or suicide; 2) percentage of the fitness difference (according to the major fitness obtained until now) to take into account in which any Ag should opt for self-destruction; 3) probability in which an Ag accepts crossing with another Ag when the quality of the applicant (who is sending the proposal) is greater or equal than that who receives the proposal; 4) probability in which an Ag accepts crossing with another Ag when the quality of the applicant is lower than that who receives the proposal; 5) probability in which an Ag decides to send a proposal for crossing; and 6) probability in which an Ag decides to mutate.

2.2 The Inter-agents Communication Protocol

As it was mentioned previously, EP-MAS.Lib includes the design of an inter-agents communication protocol (see Fig. 2). This protocol occurs in the scenario depicted in

Fig. 1-b, and as it can be seen, there are three types of messages: 1) between *Ags*; 2) between *Env*s; and, 3) between an *Ag* and an *Env*.

Based on the FIPA ACL specifications [7], agents in EP-MAS.Lib communicate with each other according to the following messages:

- INFORM (*Ag* → *Env*). For *Env* to keep some statistics of the evolutionary process and keep track of the best solution it has at any given time, any *Ag* needs to report or inform its measure of quality (fitness), as well as the structure that represents the solution to the COP (chromosome).
- PROPAGATE (*Env* → *Ag*). When an *Env* realizes that has obtained a better individual (solution) in the population of *Ag*s, it uses this message to transmit /propagate this information to the rest of *Env* (all nodes in the system keep updated).
- REQUEST (*Ag* → *Env*). In this proposal, EP parameters may change dynamically. An *Ag* should, from time to time, request with this message the *Env* to send back the parameters. Therefore, new possible values can be known.
- INFORM (*Env* → *Ag*). In response to the previous message, the *Ag* is receiving the EP parameters from the *Env*. It is important to mention that these parameters can change the way in which *Ag*s may vary to produce descendants (new *Ag*s) or self-destruction. Therefore, the population growth can be controlled.
- INFORM_REF (*Env* → *Env*). By using this message main *Env* inform the auxiliary *Env* changes in the EP parameters.
- PROPOSE (*Ag* → *Env*). In this proposal, each *Ag* is responsible for searching (selecting) a suitable partner to cross (in other proposals this is done by using a control agent [11]). Through this message the *Ag* makes a request to find a suitable partner, and sends its fitness and Id as parameters of the message. This, because they are needed when some other *Ag* accepts the request and responds appropriately to the sender. This is the selection mechanism used by our proposal.
- PROPOSE (*Env* → *Ag*). By this message the *Env* replicates the proposal sent by an *Ag* for the rest of *Ag*s. They become aware of it, so that, they can respond directly to *Ag* who made the original request (because agent Id is one parameter).
- ACCEPT_PROPOSAL (*Ag* → *Ag*). An *Ag* uses this message to respond positively to a request from another *Ag* for crossing with it. This happens only if it decides to accept the proposal (by using some probabilities and based on the fitness received). The *Ag* sends its chromosome, and the sender can use it for applying the corresponding genetic operations.
- REQUEST_WHENEVER (*Ag* → *Env*). *Ag*s are responsible for the selection mechanism and application of genetic operators to generate new *Ag*s. These new agents must be created by the *Env*. Through this message the *Env* knows that a new *Ag* must be created with the given chromosome.
- CONFIRM (*Env* → *Ag*). *Env* has the control to start and stop the evolution. When the process has to be finished, the *Env* uses this message to give to all *Ag*s the order of self-destruction.
- CONFIRM (*Ag* → *Env*): Once an *Ag* is self-destructed, either because it received an order from the *Env* or for effect of this evolutionary algorithm (by using some parameters and based on the current fitness), its uses this message to confirm his suicide.

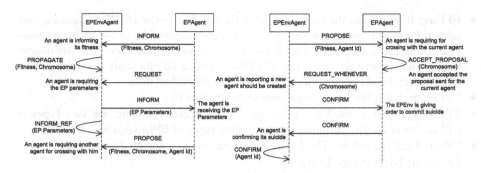

Fig. 2. Inter-agents communication protocol

- CONFIRM (*Env* → *Env*). If the receiver is the main *Env*, then some auxiliary *Env* has decided to stop its participation in the evolution process. On the other hand, if the receiver is an auxiliary *Env*, then the entire evolution process must be stopped.

2.3 The COP Specifications

Aiming to define a specific combinatorial optimization problem to be solved with EP-MAS.Lib, it is necessary to take into account the following considerations:

1. In EP-MAS.Lib a chromosome or valid solution of the problem is represented by using an *n*-vector of real values being *n* the chromosome size. Therefore, it is necessary to define the corresponding chromosome for a particular COP.
2. A valid expression to calculate the fitness value for each individual or potential solution should be given (see implementation details in Section 3). This expression should be defined so as to receive a chromosome as a parameter and return a real value as the result. The higher the calculated value is the fittest is the chromosome as a solution to the COP.
3. A valid expression to implement each genetic operation (crossover and mutation) should be given (see Section 3). A probability of occurrence is associated to each operation in order to regulate the importance of applying the operation towards the other.

Next section provides details of the EP-MAS.Lib implementation as well as some results from experiments conducted with it.

3   Implementation and Evaluation

EP-MAS.Lib was implemented by using JADE platform. JADE was selected because is FIPA-compliant, is open-source (based on Java), and it is probably the most suitable and popular agent environment for our purposes. In this matter, Fig. 3-a shows the EP-MAS.Lib class diagram with the new classes defined and the relationship of these classes with some of the JADE classes. Classes defined on the EP-MAS.Lib framework are the following:

- EPEnv: It represents the environment in the MAS; has the JADE containers; contains the *Env*, EP parameters and problem specifications. As in JADE is necessary to differentiate a main container from other containers (other nodes in the distributed system), there are two types of EPEnv: a main EPEnv (only one in the entire environment) and the auxiliary EPEnv (can be more than one).
- EPParam: It has the set of parameters to control the EP.
- EPEnvAgent: It is used to define the EPEnvAgent. As there are two types of EPEnv, there are two different corresponding types of EPEnvAgent.
- EPEnvAgentBehaviour: The JADE behaviour model associated to the *Env*.
- EPAgent: To define the EPAgent.
- EPAgentBehaviour: The JADE behaviour model associated to the *Ag*.
- EPProblem: It is an abstraction of the problem specifications (see Section 2.3).
- FitnessExp: It is an abstraction of the fitness expression to calculate the solution quality represented by the *Ags*.
- CrossoverOper: It is an abstraction of the crossover genetic operation to generate one or two new chromosomes from two given chromosomes.
- MutationOper: It is an abstraction of the mutation genetic operation to generate a new chromosome from a given chromosome.

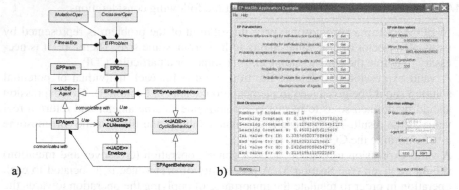

a)   b)

Fig. 3. a) EP-MAS.Lib class diagram; b) Screenshot of the main Test Application window

Algorithm 1 shows the pseudo-code related with the *Env* behaviour, specifically the communication process between the agents. Algorithm 2 shows the pseudo-code related with the *Ag* behaviour.

For evaluating the EP-MAS.Lib framework we have implemented a solution for the specific COP to find the proper setting of parameters required to configure an Artificial Neural Network (ANN) for a particular investigation that is currently being developed, and whose first results can be consulted in [12]. In this problem the fitness consists on training an ANN, configured with the parameters indicated in the individual, aiming to reduce the total average error. It is worth nothing that the calculation of this fitness is very demanding on computational capabilities.

This COP was implemented using a conventional simple genetic program as well as using EP-MAS.Lib aiming to compare the efficiency of each case to deal with the same problem. Both implementations were done with the same genetic representation and operations. The experimentation for the simple program was conducted in a PC with the following hardware platform: Intel T2600 (2.16 GHz) with 2 GB RAM. The experimentation for the distributed solution was conducted using three different nodes with the following hardware platform: 1) Intel T2600 (2.16 GHz) with 2 GB RAM; 2) Pentium 4 (3 GHz) with 1 GB RAM; 3) Pentium 4 (3.2 GHz) with 1 GB RAM.

Algorithm 1 EPEnvAgent	Algorithm 2 EPAgent
loop	loop
$\quad msg \leftarrow$ ReceiveMessage()	$\quad msg \leftarrow$ ReceiveMessage()
$\quad$ if $(\neg msg)$ then Wait()	$\quad ea \leftarrow$ the EPEnvAgent
$\quad$ if $(msg =$ INFORM) then	$\quad$ if $(\neg msg)$ then
$\quad\quad$ if $(msg.Fitness > current\ fitness)$ then	$\quad\quad$ if (has to be crossed) then
$\quad\quad\quad$ UpgradeBestSol(msg.Fitness,	$\quad\quad\quad$ SendMessage(ea, PROPOSE($current$
$\quad\quad\quad\quad msg$.Chrom)	$\quad\quad\quad\quad fitness, current\ chromo, Id$))
$\quad\quad\quad \forall x,\ x$ is EPEnvAgent	$\quad\quad$ if (has to be mutated) then
$\quad\quad\quad$ SendMessage(x,	$\quad\quad\quad Chrom \leftarrow$ ApplyMutation()
$\quad\quad\quad\quad$ PROPAGATE(msg.params))	$\quad\quad\quad$ SendMessage(ea,
$\quad$ if $(msg =$ INFORM_REF) then	$\quad\quad\quad\quad$ REQUEST_WHENEVER($Chrom$))
$\quad\quad$ UpgradeParameters(msg.EP parameters)	$\quad\quad$ Wait()
$\quad$ if $(msg =$ PROPOSE) then	$\quad$ if $(msg =$ INFORM) then
$\quad\quad \forall x,\ x$ is EPAgent	$\quad\quad$ UpgradeParameters(msg.EP parameters)
$\quad\quad$ SendMessage(x, PROPOSE(msg.params))	$\quad\quad$ if (has to be self-destructed) then
$\quad$ if $(msg =$ REQUEST) then	$\quad\quad\quad$ Delete()
$\quad\quad$ if $(status = stopped)$ then	$\quad\quad\quad$ SendMessage(ea, CONFIRM)
$\quad\quad\quad$ ReplyMessage(CONFIRM)	$\quad$ if $(msg =$ PROPOSE) then
$\quad\quad$ else	$\quad\quad$ if (accepted the proposal) then
$\quad\quad\quad$ ReplyMessage(INFORM(EP parameters))	$\quad\quad\quad$ SendMessage(msg.Id,
$\quad$ if $(msg =$ REQUEST_WHENEVER) then	$\quad\quad\quad\quad$ ACCEPT_PROPOSAL(
$\quad\quad$ CreateEPAgent(msg.Chrom)	$\quad\quad\quad\quad Current\ fitness, current\ chromo$))
$\quad\quad PopulationSize \leftarrow PopulationSize + 1$	$\quad$ if $(msg =$ CONFIRM) then
$\quad$ if $(msg =$ CONFIRM) then	$\quad\quad$ Delete()
$\quad\quad$ if $(\neg msg.Id)$ then	$\quad\quad$ SendMessage(ea, CONFIRM)
$\quad\quad\quad PopulationSize \leftarrow PopulationSize - 1$	$\quad$ if (msg = ACCEPT_PROPOSAL) then
$\quad\quad$ else	$\quad\quad C[] \leftarrow$ ApplyCrossover($current\ chromo$,
$\quad\quad\quad \forall x,\ x$ is EPAgent	$\quad\quad\quad msg$.Chr.)
$\quad\quad\quad$ SendMessage(x, CONFIRM)	$\quad\quad \forall c \in C[],\ c$ is a Chromosome
$\quad$ if $(msg =$ PROPAGATE) then	$\quad\quad$ SendMessage(ea,
$\quad\quad$ UpgradeBestSolution(msg.Fitness,	$\quad\quad\quad$ REQUEST_WHENEVER(c))
$\quad\quad\quad msg$.Chrom)	end loop
end loop	

For the experimental test with the simple genetic program, the following parameters were used: population size = 100; maximum number of evolutions = 10000; population GAP approach = 20%; crossover probability = 0.95; mutation probability = 0.25; number of repetitions of the best individual to stop the evolution = 100. Related to the EP-MAS.Lib test, the initial parameters were: probability for deciding the suicide = 0.95; percentage of fitness difference for deciding the suicide = 85%; probability of acceptance to cross when the quality of the applicant is greater or equal than that who receives the proposal = 0.95; probability of acceptance to cross when the quality of the applicant is lower than that who receives the proposal = 0.50; probability for sending a proposal for crossing = 0.95; probability for deciding to mutate = 0.95; maximum number of agent for node = 100.

Based on the results obtained we observed that the simple genetic program needed more than two days to obtain a satisfactory solution for this problem. Instead, by using EP-MAS.Lib (with the 3 nodes previously mentioned) a solution was obtained in no more than 6 hours. It is important to mention that the effectiveness and efficiency of the evolutionary process (find the optimal solution in shortest time) can be improved by changing the EP parameters as well as the evolutionary genetic operators. This is fully configurable within this proposal.

In order to compare EP-MAS.Lib with other similar related work DREAM was used. DREAN was selected because is the work that comes closer to the goal of EP-MAS.Lib. With using DREAM as a framework to implement the same experiment previously mentioned, it was necessary more than one day to obtain a satisfactory solution to configure the ANN. One of the differences between EP-MAS.Lib and DREAN is the ability of the first to change the EP parameters dynamically. Also, using EP-MAS.Lib makes easier to specify a particular COP whose complexity is similar to the one treated in this experiment.

Fig. 3-b shows a screenshot of the test application in which the main *Env* is created (node-1 and main container) as well as the initial population of *Ags* associated with this node. Upper-left hand side has the EP parameters. Upper-right hand side has some current statistics of the evolution (major and minor fitness, and current size of population). Lower-left hand side has the information of the current best chromosome or potential solution for the problem. For the other nodes (auxiliary containers) the test application window is similar to the one shown in Fig. 3-b, the only difference is that parameters cannot be changed. The best chromosome of the entire distribution system and evolution is showed in all the nodes.

4 Conclusions and Future Work

In this paper we present EP-MAS.Lib, a MAS-based framework to be used to solve a specific combinatorial optimization problem. Our proposal was implemented by using JADE and as we know, this proposal differs from other similar works in the following aspects: 1) It focuses on the communication process between agents in the system and therefore leads to the cooperation needed for solving the evolutionary algorithm, instead of the well used elements of the evolutionary algorithm (selection mechanism and genetic operations); 2) The evolutionary algorithm (selection and variation) is not controlled by a central entity, instead it's controlled by all individuals (agents) of the population actively involved in this process; 3) The needed information (EP parameters and problem specifications) is not located in a central repository, but it is replicated for all who need it; and 4) The definition of a particular COP it´s revealed by using the considerations indicated in Section 2.3 whose implementation details are shown in Section 3.

The obtained result point out that EP-MAS.Lib agents can interact with each other in a MAS-based environment to solve a particular COP within a reasonable time. However, although using a simple genetic program is more efficient than using EP-MAS.Lib for the majority COPs, problems with complex fitness calculations that require high demanding computational capabilities are more efficiently solved with the MAS-based distributed proposal presented in this paper than simple genetic programs. On the other hand, another important difference to consider between simple genetic program and EP-MAS.Lib is that the last one can dynamically change its efficiency by adding new nodes to the distributive environment.

We are working on extending this proposal so it can be used in collaborative grid environments as well as the possibility of further reducing the flow of messages and data that is required to avoid possible bottleneck.

References

1. Arenas, M.G., Collet, P., Eiben, A.E., Jelasity, M., Merelo, J.J., Paechter, B., Preub, M., Schoenauer, M.: A Framework for Distributed Evolutionary Algorithms. In: Guervós, J.J.M., Adamidis, P.A., Beyer, H.-G., Fernández-Villacañas, J.-L., Schwefel, H.-P. (eds.) PPSN 2002. LNCS, vol. 2439, pp. 665–675. Springer, Heidelberg (2002)
2. Bellifemine, F., Poggi, A., Rimassa, G.: JADE – A FIPA-compliant agent framework. Telecom Italia internal technical report. In: Proc. International Conference on Practical Applications of Agents and Multi-Agent Systems (PAAM 1999), pp. 97–108 (1999)
3. Berntsson, J.: G2DGA: an adaptive framework for internet-based distributed genetic algorithms. In: Proc. of the 2005 workshops on Genetic and Evolutionary Computation (GECCO), pp. 346–349 (2005)
4. Chmiel, K., Tomiak, D., Gawinecki, M., Kaczmarek, P., Szymczak, M., Paprzycki, M.: Testing the Efficiency of JADE Agent Platform. In: Proc. 3rd Int. Symposium on Parallel and Distributed Computing (ISPDC), pp. 49–57. IEEE Computer Society Press, Los Alamitos (2004)
5. Eiben, A.E., Smith, J.E.: Introduction to Evolutionary Computing. Springer, Heidelberg (2003)
6. Ferber, J.: Les systems multi-agents, Vers une intelligence collective, pp. 1–66. InterEditions, Paris (1995)
7. Foundation for Intelligent Physical Agents: FIPA ACL Message Structure Specification, SC00061, Geneva, Switzerland (2002), http://www.fipa.org/specs/fipa00061/index.html
8. Jain, L.C., Palade, V., Srinivasan, D.: Advances in Evolutionary Computing for System Design. Studies in Computational Intelligence, vol. 66. Springer, Heidelberg (2007)
9. Laredo, J.L.J., Eiben, E.A., Schoenauer, M., Castillo, P.A., Mora, A.M., Merelo, J.J.: Exploring Selection Mechanisms for an Agent-Based Distributed Evolutionary Algorithm. In: Proceedings Genetic and Evolutionary Computation Conference (GECCO), pp. 2801–2808. ACM, New York (2007)
10. Lee, W.: Parallelizing evolutionary computation: A mobile agent-based approach. Expert Systems with Applications 32(2), 318–328 (2007)
11. Meng, A., Ye, L., Roy, D., Padilla, P.: Genetic algorithm based multi-agent system applied to test generation. Computers & Education 49, 1205–1223 (2007)
12. Paletta, M., Herrero, P.: Learning Cooperation in Collaborative Grid Environments to Improve Cover Load Balancing Delivery. In: Proc. IEEE/WIC/ACM Joint Conferences on Web Intelligence and Intelligent Agent Technology, pp. 399–402. IEEE Computer Society, Los Alamitos (2008) E3496
13. Vacher, J.P., Galinho, T., Lesage, F., Cardon, A.: Genetic Algorithms in a Multi-Agent system. In: Proc. IEEE International Joint Symposia on Intelligent and Systems, pp. 17–26 (1998) ISBN: 0-8186-8545-4

A Framework for Dynamical Intention in Hybrid Navigating Agents

Eric Aaron and Henny Admoni

Department of Mathematics and Computer Science
Wesleyan University
Middletown, CT 06459

Abstract. As a foundation for goal-directed behavior, the reactive and deliberative systems of a hybrid agent can share a single, unifying representation of intention. In this paper, we present a framework for incorporating *dynamical intention* into hybrid agents, based on ideas from *spreading activation* models and *belief-desire-intention* (*BDI*) models. In this framework, intentions and other cognitive elements are represented as continuously varying quantities, employed by both sub-deliberative and deliberative processes: On the reactive level, representations support some real-time responsive task re-sequencing; on the deliberative level, representations support common logical reasoning. Because cognitive representations are shared across both levels, inter-level integration is straightforward. Furthermore, dynamical intention is demonstrably consistent with philosophical observations that inform conventional BDI models, so dynamical intentions function as conventional intentions. After describing our framework, we briefly summarize simple demonstrations of our approach, suggesting that dynamical intention-guided intelligence can potentially extend benefits of reactivity without compromising advantages of deliberation in a hybrid agent.

1 Introduction

Intention-guided inference is often based on propositional, deliberation-level representations, but some goal-directed intelligence can be based on dynamical, sub-deliberative representations of intention, as well. For example, consider an animated agent completing tasks, running errands in a grid world. Along with various desires and beliefs —such as, e.g., its belief that it does not have a letter to mail— the agent starts out with intentions in its cognitive system, one for each task it might perform; each intention has an associated cognitive *activation* value, representing the intensity of commitment to the corresponding task, the task's relative *priority*. As the agent begins, its plan of action is represented by its ordering of intention activations (i.e., priorities) from highest to lowest: For this example, its intention to *deposit a check* at the bank (I_{DC}) has the highest activation, so it is first in the planned task sequence; the agent's intention to *get its child* from school (I_{GC}) has slightly lower activation and is second in the planned sequence; and the remainder of the sequence follows from the agent's

E. Corchado et al. (Eds.): HAIS 2009, LNAI 5572, pp. 18–25, 2009.

other intentions, including (for illustrative purposes) its intention to mail a letter (I_{ML}), which is inconsistent with its belief that it does not have a letter.

As the agent progresses through its sequence of tasks, its cognitive state changes due to both deliberative and sub-deliberative processes. Its cognitive activations continuously evolve, for instance, unobtrusively causing I_{ML} to become negative, consistent with the agent's belief that it has no letter. Before it reaches the bank, continuous evolution also causes the activation of I_{GC} to slightly exceed that of I_{DC}; the agent then deliberates about what its highest priority task and overall task sequence should be, using geographic and task-specific knowledge to resolve uncertainty. Thus, the same representations of intentions are employed in two different contexts: Reactive processes, not conventional deliberation, resolve the mail-related inconsistency; and deliberative inference is invoked when the agent is called upon to select a new current task from two candidate tasks of essentially equivalent, maximal priority.

In this paper, we present a framework supporting such deliberative and reactive intelligence in *hybrid dynamical cognitive agents* (*HDCAs*, for short). The design of HDCAs' cognitive systems is influenced by the *belief-desire-intention* (or *BDI*) theory of intention [1]; the theory and its related implementations (e.g., [2,3] and successors) suggest that BDI elements (beliefs, desires, and intentions) are an effective foundation for goal-directed intelligence. HDCAs' sub-deliberative cognitive models are notably influenced by some *distinguishing properties* that differentiate intention from desire (noted in [1]). For examples, an intensely committed intention I diminishes impacts of other intentions on the intensity of I; the strongest intentions (i.e., intentions with the most intense commitment) need not correspond to the strongest desires; and intentions, not desires, govern HDCAs' task priorities.

In conventional deliberative agents, BDI-based intentions are represented and manipulated with mechanisms that do not emphasize *continuous-modeled* cognition, but HDCAs' cognitive models interconnect BDI elements in a continuously evolving system inspired by (though significantly different from) *spreading activation* frameworks of [4,5]. Each BDI element in an HDCA is represented by an activation value, indicating its salience and intensity "in mind" (e.g., how intensely held a belief or committed an intention), and cognitive evolution is governed by differential equations, so elements' activation values affect rates of change of other elements' activations. HDCAs employ these cognitive representations of dynamical intention on both reactive and deliberative levels, distributing the burden of goal-directed intelligence and enabling smooth hybrid integration.

2   Hybrid and Deliberative Structure

The reactive / deliberative structure of HDCAs is illustrated in Figure 1, showing sub-deliberative, continuous-modeled cognitive processes and dynamical navigation; deliberative task sequencing and path planning processes; and cognitive representations shared across levels. Each level employs cognitive representations in its own manner, but the representations fully support both levels, for

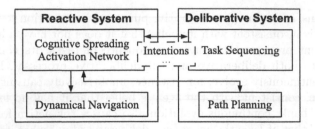

Fig. 1. System-level architecture of an HDCA, showing deliberative and reactive levels. Representations of intentions are shared by the sub-deliberative cognitive spreading activation network and the deliberative task sequencing process.

straightforward hybrid integration. The particular deliberative task sequencing and path planning processes of HDCAs in this paper are simple, although other, more complicated methods could be readily employed. Planners essentially derive "utility" values for each option, each task or path segment, based on geographic information, task-specific knowledge, and cognitive activations. Plans are then simply sequences (e.g., of tasks) in decreasing order of utility; higher commitment to an intention (i.e., higher intention activation) translates to higher utility for the associated task but is not the sole factor in determining task sequence.

Deliberation is designed to be invoked only in situations that are not well handled by fully reactive processes. In our illustrative examples, deliberation occurs only in two circumstances: if the current task is unexpectedly interrupted (e.g., by a blockaded street); or if the agent is called upon to change its current task —due to completing the previous task, evolutions of intention activations, or any other cause— and must select from multiple candidates with essentially equivalent intention activations (i.e., within a given threshold value of each other; see section 5 for an example). Unlike reactive task re-sequencing, which occurs when any intention activation values change relative ordering in their sequence and depends only on those values, HDCAs' deliberative processes also include domain-specific facts and global world knowledge. In section 5, for instance, deliberative task sequencing encodes that borrowing a book precludes buying a book, and it considers locations at which tasks are completed, making geographic distance critical to task sequencing. Deliberation also re-evaluates an agent's entire task sequence, adjusting activations of cognitive elements so that, e.g., tasks earlier in the sequence have higher activations on corresponding intentions, and precluded tasks have highly negative intentions. After deliberation, an agent simply continues with its new cognitive activation values in the reactive behavior of its new highest priority task.

In addition to being a hybrid reactive / deliberative system, an HDCA is a *hybrid dynamical system* (*HDS*, for short), a combination of continuous and discrete dynamics, modeled by a *hybrid automaton* [6,7]. A hybrid automaton is a finite state machine in which each discrete state (or *mode*) can be viewed as a continuous-modeled behavior, containing differential equations that govern system evolution in that mode. Transitions between modes (including those from

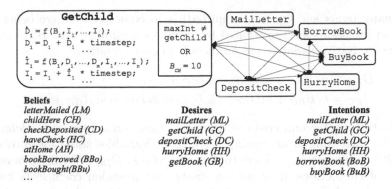

Fig. 2. Hybrid dynamical system modes and BDI elements, including abbreviations for BDI element names, for HDCAs in this paper

a mode to itself) are instantaneous, occurring when *guard* conditions are met, and may have discontinuous *side effects*, encoding discrete system dynamics. Hybrid dynamical systems are often apt models for navigating robots or animated agents (e.g., [8,9]), and HDCAs' reactive and deliberative structures naturally correspond to HDS elements: Each task of an HDCA is a reactive behavior, implemented as an HDS mode, and agents switch among these individually continuous tasks; deliberation in HDCAs only occurs during such transitions.

3   Reactive Structure

An HDCA's physical state (heading angle ϕ, position (x, y)) continuously varies as it navigates. For HDCAs in this paper, agent steering is based on [10], and simple intersection-to-intersection navigation in a grid world is similar to the method in [6], but other dynamical approaches could have been equally effective.

The continuous physical state smoothly integrates with an HDCA's cognitive system, which is based on continuously evolving activations of BDI elements (beliefs, desires, intentions); differential equations govern continuous evolutions of all elements, physical and cognitive. (Element values can also be changed discretely, as effects of mode transitions; after completing a task, for example, HDCAs' mode transitions discretely set the corresponding intention activation to the minimum possible value.) Figure 2 shows BDI elements (and abbreviations for their names) and the mode transition model for HDCAs in this paper, which is simplified to a one-to-one correspondence between intentions and actions.

In HDCAs, BDI elements are represented by *activation* values, restricted to the range $[-10, 10]$. Near-zero values indicate low salience, and greater magnitudes indicate greater salience and intensity of associated concepts, so, e.g., more active intentions represent more commitment to and urgency of the related actions. Negative values indicate salience of the opposing concept, e.g., for intentions, intention not to perform the related action. These cognitive activations are interconnected in differential equations; a partial cognitive system (with many terms and equations omitted) is in equation 1, where beliefs, desires, and intentions are represented by variables beginning with B, D, and I, and time-derivative variables are on the left in each equation:

$$\dot{D}_{DC} = a_1 B_{HC} + a_3 I_{DC} - a_5 I_{GC} + \cdots \tag{1}$$
$$\dot{I}_{DC} = b_1 B_{HC} + b_3 D_{DC} - b_6 D_{HH} + b_8 I_{DC} - b_{10} I_{GC} + \cdots .$$

This illustrates interconnectedness: Elements exert *excitatory* or *inhibitory* influence by increasing or decreasing derivatives. Variables stand for activations of cognitive elements (e.g., desire to deposit a check, D_{DC}). Coefficients encode impacts of connections; most are constants, but intention coefficients contain functions that also encode distinguishing properties of intention (see section 4).

There is also a mechanism for *perception* in HDCAs, by which proximity to an item (e.g., a UPS deposit box, a street blockade) affects agents' cognitive systems. Current HDCAs have only limited perceptual structure; potential for substantial extensions exists but is not discussed in this paper.

4   Properties of Intention

In HDCAs, there are mechanisms that ensure consistency with distinguishing properties of intention (noted in [1]) that apply to our dynamical account of intention—intentions are *conduct controlling* elements that, when salient, *resist reconsideration* and *resist conflict* with other intentions.[1]

For *reconsideration resistance*, we encode two criteria: any *high-active* intention I_a (i.e., having high activation magnitude) tends to minimize impacts on I_a from other intentions; and the magnitude of this effect grows as the activation (magnitude) of I_a grows. To enable this, for intentions I_a and I_b, for every $a \neq b$, the differential equation for $\dot{I}_a$ includes the following structure:

$$\dot{I}_a = \ldots - k_n \cdot PF(I_a) \cdot I_b \ldots . \tag{2}$$

For example, in equation 1, the coefficient of I_{GC} in the equation for $\dot{I}_{DC}$ has the form $b_{10} = k_{10} \cdot PF(I_{DC})$, with *persistence factor* PF defined as

$$PF(I_a) = 1 - \frac{|I_a|}{\sum_i |I_i| + \epsilon}, \tag{3}$$

where noise term $\epsilon > 0$ prevents division by 0, and i ranges over all intentions. For $b \neq a$, $PF(I_a)$ multiplies every intention I_b in the equation for $\dot{I}_a$, so as $PF(I_a)$ nears 0 (i.e., as I_a grows in magnitude relative to other intentions), contributions of every such I_b are diminished, and when $PF(I_a) = 1$ (i.e., $I_a = 0$), such contributions are unaffected. The denominator encodes that I_a is less reconsideration-resistant when other intentions are highly active.

---

[1] These are not the only properties of intention that are emphasized in [1]; they are, however, properties that can apply to reactive-level intention, not requiring, e.g., future-directedness incompatible with reactive implementations.

Fig. 3. Screen display of a simulation in progress. A map of the four-by-four grid world, left, shows buildings and obstacles (black boxes), targets (white squares abutting buildings), the current position of a moving blockade (gray box), and an agent's path. Numbers identify locations at which the agent invoked deliberation. On the right of the display, task sequences (including results of deliberations) are represented by sequences of color-coded circles, where radii indicate the magnitudes of activations of intentions.

Due to this and other activation-oriented mechanisms for *conduct control* and *conflict resistance* (see the supplementary website for this paper [11]), dynamical HDCA intentions are consistent with distinguishing properties of intention in [1]. Because an HDCA's deliberative level can be straightforwardly implemented to be consistent with [1], and because cross-level interconnections do not violate relevant properties, the dynamical intentions in HDCAs function as conventional intentions, rather than as some other cognitive elements inconsistent with [1].

5 Demonstrations and Experiments

Simulations of HDCAs navigating in a grid world show some features of our dynamical cognitive representations: intention-guided intelligence on both reactive and deliberative levels, extending reactive capabilities without sacrificing deliberation; and consistency of intention representations with philosophical foundations [1] of conventional BDI models. The simulations' code shows another feature, the straightforward integration of reactive and deliberative levels, united by shared cognitive representations. We briefly summarize our simulations in this section; the supplementary website for this paper [11] has animations from our simulations and more information about our code and demonstrations.

Our featured simulation of HDCA intelligence is similar to the errand-running scenario in section 1 of this paper. The HDCA navigates from intersection to intersection, eventually reaching targets in the grid world of Figure 3. That display also lists the tasks the agent might perform (*MailLetter*, ..., *BuyBook*) and indicates relative strengths of intention activations, thus showing the agent's current planned task sequence. The dynamical cognitive activations continuously evolve over time, sometimes re-sequencing tasks. The agent's simple deliberative level (see section 2) is engaged in only two circumstances: unexpected interruptions, due to an unpredictable, moving *blockade* that can block entry to any empty street; and *uncertainty* when the agent is called upon to change its current task, i.e., when there are two or more candidates for maximal priority with intention activation values differing by less than a threshold value T (see website [11] for implementation details). When deliberation is employed, the agent replans its full task sequence, incorporating task-domain and geographic knowledge.

This featured simulation contains several instances of both deliberative and reactive task sequencing. The agent in the simulation begins with maximal intention I_{DC}, but before reaching its target, activation on I_{GC} evolves to narrowly exceed that of I_{DC}. This prompts the agent to select a new current task when candidate tasks are nearly equivalent, thus invoking deliberation (at point 1 in Figure 3). The agent then generates a plan using simple methods, based on intention activations, geographic distances, and the knowledge that the agent must perform only one of *BorrowBook* and *BuyBook*. The plan is encoded by changing intention activations so that the intention corresponding to the first action in the plan gets the highest activation, etc. The agent then continues completing tasks, later deliberating when stopped by the blockade (point 2) or after completing a task when there is uncertainty about its new current task (points 3 and 4).

In addition, the agent's continuously evolving intention activations result in reactive-level task re-ordering throughout its task sequence. Most notably, as in the scenario in section 1, the agent intends to mail a letter but believes it has no letter, and it reconciles the inconsistency without deliberation: The activation of I_{ML} eventually becomes negative, consistent with its belief. Overall, the agent successfully completes tasks without superfluous activity (e.g., *BuyBook*) or action based on inconsistent intentions (I_{ML}), demonstrating the deliberative- and reactive-level intelligence enabled by its shared cognitive representations.

Other simulations directly demonstrated the consistency of HDCA intentions with distinguishing properties from [1] discussed in section 4. Some experiments simply tested HDCAs' reactive cognition, without navigation or task completion, demonstrating that terms in cognitive systems' differential equations (e.g., PF from section 4) successfully encode properties of intention. For example, when PF is present, activation values of high-active intentions are not significantly diminished by conflicting intentions, but when PF is absent, conflicting intentions can decrease high-active intention activations; similar experiments verified conflict resistance. Additionally, other experiments that did incorporate navigation demonstrated that agents' strongest desires and strongest intentions do not necessarily correspond, and intentions, not desires, serve to control HDCAs' conduct. (For more information, see this paper's supplementary website [11].)

6   Discussion and Conclusions

Hybrid dynamical cognitive agents are based on dynamical, continuous-modeled cognitive representations that extend reactive intelligence without sacrificing deliberation. For the simple implementations and explanations in this paper, HDCA deliberation is part of instantaneous HDS mode transitions, but other implementations could instead model time during deliberation—additional HDS modes could straightforwardly support this without requiring alterations to the fundamental agent model. Modeling agents as hybrid automata also supports straightforward interconnections between deliberative and reactive levels, and it potentially enables HDS analysis methods to verify some aspects of agent behavior (e.g., [12]; see [7], however, for theoretical restrictions on HDS analysis).

Because HDCAs' cognitive representations are shared by deliberative and reactive structures, intention-guided intelligence can be distributed over both levels. Moreover, HDCAs' unconventionally represented intentions are demonstrably consistent with distinguishing properties of intention noted in [1] that inform conventional BDI models, so HDCAs' intentions function as conventional intentions. Although our simple demonstrations were about a single HDCA, with limited perceptual mechanism and world interaction, they suggest more general utility of the underlying ideas: With more powerful methods for deliberative inference and perception, this dynamical intention framework could lead to intention-guided agents that rely on reactive intelligence, employing deliberation only when needed, making hybrid agents even more robust, efficient performers in dynamic multi-agent scenarios and other applications.

References

1. Bratman, M.: Intentions, Plans, and Practical Reason. Harvard University Press, Cambridge (1987)
2. Georgeff, M., Lansky, A.: Reactive reasoning and planning. In: AAAI 1987, pp. 677–682 (1987)
3. Rao, A.S., Georgeff, M.P.: Modeling rational agents within a BDI-architecture. In: Proc. of Principles of Knowledge Representation and Reasoning, pp. 473–484 (1991)
4. Collins, A.M., Loftus, E.F.: A spreading activation theory of semantic priming. Psychological Review 82, 407–428 (1975)
5. Maes, P.: The dynamics of action selection. In: IJCAI 1989, pp. 991–997 (1989)
6. Aaron, E., Ivančić, F., Metaxas, D.: Hybrid system models of navigation strategies for games and animations. In: Hybrid Systems: Computation and Control, pp. 7–20 (2002)
7. Alur, R., Henzinger, T., Lafferriere, G., Pappas, G.: Discrete abstractions of hybrid systems. Proc. of the IEEE 88(7), 971–984 (2000)
8. Axelsson, H., Egerstedt, M., Wardi, Y.: Reactive robot navigation using optimal timing control. In: American Control Conference (2005)
9. Aaron, E., Sun, H., Ivančić, F., Metaxas, D.: A hybrid dynamical systems approach to intelligent low-level navigation. In: Proceedings of Computer Animation, pp. 154–163 (2002)
10. Goldenstein, S., Karavelas, M., Metaxas, D., Guibas, L., Aaron, E., Goswami, A.: Scalable nonlinear dynamical systems for agent steering and crowd simulation. Computers And Graphics 25(6), 983–998 (2001)
11. Aaron, E., Admoni, H.: Supplementary HAIS 2009 material (2009), http://eaaron.web.wesleyan.edu/hais09_supplement.html
12. Asarin, E., Dang, T., Girard, A.: Hybridization methods for the analysis of nonlinear systems. Acta Informatica 43(7), 451–476 (2007)

Multi-agent Based Personal File Management Using Case Based Reasoning

Xiaolong Jin, Jianmin Jiang, and Geyong Min

Digital Media and Systems Research Institute, School of Informatics,
University of Bradford, Bradford, BD7 1DP, UK
{x.jin,j.jiang1,g.min}@brad.ac.uk

Abstract. Computer users have been facing a progressively serious problem, namely, how to efficiently manage computer files so as to not only facilitate themselves to use the files, but also save the scare storage resource. Although there are a lot of file management systems available so far, none of them, to the best of our knowledge, can automatically address the deletion/preservation problem of files. To fill this gap, this study explores the value of artificial intelligence techniques in file management. Specifically, this paper develops an intelligent agent based personal file management system, where Case Based Reasoning (CBR) is employed to guide file deletion and preservation. Through some practical experiments, we validate the effectiveness and efficiency of the developed file management system.

1 Introduction

Since its great invention, computer has entered people's daily life and has already fundamentally changed the ways in which people work and live. Every day, a number of new files are created and stored on each computer. If these files are not well managed, two significant problems will follow. First, since there are too many different files stored on computers, it becomes very difficult for human users to find and use desired files. Secondly, with the creation of more computer files, more storage space is needed. Therefore, how to efficiently manage the files stored on computers is a very important issue. Nowadays, a lot of file management systems have been made available [1,2]. However, there are no file management systems that are able to automatically deal with the deletion/preservation issue of files.

Recently, Intelligent Agent Technology (IAT) has emerged as a promising approach to developing intelligent software systems, as this technology enables software engineers to model applications in a natural way that resembles how humans perceive the problem domains [3]. IAT advocates designing and developing applications in terms of autonomous software entities (i.e., agents), which situate in an environment and are able to flexibly achieve their goals by interacting with one another as well as the environment. This feature is well suited to handle the complexity of developing software in modern application scenarios [4].

E. Corchado et al. (Eds.): HAIS 2009, LNAI 5572, pp. 26–33, 2009.

Therefore, IAT has been successfully applied in many industrial and commercial areas, such as, electronic commerce and human-computer interaction [5].

Cased Based Reasoning (CBR) originated from the problem solving pattern of human beings. Actually, CBR is an analogical reasoning method for solving various problems, which can adopt the knowledge of past experiences, stored in a knowledge base as cases [6,7]. In a CBR system, a new problem is solved by investigating, matching, and reusing solutions to similar cases that have been previously solved. At the highest level of generality, CBR can be formalized as a four-step procedure: *retrieve, reuse, revise,* and *retain.* CBR has gained great success in solving the problems of design, planning, classification, and advising in many different application fields [8,9], such as, image processing, molecular biology, course timetabling, spam filtering, and fault diagnosing.

This study aims to investigate intelligent file management based on the afore-mentioned artificial intelligence techniques. Specifically, we develop a multi-agent based personal file management system, where intelligent agents cooperate and coordinate with each other to manage computer files. As a core module of the management system, a CBR agent is engineered to recommend suitable actions on individual computer files using the past experience of human users.

2   Multi-agent Based Personal File Management System

Figure 1 presents a schematic diagram of the multi-agent based personal file management system. In what follows, we will briefly introduce the functionality of individual agents involved in this management system.

Human User: The human user is responsible for managing electronic files stored in the storage space, which can be a physical or logical disc driver or a logical file directory. The user can consult the CBR agent on whether a file at hand should be deleted or preserved. However, the final decision is up to the user.

CBR Agent: This agent employs the CBR mechanism to make suitable recommendation on removing or preserving computer files. A case base is provided to support the recommendation of the CBR agent, where the detailed information of deleted or preserved files are stored as past cases.

*Monitoring Agent:*This agent monitors the actions of the user on computer files and report to the CBR agent. It is also responsible for copying files at hand to the file buffer in order for the feature extraction agent and the content analysis agent to extract the metadata of individual files and analyze their contents, respectively.

Metadata Extraction Agent: This agent is designed to extract detailed metadata (e.g., size, type, and creation time) of individual files that have been temporarily copied to the file buffer. The medadata will be used by the CBR agent as the foundation to make suitable recommendation.

Content Analysis Agent: The functionality of this agent is to extract and analyze the contents of computer files. At present, it can only work on document files. It

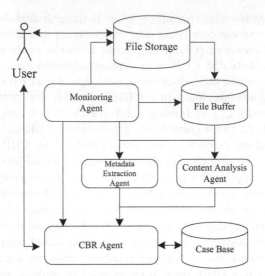

Fig. 1. A schematic diagram of the multi-agent based personal file management system

can extract their textual contents and lexically analyze them to identify a list of keywords.

3 Case Based Reasoning for File Management

First of all, it should be pointed out that for the sake of space limitation this paper focuses only on the decision making on deleting or preserving computer files. It does not cover other topics (e.g., the organization of files) of the management system. In this section, we present the key issues related to the application of CBR to file management.

3.1 Case Representation

A case is usually represented by a list of attribute-value pairs that represent the values of different features of the corresponding problem. In our CBR system, each case actually corresponds to an electronic file that has been deleted from or preserved on a computer. It consists of three types of attributes, namely, *descriptive attributes*, *solution attributes*, and *performance attributes*.

The descriptive attributes of a case refer to those attributes that represent the detailed information, including metadata and contents, of the corresponding file. In more detail, they can be listed as follows: (1) type of the file, tpf; (2) size of the file, szf; (3) creation time, crt; (4) last access time, lat; (5) last write time, lwt; (6) is a hidden file, ihf; (7) is a read-only file, irf; (8) is a system file, isf; (9) is a temporary file, itf; and (10) keywords of the file, kwd. Note that the kwd of a case is defined as an ordered set. The higher the order of a keyword in kwd, the better it can be used to describe the contents of the case.

In our study, we define two solution attributes for a case: (1) final action, act: it is the final decision made by the user on the corresponding file; (2) action time, fat: it records the specific time when the final action is performed on the file. The performance attributes of a case measures it performance on recommending suitable solutions to new cases. There are two performance attributes for a case: (1) total recommendations, trd: it is a counter that records the total times when this case is retrieved as a similar one to the case at hand; (2) correct recommendations, crd: it counts the times when the case successfully recommends a final solution to a new case.

3.2 Case Retrieval

Given a file f, in order to recommend a suitable action on it, the CBR agent will take it as a new case (without a solution) and then retrieve the n nearest neighboring cases from the case base $\aleph$, whose descriptive attributes are most similar to those of file f. In doing so, the key is to measure the similarity between the corresponding attributes and finally provide a general measure for the similarity between the new and existing cases. In this study we normalize all similarity measures into a value in $[0, 1]$. The larger the similarity value, the more similar the two cases. Later on, we suppose C_A and C_B are two cases for measuring their similarity and denote the value of a certain attribute x in any case C_X as C_X^x.

Types of Files: We use a tree structure to organize various file types. The first level (i.e., the root) is the full set of all file types. At the second level, files are classified into five types, namely, *document, image, audio, video,* and *other.* Each type has several subtypes or a sub-tree structure. Let C_A^{tpf} and C_B^{tpf} be the values of the type attribute of cases C_A and C_B, respectively, and π be their closest common ancestral node in the tree structure. We can define the similarity between C_A^{tpf} and C_B^{tpf} as follows: If π is a leaf node, $C_A^{tpf} = C_B^{tpf}$. Hence, we have $Sim(C_A^{tpf}, C_B^{tpf}) = 1$; If π is the root node, $Sim(C_A^{tpf}, C_B^{tpf}) = 0$; Otherwise, $Sim(C_A^{tpf}, C_B^{tpf}) = 2^{-(\max\{step(C_A^{tpf}), step(C_B^{tpf})\} - step(\pi))}$, where $step(x)$ counts the steps from the root of the file type tree structure to its offspring node, x.

Sizes of Files: Let C_A^{szf} and C_B^{szf} be the values of the size attribute of cases C_A and C_B, respectively. Then, we can readily give a similarity measure for the attribute, szf, as

$$Sim(C_A^{szf}, C_B^{szf}) = 1 - \frac{|C_A^{szf} - C_B^{szf}|}{\max\limits_{C_X \in \aleph} C_X^{szf} - \min\limits_{C_Y \in \aleph} C_Y^{szf}}. \tag{1}$$

Time Related Descriptive Attributes: In the case representation, we have three time related descriptive attributes, namely, $\{crt, lat, lwt\}$, to characterize cases. Let C_A^{tra} and C_B^{tra} be the values of a time related attribute corresponding to cases C_A and C_B, respectively. The similarity measure can be defined as follows:

$$Sim(C_A^{tra}, C_B^{tra}) = 1 - \frac{|(C_A^{fat} - C_A^{tra}) - (C_B^{fat} - C_B^{tra})|}{\max\limits_{C_X \in \aleph}(C_X^{fat} - C_X^{tra}) - \min\limits_{C_Y \in \aleph}(C_Y^{fat} - C_Y^{tra})}. \tag{2}$$

Binary Descriptive Attributes: As we have noticed in the previous subsection, there are four binary descriptive attributes for each case, namely, $\{ihf, irf, isf, itf\}$. The similarities between these binary attributes of two cases can also be readily measured. Let C_A^{bda} and C_B^{bda} be the values of a certain binary descriptive attribute of cases C_A and C_B, respectively. The corresponding similarity can be calculated as

$$Sim(C_A^{bda}, C_B^{bda}) = 1 - |C_A^{bda} - C_B^{bda}|. \tag{3}$$

Keywords of Files: Let $C_A^{kwd} = \{k_A^1, k_A^2, \cdots, k_A^{N_A}\}$ and $C_B^{kwd} = \{k_B^1, k_B^2, \cdots, k_B^{N_B}\}$ be the ordered sets of keywords of cases C_A and C_B, respectively, where N_A and N_B are the numbers of keywords of cases C_A and C_B. We measure the keywords similarity between C_A and C_B as follows:

$$Sim(C_A^{kwd}, C_B^{kwd}) = \frac{\sum_{\kappa \in C_A^{kwd} \cap C_B^{kwd}}^{\kappa = k_A^i = k_B^j} \left\{ 1 - \frac{|i-j|}{\max\{N_A, N_B\}} \right\}}{\max\{N_A, N_B\}}, \tag{4}$$

where κ is a common keyword in C_A^{kwd} and C_B^{kwd} and i and j are its orders in these two keyword sets, respectively. Obviously, Equation (4) fully takes into account the orders of individual keywords in the corresponding keyword sets.

There are quite a few methods available for measuring the similarity between two vectors, such as, Manhattan distance, Euclidean distance, and Mahalanobis distance [7]. In our study, we employ the Manhattan distance, namely, the weighted summation of the similarities of all descriptive attributes as the similarity function between two cases.

3.3   Case Reuse and Revision

Given a new case C_A, after the n nearest neighboring cases have been retrieved out from the case base, they will be classified according to their final action attributes. Suppose Φ be the set of the n nearest neighboring cases. Let Φ_d and Φ_p be two subsets of Φ, whose elements have final action attributes `delete` and `preserve`, respectively. Therefore, we have $\Phi_d \cup \Phi_p = \Phi$ and $|\Phi_d| + |\Phi_p| = n$. Next, the CBR agent can recommend a final action to case C_A as follows:

$$C_A^{act} = \begin{cases} \texttt{delete}, & \text{if } \sum_{C_B \in \Phi_d} \frac{C_B^{crd}}{C_B^{trd}} - \sum_{C_B \in \Phi_p} \frac{C_B^{crd}}{C_B^{trd}} > \beta \\ \texttt{preserve}, & \text{if } \sum_{C_B \in \Phi_d} \frac{C_B^{crd}}{C_B^{trd}} - \sum_{C_B \in \Phi_p} \frac{C_B^{crd}}{C_B^{trd}} \le \beta \end{cases}, \tag{5}$$

where $\frac{C_B^{crd}}{C_B^{trd}}$ is the correct recommendation rate, i.e., the ratio of the number of correct recommendations to that of the total recommendations of the case C_B; β is a positive constant indicating the preference of the file management system to preserve files. The larger the value of β, the higher the preference of the system to preserve files, and thus the more conservative the file management system.

In our CBR system, case revision mainly concerns the performance attributes of cases stored in the case base. After the final decision on a new case has been made and the corresponding action has been performed, the performance attributes of the n nearest neighboring cases in Φ will be updated. Specifically, the update will be carried out as follows:

- For all cases, increase their total number of recommendation by one;
- If the final action on the new case C_A is delete, increase the correct recommendation counter of the cases contained in Φ_d by one;
- If the final action on the new case C_A is preserve, increase the correct recommendation counters of the cases contained in Φ_p by one.

3.4 Case Retainment

Let $\hat{C}_A^{act}$ be the final action that is taken by the human user on a new case C_A. If $\hat{C}_A^{act}$ is different from the action C_A^{act} recommended by the CBR agent, the new case C_A and the corresponding final action $\hat{C}_A^{act}$ will be retained into the case base. Moreover, the two performance attributes of C_A will be initialized to be zero and the current time of the system will be taken as its action time C_A^{fat}.

4   Performance Validation

In order to validate the effectiveness and efficiency of CBR in file management, we have carried out extensive experiments. For the sake of space limitation, in this section we present only the general experimental results. In our experiments, we first created 154 cases and stored them into the case base (see Table 1 for an example case). These cases correspond two groups of deleted and preserved computer files, respectively. They can be further categorized into five types, namely,

Table 1. An example case stored in the case base

Attributes	Values
Type of the file, tpf	document/doc
Size of the file, szf	1.26 MB
Creation time, crt	22 June 2008, 21:45:55
Last access time, lat	10 January 2009
Last write time, lwt	23 June 2008, 11:38:14
Is a hidden file, ihf	No
Is a read-only file, irf	No
Is a system file, isf	No
Is a temporary file, itf	No
Keywords of the file, kwd	agent; function; requirement; metadata
Correct recommendation, crd	4
Incorrect recommendation, ird	1
Final action, act	preserve
Action time, fat	10 January 2009, 19:48:52

Table 2. The results obtained from our experiments for validating the CBR mechanism in file management

File Types	Number	Action	Correct Recom.	Incorrect Recom.	Correct Recom. Rate
Document	52	delete	8	4	67%
		preserve	30	10	75%
Image	73	delete	35	6	85%
		preserve	28	4	88%
Audio	57	delete	16	9	64%
		preserve	21	11	66%
Video	29	delete	13	5	72%
		preserve	9	2	81%
Others	25	delete	5	8	38%
		preserve	7	5	58%

document, image, audio, video, and other. Next, we request the file management system to recommend suitable actions on another group of 236 files.

In Table 2, we present the total numbers of computer files of different types. For each type of files, the numbers of correct and incorrect recommendations corresponding the two actions, `delete` and `preserve`, made by the CBR agent are listed. The correct recommendation rates are also presented. From Table 2 we can note that for the four ordinary types of files, the CBR-based file management system can provide fairly accurate recommendation. Particularly, for image files the accuracy of recommendation is considerably high. However, for other file types the performance of the system is not as good as for the ordinary ones. An interesting phenomenon is that the correct recommendation rates of all different file types corresponding to action `preserve` are higher than those of action `delete`. A potential reason is that the CBR-based file management system prefers to preserve computer files, if it cannot confidently make recommendation to delete the files.

5 Conclusions

As more and more files of different types are created and stored on computers, the efficient management of the large number of electronic files becomes an important issue to human users. Although there are a lot of file management software tools available so far, none of them is able to address the crucial deletion/preservation problem of files. Unfortunately, to the best of our knowledge this issue has also not been investigated in the open literature. To bridge this gap, in this paper we have presented a multi-agent based system for personal file management, where Case Based Reasoning (CBR) is employed to recommend suitable actions on computer files based on the knowledge learned from computer users' past behaviors on file management. Through a set of experiments and the corresponding results, we have demonstrated the effectiveness and efficiency of the developed file management system and the CBR mechanism on file management.

Acknowledgement

This work is supported by the Seventh Framework Programme of the European Union under grant (FP7-ICT-216746).

References

1. Carlton, G.H.: A critical evaluation of the treatment of deleted files in microsoft windows operation systems. In: Proceedings of the 38th Hawaii International Conference on System Sciences (HICSS 2005), pp. 1–8 (2005)
2. You, L.L., Pollack, K.T., Long, D.D.E.: Deep store: An archival storage system architecture. In: Proceedings of the 21st International Conference on Data Engineering (ICDE 2005), pp. 804–815 (2005)
3. Chmiel, K., Gawinecki, M., Kaczmarek, P., Szymczak, M., Paprzycki, M.: Efficiency of JADE agent platform. Scientific Programming 13(2), 159–172 (2005)
4. Zambonelli, F., Omicini, A.: Challenges and research directions in agent-oriented software engineering. Autonomous Agents and Multi-Agent Systems 9(3), 253–283 (2004)
5. Luck, M., McBurney, P., Shehory, O., Willmott, S.: Agent Technology: Computing as Interaction (A Roadmap for Agent Based Computing). AgentLink (2005)
6. Craw, S., Wiratunga, N., Rowe, R.C.: Learning adaptation knowledge to improve case-based reasoning. Artificial Intelligence 170(16-17), 1175–1192 (2006)
7. Khoshgoftaar, T.M., Seliya, N., Sundaresh, N.: An empirical study of predicting software faults with case-based reasoning. Software Quality Journal 14(2), 85–111 (2006)
8. Iglesias, R., Ares, F., Ferneindez-Delgado, M., Rodriguei, J.A., Bregains, J., Barrol, S.: Element failure detection in linear antenna arrays using case-based reasoning. IEEE Antennas and Propagation Magazine 50(4), 198–204 (2008)
9. Pous, C., Caballero, D., Lopez, B.: Diagnosing patients combining principal components analysis and case based reasoning. In: Proceedings of the 8th International Conference on Hybrid Intelligent Systems (HIS 2008), pp. 819–824 (2008)

Agent-Based Evolutionary System for Traveling Salesman Problem

Rafał Dreżewski, Piotr Woźniak, and Leszek Siwik

Department of Computer Science
AGH University of Science and Technology, Kraków, Poland
drezew@agh.edu.pl

Abstract. Evolutionary algorithms are heuristic techniques based on Darwinian model of evolutionary processes, which can be used to find approximate solutions of optimization and adaptation problems. Agent-based evolutionary algorithms are a result of merging two paradigms: evolutionary algorithms and multi-agent systems. In this paper agent-based evolutionary algorithm for solving well known Traveling Salesman Problem is presented. In the experimental part, results of experiments comparing agent-based evolutionary algorithm and classical genetic algorithm are presented.

1 Introduction

Evolutionary algorithms are heuristic techniques for finding (sub-)optimal solutions of continuous and discrete optimization problems [2]. They can be used in the case of highly multi-modal problems because, after using some special mechanisms for maintaining population diversity (niching techniques), the population of individuals can simultaneously detect multiple optima basins of attraction or avoid basins of attraction of local optima.

Evolutionary multi-agent systems (EMAS) are agent-based evolutionary algorithms resulting from research on merging multi-agent systems and evolutionary computations. Merging of the two paradigms leads to decentralization of evolutionary processes. In EMAS approach we usually have environment composed of computational nodes ("islands"), agents, and resources. Agents can reproduce, die (when they run out of resources) and interact with environment and other agents. There is also competition for limited resources among agents. In order to realize selection process "better" (what means that they simply better solve the given problem) agents are given more resources from the environment and "worse" agents are given less resources. This results in decentralized evolutionary processes in which individuals (agents) make independently all their decisions concerning reproduction, migration, etc., taking into consideration conditions of the environment, other agents present within the neighborhood, and resources possessed. In CoEMAS (co-evolutionary multi-agent system) approach we additionally have many species and sexes of agents which interact with each other [3]. Thus we can model co-evolutionary interactions and sexual selection mechanisms, which promote diversity of population.

Traveling Salesman Problem (TSP) is old and well known NP-hard problem. Many real-life problems can be modeled as TSP or some variants of it. The TSP problem has

E. Corchado et al. (Eds.): HAIS 2009, LNAI 5572, pp. 34–41, 2009.

quite simple concept. Traveling salesman has to visit all cities just one time, and in the end go back to the starting city. He knows cost of traverse between cities and has to plan the shortest route (with minimal cost). Because TSP is NP-hard problem we must rather try to find satisfying sub-optimal solution using some heuristic technique (simulated annealing, ant colony optimization, evolutionary algorithm) because search for global optima would be too expensive. Evolutionary algorithms are well suited to solve such problems because they usually find approximate solutions, or saying this in other words—they locate basins of attraction of optima. An introduction to solving TSP with the use of evolutionary algorithms may be found for example in [2,4]. Good results are usually obtained when some heuristic technique is combined with a local search technique (hybrid approaches).

When we try to solve TSP with evolutionary algorithm we will realize that it has quite simple fitness function. Each of solutions is a permutation of cities. To compare the quality of each solution, it is sufficient to calculate length of the route for each permutation of cities and choose the shortest one. But the choice of route representation and genetic operators is usually not so obvious. It's rather not recommended to use binary representation in TSP. In the last few years, mainly three representation techniques were used: adjacent method, order method and path method. Each of them has its own appropriate genetic operators, but the route is connected sequence of cities in all techniques.

The main goal of this paper is to apply agent-based model of evolutionary computations to TSP and to compare the obtained results with those obtained by standard genetic algorithm. In the following sections we will present the architecture of the agent-based evolutionary system for solving TSP, and results of preliminary experiments.

2 Agent-Based Evolutionary System for Solving TSP

In this section the architecture and algorithms of the agent-based evolutionary system for solving TSP are presented. The system was implemented with the use of agent-based evolutionary platform JAgE [1]. This platform has all necessary mechanism for the realization agent-based models of (co-)evolutionary computations: computational environment composed of nodes (islands), agents, agent-agent and agent-environment communication mechanisms, and mechanisms for distributed computations.

The platform has component architecture, what means that every component is responsible for selected tasks of the system. There can be distinguished four basic components [1]:

- *Workplace*—main component which controls all computation processes. Every computation node has its own instance of *Workplace*. It includes one (main) agent which controls all descendant agents. *SimpleWorkplace* is a basic implementation of *Workplace*. It stores addresses which are needed for exchanging information between agents.
- *ConnectionManager* is responsible for connections, communication, and transmission of objects between computational nodes.
- *ControlManager*—its main task is monitoring and controlling computation processes.

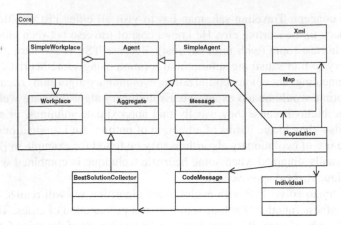

Fig. 1. Classes of the agent-based evolutionary system for solving TSP and their relations with jAgE core classes

- *ResourceManager* is a component which includes methods for system resource management.

Agent-based evolutionary system for solving TSP was implemented as a component of jAgE platform (see fig. 1). *XML* class is responsible for supplying external data to evolutionary algorithm solving TSP. *Individual* class includes definition of chromosome, methods for route length calculation and stores auxiliary information about individuals. The core of computations—where evolutionary process is realized—is the *Population* class. In this class evolutionary algorithm solving TSP was implemented.

Algorithm 1. Genetic algorithm
1 Generate random population;
2 Calculate length of route for each individual;
3 Calculate their absolute and relative fitness;
4 **while** *stopping condition is not fulfilled* **do**
5   Send the best individual to best solution collector class;
6   **for** $i \leftarrow 0$ **to** *NumberOfIndividuals* **do**
7   Add to the base population individual which was selected via roulette selection
8   **end**
9   **for** $i \leftarrow 0$ **to** *NumberOfIndividuals*/2 **do**
10   crossover two randomly selected individuals;
11   **end**
12   Perform mutation of the offspring;
13   Calculate length of route for each individual;
14   Calculate their absolute and relative fitness;
15 **end**

In the system described, two algorithms were implemented: agent-based evolutionary algorithm—EMAS (see alg. 2) and genetic algorithm—GA (see alg. 1). Genetic algorithm served as a reference point during experiments.

Algorithm 2. Agent-based evolutionary algorithm

```
 1  Generate random population of agents;
 2  Calculate length of route for each agent;
 3  Calculate their absolute and relative fitness;
 4  Distribute resources between agents;
 5  while stopping condition is not fulfilled do
 6  |   if agent enters the given population as a result of migration then
 7  |   |   Add it to the population;
 8  |   end
 9  |   Find the best agent in population;
10  |   if it is better than previous best one then
11  |   |   Send him to the best solution collector class;
12  |   end
13  |   if if the amount of resources of the agent that wants to crossover is above the
    |      minimum then
14  |   |   Select partner and perform crossover;
15  |   |   Assign children to temporary population;
16  |   end
17  |   Mutate offspring;
18  |   Merge temporary population with base population;
19  |   Give back the rest of agents' resources to the environment;
20  |   if agent's age exceeds limit then
21  |   |   Remove it from the system;
22  |   end
23  |   Calculate length of route for each agent;
24  |   Calculate their absolute and relative fitness;
25  |   Distribute resources between agents;
26  |   if agent can not perform any action then  /* agent has too few resources */
27  |   |   Remove it from the system;
28  |   end
29  |   if k steps of the algorithm passed from the previous modification then
30  |   |   Modify parameters of mutation and crossover;
31  |   end
32  end
```

3 Experimental Results

In this section results of experiments, which goal was to compare implemented agent-based evolutionary system (EMAS) and genetic algorithm (GA), are presented. The basic features of algorithms which were compared during experiments include the working time and the number of steps needed to find an (sub-)optimal solution. The results of experiments presented will also answer the question of how each parameter influence the course of evolutionary processes.

Input data should be identical for all algorithms in order to correctly estimate the quality of each technique. The problem consisted of 30 cities, identical for all techniques, was used during experiments. Certain number of cities from this problem was chosen to change the number of cities in consecutive experiments. In experiments 10, 15, 20, 25 and 30 cities samples were used. During experiments also Euclidean problem with 92 cities from TSP library [5] was used.

During preliminary experiments many different parameters' values were tested. It was observed that some configurations give better results, but others result in shorter working time of the algorithm. In the case of GA there were not too many possibilities:

Table 1. Values of parameters used during experiments

	EMAS	GA
Number of individuals	100 on each of 5 islands	100
Minimal number of individuals	10	-
Probability of crossover	0.8	0.7
Probability of mutation	0.01	0.1
Migration cost	0.3	-
Crossover cost	0.18	-
Mutation cost	0.15	-
Environment resources	equal to the initial number of individuals	-
Minimal amount of resources	0.04	-
Frequency of migration	after each 500 steps	-
Individual's life time limit	15 steps	-

only the number of individuals and parameters used during crossover and mutation. EMAS gives much more options. Optimal values of parameters used during main part of experiments for each technique are shown in table 1. During experiments the maximal number of steps (stopping condition) was set on the basis of improvements of average fitness of the population.

Number of individuals. It is very important parameter of evolutionary algorithm. In the case of multi-modal optimization problems it affects the number of located basins of attraction of local and global optima. In the case of multi-objective optimization problems it affects the coverage of Pareto frontier. Relatively, greater number of individuals is desirable in the case of hard computational problems, with many local optima, but the size should be carefully chosen because too great value can considerably slow down the computational process.

In the case of GA, the number of individuals is the same in all stages of algorithm. Each base and descendant population has the same size, so the number of individuals never changes. Results of experiments show that the number of individuals in EMAS grows during initial faze but then stabilizes at the given level, which depends on the total amount of resources present in the system. Number of individuals oscillates around this value during the consecutive steps of algorithm (see fig. 2).

There was also observed the fact that some parameters' values influence the number of individuals in the EMAS. One of them is minimal resource amount, which agent must possess in order to perform some activities in the system. Low minimum amount needed for crossover and mutation causes that average number of individuals grows and evolutionary process runs faster (and less precisely as well). It's connected with the fact that resources of the environment are limited and dependent on the initial number of individuals. When the population size is very large it results in limited activities of the agents because each of agents receives much less resources from the environment than in the case when there are less agents in the population. Other parameter which influences the population size is crossover cost. When it is too high it causes that less agents are able to initialize the process of reproduction and crossover, and as a result less new agents enter the population.

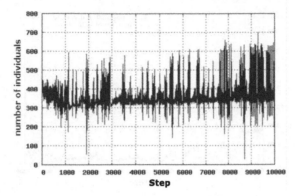

Fig. 2. Number of individuals in EMAS (92 cities, average from 5 experiments)

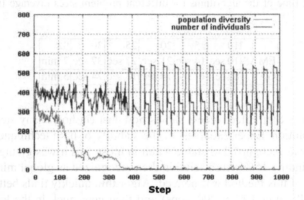

Fig. 3. Population diversity in EMAS (25 cities, average from 5 experiments)

Population diversity. In our experiments population diversity was measured as the number of different solutions present within the population in the given step.

In the case of EMAS it's easy to observe the tendency to slowly reduce the diversity during the experiment (see fig. 3). This phenomena is connected with the process of approaching the areas where optima are located—the search is then limited only to the very small areas. At the time of reducing diversity algorithm usually found good solution for 25 and 30 cities. Interdependence of diversity and the number of individuals present in the population can also be observed. When the diversity goes down then the number of individuals oscillates with greater amplitude (see fig. 3). This is due to the fact that there are more very similar individuals (with the same or almost the same fitness) in the population, and as a consequence resources from the environment are more evenly distributed between agents. Dying off and reproduction of agents is realized in cyclic way.

GA with roulette selection shows tendency to significant loss of diversity during first 100 steps (see fig. 4). In the next steps, diversity keeps on safe stage. Contrary to the EMAS, increasing the number of cities does not result in differences in the diversity of population.

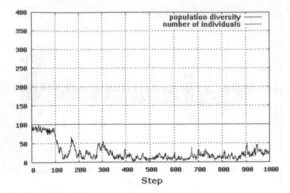

Fig. 4. Population diversity in GA (25 cities, average from 5 test). Genetic algorithm

Table 2. Working time of the algorithms for different problem sizes (average from 10 experiments)

	10	15	20	25	30	92
EMAS	10 sec	14 sec	21 sec	27 sec	37 sec	7 min
GA	8 sec	11 sec	14 sec	18 sec	24 sec	5 min

Step effectiveness. Step effectiveness shows how fast (in how many steps) algorithm finds the best solution. Experiential test of the stability of both techniques was carried out. Algorithm was recognized as stable when in most runs it was giving similar solutions. GA was finding 15% worse solution then recognized as global minimum. Some specific schema of this method was noticed. Algorithm quickly finds better solution in first steps (usually after 100 or 200 steps) and then gets stuck in the local minimum or during the rest time of experiment only slightly improves founded solution. Tests with 25 and 30 cities showed advantage of EMAS over GA. First method still returned good solutions, 10% worse than recognized as global minimum. GA went considerably poorer, and found 30% worse solution for 25 cities and 35% worse for 30 cities.

Time effectiveness. During experiments we also tried to compare GA and EMAS on the basis of time needed to find good solution. Unfortunately, it turned out to be the very hard task to prove superiority of some method over other looking only on the algorithms' working time. Presented techniques differ from each other on various levels and in so many details that comparison is not easy. First significant difference is population. GA has constant population size and it is equal to 100. In EMAS there are 5 islands with separated sub-populations. Initially, each island has 100 or 300 individuals. If we compare it with GA, the EMAS has five times more individuals in the initial step than GA. If we add the fact that in EMAS the number of individuals changes along the time, the difference in system load in the case of each algorithm is significant. These facts cause that EMAS should be, and in fact is, slower (see tab. 2). Other side effect of the fact that EMAS is more complex algorithm than GA is that it is much more configurable than GA, what gives more possibilities to tune the algorithm on the basis of the problem being solved.

4   Summary and Conclusions

In this paper agent-based evolutionary system for solving TSP was presented. There were two evolutionary algorithms compared during experiments: agent-based one and classical one. Experiments were carried out with the use of two problems with varying number of cities.

Results of preliminary experiments showed that EMAS obtained better results than GA, although the time needed to solve the problem was longer than in the case of GA. Longer run time mostly resulted from the fact that EMAS used larger initial population.

This was the first attempt to apply agent-based evolutionary algorithm to TSP— previously such approach was used mostly to solve multi-modal and multi-objective problems. Future research will certainly include further experimental verification of the proposed approach and the comparison to other "classical" and modern versions of evolutionary algorithms for solving TSP. Also, the application of EMAS to solving more complex problems like Vehicle Routing Problem (VRP) or VRP with Time Windows (VRPTW) is included in future research plans.

References

1. Agent-based evolution platform (jAgE), http://age.iisg.agh.edu.pl
2. Bäck, T., Fogel, D., Michalewicz, Z. (eds.): Handbook of Evolutionary Computation. IOP Publishing/ Oxford University Press (1997)
3. Dreżewski, R.: A model of co-evolution in multi-agent system. In: Mařík, V., Müller, J.P., Pěchouček, M. (eds.) CEEMAS 2003. LNCS, vol. 2691, pp. 314–323. Springer, Heidelberg (2003)
4. Michalewicz, Z.: Genetic Algorithms + Data Structures = Evolution Programs. Springer, Heidelberg (1996)
5. Moscato, P.: TSPBIB home page,
   http://www.ing.unlp.edu.ar/cetad/mos/TSPBIB_home.html

A Vehicle Routing Problem Solved by Agents

Mª Belén Vaquerizo García

Languages and Systems Area
Burgos University, Burgos
belvagar@ubu.es

Abstract. The main purpose of this study is to find out a good solution to the vehicle routing problem considering heterogeneous vehicles.

This problem tries to solve the generation of paths and the assignment of buses on these routes. The objective of this problem is to minimize the number of vehicles required and to maximize the number of demands transported.

This paper considers a Memetic Algorithm for the vehicle routing problem with heterogeneous fleet for any transport problem between many origins and many destinations. A Memetic Algorithm always maintains a population of different solutions to the problem, each of which operates as an agent. These agents interact between themselves within a framework of competition and cooperation.

Extensive computational tests on some instances taken from the literature reveal the effectiveness of the proposed algorithm.

Keywords: Vehicle Routing Problem, Heterogeneous Fleet, Evolutionary Algorithms, Memetic Algorithms, Agents.

1 Introduction

A vehicle routing problem with heterogeneous fleet is one of the most important problems in distribution and transportation. So, this generic problem and its practical extensions are discussed in great detail in the literature [2].

This paper considers a Memetic Algorithm for a vehicle routing problem with heterogeneous fleet for any transport problem between many origins and many destinations. The objective of this problem is to minimize the number of vehicles required and to maximize the number of demands transported.

A good example of this type of problem is the urban and interurban transport. In this problem is considered the heterogeneous fleet and networks symmetric or asymmetric. In the heterogeneous fleet, different types of vehicle with different capacities may be considered, and, furthermore, only in symmetric networks, the distance between two nodes is considered as the same in two directions.

More precisely, this problem can be defined as a problem of determining a set of routes and the best assignment of the available vehicles over these routes. And, besides, this problem must cover all passenger demand subject to the next constraints:

- Each vehicle performs just one route.
- For each route, the total number of passengers to be covered in each period of time considered, should not exceed the maximum capacity of the transport vehicle, so it is necessary to determine the frequency of vehicles to take this route in each time period.

E. Corchado et al. (Eds.): HAIS 2009, LNAI 5572, pp. 42–49, 2009.

- The total transportation cost should be minimized.
- Tours with setbacks are not allowed on routes.

According to this, the planning of this problem is as it is shown in the next figure.

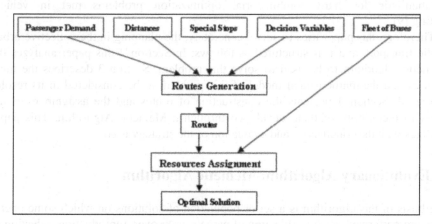

Fig. 1. Planning of an Urban Public Transport

By other hand, in the theory of computational complexity, this type of problem belongs to the class of NP-complete problems. Thus, it is assumed that there is no efficient algorithm for solving it, and in the worst case the running time for an algorithm for this problem depends exponentially on the number of stops, so that even some instances with only dozens of stops cannot be solved exactly.

In artificial intelligence, evolutionary algorithms are a large set of stochastic optimization algorithms inspired by biology [1]. They are based on biological processes that allow populations of organisms adapt to their environment. Also they can be considered for searching solutions.

Memetic Algorithms belong to the class of evolutionary algorithms that hybridize local search within a classical Genetic Algorithm framework. They are an optimization paradigm based on the systematic knowledge about the problem to be solved, and the combination of ideas taken from different metaheuristics, both based on population as based on local search. It is an Algorithm that can be viewed as a variant of Genetic Algorithms. Basically, they combine local search heuristics with crossover operators. For this reason, some researchers have viewed them as Hybrid Genetic Algorithms [12]. However, combinations with constructive heuristics or exact methods may also belong to this class of metaheuristics.

A Memetic Algorithm always maintains a population of different solutions to the problem, each of which operates as an agent. These agents interact between themselves within a framework of competition and cooperation. It is very similar to what occurs in nature between individuals of the same species. Each generation consists of updating the population of agents [3].

Otherwise, it is necessary to adapt the algorithm to the characteristics of the problem to be solved [7]. Thus, it needs to use a guide function that quantifies how good are the agents to can solve this problem.

In conclusion, the framework of this research is the development of and effective metaheuristic for hard combinatorial optimization problems met in vehicle routing [9].

Therefore, this paper tries to solve, concretely, the planning in the collective urban public transport, and it is structured as follows: In Section 2 this paper analyzes the Memetic Algorithm to be used to solve this problem. Section 3 describes the main features and the mathematical model of this problem to be considered in its resolution. And, Section 4 presents the construction of routes and the assignment of the buses to them, both of them developed through a Memetic Algorithm. This paper finalizes with the conclusions and evaluation of the strategy used.

2 Evolutionary Algorithm: Memetic Algorithm

The basis of this algorithm is a set or population of solutions on which some operations are performed, such as selection of parents, crossover, reproduction, selection of survivors or mutation, and each of the iterations in which this sequence is performed, are called generation [4].

Both, the selection and the replacement processes, are purely competitive, and in they only vary the distribution of the existing agents. The reproduction of agents is usually done through two operators: recombination and mutation.

Thus, the main feature of this algorithm is the use of crossover or recombination operator as the primary mechanism for search, and to build descendants of chromosomes that have the same features as the chromosomes that are crossed. Its usefulness is the assumption that different parts of the optimal solution can be discovered independently and, then, they can be combined to form better solutions. It ends when a certain number of generations without improvement are passed, or when other predetermined criteria are met.

By other hand, the basic idea of this algorithm is to incorporate as much knowledge to the problem domain as it is possible during the process of generating a new population. In order to this, this knowledge can be incorporated in the selection of the attributes of the parents to be transmitted to children, and in the selection of the attributes that are not parents to be transmitted to children. Although in a local search algorithm it was defined a neighbourhood for a single individual, in a population of individuals the vicinity of that population can be obtained through the composition of the individuals.

However, the objective is to build new solutions from the existing ones, and this can be done by identifying and combining attributes of the current solutions.

Otherwise, this algorithm requires two things to be defined: A genetic representation of the solution domain, and a fitness function to evaluate the solution domain.

According to this, the framework of the proposed Memetic Algorithm used in this work is the next [11]:

```
Procedure MA;
{
    for (j=1;j++; j<=popsize)
    {
        i=GenerateSolution();
        i=Local-Search(i);
        add individual i to P;
    }
    while terminate!=true
    {
        for (i=1;i++;i<= #recombination)
        {
            select two parent iₐ,iᵦ ∈ P Randomly;
            i_c =Recombine(iₐ,iᵦ);
            i_c =Local-Search(i_c);
            add individual i_c to P;
        }
        for (i=1;i++;i<= #mutations)
        {
            select an individual i∈ P Randomly;
            iₘ=Mutate(i);
            iₘ=Local-Search(iₘ);
            add individual iₘ to P;
        }
        P=select(P);
        if (P converged) P=mutateAndLS(P);
    }
}
```

Fig. 2. Framework of the Memetic Algorithm used

3   The Vehicle Routing Problem. Public Transport Routes

As mentioned above, this work is done to try to improve the urban public transport in any mid-size city, and, for that, previously has to be analyzed the current situation in that city, in order to obtain a detailed knowledge of its structure and function with regard to the following points:

- The number of lines and characteristics of their stops.
- The frequency of each line.
- The estimated speed at which buses can go through the city.
- The planning schedules of the lines.
- The estimation of the passengers getting on and off at each stop and in each time period considered within the timeframe set.
- The number of buses available. Different types of buses available and their restrictions.
- Special passengers demand in specific stops and in specific periods of time.
- The network topography of the city in order to know the limitations provided by the city streets as a cut street, a narrow street, a one-way street, etc.

Therefore, the identification of all these data (input data and decision variables) will allow subsequently to formalize the problem to solve.

As discussed before, this problem is considered a complex problem of combinatorial optimization, which belongs to the class of problems known as NP-hard, those whose exact resolution requires specific models and algorithms with great difficulty and much processing time to obtain the optimal solution. For this reason, when a problem about routes is discussed, it is often addressed through the formulation of models whose decisions can be obtained through algorithms to solve instances of suitable size in an optimal way and in a reduced time.

By other hand, it should be noted that the evaluation of the solution implies calculating the objective function of the model, and that, moreover, it implies to choose a resolution method, in which there must be a commitment between efficiency and quality of the solution obtained.

Thus, the mathematical model, that represents the problem studied, can be seen as a single objective function (maximize the level of service, considering the number of buses available as a restriction). In this case, it should be noted that the approach to maximize the level of service is considered as minimizing travel time and waiting time of passengers at stops. In the next table, the mathematical model, for the geographical network of a city to be considered, is presented detailing the notation as follows:

Table 1. Mathematical Model for the Planning of a Urban Public Transport

$G = (A, N)$ where $N = \{1, 2,...,n\}$ *(set of nodes)* and $A = \{a1, a2 ...,am\}$ *(set of arcs)*
$P \subset N, P = \{i1, i2,..., ip\}$ *(set of stops)* $\forall\ i, i' \in P$ *demii' = Passenger demand from stop i to stop i'*
$RT = \{RR1, RR2,... RRs, ...,RRntrc\}$ *set of Full Routes*
$\sigma(i, i') = \{Routes\ with\ this\ pair\ of\ stops\ in\ their\ path\}$ Xs = Number of vehicles assigned to the route RRs
tpii',s = Travel time from the stop *i* to the stop *i'* in the Route *RRs*
ntv = Number of different types of vehicles $\lambda(l)$ = {*Routes covered by vehicles of the type l*}
freqs = *number of times you can start the route s in that period of time*
The Objective Function would be to Minimize Waiting Times and Travel Times, with the restrictions (3, 4) :
(1)   Waiting Times at Stops:
$$\sum_{i,i' \in P} \frac{60}{\sum_{s \in \sigma(i,i')} freq_s} \cdot dem_{i,i'}$$
(2)   Travel Times:
$$\sum_{i,i' \in P} \left[\left(\sum_{s \in \sigma(i,i')} tp_{ii',s} \cdot freq_s \right) \Big/ \sum_{s \in \sigma(i,i')} freq_s \right] \cdot dem_{ii'}$$
(3) Number of Vehicles for each type: $\forall\ l = 1... ntv$ $\sum_{s \in \lambda(l)} X_s \leq nv_l$
(4) All demand covered: $\forall\ i, i' \in P\ /\ demi,i' > 0$ $\sum_{s \in \sigma(i,i')} X_s > 0$

4   Solving the Problem Using a Memetic Algorithm

This work has implemented the following general layout of a Memetic Algorithm: Once the genetic representation and the fitness function are defined, the algorithm proceeds to initialize a population of solutions randomly, then it is improved through a repetitive application of mutation, crossover, inversion and selection operators [5,6,10].

The evolution starts from a population of randomly generated individuals and happens in generations. In each generation, the fitness of every individual in the population is evaluated, multiple individuals are stochastically selected from the current population (based on their fitness), and modified (recombined and possibly randomly mutated) to form a new population. The new population is then used in the next iteration of the algorithm. Thus, each generation consists on updating the population of agents, using for it a new population obtained by recombination of the characteristics of some selected agents [13]. The selection operation choose a selection of the best agents in the current population, it requires the use of a function for measuring the quality of each agent in the resolution of the problem. Then, the competition between agents is through the selection and update operations.

This algorithm terminates when either a maximum number of generations has been produced, or a satisfactory fitness level has been reached for the population. If the algorithm has terminated due to a maximum number of generations, a satisfactory solution may or may not have been reached.

By other hand, is very important to fix some features in the representation of the objective functions, which are efficiently exploited by a memetic approach. The first element to be determined is the representation of solutions to be used. It is important to clarify here that the representation should not be built as merely coding, for which the relevant considerations are related to memory consumption, complexity handling, etc. On the contrary, the representation refers to the abstract formulation of the solutions from the perspective of the algorithm. In this case, arrays of different structures have been used, and in this problem it has been determined that the most appropriate way to store the information have been by adjacency.

Besides, the fitness function is defined over the genetic representation and measures the quality of the represented solution. Candidate solutions to the optimization problem play the role of individuals in a population, and the fitness function determines the environment within which the solutions are considered valid.

The computational evaluation was performed on little instances of about 60 stops. Thus, considering the variables and restrictions listed before and for the following network, as an example in Figure 3, the system would provide the next designs of routes shown below in Table 2.

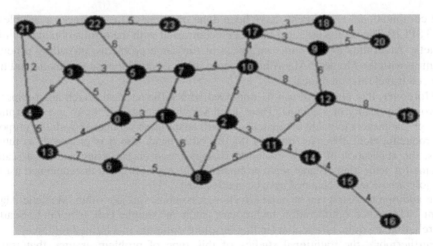

Fig. 3. Topographical network considered as example. Stops, arcs and distances.

Table 2. A solution about generation of routes and their buses using a Memetic Algorithm

Routes	Path										Demand covered	Travel Time	Buses
1:													3,7
→:		20	18	17	10	7	5	0	13		41	26	
←:	13	0	1	6	8	2	10	17	18	20	81	37	
2:													4,6,8
→:		16	15	14	11	2	1	7	5	22	71	31	
←:	22	21	3	0	1	2	11	14	15	16	107	34	
3:													1
→:		4	13	6	1	7	10	17	9		43	30	
←:		9	17	10	7	5	3	4			23	22	
4:													5
→:		4	3		5	7	10	12	19		31	31	
←:		19	12	10	7	5	3	4			29	31	
5:													2
→:		8	11	12	9	17	23	22			37	31	
←:			22	21	3	0	1	8			22	21	

These feasible routes satisfy specific conditions based on the number of stops, total distance, maximum time travel, passenger demand satisfied, etc.

To obtain this solution, it has been necessary to define in the algorithm the following points, among others: Number of individuals to cross, how selecting the parents, the type of selection, the type of combination, the type of crossing, the number of children.

Moreover, in this problem, routes are going to be exchanged and the individuals which are going to be crossed are solutions of the current population. And new individuals called children are obtained by crossing. The value of the solution, it is used to sort the solutions in the population to be the best solution which holds less value. This value is based on the ratio between the demand and travel time of the solution.

5 Conclusions and Future Lines

The computational experiments have been done on some of the instances available in the TSPLIB.The results obtained have been contrasted with the results obtained with a Genetic Algorithm for the same instances of the same problems, providing a better solution with the Memetic Algorithm because it has covered a greater demand and has reduced travel time on routes.

Moreover, this algorithm can be improved with different local search algorithms, to improve the quality of solutions. These local search methods will make use of explicit strategy parameters to guide the search, and will adapt these parameters with the purpose of producing more effective solutions [8]. By other hand, also it is proposed, as future lines, the application of a new evolution strategy based on clustering and local search scheme for some kind of large-scale problems of this type [14], and the combined use of two jobs whose results obtained have been effective [2,11].

In addition to this, it can be quite effective to combine this algorithm, Memetic Algorithm, with other optimization techniques, such as simple Hill Climbing, Scatter Search, that are quite efficient at finding absolute optimum in a limited region.

Furthermore, the traditional studies of this type of problem assume that travel speed of the vehicle is constant, which is an approximation of the real-world conditions in order to simplify the model of computing, but in the real-world this assumption is not very reasonable, especially in urban city conditions. By this reason, it is ssugested in a future work to consider the study on time-dependent vehicle routing problems in regard to congestion during peak time in urban areas, accidents like vehicle breakdown, severe weather factors and so on.

Finally, among the main difficulties of this work, to get the service previously established, is the need to know the number of people using each line for each time period. For these reasons, for a better treatment of this problem, it is suggested in a future work the use of the Fuzzy Sets Theory, and the combination of a Memetic Algorithm with others efficient algorithms such as mentioned above.

References

1. Alba, E., Cotta, C., Herrera, F.: Computación Evolutiva. Inteligencia Artificial, Revista Iberoamericana de Inteligencia Artificial (1998)
2. Buriol, L., Franca, P.M., Moscato, P.: A new memetic algorithm for the asymmetric traveling salesman problem. Journal of Heuristics 10 (2004)
3. Carrascosa, C., Bajo, J., Julian, V., Corchado, J.M.L., Botti, V.J.: Hybrid multi-agent architecture as a real-time problem-solving model. Expert Syst. Appl. (ESWA) 34(1), 2–17 (2008)
4. Cotta, C.: Una visión general de los algoritmos meméticos. Rect@ 3, 139–166 (2007)
5. García, S., Cano, J.R., Herrera, J.R.: A memetic algorithm for evolutionary prototype selection: A scaling up approach. Pattern Recognition (PR) 41(8), 2693–2709 (2008)
6. Gutin, G., Karapetyan, D.: A Memetic Algorithm for the Generalized Traveling Salesman Problem. CoRR abs/0804.0722 (2008)
7. Handa, H., Chapman, L., Yao, X.: Robust Salting Route Optimization Using Evolutionary Algorithms. Evolutionary Computation in Dynamic and Uncertain Environments (2007)
8. Molina, D., Lozano, M., García-Martínez, C., Herrera, C.: Memetic Algorithm for Intense Local Search Methods Using Local Search Chains. In: Hybrid Metaheuristics 2008, pp. 58–71 (2008)
9. Prins, C.: A simple and effective evolutionary algorithm for the vehicle routing problem. Computers and Operations Research (2004)
10. Rodrigues, A.M., Soeiro Ferreira, J.: Solving the rural postman problem by memetic algorithms. In: de Sousa, J.P. (ed.) Proceedings of the 4th Metaheuristic International Conference (MIC 2001), Porto, Portugal (2001)
11. Tavakkoli-Moghaddam, R., Saremi, A.R., Ziaee, M.S.: A memetic algorithm for a vehicle routing problem with backhauls. Applied Mathematics and Computation 181 (2006)
12. Wei, P., Cheng, L.X.: A hybrid genetic algorithm for function optimization. Journal of Software 10(8), 819–823 (1999)
13. Jin, X., Abdulrab, H., Itmi, M.: A multi-agent based model for urban demand-responsive passenger transport services. In: IJCNN 2008, pp. 3668–3675 (2008)
14. Wang, Y., Qin, J.: A Memetic-Clustering-Based Evolution Strategy for Traveling Salesman Problems. In: Yao, J., Lingras, P., Wu, W.-Z., Szczuka, M.S., Cercone, N.J., Ślezak, D. (eds.) RSKT 2007. LNCS, vol. 4481, pp. 260–266. Springer, Heidelberg (2007)

MACSDE: Multi-Agent Contingency Response System for Dynamic Environments

Aitor Mata[1], Belén Pérez[1], Angélica González[1], Bruno Baruque[2], and Emilio Corchado[2]

[1] University of Salamanca, Spain
aitor@usal.es, lancho@usal.es, angelica@usal.es
[2] University of Burgos, Spain
bbaruque@ubu.es, escorchado@ubu.es

Abstract. Dynamic environments represent a quite complex domain, where the information available changes continuously. In this paper, a contingency response system for dynamic environments called MACSDE is presented. The explained system allows the introduction of information, the monitoring of the process and the generation of predictions. The system makes use of a Case-Based Reasoning system which generates predictions using previously gathered information. It employs a distributed multi-agent architecture so that the main components of the system can be accessed remotely. Therefore, all functionalities can communicate in a distributed way, even from mobile devices. The core of the system is a group of deliberative agents acting as controllers and administrators for all functionalities. The system explained includes a novel network for data classification and retrieval. Such network works as a summarization algorithm for the results of an ensemble of Self-Organizing Maps. The presented system has been tested with data related with oil spills and forest fire, obtaining quite hopeful results.

Keywords: Dynamic environments; Case-Based Reasoning; oil spill; forest fire; Self Organizing Memory; summarization.

1 Introduction

Dynamic systems represent a quite difficult knowledge field where new techniques and architectures can be applied to check their useful applications. In this paper, a multi-agent architecture is applied to natural dynamic systems, which use a large number of parameters and contain a high level of system variability.

This paper presents the MACSDE architecture. It is a new approach to open systems, where the different parts of the applications that are implemented communicate with each other by an internal architecture that allows different elements to connect to the system or to eventually disconnect from the system if they are not going to be operative for a time.

The system explained in this work implements the Case-Based Reasoning methodology as the core of the system, and generated the predictions for solving the problems presented. The current system has been expanded from a previously local

E. Corchado et al. (Eds.): HAIS 2009, LNAI 5572, pp. 50–59, 2009.

version to one that is distributed, where different users with different roles communicate with each other through the MACSDE architecture, and use the CBR inner methodology to solve problems and generate predictions. The main phases of the CBR cycle are implemented in this application since services are already part of the MACSDE structure.

This paper presents the problems that the architecture faces with regards to predictions in natural environments specifically oil spills and forest fires. In both cases, the system must generate predictions in order to know, first of all, where the oil slicks that generated after a spill will be, and secondly, where the fire will be at a specific time.

The remainder of this paper is structured as follows: section 2 provides a brief explanation of the two basic methodologies used in MACSDE; section 3 presents the proposed model as well as a description of the services offered; section 4 shows the results obtained with MACSDE, being applied to oil spills and forest fires, and finally section 5 includes the conclusions obtained and future lines of work.

2 Combining a Multi-agent Architecture and Case-Based Reasoning Systems

Agents and multi-agent systems have been successfully applied to several scenarios, such as education, culture, entertainment, medicine, robotics, etc. [1]. Agents have a set of characteristics, such as autonomy, reasoning, reactivity, social abilities, proactivity, mobility, organization, etc. which allow them to cover several needs for developing contingency response systems [2].

The agents' characteristics make them appropriate for developing dynamic and distributed systems, as they possess the capability of adapting themselves to the users and environmental characteristics [3]. In addition, the continuous advances in mobile computing make it possible to obtain information about the environment and also to react physically over it in more innovative ways. The agents in MACSDE multi-agent system are based on the deliberative (Belief, Desire, Intention - BDI) model [4], where the agents' internal structure and capabilities are based on mental aptitudes, using beliefs, desires and intentions for solving problems. However, modern developments need higher adaptation, learning and autonomy levels than pure BDI model [4]. This can be achieved by modeling the agents' characteristics for providing them with mechanisms that allow them solving complex problems and achieving autonomous learning. Some of these mechanisms are Case-Based Reasoning (CBR) systems [5], where problems are solved by using solutions to similar past problems [1]. The origin of the CBR methodology is in knowledge based systems and has been used on several scenarios such as health care, e-Learning, ubiquitous computing, oceanography, etc. [6]. Solutions are stored into a case memory which the system can consult in order to find better solutions for new problems. Deliberative agents can make use of these systems to learn from past experiences and to adapt their behavior according each situation.

The main element of CBR is the case base which is a structure that stores problems, elements (cases) and its solutions. A case base can be visualized as a database where a collection of problems is stored keeping a relationship with the solutions to every problem stored. This gives the system the ability to generalize in order to solve new problems. In the case of the present work the case base is implemented by means of a topology preserving meta-algorithm called Weighted Voting Superposition of Self-Organizing Maps (WeVoS-SOM). The learning capabilities of CBR are due to its own structure composed of four main phases [5]: retrieval, reuse, revision and retention.

The retrieve phase consists on finding the most similar cases to the current problem from the case base. Once a series of cases are extracted they can be reused. The reuse phase adapts the selected cases for solving the current problem through a new solution. The revision phase revises the solution to check if it is an adequate solution to the current problem. If the solution is confirmed, then it is retained by means of the retain phase. This new solution could eventually serve as a solution for future problems. In most cases, CBR should not been used alone but combined with artificial intelligence techniques on each phase. For example, Growing Cell Structures has been used with CBR to automatically create the intern structure of the case base from existing data and it has been combined with multi-agent applications [7] to improve its results. ART-Kohonen neural networks, artificial neural networks, genetic algorithms and fuzzy logic [8] have been used to complement the capabilities of CBR systems. These techniques enhance the way CBR systems create better solutions.

3 Presenting a New Multi-agent Contingency Response System for Dynamic Environments

Cases are the key to obtain solutions to future problems through a CBR system. The functionalities of MACSDE (Multi-Agent Contingency response System for Dynamic Environments) can be accessed using different interfaces executed on PCs or PDAs (Personal Digital Assistant). Users can interact with the system by introducing data, requesting a prediction or revising a solution generated (i.e. prediction). The interface agents communicate with the services through the agents' platform and vice versa. The interface agents perform all the different functionalities which users can make use for interacting with MACSDE. The different phases of the CBR system have been modeled as services, so each phase can be requested independently. For example, one user may only introduce information in the system (e.g. a new case), while another user could request a new prediction.

All information is stored in the case base and MACSDE is ready to predict future situations. A problem situation must be introduced in the system for generating a prediction. Then, the most similar cases to the current situation are retrieved from the case base. Once a collection of cases are chosen from the case base, they must be used for generating a new solution to the current problem. Growing Radial Basis Functions (RBF) Networks [9] are used in MACSDE for combining the chosen cases in order to obtain the new solution.

MACSDE determines future probabilities in a certain area. It divides the area to be analyzed in squares of approximately half a degree side for generating a new prediction. Then, the system determines the amount of slicks in each square. The squares are colored with different gradation depending on the probability of finding the analyzed element.

Fig. 1 shows the structure of MACSDE, where the different elements of the system are related and where the communication paths between the elements are also shown.

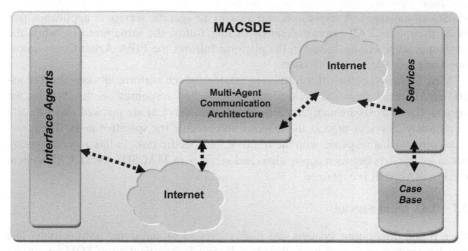

Fig. 1. MACSDE Architecture

There are four basic blocks in MACSDE: Applications, Services, Agent Platform and Communication Protocol. These blocks provide all the system functionalities:

Interface Agents. These represent all the programs that users can use to exploit the system functionalities. Applications are dynamic, reacting differently according to the particular situations and the services invoked. They can be executed locally or remotely, even on mobile devices with limited processing capabilities, because computing tasks are largely delegated to the agents and services.

Services. These represent the activities that the architecture offers. They are the bulk of the functionalities of the system at the processing, delivery and information acquisition levels. Services are designed to be invoked locally or remotely. Services can be organized as local services, web services, GRID services, or even as individual stand alone services. MACSDE has a flexible and scalable directory of services, so they can be invoked, modified, added, or eliminated dynamically and on demand. It is absolutely necessary that all services follow a communication protocol to interact with the rest of the components. Services make use of the information stored in the case base to solve the proposed problems.

Multi-Agent Communication Architecture. This is the core of the system, integrating a set of agents, each one with special characteristics and behavior. An important feature in this architecture is that the agents act as controllers and administrators for all applications and services, managing the adequate functioning of the system, from services, applications, communication and performance to reasoning and decision-making. In MACSDE, services are managed and coordinated by deliberative BDI agents. The agents modify their behavior according to the users' preferences, the knowledge acquired from previous interactions, as well as the choices available to respond to a given situation. The communication protocol allows applications and services to communicate directly with the Agents Platform. This protocol is based on SOAP specification to capture all messages between the platform and the services and applications [10]. Services and applications communicate with the *Agents Platform* via SOAP messages. A response is sent back to the specific service or application that made the request. All external communications follow the same protocol, while the communication among agents in the platform follows the FIPA Agent Communication Language (ACL) specification.

MACSDE also defines three different services which perform all tasks that the users may demand from the system. All requests and responses are handled by the agents. The requests are analyzed and the specified services are invoked either locally or remotely. Services process the requests and execute the specified tasks. Then, services send back a response with the result of the specific task. In this way, the agents act as interpreters between applications and services in MACSDE. Next, CBR system used in MACSDE is explained.

3.1 Data Input Service

Case-Based Reasoning systems are highly dependent on stored information. The novel algorithm presented here, *Weighted Voting Summarization of SOM ensembles (WeVoS-SOM)* [11, 12] is used to organize the data that is accumulated in the case base. It is also used to recover the most similar cases to the proposed problem.

The main objective of the WeVoS-SOM is to generate a final map processing several other similar maps unit by unit. Instead of trying to obtain the best position for the units of a single map trained over a single dataset, it aims to generate several maps over different parts of the dataset. Then, it obtains a final summarized map by calculating by consensus which is the best set of characteristics vector for each unit position in the map. To do this calculation, first this meta-algorithm must obtain the "quality" [13] of every unit that composes each map, so that it can relay in some kind of informed resolution for the fusion of neurons.

The final map obtained is generated unit by unit. The units of the final map are first initialized by determining their centroids in the same position of the map grid in each of the trained maps. Afterwards, the final position of that unit is recalculated using data related with the unit in that same position in every of the maps of the ensemble. For each unit, a sort of voting process is carried out as shown in Eq. 1:

$$V_{p,m} = \frac{\sum b_{pm}}{\sum_1^M b_p} \cdot \frac{q_{pm}}{\sum_1^M q_p} \tag{1}$$

where, V_{pm} is the weight of the vote for the unit included in map m of the ensemble, in its position p; M is the total number of maps in the ensemble; b_{pm} is the binary vector used for marking the dataset entries recognized by the unit in position p of map m; and, q_{pm} is the value of the desired quality measure for the unit in position p of map m.

The final map is fed with the weights of the units as it is done with data inputs during the training phase of a SOM [14], considering the "homologous" unit in the final map as the BMU. The weights of the final unit will be updated towards the weights of the composing unit. The difference of the updating performed for each "homologous" unit in the composing maps depends on the quality measure calculated for each unit. The higher the quality (or the lowest error) of the unit of the composing map, the stronger the unit of the summary map will be updated towards the weights of that unit. The summarization algorithm will consider the weights of a composing unit "more suitable" to be the weights of the unit in the final map according to both the number of inputs recognized and the quality of adaptation of the unit (Eq. 1). With this new approach it is expected to obtain more faithful maps to the inner structure of the dataset.

3.2 Prediction Generation Service

When a prediction is requested by a user, the system starts recovering from the case base the most similar cases to the problem proposed. Then, it creates a prediction using artificial neural networks. Once the most similar cases are recovered from the case base, they are used to generate the solution. Growing RBF networks [15] are used to obtain the predicted future values corresponding to the proposed problem. This adaptation of the RBF networks allows the system to grow during training gradually increasing the number of elements (prototypes) which play the role of the centers of the radial basis functions. The creation of the Growing RBF must be made automatically which implies an adaptation of the original GRBF system. The error for every pattern is defined by (Eq. 2).

$$e = \frac{1}{n} \sum_{k=1}^{p} \left\| t_{ik} - y_{ik} \right\| \tag{2}$$

where t_{ik} is the desired value of the k_{th} output unit of the i_{th} training pattern, y_{ik} the actual values of the k_{th} output unit of the i_{th} training pattern.

Once the GRBF network is created, it is used to generate the solution to the proposed problem. The solution proposed is the output of the GRBF network created with the retrieved cases. The GRBF network receives, as input, the values stored in the case base. With those values, the network generates the proposed solution, using only the data recovered from the case base in previous phases.

3.3 Revision Service

After generating a prediction, the system needs to validate its correction. MACSDE can also query an expert user to confirm the automatic revision previously done. The system also provides an automatic method of revision that must be also checked by an expert user which confirms the automatic revision.

Explanations are a recent revision methodology used to check the correction of the solutions proposed by CBR systems [16]. Explanations are a kind of justification of the solution generated by the system. To obtain a justification to the given solution, the cases selected from the case base are used again. As explained before, a relationship between a case and its future situation can be established. If both the situations defined by a case and the future situation of that case are considered as two vectors, a distance between them can be defined, calculating the evolution of the situation in the considered conditions. That distance is calculated for all the cases retrieved from the case base as similar to the problem to be solved. If the distance between the proposed problem and the solution given is not greater than the average distances obtained from the selected cases, then the solution is a good one, according to the structure of the case base. If the proposed prediction is accepted, it is considered as a good solution to the problem and can be stored in the case base in order to solve new problems. It will have the same category as the historical data previously stored in the system.

4 Results

MACSDE has been tested in two different fields, both related with natural dynamic environments. It has been checked with a subset of the available data that has not been previously used in the training phase. The predicted situation was contrasted with the actual future situation as it was known (historical data was used to train the system and also to test its correction). The proposed solution was, in most of the variables, close to 90% of accuracy.

Table 1. Percentage of good predictions obtained with different techniques

Number of cases	RBF		CBR		RBF + CBR		MACSDE	
	Oil spill	Forest fires	Oil spill	Forest fires	Oil spill	Forest fires	Oil spill	Forest fires
500	43 %	38 %	40 %	41 %	43 %	44 %	46 %	47 %
1000	47 %	44 %	48 %	47 %	52 %	51 %	62 %	66 %
3000	54 %	51 %	56 %	53 %	64 %	65 %	78 %	73 %
5000	61 %	58 %	63 %	61 %	74 %	70 %	87 %	84 %

Table 1 shows a summary of the obtained results. In this table different techniques are compared. The evolution of the results is shown along with the augmentation of the number of cases stored in the case base. All the techniques analyzed improve their results at the same time the number of stored cases is increased. The solution proposed do not generate a trajectory, but a series of probabilities in different areas, what is far more similar to the real behaviour of the oil slicks. The left column of each technique is referred to the oil spill problem, while the right column is referred to the forest fires.

The *"RBF"* column represents a simple Radial Basis Function Network that is trained with all the data available. The network gives an output that is considered a solution to the problem. The *"CBR"* column represents a pure CBR system, with no additional techniques included. The cases are stored in the case bases and recovered considering the Euclidean distance. The most similar cases are selected and after applying a weighted mean depending on the similarity, a solution is proposed. It is a *mathematical* CBR. The *"RBF + CBR"* column corresponds to the possibility of using a RBF system combined with CBR. The recovery from the CBR is done using the Manhattan distance to determine the closest cases to the introduced problem. The RBF network works in the reuse phase, adapting the selected cases to obtain the new solution. The results of the *"RBF+CBR"* column are, normally, better than those of the *"CBR"*, mainly because of the elimination of useless data to generate the solution. Finally, the *"MACSDE"* column shows the results obtained by the proposed system, being better than the three previous solutions analyzed.

Several tests have been done to compare the overall performance of MACSDE. The tests consisted of a set of requests delivered to the Prediction Generation Service (PGS) which in turn had to generate solutions for each problem. There were 50 different data sets, each one with 10 different parameters. The data sets were introduced into the PGS through a remote PC running multiple instances of the Prediction Agent. The data sets were divided in five test groups with 1, 5, 10, 20 and 50 data sets respectively. There was one Prediction Agent for each test group. 30 runs for each test group were performed. First, all tests were performed with only one Prediction Service running in the same workstation on which the system was running. Then, five Prediction Services were replicated also in the same workstation. For every new test, the case base of the PGS was deleted in order to avoid a learning capability, thus requiring the service to accomplish the entire prediction process.

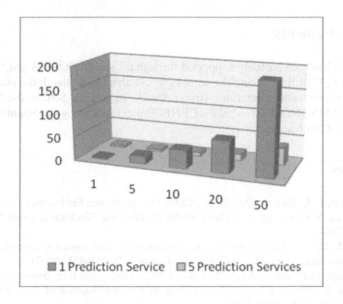

Fig. 2. Average time needed to generate all solutions

Fig. 2 shows the average time needed by MACSDE for generating all solutions for each test group. The time exponentially increases when there is only one PGS running. This is because the service must finish a request to start the next one. So, for the last test group (50 data sets) the service was overcharged. On the other hand, with five replicated services, the system can distribute the requests among these services and optimize the overall performance. The system performed slightly faster when processing a single request, but the performance was constantly reduced when more requests were sent to the service.

5   Conclusions and Future Work

As conclusions, a novel hybrid model called MACSDE, which aim is to provide the users with predictions to asses their actions regarding the contingency response in dynamic environments, is presented in this work.

It has been tested under the frame of two different real cases and compared with other previous models. From this comparison it can be concluded that the novel hybrid model presented outperforms previous simpler models when used for this same purpose.

Future work will be focused on the independent enhancement of different parts of the system. For example, for the case base, other topology preserving models can be used in combination with the ensemble meta-algorithm to improve the organization of the different cases. Other techniques in the field of CBRs and multi-agent systems will also be integrated into the system to try to improve the currently obtained results.

Generalization is also a future objective of this model, which aim is to be applied to different knowledge fields even if tuning modifications are needed to adapt the model to the new circumstances.

Acknowledgements

This research has been partially supported through project BU006A08 and SA071A08 by the Junta de Castilla y León. The authors would also like to thank the manufacturer of components for vehicle interiors, Grupo Antolin Ingeniería, S.A. in the framework of the project MAGNO 2008 – 1028 – CENIT Project funded by the Spanish Ministry of Science and Innovation.

References

[1] Corchado, J.M., Bajo, J., De Paz, Y., Tapia, D.I.: Intelligent Environment for Monitoring Alzheimer Patients, Agent Technology for Health Care. Decision Support Systems (in press) (2008)

[2] Yang, J., Luo, Z.: Coalition formation mechanism in multi-agent systems based on genetic algorithms. Applied Soft Computing Journal 7(2), 561–568 (2007)

[3] Jayaputera, G.T., Zaslavsky, A.B., Loke, S.W.: Enabling run-time composition and support for heterogeneous pervasive multi-agent systems. Journal of Systems and Software 80(12), 2039–2062 (2007)

[4] Bratman, M.E., Israel, D., Pollack, M.E.: Plans and resource-bounded practical reasoning. Computational Intelligence 4, 349–355 (1988)

[5] Aamodt, A., Plaza, E.: Case-Based Reasoning: foundational Issues, Methodological Variations, and System Approaches. AI Communications 7(1), 39–59 (1994)

[6] Corchado, J.M., Lees, B., Aiken, J.: Hybrid instance-based system for predicting ocean temperatures. International Journal of Computational Intelligence and Applications 1(1), 35–52 (2001)

[7] Carrascosa, C., Bajo, J., Julian, V., Corchado, J.M., et al.: Hybrid multi-agent architecture as a real-time problem-solving model. Expert Systems With Applications 34(1), 2–17 (2007)

[8] Fdez-Riverola, F., Iglesias, E.L., Díaz, F., Méndez, J.R., et al.: Applying lazy learning algorithms to tackle concept drift in spam filtering. Expert Systems With Applications 33(1), 36–48 (2007)

[9] Karayiannis, N.B., Mi, G.W.: Growing radial basis neural networks: merging supervised andunsupervised learning with network growth techniques. IEEE Transactions on Neural Networks 8(6), 1492–1506 (1997)

[10] Cerami, E.: Web Services Essentials Distributed Applications with XML-RPC, SOAP, UDDI & WSDL. O'Reilly & Associates, Inc. (2002)
[11] Baruque, B., Corchado, E., Rovira, J., González, J.: Application of Topology Preserving Ensembles for Sensory Assessment in the Food Industry. In: Intelligent Data Engineering and Automated Learning - IDEAL 2008, pp. 491–497 (2008)
[12] Corchado, E., Baruque, B., Yin, H.: Boosting Unsupervised Competitive Learning Ensembles. In: de Sá, J.M., Alexandre, L.A., Duch, W., Mandic, D.P. (eds.) ICANN 2007. LNCS, vol. 4668, pp. 339–348. Springer, Heidelberg (2007)
[13] Pölzlbauer, G.: Survey and Comparison of Quality Measures for Self-Organizing Maps. In: Paralic, J., Pölzlbauer, G., Rauber, A. (eds.) Fifth Workshop on Data Analysis (WDA 2004), pp. 67–82. Elfa Academic Press, London (2004)
[14] Kohonen, T.: The Self-Organizing Map. Neurocomputing 21, 1–6 (1998)
[15] Ros, F., Pintore, M., Chrétien, J.R.: Automatic design of growing radial basis function neural networks based on neighboorhood concepts. Chemometrics and Intelligent Laboratory Systems 87(2), 231–240 (2007)
[16] Plaza, E., Armengol, E., Ontañón, S.: The Explanatory Power of Symbolic Similarity in Case-Based Reasoning. Artificial Intelligence Review 24(2), 145–161 (2005)

Measuring and Visualising Similarity of Customer Satisfaction Profiles for Different Customer Segments

Frank Klawonn[1], Detlef D. Nauck[2],
and Katharina Tschumitschew[1]

[1] Department of Computer Science University of Applied Sciences BS/WF
Salzdahlumer Str. 46/48, D-38302 Wolfenbuettel, Germany
[2] BT Group, Chief Technology Office, Research and Venturing Intelligent Systems
Research Centre Adastral Park, Orion Building pp1/12, Ipswich IP5 3RE, UK

Abstract. Questionnaires are a common tool to gain insight to customer satisfaction. The data available from such questionnaires is an important source of information for a company to judge and improve its performance in order to achieve maximum customer satisfaction. Here, we are interested in finding out, how much individual customer segments are similar or differ w.r.t. to their satisfaction profiles. We propose a hybrid approach using measures for the similarity of satisfaction profiles based on principles from statistics in combination with visualization techniques. The applicability and benefit of our approach is demonstrated on the basis of real-world customer data.

Keywords: customer satisfaction; rank correlation; MDS; cluster analysis.

1 Introduction

Customer satisfaction is a key issue for a company to maintain and improve its position in the market. Questionnaires filled in by customers via telephone interviews, direct interviews, mail or the Internet provide a very important source of information on customer satisfaction. Such questionnaires usually contain questions concerning different aspects of customer satisfaction as well as other questions regarding other general or specific information like age of the customer or which item or service they have purchased from the company. There are, of course, many ways to analyse the data available from such questionnaires, depending on the kind of question or information the company is interested in [3,4,8]. This paper focuses on the following aspect. Customers are usually grouped into different customer segments. Customer satisfaction might be similar or differ among these customer segments. When significant differences among customer segments can be identified, the company can use this information to take appropriate actions in order to improve the customer satisfaction, especially for those customer segments where the satisfaction is lower. Differences found in customer satisfaction can also help to estimate the impact of possible company campaigns or actions on the customer segments.

E. Corchado et al. (Eds.): HAIS 2009, LNAI 5572, pp. 60–67, 2009.

2   Problem Description

Here, the only questions of interest within a questionnaire are those asking customers directly about their satisfaction. Nevertheless, besides the overall satisfaction a questionnaire will usually contain questions regarding customer satisfaction with respect to different criteria, for instance concerning different services, quality of products, prices or information provided by the company. The answers to a question regarding customer satisfaction are usually limited to a specific ordinal scale with varying granularity for different criteria. In the simplest case the ordinal scale might only contain two answers, i.e. "Are you satisfied with . . . ? Yes/No". But in most cases a set of more refined answers is provided, for instance, extremely satisfied, very satisfied, satisfied, . . . , extremely dissatisfied.

In addition to the ordinal scale that allows specifying the degree of customer satisfaction explicitly, there are usually additional answers like "don't know", "not applicable" or "refuse to answer". It can also happen that a customer does not answer a question. Here, all these cases that do not provide an explicit evaluation of the customer satisfaction, are considered in the same way as a null answer. In the following, we will refer to a null answer as a missing value.

Finding similarities and especially differences concerning customer satisfaction for different customer segments is the focus of this paper. We assume that the customer segments are given. The segmentation might depend on the customer's age, income, area of residence and other aspects. How the customer segmentation is defined exactly is not relevant for this paper.

We assume that altogether a number of q different customer satisfaction questions are considered. Each question has an individual ordinal scale of possible answers plus a specific category for "missing value" as described above. We also consider a number of c different customer segments. We assume that statistics for each customer segment and each customer satisfaction question are available. This means for each customer segment and each question, we know either the absolute or the relative frequencies of the possible answers, including missing values, to the question. It is not required that exactly the same questionnaire was presented to customers from different customer segments. It is only necessary that at least the same q questions concerning customer satisfaction were contained in each questionnaire. Furthermore, all questionnaires must use the same granularity for the ordinal scale of corresponding customer satisfaction questions while the ordinal scales for different questions may vary.

3   Measuring Similarities between Customer Satisfaction Profiles

In order to compare the satisfaction profiles of different customer segments with respect to any of the questions, we first have to compare the corresponding distributions over the possible answers. Initially, we restrict the comparison of two customer segments to a single question. The combination of a number of customer satisfaction questions will be considered later in this section.

A very naive approach for comparing two segment on one question would be to simply compare the distributions on all answers including the missing values. However, this can be misleading if the proportions of missing values are not identical in the two customer segments. To illustrate this effect, consider the following artificial example. Assume that in both customer segments all customers who have provided an answer on the ordinal scale have voted for the same degree of satisfaction. However, in the first customer group there are 20% missing values, whereas there are no missing values in the second group. The difference in customer satisfaction for these two groups lies only in the proportion of missing values, but not in the distribution of those who have provided an evaluation of their customer satisfaction. For this reason, we consider the distributions on the ordinal scale and the proportions of missing values separately.

Ignoring the missing values means we first have to normalise the two distributions over the values of the ordinal scale, so that the frequencies add up to 100%. In the above simple example, this would mean that the two distributions over the non-missing answers would be identical after normalisation.

The similarity or difference of two probability distributions over an ordinal scale could be measured on the basis of the differences of the frequencies or in terms of the Kullbach-Leibler entropy (see for instance [2]). However, in this way the ordinal scale would be considered as a finite set of discrete values without any ordering structure. As an extreme example consider three distributions. For the first distribution 100% of the probability mass is concentrated on the largest value of the ordinal scale, for the second one 100% of the probability mass is concentrated on the second largest value, for the third 100% of the probability mass is concentrated on the smallest value. Comparing these distributions in terms of frequency differences or in terms of the Kullbach-Leibler entropy would tell us that they differ in the same way. However, it is obvious that the the first distribution is more similar to the second one than to the last one, for example.

Therefore, we propose to compare the cumulative distribution functions over the ordinal scale in a manner not identical, but similar to the Wilcoxon rank test, also called Mann-Whitney-U-test, known from statistics (see for instance [5,7,11]). When the ordinal scale for the question X has the values (possible answers) $v_1, \ldots, v_h$ and the probability distribution is given as $P(X = v_k) = p_k$, $(k = 1, \ldots, h)$, then the cumulative distribution function is $P(X \leq v_k) = F_k = \sum_{i=1}^{k} p_i$. The pointwise difference

$$d_0\left(P^{(1)}, P^{(2)}\right) = \sum_{k=1}^{h} \left| F_k^{(1)} - F_k^{(2)} \right| = \sum_{k=1}^{h-1} \left| F_k^{(1)} - F_k^{(2)} \right| \tag{1}$$

between the cumulative distribution functions seems to be more appropriate to measure the difference between two probability distributions on an ordinal scale. Note that $F_h^{(1)} = F_h^{(2)} = 1$ always holds.

The distance measure d_0 will have a tendency to higher values, when the ordinal scale has more values, i.e. h is large. This means that questions with finer granularity tend to contribute much more to the difference between satisfaction profiles. We take this effect into account as follows. Consider the two cases:

1. Assume an ordinal scale with just two values (i.e. $h = 2$, for instance, when a question with the only answers yes and no is considered). For the two extreme distributions where 100% of the probability is put on one answer and 0% on the other, the distance measure (1) will yield the value $d_0 = 1$.

2. Now assume an ordinal scale with $h = 2r$ values. Consider the distribution $P^{(1)}$ where the answers are uniformly distributed over the first r values, i.e. $p_1^{(1)} = \ldots = p_r^{(1)} = 1/r$ and $p_{r+1}^{(1)} = \ldots = p_{2r}^{(1)} = 0$, and the distribution $P^{(1)}$ where the answers are uniformly distributed over the last r values, i.e. $p_1^{(2)} = \ldots = p_r^{(2)} = 0$ and $p_{r+1}^{(2)} = \ldots = p_{2r}^{(2)} = 1/r$. Then we obtain $d_0\left(P^{(1)}, P^{(2)}\right) = r = \frac{h}{2}$.

We require that the distance (dissimilarity) between the distributions in the first case should be the same as the distance between the two distribution in the second case. Therefore, we introduce a correction factor and use $d_{\mathrm{ord}}\left(P^{(1)}, P^{(2)}\right) = \frac{2}{h} \cdot d_0\left(P^{(1)}, P^{(2)}\right)$ as the distance between two probability distributions on an ordinal scale with h values (possible answers).

So far we have only compared probability distributions over an ordinal scale ignoring missing values. In order to take the missing values into account, we compute the difference $d_{\mathrm{miss}}\left(p_{\mathrm{miss}}^{(1)}, p_{\mathrm{miss}}^{(2)}\right) = \left|p_{\mathrm{miss}}^{(1)} - p_{\mathrm{miss}}^{(2)}\right|$ between the relative frequencies of the missing values. The overall distance between the (normalised) probability distributions $P^{(1)}$ and $P^{(2)}$ with a proportion (relative frequency) of missing values $p_{\mathrm{miss}}^{(1)}$ and $p_{\mathrm{miss}}^{(2)}$, respectively, is a convex combination of the distances d_{ord} and d_{miss}: $d_{\mathrm{ord+miss}}\left(\left[P^{(1)}, p_{\mathrm{miss}}^{(1)}\right], \left[P^{(2)}, p_{\mathrm{miss}}^{(2)}\right]\right) =$

$$\left(1 - \max\{p_{\mathrm{miss}}^{(1)}, p_{\mathrm{miss}}^{(2)}\}\right) \cdot d_{\mathrm{ord}}\left(P^{(1)}, P^{(2)}\right) + \max\{p_{\mathrm{miss}}^{(1)}, p_{\mathrm{miss}}^{(2)}\} \cdot d_{\mathrm{miss}}\left(p_{\mathrm{miss}}^{(1)}, p_{\mathrm{miss}}^{(2)}\right).$$

This way, the influence of the difference between the probability distributions on the ordinal scale is reduced when at least one of them has a high proportion of missing values.

So far, we have only discussed measuring the difference between two distributions over an ordinal scale incorporating missing values. In the case of customer satisfaction profiles, we may use this approach for comparing the distributions of answers in two customer segments with respect to one question. For comparing two customer segments with respect to a number of questions we simply add up the distances obtained for the single questions.

4   Visualisation

The previous section provides a method to compute the dissimilarity between two customer segments with respect to their answers to selected questions. Although a pairwise comparison of customer satisfaction profiles will already provide important insights in the relation between customer segments and customer satisfaction, it is even more interesting to have an overall overview about how similar or different the satisfaction profiles of a collection of customer segments are. In order to provide this overview, we visualise the distances (dissimilarities) in the plane. We compute the pairwise distances between the customer segments on the basis of the considerations described in the previous section. Then we represent each customer segment by a point in the plane. These points should be positioned in the plane in such a way that the distances between them are as close to the computed dissimilarities of the customer satisfaction profiles as possible. In general, it will not be possible to place the points so that the computed dissimilarities exactly coincide with the geometric distances. It is, for instance, impossible to place four points in the plane with the same (non-zero) distance between each of them.

Nevertheless, there are well-known techniques to position points in the plane so that the distances between the points approximate given (abstract) distances (or distances in a higher-dimensional space). Multidimensional scaling (MDS) and especially Sammon mappings (see [1]) belong to these techniques.

Given a collection of customer segments and a set of satisfaction questions, we compute the pairwise differences (dissimilarities) between the customer segments according to the method described in Section 3. Then we apply MDS based on these pairwise distances to visualise the dissimilarities in satisfaction profiles between all customer segments.

5   Results

The proposed approach was tested on data from over 10,000 customer questionnaires from eight different customer segments marked by the numbers 0,...,7. The customer segments have been found by a typical marketing analysis where demographic and product data are run through a cluster analysis and the identified clusters are later identified and labelled by marketing experts. The actual meaning of the segments is confidential. Each of the customer segments contains between 1300 and 1600 customers. The satisfaction profile for the customer segments is defined on the basis of four questions concerning satisfaction with different ordinal scales with 6-8 values plus a null answer, comprising no answer given and "don't know's".

Figure 1 shows the result of MDS applied to the computed dissimilarities for the eight customer segments on the left hand side. Each spot represents a customer segment. The closer two spots are the more similar are the satisfaction profiles of the corresponding customer segments. An alternative to the MDS approach is hierarchical clustering as it is shown in figure 2.

Figures 3 and 4 show the distributions over the answers to questions 0 and 3 (the distributions over the answers to questions 1 and 2 are not shown in this paper for rasons of limited space). Missing values are ignored for these distributions. Only the distribution over the ordinal scales are shown as relative frequencies. Each figure contains the distribution of all customer segments with respect to one question. Eight neighbouring bars represent the freqencies of the eight customer segments for one ordinal value of the corresponding question.

It is also interesting to see how the satisfaction profiles of customer segments change over time. We apply the same technique as above, but simultaneously

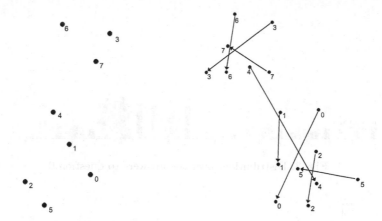

Fig. 1. Visualisation of the similarities of customer profiles (left) and changes of customer profiles (right) based on MDS

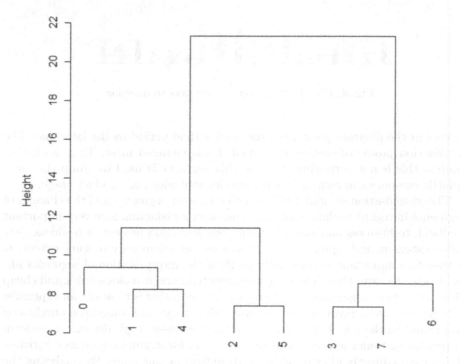

Fig. 2. Visualisation of the similarities of customer profiles based on hierarchical clustering

for different time periods. Here we consider again eight customer segments with questionnaire result from two time periods. Therefore, instead of eight points in the MDS scatterplot, we now have 16 points, two points for each customer segment. Figure 1 shows the result of this analysis on the right hand side. The

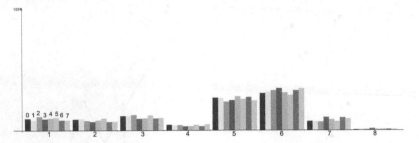

Fig. 3. Distributions over the answers to question 0

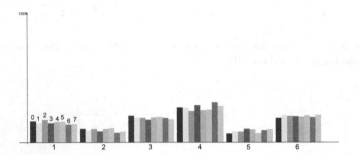

Fig. 4. Distributions over the answers to question 3

arrows in the diagram point from the earlier time period to the later one. The satisfaction profile of customer segment 4 has changed most. To a marketing analyst this is not surprising, because this segment is used to group the most volatile customers in terms of their lifestyle, attitudes and product usage.

The visualisation of similarity between customer segments and the changes of segments in regard to dimensions like customer satisfaction give very important feedback to business analysts. Customer segmentation projects a business view onto customers and represents true customer behaviour only to some extent. It is therefore important to constantly verify if the interpretation of segments has to be adjusted over time. The analysis presented here provides a quick and cheap alternative to re-segmentation. Customer segmentation is typically an expensive activity because it involves running in-depth surveys and sometimes purchasing additional marketing or demographic data. Businesses typically run some form of regular customer analysis on a smaller scale, for example, customer satisfaction surveys directly after engaging with individual customer. By analysing the similarity of customer segments in relation to available process or survey data we can quickly establish relationship and their changes between segments. This analysis can reveal interesting, previously unknown information or prompt a required re-segmentation because discovered relationships no longer align with the interpretation of segments.

6   Conclusions

The proposed approach to analysing similarities between customer satisfaction profiles of different customer segments has shown interesting results and justifies further investigation. The visualisation technique can also be used to track historical or hypothetical (what-if-analysis) changes in the satisfaction profiles of customer segments. For tracking purposes, instead of MDS more sophisticated methods like NeuroScale [6], MDS_{polar} [9] or its extensions [10] that construct an explicit mapping from the high-dimensional space to the visualisation plane could be used.

References

1. Borg, I., Groenen, P.: Modern multidimensional scaling: Theory and applications. Springer, Berlin (1997)
2. Cover, T., Thomas, J.: Elements of information theory. Wiley, New York (1991)
3. von Hagen, F., Baaken, T., Holscher, V., Plewa, C.: International research customer satisfaction surveys (Germany and Australia) and research provider surveys (Germany and Europe). Int. Journ. of Technology Intelligence and Planning 2, 210–224 (2006)
4. Hwang, H.-G., Chen, R.-F., Lee, J.M.: satisfaction with internet banking: an exploratory study. Int. Journ. of Electronic Finance 1, 321–335 (2007)
5. Lehmann, E.L.: Nonparametrics: Statistical methods based on ranks. Springer, Berlin (2006)
6. Lowe, D., Tipping, M.E.: Feed-forward neural networks topographic mapping for exploratory data analysis. Neural Computing and Applications 4, 83–95 (1996)
7. Mann, H.B., Whitney, D.R.: On a test of whether one of two random variables is stochastically larger than the other. Annals of Mathematical Statistics 18, 50–60 (1947)
8. Nauck, D.D., Ruta, D., Spott, M., Azvine, B.: A Tool for Intelligent Customer Analytics. In: Proceedings of the IEEE International Conference on Intelligent Systems, pp. 518–521. IEEE Press, London (2006)
9. Rehm, F., Klawonn, F., Kruse, R.: MDS_{polar}: A new Approach for Dimension Reduction to Visualize High Dimensional Data. In: Famili, A.F., Kok, J.N., Peña, J.M., Siebes, A., Feelders, A. (eds.) IDA 2005. LNCS, vol. 3646, pp. 316–327. Springer, Heidelberg (2005)
10. Rehm, F., Klawonn, F., Kruse, R.: POLARMAP – Efficient visualisation of high dimensional data. In: Banissi, E., Burkhard, R.A., Ursyn, A., Zhang, J.J., Bannatyne, M., Maple, C., Cowell, A.J., Tian, G.Y., Hou, M. (eds.) Information visualization, pp. 731–740. IEEE, London (2006)
11. Wilcoxon, F.: Individual comparisons by ranking methods. Biometrics Bulletin 1, 80–83 (1945)

Development of a Decision-Maker in an Anticipatory Reasoning-Reacting System for Terminal Radar Control

Natsumi Kitajima, Yuichi Goto, and Jingde Cheng

Department of Information and Computer Sciences,
Saitama University, Saitama, 338-8570, Japan
{kitajima,gotoh,cheng}@aise.ics.saitama-u.ac.jp

Abstract. Terminal radar control is more and more complex in recent years. To reduce human errors in terminal radar control, an automatic system to support conflict detection and conflict resolution is required for reliable and safe terminal radar control. An anticipatory reasoning-reacting system for terminal radar control is a hopeful candidate for such systems. This paper proposes a methodology of decision-making in an anticipatory reasoning-reacting system for terminal radar control, presents a prototype of decision-maker, and shows that it can make appropriate decisions in anticipatory reasoning-reacting system for terminal radar control.

Keywords: Terminal radar control, Anticipatory reasoning-reacting system, Decision-making, Reasoning about actions.

1   Introduction

Terminal radar control is an air traffic control facility which controls approaching and departing of aircraft in an airport with radar systems. Recently, air traffic has rapidly increased [1], and the growth of air traffic will continue for the long term [2]. Terminal radar control is more and more complex as the growth of air traffic. Conflict detection and conflict resolution are especially important tasks for reliable and safe terminal radar control. However, conflict detection and conflict resolution by humans are not so reliable because humans sometime cause errors. Furthermore, conflict detection and conflict resolution in complex situations are very difficult for humans, and such tasks cause stress and tiredness that increase probability of human errors. To make terminal radar control more reliable and safe, it is necessary to develop an automatic system to support conflict detection and conflict resolution in terminal radar control.

On the other hand, anticipatory reasoning-reacting systems (ARRSs for short) was proposed as a new generation of reactive systems with high reliability and high security [3,4]. An ARRS detects and predicts omens of attacks and failures anticipatorily, takes some actions to inform its users, and performs some operations to defend attacks and failures by itself. A prototype of ARRS for elevator control has been developed, and it shows usefulness of ARRSs [5].

E. Corchado et al. (Eds.): HAIS 2009, LNAI 5572, pp. 68–76, 2009.

We consider that an ARRS is a hopeful candidate for implementing an automatic system to support conflict detection and conflict resolution. The functions of prediction and decision-making of an ARRS may be effective to conflict detection and conflict resolution. Terminal radar control with an ARRS may be free from problems of traditional terminal radar control. As the first step to develop a practical ARRS for terminal radar control, to establish a methodology of decision-making in an ARRS is indispensable.

The rest of paper is organized as follows: Section 2 explains terminal radar control, Section 3 proposes an ARRS for terminal radar control and its decision-maker, Section 4 presents a prototype of the decision-maker and shows some experiments, and Section 5 gives contributions and future works.

2   Terminal Radar Control

Terminal radar control is performed in fixed terminal area of an airport. An automated radar terminal system receives data about aircraft from radar systems, and provides information of aircraft, such as their discrete ID, position, altitude, direction, speed, and so on, for controllers. A controller receives the information of aircraft, predicts future conflicts of aircraft, and decides next actions to avoid the conflicts. A controller performs prediction or decision-making based on his/her empirical knowledge and rules, and gives instructions to pilots. Pilots operate aircraft according to the instructions.

In terminal radar control, the most important tasks of controllers are conflict detection and conflict resolution. A conflict is a situation involving aircraft and hazards (e.g., other aircraft, weather, and so on) in which the applicable separation minima may be compromised [6]. Conflict detection is to predict future possible conflicts, and conflict resolution is to avoid conflicts by changing actions of aircraft. In Japan, vertical separation minima is $1,000$ feet (if an altitude is not over than $41,000$ feet) or $2,000$ feet (if an altitude is over than $41,000$ feet). Horizontal separation minima is 5NM (nautical mile) [7]. A conflict may cause serious accidents, therefore conflict detection and conflict resolution are very responsible tasks.

3   An Anticipatory Reasoning-Reacting System for Terminal Radar Control

3.1   An Overview of the System

A practical ARRS may consist of a reactive subsystem, which is a traditional reactive system, and some additional components such as a predictor, a decision-maker, and a coordinator. The predictor receives sensory data of the system itself and its external environment, and predicts about possible future events/situations based on some predictive models. The decision-maker receives results of prediction, and decides next actions. The coordinator controls the work flow among components, receives sensory data from the reactive subsystem, and sends next actions

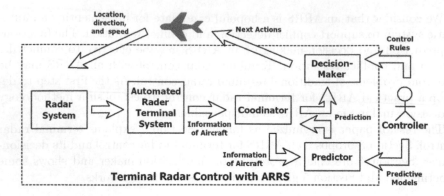

Fig. 1. An Overview of Terminal Radar Control with an ARRS

to the reactive subsystem. Until now, as a methodology of prediction, anticipatory reasoning based on temporal relevant logic was proposed [4], and an anticipatory reasoning engine which is an essential component of the predictor has been developed [8].

We propose an ARRS for terminal radar control which predicts conflicts, and decides actions to avoid them. Fig. 1 shows an overview of terminal radar control with an ARRS. An ARRS provides functions of prediction and decision-making, therefore controllers do not have to perform prediction and decision-making. On the other hand, these components receive empirical knowledge of controllers such as control rules, and predictive models. In an ARRS for terminal radar control, the roles of each component are as follows: the predictor predicts future conflicts from the present state of aircraft; the decision-maker receives the results of prediction, and decides next actions of the aircraft to avoid the conflicts based on control rules; the coordinator controls work flow of the predictor and the decision-maker. In terminal radar control with an ARRS, controllers give predictive models and control rules to the ARRS. A predictive model is a description to predict future events/situations such as conflicts between aircraft, and control rules are rules of how to control aircraft. In terminal radar control, there are some predictive model and control rules according to situations of terminal area or policies of control. Therefore, an ARRS for terminal radar control should deal with various predictive models and control rules.

3.2 Development of a Decision-Maker

In this paper,we focus our mind on decision-making in an ARRS for terminal radar control. To develop a decision-maker of an ARRS for terminal radar control, we defined requirements of the decision-maker as follows:

R1. A decision-maker must decide next actions of aircraft according to the present state of aircraft and given predictions.

R2. A decision-maker must decide next actions of aircraft based on various control rules given by controllers.

R3. A decision-maker must decide appropriate actions of aircraft based on control rules if control rules are appropriate.

R4. A decision-maker must decide next actions of aircraft before the point of time when conflicts will occur.

In our previous work, we have proposed an approach for reasoning about actions based on deontic relevant logics [9]. Reasoning about actions is the process of draw new actions from some given premises. For satisfying the above requirements, reasoning about actions is an indispensable process in decision-making because candidates of the next actions are previously unknown or unrecognised, and reasoning is only way to find candidates of the next actions from given premises and control rules. It is necessary right fundamental logic systems underlying reasoning about actions to ensure that appropriate candidates of next actions are reasoned out from appropriate premises. We adopted deontic relevant logics (DRLs for short) [10,11] as the fundamental logics. DRLs are suitable for logic systems underlying reasoning about actions [9]. DRLs are obtained by introducing deontic operators into strong relevant logics. Deontic operators are useful to represent 'what should do' and 'what must not do.' Reasoning underlying strong relevant logics or their extensions is valid reasoning in contrast with reasoning underlying classical mathematical logic or its traditional extensions [11]. Therefore, DRLs are suitable for representation of rules and reasoning about actions appropriately. We have also constructed an action reasoning engine (AcRE for short) for general purpose [9]. The AcRE can perform reasoning about actions based on DRLs, and deduce candidates of the next actions according to given control rules.

Reasoning about actions based on DRLs is useful while it has an issue that number of premises increases with complexity of a situation, i.e., scaling up to larger problem. To resolve this issue, we adopt qualitative reasoning approach for reasoning about actions. Qualitative reasoning concerns representations for continuous aspects of the world, such as space, and quantity, and reasoning with very little information [12]. Qualitative reasoning deals with information which represents qualitatively. We already showed qualitative reasoning is suitable for reducing scale of reasoning and fast reasoning [13]. To perform qualitative reasoning, the AcRE should deal with only qualitative information. Therefore, additional functions which deal with quantitative information are needed for before or after qualitative reasoning because only qualitative information is not enough to make decisions.

We defined data and functions of a decision-maker by using an AcRE and qualitative reasoning techniques. Fig. 2 shows data and functions of the decision-maker. Each data and functions are as follows:

Data:

- **Conflicts** are possible conflicts predicted by the predictor, and represented as logical formulas in the form of $CT(x, y)$ where CT is conflict type, and x, y are aircraft IDs which will conflict. Conflict type is classified into three types: $Merge$, $Cross$ and $RearEnder$.

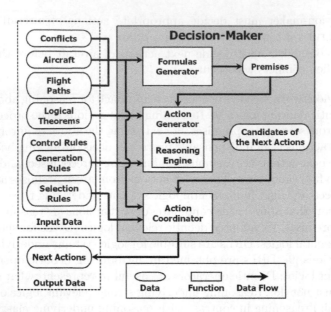

Fig. 2. Data and Functions of a Decision-Maker

- **Aircraft** is the present state of aircraft at a certain time, and represented as 5-tuples in the form of (id, x, y, z, v, r) where id is the ID number of each aircraft, x is the current longitude of the aircraft, y is the current latitude of the aircraft, z is the current altitude of the aircraft, v is the current speed of the aircraft, and r is the current segment of the aircraft in the flight path.
- **Flight Paths** are standard flight paths to the airport. This data consists of point and segment. Data of point are represented as 3-tuples in the form of $(name, x, y)$ where $name$ is a name of each fixed point in the path, and x is the longitude of the point, y is the latitude of the point. Data of segment are represented as 3-tuples in the form of $(segname, previous, next)$ where $segname$ is a name of each segment, $previous$ is a name of previous segment, and $next$ is a name of next segment.
- **Control Rules** are classified into two classes: generation rules and selection rules. Generation rules are rules of what to do next. A generation rule is represented by logical formulas of the first-order predicate formal language based on DRLs. Selection rules are rules of how to select actions, and represented as 2 tuples in the form of $(name, w)$ where $name$ is a name of estimation function for estimation of next actions, and w is the weight of estimation which is showed from 0 to 5.
- **Logical Theorems** are logical theorems of DRLs which is used for reasoning about actions.
- **Premises** are facts in the control are of terminal radar control, and represented by atomic formulas. For example, '$Following(x, y)$' is one of premises which represents 'aircraft x is followed by aircraft y'.

- **Candidates of the Next Actions** are actions which each aircraft should take, and represented by atomic formulas. For example, '$O(Deceleration(x))$' is one of candidates which represents 'aircraft x should decelerate'.
- **Next Actions** are selected actions to avoid a possible conflict, and represented as 3-tuples in the form of $(type, aircraft, degree)$ where $type$ is action type, $aircraft$ is the ID of aircraft which takes the action, and $degree$ is a quantity of the action such as angle, distance, and speed. Action type is classified into three types: $Deceleration$, $TurnRight$, and $TurnLeft$.

Functions:

- **Formulas Generator** executes programs which generate premises of reasoning about actions.
- **Action Generator** generates candidates of the next actions. This function contains the AcRE and converters of input/output data of the AcRE.
- **Action Coordinator** selects next actions, and decides detail of each action.

An decision-maker which we designed receives data of conflicts, aircraft, flight paths, control rules, and logical theorems. Then, the decision-maker decides next actions based on these input data. In the decision-maker, first, formulas generator receives data of conflicts, aircraft, and flight paths, and generates premises which are used for action generator. Second, action generator receives premises generated by formulas generator, generation rules and logical theorems of DRLs, performs reasoning about actions with the AcRE, and generates candidates of the next actions. Finally, action coordinator receives candidates of the next actions and selection rules, selects next actions according to given selection rules, and calculates degree of each action. Degree of an action represents quantity such as speed, angle, or length.

4 A Prototype of Decision-Maker and Experiments

4.1 Development of a Prototype of the Decision-Maker

We developed a prototype of the decision-maker which supports essential processing to show it can make decisions in actual situations. The prototype is developed according to our methodology of decision-making while detail processing of formulas generator and action coordinator are implemented based on very simple algorithm.

As data of conflicts, the prototype receives only one conflict and decides next actions for avoidance of the conflict. As generation rules, the prototype supports a generation rules which is represented by defined predicates. Predicates which the prototype supports are '$High(x,y)$' (aircraft x is higher than y), '$Following(x,y)$' (aircraft x is followed by aircraft y), '$NoFollowing(x)$' (aircraft x has no following aircraft), '$EnoughInterval(x)$' (aircraft x has enough interval between following aircraft to take an action), '$EnoughRightSpace(x)$' (aircraft x has enough space on the right), and '$EnoughLeftSpace(x)$' (aircraft x has enough space on the left). As selection rules, the prototype supports following two elimination type: the number of aircraft and distance. As next actions, the prototype outputs next actions represented by following action type: Deceleration, Turn Right, Turn Left, Up and Down.

In the prototype, we uses logical theorems of DEc, which is one of deontic relevant logic systems [10]. Selected theorems are as follows: $((A \Rightarrow (\neg B)) \Rightarrow (B \Rightarrow (\neg A)))$, $(((A \Rightarrow B) \wedge (A \Rightarrow C)) \Rightarrow (A \Rightarrow (B \wedge C)))$, $((A \wedge B) \Rightarrow (B \wedge A))$, $(((A \Rightarrow B) \wedge (B \Rightarrow C)) \wedge (A \Rightarrow C))$, $((\neg(A \wedge (\neg A))))$, $O(A \Rightarrow B) \Rightarrow (O(A) \Rightarrow O(B))$, $O(A) \Rightarrow (\neg O(\neg A))$, $\neg(O(A) \wedge O(\neg A))$, $O(A \wedge B) \Rightarrow (O(A) \wedge O(B))$, and $(\neg O(\neg(A \wedge B))) \Rightarrow ((\neg O(\neg A)) \wedge (\neg(O(\neg B))))$. In above formulas, O shows obligation operator of deontic relevant logic.

A decision-maker in an ARRS receives input data from other component, therefore we developed a simple simulator. The simulator generates input data from prepared files, and executes the decision-maker.

4.2 Experiments

We made some experiments of decision-making by our prototype of the decision-maker. We adopted an example of terminal radar control in Tokyo International Airport which is one of the largest airports in Japan. We assumed conflict in terminal area of Tokyo International Airport, and prepared input data with different parameters by using actual flight data of aircraft in the airport. Terminal radar control is performed in fixed terminal area by each airport. In Tokyo National Airport, current capacity of aircraft which is under terminal radar control at the same time is about 10 [14].

We performed some decision-making with different number of aircraft (N_a) as follows: $N_a = 2$ (only two aircraft which will conflict), $N_a = 5$ (light air traffic), and $N_a = 10$ (heavy air traffic). As a conflict, we set merge between aircraft $A004$ and aircraft $A005$. As a runway which aircraft arrives, we set 32R which is one of runway of the airport. A flight path is fixed according to arrival runway. Furthermore, we prepared the control rules to takes actions with applicable separation. In this control rules, the prototype decides several actions to keep applicable separation when an action will affect other aircraft. In each decision-making, we also tried to measure execution time of the prototype.

Table 1 shows results of selected actions by the prototype and execution time of its decision-making. From results of selected actions in the table, selected actions are changed by given situations of aircraft. A predicted conflict is avoid by selected actions, and applicable separation between aircraft is kept by selected actions. Therefore, we can say that the prototype can decide actions from given situations and control rules. From result of execution time in the table, we can also say that number of candidates and execution time increase according to number of aircraft.

The above experimental results show that our prototype of decision-maker can decide next actions from given data of a conflict, aircraft, and flight paths, and the decided next actions follow given control rules. In this experiment, we can say that the execution time of our prototype is acceptable if a conflict is predicted appropriately. Current techniques of conflict detection can predict a conflict at

Table 1. Results of Selected Actions and Execution Time

N_a	selected actions	time
2	A005 decelerates 20kn.	1.506s
5	A005 decelerates 20kn.	5.601s
	A007 decelerates 20kn.	
10	A005 decelerates 20kn.	25.061s
	A007 decelerates 20kn.	
	A009 decelerates 20kn.	

three minutes before [15]. However processing of our prototype is simplified, and we guess the prototype cannot decide next actions in time if advanced processing is implemented. We have to perform some additional experiments with different parameters to evaluate out decision-maker from various aspects, and have to improve the prototype of the decision-maker to perform more efficient decision-making in terminal radar control.

5 Concluding Remarks

We have proposed an anticipatory reasoning-reacting system for reliable and safe terminal radar control. We also designed a decision-maker with reasoning about actions based on deontic relevant logics, presented a prototype of the decision-maker, and showed some experiments. Experimental results showed that our decision-maker can make appropriate decisions in actual terminal radar control. As a result, we can expect to develop a practical anticipatory reasoning-reacting system for reliable and safe terminal radar control.

Future works are as follows: to make more experiments with various situations in terminal radar control, to improve performance of the decision-maker, and to improve the decision-maker to decide the most appropriate actions.

References

1. Boeing Commercial Airplanes: Statistical summary of commercial jet airplane accidents (2008), http://www.boeing.com/news/techissues/
2. International Civil Aviation Organization: Growth in ATR traffic projected to continue. Technical report (2001)
3. Cheng, J.: Anticipatory reasoning-reacting systems. In: Proc. International Conference on Systems, Development and Self-organization, Beijing, China, pp. 161–165 (2002)
4. Cheng, J.: Temporal relevant logic as the logical basis of anticipatory reasoning-reacting systems. In: Dubois, D.M. (ed.) Computing Anticipatory Systems: CASYS 2003 - Sixth International Conference, Liege, Belgium, August 11-16, 2003. AIP Conference Proceedings, vol. 718, pp. 362–375. American Institute of Physics (2004)

5. Shang, F., Nara, S., Omi, T., Goto, Y., Cheng, J.: A prototype implementation of an anticipatory reasoning-reacting system. In: Dubois, D.M. (ed.) Computing Anticipatory Systems: CASYS 2005 - Seventh International Conference, Liege, Belgium, August 8-13, 2005. AIP Conference Proceedings, vol. 839, pp. 401–414. American Institute of Physics (2006)
6. International Civil Aviation Organization: Global air traffic management operational concept (2005), http://www.icao.int/
7. Senoguchi, A., Fukuda, Y.: An idea of altitude prediction model for conflict detection. Technical report, IEICE (2006) (in Japanese)
8. Nara, S., Shang, F., Omi, T., Goto, Y., Cheng, J.: An anticipatory reasoning engine for anticipatory reasoning-reacting systems. International Journal of Computing Anticipatory Systems, (CHAOS), 225–234 (2006)
9. Kitajima, N., Nara, S., Goto, Y., Cheng, J.: A deontic relevant logic approach to reasoning about actions in computing anticipatory systems, CHAOS. International Journal of Computing Anticipatory Systems 20, 177–190 (2008)
10. Tagawa, T., Cheng, J.: Deontic relevant logic: A strong relevant logic approach to removing paradoxes from deontic logic. In: Ishizuka, M., Sattar, A. (eds.) PRICAI 2002. LNCS, vol. 2417, pp. 39–48. Springer, Heidelberg (2002)
11. Cheng, J.: Strong relevant logic as the universal basis of various applied logics for knowledge representation and reasoning. In: Kiyoki, Y., Henno, J., Kangassalo, H. (eds.) Information Modelling and Knowledge Bases XVII, Frontiers in Artificial Intelligence and Applications, vol. 136, pp. 310–320. IOS Press, Amsterdam (2006)
12. Forbus, K.: Qualitative reasoning. In: CRC Handbook of Computer Science and Engineering, pp. 715–733. CRC Press, Boca Raton (1996)
13. Kitajima, N., Goto, Y., Cheng, J.: Fast qualitative reasoning about actions for computing anticipatory systems. In: Proc. Third International Conference on Availability, Reliability and Security, pp. 171–178 (2008)
14. Ministry of Land, Infrastructure, Transport and Tourism: Current states and issues of capacity of departure and arrival in tokyo international airport (in japanese) (2000), http://www.mlit.go.jp/
15. Fukuda, Y., Senoguchi, A.: Research and development on conflict alert function of air traffic control workstation (in Japanese). Technical report, IEICE (2006)

Study of Outgoing Longwave Radiation Anomalies Associated with Two Earthquakes in China Using Wavelet Maxima

Pan Xiong[1], Yaxin Bi[2], and Xuhui Shen[1]

[1] Institute of Earthquake Science, China Earthquake Administration,
Beijing, 100036, China
xiong.pan@gmail.com, shenxh@seis.ac.cn
[2] School of Computing and Mathematics, University of Ulster,
Co. Antrim, BT37 0QB, United Kingdom
y.bi@ulster.ac.uk

Abstract. The paper presents an analysis of the continuous outgoing longwave radiation (OLR) based on time and space by using the wavelet-based data mining techniques. The analyzed results reveal that the anomalous variations exist prior to the earthquakes. The methods studied in this work include wavelet transformations and spatial/temporal continuity analysis of wavelet maxima. These methods have been applied to detect singularities from OLR data that correspond to seismic precursors, particularly to a comparative study of the two earthquakes of Wenchuan and Pure recently occurred in China.

1 Introduction

By studying remote sensing data, researchers have found various abnormal activities in earth, atmosphere and ionosphere prior to large earthquakes, which are reflected in anomalous thermal infrared (TIR) signals [1], outgoing longwave radiation [5] and surface latent heat flux (SLHF) [2,3,8] and anomalous variations of the total electron content (TEC) [6,7] prior to the earthquake events. The latest advancements in litho-sphere – atmospheric – ionospheric models provide a possible explanation to the origin of these phenomena [6,7], and also permit us to explore possible new studies on the spatial and temporal variability of remote sensing data before and during major earthquakes.

Several studies have recently been carried out to analyze thermal infrared anomalies appearing in the area of earthquake preparation a few days before the seismic shock [9, 10]. These studies analytically compare a single image of pre (vs. post) earthquake satellite TIR imagery [10]; analyze the pixel temperature variance from long term scene threshold temperatures to identify "hot" areas [11]; perform a multi-spectral thermal infrared component analysis on the Moderate Resolution Imaging Spectroradiometer (MODIS) on Terra and Aqua satellites by using Land Surface Temperature (LST) [4]; assess the anomalous SLHF peaks a few days prior to the main earthquake event in the case of coastal earthquakes [8]; and study OLR data and discover anomalous variations prior to a number of medium to large earthquakes [5].

E. Corchado et al. (Eds.): HAIS 2009, LNAI 5572, pp. 77–87, 2009.

OLR is the thermal radiation flux emerging from the top of the atmosphere and being connected with the earth–atmosphere system. It is often affected by cloud and surface temperature. Due to OLR resulting from infrared band telemetry, not only OLR data is continuous, stable and commeasurable, but also it is sensitive to the sea surface layer and the near ground temperature change. It can be therefore regarded as an ideal device for monitoring the symptoms of some natural disasters linking to "hot" origin of phenomena, like earthquakes.

Precisely detecting seismic precursors within OLR data prior to earthquakes is vitally important to sufficiently make use of OLR resources to monitor stable conditions of active faults beneath the earth and to identify the potential earthquake zones. A possible solution to these problems is to employ computer-assisted intelligent analysis methods – advanced data mining methods – to detect abnormal events embedded in OLR data. The key challenge facing data mining research is to properly and rapidly digest massive volumes of OLR data in order to detect abnormal events. More recently Cervone et al. have developed a new data mining method based on wavelet analyses to detect anomalous SLHF maxima peaks associated with four coastal earthquakes [2, 3].

In this paper we propose to use wavelet transformations as a data mining tool to detect seismic anomalies within OLR data. We have performed an assessment on a number of wavelet methods and selected two real continuous Daubechies Wavelet and Gaussian Derivative Wavelet. The distinguishing feature of our method is that we calculate the wavelet maxima that propagate from coarser to finer scales over the defined grids and then identify strong anomalies in the maxima lines distributing on the grids by only accounting for those phenomena that show continuity in both time and space. The identified anomalies are regarded as potential precursors prior to the earthquakes. In this context, the time continuity means that the detected anomalies occur at the same time or with a short delay of each other, while the space continuity means that the detected anomalies are distributed in space according to a precise geometry conforming to the geological settings of the region. The proposed method could be applied to different types of spatial and temporal data, which is not restricted to a particular resolution or time sampling.

The proposed method has been applied to analyze the OLR data associated with the two earthquakes recently occurred in Wenchuan and Puer of China, respectively. Combining with the tectonic explanation of spatial and temporal continuity of the abnormal phenomena, the analyzed results have shown seismic anomalies to be as the earthquake precursors.

2   Earthquakes and Data

In this study, two earthquakes are selected for evaluating the proposed method. The first one is the Wenchuan earthquake of magnitude 8.0 that is the largest earthquake in China in the past thirty years. It occurred on 12th May 2008, the location of the epicenter is at 30.986°N, 103.364°E, and the depth is 19 km. The main earthquake was followed by a series of smaller aftershocks. The second is the Puer earthquake of magnitude 6.4 that occurred on 3rd June 2007. The earthquake's epicenter is at 23°N, 101.1°E.

The OLR energy flux is characterized by a number of parameters, such as the emission from the ground, atmosphere and clouds formation, which have been being observed on the top of the atmosphere by National Oceanic and Atmosphere Administration (NOAA) satellites [15]. These OLR data have been recorded twice-daily by the several polar-orbiting satellites for more than eight years, forming time series data across the different periods of time along with the spatial coverage of the entire earth. The original OLR data are processed by the interpolation technique to minimize the distance in space or time over which a value is interpolated. The detail of the interpolation technique has been given by Liebmann and Smith [13].

The data used for this study are twice-daily means from the NOAA-18 satellite. Their spatial coverage is 1×1 degree of latitude by longitude covering the area of 90°N – 90°S and 0°E – 357.5°E, and the time range is from 3rd September 2006 to 28th September 2008, forming time series data over the specified region.

3 Methodology

There are several wavelets to choose in the analysis of OLR data. The best one for our application depends on the nature of OLR data and what we require in detecting seismic anomalies. We have undertaken an empirical analysis on several wavelet methods and selected two for our study. The first method is one of the Daubechies Wavelets, called a db1, and the second is the Gaussian Derivative Wavelets called a gaus3. Both of these methods employ one dimensional continuous wavelet transformations. We use these two methods to analyze the continuity of modulus maximum in time and space and to detect singularities within the OLR data of the defined grids covering the two earthquakes.

3.1 Wavelet Transformation

The formalism of the continuous wavelet transform (CWT) was first introduced by Grossmann and Morlet [18]. Formally it is written as:

$$\gamma(s,\tau) = \int f(t)\psi^{*}_{s,\tau}(t)dt \qquad (1)$$

where * denotes complex conjugation, s is the scale factor and τ is the translation factor which are the new dimensions after the wavelet transform. This equation shows how a function $f(t)$ is decomposed into a set of basis functions, called the wavelets. For the sake of completeness, formula (2) gives the inverse wavelet transform.

$$f(t) = \iint \gamma(s,\tau)\psi_{s,\tau}(t)d\tau ds \qquad (2)$$

The wavelets are generated from a single basic wavelet (t), the so-called *mother wavelet*, by scaling and translation:

$$\psi_{s,\tau}(t) = \frac{1}{\sqrt{s}}\psi\left(\frac{t-\tau}{s}\right) \qquad (3)$$

where the factor $1/\sqrt{s}$ is for energy normalization across the different scales. It is important to note that the theory of wavelet transforms not only deals with the general properties of the wavelets and wavelet transforms, but it also defines a framework for designing other types of wavelets.

The methods used in this study are Daubechies Wavelets and Gaussian Derivative Wavelet. The general characteristics of Daubechies Wavelets can be compactly supported with external phase and highest number of vanishing moments for a given support width. Associated scaling filters are minimum-phase filters [16]. The Gaussian Wavelets can be derived from the Gaussian probability density function. Gaussian functions are optimal in terms of their time-frequency localisation. The time-frequency localisation property of the Gaussian wavelet make it possible to design Glters with very narrow frequency band.

Mallat and Hwang [14] introduced a method for processing and detecting singularities using wavelets, in which detection singularities are initially through calculating the local maxima of the wavelet transform modulus. It is proved that modulus maxima detect all singularities.

In this work, we have experimented several one dimensional wavelet transformations and selected better performed wavelet functions db1 (Daubechies Wavelets) and gaus3 (Gaussian Wavelets) to calculate maxima lines. The basic idea is that using the wavelet functions db1 and gaus3, we can calculate the wavelet coefficients of the time series OLR data, and then we perform a further calculation on these wavelet coefficients, resulting in a sets of numeric values called wavelet maximas.

3.2   Calculating Singularities

For any given real valued function ϕ with zero average $\int_{-\infty}^{\infty}\phi(t)dt = 0$, let

$$Wf(u,s) = \int \frac{f(t)}{\sqrt{(s)}}\phi(\frac{t-u}{s})dt$$

be the real continuous wavelet transform of a function f. Since ϕ has zero mean, the previous integral measures the variation of f in a neighborhood of time u of size proportional to the so called scale factor $s > 0$. (u_0, s_0) is defined to be a modulus maximum if $|Wf(u_0, s_0)|$ is a local maximum, i.e. if

$$\frac{\partial Wf(u_0, s_0)}{\partial u} = 0$$

and if $Wf(u_0,s_0)$ is strictly increasing to the left of u_0 or strictly decreasing to the right of u_0. In other words, the goal is to identify the isolated local maxima of the wavelet transform $Wf(u_0,s_0)$ for each $s>0$. A connected curve γ in the scale-time plane is called "a maxima line" if $(u,s)\in \gamma$ implies (u,s) is a modulus maximum.

Modulus maxima carries a significant degree of information about the position of singularities, particularly, it is possible to prove that for every singularity t_0 of f one finds "a sequence" of modulus maxima (u_i, s_i) such that $s_i \to 0$ and $u_i \to t_0$.

3.3 Continuity in Space and Time

By considering the tectonic background, continental boundaries and fault lines, we define the study area and divide it into a set of grids. The analyzed results on the different grids are combined into a $n \times m$ matrix, in which the rows n correspond to the selected scales at each of the grids in which the wavelet analysis has been performed, the columns m correspond to time, and the entry values of the matrix are either the propagation lengths of significant maxima lines or zero if none has been detected at this particular point in space/time. As such, the time-series measurements of wavelet maxima are continuous in space and time of the grid path.

To effectively visualize maxima lines, we use different colors to represent the degree of maxima magnitudes, in which the dark color indicates the most significant singularity. By plotting the sequences of modulus maxima lines, we can get the continuous curves of maxima both in space and time.

The third part of Fig. 2 shows an example of a set of maxima lines. Fig. 3 shows the curves of the sequences of modulus maxima and the Fig.4 indicates the identified singularities.

3.4 Experimental Procedure

This section describes the experimental procedure and analysis method through an example of the Wenchuan earthquake.

First, we define an experimental area. The Wenchuan earthquake is postulated as the result of motion on a northeast striking reverse fault or thrust fault on the northwestern margin of the Sichuan Basin. By taking into account the tectonic background, continental boundaries and active faults, we define an experimental area and divide it into a set of grids as shown in Fig.1 [17].

Secondly, based on the defined grids, OLR daily data, from 28th September 2007 to 28th September 2008, are downloaded from the NOAA Climate Prediction Center. After pre-processing, we employ the wavelet methods db1 and gaus3 to analyze the data and generate wavelet maxima values. The singularities detected from these values on each o the grids are then visualized. For example, Fig.2 shows the resulting maxima curves in the first grid. The figure consists of three components: a) the original time series OLR data, b) a pseudo three dimensional representation of the wavelet coefficients, and c) significant wavelet maximas detected over time. The colors in the legend indicate the degrees of maxima magnitudes from the largest degree to the smallest one.

Thirdly, we take every maxima curve in each of the grids and rearrange them onto one diagram as shown in Fig.3. In the figure the x-axis represents time in day units, and the y-axis represents the grids in a sequential order. The magnitudes of maxima represent the degrees of seismic anomalies, where the larger the magnitude, the higher the degree of seismic anomalies. The figure heading lists the earthquake name, the period of selected data, data type, region of data, grid path and the wavelet method used. The red line indicates the day when the earthquake occurred.

82   P. Xiong, Y. Bi, and X. Shen

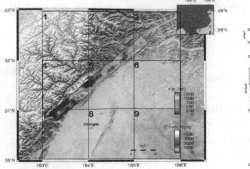

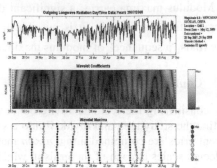

Fig. 1. Grids defined for the Wenchuan earthquake (the map is adapted from)

Fig. 2. An example of analysis results and corresponding maxima curves

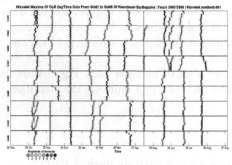

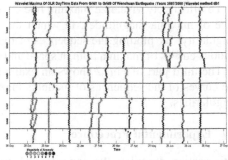

Fig. 3. The curves of wavelet maxima computed wavelet transformation

Fig. 4. Identified singularities from the curves in the modulus maxima

Final stage is to identify singularities from the maxima curves. The key feature of singularities is that they form a continuous maxima line with a large magnitude. In Fig. 3, we can find several singularities, which can be grouped into three types – pre- and post-earthquake and when the earthquake occurred, corresponding to 1) singularities prior to the earthquake, which may be caused by the large energy flux before the earthquake; 2) singularities in the time when the earthquake occurred, which may be caused by the release of a large amount of energy; 3) singularities after the earthquake, perhaps caused by many aftershocks after the earthquake. From Fig.4, three obvious singularities prior to the earthquake are highlighted with the dashed red ovals.

4   Results and Discussion

4.1   Wenchuan Earthquake

We have done two experiments with the Wenchuan earthquake based on time and space.

1) Based on historical seismic activities and tectonic characteristics, we select three areas, namely regions 1 and 2, and the Wenchuan earthquake region called the main

region to carry out comparative studies on the three regions from time and space. The main region covers the earthquake. Region 1 is adjacent to the active fault line and Region 2 is far from the fault zones and there are no historical earthquakes recorded. The duration of the OLR data used for three regions is from 28th September, 2007 to 28th September, 2008. The location of the main region is from 30°N, 103°E to 33°N, 106°E, Region 1 is from 28°N, 105°E to 31°N, 108°E, and Region 2 is from 44°N, 113°E to 47°N, 116°E.

The following figures (Fig.5, Fig.6 and Fig.7) show the wavelet maxima curves of the three regions produced by using gaus3. The red line indicates the day when the Wenchuan earthquake occurred.

In Fig.5 several continuous singularities are identified, some of them are around the Wenchuan earthquake. These singularities may be caused by the large amount of energy generated by the Wenchuan earthquake. Compared with Fig.6, the maxima curves are more disorder, but one continuous singularity can be clearly observed. Looking at Fig.7, the maxima lines are complete disorder.

The distribution of the singularities in Fig.6 is similar to that in Fig. 5. However in Fig.5 the maxima lines of singularities are more continuous with larger magnitudes and a clear singularity appears on the day when the earthquake occurred. Although a similar distribution appears in Fig.7, the maxima lines are disorder and the magnitudes of the maxima are also smaller. Considering the factors of geographic region and tectonic background of the earthquake, we could conclude that the singularities from the wavelet maxima curves of the Wenchuan region are more informative and regular than those in the other two regions. In particular, the singularities in Region 2 are completely in disorder since the region is stable and there are almost no earthquakes in this area in past decades. The singularities can also be discovered in Region 1 on the day when the Wenchuan earthquake occurred, these could be due to that Region 1 is close to the active fault line and the epicenter of the Wenchuan earthquake.

2) In order to investigate the period of the occurrence of singularities, we selected OLR data in a different year based on the same grids, i.e. from 28th September, 2006 to 28th September, 2007, and carry out another experiment using the wavelet method db1. The comparative results are illustrated in Fig. 8 and Fig.9, respectively.

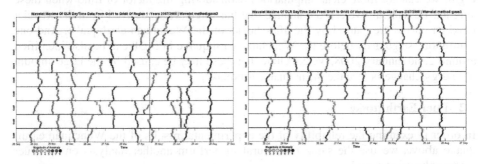

Fig. 5. Maxima curves of the Wenchuan region   **Fig. 6.** Maxima curves of the Region 1

84   P. Xiong, Y. Bi, and X. Shen

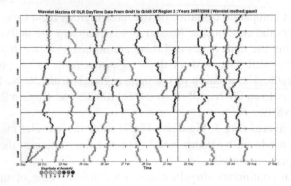

Fig. 7. Wavelet maxima analysis curves of the Region 2

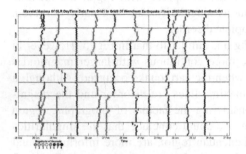

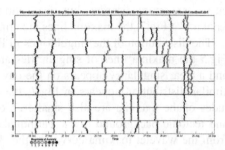

Fig. 8. Maxima curves of the Wenchuan from 28th Sep, 2007 to 28th Sep, 2008

Fig.9. Maxima curves of the Wenchuan 28th Sep, 2006 to 28th Sep, 2007

In Fig. 9, the distribution of singularities is discontinuous and disorder, from Fig. 8, we can find more continuous maxima lines and more singularities, the Wenchuan earthquake could be the main reason to cause such effect, thus the singularities in the period when the Wenchuan earthquake occurred (from 28 Sep, 2007 to 28 Sep, 2008). However, the similar distribution can be found from the above figures, the cause may be the OLR data and its similar variation rule every year. From Fig.8 and Fig.9 we can also observe several singularities after the Wenchuan earthquake, especially in Fig.8, we can get more clear singularities. Considering the factor of time when these singularities appear, we could conclude that the singularities derived from the wavelet maxima curves are caused by the aftershocks of the Wenchuan earthquake. And one continuous singularity before the Wenchuan earthquake in Fig.8 can be also identified around 22nd January 2008, we conclude that these may be caused by the large amount of energy change before the Wenchuan earthquake.

4.2 Puer Earthquake

In order to examine the reliability of the experimental results of the Wenchuan earthquake above, we use the same procedural to perform another analysis on the Puer earthquake using gaus3.

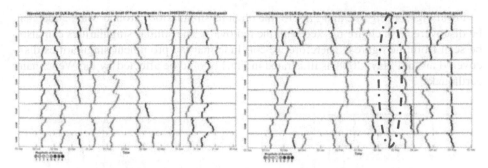

Fig. 10. Maxima curves of the Puer Sep, 2006 **Fig. 11.** Maxima curves of the Puer from 3rd
to 3rd Sep, 2007   Sep, 2007 to 3rd Sep, 2008

Since the Puer earthquake occurred on 3rd June, 2007, thus we selected two periods of OLR data, one is from 3rd September, 2006 to 3rd September, 2007, and another is from 3rd September, 2007 to 3rd September, 2008. The location of the Puer Region girds is from 21°N, 100°E to 24°N, 103°E.

From Fig. 10, it can be seen that the distribution of singularities is discontinuous and disorder, all of the singularities are not obvious except one before the Puer earthquake. It is observed that the singularities prior to the Puer earthquake are not as clear and regular as before the Wenchuan earthquake. A possible reason for this is the magnitude of the Puer earthquake (Mw=6.4) is smaller than that of the Wenchuan earthquake, resulting in a smaller amount of energy release.

In Fig. 11, though the distribution of singularities is disorder, the singularities are clearer. Especially around the red line, we can identify one continuous singularity around 2nd May 2008 and another one around 3rd August 2008.

However, comparing Fig.10 with Fig.11, the prominent singularities can be observed from these maxima curves, which are highlighted with the dashed red ovals. A possible factor of causing these singularities is the time of the singularity occurred. Thus we postulated that such effect would result from the influence of the Wenchuan earthquake, which occurred on 12th May, 2008.

In order to examine the possibility of the influence of the Wenchuan earthquake to the adjacent regions, we select the other two regions, Kunming (Yunnan province) and Panzhihua (Sichuan province). The location of Kunming region girds defined is from 23°N, 102°E to 26°N, 105°E and the location of Panzhihua region girds is from 24°N, 100°E to 27°N, 103°E. We divided these regions into two sets of grids and downloaded the same period of OLR data and carried out a further experiment. Two wavelet maxima curves are generated over the OLR daytime data from 3rd September, 2007 to 3rd September, 2008 as shown in Fig.12 and 13 (db1).

The singularities in Fig.12 are not obvious but some are around the red line. Around 2nd May 2008, there are some obvious but they are not continuous singularities. From Fig.13 we can observe more obvious and continuous singularities, the singularities identified around the red line are as same as those in Fig.12. But there is one continuous singularity before the red line. Around 4th June 2008, we can observe some singularities but in discontinuity. After the red line, two continuous singularities can be identified but the magnitudes of the singularities are relatively smaller.

From Figs.11, Figs.12 and Figs.13, the obvious singularities around May, 2008 can be discovered, which are marked with the red ovals. The similar singularities can also be seen from the figures. Therefore we conclude that the prominent singularities discovered around May, 2008 as illustrated in Fig.11 could be affected by the Wenchuan earthquake.

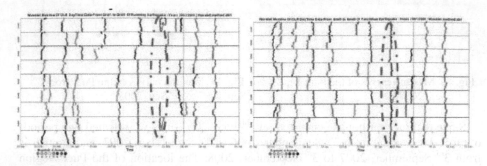

Fig. 12. Maxima curve of the Kunming from 3rd Sep, 2007 to 3rd Sep, 2008

Fig. 13. Maxima curves of the Panzhihua from 3rd Sep, 2007 to 3rd Sep, 2008

5   Conclusion

This paper presents an analysis on the selected OLR data associated with the Wenchuan and Pure earthquakes and explains how the OLR singularities discovered are related to the earthquakes. The methodology discussed in the present paper uses data mining techniques, including wavelet transformations and spatial/temporal continuity analysis of the wavelet maxima to identify singularities before the earthquakes. The numerous comparative experiments and analyses on time and space of the Wenchuan and Pure earthquakes conclude that the prominent OLR singularities could be found prior to the earthquakes in the wavelet maxima curves, which follow continuity both in space and time. Therefore our studies conclude that singularities discovered within OLR data could be regarded as an effective indicator to detect seismic anomalies. This finding will be further validated by using more earthquake data in the future.

Acknowledgements

This work is supported by the project of "Data Mining with Multiple Parameters Constraint for Earthquake Prediction (founded by the Ministry of Science and Technology of China, Grant No.:2008BAC35B05)". The authors would like to acknowledge the School of Computing and Mathematics at the University of Ulster for hosting the first author's visit and NOAA for making OLR data available for various research communities.

References

1. Carreno, E., Capote, R., Yague, A., et al.: Observations of thermal anomaly associated to seismic activity from remote sensing. In: General Assembly of European Seismology Commission, Portugal, September 10–15, pp. 265–269 (2001)
2. Cervone, G., Kafatos, M., Napoletani, D., et al.: Wavelet Maxima Curves Associated with Two Recent Greek Earthquakes. Nat. Hazards Earth Syst. Sci. 4, 359–374 (2004)
3. Cervone, G., Singh, R.P., et al.: Wavelet maxima curves of surface latent heat flux anomalies associated with Indian earthquakes. Nat. Hazards Earth Syst. Sci. 5, 87–99 (2005)
4. Ouzounov, D., Freund, F.: Mid-infrared emission prior to strong earthquakes analyzed by remote sensing data. Advances in Space Research 33(3), 268–273 (2004)
5. Ouzounov, D., Liu, D., et al.: Outgoing long wave radiation variability from IR satellite data prior to major earthquakes. Tectonophysics 431, 211–220 (2007)
6. Pulinets, S.A., et al.: Ionospheric Precursors of Earthquakes, p. 316. Springer, Berlin (2004)
7. Hayakawa, M., Molchanov, O.A.: Seismo Electromagnetics:Lithosphere-Atmosphere-Ionosphere Coupling, TERRAPUB,Tokyo, p. 477 (2002)
8. Dey, S., Singh, R.P.: Surface latent heat flux as an earthquake precursor. Nat. Haz. Earth Sys. Sc. 3, 749–755 (2003)
9. Tronin, A.A.: Satellite thermal survey application for earthquake prediction. In: Hayakawa, M. (ed.) Atmospheric and Ionospheric Phenomena Associated with Earthquakes, TERRAPUB, Tokyo, pp. 717–746 (1999)
10. Tronin, A.A., Hayakawa, M., Molchanov, O.A.: Thermal IR satellite data application for earthquake research in Japan and China. J. Geodyn. 33, 519–534 (2004)
11. Tramutoli, G., Di Bello, N., Pergola, S.: Robust satellite techniques for remote sensing of seismically active areas. Annals of Geophysics 44(2), 295–312 (2001)
12. Qiang, Z.: Thermal Infrared Anomoly Precursor of Impending Earthquakes. Pure Pur. A. Geoph. 149, 159–171 (1997)
13. Liebmann, B., et al.: Description of a Complete (Interpolated) Outgoing Longwave Radiation Dataset. Bulletin of the American Meteorological Society 77, 1275–1277 (1996)
14. Mallat, S., Hwang, W.L.: Singularity Detection And Processing With Wavelets. IEEE Transactions on Information Theory 38, 617–643 (1992)
15. NCAR and NOAA, ftp ftp.cpc.ncep.noaa.gov; cd precip/noaa* for OLR directories (2008)
16. Misiti, M., Misiti, Y., et al.: Wavelet ToolboxTM 4 User's Guide (2008)
17. The Science Behind China's Sichuan Earthquake (2008), http://www.tectonics.caltech.edu
18. Grossmann, A., Morlet, J.: Decomposition of Hardy functions into square integrable wavelets of constant shape. SIAM J. Math. 15, 723–736 (1984)

A Hybrid Approach for Designing the Control System for Underwater Vehicles

A. Lamas, F. López Peña, and R.J. Duro

Grupo Integrado de Ingenieria, Universidad de La Coruña
Mendizábal s.n. 15403 Ferrol, Spain
Phone: +34 981142364, Fax: +34 981142364
flop@udc.es, richard@udc.es, alamas@udc.es

Abstract. An approach in the form of an automatic evolutionary design environment for obtaining any type of control systems for underwater vehicles is presented. A specific case is considered in which this strategy is hybridized with Artificial Neural Networks. The design procedure is carried out by means of evolutionary techniques from a set of specifications using as a fitness evaluator an ad-hoc hydrodynamic simulator which includes the estimation of added mass and added inertia coefficients. The resulting design environment was used to construct the neural network based controllers of a submersible catamaran. Results of the application of the automatic design procedure and of the operation of the controllers thus obtained are presented.

Keywords: Evolutionary Design, Artificial Neural Networks, AUVs.

1 Introduction

Modeling and controlling underwater vehicles is a complex problem due to their dynamic characteristics, which are usually described by highly non linear equation systems with uncertainties in their coefficients. In general, submarine controllers must verify several simultaneous maneuvering constraints depending on the type of vehicle and mission to be accomplished; thus increasing the complexity of their design.

Different approaches to submarine vehicle control can be found [1-3], but it is hard to find references addressing control design on the six degrees of freedom. It is common to design linear controllers as autopilots starting from a linear model of the plant. This initial model, as in the case of Feng and Allen [1], is usually developed starting from a stationary operation point and obtaining the transfer function. However, to obtain and assess a linear model is hard and its validity is, at least, questionable in many cases which are highly non linear and comprise different coupled processes. For this reason, some authors have built prototypes of ROV's or AUV's in order to test their control systems [6].

Another type of control system for underwater vehicles that display better results than traditional ones are those based on non linear models. They consist of a dynamic plant model and a reference model. The finite dimensional approximations to the dynamics of the vehicle are structurally similar to the rigid body equations. Different control techniques can be applied using these models and their stability analytically obtained. This is the approach followed by Smallwood and Whitcomb [2].

E. Corchado et al. (Eds.): HAIS 2009, LNAI 5572, pp. 88–95, 2009.
© Springer-Verlag Berlin Heidelberg 2009

The design process becomes very complicated when trying to obtain controllers for highly non linear systems. To address this problem using intrinsically non linear modelers/controllers, some authors have resorted to neural networks or neuro fuzzy based systems [3-8]. Kodogiannis [6], describes the application of ANNs to the control of a remotely operated underwater vehicle. In his work, an adaptive control is combined with neural network based architectures in order to approximate the non linearity of the system and its dynamics in the simulation of the vehicle. Other authors, such as Mills y Harris [7] or Suto and Ura [8] use neuro-fuzzy systems or simply sets of neural networks that have been designed ad hoc for their particular characteristics. These approaches required a training set for each controller, which is very difficult to obtain when the different controllers are coupled.

In the work presented here we have addressed the problem not only from the point of view of the automation of the operational control but also from that of the automation of the process of designing the optimal controller set starting from a group of specifications provided as an appropriate definition of the fitness of the results. To this end we have had to develop two different tools, first an evolutionary environment for automatic design [9] and, additionally, a computationally efficient hydrodynamic simulator that can provide evaluations of the fitness of the controller sets by simulating the behavior of the whole submarine in a mission sequence.

2 Evolutionary Based Automatic Design System

The three main pillars of our work that are hybridized into a single environment are the capabilities of Artificial Neural Networks to generalize complex non linear functions from a discrete set of samples, the efficient exploration of the design space provided by Evolutionary techniques and the development of a reliable hydrodynamic simulator with contained computational cost characteristics. All of these elements are linked in order to obtain the appropriate set of controllers.

The evolutionary based automatic design system is basically an evolutionary environment using genetic algorithms that permits multiple encoding mechanisms for the structures to be evolved and, more importantly, permits a modular and fluid interaction with mechanical, hydrodynamic and control simulators so that the individuals in a population may be evaluated in conditions that are as similar as possible to reality. Obviously, the more precise the simulators used to evaluate phenotypes, the better the results. On the other hand, simulation fidelity usually results in a high computational cost. For this reason, we have implemented the system as a distributed environment capable of operating in computer clusters using MPI.

The environment consists of three basic modules or components: a decision module made up of a set of simulators and human-machine interfaces that are integrated through a coordinator. The outputs of the simulators are inputs to a fitness function, which is also a part of the module and which provides a fitness value or vector. A search module implemented as a genetic algorithm in this particular case, although any other evolutionary technique can be easily inserted. The elements to evolve are Artificial Neural Networks (ANNs) that will be used as controllers for the submarine. The inputs to the networks are the data provided by the angular and linear position, velocity and acceleration sensors and their outputs the commands to the control planes assigned to each one of the controllers.

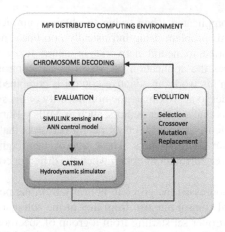

Mass: 26373 Kg
Maximum Thrust: 9090 Nw.
Power: 80 Kw diesel.
Electrical propulsion: 10 Kw.
Surface speed: 9.5 kn.
Underwater speed: 2 knots.
Maximum depth: 30 m.

Fig. 1. Evolutionary Design Environment (left). Catamaran of the example (right).

Finally a computation distribution module provides the distribution capabilities in order to parallelize the processing.

For the Genetic Algorithm to operate, the fitness of the individuals is obtained by means of set of simulators in the decision module. In the case of the application presented here, it uses a hydrodynamic simulator that was developed ad hoc. The controller simulator sends the control signals obtained from the simulation of the artificial neural networks resulting from the decoding of the genotype to the actuators in the submarine model specified in the hydrodynamic simulator. The hydrodynamic simulator returns the values for the angular and linear position, speed and acceleration, which are taken (after the addition of a level of noise for robustness) as sensor values and which are the ones used by the networks as inputs in order to calculate the actuation the next instant of time.

To calculate the fitness of a given controller set that makes up an individual it is necessary for the phenotype (the vehicle with the controllers) to undergo several trial runs in the hydrodynamic simulator. This implies a large number of simulations per generation of the evolutionary process. Consequently, there is a need for the last module, that is, the computation distribution module, which is in charge of coordinating the execution of the simulations distributing them among whatever computational resources it has available. In this case, the computations are distributed over a cluster of PCs using MPI (Message Passing Interface) to make the execution times manageable.

3   Hydrodynamic Simulator

The dynamical model has been developed from scratch independently of the ones used in other studies. Being aware that submarines and airships are governed by the same equations of motion and that the similitude parameters and geometries are quite close in both cases, we have made use of the theoretical and experimental developments carried out on airships in order to provide simple analytical methods for the determination of added mass, added inertia, and force coefficients in the equations of motion. In general, these coefficients are external data provided as input to the hydrodynamic simulator configured by these equations of motion, which are the conservation laws of momentum and angular momentum. Expressed in a coordinate system fixed to a submarine of mass m moving at speed $\mathbf{V}$ these are:

$$\frac{\mathbf{F}}{m} = \dot{\mathbf{V}} + \omega \otimes \mathbf{V}$$

$$\mathbf{M} = \mathbf{I} \cdot \dot{\omega} + \omega \otimes (\mathbf{I} \cdot \omega) \tag{1}$$

The force $\mathbf{F}$ and angular momentum $\mathbf{M}$ vectors account for the forces and moments directly applied on the vehicle. The inertia matrix $\mathbf{I}$ is a function of the vehicle's geometry and its weight's distribution. The vehicle's angular velocity is ω.

The forces exerted on the submarine can be written as:

$$\mathbf{F} = \mathbf{F_H} + \mathbf{F_T} + m(\mathbf{g} - \mathbf{b}) \tag{2}$$

Where $\mathbf{F_H}$ are the hydrodynamic forces, $\mathbf{F_T}$ the thrust and $\mathbf{b}$ is the buoyancy force divided by the mass m of the vehicle. Similarly, the moments on the center of gravity of the forces exerted on the submarine can be written as:

$$\mathbf{M} = \mathbf{M_H} + \mathbf{M_T} + md\mathbf{b} \tag{3}$$

Where $\mathbf{M_H}$ is the vector of the moments of hydrodynamic forces, $\mathbf{M_T}$ the moment produced by thrust and d is the distance between the center of gravity and the center of buoyancy. Notice than the thrust $\mathbf{F_T}$ and its moment $\mathbf{M_T}$ are known while the hydrodynamic forces $\mathbf{F_H}$ are given in the following general form:

$$\mathbf{F_H} = -\tfrac{1}{2} \rho V^2 V^{\frac{2}{3}} \mathbf{C_F} + \rho V \dot{\mathbf{V}} \cdot \mathbf{k} \tag{4}$$

Where ρ is the density of the fluid, V is the volume of the vehicle, $\mathbf{C_F} = (C_D, C_Q, C_L)$ is the vector formed with the drag, lateral force and lift coefficients, while $\mathbf{k}$ is a diagonal matrix containing the added mass coefficients. Similar coefficients, called inertia moment coefficients, appear in the determination of the momentum exerted by the fluid on the vehicle:

$$\mathbf{M_H} = \tfrac{1}{2} \rho V^2 V \mathbf{C_M} + \rho \mathbf{I}' \cdot \dot{\omega} \tag{5}$$

Where $\mathbf{I}'$ represent the volumetric inertia matrix of the volume occupied by the vehicle. The terms of added mass or added inertia moments are relevant when the densities of fluid and vehicle are similar, as in cases of submarines or airships.

Force, mass, and inertia moment coefficients can be determined experimentally, numerically or analytically. In principle they should be known as input for our simulator. Some numerical and analytical methods can be found in the literature to estimate then, as the analytical method by Tuckerman [10] originally developed to be used in airship design. The later one has been used in the present investigation. Details of this implementation can be found elsewhere [11][12].

In addition, the forces and moments exerted by the fins and rudders should be included. It's modeling is fundamental in designing the control system for the vehicle. In general, considering these forces as a result of hydrodynamic reactions, they can be expressed as:

$$F_{Rudder} = \frac{1}{2} \cdot C_{L.Rudder} \cdot S_{Rudder} \cdot \mathbf{V}^2 \cdot (\xi + \delta_p + \tau) \tag{6}$$

Where the incidence angle of the flow arriving to the rudder is taken as the sum of the vehicle's general incidence angle ξ in the plane normal to the rudder, plus the deflection of the rudder δ, plus the deviation angle τ due to the vehicle's angular velocity. Fernandez Ibarz et Al. [11][12] shows details about this simulator and the method for obtaining the different coefficients used.

4 Experiments

In order to test the evolutionary environment for the automatic design of submarine controllers using the hydrodynamic simulator described above, some numerical experiments consisting in obtaining a maneuver control system for a submersible catamaran in an automatic manner have been performed. The submersible catamaran considered as a test case is aimed at passenger transportation is shown in figure 1:

The elements that can be acted on to perform the control are the depth and direction rudders as well as the values of the thrust. A model with two perpendicular control surfaces on the stern is considered to study the control of this vehicle. The objective is to obtain a set of controllers that permit following trajectories in three dimensions as well as performing speed changes in such a way that the people inside are comfortable. The different types of maneuvers will be specified through pitch, speed, depth, roll or yaw commands or setpoints.

The complexity of the problem is evident, on one hand due to the coupling between the actuations of the controllers and, on the other, due to the fact that the response of the system is not instantaneous because of its inertia and the reactions of the medium in which it is immersed. Two sets of variables were determined to be very influential during the runs: the attack and sliding angles on one hand and the pitch, yaw and roll angles on the other. Depending on whether the catamaran is ascending or descending, the attack angles, the moments and the hydrodynamic forces on the vehicle change very rapidly. This variation and the difference between wind axes and body axes make the direction of motion of the catamaran different from that to which it is pointing. In addition, a roll angle appears when turning due to the asymmetry of the vehicle.

In the examples presented the capacity of the controllers obtained in following step type commands and their response to perturbations is analyzed. Whenever a setpoint is provided, the controller must be able to make the vehicle adapt to this command with the minimum error possible and by means of a response that applies speeds and accelerations that allow passengers a pleasant voyage. This implies that the pitch and roll angles must always be smaller than 15° and the accelerations of the system must be smooth and below 1m/s2.

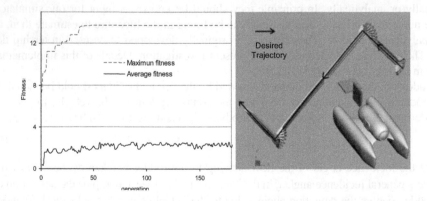

Fig. 2. Left. Evolution of fitness during the process of obtaining the pitch controller. Right. Catamaran following the trajectory indicated using the yaw controller.

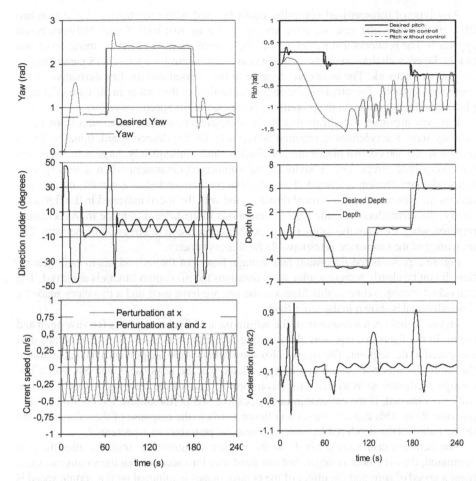

Fig. 3. Left column: Response of the submarine´s yaw control when it suffers a tridimensional perturbing current. Center: Actuation of the direction rudder. Bottom: Perturbing current at the direction x, y and z. Top right: Angular pitch behavior of the submarine for different commands. The dashed line indicates the behavior of the submarine without pitch control. Bottom right two: Actuation of the depth controller and acceleration induced on the passengers.

Regarding the evolutionary part of the system, during these experiments the mutation rate was 10% using a non linear mutation probability that favored small mutations, crossover was a typical two point crossover operator, tournament selection was chosen and the fitness function was a relatively simple one where each set of controllers in the population was executed three times with different objective trajectories for a given interval of time. The evaluation was carried out taking the worst run of the controller in the simulator by combining in a proportion of 3 to 1 the mean squared error of the resulting motion with respect to the desired one and the time it took the controller to be stable on the desired trajectory after each actuation. The number of individuals participating were 200 and in order to achieve reasonable execution times, the computation was distributed over a 50 PC cluster.

Any type of parametrized controller could be used with this strategy, i.e. a standard PID. In this particular case we have chosen to use an Artificial Neural Network based approach. The networks that model each of the controllers within a set are made up of two hidden layers with three neurons, two inputs and one output for a total of 18 parameters or weights per network. The inputs are distance to the setpoint and the first derivative of the controlled variable. The output is the angular actuation on the rudder in the case of control planes and the actuation on the propulsion in the case of the speed controller.

It is very important to indicate that all the maneuvers of the catamaran start from a zero velocity state. It accelerates as required until it achieves the desired speed. Initially, due to its low speed, the control planes are not efficient, and, consequently, large actuations produce very little effect. The behavior of the submerged catamaran when it undergoes a sinusoidal perturbation is studied. This perturbation is simulated as a variable current that acts on the submarine in the vertical direction and with the speed indicated in the figure. In theory, the controllers are quite hard to obtain due to the large difference in the actuation required when we want the submarine to perform immersion or emersion maneuvers. The buoyancy of the submarine is responsible for this asymmetry.

Figure 2 presents the maximum and average fitness of the individuals in each generation. It can be clearly seen that after 80 generations the maximum fitness is achieved. The individual corresponding to this fitness is the one we have used and a trajectory made by this individual is shown in the same figure.

Figure 3 displays the response of the submarine to different setpoints for yaw pitch and depth. In the figure we represent the resulting actuation in terms of the value obtained as compared to the setpoint. The figures also show what would happen with the catamaran if the controller was not used. It can be appreciated that the response is fast but with a low enough oscillation so as to allow for a comfortable ride for the passengers. Once the controller is obtained, if the catamaran undergoes external perturbations, the controller compensates them. This can also be seen in figure 4, where the response of the submarine and the actuation of the depth rudder under a sinusoidal perturbation is presented.

The behavior conforms perfectly to the motion requirements. Initially, after the first command, the overshoot is larger, but one must take into account that the catamaran starts from a speed of zero and the effect of the control planes is minimal until a certain speed is achieved. As controllability increases the way the catamaran follows the desired trajectory clearly improves. The effect of the perturbations on the system is smoothly compensated by the depth rudder. Finally, as indicated we have sought a low value for the acceleration in order to respect the comfort of the ride. If one looks at the figure it is easy to see that the acceleration never surpasses 0.9 m/s2.

The different actuations of the controllers are coupled. The best indication of the quality of the results obtained is when all of the controllers are acting together. This is the reason why in all of the figures we have presented in this section the submarine was operating using all of the controllers simultaneously. In every case, the submersible started from an initial velocity of zero.

5 Conclusions

A hybrid automatic design environment that integrates evolution, artificial neural networks and an efficient hydrodynamic simulation strategy for submarines has been presented in this paper. Thus, neural network based controllers for operating a submersible catamaran whose maneuvering capacity is based on control surfaces were automatically designed to compensate perturbing forces acting on the vehicle while taking it to the assigned position and orientation setpoints through the actuation over a propulsion system and depth and direction rudders. The objective was to control a 20 ton submersible catamaran aimed at tourist transportation and this imposed very severe constraints in terms of security and comfort parameters, limiting allowed angles and accelerations during the motion of the submersible. The results obtained clearly meet all the design parameters established. Very precise behaviors were obtained without having to resort to complicated non linear models of the plant in the controller.

Acknowledgments. This work was partially funded by the MEC of Spain through projects DEP2006-56158-C03-02 and DPI2006-15346-C03-01.

References

1. Feng, Z., Allen, R.: Modeling of Subzero II, ISVR Technical Memorandum 880, Southampton, University of Southampton (2001)
2. Smallwood, D.A., Whitcomb, L.L.: Toward Model Based Trajectory Tracking of Underwater Robotic Vehicles: Theory and Simulation. In: 21th International Symposium on Unmanned Untethered Submersible Technology, Durham, NC, USA (2001)
3. Kim, T.W., Yuh, J.: A novel neuro-fuzzy controller for autonomous underwater vehicles. In: Proceedings of the IEE Int. Conf.on Robotics and Automation, May 21-26, pp. 2350–2355 (2001)
4. Yuh, J.: A neural net controller for underwater robotic vehicles. IEEE Journal of Oceanic Engineering 15(3), 161–166 (1990)
5. Choi, S.K., Yuh, J., Takeshige, G.: Development of the Omnidirectional Intelligent Navigator. IEEE Robotics & Automation Mag. 2, 44–53 (1995)
6. Kodogiannis, V.S.: Neural Network Adaptive Control for Underwater Robotic Systems. Mechatronics Group, Dep. of Computer Science. U Westmister, London U.K (2001)
7. Mills, D., Harris, C.J.: Neurofuzzy modeling and control of a six degree of freedom AUV. Prentice Hall, Helmel Hapmstead (1994)
8. Suto, T., Ura, T.: Unsupervised Learning System for Vehicle Guidance Constructed with Neural Networks. In: Proc. 8th International Symposium on Unmanned Untethered Submersible Technology, Durham, New Hampshire, pp. 222–230 (1993)
9. Lamas, A., Duro, R.J.: ADEUS: Integrating Advanced Simulators, Evolution and Cluster Computing for Autonomous Robot Design. In: Proceedings of the International Conference on Computational Intelligence, Robotics and Autonomous Systems, pp. 398–404 (2001)
10. Tuckerman, L.B.: Inertia Factors of Ellipsoid for Use in Airship Design. Report NACA N°.210. National Advisory Committee for Aeronautics (1926)
11. Fernández Ibarz, J.: Modelo de simulación y control dinámico para un catamarán submarino con seis grados de libertad. PhD Thesis, Universidade Coruña, Spain (April 2003)
12. Fernández Ibarz, J., Lamas, A., López Peña, F.: Study of the Dynamic Stability of an Underwater Vehicle. In: Proceedings of the 8th International Conference on the Stability of Ships and Ocean Vehicles, Madrid, September 15-19, pp. 129–140 (2003)

Hydrodynamic Design of Control Surfaces for Ships Using a MOEA with Neuronal Correction

V. Díaz-Casás, Francisco Bellas, Fernando López-Peña,
and Richard Duro

Integrated Group for Engineering Research,
Universidade da Coruña, 15403, Ferrol, Spain
http://www.gii.udc.es

Abstract. In this paper we present a hybrid intelligent system for the hydrodynamic design of control surfaces on ships. Our main contribution here is the hybridization of Multiobjective Evolutionary Algorithms (MOEA) and a neural correction procedure in the fitness evaluation stage that permits obtaining solutions that are precise enough for the MOEA to operate with, while drastically reducing the computational cost of the simulation stage for each individual. The MOEA searches for the optimal solutions and the neuronal system corrects the deviations of the simplified simulation model to obtain a more realistic design. This way, we can exploit the benefits of a MOEA decreasing the computational cost in the evaluation of the candidate solutions while preesrving the reliability of the simulation model. The proposed hybrid system is successfully applied in the design of a 2D control surface for ships and extended to a 3D one.

Keywords: Automatic Design, Multiobjective Evolutionary Algorithms, Artificial Neural Networks, Control Surfaces.

1 Introduction

Naval designers have to take many aspects into account in the process of designing a ship, encompassing both the design of the shapes and that of the structure. One of the main aspects to consider in the external design of a ship are the control surfaces. Examples of control surfaces are the rudders, the daggerboards, the masts, the fairings and the hulls. Formally speaking, control surfaces are moving parts that cause various effects during the process of navigating, and are regarded as very relevant elements due to their role (governing the vessel, reducing drift, improving stability, ...). It is, therefore, necessary to obtain their design with great precision to achieve optimal surfaces.

The real modeling of control surfaces on a ship is governed by hydrodynamic criteria, leading to analytical equations difficult to solve and forcing engineers to rely on numerical solution software tools to produce the final design parameters that provide a reliable and efficient final product. Typically, these applications are used to check the results obtained with different control surfaces by simulating them. However, there is no general software tool for the design and

E. Corchado et al. (Eds.): HAIS 2009, LNAI 5572, pp. 96–103, 2009.

optimization of control surfaces. This was the main objective of the research line this work belongs to: developing and implementing a computational procedure for the automatic design and optimization of control surfaces.

To this end, we must point out two features of our problem domain: the high dimensionality of the search space (with many parameters) and the existence of multiple objective functions and several constraints. As a result, traditional optimization methods, such as gradient descent or simple heuristics, are not effective in this kind of engineering problems because they require a priori knowledge of the search space that, in most cases, is not available. This is the reason that such problems have begun to be addressed by computational metaheuristic techniques based on the combination of evolutionary algorithms (EA) and multiobjective techniques (MO), known as multiobjective evolutionary algorithms (MOEA). Although there are several MOEA algorithms [1] and they are increasing in specialty and effectiveness in real applications [2], recent comparison contests [3] have shown that the NSGA-II algorithm [4] provides, in most cases, the best results in optimization tests and real applications [5]. In this work, we are not interested in developing a new MOEA or improving the existing ones, but rather on producing an integrated system, consequently, we have selected the NSGA-II for the search process of our computational procedure.

In the particular case of naval engineering applications, one of the first works we can find is that by Quagliarella and Vinici [6] where the authors present a system for designing multicomponent airfoils for high-lift applications using a MOEA with two coupled flow solvers. To investigate the feasibility of full stern submarines, Thomas [7] has applied a MOEA with three objectives: the maximization of internal volume, the minimization of the power coefficient for propulsors and the minimization of the cavitation index. Mierzwicki [8] presented in his thesis a simplified metric and methodology for measuring the risk of ship design concepts as part of a MOEA system. Olcer [9] has developed a two-stage hybrid approach for solving a particular multiobjective problem in ship design, the subdivision arrangement of a ROPAX vessel. In [10], the authors optimize a hydrodynamic shape using the NSGA-II algorithm, specifically a Wigley hull.

The main difference with the tool we are presenting here is that most of these works provide an "ad-hoc" approach for design and implementation involving the user in much of the process. In addition, the hydrodynamic models are typically simplified to reduce the computational cost with the consequent deviation with respect to reality. To solve it, we propose a hybrid system that uses a MOEA for the optimization with the addition of a neuronal correction procedure to compensate for the deviations of the simplified simulation model.

2   The Automatic Design Procedure

The computational procedure we are presenting here has been developed from a very general perspective, becoming a generic shape optimization system that, in this particular case, we are going to apply to the design of control surfaces for ships. The system has been implemented as a set of independent blocks which can be substituted and relinked depending on the particular design process. The structure of the design system can be viewed in Fig. 1 and comprises three modules: Search Module, Decoder Module and Evaluation Module.

The Search Module contains the MOEA that performs the search process in the solution space. As commented in the introduction, we have selected the NSGA-II algorithm due to its successful performance in comparison tests. The constraint handling is achieved using the mechanism proposed by Deb et al. in [4]. The specific encoding of the chromosomes, reproduction operators and computational cost depends on the particular application, so they will be commented in the results section. Every time a new generation of solutions is obtained in the Search Module, a decoding process is necessary. This process is performed by the Decoder Module which conforms an interface between the parametric definition of the model used by the search strategy and the fluid dynamics solver. Finally, the Evaluation Module calculates the quality of each alternative through a fluid-dynamic simulation. The output of this module is a fitness value (or vector) used as input by the Search Module to evaluate the solutions. The system execution finishes after a fixed number of iterations.

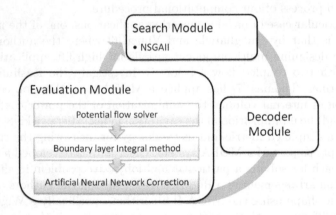

Fig. 1. Automatic Design Procedure

2.1   Hydrodynamic Simulation Model

Fluid dynamics problems are described by the Navier-Stokes equations. However, the complete system generated by the direct application of these equations is not suitable for any design procedure in real problems due to the high computational cost required. Thus, in each design problem this simulation model must be adapted in order to reduce its computational cost. This is especially true when using evolutionary algorithms which, as commented before, imply a large number of simulations.

To evaluate the fitness of control surfaces, under certain conditions we can transform the three-dimensional problem into a set of two-dimensional ones and thus compute the performance of 2D profiles. To minimize the computational cost of the simulator, viscous forces and turbulence effects are initially neglected following a customary approach in this type of problems. Hence, a basic fluid dynamics principle indicates that the corresponding 2D velocity field can be obtained as the gradient of a potential function. Thus, the pressure acting on the profile surface may be obtained according to the classical Bernouilli principle by calculating the flow kinetic energy variation along the airfoil. Consequently, the resulting lift (C_L) and drag forces (C_D) are calculated by integration of the pressure distribution along the surface.

As fluid viscosity is being neglected an important component of drag forces is being ignored. The effects of viscous processes appearing on the surfaces can be accounted for by considering the boundary layer. An effective way to model this layer is by using a classical integral method. In our case the ones by Thwaites [11], Michel [12] and Head [13], are used in laminar, transitional and turbulent boundary layers, respectively. After that, the airfoil drag coefficient is calculated by using the Squire-Young formula[14].

However, when simulating cases where the profile is set at large angles of attack α to the incoming flow, processes like the boundary layer detachment are not considered by this classical approach, and thus the error of the simulation increases. This implies that additional corrections must be included in the simulator in order to achieve realistic results. The effective torque and force that each surface can produce was calculated as the integral of the effects of each airfoil. To extrapolate 2D airfoil results to 3D blade element, it was necessary to introduce the effect of fluid circulation around the blade in the calculus procedure, that was achieved using the Prandtl[15] correction.

2.2 Neuronal Correction

The large range of angles of attack α that can appear in the 2D profiles imply that, in some of them, especially in those of higher value, the detachment of the boundary layer can occur. When this happens, the potential flow condition is not valid anymore in a large portion of the flowfield, and large errors are introduced. To correct these deviations, we have developed a non-linear correlation profiling method through the application of an artificial neural network based model. As an initial approach, we have used a multilayer perceptron neural network that has six input neurons, two hidden layers of eight neurons and two outputs. The inputs are the lift coefficient (C_L), the drag coefficient (C_D), the Reynolds number (R_e), the angle of attack (α), the detachment point on the suction side and the detachment point on the pressure side.

The network parameters have been trained using a classical backpropagation algorithm. We have considered a training set of 450 values obtained from published results [16] of NACA 2410, NACA 0009, NACA 2415, NACA 2541, CLARKY, ClarkySM and DAE11 profiles. Fig. 2 shows the real, calculated and corrected lift coefficient for the NACA 6409 profile. After 1000 training steps, the initial average error was reduced from 14,5% to 5%, providing a suitable precision in the results with a much lower computational cost than a turbulent flow

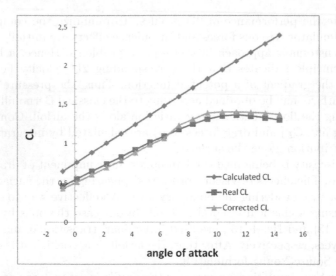

Fig. 2. Real, calculated and corrected lift coefficient for the NACA 6409 profile. The correction is made using the neural network.

solver. The network obtained has been applied in the final step of the evaluation of the model as shown in Fig. 1.

3   Design Example

To test the behavior of the automatic design procedure, we have created an experiment to solve the optimization problem of a hydrodynamic aerodynamic 2D profile based on a modified 4-digit NACA shape. To achieve it, we define 4 parameters that must be optimized:

- **Thickness**, limited from 2% to 36% of the chord.
- **Point of maximum deformation**, located in the region between 20% and 80% of the chord.
- **Maximum value and sense of the arrow in the middle section**, from 0% to 8% of the chord.
- **Displacement of the maximum chord thickness that defines the profile**, limited from -10% to 10% of the chord with respect to the original value in the NACA profile.

With the aim of analyzing the advantages and difficulties derived from the minimization or maximization of several objectives simultaneously, three different cases have been considered:

- **Two goals**: the objectives are the maximization of the lift coefficient for angles of attack of 5° and -5°.

- **Three goals**: we add the minimization of the drag coefficient for a 0° angle of attack to the previous objectives
- **Three goals and one constraint**: we introduce an additional requirement on the maximum thickness of the profile. We consider two cases:
  - Maximum 1% of the chord.
  - Maximum 0.5% of the chord.

In addition, we have decided to define a sense in the curvature of the profile, so that the algorithm can select the one that best suites the different angles of incidence studied. This sequence of problems involves a gradual increase in complexity. The gradual incorporation of new objectives and constraints allows us to analyze their individual influence on the shape of the profile.

In a first phase of the experiment, we have analyzed the influence of the parameters of the NSGA-II algorithm on the final result which has defined the most suitable population size and number of generations for the problems. We have sought a compromise between diversity of the population, quality of results and computational requirements. With the aim of comparing the results of the three problems, the parameters were fixed for all of them. At this point it is necessary to indicate that the computational cost of the evaluation is identical in all cases. The tests were carried out using a population of 20 individuals and 1000 generations of evolution. With these parameters, the tests were performed in a computer cluster with *Dual-Core AMD Opteron 2212 Processor*, obtaining an average time length for each test of 17.28 minutes, with a standard deviation of 1.43 minutes.

Fig. 3 left shows the Pareto fronts for the tests carried out with respect to the lift coefficients for 5° and -5°. The need to maximize these conflicting objectives has generated three population groups: one with positive curvature, adapted to the 5° angle of attack, and negative lift coefficients for the negative -5° angle of attack; an opposite case with negative curvature and negative lift values for 5° angle of attack, and a third group of symmetric profiles with moderate lift values for both angles. The effects of the constraints on this problem appear as a limitation on the maximum values of the lift coefficient.

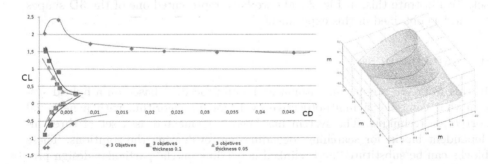

Fig. 3. Left image: Pareto fronts for the lift CL and drag CD coefficients. Right image: 3D rudder obtained.

The effects of minimizing the aerodynamic resistance coefficient with zero angle of attack are shown in Fig. 3 left. The different fronts that are generated due to the introduction of constraints on the thickness can be distinguished. This way, by limiting the thickness of the profile, its aerodynamic resistance is limited and, consequently, parallel fronts of least drag and lift are generated.

In these problems we are trying to maximize the lift at 5° and -5° and minimize the drag when the angle of attack is zero. In Fig. 4 we can see the evolution of the average value of the drag, and reflects the increase of the average value during the optimization process which runs opposite to the last objective. Fig. 3 left shows the effects of the constraint on minimum thickness. As shown, when the thickness is not constrained larger lift values are obtained at the expense of significantly increasing the drag.

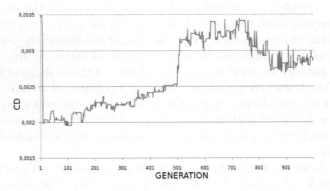

Fig. 4. Evolution of the average drag coefficient

Finally, we must take into account that, as the objectives are conflicting, at the end of each optimization process, we do not obtain a single solution but a set of compromise solutions. Thus, depending on the weighing the designer determines for the problem, the most appropriate one must be selected from the set. To illustrate this, in Fig. 3 right we have represented one of the 3D shapes (a rudder) obtained in this experiment.

4 Conclusions

In this work, we present an automatic design procedure based on a hybrid system that combines a multi-objective evolutionary algorithm with a neuronal correction technique. The system has been implemented as a set of three independent blocks for searching, decoding and evaluating the solutions. These blocks can be substituted and relinked depending on the particular design process. The search module implements a NSGA-II algorithm that tries to optimize the design parameters. The evaluation module calculates the quality of each candidate solution through a simplified fluid-dynamic simulator, the results of which are corrected using a artificial neural network based technique. The system has been applied to the optimization of control surfaces for ships, using 4 parameters to define the problem and three different combinations of objectives and constraints. The results obtained show the successful behavior of the automatic design procedure and correction mechanism.

Acknowledgments. This work was partially funded by the MEC through projects DEP2006-56158-C03-02 and DPI2006-15346-C03-01 and FPU program.

References

1. Coello, C., Van Veldhuizen, D., Lamont, G.: Evolutionary Algorithms for Solving Multi-Objective Problems, 2nd edn. Kluwer Academic Publishers, Dordrecht (2007)
2. Coello, C., Lamont, G.: Applications of Multi-Objective Evolutionary Algorithms. World Scientific, Singapore (2004)
3. Huang, V.L., Qin, A.K., Deb, K., Zitzler, E., Suganthan, P.N., Liang, J.J., Preuss, M., Huband, S.: Problem Definitions for Performance Assessment of Multi-objective Optimization Algorithms Special Session on Constrained Real-Parameter Optimization, Technical Report, Nanyang Technological University, Singapore (2007)
4. Deb, K., Pratap, A., Agarwal, S., Meyarivan, T.: A Fast and Elitist Multi-Objective Genetic Algorithm. NSGA-II. IEEE Transactions on Evolutionary Computation 6(2), 182–197 (2000)
5. Suganthan, P.N.: Special Session on Constrained Real-Parameter Optimization, CEC (2007),
   http://www.ntu.edu.sg/home/EPNSugan/index_files/CEC-07/
   CEC-07-Comparison.pdf
6. Vinici, A., Quagliarella, D.: Inverse and Direct Airfoil Design using a Multiobjective Genetic Algorithm. AIA A Journal 35(9), 1499–1505 (1997)
7. Thomas, M.W.: Multi-Species Pareto Frontiers in Preliminary Submarine Design. Foundations of Computing and Decision Sciences 25(4), 273–289 (2000)
8. Mierzwicki, T.S.: Risk Index for Multi-Objective Design Optimization of Naval Ships MSc thesis, Virginia Polytechnic Institute and State University, (2003)
9. Olcer, A.I.: A hybrid approach for multi-objective combinatorial optimisation problems in ship design and shipping. Computers&Operations Research 35(9), 2760–2775 (2008)
10. Clemente Fernández, J.A., Pérez Rojas, L., Pérez Arribas, F.: Optimización de una forma hidrodinámica mediante algoritmos evolutivos XIX Congreso Panamericano de IngenierPía Naval, Transporte Marítimo e Ingeniería Portuaria (2005)
11. Thwaites, B.: On the momentum equation in laminar boundary layer flow. ARC RM 2587 (1952)
12. Cebeci, T., Smith, A.: Analysis of turbulent Boundary layers, pp. 332–333 (1974)
13. Head, M.R.: Entrainment in the turbulent boundary layers. ARC R&M 3152 (1958)
14. Martin, O.L., Hansen: Aerodynamics of Wind, pp. 11–66. Turbines Science Publishers. Ltd. (2000)
15. Moores, J.: Potential Flow 2-Dimensional Vortex Panel Model: Applications to Wingmills, University of Toronto (2003)
16. Abbot, H., Von Doenhoff, A.E.: Theory of Wing Sections. Dover, New York (1949)

Closures of Downward Closed Representations
of Frequent Patterns

Marzena Kryszkiewicz

Institute of Computer Science, Warsaw University of Technology
Nowowiejska 15/19, 00-665 Warsaw, Poland

Abstract. The discovery of frequent patterns is one of the most important issues in the data mining area. A major difficulty concerning frequent patterns is huge amount of discovered patterns. The problem can be significantly alleviated by applying concise representations of frequent patterns. In this paper, we offer new lossless representations of frequent patterns that are derivable from downward closed representations by replacing the original elements and eventually some border ones with their closures. We show for which type of downward closed representations the additional closures are superfluous and for which they need to be stored. If the additional closures are not stored, the new representations are guaranteed to be not less concise than the original ones.

Keywords: frequent patterns, downward closed sets, closures, closed sets.

1 Introduction

The problem of discovering frequent patterns in large databases is one of the most important issues in the area of data mining. It was introduced in [1] for a sales transaction database. Frequent patterns were defined there as sets of items that co-occur more often than a given threshold. Frequent patterns are commonly used for building association rules. For example, an association rule may state that 30% of sentences with the term "study" have also a term "research".

The number of discovered patterns is usually huge, which makes analysis of the discovered knowledge infeasible. It is thus preferable to discover and store a possibly small fraction of patterns, from which one can derive all other significant patterns whenever required. Lossless representations of frequent patterns were discussed e.g., in [2-16,18]. First representations of frequent patterns were based on closed sets [18] and generators [7]. At present, the representations using generalized disjunctive rules for reasoning about patterns are regarded as most concise ones. The generalized disjunction-free set representations [8-12] and the non-derivable itemset representation [4-5] are typical instances of such representations. They are by up to a few orders of magnitude more concise than the frequent patterns, as well as the generators and closed itemsets representations [5,8,11]. This is particularly evident for the representation offered in [11], which represents not only frequent positive patterns, but also patterns with negation. The representation of frequent patterns based on essential patterns, proposed in [6], was reported as by up to two orders of magnitude more concise, than the closed itemsets representation.

E. Corchado et al. (Eds.): HAIS 2009, LNAI 5572, pp. 104–112, 2009.
© Springer-Verlag Berlin Heidelberg 2009

The representations of frequent patterns find practical applications in data discovering tasks, such as mining grammatical patterns [19], hierarchical clustering of the sets of documents [14], discovering synonyms or homonyms [19,20], as well for discovering dominated meanings from WEB [17].

In this paper, we offer new lossless representations of frequent patterns that are derivable from downward closed representations by replacing the original elements and eventually some border ones with their closures. The first trial of using closures instead of a downward closed representation was reported in [16] for frequent non-derivable itemsets. In our paper, we do not concentrate on a particular representation. Instead we propose a general framework of transforming any downward closed representation into its closure variant. We show for which type of downward closed representations the additional closures are superfluous and for which ones they need to be stored. In the cases when the additional closures are not stored, the new representations are guaranteed to be not less concise than the original representations.

The layout of the paper is as follows: Section 2 recalls the notions of frequent itemsets, downward closed sets, closures, closed sets, generators and key generators. Our new contribution is presented in Section 3. There, we identify types of downward closed representations available in the literature and propose new closure representations for each type. Section 4 contains the summary and conclusions.

2 Basic Notions

2.1 Itemsets, Supports of Itemsets, Frequent Itemsets, Downward Closed Sets

Let $I = \{i_1, i_2, ..., i_m\}$, $I \neq \emptyset$, be a set of distinct *items*. In the case of a transactional database, a notion of an item corresponds to a sold product, while in the case of a relational database an item is an attribute-value pair. Any set of items is called an *itemset*. Without any loss of generality, we will restrict further considerations to transactional databases. Let $\mathcal{D}$ be a set of transactions, where each transaction is a subset of I. *Support* of itemset X is denoted by $sup(X)$ and is defined as the number (or percentage) of transactions in $\mathcal{D}$ that contain X.

Property 2.1.1 [1]. Let $X \subseteq I$. If $Y \subset X$, then $sup(Y) \geq sup(X)$.

An itemset X is called *frequent* if its support is greater than some user-defined threshold *minSup*, where $minSup \in [0, |\mathcal{D}|]$. The set of all frequent itemsets will be denoted as $\mathcal{F}$; i.e., $\mathcal{F} = \{X \subseteq I | sup(X) > minSup\}$.

A set $X \subseteq 2^I$ is defined as *downward closed*, if for each itemset in X, all its subsets are also in X; that is, if

$$\forall X \in \mathcal{X}, Y \subset X \Rightarrow Y \in \mathcal{X}.$$

Please note that all supersets of an itemset which does not belong to a downward closed set X do not belong to X either.

Property 2.1.2. Let X be a downward closed set. $\forall X \notin \mathcal{X}, Y \supset X \Rightarrow Y \notin \mathcal{X}$.

Property 2.1.3 [1]. $\mathcal{F}$ is downward closed.

2.2 Closures, Closed Itemsets

A *closure* of an itemset X is denoted by $\gamma(X)$ and defined as the greatest (w.r.t. set inclusion) itemset in 2^I that occurs in all and only transactions in $\mathcal{D}$ in which X occurs; that is,

$$\gamma(X) = I \cap (\cap\{T \in \mathcal{D} | T \supseteq X\}).$$

Property 2.2.1. Let $X, Y \subseteq I$.

a) $|\gamma(X)| = 1$.
b) $sup(\gamma(X)) = sup(X)$.
c) If $X \subset Y$, then $\gamma(X) \subseteq \gamma(Y)$.
d) If $X \subset Y \subseteq \gamma(X)$, then $sup(Y) = sup(X)$.

Sometimes the closure of an itemset X is defined equivalently as the greatest superset of X that has the same support as $sup(X)$.

Example 2.2.1. Let $\mathcal{D}$ be the database from Table 1. Table 2 shows the supports and closures of sample itemsets in $\mathcal{D}$.

<table>
<tr><td colspan="2" align="center">Table 1. Sample database $\mathcal{D}$</td></tr>
</table>

Table 1. Sample database $\mathcal{D}$

Id	Transaction
T_1	$\{abcde\}$
T_2	$\{abcdef\}$
T_3	$\{abcdehi\}$
T_4	$\{abe\}$
T_5	$\{bcdehi\}$

Table 2. Supports and closures of itemsets

Itemset X	Transactions supporting X	$sup(X)$	$\gamma(X)$
$\varnothing$	$\{T_1,T_2,T_3,T_4,T_5\}$	5	$\{be\}$
$\{a\}$	$\{T_1,T_2,T_3,T_4\}$	4	$\{abe\}$
$\{c\}$	$\{T_1,T_2,T_3,T_5\}$	4	$\{bcde\}$
$\{h\}$	$\{T_3,T_5\}$	2	$\{bcdehi\}$
$\{i\}$	$\{T_3,T_5\}$	2	$\{bcdehi\}$
$\{chi\}$	$\{T_3,T_5\}$	2	$\{bcdehi\}$

The set of the closures of all itemsets in X, $X \subseteq 2^I$, will be denoted by $C(X)$; that is,

$$C(X) = \{\gamma(X) | X \in X\}.$$

For $X = 2^I$, $C(2^I)$ will be denoted briefly as C. Please note that $C(X)$, and in particular C, may not be downward closed. Itemset X is defined as *closed* if $\gamma(X) = X$.

Property 2.2.2 [8]. Let $X \subseteq I$.

$X \in C$ iff $\gamma(X) = X$ iff $\forall Y \supset X$, $sup(X) > sup(Y)$ iff $sup(X) \neq \max\{sup(X \cup \{a\}) | a \in I \setminus X\}$.

The knowledge of the supports of all itemsets in C is sufficient to determine the support of each itemset X in 2^I as the support of either the closed itemset being the closure of X (see Property 2.2.1b and Property 2.2.3) or an itemset with maximal support among closed supersets of X (see Property 2.2.4).

Property 2.2.3 [18]. Let $X \subseteq I$. $\gamma(X)$ is a least superset of X in C.

Property 2.2.4 [7,8]. Let $X \subseteq I$. $sup(X) = \max\{sup(Y) | Y \in C \wedge Y \supseteq X\}$.

The closures of all frequent itemsets are all frequent closed itemsets (i.e. $C(\mathcal{F}) = \mathcal{F} \cap C$. An itemset is frequent if and only if it has a closure in $C(\mathcal{F})$. Clearly, frequent closed itemsets stored altogether with their supports enable correct determination of the supports of all frequent itemsets [7,8].

2.3 Generators

Let $X \subseteq I$. A minimal itemset $Y \subseteq X$ satisfying $\gamma(Y) = \gamma(X)$ is called a *generator* of X. By $\mathcal{G}(X)$ we denote the set of all generators of X, i.e. $\mathcal{G}(X) = \text{MIN}\{Y \subseteq X | \gamma(Y) = \gamma(X)\}$.

Property 2.3.1. Let $X \subseteq I$.

a) $|\mathcal{G}(X)| \geq 1$.

b) $sup(X) = sup(Y)$ for each $Y \in \mathcal{G}(X)$.

The set of the generators of all itemsets in $\mathcal{X}$, $\mathcal{X} \subseteq 2^I$, will be denoted by $\mathcal{G}(\mathcal{X})$, i.e.,

$\mathcal{G}(\mathcal{X}) = \cup_{X \in \mathcal{X}} \mathcal{G}(X)$. For $\mathcal{X} = 2^I$, $\mathcal{G}(2^I)$ will be denoted briefly as $\mathcal{G}$.

X is defined as a *key generator* if $\mathcal{G}(X) = \{X\}$.

Property 2.3.2 [8]. Let $X \subseteq I$.

$X \in \mathcal{G}$ iff $\mathcal{G}(X) = \{X\}$ iff $\forall Y \subset X$, $sup(X) < sup(Y)$ iff $sup(X) \neq \text{min}\{sup(X \setminus \{a\}) | a \in X\}$.

Property 2.3.3 [2,7,8]. $\mathcal{G}$ and $\mathcal{G}(\mathcal{F})$ are downward closed.

Property 2.3.4 [7,8]. Let $X \subseteq I$. $sup(X) = \text{min}\{sup(Y) | Y \in \mathcal{G} \wedge Y \subseteq X\}$.

The lossless generators representation of frequent itemsets introduced in [7] consists of all frequent key generators ($\mathcal{G}(\mathcal{F})$) stored altogether with their supports and of all infrequent key generators all proper subsets of which are in $\mathcal{G}(\mathcal{F})$. Clearly, the generators representation is downward closed.

3 Representing Patterns with Closures of Downward Closed Sets

In this section, we will consider an arbitrary downward closed representation $\mathcal{R}$ stored altogether with the supports of their elements.[1] We identify the following two types of downward closed representations that occur in the literature:

- type 1: for each itemset X in 2^I, it can be determined if X belongs to the representation based on the supports of only proper subsets of X; the frequent non-derivable itemsets are of this type;

- type 2: for each itemset in 2^I, it can be determined if X belongs to the representation based on the supports of only X and its proper subsets; examples of such representations are: the generators representation, the generalized disjunction-free representation, generalized disjunction-free representation of patterns admitting negation, and the union of frequent essential patterns, $\{\varnothing\}$ and minimal infrequent essential patterns[2].

For both types of downward closed representations, we will examine the consequences of replacing the elements of $\mathcal{R}$ with their closures, i.e. with $C(\mathcal{R}) = \{\gamma(X) | X \in \mathcal{R}\}$. Each element of $C(\mathcal{R})$ will be stored with its support value. Since each

---

[1] This restriction may be relaxed in the case of some downward closed representations, where the supports of infrequent itemsets do not need to be stored.

[2] Originally, the representation based on essential patterns was defined in [6] as the union of all frequent essential (non-empty) itemsets and maximal frequent itemsets. As this representation is not downward closed, we modified it here to obtain its downward closed variant.

itemset has exactly one closure, which may be also the closure of other itemsets, the cardinality of $C(R)$ does not exceed the cardinality of R.

Proposition 3.1. Independently of type of R, $C(R)$ is sufficient to determine the closure and support for each itemset in R correctly.

Proof. $C(R)$ contains the closures of all itemsets in R. Hence for any itemset X in R, $\gamma(X) \in C(R)$ and by Property 2.2.3, $\gamma(X)$ is a least superset of X in $C(R)$. Clearly, $sup(X) = sup(\gamma(X))$.   □

Theorem 3.1. If downward closed representation R is of type 1, then $C(R)$ is sufficient to reconstruct R and the supports of all its elements correctly.

Proof (constructive). If $C(R)$ is empty, then there is no itemset in R. Otherwise, $\emptyset$ belongs to R and its support equals the support of its closure $\gamma(\emptyset)$ in $C(R)$. From now on, consecutive elements of R are determined in an iterative way. During each ith iteration, itemsets of cardinality i are considered. Each itemset having at least one proper subset not belonging to R does not belong to R (by Property 2.1.2). Each remaining itemset is classified to R or discarded based on the supports of only its proper subsets. The support of each newly found element of R is equal to the support of its closure in $C(R)$. If among all itemsets of the considered cardinality i, there is none having all proper subsets in R, then all itemsets of the cardinality $k \geq i$ do not belong to R.   □

Later on in this section, we will consider the representation R assuming it is of type 2. Before we proceed, let us consider the following example.

Example 3.1. Let us assume that $R = \{\emptyset_5, \{a\}_4, \{h\}_2, \{i\}_2\}$ (values in square brackets in the subscript denote supports of itemsets) is a downward closed representation of type 2 found in Table 1. Clearly R is downward closed. Then $\gamma(\emptyset)=\{be\}$, $\gamma(\{a\})=\{abe\}$, $\gamma(\{h\})=\gamma(\{i\})=\{bcdehi\}$. Hence, $C(R) = \{\{be\}_5, \{abe\}_4, \{bcdehi\}_2\}$.

Now let us assume that we do not know which itemsets are elements of R, and want to reconstruct R based on $C(R)$. By Proposition 3.1, the closures and supports of all elements of R will be determined correctly; e.g. $\{be\}$ will be found a closure of $\emptyset$ as a least closed superset of $\emptyset$ (see Property 2.2.3). Similarly, the closures and supports of the remaining elements of R will be identified correctly.

Now, we will try to determine the closure and support of $\{c\}$. Again using Property 2.2.3, we will find that $\gamma(\{c\})=\{bcdehi\}_2$, so $sup(\{c\})=2$. This time both the closure and the support were determined incorrectly. The real closure of $\{c\}$ is $\{bcde\}_4$ and the real support equals 4. The reason of the problem lies in the fact that $C(R)$ does not contains the real closure of $\{c\}$, but contains the closure $\{bcdehi\}$ of another element of R (here: of $\{h\}$ and of $\{i\}$) that is a least superset of $\{c\}$ in $C(R)$.

Now, let us consider itemset $\{f\}$. We try to use Property 2.2.3, but we do not find any closed superset of $\{f\}$ in $C(R)$. In this case, we may conclude correctly that $\{f\}$ does not belong to R, since its closure is not present in $C(R)$.   □

Let us conclude our findings from Example 3.1:

F1. $C(\mathcal{R})$ does not guarantee the correct reconstruction of the downward closed representation $\mathcal{R}$ of type 2, although the closures and supports of the elements of $\mathcal{R}$ are properly derivable.

F2. Itemsets that do not have supersets in $C(\mathcal{R})$ certainly do not belong to $\mathcal{R}$.

F3. Itemsets that have supersets in $C(\mathcal{R})$ are not guaranteed to belong to $\mathcal{R}$. An itemset X not belonging to $\mathcal{R}$ the closure of which is a subset of an itemset in $C(\mathcal{R})$, may be assigned an incorrect closure and support in the case when there is exactly one minimal (least) itemset among X's supersets in $C(\mathcal{R})$. Otherwise, X will be identified correctly as not belonging to $\mathcal{R}$ since an itemset cannot have more than one closure.

It follows from conclusion F2 that within the process of reconstructing $\mathcal{R}$ from $C(\mathcal{R})$, only the itemsets that have supersets in $C(\mathcal{R})$ are likely to be elements of $\mathcal{R}$.

In the sequel, by $Cover(X)$, where $X \subseteq 2^I$, we will denote the set of all itemsets each of which has a superset in X; that is, $Cover(X) = \{Y \subseteq X | X \in \mathcal{X}\}$. Hence, $Cover(C(\mathcal{R}))$ contains all itemsets having supersets in $C(\mathcal{R})$.

Corollary 3.1. If downward closed representation $\mathcal{R}$ is of type 2, then only elements of $Cover(C(\mathcal{R}))$ may belong to $\mathcal{R}$.

Corollary 3.2. $Cover(C(\mathcal{R})) \supseteq \mathcal{R}$.

Corollary 3.3. The set $C(Cover(C(\mathcal{R})))$ stored altogether with the supports of its elements is sufficient to determine correctly the closure and support of each itemset in $Cover(C(\mathcal{R}))$, and by this in $\mathcal{R}$.

Eventually, we are ready to address issue F1 and discuss the correctness of reconstruction of $\mathcal{R}$ with $C(Cover(C(\mathcal{R})))$.

Theorem 3.2. If downward closed representation $\mathcal{R}$ is of type 2, then $C(Cover(C(\mathcal{R})))$ is sufficient to reconstruct $\mathcal{R}$ and the supports of all its elements correctly.

Proof (constructive). If $C(Cover(C(\mathcal{R})))$ is empty, then there is no itemset in $\mathcal{R}$. Otherwise, $\varnothing$ belongs to $\mathcal{R}$ and its support equals the support of its closure $\gamma(\varnothing)$ in $C(Cover(C(\mathcal{R})))$. From now on, consecutive elements of $\mathcal{R}$ are determined in an iterative way. During each ith iteration, itemsets of cardinality i are considered. Each itemset having at least one proper subset not belonging to $\mathcal{R}$ does not belong to $\mathcal{R}$ (by Property 2.1.2). For the remaining itemsets, their supports are calculated as the supports of their closures in $C(Cover(C(\mathcal{R})))$. Each remaining itemset is classified to $\mathcal{R}$ or discarded based on only its support and the supports of its proper subsets. If among all itemsets of the considered cardinality i, there is none having all proper subsets in $\mathcal{R}$, then all itemsets of the cardinality $k \geq i$ do not belong to $\mathcal{R}$.   □

Now, we will consider the ways of reducing $C(Cover(C(\mathcal{R})))$ that will not lead to incorrect reconstruction of $\mathcal{R}$. Clearly, potential candidates for reduction are closures of itemsets that do not belong to $\mathcal{R}$. First we will want to make use of the property of downward closed sets stating that all supersets of an itemset not belonging to downward closed set $\mathcal{R}$ do not belong to $\mathcal{R}$.

Let us define a *negative border* of $\mathcal{R}$, denoted as $Bd^-(\mathcal{R})$, as the set of all minimal itemsets in 2^I that do not belong to $\mathcal{R}$, i.e.:

$$Bd^-(\mathcal{R}) = \{X \subseteq I \mid X \notin \mathcal{R} \wedge \forall Y \subset X, Y \in \mathcal{R}\}.$$

Clearly, $C(\mathcal{R}) \cup C(Bd^-(\mathcal{R}))$ is sufficient to determine both all itemsets belonging to $\mathcal{R}$ and all (minimal) itemsets not belonging to $\mathcal{R}$. We already know that $C(\mathcal{R})$ is sufficient to identify those (minimal) itemsets not belonging to $\mathcal{R}$ that do not have closures in $Cover(C(\mathcal{R}))$ (see F2). Thus, in order to correctly identify $\mathcal{R}$ and the remaining (minimal) itemsets not belonging to $\mathcal{R}$, we need to extend $C(\mathcal{R})$ with the closures of those itemsets in $Bd^-(\mathcal{R})$ that are contained in $Cover(C(\mathcal{R}))$.

Theorem 3.3. If downward closed representation $\mathcal{R}$ is of type 2, then $C(\mathcal{R}) \cup C(\{X \in Bd^-(\mathcal{R}) \mid X \in Cover(C(\mathcal{R}))\})$ is sufficient to reconstruct $\mathcal{R}$ and the supports of all its elements correctly.

Taking into account that an itemset belongs to $Cover(C(\mathcal{R}))$ if and only if it has a superset among itemsets in $C(\mathcal{R})$, allows us to infer Corollary 3.4.

Corollary 3.4. If downward closed representation $\mathcal{R}$ is of type 2, then $C(\mathcal{R}) \cup C(\{X \in Bd^-(\mathcal{R}) \mid X \subseteq Y \wedge Y \in C(\mathcal{R})\})$ is sufficient to reconstruct $\mathcal{R}$ and the supports of all its elements correctly.

Further reduction of a closure representation of $\mathcal{R}$, which we propose beneath, follows from the observation F3 stating that an itemset for which there are more than one minimal superset in $C(\mathcal{R})$ does not belong to $\mathcal{R}$.

Theorem 3.5. If downward closed representation $\mathcal{R}$ is of type 2, then $C(\mathcal{R}) \cup C(\{X \in Bd^-(\mathcal{R}) \mid |MIN\{Y \in C(\mathcal{R}) \mid X \subseteq Y\}| = 1\}$ is sufficient to reconstruct $\mathcal{R}$ and the supports of all its elements correctly.

When building a closure variant of $\mathcal{R}$ of any type, it is worth to remember that each itemset has the same closure as its generator, so $C(\mathcal{R})$ can be determined as the closures of only the generators in $\mathcal{R}$, i.e. as $C(\mathcal{G}(\mathcal{R}))$.

4 Summary and Conclusions

In this paper, we proposed a general framework of transforming a downward closed representation into its closure variant. We identified two types of downward closed representations that are available in the literature. A downward closed representation of type 1 enables determining for each itemset X in 2^I, if X belongs to the representation based on the supports of only proper subsets of X. A downward closed representation of type 2 enables determining for each itemset X in 2^I, if X belongs to the representation based on the supports of only X and its proper subsets.

We have proved that for any downward closed representation $\mathcal{R}$ of type 1, the closures of all elements of $\mathcal{R}$ (i.e. $C(\mathcal{R})$) are sufficient to reconstruct $\mathcal{R}$. If an itemset in a new closure representation is a closure of on average n itemsets in $\mathcal{R}$, then the new representation is n times more concise than $\mathcal{R}$.

We have shown that $C(\mathcal{R})$ may not be sufficient to reconstruct $\mathcal{R}$ in the case of a downward closed representation $\mathcal{R}$ of type 2. We have proved, however, that $C(\mathcal{R})$ extended with the closures of each minimal itemset not belonging to $\mathcal{R}$ that has exactly one minimal superset in $C(\mathcal{R})$ is sufficient to reconstruct $\mathcal{R}$.

References

1. Agrawal, R., Imielinski, T., Swami, A.: Mining Associations Rules between Sets of Items in Large Databases. In: ACM SIGMOD, Washington, USA, pp. 207–216 (1993)
2. Bastide, Y., Taouil, R., Pasquier, N., Stumme, G., Lakhal, L.: Mining Frequent Patterns with Counting Inference. ACM SIGKDD Explorations 2(2), 66–75 (2000)
3. Bykowski, A., Rigotti, C.: A Condensed Representation to Find Frequent Patterns. In: PODS 2001. ACM SIGACT-SIGMOD-SIGART, USA, pp. 267–273 (2001)
4. Calders, T., Goethals, B.: Mining All Non-Derivable Frequent Itemsets. In: Elomaa, T., Mannila, H., Toivonen, H. (eds.) PKDD 2002. LNCS, vol. 2431, pp. 74–85. Springer, Heidelberg (2002)
5. Calders, T., Goethals, B.: Non-Derivable Itemset Mining. In: Data Mining and Knowledge Discovery, vol. 14, pp. 171–206. Kluwer Academic Publishers, Dordrecht (2007)
6. Casali, A., Cicchetti, R., Lakhal, L.: Essential Patterns: A Perfect Cover of Frequent Patterns. In: Tjoa, A.M., Trujillo, J. (eds.) DaWaK 2005. LNCS, vol. 3589, pp. 428–437. Springer, Heidelberg (2005)
7. Kryszkiewicz, M.: Concise Representation of Frequent Patterns Based on Disjunction–Free Generators. In: ICDM 2001, San Jose, California, USA, pp. 305–312 (2001)
8. Kryszkiewicz, M.: Concise Representations of Frequent Patterns and Association Rules. Publishing House of Warsaw University of Technology, Warsaw (2002)
9. Kryszkiewicz, M.: Reducing Infrequent Borders of Downward Complete Representations of Frequent Patterns. In: The First Symposium on Databases, Data Warehousing and Knowledge Discovery, Baden-Baden, Germany, pp. 29–42 (2003)
10. Kryszkiewicz, M.: Reducing Borders of k-Disjunction Free Representations of Frequent Patterns. In: ACM Symposium on Applied Computing (SAC), pp. 559–563. ACM, Nikosia (2004)
11. Kryszkiewicz, M.: Generalized Disjunction-Free Representation of Frequent Patterns with Negation. J. JETAI, 63–82 (2005)
12. Kryszkiewicz, M., Gajek, M.: Concise Representation of Frequent Patterns Based on Generalized Disjunction-Free Generators. In: Chen, M.-S., Yu, P.S., Liu, B. (eds.) PAKDD 2002. LNCS, vol. 2336, pp. 159–171. Springer, Heidelberg (2002)
13. Kryszkiewicz, M., Rybiński, H., Gajek, M.: Dataless Transitions between Concise Representations of Frequent Patterns. J. Int. Inf. Systems 22(1), 41–70 (2004)
14. Kryszkiewicz, M., Skonieczny, Ł.: Hierarchical Document Clustering Using Frequent Closed Sets. In: Advances in Soft Computing, pp. 489–498. Springer, Heidelberg (2006)
15. Mannila, H., Toivonen, H.: Multiple Uses of Frequent Sets and Condensed Representations. In: KDD 1996, Portland, USA, pp. 189–194 (1996)

16. Muhonen, J., Toivonen, H.: Closed Non-derivable Itemsets. In: Fürnkranz, J., Scheffer, T., Spiliopoulou, M. (eds.) PKDD 2006. LNCS, vol. 4213, pp. 601–608. Springer, Heidelberg (2006)
17. Nykiel, T., Rybinski, H.: Word Sense Discovery for Web Information Retrieval. In: MCD Workshop (ICDM) 2008, Piza (2008)
18. Pasquier, N., Bastide, Y., Taouil, R., Lakhal, L.: Efficient Mining of Association Rules Using Closed Itemset Lattices. J. Inf. Systems 24(1), 25–46 (1999)
19. Rybinski, H., Kryszkiewicz, M., Protaziuk, G., Jakubowski, A., Delteil, A.: Discovering Synonyms based on Frequent Termsets. In: Kryszkiewicz, M., Peters, J.F., Rybinski, H., Skowron, A. (eds.) RSEISP 2007. LNCS (LNAI), vol. 4585, pp. 516–525. Springer, Heidelberg (2007)
20. Rybinski, H., Kryszkiewicz, M., Protaziuk, G., Kontkiewicz, A., Marcinkowska, K., Delteil, A.: Discovering Word Meanings Based on Frequent Termsets. In: Raś, Z.W., Tsumoto, S., Zighed, D.A. (eds.) MCD 2007. LNCS, vol. 4944, pp. 82–92. Springer, Heidelberg (2008)

Transductive-Weighted Neuro-Fuzzy Inference System for Tool Wear Prediction in a Turning Process

Agustín Gajate[1], Rodolfo E. Haber[1], José R. Alique[1], and Pastora I. Vega[2]

[1] Instituto de Automática Industrial , Spanish Council for Scientific Research,
Ctra. Campo Real Km. 0.200, Arganda del Rey, 28500 Madrid, Spain
{agajate,rhaber,jralique}@iai.csic.es
[2] Departamento de Informática y Automática, Universidad de Salamanca,
Pza. de los Caídos s/n, 37008 Salamanca, Spain
pvega@usal.es

Abstract. This paper presents the application to the modeling of a novel technique of artificial intelligence. Through a transductive learning process, a neuro-fuzzy inference system enables to create a different model for each input to the system at issue. The model was created from a given number of known data with similar features to data input. The sum of these individual models yields greater accuracy to the general model because it takes into account the particularities of each input. To demonstrate the benefits of this kind of modeling, this system is applied to the tool wear modeling for turning process.

Keywords: Transductive reasoning, Neuro-fuzzy inference system, Modeling, Tool wear.

1 Introduction

Nowadays, new strategies for modeling dynamic systems have become essential to qualitatively improve complex and large-scale processes. A model is considered to be a depiction of the physical properties of an object or process, so if we have a model of the process, we actually have a tool for planning, recognition and depiction of the information involved in the process. Moreover, we can infer properties from the process that can be used to understand, to predict and to control the process itself.

There are two ways of dealing with process modeling. The first approach ("white box") is based on the assumption that the process can be fully described by mathematical equations representing the corresponding physical laws [1]. The second approach ("black box") is based on the idea that the process is completely unknown and there is no *a priori* knowledge of the model structure to reflect the process's physical behavior. The unknown model parameters are estimated using experimental data to obtain an input/output relationship. The main advantage of the "black box" approach when using accurately measured data is that it is possible to develop a model without requiring physical process knowledge [2]. The main drawbacks of this method lie in the structure of the model, which is unable to offer any physical meaning. In practice, the best way is to combine the two approaches, if possible, so that the more thoroughly-known parts can be modelled using physical knowledge and the less-known

E. Corchado et al. (Eds.): HAIS 2009, LNAI 5572, pp. 113–120, 2009.

ones can be approximated through the "black box" approach. This is what is known as the "gray box" approach [3].

Sometimes the process is extremely complex and fraught with uncertainty, and its behavior is practically impossible to describe exactly by conventional modelling tools. In that case approaches based on artificial intelligence techniques like Fuzzy Logic and Neural Networks are the best way to cope with this problem. Recent years have been characterized by the development of new paradigms in the field of artificial intelligence (AI). Nowadays, the hybridization of fuzzy logic with neural networks is the most well-established and best-known method.

By the late nineties, several hybrid neuro-fuzzy systems have already been developed, which may be separated into two major groups: neural networks endowed with the ability to handle fuzzy information [fuzzy-neural networks (FNN)] [4,5], and fuzzy systems combined with neural networks in order to enhance certain desirable characteristics [neural-fuzzy systems (NFS)] [6,7]. A deep review of available neural-fuzzy strategies goes beyond the scope of the paper [8,9].

The simplest and easiest way to obtain a neuro-fuzzy model is to create its knowledge base using verbalization techniques. Frequently, a complete verbal description of how a complex process behaves is quite difficult to obtain. In such situations an identification procedure is required. So then, a neuro-fuzzy model can be built from measured input/output ("black box") data using engineering knowledge about the process variables, goals and disturbances ("white box") by applying a recursive identification technique.

From the viewpoint of systems theory and system modeling, transductive methods generate a model at a single point of the workspace. For each new datum that has to be processed, the closest examples are selected among the known data, with the goal of creating a new local model that dynamically approximates the process in its new state as close as possible. The main issue is therefore how to assign more weight to the specific information related with the datum to be processed than to the general information provided by the entire training set [10].

Machining processes are widely used in manufacturing. Four basic types of operations are turning, drilling, milling, and grinding, performed by different machine tools. Indeed, the importance of maximising a tool's working time and doing the utmost to keep tools from breaking is directly related with turning-process optimisation. The key issue is to find an appropriate trade-off between tool wear and productivity considering the tool's cost, its replacement cost, the cost of writing off the machine's idle time, and so forth. Avoiding breakage derived from a excessive tool wear is another capital factor, because replacing the tool after it breaks means increased costs, since the post-breakage stage is one of the trickiest, most unpredictable times, aside from the damage that may be done to the part and, not unusually, to the whole machine itself.

From the best of authors' knowledge, transductive reasoning methods have not been applied yet for modeling of machining processes. Moreover, this work proposes the application of a Transductive-Weighted Neuro-Fuzzy Inference System to obtain local models for predicting tool wear in a turning process [11]. Furthermore, the proposed strategy for modeling is simpler, faster and more accurate than other neuro-fuzzy inference systems to deal with complex processes. The main contributions are therefore not only a modified strategy of an state-of-the-art neuro-fuzzy inference system for modeling, but also its application for monitoring tool wear in a turning process.

2 Neuro-fuzzy Inference Systems and Transductive Reasoning

In terms of learning procedures, most of evolutionary neuro-fuzzy strategies apply inductive reasoning systems. In inductive reasoning the key issues is to find a general model drawn from the entire set of input/output data representing the whole system. The model is later used for designing the required control system. In contrast, there are transductive reasoning methods that generate a model at a single point of the workspace. Transductive methods have some advantages over inductive methods, because sometimes creating a valid model for the entire space or region of operation is a difficult task, yielding insufficient performance in some cases. The dynamic generation of local models enables to easily expand the represented knowledge as the set of known data facilitating incremental on-line learning. In addition, these strategies are capable of functioning correctly with a small training set.

Transductive reasoning methods have been applied to text recognition applications [12], time series prediction and medical diagnosis applications [11]. However, on the basis of reviewed literature, applications in the field of manufacturing processes have not been previously reported.

Neuro-fuzzy inference techniques combine the paradigms of fuzzy logic and neural networks in order to take advantage of both techniques achieving the simplicity of modelling (neural networks), while providing knowledge explicitly expressed in a set of "if-then" rules (fuzzy logic).

2.1 Transductive-Weighted Neuro-Fuzzy Inference System

This work is inspired on a new paradigm of neuro-fuzzy inference systems to obtain local models of the process. Transductive-Weighted Neuro-Fuzzy Inference System (TWNFIS) is a relatively new transductive reasoning system that consists in a dynamic neuro-fuzzy inference system with local generalization [11]. TWNFIS is endowed with three important characteristics: Neural, Fuzzy and Trasductive.

- Neural: Excellent ability to model any nonlinear function with a high accuracy in addition to possessing a high learning capacity.
- Fuzzy: Semantic transparency, ability to represent human thought as well as excellent behavior before uncertainty and imprecision.
- Transductive: Estimation of the model in a single input/output set of the space, using only information related with the corresponding set.

The relevant steps for modeling on the basis of TWNFIS are given as follows. The system's inputs can be treated in different kinds of physics units but the normalization is recommended. In this paper, each input data x' is normalized according to (1):

$$x = \frac{x' - \mu_x}{\sigma_x} \tag{1}$$

where μ_x is the mean and σ_x is the standard deviation of the set of known data or training set.

A local model is created using data from the training set that are the closest to each new input datum. The weighted Euclidean distance is used for selecting each data subset (2). The size of the subset (N_q) is one parameter of the algorithm. Weights (w_j) of each element of the input vector ($w_j \in [0,1]$) are computed in a *posteriori* model-adjusting process, reflecting the importance of each variable. Initially they all have unitary value.

$$\|\overline{x} - \overline{y}\| = \left[\frac{1}{P} \sum_{j=1}^{P} w_j \left| x_j - y_j \right|^2 \right]^{\frac{1}{2}} \tag{2}$$

where P is the number of elements in the input data vector, $\overline{x}$ is the input data vector, and $\overline{y}$ is each one of the vectors in the training set.

Membership functions are built iteratively on the basis of the closest data. TWNFIS uses the Evolving Clustering Method (ECM) to create these functions. The main difference with regard to the approach here-in proposed is the use of a clustering algorithm more suitable for real-time modelling dynamic systems instead of ECM. A clustering strategy called Quality Cluster Algorithm (QT_Clust) is then applied [13]. This algorithm utilizes two parameters: a threshold to indicate a maximum diameter of the clusters and a minimum number of elements (data) in a cluster. A candidate cluster is created using the first datum. The other elements are iteratively added without exceeding the maximum diameter. A second candidate cluster is formed starting with the second data and repeating the procedure. The number of candidate clusters is equal to the number of closest data. At this point, the largest candidate cluster is selected and retained. Data are removed from consideration and the entire procedure is repeated on the smaller set. The resulting clusters are ellipsoids. The center and the radio of the clusters set the center and the width of the Gaussian membership functions, respectively.

The weight of each variable is adjusted according to its relevance within each subspace. A gradient-descent algorithm to optimize the weights and parameters of the fuzzy rules is then applied. If the closest neighbors do not change due to the new adjusted weights of the variables, a new model is created setting the weights obtained in the previous iteration. Finally, the model is used to predict the system output.

Defuzzification is done using a modified center of area method. The resulting error function is stated as a weighted quadratic error function that is derivable:

$$E = \frac{1}{2} v_i \left[f(x_i) - d_i \right]^2 \tag{3}$$

where $f(x_i)$ is the defuzzification function that yields the output, d_i are the target values, and v_i indicates the proximity of each target to the expected prediction. A Gradient-descent algorithm is then applied to after deriving (3).

2.2 Tool Wear Model in Turning Process

Tool wear is generally caused by a combination of various phenomena, although it is an event inherent to the cutting process. Tool wear can occur gradually or in drastic breakdowns. Gradual wear may occur by adhesion, abrasion or diffusion, and it may appear in two ways: wear on the tool's face or wear on its flank. Contact with the chip produces a crater into the tool's face. Flank wear, on the other hand, is commonly due to friction between the tool and the work piece material.

Tool wear is not a physical variable whose value may be measured by any specific method, but rather a subjective estimate a specialist can make, depending on the condition of the tool's edges and surfaces. Since there is no single criterion for deciding when a tool needs sharpening, different lifetimes may be predicted for the same tool employed in the same process. Two widely used criteria are catastrophic failure and changes in tool geometry. Other criteria that are sometimes used are a degraded tool-surface finish, deviation in cutting forces, increased power consumption, overheating, non-tolerant pieces and the appearance of chattering. This paper deals the tool wear modeling of the turning process. Data supplied by [14] coming from the measurement of different sensors such as acoustic emissions signals, vibrations (accelerations) and cutting forces (Fig. 1). Therefore, it is a multiple-input /single-output model to predict tool wear.

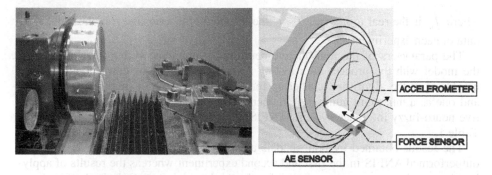

Fig. 1. Turning process and sensors related to the model

The choice of cutting force, vibration and acoustic emission signals is because they are the most affected by the tool wear. Among the cutting force signals, it has become experimentally that the cutting force in the z axis is more sensitive to changes in the wear compared to the cutting forces in the components on the x and y axes.

3 Results

Modeling of tool wear for turning processes is based on the experimental data provided in [14]. For the sake of simplicity the table with experimental data is shown in Appendix I and a brief description is given as follows.

The experimental study was carried out for turning operations on cast iron work piece material and uncoated coated carbide insert tool material The process parameters were: cutting speed (94 m/min and 188 m/min), feed: 0.06 mm/rev and 0.08 mm/rev, depth: 0.7 mm (constant). The turning operations were carried out on a high precision lathe machine. Acoustic emission signals (ring down count), vibrations (acceleration), cutting forces and tool wear were recorded for each operation on the machine. The authors stated that tool wear (flank wear) was measured off-line using a microscope.

The tool wear T_w' was modeled through the time t, the cutting force in the direction of the cutting speed F_z, the vibrations (accelerations) of the tool a_t, and the acoustic emission signals AES. Therefore, depending on the model, the tool wear was estimated as follows:

$$T_w' = \hat{H}(t, F_z, a_t, AES) \qquad (4)$$

where $\hat{H}$ represents the corresponding neuro-fuzzy system.

The Total Average Error TAE (5) is used to assess the accuracy of the model. Additionally, another measure of accuracy is the number of data that cross an individual error of 10%.

$$TAE = \frac{1}{n} \sum_n \frac{\left| T_w - T_w' \right|}{T_w} \cdot 100 \qquad (5)$$

where T_w is the real tool wear, T_w' is the modeled tool wear and n is the number of data of each experiment.

The parameters of the transductive neuro-fuzzy inference system that best match the model with the process after testing several configurations correspond to: three neighbors in the algorithm, clustering threshold value of 2.53 (maximum diameter) and one as a minimum number of elements. The parameters of TWNFIS and inductive neuro-fuzzy inference system (ANFIS) [14] for modeling tool wear are shown in Table 1.

The results obtained by both models are shown in Fig. 2 and Table 2. TWNFIS outperformed ANFIS in the first and second experiment whereas the results of applying these techniques are very similar for the third experiment. Only in the fourth experiment ANFIS showed better accuracy than TWNFIS.

Table 1. Neuro-fuzzy algorithms for modeling tool wear

Algorithm	Inductive System [14]	Transductive System
System	MISO	MISO
Clustering	Substractive	Quality Algorithm
Membership functions type	Gaussian	Gaussian
Inference system	Takagi-Sugeno	Mamdani
Number of membership functions	7	Variable each run (max. 3)
Number of rules	7	Variable each run (max. 3)
Training algorithms	BP + Least Square Method	Back Propagation
Iterations	3	3
Learning rate	10^{-3}	10^{-3}
Error tolerance	0	0
Training data set	24 samples	24 samples
Validation data set	47 samples	47 samples

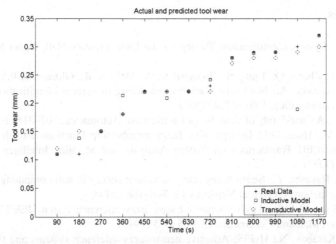

Fig. 2. Real tool wear and obtained with the models for experiment 1

Table 2. Average errors of the created models

Experiment	Inductive Model [14]	Point with AE >10%	Proposed Model	Point with AE >10%
1	7.12 %	3	5.37 %	1
2	40.4 %	4	5.30 %	3
3	3.46 %	1	3.98 %	0
4	1.97 %	0	7.19 %	3

4 Discussion

The first application of a paradigm for modeling a complex dynamical process inspired in TWNFIS is proposed in this paper. The introduction of a more efficient clustering technique to this strategy in order to deal with computing and real-time constraints is a novelty with regard to the original approach proposed in [11].

Likewise, from the best of authors knowledge, the application for modeling tool wear in a real manufacturing process (a turning process), has not been previously reported in the literature. The comparison of this technique with ANFIS according with the work reported in [14] demonstrated the superiority of the suggested approach..

From technical viewpoint an efficient computational model for predicting tool wear is essential. The importance of maximising a tool's working time and doing the utmost to keep tools from breaking is directly related with cutting-process optimisation. The key issue is to find an appropriate trade-off between tool wear and productivity considering the tool's cost, its replacement cost, the cost of writing off the machine's idle time, and so forth.

Acknowledgments. This work was supported by DPI2008-01978 COGNETCON and CIT-420000-2008-13 NANOCUT-INT projects of the Spanish Ministry of Science and Innovation.

References

1. Ljung, L.: System Identification: Theory for the User. Prentice-Hall, Upper Saddle River (1999)
2. Sjoberg, J., Zhang, Q., Ljung, L., Benveniste, A., Delyon, B., Glorennec, P.Y., Hjalmarsson, H., Juditsky, A.: Nonlinear black-box modeling in system identification: a unified overview. Automatica, 1691–1724 (1995)
3. Bohlin, T.: A Case-Study of Gray Box Identification. Automatica, 307–318 (1994)
4. Keller, J.M., Hunt, D.J.: Incorporating fuzzy membership functions into the perceptron algorithm. IEEE Transactions on Pattern Analysis and Machine Intelligence, PAMI 7, 693–699 (1985)
5. Mitra, S., Hayashi, Y.: Neuro-fuzzy rule generation: survey in soft computing framework. IEEE Transactions on Neural Networks 11, 748–768 (2000)
6. Jang, J.-S.R.: ANFIS: adaptive-network-based fuzzy inference system. IEEE Transactions on Systems, Man and Cybernetics 23, 665–685 (1993)
7. Kim, J., Kasabov, N.: HyFIS: Adaptive neuro-fuzzy inference systems and their application to nonlinear dynamical systems. Neural Networks 12, 1301–1319 (1999)
8. Nauck, D., Klawonn, F., Kruse, R.: Foundations of Neuro–Fuzzy Systems. Wiley, Chichester (1997)
9. Wang, L.X.: A course in fuzzy systems and control. Prentice-Hall, Inc., Upper Saddle River (1996)
10. Vapnik, V.: Statistical Learning Theory. John Wiley & Sons, Inc., New York (1998)
11. Song, Q., Kasabov, N.: TWNFI - a transductive neuro-fuzzy inference system with weighted data normalization for personalized modeling. Neural Networks 19, 1591–1596 (2006)
12. Joachims, T.: Transductive inference for text classification using support vector machines. In: Machine Learning: Proceedings of the Sixteenth International Conference (1999)
13. Heyer, L.J., Kruglyak, S., Yooseph, S.: Exploring expression data identification and analysis of coexpressed genes. Genome Research 9, 1106–1115 (1999)
14. Sharma, V.S., Sharma, S.K., Sharma, A.K.: Cutting tool wear estimation for turning. Journal of Intelligent Manufacturing 19, 99–108 (2008)

Review of Hybridizations of Kalman Filters with Fuzzy and Neural Computing for Mobile Robot Navigation

Manuel Graña, Iván Villaverde, Jose Manuel López Guede,
and Borja Fernández*

Grupo de Inteligencia Computacional, UPV/EHU
www.ehu.es/ccwintco

Abstract. Kalman Filters (KF) are at the root of many computational solutions for autonomous systems navigation problems, besides other application domains. The basic linear formulation has been extended in several ways to cope with non-linar dynamic environments. One of the latest trend is to introduce other Computational Intelligence (CI) tools, such as Fuzzy Systems or Artificial Neural Networks inside its computational loop, in order to obtain learning and advanced adaptive properties. This paper offers a short review of current approaches.

1 Introduction

Navigation, defined as the the process of determining and maintaining a course or trajectory to a goal location[8], has been considered the most basic and key problem in mobile robotics since the beginning of the research on that area. Opposite to a "wandering" behavior of the robot, in which it simply "moves around" by just avoiding obstacles in its path, navigation implies a knowledge of the structure of its environment and the ability to self-locate itself in it and in respect to its goal. That first condition requires that the robot has some a priori information about its surroundings. Autonomy increases inversely withe a priori information needed by the system to navigate in its environment. Maps[23] are the world representation used to guide navigation. The most autonomous form of navigation is Simultaneous Localization and Mapping (SLAM)[1,6,7], in which the robot builds a map while localizing itself in it, and uses this position estimation to integrate new measurements from the environment into the previous map. This basic dynamic models in SLAM are probabilistic, and the computational solutions are stochastic ones such as Expectation Maximization algorithms, Monte Carlo particle filters and Kalman Filters (KF). The latter were developed in the early 60's by R. E. Kalman and R. S. Bucy[11,24]. The origin of the use of Kalman Filters in mobile robotics can be traced back to the 80's, when the actual formulation was introduced by Smith, Cheeseman and Self [21,22].

---

* The Spanish Ministerio de Educacion y Ciencia supports this work through grant DPI2006-15346-C03-03.

E. Corchado et al. (Eds.): HAIS 2009, LNAI 5572, pp. 121–128, 2009.

Because the KF is in essence a linear method, the classical extensions found in the literature to cope with non-linear phenomena are linearization strategies. In this paper we explore the diverse hybridizations of the KF with of Computational Intelligence (CI) tools, which follow an intrinsic non linear modelling approach. These hybridizations may help to improve the accuracy and tracking power of the KF. This survey is intended to help opening avenues for research n Hybrid Systems and their applications in mobile robotics. We have found that the main KF hybridizations are: (1) using KF as an estimation algorithm to train computational systems (Artificial Neural Networks, Fuzzy Systems), instead of the simple gradient descent algorithms, (2) using CI tools to model the KF elements more realistically, (3) mixing EKF with other (fuzzy) representations.

The structure of the paper is as follows: We give a short review of the Kalman filters (KF) and their variants that have found application in mobile robotics in section 2. In section 3 we present the approaches to hybrid KF. Finally, section 4 gives final conclusions and comments.

2   Kalman Filters and Non-linear Variants

In a typical mobile robot SLAM navigation, the robot moves around taking measurements of its environment trying to self-localize in the world, and uses them also to build a map to refer this localization. Uncertainties in the measurement come from the sensor noise, and from the inaccurate realization of the commanded actions (i.e. wheel skid). Formally, at each time instant k the robot would be at state $x_k \in \Re^n$, and its motion is assumed to follow a given model $x_k = f(x_{k-1}, u_{k-1}, w_{k-1})$, which becomes $x_k = A_k x_{k-1} + B_k u_k + w_k$ in the linear case, where w_{k-1} is the motion noise, and u_{k-1} the motion command issued to the system at time $k - 1$. At each time instant the robot makes some environment measurements z_k that can be assumed to follow a model $z_k = h(x_k, v_k)$ of the system state x_k and observation noise v_k, which becomes $z_k = H_k x_k + v_k$ in the linear case.

The KF is a recursive method to estimate the state of a system. The knowledge of state of the system at time k will be represented by its estimation $\hat{x}_k$ and its related error covariance matrix P_k. This error covariance is a measure of the accuracy of the state estimation, defined as $P_k = E[e_k e_k^t]$, being $e_k = x_k - \hat{x}_k$ the estimation error. State estimation is performed by means of a two step prediction-correction feedback process. In the first step, the Prediction step, an *a priori* estimation of the state ($\hat{x}_k^-$ and P_k^-) is done. Then, measurements of the environment are taken into account in a second step, the Correction step, to compute the *a posteriori* estimation ($\hat{x}_k^+$ and P_k^+), which becomes the sytem estimation ($\hat{x}_{k+1}$ and P_{k+1}) at time $k + 1$. This recursive computation schema is the core of the KF, and all their variants.

The Discrete Kalman Filter (DKF) is the classical formulation of the linear KF for linear problems with known noise covariance matrices, the well know Prediction-Correction equations are as follows:

$$\hat{x}_k^- = A\hat{x}_{k-1} + Bu_{k-1}; \; P_k^- = AP_{k-1}A^T + Q, \tag{1}$$

$$\hat{x}_k^+ = \hat{x}_k^- + K_k(z_k - H\hat{x}_k^-); \quad P_k^+ = (I - K_k H)P_k^-, \tag{2}$$

where K_k is known as the *Kalman Gain*, and is a factor that relates the uncertainty between the state estimation and the residual prediction error.

The Extended Kalman Filter (EKF) is an improvement of the DKF to deal with non-linear problems. Assumed that the non-linear model is known, the state is predicted according to its model, and aPredictive step and the performed observation follows the bare application of this model:

$$\hat{x}_k^- = f(\hat{x}_{k-1}, u_{k-1}, 0); \quad \hat{z}_k^- = h(\hat{x}_k^-, 0). \tag{3}$$

A linear approximation of the non-linear model is performed at the point $\left(\hat{x}_k^- \hat{z}_k^-\right)$:

$$x_k \approx \hat{x}_k^- + A(x_{k-1} - \hat{x}_{k-1}) + W w_{k-1}; \quad z_k \approx \hat{z}_k^- + H(x_k - \hat{x}_k^-) + V v_k, \tag{4}$$

where A and W are the Jacobian matrices of f relative to x and w, respectively, and H and V are the Jacobian matrices of h relative to x and v, respectively. From this point the EKF is equal to a DKF, it applies the coorresponding equations adapted from 1 and 2. EKF is not an optimal estimator (as the DKF was) and it is very sensitive to wrong initial condition estimation and bad process models. The way it linearizes the functions it is also source of inefficiency. As second and higher terms of the Taylor series are discarded, if those terms are not negligible, it introduces error and bias in the transformation, as the local linearity assumption is broken. Also, the derivation of the Jacobian matrices is not usually trivial and can be very difficult in some applications.

The Unscented Kalman Filter (UKF) [9] replaces the linear approximation of the EKF with the *unscented transformation*. The aim of this transformation is to calculate the statistics of a random variable which undergoes a non-linear transformation. The basic idea is that is easier to approximate a probability distribution than an arbitrary non-linear function, and it does so by applying the non-linear function to a set of points $\{\chi_i\}$ sampled deterministically from a hypothetical population of points with mean $\bar{x}$ and spatial covariance P_x and calculating the mean and covariance of the transformed points. The non-linear dynamic system model is applied to these points $\gamma_i = f[\chi_i]$ and first and second order statistics are computed (applying convenient weights $\{W_i\}$):

$$\bar{y} = \sum_{i=0}^{2n} W_i^{(m)} \gamma_i; \quad P_y = \sum_{i=0}^{2n} W_i^{(c)} (\gamma_i - \bar{y})(\gamma_i - \bar{y})^T. \tag{5}$$

The UKF uses these estimations as the a priori estimation computed at the Prediction step, $\hat{x}_k^- = \bar{y}$ and $P_k^- = P_y$ as computed in 5. The system observation prediction $\hat{z}_k^-$ and its covariance matrix P_{z_k} is computed in an analogous way. Finally, the correction step is computed:

$$\hat{x}_k = \hat{x}_k^- + K_k\left(z_k - \hat{z}_k^-\right); \quad P_k = P_k^- - K_k P_{z_k} K_k^T, \tag{6}$$

where the gain is computed as $K_k = P_{x_z y_z} P_{z_k}^{-1}$.

Other variants proposed in the literature try to cope with non-stationary processes by estimating the noise covariance matrix at each time step, or work with agregates instead of the original variables to reduce the dimensionality of the state vector [12], or merge KF with Monte Carlo techniques [15].

3 Hybrized Approaches

We will consider three kind of hybridizations found in the literature: The use of KF as estimation (learning, tuning) algorithm, the enhancement of KF elements with nonlinear modeling abilities, and the mixture of representations. Although the first one is not greatly relevant to mobile robotics navigation, we discuss it for the sake of completeness.

3.1 KF as an Estimation Algorithm

The KF can be viewed as a parameter estimation algorithm with the added ability to take into account the uncertainty of the estimation to provide an adaptive gain in the learning steps. It can be, thus, used as the training algorithm for the CI leaning approaches, such as Artificial Neural Networks or Fuzzy Systems. Following this philosophy, the EKF has been applied as the basic learning algorithm for the estimation of Radial Basis Functions (RBF) parameters [14,20] inside a growing architecture. The Node Decoupled learning applies an EKF to each network node independently, the system state corresponds to the RBF weights, both hidden units mean and variance and the hidden to output weights, the observation being the desired output. The Kalman gain is computed on an estimation of the network prediction error covariance, assuming a constant noise distribution. EKF improves over gradient descent because it introduces an adaptive learning gain (the Kalman gain) depending on the uncertainty of the network prediction, thus ensuring smooth convergence. The EKF has been also applied to tuning the membership functions in Mamdani type of fuzzy systems with correlation inference [19]. There the estimation of the optimal membership parameters is found equivalent to nonlinear dynamic system identification problem, solved applying the EKF.

3.2 Non-linear Enhancement of KF

There have been some attemps to embedd CI algorithms to relax the strong assumptions of the KF and its extensions (EKF, UKF). Superseding the linear dynamical and observation model, relaxing the constraints on the assumed noise distribution or the computation of the Kalman gain are some of the KF algorithm elements that have beed influenced by CI techniques.

Because EKF relies heavily on the assumption of white noise, it can ben enhanced by introducing a noise covariance estimation algorithm that can cope with coloured noise or systematic error bias. The work in [4] uses Artificial Neural Networks to model the noise in the motion and sensing of a mobile robot, showing improvements over the accuracy of positioning. The main handicap of their approach is that the ANN training must be performed off-line, so the environment has to be well defined, as well as the robot characteristics. For SLAM applications it is desirable that the noise modeling is done on-line with the state updating and control. This endeavor is taken in [3] where they apply a Neuro-Fuzzy System (NFS) approach to perform the instantaneous estimation of the observation noise covariance matrix parameters. Independent NFS are used for each parameter, consisting of thre IF-THEN fuzzy rules defined on the covariances of the innovation sequences, which in turn are estimated by a moving average filter. They perform simulations of SLAM navigation on a 2D simplified synthetic world, where they experiment with induced systematic errors, so their advance to real world conditions is a compelling challenge. In [17] a Takagi-Sugeno approach is applied to approximate the local behavior of the system, then it is decomposed in a set of linear systems that are dealt with by conventional DKF; the output of the decoupled systems is linearly combined to obtain an estimate of the global system state. Again, results are shown on simulation environments and the T-S system parameters must be trained off-line on accurate sampling of the data that the system will encounter. To approach on-line performance some authors built their fuzzy inference system based on theoretical physical understanding of the robot-environment-sensing relations, such as in the work reported in[18] where the GPS measurements noise covariance is estimated by this means.

Another way to mix Fuzzy Systems and EKF is performing the Kalman gain estimation by Fuzzy techniques. This approach is demonstrated in [16] on a prototype land mine detector robot. One critical feature of this robot is an accurate map of the terrain undulations, needed to align the sensor with terrain. The DKF is used to maintain an accurate estimation of the terrain model integrating the laser rangefinder readings. To avoid non-linear modelling, the Kalman gain is set according to the terrain classification performed by a Takagi-Sugeno system, trained on measurement samples from an in-house terrain database.

Finally, a way to improve convergence of the EKF is manipulate the state covariance predictor multiplying it by a diagonal matrix of "fading memory factors", which is equivalent so a low pass FIR applied to the state covariance predictor. A suboptimal algorithm to set the filter coefficients has been called the Strong Tracking Kalman Filter (STKF). In [10] an adaptive method to set the STKF parameters has been proposed, based on a Takagi-Sugeno fuzzy system defined on the innovation (the difference between the predicted and observed sensor information) divergence. However, the rules of the Takagi-Sugeno are built beforehand, with membership functions defined arbitrarily. Results are shown on simulations of GPS based localization of mobile systems.

3.3   Information Fusion

The state representation and dynamics of the KF type algorithms can be combined with the results of other CI techniques, allowing for the increase in robustness and a kind of symmetric validation in some cases. EKF has been used

along with Fuzzy Occupancy Maps (FMO) to perform localization of legged
robots in the context of Robot Soccer [13]. The FMO is a grid representation
of the playing field, each cell in the grid has a fuzzy ocuppancy value which is
updated when the robot performs a movement. Updating is done by blurring
the FMO in the direction of commanded movement, and correcting the position estimate on the basis of visual detection of known landmarks. The state
definition of the EKF is the robot position and orientation, and an analytical
expression for the Jacobian matrix is given. The fusion of both methods allows
the easy initialization of the EKF, which performs a computationally efficient
local process of continued localization of the robot, increasing the accuracy of
the FMO with large cell sizes. One way to realize the fusion of both methods
consists in running the EKF and the FMO in parallel, using each other to confirm the results, being the FMO the main reference. When occupancy fuzzy
values are low, the system tends to confuse. Then a population of EKF's are
running in parallel, maintaining competing hypothesis about the robot position.
The ones that show lower agreement with the FMO can be removed. Another
way to perform the representation fusion is reported in [2], where a collection
of KF is used with diverse noise covariances, the resulting estimation is fused
by a Takagi-Sugeno Fuzzy System, providing an improved traking in mobile
localization.

4   Conclusions

In summary KF approaches have used as estimation methods for learning processes as well as they have used learning processes to estimate some of their
computational elements, namely the noise covariance, the Kalman gain or the
non-linear model linear approximation. The KF aproach provides a convenient
way to adapt the estimation process to improve convergence, while, on the other
hand, its linear assumptions are too strong for most real-life applications, so
that non-linear learning algorithms help to make the KF approaches applicable
to applications such mobile robot navigation.

There are a number of approaches that will be worth to explore, like em-
bedding DKF or EKF into Evolutionary Strategies (ES) in order to use the
uncertainty management of KF approaches to improve the self-tuning of the ES
parameters, or to deal with uncertain fitness functions. Also, there are few works
on the non-linear modelling of the sensor information, so that non-linear observation functions could be learned from the data, even on-line. Training Neural
Networks or other Computational Intelligence tools for this functionality can be
a promising line of research.

One avenue for further research in the application of the KF and hybridized
approaches is that of Multi-Robot Systems, such as swarms of modular robots.
For instance, it would be appealing to explore the application of distributed
KF approaches to task like the cooperative map building among swarms of
autonomous robots [5].

References

1. Bailey, T., Durrant-Whyte, H.: Simultaneous localization and mapping (slam): part ii. Robotics & Automation Magazine 13(3), 108–117 (2006)
2. Carrasco, R., Cipriano, A.: Fuzzy logic based nonlinear kalman filter applied to mobile robots modelling. In: 2004 IEEE international conference on fuzzy systems, vol. 3, pp. 25–29 (2004)
3. Chatterjee, A., Matsuno, F.: A neuro-fuzzy assisted extended kalman filter-based approach for simultaneous localization and mapping (slam) problems. IEEE Transactions on Fuzzy Systems 15(5), 984–997 (2007)
4. Choi, M., Sakthivel, R., Chung, W.K.: Neural network-aided extended kalman filter for slam problem. In: IEEE International Conference on Robotics and Automation, pp. 1686–1690 (2007)
5. Di Marco, M., Garulli, A., Giannitrapani, A., Vicino, A.: Simultaneous localization and map building for a team of cooperating robots: a set membership approach. IEEE Transactions on Robotics and Automation 19(2), 238–249 (2003)
6. Dissanayake, M.W.M.G., Newman, P., Clark, S., Durrant-Whyte, H.F., Csorba, M.: A solution to the simultaneous localization and map building (slam) problem. IEEE Transactions on Robotics and Automation 17(3), 229–241 (2001)
7. Durrant-Whyte, H., Bailey, T.: Simultaneous localization and mapping: part i. Robotics & Automation Magazine 13(2), 99–110 (2006)
8. Franz, M.O., Mallot, H.A.: Biomimetic robot navigation. Robotics and Autonomous Systems 30(1-2), 133–153 (2000)
9. Julier, S.J., Uhlmann, J.K.: A new extension of the kalman filter to nonlinear systems. In: Proceedings of AeroSense: The 11th International Symposium on Aerospace/Defence Sensing, Simulation and Controls (1997)
10. Jwo, D.-J., Wang, S.-H.: Adaptive fuzzy strong tracking extended kalman filtering for gps navigation. Sensors Journal 7(5), 778–789 (2007)
11. Kalman, R.E.: A new approach to linear filtering and prediction problems. Transaction of the ASME - Journal of Basic Engineering, 35–45 (1960)
12. Mandel, J.: A brief tutorial on the ensemble kalman filter. Technical Report CCM Report 242, University of Colorado at Denver and Health Sciences Center (February 2007)
13. Martin, F., Matellan, V., Barrera, P., Cañas, J.M.: Localization of legged robots combining a fuzzy-markov method and a population of extended kalman filters. Robotics and Autonomous Systems 55(12), 870–880 (2007)
14. Meng, Q., Lee, M.: Error-driven active learning in growing radial basis function networks for early robot learning. Neurocomputing 71(7-9), 1449–1461 (2008)
15. Murphy, K., Russell, S.: Rao-Blackwellised particle filtering for dynamic bayesian networks. In: Sequential Monte Carlo Methods in Practice. Springer, Heidelberg (2001)
16. Najjaran, H., Goldenberg, A.: Real-time motion planning of an autonomous mobile manipulator using a fuzzy adaptive kalman filter. Robotics and Autonomous Systems 55(2), 96–106 (2007)
17. Pathiranage, C.D., Watanabe, K., Izumi, K.: Simultaneous localization and mapping (slam) based on pseudolinear measurement model with a bias reduction approach. In: International Conference on Industrial and Information Systems. ICIIS 2007, pp. 73–78 (2007)
18. Reina, G., Vargas, A., Nagatani, K., Yoshida, K.: Adaptive kalman filtering for GPS-based mobile robot localization. In: IEEE International Workshop on Safety, Security and Rescue Robotics. SSRR 2007, pp. 1–6 (2007)

19. Simon, D.: Training fuzzy systems with the extended kalman filter. Fuzzy Sets and Systems 132(2), 189–199 (2002)
20. Simon, D.: Training radial basis neural networks with the extended kalman filter. Neurocomputing 48, 455–475 (2002)
21. Smith, R., Self, M., Cheeseman, P.: Estimating uncertain spatial relationships in robotics. In: Autonomous Robot Vehicles, pp. 167–193. Springer, Heidelberg (1990)
22. Smith, R.C., Cheeseman, P.: On the representation and estimation of spatial uncertainty. Technical Report TR 4760 & 7239, SRI (1985)
23. Thrun, S.: Robotic Mapping: A Survey. In: Exploring Artificial Intelligence in the New Millenium (2002)
24. Welch, G., Bishop, G.: An introduction to the kalman filter. In: SIGGRAPH 2001 In Computer Graphics, Annual Conference on Computer Graphics & Interactive Techniques. ACM Press, New York (2001)

A Real-Time Person Detection Method for Moving Cameras*

Javier Oliver[1], Alberto Albiol[2], Samuel Morillas[1], and Guillermo Peris-Fajarnés[1]

[1] Centro de Investigación en Tecnologías Gráficas (CITG)
jaolmol@upvnet.upv.es
[2] Dep. de Comunicaciones (DCOM)
Universidad Politécnica de Valencia Spain

Abstract. In this paper, we introduce an advanced real-time method for vision-based pedestrian detection made up by the sequential combination of two basic methods applied in a coarse to fine fashion. The proposed method aims to achieve an improved balance between detection accuracy and computational load by taking advantage of the strengths of these basic techniques. Boosting techniques in human detection, which have been demonstrated to provide rapid but not accurate enough results, are used in the first stage to provide a preliminary candidate selection in the scene. Then, feature extraction and classification methods, which present high accuracy rates at expenses of a higher computational cost, are applied over boosting candidates providing the final prediction. Experimental results show that the proposed method performs effectively and efficiently, which supports its suitability for real applications.

1 Introduction

The CASBliP Project [1] aims at developing a navigation assistance system for blind and visually impaired users. This system aids the users by means of the generation of real-time simplified acoustic maps that describe the scene. These maps consist of acoustic representations of the main elements that appear in the scene by means of spatially localized sounds associated to each object. Persons, free paths, moving objects and other hazards are the main objects in the scene to be included in the acoustic maps. In this work, we focus on the task of person detection.

Real-time person detection in video sequences is a challenging task since humans may appear in a great variability of appearances, poses and illumination conditions, and in variable scenarios such as urban, traffic and cluttered environments. Furthermore, the problem significantly increases when the acquisition system works onboard a person.

There exists extensive literature on vision-based object detection and in particular on human detection. The problem has been handled by authors combining different feature extraction methods with different classification machines for the retrieval process, which

* This work is supported by CASBLIP project 6-th FP [1]. The authors acknowledge the support of the Technological Institute of Optics, Colour and Imaging of Valencia - AIDO. Dr. Samuel Morillas acknowledges the support of Generalitat Valenciana under grant GVPRE/2008/257 and Universitat Politècnica de València under grant Primeros Proyetos de Investigación 2008/3202.

E. Corchado et al. (Eds.): HAIS 2009, LNAI 5572, pp. 129–136, 2009.

may be carried out either using single detection window [4] or a parts-based approach [6]. Haar-like features combined with Boosting classifiers have been commonly used for face detection by Viola et al [2] [3] yielding excellent results. Similar combination of methods has been used for human detection by other authors: Oren, Papageorgiou and Poggio [11] use Haar-like features combined with Boosting, where SVM are used as weak classifiers [15]. Viola et al. [12] extend Haar-Boosting-based methods to handle space-time information for moving-human detection. During the last years, feature extraction methods like Histograms of Oriented Gradients (HOG) in combination with SVM classifiers have been widely used by several authors [4] [6]. Other authors use HOG features combined with Boosting-based classifiers, where SVM are used as weak classifiers in each stage of the cascade [13]. In this context, previous works on different classification problems have shown that Boosting-based methods are time-efficient methods [2] whereas, on the other hand, SVM-based techniques provide more accurate results at the expense of an increase in computation cost. For our application [1], we need an effective real-time processing method. Therefore, in order to obtain an improved trade-off between detection accuracy and computational efficiency we propose an advanced pedestrian detection method which is made up by a coarse-to-fine combination of these two basic approaches.

The rest of the paper is organized as follows: Section 2 gives a description of the method and the training methodology. Section 3 shows the experimental evaluation. The main conclusions are summarized in Section 4.

2   System Description

The introduced method, which we name Haar-Boosting-HOG-SVM detection method (HBHS), performs the pedestrian identification task by dividing the processing chain into two stages (see Fig. 1). In the first stage, a coarse selection of candidates in the scene is provided by a Haar-Boosting method, where Haar-like features extracted from the input image and classified using a Boosting-based method provide a preliminary candidate selection. In the second stage, single-window HOG descriptors are computed on each preliminary candidate and then classified using SVM in order to refine the initial findings. Predictions yielded by SVM are the final labelling of the method.

A set of Boosting classifiers composed by 20 to 45 stages and a set of *Polynomial* and *Gaussian* SVM kernels have been considered in order to find out the optimal balance regarding the computation cost and the accuracy of the HBHS method. MIT-Pedestrian [17] and INRIA [18] datasets have been used to train the Boosting and the SVM classification machines.

The HBHS performance of has been obtained from our particular set of test sequences used in the CASBliP, that gather different scenarios, positions of the camera, backgrounds and light conditions.

General description and training methodology for each HBHS module is detailed next.

2.1   Coarse Candidate Selection Module

A fast preliminary coarse candidate selection is based on the classification of the Haar-like features as in [2]. These features, which are based on differences of mean intensity

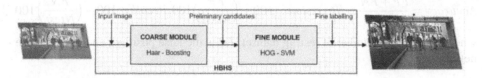

Fig. 1. Flowchart of the HBHS method

among adjacent rectangular groups of pixels, are classified by means of a Boosting based classifier, which consist of a cascade of weak stage classifiers that have been trained using the Adaboost [16] method.

The Boosting classifier has been trained using 3365 positive and 200 negative examples selected from the image databases.

2.2 Fine Labelling Module

This module works on the preliminary set of candidates retrieved by the coarse module and provides a refined labelling by using SVM for classification of a HOG descriptor of each preliminary candidate. SVM classification of HOG features provides an accurate identification but it is also much more computationally demanding. However, since this refined classification is performed only for the preliminary candidates, the increased computational load can be assumed within the required real-time processing.

HOG features, which are computed using a simplification of Lowe's SIFT algorithm [10], is a vector containing the accumulated histogram of the oriented gradients of the image. Gradients, which are represented by its magnitude and orientation, are computed over a dense cell grid, where cells are grouped into overlapped blocks of NBP (Number of Spatial Bins) cells. The gradient module is normalized by the highest value in the block and is spatially weighted by a Gaussian function. The gradient orientation information is discretized into NBO bins (Number of Orientation Bins). The accumulated histogram of all blocks shapes the image feature vector.

In the last step of the fine labeling module, HOG feature vectors are classified using SVM. Several *Polynomial* and *Gaussian* kernels [19] have been considered, varying the degree *d* of the *Polynomial* function and the exponential term *gamma* for the *Gaussian*, as defined in [14].

The two imagery datasets have been fully used for the training stage considering the left-right reflections, obtaining a global dataset with 4991 positive and 16630 negative examples.

3 Experimental Results

The method has been implemented in C using OpenCV [20] and libsvm [14]libraries. Training and testing images for Boosting and SVM classification machines were 64x128 pixels size. Scanning images for the final test of the HBHS method were 320x240 pixels. The method performance is measured as the trade-off among *time cost* and *Accuracy, Precision, Recall*, given as

$$Accuracy = \frac{FP+FN}{Ntest} \quad , Precision = 100 - \left(\frac{FP}{Ntest}\right)100, Recall = 100 - \left(\frac{FN}{Ntest}\right)100$$

where Ntest is the number of testing dataset, FP is the false positives and FN the false negatives.

3.1 Coarse Module Performance

A study of the Boosting classifier performance depending on the number of stages has been carried out using our personal video-sequences collection for the test. Table 1 shows that the results of boosting method by itself depends on the number of stages of the classifier. We find that for those classifiers with low number of stages the method provides high number of false alarms, whereas classifiers composed by more stages provide better results in terms of *Precision*. Regarding the *Recall*, we find that the highest value is obtained for 30 stages with a value of 80.60%. However, for 35 stages we obtain nearly the same *Recall* and *Precision* improves considerably. Moreover, the last row in Table 1 shows that computation cost is less than 50 ms per frame and is held constant for all the different stages of the Boosting classifier. This low time is due to the fact that features are extracted from groups of pixels rather than from single pixels. Besides, Boosting-based classifiers discard most of the background patches in earlier stages of the cascade, speeding up the process and significantly increasing the detection rate.

3.2 Fine Module Performance

HOG parameters discussion. There are several parameters in the HOG computation, such as *cell size* (measured in pixels), *block size* (measured in cells), NBO and *block overlapping*, that determine the description capability of an image descriptor vector.

First, we analyzed the effect of considering boundary blocks in the image descriptor computation. Results showed that boundary blocks reduction decreased the SVM accuracy in 0.19%. Therefore, such reduction must be considered to speed up the process at expense of the insignificant decrease in the SVM accuracy.

Second, we analyzed the effect of the block overlapping, considering $\frac{1}{2}$, $\frac{1}{4}$ and $\frac{3}{4}$ overlapping, values which are proposed by Dalal [4]. The best trade-off between detection rate and computational cost was obtained for $\frac{1}{4}$ *block overlapping*, where detection rate was optimal and time cost per sample, measured in seconds, dropped from 0.020 in the case of $\frac{3}{4}$ overlapping to 0.013 in the $\frac{1}{2}$ and 0.006 in the $\frac{1}{4}$ case.

Furthermore, we analyzed the effect of the block size, being $NBP = 4, 8$. The optimal values regarding detection rate and time cost were found for $NBP = 4$, where accuracy on test raised from 99.20 to 99.32 and computation cost per sample dropped from 0.0058 to 0.0052.

SVM kernel selection. For the best SVM kernel selection, we trained and tested two different kernels: *Polynomial* and *Gaussian*. Optimal kernels have been obtained for the *Gaussian* kernel, for $gamma = [0.05 - 0.25]$. Kernels with relative higher values of gamma ($g = 0.25$) achieve *Precision* values of 100% but *Recall* slightly drops and the computation cost takes more than 6 ms with respect to the optimal value. On other hand,

Table 1. Boosting method performance depending on the number of stages of the classifier. Optimal values in terms of computational cost and recall are found for 30 and 35 stages.

	Nstg 20	Nstg 25	Nstg 30	Nstg 35	Nstg 40	Nstg 45
Prec(%)	10.53	15.29	22.86	34.28	50.07	69.14,
Rec(%)	59.94	79.33	80.60	79.86	75.52	72.95
Tcost(ms)	48.8	48.4	49.0	47.3	49.2	48.2

lower values of gamma ($g = 0.05$) yield the best *Recall* value, 93.9%, but *Precision* decreases to 99.25%. The best balance among these parameters is obtained for $g = 0.2$, where *Precision* = 99.93%, *Recall* = 93.70%, *Accuracy* = 99.6% and the runtime is 11ms. It is important to remark that we prefer higher *Precision* values rather than higher *Recall* values since SVM will work in the second stage of the process and will have to yield low number of false positive images. From here on we will use radial basis kernel with *gamma* = 0.25.

3.3 HBHS Performance

In this section we test the performance of the HBHS method. Several stages have been studied for the boosting classifier. HOG parameters have been set to: $NBP = 4$, $NBO = 4\{0 - 180\}$, *block overlapping* = $1/4$, *feature vector* = 512 components for a 64x128 window. A *Radial Basis kernel* with *gamma* = 0.2 has been used for the SVM classifier, and the optimal threshold in the classification has been studied. The following figure shows the method performances combining different boosting classifiers, varying the number of stages, and studying the effect of the detection threshold in the SVM contribution. Figure 2 shows the relationship between *Precision* and *Recall* of the HBHS for different boosting classifiers.

Boosting results involve a constraint in the maximum performance achievable in the *Recall* of the HBHS method. On the other hand, *Precision* is highly improved with SVM contribution, achieving values of 100% of *Precision* for *Recall* values under 50%. For *Recall* values higher than 50%, the *Precision* starts to drop off. The best balance between *Precision* and *Recall* is obtained for the 35-stage classifier with a threshold near -0.5.

A study of the computation cost of each of these methods is necessary to complete the study. Table 2 shows mean values of the HBHS time cost for the same scenario. Row 1 shows the number of candidates per image selected by the boosting method and Rows 2-4 show detailed averaged time information of each processing module during the scanning of the 320x240 input images. It is easy to find that SVM method is the bottle neck of the process in terms of computational cost. Therefore, the lower number of candidates provided by Boosting positively affects the SVM computational cost. Note that for lower number of stages, more candidates are yielded by boosting and therefore higher computational cost is needed by SVM. However, for those situations with more stages in Boosting classifier, the necessary time for SVM significantly decreases. See that there is a significant step in the final temporal cost for a boosting classifier with 30 and 35 stages. The total cost decreases in 50 ms for 35 stages, which is more than

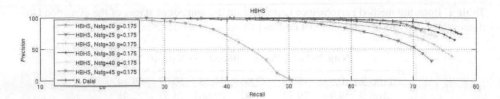

Fig. 2. Comparison among several HBHS classifiers varying the number of stages in Boosting and the SVM classification threshold. The SVM has been trained with a Gaussian kernel with *gamma* = 0.175. In coloured lines, HBHS performance. In black, N. Dalal performance.

Table 2. HBHS temporal analysis for different number of Boosting stages and a *Gaussian* kernel with *gamma* = 0.175. Mean values per window.

	Nstg 20	Nstg 25	Nstg 30	Nstg 35	Nstg 40	Nstg 45
Boosting candidates	8.84	4.80	3.24	2.27	1.97	1.64
Boosting time (ms)	48.8	48.4	49.0	47.3	49.2	48.2
HOG time (ms)	141.6	78.1	52.8	36.4	31.9	25.8
SVM time (ms)	314.9	169.7	116.8	80.4	71.4	57.9
HBHS time (ms)	505.5	296.3	218.7	164.2	152.6	131.9

25%. However, a classifier based on 40 stages does insignificantly reduce the cost in comparison to the 35 stages. According to the figure 2, the best results were found for 30, 35 and 40 stages. On the other hand, results on time study reveal that a classifier based on 35 stages is preferred rather than the 30 one. Although classifiers with 40 and 45 stages provide good temporal results, *Precision* and *Recall* drops considerably. Therefore, the best combination of *Accuracy* and *Temporal Cost* is reached for a boosting classifier with 35 stages and a threshold set to -0.4. Figure 3 shows some examples of the HBHS method performances over our collected dataset for different stages of boosting method. Boosting candidates are surrounded by white boxes. SVM contribution and therefore the final prediction is represented in grey colour. The visual analysis reveals that sometimes the positive boosting candidates are not well framed. Then, the feature extraction is carried out over an image where the candidate is not centred in the window. SVM is more sensitive to changes than boosting, and therefore a slight drift in the position implies a wrong classification. The problem increases for high stages of boosting classifiers. For that reason, there are several occasions where SVM fails in the labelling. However, the final labelling is accurate and fast enough for real-time processing of complex scenarios.

3.4   HBHS against Other Methods

In this subsection we present a comparison of our method with two reference methods, the one based on Viola and Jones [2] and the other based on Dalal [4]. For the first method, we have developed our own implementation according to their theoretical basis. For Dalal's method, we have straight downloaded the source code from his official

Fig. 3. HBHS performance over our particular imagery collection, for different stages of the coarse classifier. From left to right, classifiers comprise from 25 to 45 stages. In blue, discarded coarse predictions. In green, SVM candidates and therefore the final prediction of HBHS.

Table 3. Comparison of HBHS among other methods

	Precison	Recall	Time(ms)
Viola and Jones	34.28	79.86	47
Dalal	85.69	72.61	1990
HBHS	79.36	72.77	164

website [5] and compiled it under linux using gcc. Testing has been carried out using our personal imagery dataset.

Table 3 shows a comparison among Viola and Jones approach, Dalal person detector and HBHS. This table shows that our method presents a good balance among *time cost*, *Precision* and *Recall*, achieving high temporal performance providing high *Precision* and *Recall* rates as well. The best rates of *Precision* and *Recall* are achieved by Dalal, but temporal requirements are significant, too. On the other hand, Viola and Jones approach needs the lowest computational cost but poorer detection rates are achieved. HBHS, however, provides similar detection rates to Dalal and still presents low computational cost, 10 times lower than Dalal's method requirements, achieving real-time detection at a high detection rates.

4 Conclusions and Future Work

In this paper we present a new real-time method to detect persons in image sequences of diverse nature. This method has been devised with the aim of improving the balance detection accuracy and computational cost of the state-of-the-art methods. The proposed method comprised the sequential combination of two basic methods performing in a coarse-to-fine fashion. A fast coarse preliminary candidate selection is provided by boosting techniques. Then, SVM-based classification of HOG features extracted from the preliminary candidates refines the selection and provides the final detection. Extensive experimental results using real sequences have been carried out both for parameter adjustment and performance assessment. It can be seen that the proposed method obtains an improved trade-off between computational cost and detection accuracy.

Several modifications that may enhance the overall performance of the method might be the following: First, tracking techniques may be considered in order to improve the high miss rate and the slightly deviations of the candidate bounding box that Adaboost produces; Second, other implementations of SVM may be used to reduce the high computational cost of the fine classification. Third, stereo-cameras may be used in order to perform the preliminary candidate selection and intelligent scan considering the expected size of a person at each depth.

References

1. CASBLIP Project (FP6-2004-IST-4), http://www.casblip.upv.es
2. Viola, P., Jones, M.: Rapid object detection using a boosted cascade of simple features. IEEE Computer Vision and Pattern Recognition 25, 29 (2001)
3. Mohan, A., Papageorgiou, C., Poggio, T.: Example-based object detection in images by components. IEEE Trans. Pattern Anal. Mach. Intell. 23(4), 349–361 (2001)
4. Dalal, N., Triggs, B.: Histograms of Oriented Gradients for Human Detection. In: Proceedings of IEEE Conference Computer Vision and Pattern Recognition, San Diego, USA, pp. 886–893 (June 2005)
5. Dalal, N.: Pedestrian Detector source code,
   http://www.navneetdalal.com/software
6. Parra, I., Fernandez, D., Sotelo, M.A., Bergasa, L.M., Revenga, P., Ocaña, M., Garcia, M.A.: Combination of Feature Extraction Methods for SVM Pedestrian Detection. IEEE Trans. Intell. Trans. Sys. 8(2) (June 2007)
7. Gavrila, D., Philomin, V.: Real-time object detection for "smart" vehicles. In: Proc IEEE Int. Conf. Comput. Vis, pp. 87–93 (1999)
8. Broggi, A., Bertozzi, M., Fascioli, A., Sechi, M.: Shape-based pedestrian detection. In: Proc. IEEE Intell. Veh. Symp. Dearborn, MI, pp. 215–220 (October 2000)
9. Bertozzi, M., Broggi, A., Chapuis, R., Chausse, F., Fascioli, A., Tibaldi, A.: Shape-based pedestrian detection and localization. In: Proc IEEE ITS Conf., Shanghai, China, pp. 328–333 (October 2003)
10. Lowe, D.: Distinctive image features from scale-invariant keypoints. International Journal of Computer Vision 60(2), 91–110 (2004)
11. Oren, M., Papageorgiou, C., Sinha, P., Osuna, E., Poggio, T.: Pedestrian Detection Using Wavelet Templates. Computer Vision Pattern Recognition (1997)
12. Viola, P., Jones, M., Snow, D.: Detecting pedestrians using patterns of motion and appearance. In: International Conference on Computer Vision (2003)
13. Carmen, A., Albiol, A.: Authomatic Pedestrian Detection using boosting techniques. Final Degree Project. Technical University of Valencia (2006)
14. An implementation of Support Vector Machines (SVM) in C,
    http://svmlight.joachims.org/
15. Yuan, Z., Yang, L., Qu, Y., Liu, Y., Jia, X.: A Boosting SVM Chain Learning for Visual Information Retrieval. In: Wang, J., Yi, Z., Żurada, J.M., Lu, B.-L., Yin, H. (eds.) ISNN 2006. LNCS, vol. 3971, pp. 1063–1069. Springer, Heidelberg (2006)
16. Freund, Y., Schapire, R.E.: A Short Introduction to Boosting. Journal of Japanese Society for Artificial Intelligense 14(5), 771–780 (1999)
17. Pedestrian Dataset from MIT,
    http://cbcl.mit.edu/software-datasets/PedestrianData.html
18. Pedestrian Dataset from INRIA, http://pascal.inrialpes.fr/data/human/
19. Vapnik, V.: The Nature of Statistical Learning Theory, p. 314. Springer, Heidelberg (2000)
20. OpenCV library, http://sourceforge.net/projects/opencvlibrary/

Unsupervised Methods for Anomalies Detection through Intelligent Monitoring Systems

Alberto Carrascal, Alberto Díez, and Ander Azpeitia

Fundación Fatronik-Tecnalia, Paseo Mikeletegi 7, Parque
Tecnológico, 20009 Donostia, Spain

Abstract. The success of intelligent diagnosis systems normally depends on the knowledge about the failures present on monitored systems. This knowledge can be modelled in several ways, such as by means of rules or probabilistic models. These models are validated by checking the system output fit to the input in a supervised way. However, when there is no such knowledge or when it is hard to obtain a model of it, it is alternatively possible to use an unsupervised method to detect anomalies and failures. Different unsupervised methods (HCL, K-Means, SOM) have been used in present work to identify abnormal behaviours on the system being monitored. This approach has been tested into a real-world monitored system related to the railway domain, and the results show how it is possible to successfully identify new abnormal system behaviours beyond those previously modelled well-known problems.

Keywords: Unsupervised Anomaly Detection, Unsupervised Classification, Intelligent Monitoring Systems, Clustering.

1 Introduction

Because of the recent technological revolution occurred in industrial sector, it turns increasingly difficult to raise any appropriate manual maintenance process. Thus, the amount of information about the state of the system is being monitored is continuously increasing, exceeding the capacity of maintenance technicians. While the industry is undergoing a technological revolution, new reactive, proactive and predictive maintenance approaches are being developed.

The success of the majority of the monitoring and intelligent diagnosis systems relies on the use of the knowledge regarding existing domains (Knowledge Based Systems, or KBS) [1]. In this kind of domains, the main difficulty regarding failure and anomaly detection is how to make the expert knowledge explicit and how to model it. Knowledge modelling based on rules is one of the most common approaches [2]. Nevertheless, there exist domains where this approach can not be applied, due to either, non-existing previous expert knowledge or overly complex knowledge base management [3].

Supervised learning models do not successfully resolve this problem as they require previous knowledge about which should be the system output when new data come in, and also a high external support will be needed [4]. On the contrary, unsupervised learning models classify monitored system data by means of some similarity measure

E. Corchado et al. (Eds.): HAIS 2009, LNAI 5572, pp. 137–144, 2009.
© Springer-Verlag Berlin Heidelberg 2009

without any external support. Failure detection is achieved by comparing and identifying new cases with past breakdowns, whilst anomalies are detected whenever there is no mapping with any previous case. Therefore, the problem of identifying failures and anomalies can be transformed into an unsupervised classification problem [5].

The monitoring domain presented in this work concerns the railway domain. It is an especially critical domain in which is really important to assure the safety for every journey, for both passengers and cargo; which implies that all the components embedded into the train accomplish some reliability standards. In such domain, an exhaustive control of life cycle parameters of train components has to be carried out, guaranteeing correct operation working for all of them throughout their service life-time.

2 Unsupervised Methods

Unsupervised methods have been used in many contexts and domains, involving different unsupervised learning problems. The main goal of these techniques is to perform a clustering of similar datasets, which are supposed to have the same pattern. Such pattern could be very significant in order to classify or to identify behaviours linked to the data, or in order to detect or to infer possible failures or anomalous conditions; different from supervised learning (and reinforcement learning), where the learner is only provided with unlabelled examples.

In the study presented in this paper, the performance of different techniques have been tested, to illustrate the differences between them and to analyze their behavior in a real monitoring system. It is hardly important to underline that, methods selected for this study (HCL, K-Means and SOM) are a representative subset of the unsupervised classification approaches.

HCL (Hierarchical Clustering) is an algorithm that builds clusters iteratively in a hierarchical structure. The iterative process can be either agglomerative or divisive. Normally, agglomerative strategy is more commonly used [6]. Differences between methods arise because of the different ways of defining similarity (distance) between clusters. Several agglomerative techniques such as complete linkage, single linkage, average linkage, weighted average linkage, median linkage, centroid linkage and Ward's linkage are possible.

HCL graphic results (dendograms) are complex and confusing when the amount of data is considerable. This fact, together with the fact that the complexity of this algorithm is on the order of O(n2), depending on the configuration, makes difficult to employ this technique when the amount of data is over thousands of samples. This is the main reason why in those cases it is more advisable to apply other approaches that are easier to interpret and less computationally expensive, such as K-Means and SOM [7], [8].

K-Means classification algorithm performs a partition of data space into k clusters. Each cluster is represented by an element, the centroid or a mean point, whose initial value can be randomly set or estimated by applying some kind of heuristic. In an iterative process, the elements are assigned to the partition with the least distance between them and the centroid of the partition. After elements assignation into clusters, cluster centroids are recalculated with the elements that belong to its partition. The process converges to a solution with a linear complexity $O(n)$, which is not always the global optimum. The success of K-Means approach is strongly determined by the choice of the k value, the metric employed and the initial centroids values.

The Self-Organizing Maps (SOM) approach allows representing into a low-dimensional map a high-dimensional data set, so that the similarity between analyzed data can be easily identified [9]. SOM map is composed by neurons grouped according to a topology (hexagonal and rectangular topologies are the most common ones). Each neuron has associated a weighted vector that allows mapping entry data into each neuron on the basis of a given measure. To achieve this, the data are presented to the network iteratively, so that at each time step, the winning neuron, or the neuron that has associated the weighted vector most similar to the sample, modifies its associated vector to increase its similarity with given data. The vector associated to the winning neuron, and to the neighbouring neuron vectors according to the topology used, is modified by means of a decreasing function of the distance between nodes on the map grid. Neighbourhood functions most commonly used are Gaussian and Bubble functions.

SOM provides a non-linear, ordered, smooth classification of high-dimensional input data, preserving neighbourhood relations. This capacity for managing high dimensional data with good results and performance, makes this approach possible to be applied in many complex domains, such us engineering, bioinformatics and genetics, communications, etc. [10], [11], [12].

3 Monitored System

Monitored system based on intelligent diagnostics that has been used for present study, is related to one of the most critique last generation train component, made by CAF [13] company: self-propelled, dual voltage electric train units with a variable gauge system (ATPRD). As safety measure, ATPRD incorporate ATMS (Acceleration and Temperature Monitoring System) equipment, developed by CAF; which allows knowing temperatures and accelerations at any time inside the train motion units, called bogies. The importance of these component measurements is critical, since the failures that can occur on the trains are mainly associated to anomalous behaviours inside the bogies.

There are several sensors to monitor the acceleration and temperature of the bogies, strategically replicated and placed over the train. Every 5 minutes during a train journey, sensors acquire readings of those parameters which are forwarded by means of a GSM connection with the train. Such information is registered and stored in a database to provide the needed data input to our approach.

Sensors are distributed in 8 bogies per train, as it is showed in Figure 1. Each bogie has 32 sensors which can be divided in five groups: internal and external wheel groups with 16 temperature sensors installed on the wheels (4 per wheel), cylindrical and conic hollow shaft groups with 8 temperature sensors installed on the hollow shafts (HS hereafter), and finally, reduction gear group with 5 temperature sensors.

3.1 Derived Variables

In order to easily identify journey anomalies, a model representation that easily allows comparing the main characteristics of different journeys as simple data vectors was needed. To that, due to the lack of any thermodynamic model of the train, a set of quantitative variables has been defined to characterize the bogies behaviour each journey.

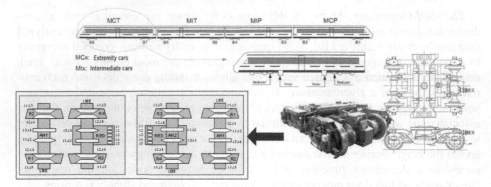

Fig. 1. Bogies and sensors distribution in the ATPRD unit

Derived parameters also eliminate temporal dependency of collected sensor data; so, for each journey, each derived parameter has an associated value. This representation also allows comparing journeys of different duration, since each journey is identified by a vector of values of constant dimension. Derived parameters used in this study are the following:

- Correlation of a sensor with its pair: Pair sensors are sensors physically located very close, so that the correlation between their readings is used to validate its correct operation.
- Volatility of a sensor: this variable measures the variability of a certain sensor as the absolute differences mean.
- Sensors mean group.
- Mean square of sensors group: the mean of a group of sensors squared.
- Maximum absolute value of a subgroup of sensors: this derived parameter allows identifying atypical values measured by the sensors.
- Percentiles (10, 20, 80, and 90) of a group of sensors: P_{10} and P_{20} allow identifying low sensor values, whereas P_{80} and P_{90} are intended to identify high sensor values.

Considering the number of sensors and groups the total number of derived parameters obtained is 65. This way, different bogie behaviours are characterized by 65-dimensional real vectors.

4   Results

Test data used in this study have been collected from 12 ATPRD units, monitored during eleven months (from January to October 2008), obtaining a total of 9.100 vectors that represent the behaviour of the different bogies over units monitored. Euclidean distance has been adopted as similarity measure used by every analysed unsupervised classification method. Gaussian normalization has been applied in order to minimize the impact related to the different domains, means and variances over each component vector.

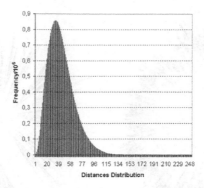

Fig. 2. Whole data distances distribution

A first similarity analysis on existing data shows a high level of regularity regarding the data patterns. This regularity is shown in figure 2 where the whole data distances distribution can be shown. There exists an expected non-stochastic behaviour in the distances distribution, with a clear deviation to little distances. This confirms that rail lines regularity (concerning journey duration, velocity profile, et cetera) is reflected by the regularity of the vectors that represent train bogies behaviours.

Classification of the different bogie behaviours dealing with the similarity of the vectors that represent them has been carried out. HCL method has been applied with an agglomerative strategy and average linkage. As shown in Figure 3, the complexity of the graphic representation of the output obtained by applying this algorithm is an important handicap when interpreting the results. Only the most clearly divergent cases can be easily isolated. A more detailed analysis of marginal cases identified by HCL technique clearly shows a small group of five anomalous bogie behaviours. These cases have been contrasted with the sensor values in those dates, concluding that they are related to anomalous behaviours on different group of sensors: Wheels, HS and Reduction gear group. Figure 4 (A) shows an example of this anomalous behaviour on Wheels group. The graphic illustrates the signal related to train speed (bottom signal), in relation with the signals related to the sensors of current group. The irregular signal (upper signal) that is uncorrelated with the other ones is obviously associated to a malfunction of the sensor that represents. Regarding these five anomalous samples, the calculated derived parameter values are notably different from the other cases analyzed by the HCL method. As has been checked, the main causes of the sensors failures are wrong connections and water invasion.

In contrast to HCL approach, unsupervised classification methods, such as K-Means and SOM, allow a more intuitive interpretation of generated clusters. Both techniques require the specification of the maximum number of clusters to be considered. After some experimental tests, this maximum value was fixed to 225 (15x15-dimensional SOM map).

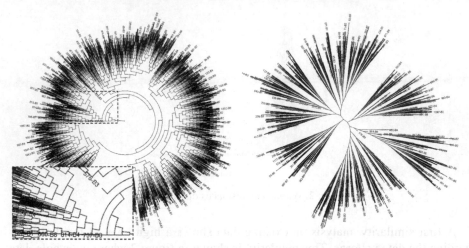

Fig. 3. HCL graphical output

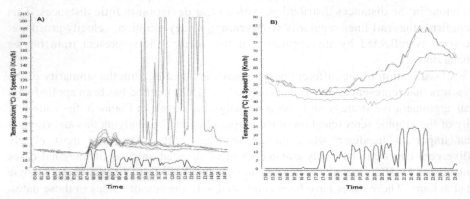

Fig. 4. (A) Bogie anomalous behaviour example related to wheels sensors group failure. **(B)** Example of Bogie anomalous temperature profile.

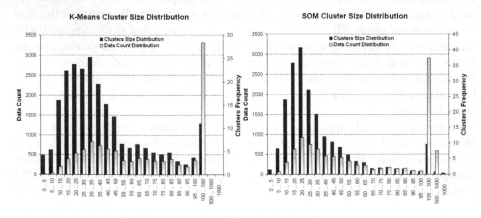

Fig. 5. K-Means and SOM clusters size distribution

A rectangular topology, Gaussian neighbouring function and a learning rate value of 0.5 were chosen to configure the Kohonen network. In the same way, clusters medium size was determined after ten test executions of each method. This executions rate was considered as enough in order to obtain a reliable clusters size distribution. As illustrated in the figure 5, around 30% of the data is grouped into clusters with size between 100 and 500 elements. The rest of the data is distributed into lower size clusters, with a more uniform behaviour in the case of K-Means. Small clusters (with a size between 0 and 5), are more frequent in K-Means approach, showing more sensibility to little variations on the data.

In order to identify failures on bogies behaviour, a more detailed analysis of smallest cluster was performed, realizing that in the case of SOM approach, data classified into small clusters are severely modified with each method execution. Nevertheless, K-Means results are more homogeneous. Anomalous elements identified by HCL technique are all located into lower size clusters obtained by applying K-Means. On the contrary, this clusters distribution was not achieved with SOM.

Regarding small clusters given by SOM or mainly by K-Means, the results have been really interesting, detecting other anomalous situations (see Figure 4.B). These behaviours are related to sensor failures but also to anomalous journeys caused by external conditions such as journey sections with unusual speed, exceptional climatic conditions, non-scheduled stops, etc.

5 Conclusions

In order to address the problem related to the process of anomaly identification in intelligent monitored systems, unsupervised methods are a very useful alternative when there is no previous expert knowledge about the application domain. Tests performed highlighted how anomalous situations of interest were detected by means of classic unsupervised methods: HCL, SOM and K-Means.

The railway sector analyzed in present paper shows a very regular behaviour. A significant set of cases is enough to cover all possible normal behaviours associated to the data. Anomalous behaviours that imply failures can be detected by means of unsupervised classification techniques.

In experiments performed, HCL model shows very good results regarding detection of anomalous cases. However, HCL is computationally expensive and the interpretation of its results demands increasing effort as the size data grows. SOM model is computationally less expensive and its graphical output is more intuitive. Nevertheless, owing to the input noise tolerance associated to artificial neural networks approaches and to the topology preservation, the obtained classifications show less accuracy level than those obtained when using K-Means approach. Further, the K-Means method improves the detection of slight data variations related to bogies behaviours.

Train bogies anomaly behaviours detected in this paper have been contrasted with maintenance information. From the analysis, these anomalies have been found to match failures in sensors or train journeys strongly influenced by external conditions, which confirms the validity of our approach.

Acknowledgments. Present work has been financed by Fatronik-Tecnalia. The data used for the study has been provided by NEM Solutions and CAF.

References

1. Gonzalez, A.J., Dankell, D.D.: Engineering of knowledge-based systems. Prentice-Hall, Englewood Cliffs (1993)
2. Brachmand, R.J., Levesque, H.J.: Knowledge Representation and Reasoning. MIT Press, Cambridge (2003)
3. Preece, A.D.: Validation of Knowledge-Based Systems: The State-of-the-Art in North America. J. Study of Artificial Intelligence Cognitive Science and Applied Epistomology 11(4) (1994)
4. Alpaydin, E.: Introduction to Machine Learning. In: Adaptive Computation and Machine Learning. MIT Press, Cambridge (2004)
5. Eskin, E., Arnold, A., Prerau, M., Portnoy, L., Stolfo, S.: A Geometric Framework for Unsupervised Anomaly Detection: Detecting Intrusions in Unlabeled Data. In: Barbara, D., Jajodia, S. (eds.) Matrix Eigensystem Routines - EISPACK Guide. LNCS, vol. 6(4) (2002)
6. Williams, C.: A MCMC approach to hierarchical mixture modeling. Advances in Neural Information Processing Systems 12, 680–686 (2000)
7. Garrett-Mayer, E., Parmigiani, G.: Clustering and Classification Methods for Gene Expression Data Analysis. Johns Hopkins University, Dept. of Biostatistics Working Papers. Working Paper 70 (2004)
8. Yin, L., Huang, C.H., Ni, J.: Clustering of gene expression data: performance and similarity analysis. BMC Bioinformatics 7(suppl. 4), 19 (2006)
9. Kohonen, T.: Self-organizing maps. Springer, Berlin (1997)
10. Carrascal, A., Couchet, J., Ferreira, E., Manrique, D.: Anomaly Detection using prior knowledge: application to TCP/IP traffic. Artificial Intelligence in Theory and Practice - IFIP International Federation for Information Processing 217, 139–148 (2006)
11. Kohonen, T., Oja, E., Simula, O., Visa, A., Kangas, J.: Engineering Applications of the Self-Organizing Map. Proceedings of the IEEE 84(10), 1358–1384 (1996)
12. Huang, S., Ward, M.O., Rundensteiner, E.A.: Exploration of dimensionality reduction for text visualization. Technical Report TR-03-14, Worcester Polytechnic Institute, Computer Science Department (2003)
13. Construcciones y Auxiliar de Ferrocarriles,
    http://www.caf.net/caste/home/index.php

Architecture for Hybrid Robotic Behavior

David Billington[1], Vladimir Estivill-Castro[2], René Hexel[1], and Andrew Rock[1]

[1] ICT/IIIS, Griffith University, Nathan, QLD, 4111, Australia
{d.billington,v.estivill-castro,r.hexel,a.rock}@griffith.edu.au
www.griffith.edu.au/mipal
[2] Visiting Scholar, Universitat Popeu Fabra, Barcelona, Spain

Abstract. Software architectures for agent technology and robots have been polarized between reactive architectures and architectures based on planning and reasoning. Although hybrid architectures have been shown to offer benefits from both, these seem complicated to integrate. In this paper we integrate the reactive nature of finite state machines and the reasoning capabilities of non-monotonic logics to produce intelligent autonomous robots. In particular, we demonstrate this with a robotic poker player. The robotic player integrates vision, sound recognition, motion control and the reasoning to perform competitively as a player in a complex game with incomplete information.

Keywords: Non-monotonic logics, finite state machines, software patterns, software engineering, software architecture.

1 Introduction

The implementation of the behavior of an autonomous robot is a delicate and sophisticated engineering task. Typically, one would like to produce an architecture that combines a reactive architecture (considered suitable for unknown but simple environments and tasks) with a planning/reasoning approach (suitable for complex worlds which need sophisticated knowledge about the domain and the environment). We present a software architecture that enables behavior designers to specify behaviors using the reactive modeling tool of finite state machines; however, we enable predicates of non-monotonic logics to label transitions. This simplifies the design task because the reasoning component of the logic will resolve conflicts in the description, while the descriptive nature of non-monotonic logics relieves the designer from many of the concerns regarding implementation or the burdens and pitfalls of procedural implementation.

We enable the design of robotic behaviors in terms and formalisms that are accessible to humans. This will become significantly more relevant as collaborative applications for teams of autonomous robots in human environments emerge in the near future. Robots have penetrated carefully controlled industrial environments where they perform well-defined, repetitive tasks. However, the emergence of agent technology and reduced costs of hardware for sensors, batteries, networking, and computational power, suggest robots will be deployed in much

E. Corchado et al. (Eds.): HAIS 2009, LNAI 5572, pp. 145–156, 2009.
© Springer-Verlag Berlin Heidelberg 2009

more challenging environments that continuously change, often in unpredictable ways. Today's technologies only handle the complexity of human environments to a very limited extent, but it is expected that in the near future, intelligent integrated systems with the capacity to act within such a complex environment will collaborate with their users in many tasks [1].

There is now an emerging line of research where the human-machine interaction anticipates the ability of all parties to act and co-exist with the environment both in cooperation but also in competition with each other and other collaborative teams [2,3]. Therefore, the area of social robots that interact with people is gaining prevalence [4], together with the area of human-robot interaction [5].

Cognitive Robotics aims at programing robots using only high-level actions and relations among actions described by a formal logic. With the situation calculus as foundation, Golog is arguably the most studied high level logical language in this direction [6], and many extensions have appeared in the literature. However, little exists in terms of comparisons and implementations in robots [7]. Non-monotonic logics can and should be incorporated into formalisms for the specification, analysis, and design of behavior [8]. We incorporate them into the central behavioral artifacts provided by state machines. This paper describes how we implemented this approach. A specific non-monotonic logic (*Plausible Logic* [9,10,11] (PL), which is the only non-monotonic logic with an efficient non-looping algorithm [11], was used. This paper describes the infrastructure that enables programmers to design, validate, and deploy graphical models of behavior in autonomous mobile robots. We describe the generic architecture that solves issues of control, interaction with the environment and knowledge representation, and how developers define behaviors using this infrastructure elsewhere [12]. Here, we give an illustration with an application where robots have multi-modal interactions with humans in a game of poker. We believe this will demonstrate the benefits of our approach for intelligent integrated systems that combine capabilities such as reasoning and planning, voice recognition, image analysis, and motion control. Games are considered a suitable methodology for evaluating robot-human interaction [13] while general game playing is the new frontier of artificial intelligence and agent technology [14]. Our robot acts as a multi-modal interface that perceives multiple aspects of the environment and produces diverse types of outputs, such as sounds and gestures, and even acts on the environment. It interacts with a human in a competitive environment with incomplete information.[1] However, the architecture has also been applied to other scenarios where decision-making is complicated because there is incomplete information about a dynamic environment, for example robotic soccer [15] and robots for multi-modal interfaces [16]. Algorithms for signal analysis, computer vision, image processing, and gesture recognition are all involved to capture information from the environment including human speech and human actions.

---

[1] In game theory, making a decision without knowledge of all the values of variables that determine the state of the environment is labeled as *incomplete information,* but also, in the literature of agents this is referred to as an *inaccessible environment.*

2   Software Architecture

The most general architecture for an agent interacting with its environment presumes an execution cycle consisting of a phase where the agent collects information through its sensors, decides on an action, and then applies this action [17]. This provides a preliminary answer to the first problem the architecture is to solve — *the interaction problem*, i.e. how does the robot/agent receive information from the environment, and how does it act on it. We use a global architecture that provides a series of services that enable high-level PL descriptions to be compiled and executed directly on board a robot.

2.1   Format of the Generic Software Architecture

Our generic software architecture shares many fundamental and structural components with other proposed software architectures for robotics [18]. This will illustrate that our incorporation of non-monotonic reasoning and its tools for visual description and for designing behavior are also applicable to many other architectures. Our architecture has also proven to be a solid framework[2] for development from the software engineering perspective in two important aspects. Modules and subcomponents can be developed by a team of programmers working relatively independently of each other. The architecture facilitates integration and supports a development cycle that consists of regular version refinement and improvement, almost like *Extreme Programming* [20].

How the robot encodes all information collected about the environment, including domain knowledge provided a priori, is the *knowledge representation problem*. Our architecture proposes to have this at several levels of detail. At one level, we use what we call a *whiteboard* where almost all modules write information they have come across, mainly to facilitate module communication. From the perspective of knowledge representation with logics, the *whiteboard* comprises all the facts (including a time-stamp and an author), allowing reasoning associating agency and negotiation. E.g., in soccer we can interpret *whiteboard* messages as "vision believes the ball is in front" and "sonar believes something is ahead", and fuse this to increase our confidence that the ball is ahead.

This basic knowledge representation is complemented with formal logics (and in particular PL) to represent significant issues regarding the domain of operation. Eg. the rules of poker and the strategies to make decisions should be expressed in logic. While this would enable reasoning regarding the action to be performed in a certain state of the environment, the other important aspect is a software engineering concern. Robot control software becomes rather large very quickly, and the analysis of certain situations would be better expressed in a descriptive language close to human understanding enabling much easier analysis of the validity of the knowledge implicitly determining robot's actions.

The nesting, presentation, and meaning of rules encoded in, e.g., C++ or Java becomes complicated and beyond human comprehension. Logic models can be

---

[2] In the sense of 'framework' defined by Larman [19].

created and evolved with mechanisms that abstract the logical inference algorithms, but will facilitate the validation and testing. They should also facilitate their improvement through iterative refinement by humans; much in the way the theory of expert systems approached knowledge elicitation [21].

When addressing the *control problem*, the architecture determines what takes control of the robot's actuators, how a decision on the next action is made, and how progress and the environment is being monitored to adapt the course of action if the current setting is not what we hoped for? There have been many debates between reactive systems and reasoning agents [17]. Reactive agents typically do not build plans, carry out no reasoning, and rarely represent knowledge in any formal logic. Reasoning architectures try to build sophisticated knowledge representations of the environment in order to perform high-level reasoning and to conclude what should be the best action. They may identify goals, build plans, evaluate plan feasibility and monitor progress. Some argue that it is not possible to control a robot in a complex environment unless one applies variants of the subsumption architecture proposed by Brooks. Others [22] suggest a combination of non-monotonic reasoning in reactive systems, placed as "knowledge middleware" to bridge the reactive sensor-based approach and reasoning.

We have a hybrid architecture for the control problem. Like subsumption, it uses priority discrimination. Behaviors are organized to provide a hierarchical structure to the type of behaviors or actions suitable for a certain setting. *External States* characterize some high-level settings in the environment.

Consider a reactive system with behaviors that control the robot and described with state-machines. A behavior consists of *behavior states* and transitions between them. For example, in robotic soccer, a simple ball-chaser behavior could be defined by two states and two transitions. In the state of FOLLOW the robot follows the ball. If the ball goes out of sight, a BALL_NOT_VISIBLE transition moves the behavior to a state of SEARCH where the robot spins around. When the ball becomes visible again, (BALL_VISIBLE transition), the robot changes state back to the state FOLLOW (Fig. 1(a). This is reactive behavior, because as soon as the conditions that label a transition become effective, the system reacts by performing the actions in the new state.

However, there are elements of planning and reasoning in our agents (robots), as formal logic statements label the transitions between states. Reasoning is performed to establish if a transition should take effect. In the example above, the change to the ball searching state may not be simply the fact that one frame has not identified the ball, but involve a more complex evaluation of other aspects such as path calculation, ball speed and/or distance, as well as the vision module error rate, recent changes of state, or minimum time intervals.

Our architecture has the capability to develop and incorporate complex behaviors while maintaining some grasp on the correctness of them. For this, we enable the programming of a robot in a high-level descriptive language (not just procedural or object-oriented coding in C++ or Java). We describe the knowledge base with a formal logic, and the behaviors are described by a hierarchy of finite state machines. Both of these descriptions are manipulated by

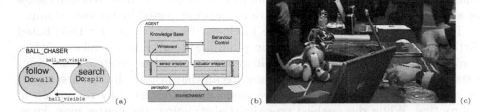

Fig. 1. (a) Illustration of simple behavior. (b) Illustration of the architecture operating under an action-perception cycle. (c)The prototype interacting with humans.

programming tools that allow visualization and development by software engineers, while enabling a significant amount of testing in simulators.

2.2   Environmental Interaction – The Action-Perception Cycle

The action-perception cycle is perhaps an issue of control, but we discuss it further since it defines the interface with the environment. More importantly, it defines the possibility of the same agent to operate on different hardware (and possibly a different environment). This is because the specific sensors interact with behavior control and the knowledge base through wrappers[3] that can have common interfaces. Consider the diagram in Fig. 1(b). Sensors collect information about the environment and deposit this on the *whiteboard*. The behavior control deposits messages on the *whiteboard* that actuator wrappers grab if the message belongs to them and in turn operate their actuator. A particular sensor (e.g. a camera, gyroscope, sonar sensor, or microphone) may in fact be accessible to a series of software layers. Also, the sensor wrapper may be in itself a sophisticated layered or pipelined architecture (e.g. a vision module representing a pipeline for image segmentation, edge detection and object recognition).

The wrapper allows replacing the physical environment with a virtual environment (e.g. through connecting to a networking socket instead of a real sensor) or portability between different hardware platforms in line with considerations by Kim[24]. In this way, the intelligence (and arguably the personality) of the agent remains the same, operating unaware of a virtual or a physical environment.[4] This idea also illustrates that the majority of the architecture is isolated from the actual physical platform it operates on. Thus, from the software engineering perspective, this means that we can test and verify a large portion of our internal modules by providing simulators for the sensors and the actuators.

2.3   Knowledge Representation – The Whiteboard

The *whiteboard* is an abstract data structure where a module can deposit a message. Each message has a type and a time stamp, and is signed by the module

---

[3] The design pattern *wrapper* [19, Page 418] is also named *facade* [23, Page 185].

[4] We have denoted this capability *the matrix* in honor of the series of 3 science-fiction films where the sensors of humans are bridged to a virtual environment, but in the literature on robotics and agents there are other names for this.

that deposits the message. Modules can read all or just the most recent message of a given type. Modules can also retrieve messages sorted by the time-stamp.

The *whiteboard* is inspired by the blackboard architectures for Distributed Artificial Intelligence [25] and by the publish/subscribe and similar software engineering patterns [26,19]). This eliminates the need to create a more complex module communication mechanism. Recall that historically, the first model of communication was a master-slave model best illustrated by the notion of subroutine. While this enabled procedural abstraction, the master must know how and when to call the subroutine. The client-server model provides a step forward due to the independence of flow of control to each module. Nevertheless, the client must be aware of the server interface. The *whiteboard* allows a further level of decoupling. The provider may supply information for unknown consumers who may not even be active. There is no need to be aware of the consumer's interface, only the interface to the *whiteboard* is necessary.

Sensory information in robotic applications is noisy, and may lead to false beliefs about the ground truth. Information with the same time-stamp in the *whiteboard* may in fact not be synchronous in the environment. A typical example of this is the challenge of reading angles for joints in the head of a SONY Aibo and associate them with an image from the camera. A moving head can result in images whose associated angles for head-joints are more than 12 degrees off. Of course this can be eliminated by commanding the head to stay still and then grabbing the image, but in robotic soccer, this would slow down participation in the game beyond any level of competitive performance. Other software modules need to perform the corresponding sensor fusion (data fusion) to build a reasonable picture of the environment. The *whiteboard* can be considered a series (a table) of facts of the form "at time X, module Z believed Y". Every sensor wrapper will deposit into the *whiteboard* whatever information it can report. Using non-monotonic logic solves the sensor fusion issues by integrating the different beliefs about the environment.

However, there are some challenges with the *whiteboard*. For efficiency reasons, some message may hold pointers (references) to other objects[5] that were not replicated (cloned). E.g. with the images from the vision module, it is too costly (in terms of memory and CPU time) to copy the image pixels. Therefore, the potential exists for the publisher to have deleted the object(s) pointed to by the message. Vision may need to free memory more frequently than other modules as it handles much larger objects. Reference counting offers a solution, but in C++ requires retro-fitting into all affected classes. Nonetheless, the *whiteboard* offers an interface to test references that no longer have a footprint.

2.4   Reasoning and Planning

The large list of facts that are available on the *whiteboard* need further processing by reasoning. We have shown that non-monotonic logic can reason about the landmarks reported by vision [27,28]. We have also used this to construct

---

[5]   Instances of classes in the sense of the object-oriented model.

behaviors for triggering alarms if certain conditions are observed in the environment [29]. The first point that the above examples of reasoning illustrate is that the reasoning engine can sit independently of the perception-action cycle.

1. It may sit somewhere in the list of activities that are performed when we respond to the sensor that has brought us to the start of the cycle (e.g. analyzing sightings after vision has processed an image).
2. It may sit as part of some behavior. That is, the conditions that label the transitions in the finite state machine that describes the behavior.

The implementation of PL required the development of a logic programing language (DPL) and a reasoning engine in C. The DPL representation of a set of rules is parsed into a binary representation that can be uploaded onto the robot or used in a simulator for testing. The very same reasoning engine runs on both the robot and the simulator, allowing us to test and debug logic models in a controlled environment outside the robot.

Glue code between the C++ implementation on the robot and the reasoning engine allows us to use predicates that we know can be implemented more efficiently in the native programming language than in logic itself. For example, it may be very laborious to describe integer arithmetic in logic. Asking if an integer is larger than another (which may be necessary for testing if an object is perceived above another) or computing the angle between to perceived objects is best performed in the SONY Aibo in C++. Therefore, we expect to naturally construct logic models for which some predicates consist of collecting a bunch of facts from the *whiteboard* (and thus perhaps from many other modules) and formulating a new fact by posting a new message onto the *whiteboard*.

We have a direct mechanism to place automatically generated C++ code (from a PL) in a *Framework* using a *Template Method*[6]. The template method has three main phases. The first phase (initialization) sets all the variables of the logic to false, ensuring that, by default, all facts are unknown. Then, those predicates that are calculated in the native language are evaluated once using the information on the *whiteboard* (in the simulator, they are set by the operator of the simulator). The resulting values modify the Boolean variables that were initialized in the first step. The logic engine can now be invoked with one or more formulas, and as a result, some new facts can be posted to the *whiteboard*, with the signature of the module using the engine.

2.5 Modeling Behaviors

We now provide a description of the tools and our approach to the implementation and the programming of behaviors. As we already mentioned, this was originally implemented on a SONY Aibo for several applications. Easy migration to other platforms has been demonstrated by a Mac OS X (Cocoa) simulator implementation, and a recent port to Aldebaran's Nao robot.

---

[6] Larman describes frameworks using the design pattern *Template Method* [19].

Our sense of subsumption is not strictly in the sense of its proponent [30,31]. The highest level in the hierarchy is what we call *external states*. Actions issued by external states have the highest priority. They are modeled by finite state machines (a programmer can define the external states and its transitions with a finite state machine). We use the term state to reflect again the notion of state in finite state machines, or OMT/UML state diagrams [26,19] but in our nomenclature external state refers to overarching states where *behaviors* take place. An external state is a general top-level, *easily perceived*, long lasting condition of the robot. The robot changes from these external states because an external event produces some stimulus. Our external states correspond to the states designed for the application at hand. These are also what external agents will believe are states of the application (this is what we mean by "easily perceived"). Eg., for playing soccer, the external states were `ready`, `playing`, `off-field`, `booting`, `set-team`, `getting-up`, and `returning-home` (perhaps a better name is "meta-states" or "modes"). External states have the highest controller priority. For example, if gyroscopes tell the robot that it has fallen over, such meta-state components may send instructions and issue joint commands to recover from falling over. These actions may dispute the control over "behaviors" under that meta-state (this is where meta-states take priority over behaviors and is our analogy for "subsumption" [30]). For poker and dominoes, the indication that a player's turn has arrived, will constitute a transition of external states.

`Behaviors` can be implemented under the assumption that pre-conditions of the `State` are met. For example, the designer of a `Behavior` may assume to a significant extent that the robot is standing (because if not the meta-state would have taken over and performed the transition to get it standing, despite the fact that the `Behavior` will still execute if the robot is up-side-down; however, none of the commands will be able to overrule the commands issued by the external state). This point is important because the programmer of the behavior cannot totally assume for example that it will always have images as if standing up. In fact, some variables may have values from an up-side-down robot, and this may create a programming bug in the behavior. The true assumption for behavior designers is that their command would have no effect if the robot is up-side-down. However, programming with the aid of non-monotonic reasoning facilitates operation under these contexts, because again, any sensor information can be incorporated in the from of "sensor X believes Y" and thus, it does not matter that it is inconsistent with other information. The logic model should account for this, we will illustrate this point further with an example.

A `Behavior` is a long lasting activity, (however, the code will only do a little bit of work in every call), a personality that defines what the robot does under this behavior. We must emphasize that a behavior must make sense for each of the possible external states. Behaviors are constructed from actions (or commands) to the wrappers of actuators (if they are basic behaviors) or by sub-behaviors (if they are composite behaviors). Commands and actions are thus detailed motions, while behaviors are intermediate between commands and external states. An example of the benefits of this design is our RoboCup soccer implementation, by which the state component maintains whether the robots are playing on the blue team or on the red team. Mi-Pal played both halves of a match with the same memory stick even before the league considered this.

Behaviors operate under a `BehaviorControl` container (a singleton[7] object). `BehaviorControl` has different Behaviors and decides which to run based on the current external `State`. If the external `State` has changed, then we will call `SwitchBehavior` in `BehaviorControl`; this will select the `Behavior` for that `State`. Thus, fundamentally, states are aware of behaviors, but behaviors should be unaware of states. This should enable the same behavior to be used in a different state. However, behaviors will be aware that an external state transition has occurred. Each external state should have at least one behavior responsible for controlling the robot in that external state. Therefore, behaviors act in parallel. When the external state is established (because of a external-state switch) the method `DoBehavior` in `BehaviorControl` will perform what we call an update of all behaviors in all states, namely `UpdateBehavior()` is called. This makes every behavior aware of changes in the environment, even if they are not "active".

3   Illustration and Conclusions

Interpreting the information from sensors (or from other reasoning modules) as beliefs for combining them for common-sense decision making has several benefits. We use the example of a poker player to illustrate that both backward chaining and forward chaining are possible. It is also possible to consider a model of (multiple) agency within a single robot. We illustrate first the use modalities to fuse inputs or expertise from different sub-modules. For example, the ranking of the 169 possible initial hands of Texas-Hold'em Poker is subject to contention. Therefore, perhaps it is better, after recognizing its hold, that the robot represents this fact as "vision believes our hand-strength pre-flop is in the top 50% of all possible hands". Sensors supply contradictory information. In soccer, vision on the Aibo regularly reports a distance to the ball different from what the chest-sonar sensor reports. Thus, a fact such as "sensor X says temperature is above 20" can be modeled as a simple belief using the power of plausible rules. That is, the temperature of the environment in unknown, but we can have a rule that says "if sensor X says temperature is above 20, then usually the temperature is above 20". So evaluations and measurements of the environment, and comparisons for specific constants and values, can be handled with more flexibility. We incorporate sensor information on the same aspect of the environment from more than one sensor by plausible rules and a priority relation on the rules. Similarly, we can incorporate two or more personalities into a poker player that give potentially contradictory advice on the next action. That is, our poker strategy is by default a tight aggressive strategy. However, we monitor the opponent's moves. If we learn that the opponent is tight and passive, we become even tighter and more aggressive.

---

[7] A singleton is the only object of this class in the system [19, Page 413][23, Page 125].

In many interactions for games it is not uncommon to have a state diagram leading to the same set of outputs. For example, the design demands a behavior that has few final options as its completion. This is very common in games like poker or dominoes, where the *external state* of IN_MY_TURN has very few ways to finish the behavior. For example, an active poker player can only *call, raise,* or *fold* (and in some situations, only a subset of these options). The corresponding state-diagrams do not have transitions back. When this happens (no transitions back), our tools that evaluate the large system of PL modules can operate in two formats that are analogous to the two types of automatic reasoning (namely forward-chaining and backward chaining). In the forward chaining approach, the execution follows the execution that the finite state machine would follow. That is, the PL expression guarding the initial state is evaluated, and a new state is determined as a result. Once again, the new state is taken as the current state and the expressions guarding transitions are used to determine the next state. This is repeated until a final state is determined. In the poker player, when it is its turn, the robot runs the PL module that determines if the type of opponents is to be changed, and according to the outcome we may now be in a new state to play tighter, it tosses a coin to perhaps bluff, and finally, after deciding on the personality it is going to go with, it uses the PL of the personality it has decided upon to make the final decision whether to call, fold or raise.

Backward chaining is also possible, that is, to execute all the PL modules leading to final states asynchronously (but completely). Then the ones on the previous level synthesize from the current level, and so on, traversing backwards in the transition chain until the initial state is reached. In the the poker player, we may run all styles of play, and get "suggestions" on how to play (as if these were experts on the next move). Then, we synthesize back, so that now we can take the advice from the strategy that best counters the type of opponent we believe we are facing, and directly select among the suggested actions to call, raise, or fold. While backward chaining may seem more wasteful, we have found that typically the run time of the embedded PL reasoning engine is negligible within the action cycle of the robot. We can therefore easily execute all PL modules, rather than selectively execute them. But, we have left this as an optional alternative as we believe both forms may be useful in different behaviors.

Moreover, this also illustrates that the architecture supports a multi-agent model, which is proposed as a post-object-orientation paradigm for software development [17]. The agent model suggests negotiation, perhaps interaction through auctions or regulators. Since behaviors are composed of loosely coupled modules capable of non-monotonic reasoning demonstrates that we can model and support several agents who may arrive at rather contradictory conclusions or bid for possibly incompatible actions. The overall system will mediate between them for a global, well-defined behavior. This was illustrated with the example of a poker player modeled as several personalities that would not necessarily suggest the same action on a particular scenario for the game. Nevertheless, all can execute and are mediated by a non-monotonic reasoning regulator.

References

1. Wichert, G.V., Lawitzky, G.: Man-machine interaction for robot applications in everyday environments. In: IEEE Int. Workshop on Robot and Human Interactive Communications, Bordeaux, Paris, pp. 343–346 (2001)
2. Clarkson, J., Dowland, B., Cipolla, R.: The use of prototypes in the design of interactive machines. In: Int. Conf. Engineering Design ICED 1997, Tampere (1997)
3. Tzafestas, S., Tzafestas, E.: Human-Machine interaction in intelligent robotic systems: A unifying consideration with implementation examples. J. Intelligent and Robotic Systems 32(2), 119–141 (2001)
4. Breazeal, C., Takanishi, A., Kobayashi, T.: Social robots that interact with people. Springer Handbook of Robotics, pp. 1349–1369. Springer, Berlin (2008)
5. Sankar, N. (ed.): Human-Robot Interaction. I-Tech Education, Vienna (2007)
6. Vassos, S., Levesque, H.: Progression of situation calculus action theories with incomplete information. In: 20th IJCAI, Hyderabad, India, pp. 2024–2029 (2007)
7. Trevizan, F.W., de Barros, L., Corrêa da Silva, F.: Designing logic-based robots. Inteligencia Artificial 10(31), 11–22 (2006)
8. Antoniou, G.: Nonmonotonic Reasoning. MIT Press, Cambridge (1997)
9. Billington, D., Rock, A.: Propositional plausible logic: Introduction and implementation. Studia Logica 67, 243–269 (2001)
10. Rock, A., Billington, D.: An implementation of propositional plausible logic. In: 23rd Australasian Computer Science Conf. of Australian CSC, vol. 22(1), pp. 204–210 (2000)
11. Billington, D.: The proof algorithms of plausible logic form a hierarchy. In: Zhang, S., Jarvis, R. (eds.) AI 2005. LNCS (LNAI), vol. 3809, pp. 796–799. Springer, Heidelberg (2005)
12. Billington, D., Estivill-Castro, V., Hexel, R., Rock, A.: Plausible logic facilitates engineering the behavior of autonomous robots (submitted)
13. Xin, M., Sharlin, E.: Playing games with robots — a method for evaluating human-robot interaction. In: Sankar, N. (ed.) Human-Robot Interaction, Vienna, Austria, ch. 26, pp. 469–480. I-Tech Education and Publishing (2007)
14. Schiffel, S., Thielscher, M.: Fluxplayer: A successful general game player. In: Twenty-Second AAAI Conf. on Artificial Intelligence, pp. 1191–1196. AAAI Press, Menlo Park (2007)
15. Lovell, N.: Machine Vision as the Primary Sensory Input for Mobile, Autonomous Robots. PhD thesis, School of ICT, Griffith University, Nathan, QLD (2006)
16. Estivill-Castro, V., Seymon, S.: Mobile robots for an e-mail interface for people who are blind. In: Lakemeyer, G., Sklar, E., Sorrenti, D.G., Takahashi, T. (eds.) RoboCup 2006: Robot Soccer World Cup X. LNCS, vol. 4434, pp. 338–346. Springer, Heidelberg (2007)
17. Wooldridge, M.: An Introduction to MultiAgent Systems. John Wiley, NY (2002)
18. Liu, T.X.W., Baltes, J.: An intuitive and flexible architecture for intelligent mobile robots. In: 2nd Int. Conf. Autonomous Robots and Agents, NZ, pp. 52–57 (2004)
19. Larman, C.: Applying UML and Patterns: An Introduction to Object-Oriented Analysis and Design and Iterative Development. Prentice-Hall, NJ (1995)
20. Jeffries, D., Anderson, A., Hendrickson, C.: Extreme Programming Installed. Addison-Wesley, MA (2001)
21. Compton, P., et al.: Ripple down rules: possibilities and limitations. In: 6th Banf AAAI Knowledge Acquisiiton for Knowledge Based Systems Workshop (1991)

22. Heintz, F., Rudol, P., Doherty, P.: Bridging the sense-reasoning gap using dyknow: A knowledge processing middleware framework. In: Hertzberg, J., Beetz, M., Englert, R. (eds.) KI 2007. LNCS, vol. 4667, pp. 460–463. Springer, Heidelberg (2007)

23. Gamma, E., Helm, R., Johnson, R., Vlissides, J.: Design Patterns: Elements of Reusable Object-Oriented Software. Addison-Wesley, MA (1995)

24. Kim, J.H., Lee, K.H., Kim, Y.D.: The origin of artificial species: Generic robot. Int. J. Control, Automationa, and Systmes 3(4), 564–570 (2005)

25. Hayes-Roth, B.: A blackboard architecture for control. In: Distributed Artificial Intelligence, pp. 505–540. Morgan Kaufmann, San Francisco (1988)

26. Rumbaugh, J.R., Blaha, M.R., Lorensen, W., Eddy, F., Premerlani, W.: Object-Oriented Modeling and Design. Prentice-Hall, NJ (1991)

27. Billington, D., Estivill-Castro, V., Hexel, R., Rock, A.: Non-monotonic reasoning for localisation in robocup. In: Australasian Conf. on Robotics and Automation, Sydney, Australian Robotics and Automation Association (2005)

28. Billington, D., Estivill-Castro, V., Hexel, R., Rock, A.: Using temporal consistency to improve robot localisation. In: Lakemeyer, G., Sklar, E., Sorrenti, D.G., Takahashi, T. (eds.) RoboCup 2006: Robot Soccer World Cup X. LNCS, vol. 4434, pp. 232–244. Springer, Heidelberg (2007)

29. Billington, D., Estivill-Castro, V., Hexel, R., Rock, A.: Non-monotonic reasoning on board a sony AIBO. In: Lima, P. (ed.) Robotic Soccer, ch. 3, Vienna, Austria, pp. 45–70. I-Tech Education and Publishing (2007)

30. Brooks, R.: Intelligence without reason. In: 12th ICJAI, Sydney, Australia, pp. 569–595. Morgan Kaufmann, San Francisco (1991)

31. Brooks, R.: How to build complete creatures rather than isolated cognitive simulators. In: Architectures for Intelligence, pp. 225–239. Lawrence Erlbaum, Mahwah (1991)

A Hybrid Solution for Advice in the Knowledge Management Field

Álvaro Herrero[1], Aitor Mata[2], Emilio Corchado[1], and Lourdes Sáiz[1]

[1] Department of Civil Engineering, University of Burgos
C/ Francisco de Vitoria s/n, 09006 Burgos, Spain
{ahcosio,escorchado,lsaiz}@ubu.es
[2] Department of Computing Science and Automatic, University of Salamanca
Plaza de la Merced, s/n, 37008 Salamanca, Spain
aitor@usal.es

Abstract. This paper presents a hybrid artificial intelligent solution that helps to automatically generate proposals, aimed at improving the internal states of organization units from a Knowledge Management (KM) point of view. This solution is based on the combination of the Case-Based Reasoning (CBR) and connectionist paradigms. The required outcome consists of customized solutions for different areas of expertise related to the organization units, once a lack of knowledge in any of those has been identified. On the other hand, the system is fed with KM data collected at the organization and unit contexts. This solution has been integrated in a KM system that additionally profiles the KM status of the whole organization.

1 Introduction

Knowledge Management (KM) enables organizations to capture, share, and apply the collective experience and know-how (knowledge) of their staff. Ever-growing volumes of data are increasingly viewed as important and essential sources of information that may eventually be turned into knowledge.

KM can be successfully applied in organizations by developing and implementing knowledge infrastructures [1]. These knowledge infrastructures consist of three main dimensions: people, organizational and technological systems.

In recent years, the deployment of information technology has become a crucial tool for enterprises to achieve a competitive advantage and organizational innovation [2]. In keeping with this idea, Artificial Intelligence (AI) can be applied in KM systems in order to speed up processes, classify unstructured data formats that KM is unable to organize, visualize the intrinsic structure of data sets, and select employee-related knowledge from large amounts of data, among other processes.

This paper proposes the application of the Case-Based Reasoning (CBR) [3] and connectionist paradigms to profile the KM status of an enterprise and then automatically generate improvement proposals. The underlying idea is to produce specific and customized suggestions without human intervention to improve the KM status of the analyzed organization. The inputs of this hybrid advising solution are KM data gathered from the analyzed organization by surveys [4]. To process such data, a connectionist projection model (See Section 2) is used.

E. Corchado et al. (Eds.): HAIS 2009, LNAI 5572, pp. 157–168, 2009.

The paper is structured in the following way. Section 2 introduces the unsupervised neural projection model applied in this work, while section 3 describes the CBR paradigm. Section 4 presents the proposed hybrid solution applying the two previously introduced AI paradigms. Section 5 describes the application of the proposed solution to a KM system. Finally, Section 6 presents the conclusions and some proposals for future work in the same field.

2 Connectionist Projection Model

The identification of patterns that exist across dimensional boundaries in high dimensional data sets is a challenging task. Such patterns may become visible if changes are made to the spatial coordinates. Projection models perform such change by projecting high-dimensional data onto a lower dimensional space in order to identify "interesting" directions in terms of any specific index or projection. Such indexes or projections are, for example, based on the identification of directions that account for the largest variance of a data set –as is the case of Principal Component Analysis (PCA) [5], [6], [7] or the identification of higher order statistics such as the skew or kurtosis index -as is the case of Exploratory Projection Pursuit (EPP) [8]. Having identified the most interesting projections, the data is then projected onto a lower dimensional subspace plotted in two or three dimensions, which makes it possible to examine its structure with the naked eye.

The combination of this technique together with the use of scatter plot matrixes constitutes a very useful visualization tool to investigate the intrinsic structure of multidimensional data sets, allowing experts to study the relations between different components, factors or projections, depending on the applied technique.

The solution proposed in this paper applies an unsupervised neural model called Cooperative Maximum Likelihood Hebbian Learning (CMLHL) [9], [10]. It is based on Maximum Likelihood Hebbian Learning (MLHL) [9], [11]. Considering an N-dimensional input vector (x), and an M-dimensional output vector (y), with W_{ij} being the weight (linking input j to output i), then CMLHL can be expressed as:

1. Feed-forward step:

$$y_i = \sum_{j=1}^{N} W_{ij} x_j, \forall i .$$ (1)

2. Lateral activation passing:

$$y_i(t+1) = \left[y_i(t) + \tau(b - Ay) \right]^+ .$$ (2)

3. Feedback step:

$$e_j = x_j - \sum_{i=1}^{M} W_{ij} y_i, \forall j .$$ (3)

4. Weight change:

$$\Delta W_{ij} = \eta . y_i . sign(e_j) | e_j |^{p-1} .$$ (4)

Where: η is the learning rate, τ is the "strength" of the lateral connections, b the bias parameter, p a parameter related to the energy function [9], [10], [11] and A a symmetric matrix used to modify the response to the data [10]. The effect of this matrix is based on the relation between the distances separating the output neurons.

3 CBR Paradigm

Case-Based Reasoning [12] origins are in knowledge based systems. CBR systems solve new problems acquiring the needed knowledge from previous situations [13]. The main element of a CBR system is the case base, a structure that stores the information used to generate new solutions.

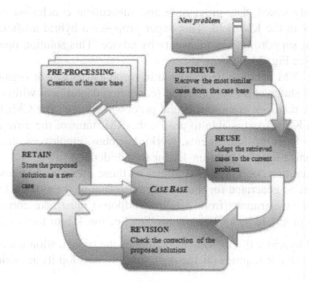

Fig. 1. Basic structure of a CBR system, including the four main stages of the CBR cycle

The learning capabilities of CBR systems are due to its own structure, composed of four main stages [14]: *retrieve*, *reuse*, *revision* and *retain*. These stages are depicted in Fig. 1. The first stage is called *retrieve*, and consists in finding the cases (from the case base) that are most similar to the new problem. Once a set of cases is extracted from the case base, they are *reused* by the system. In this second stage (*retain*), the selected cases are adapted to fit in the new problem. After applying the new solution to the problem, that solution is *revised* to check its performance. If it is an acceptable solution, then it is *retained* by the system and could eventually serve as a solution to future problems.

As a methodology [12], CBR has been used to solve a great variety of problems. It is a cognitive structure that can be easily applied to solve problems such as those related with soft computing, since the procedures used by CBR are quite easy to assimilate by soft computing approaches. CBR has also helped to create applications related to quite different environments, such as health sciences [15], eLearning [16], planning [17], Intrusion Detection Systems [18] and oceanographic matters [19]. Applying CBR to solve a problem usually requires other AI techniques. It is not only a simple way of structuring the reutilization of information, but also a model that can combine different techniques to improve their individual results. Different kinds of neural networks such as ART-Kohonen [20] or Growing Cell Structures [21] have been combined with CBR to automatically create the inner structure of the case base. Some effort has also been devoted to the case-based maintenance issue [22].

4   A Hybrid Advising Solution

As it is previously stated, data processing and subsequent conclusion extraction are challenging tasks in the KM field. This paper proposes a hybrid artificial intelligent solution aimed at supporting KM managers by advice. This solution operates in two different steps (See Fig. 2 below):

- KM Profiling: KM data can be available in some enterprises or organizations. To know the KM status of an organization (company, department within a company, etc) from such data, we propose a neural processing phase. The CMLHL model in unison with a KM expert is able to profile the KM status of the different elements (specific knowledge of departments, working groups, employees, etc.) within the analyzed organization (See Section 4.1 for further details).
- Proposal Generation: once the KM status of those elements is known, coherent proposals must be generated for the worst cases. That is, the elements whose situation is critical or alarming (from a KM standpoint) must take corrective action. That action is proposed by a CBR system (See Section 4.2 for further details).

The proposed hybrid artificial system facilitates the organization under analysis to access the knowledge it requires (at the right time) to develop its activities in a satisfactory way.

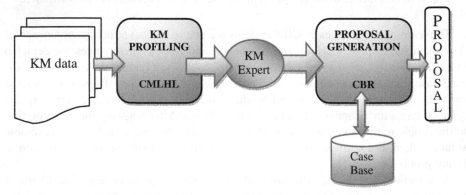

Fig. 2. Schema of the hybrid solution

4.1 KM Profiling

High-volume KM data can be gathered in an organization. This does not guarantee that an in-depth profitable analysis could be subsequently performed. To do so, the application of the CMLHL model is proposed in this paper.

In order to show the KM status of the different units (departments, working groups, employees, etc.) within an organization, some data must be collected. KM experts must choose and define the procedures and methodologies to do it according to the context. Some different techniques can be applied: interviews, surveys, database mining, and so on.

Once these data are captured, CMLHL is applied to visualize the KM status of the different units. CMLHL, as a topology preserving mapping due to the use of lateral connections, clusters projected similar samples of the dataset. The projection is then interpreted by the KM expert to generate a KM profile matrix. According to this matrix, every single department is characterized by 2 features:

- *Temporal restrictions*: urgency of improving the situation.
- *Required knowledge*: amount of knowledge needed to get to a proper situation.

These two features are included in the case definition of the Proposal Generation as described below.

4.2 Proposal Generation

The case base stores information about the previously solved problems together with the decision taken by the expert (supervised phase) or by the system (unsupervised phase) to solve a new problem. This is the main idea underlying the CBR methodology: it is possible to reuse past information in order to solve new problems.

The proposed CBR system is trained until the case base grows up to a big and valid enough state. Then it will generate customized solutions without the intervention of the KM expert. Information about problems and related solutions is stored in the case base that follows the structure described in Table 1. The first eleven parameters refer to the problem to be solved, including data related to the analyzed organization. The other two parameters defining a case (KM Profiling Features) are obtained in the first step of this solution (KM Profiling). All these thirteen features were selected to define cases as the proposal to be generated strongly depends on them. The values of these parameters compose a vector that determines the location of the cases in the case base. When a new case is introduced in the case base, a vector is assigned to that new case. Values of vectors will be similar if the cases they represent are similar. Additionally, these vectors will determine the cases to be recovered from the case base in order to reuse them for a solution. When a new problem comes to the system, those cases that are most similar to the new problem are recovered from the case base. This similarity is calculated by the vector identifying the new problem and those associated to stored cases.

The final solution (proposal) to the problem must be in the set of solutions shown in Table 2. There may be one or several solutions for a certain problem, depending on the parameters of the organization under analysis. As the case base is enlarged by introducing the decisions taken by the expert, the CBR system generates its own automatic decisions. To know when the human expert is not needed, the solution (set of proposals) given by the expert is compared with the eventual solution given by the CBR system from the data available until the moment. When the number of cases in which there is any difference between them is under a certain threshold (never greater than 10% of situations; i.e. the system chooses the right solution in 90% of the cases at least), human supervision is not required. Then, the expert may only be consulted in special situations. If the system is not able to produce a proper solution and it is necessary to check the human expert opinion, the expert will be offered a set of available solutions (those collected in Table 2), if none of them fits the problem, then the expert will introduce in the system a proper solution.

Table 1. Description of the parameters defining a case. The table shows the different parameters (organized in three different levels: Initial Parameters, KM Profiling Features, Solution) and their possible values.

Level	Parameter	Possible Values
Initial Parameters	General Environment – Stability	Stable (1), dynamic (0).
Initial Parameters	General Environment – Complexity	Simple (1), complex (0).
Initial Parameters	General Environment – Hostility	Favourable (1), hostile (0).
Initial Parameters	General Environment – Diversity	Integrated (1), diverse (0).
Initial Parameters	Economic sector	Emerging (4), growing (3), mature (2), declining (1).
Initial Parameters	Developing methods	Internal (1), external (2), cooperation (3), internalization (4).
Initial Parameters	Organizational structure	Simple (1), functional (2), divisional (3), matrix structure (4).
Initial Parameters	Number of employees	Number of employees of the analysed organization.
Initial Parameters	Employees average age	Average age of all the employees.
Initial Parameters	Type of work	Individual (1), collaborative (2).
Initial Parameters	Scope of application	Local (1), regional (2), national (3), continental (4), international (5).
KM Profiling Features	Temporal restrictions	A lot of urgency (3), during this year (2), later (1).
KM Profiling Features	Level of required knowledge	Wide (3), medium (2) or basic (1).
Solution	Proposal	One or more of the solutions in Table 2.

Table 2 shows a predefined sample set of possible solutions that may be applied to an organization after being analyzed by the system. The solutions range from easy internal collaboration solutions (first one) to intensive abstract working (last one). The system will be adapted to the specific characteristics of the analyzed organizations. To do so, the expert decisions will increase this set by proposing new solutions (when the existing ones are not appropriate enough). Already generated solutions can be used in the future to different organizations in new proposals, even if they have never been used in such a knowledge field.

Table 2. Solutions that may be proposed by the system

Solution	Description
1	Experimented employees should collaborate with the new ones.
2	Staff should be trained in their specialized knowledge in deep.
3	Staff should swap their responsibilities from time to time.
4	Experts in a certain area should be employed.
5	Collaborative work should be done.
6	Collaborate with other enterprises in the same field.
7	Take into account suggestions and opinions of clients and providers.
8	Temporally hire an expert in some specific area.
9	Encourage the proposition of new ideas and solutions.
10	Enrol in external knowledge communities.
11	Report detected successes, fails and mistakes.
12	Identify the external and internal knowledge networks.
13	Identify the knowledge experts.
14	Evaluate the learning time in a certain field.
15	Estimate the probability of losing an expert.
16	Estimate the needed time to transfer the knowledge from an expert.
17	Describe the potential uses of the available knowledge.
18	Define internal knowledge communities.

4.3 A Real-Life Case Study

The proposed hybrid solution was applied to a real-life case study: companies from the wall painting sector in the Spanish autonomous region of Castilla y León [4]. The data selected for the KM profiling were taken from a staff survey. A total of 68 records from 39 different companies were generated. The information contained in the 88-feature data set relates to 21 painting techniques (brush painting, spray varnishing, plaster or stucco work, etc...). For each one of these techniques, the survey measured the 4 following factors:

- Knowledge level held: taking values from 2 (lowest level of knowledge) to 8 (highest level of knowledge).
- Willingness to acquire new knowledge: binary value.
- Interest in updating the knowledge held: binary value.
- Interest in sharing the knowledge held: binary value.

As an example, the application of the proposed solution to two employees (E_1 and E_2) is described in the following paragraphs.

4.3.1 A KM Profiling Example
As previously stated, 88 questions were answered by employees E_1 and E_2. Thus, it is not possible to present all the acquired data, although some information may be supplied. The answers from the questions on spray varnishing (one of the 21 painting techniques) were as shown in Table 3.

Table 3. Sample data for the KM Profiling step. The table shows some questions included in the survey and the answers from the two sample employees.

	Value of the answer	
Question	Employee E_1	Employee E_2
Knowledge level held	6	6
Willingness to acquire new knowledge	1	0
Interest in updating the knowledge held	0	0
Interest in sharing the knowledge held	0	0

The 88-feature data were projected by means of CMLHL. The obtained projection is shown in Fig. 3.

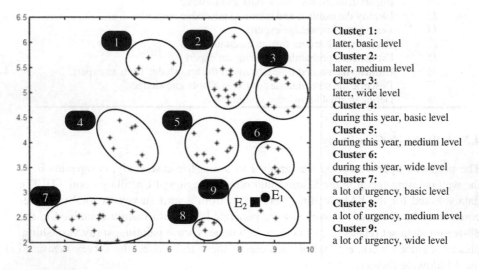

Cluster 1:
later, basic level
Cluster 2:
later, medium level
Cluster 3:
later, wide level
Cluster 4:
during this year, basic level
Cluster 5:
during this year, medium level
Cluster 6:
during this year, wide level
Cluster 7:
a lot of urgency, basic level
Cluster 8:
a lot of urgency, medium level
Cluster 9:
a lot of urgency, wide level

Fig. 3. a) CMLHL factor pair 1-2 projection of the wall-painting case study. b) KM profiling values assigned (temporal restrictions and required knowledge) by the KM expert to each cluster.

Employees E_1 and E_2 were diagnosed as having a "Knowledge Deficit" that is borne out by their belonging to one of the worst clusters. The following values were given to the KM profiling features:
- Temporal restrictions: a lot of urgency (3).
- Required knowledge: wide level (3).

For the sake of brevity, the rest of this section is only describing the E_1 case.

4.3.2 A Proposal Generation Example

Once the KM profiling features of E_1 were known, values were given to the other case parameters as shown in Table 4. These values allow the system to recover (from the case base) those cases that have similar values in their parameters. There is at least one proposal associated to any of the cases recovered. The system measures the

Table 4. E_1 case definition

Parameter	Discrete Values of Parameters
General Environment – Stability	Dynamic (0)
General Environment – Complexity	Simple (1)
General Environment – Hostility	Favourable (1)
General Environment – Diversity	Diverse (0)
Economic sector	Mature (2).
Developing methods	Internal (1).
Organizational structure	Functional (2).
Number of employees	6
Employees average age	49
Type of work	Collaborative (2)
Scope of application	Regional (2)
Temporal restrictions	A lot of urgency (3)
Required knowledge	Wide (3)

similarity between the values of the recovered cases and the analyzed values of E_1 and consider those most similar and their solutions. If there is a common proposal (or even a mainly common proposal) for those cases, then that is the proposal offered by the system as solution to E_1.

Taking this information into account, the CBR system generates the following proposals for the E_1 case:

- 2.- Staff should be trained in their specialized knowledge in deep.
- 11.- Report detected successes, fails and mistakes.

These solutions were also checked by the KM expert, being verified that they are the best solutions for such situation.

5 Enhancing DIPKIP

DIPKIP [4] is a KM system that responds to the need for information management and knowledge flows within a KM organization. It proved itself to be a robust tool for the analysis and identification of critical situations that enable companies to take decisions in the field of KM. It is named after its 4 steps: Data acquisition, Intelligent Processing, Knowledge Identification and Proposal that can be briefly described in the following way:

- **First Step - Data Acquisition**: it aims to capture information about the organization in which DIPKIP is to be applied. Information can be acquired through interviews, surveys, database mining, a combination of these, and so on.
- **Second Step - Intelligent Processing**: the data obtained in the first step is analyzed through CMLHL. This model provides a visualization of the internal structure of the data set. CMLHL was selected as it provided the clearest projections of the case studies for subsequent expert analysis.
- **Third Step - Knowledge Identification**: a KM expert, based on the data projection generated in the second step, catalogues the analyzed organization into one of three classes, according to the situations that can arise in the field of strategic knowledge - knowledge deficit, partial knowledge deficit and no knowledge deficit.

- **Fourth Step - Proposal**: based on the previous step, DIPKIP sets out proposals relating to the following KM processes: creation/acquisition, transference/ distribution and putting into practice/updating. Once the required KM processes have been identified, the KM expert has to decide on the specific actions to implement these processes. DIPKIP outputs must be customized by considering the situation of the analyzed organization. The purpose of DIPKIP is to support decision making that relates to knowledge acquisition, sharing and updating processes that are key to KM processes in the company.

In the original version of DIPKIP, the KM expert was in charge of analyzing the data projection generated in the third step: Knowledge Identification. This analysis was intended to determine the DIPKIP proposals for the analyzed organization. In the proposed enhanced version of DIPKIP presented in this paper, it is extended by means of the CBR paradigm as previously described in section 4: a CBR system is included in DIPKIP last step in order to automate the proposal generation. Thus, the KM expert intervention is only required in the most difficult situations.

The initial version of the DIPKIP system required expert intervention to identify the knowledge (third step) and to generate a proposal (fourth step). In this version, the fourth step is split in two different phases to eliminate the need of a human expert. The first phase relies on supervised training: the CBR system is trained by storing in the case base the relation between the clustered data and the decision taken. To store information in the case base, it is necessary to analyze the output of the third step.The case base stores information related to the previously solved problems together with the decision taken by the expert (or by the system on its own) to solve the proposed situation.

A schema about the first (supervised) phase of the extended fourth step is depicted in Fig. 4.a. The decisions taken by the KM expert make the case base grow, until the CBR system can work in an independent way. After that, the CBR system is then autonomous and the KM expert is only consulted, in this fourth step, when the CBR

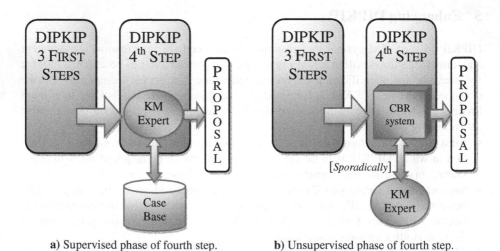

a) Supervised phase of fourth step.   b) Unsupervised phase of fourth step.

Fig. 4. Two phases of the proposed DIPKIP system extension

system is not sure enough to propose a good solution (Fig. 4.b). If the expert is consulted by the system, when working in an autonomous way, new solutions may be stored in the system.

6 Conclusions and Future Work

This study presents a novel KM hybrid artificial intelligent solution that automates the proposal generation for overcoming KM deprived situations. To do so, the Case-based Reasoning and connectionist paradigms have been applied. Additionally, this solution has been applied to a four-step KM system called DIPKIP, in which the last step has been split into two different phases. This upgrading has been performed by means of a CBR system allowing an automatic proposal generation. The new model has been applied successfully to a real-case study.

Future work will focus on the application of this hybrid solution to different case studies.

Acknowledgments. This research has been partially supported by the projects BU006A08 and SA071A08 of the JCyL. The authors would also like to thank the manufacturer of components for vehicle interiors, Grupo Antolin Ingeniería, S.A. in the framework of the project MAGNO 2008 - 1028.- CENIT Project funded by the Spanish Ministry of Science and Innovation.

References

1. Sivan, Y.Y.: Nine Keys to a Knowledge Infrastructure: A Proposed Analytic Framework for Organizational Knowledge Management. In: Proceedings of WebNet 2000 - World Conference on the WWW and Internet, pp. 495–500. AACE (2000)
2. Shu-Mei, T.: The Effects of Information Technology on Knowledge Management Systems. Expert Systems with Applications: An International Journal 35(1-2), 150–160 (2008)
3. Aamodt, A., Plaza, E.: Case-Based Reasoning - Foundational Issues, Methodological Variations, and System Approaches. AI Communications 7(1), 39–59 (1994)
4. Herrero, A., Corchado, E., Sáiz, L., Abraham, A.: DIPKIP: A Connectionist Knowledge Management System to Identify Knowledge Deficits in Practical Cases. Computational Intelligence (in press, 2009)
5. Hotelling, H.: Analysis of a Complex of Statistical Variables into Principal Components. Journal of Education Psychology 24, 417–444 (1933)
6. Pearson, K.: On Lines and Planes of Closest Fit to Systems of Points in Space. Philosophical Magazine 2(6), 559–572 (1901)
7. Oja, E.: Neural Networks, Principal Components, and Subspaces. Int. Journal of Neural Systems 1, 61–68 (1989)
8. Friedman, J.H., Tukey, J.W.: A Projection Pursuit Algorithm for Exploratory Data-Analysis. IEEE Transactions on Computers 23(9), 881–890 (1974)
9. Corchado, E., MacDonald, D., Fyfe, C.: Maximum and Minimum Likelihood Hebbian Learning for Exploratory Projection Pursuit. Data Mining and Knowledge Discovery 8(3), 203–225 (2004)

10. Corchado, E., Fyfe, C.: Connectionist Techniques for the Identification and Suppression of Interfering Underlying Factors. Int. Journal of Pattern Recognition and Artificial Intelligence 17(8), 1447–1466 (2003)

11. Fyfe, C., Corchado, E.: Maximum Likelihood Hebbian Rules. In: Proc. of the 10th European Symposium on Artificial Neural Networks (ESANN 2002), pp. 143–148 (2002)

12. Watson, I.: Case-based Reasoning is a Methodology Not a Technology. Knowledge-Based Systems 12(5-6), 303–308 (1999)

13. Aamodt, A.: A Knowledge-Intensive, Integrated Approach to Problem Solving and Sustained Learning. Knowledge Engineering and Image Processing Group. University of Trondheim (1991)

14. Aamodt, A., Plaza, E.: Case-Based Reasoning: Foundational Issues, Methodological Variations, and System Approaches. AI Communications 7(1), 39–59 (1994)

15. Corchado, J.M., Bajo, J., Abraham, A.: GERAmI: Improving the Delivery of Health Care. IEEE Intelligent Systems. Special Issue on Ambient Intelligence (2008)

16. Althoff, K.D., Mänz, J., Nick, M.: Maintaining Experience to Learn: Case Studies on Case-Based Reasoning and Experience Factory. In: Proc. 6th Workshop Days of the German Computer Science Society (GI) on Learning, Knowledge, and Adaptivity (LWA 2005) (2005)

17. Cox, M.T., Muñoz-Avila, H., Bergmann, R.: Case-based Planning. The Knowledge Engineering Review 20(3), 283–287 (2006)

18. Herrero, A., Corchado, E., Pellicer, M.A., Abraham, A.: MOVIH-IDS: A Mobile-Visualization Hybrid Intrusion Detection System. Neurocomputing (in press, 2009)

19. Fdez-Riverola, F., Corchado, J.M.: FSfRT: Forecasting System for Red Tides. Applied Intelligence 21(3), 251–264 (2004)

20. Yang, B.S., Han, T., Kim, Y.S.: Integration of ART-Kohonen Neural Network and Case-based Reasoning for Intelligent Fault Diagnosis. Expert Systems With Applications 26(3), 387–395 (2004)

21. Diaz, F., Fdez-Riverola, F., Corchado, J.M.: Gene-CBR: A Case-based Reasoning Tool for Cancer Diagnosis using Microarray Data Sets. Computational Intelligence 22(3/4), 254–268 (2006)

22. Liu, C.-H., Chen, L.-S., Hsu, C.-C.: An Association-based Case Reduction Technique for Case-based Reasoning. Information Sciences 178(17), 3347–3355 (2008)

A Cluster-Based Feature Selection Approach

Thiago F. Covões[1], Eduardo R. Hruschka[1], Leandro N. de Castro[2],
and Átila M. Santos[2]

[1] University of São Paulo (USP), Brazil
tcovoes@icmc.usp.br, erh@icmc.usp.br
[2] Catholic University of Santos (UniSantos), Brazil
lnunes@unisantos.br, atila_msantos@yahoo.com.br

Abstract. This paper proposes a filter-based method for feature selection. The filter is based on the partitioning of the feature space into clusters of similar features. The number of clusters and, consequently, the cardinality of the subset of selected features, is automatically estimated from the data. Empirical results illustrate the performance of the proposed algorithm, which in general has obtained competitive results in terms of classification accuracy when compared to a state of the art algorithm for feature selection, but with more modest computing time requirements.

Keywords: Feature Selection, Filters, Clustering and Classification Problems.

1 Introduction

Feature selection involves choosing a subset of original variables (attributes) by eliminating the redundant, uninformative, and noisy ones. This issue has been broadly investigated in supervised learning tasks for which datasets with many features are available, like in text mining and gene expression data analysis. A comprehensive survey of feature selection algorithms is presented in [12]. In brief, there are two fundamentally different approaches for feature selection [11][9]: *wrapper* and *filter*. The former evaluates the subset of selected features using criteria based on the results of learning algorithms, while the latter selects features based on intrinsic properties of the data, being independent of the learning algorithm to be used. Wrappers are often criticized because they require massive amounts of computation [5]. In data mining applications one usually faces large datasets, and thus methods called filters, which are commonly faster than wrappers, are often more interesting [11]. The readers interested in filtering methods are referred to references [3][5][11][13][16][17] and the bibliography therein. While some filters may involve some kind of transformation of the feature space (e.g., principal component analysis and factor analysis), the present work focuses on finding subsets of features of the original space, mainly because this often allows much simpler and comprehensible results, maintaining the physical interpretation of the selected features.

The feature selection method proposed in this paper is based on the filter described by Mitra et al. [13]. Such filter is considered a state of the art method for feature selection in the pattern recognition field [17]. In brief, the filter described in [13] removes redundant features by partitioning the original feature set into some distinct

E. Corchado et al. (Eds.): HAIS 2009, LNAI 5572, pp. 169–176, 2009.

clusters, which are formed by similar features. Then, a single feature from each cluster is selected, allowing dimensionality reduction. The size of the feature subset depends on the initial choice of a parameter that represents the number of neighboring features. The approach proposed here is also based on the clustering of similar features. However, we use a clustering algorithm that can estimate, according to a numeric criterion, the number of clusters, which also corresponds to the number of features to be selected. From this perspective, the proposed method can find a subset of non-redundant features in a more automatic fashion.

2 Related Work

The unsupervised filter proposed by Mitra et al. [13] involves two main steps, namely: (i) the partitioning of the complete feature set into clusters and; (ii) the selection of a representative feature from each cluster. Linear dependency is used to assess similarities between two features and, consequently, to induce clusters. In particular, it is shown that the proposed *maximal information compression index* - $\lambda(X_i,X_j)$ - in Eq. (1) is a measure of the minimum amount of information loss (or the maximum amount of information compression) possible. Hence, it is a dissimilarity measure that may be suitably used for redundancy reduction. Let X_i and X_j be two random variables (here called features). $\lambda(X_i,X_j)$ is defined as:

$$\lambda(X_i, X_j) = \xi - \sqrt{\xi^2 - 4\,\mathrm{var}(X_i)\,\mathrm{var}(X_j)(1 - \rho(X_i, X_j)^2)} \tag{1}$$

where $\xi = var(X_i) + var(X_j)$, $var(X_i)$ denotes the variance of X_i, and $\rho(X_i,X_j)$ is the Pearson correlation coefficient between X_i and X_j.

Clusters of features are obtained via the well-known k-nearest neighbors (k-NN) principle. Initially (first iteration of the algorithm), the k nearest neighbors (k_{NN}) of each feature are computed. Among them, the feature that has the most compact cluster is selected, and its k_{NN} neighboring features are discarded. The distance of a given feature to its farthest neighbor measures the *lack of compactness* of a given cluster. The process is repeated for the remaining features, iterating until all of them are classified as either selected or discarded. During the execution of the algorithm, the k_{NN} value is indirectly controlled by a parameter called *constant error threshold*, ε, which is set equal to the distance of the kth nearest-neighbor of the feature selected in the first iteration. In subsequent iterations, the *lack of compactness* value is checked to verify whether it is greater than ε or not. If that is true, the k_{NN} value is decreased. It is important to note that the initial value of k_{NN} is chosen by the user, and it controls the cardinality of the subset of selected features. As claimed by the authors [13], on the one hand it may be useful to control the representation of the data at different levels of details, performing some kind of exploratory data analysis. On the other hand, the choice of the value of such a parameter may be hard to be accomplished in practice, for the user is left to estimate a critical parameter of the algorithm. The method proposed in this paper is aimed at further improving the algorithm proposed in [13] by making it capable of selecting features in a more automatic fashion with regards to the number of selected features. Moreover, the proposed algorithm has been designed to be at least as computationally efficient as the Mitra et al.'s algorithm - whose overall computational complexity is estimated as $O(M^2 \cdot N)$ if the value of the k_{NN} parameter is a priori known, where M and N are the number of features and instances of the dataset, respectively.

3 Simplified Silhouette Filter (SSF)

After partitioning a set X of features $X=\{X_1, X_2, ..., X_M\}$, $X_j \in \Re^N$, into a collection $C^X=\{C_1,C_2,...,C_k\}$ of k mutually disjoint subsets of correlated features C_i of X, it is expected that features that belong to the same cluster should be more similar (correlated) to each other than features that belong to different clusters. Therefore, it is necessary to devise means of evaluating similarities (correlations) between feature sets. This problem is often tackled indirectly, i.e. distance measures can be used to quantify dissimilarities (*lack of correlation*) between features. We employ the maximal information compression index – Eq. (1) – to find clusters of features.

Attempting to find a globally optimum solution for clustering problems is usually not computationally feasible [1]. This difficulty has stimulated the search for efficient approximate algorithms. This work follows this trend, employing a heuristic procedure, which is based on the simplified silhouette criterion [6], for finding the number of clusters and the corresponding feature partitions.

To define the simplified silhouette (SS) [6], consider a feature X_i belonging to cluster C_a. The dissimilarity of X_i to the medoid of C_a is denoted by $a(i)$. Now let us take into account cluster C_j. The dissimilarity of X_i to medoid of C_j will be called $d(X_i,C_j)$. After computing $d(X_i,C_j)$ for all clusters $C_j \neq C_a$, the smallest one is selected, i.e. $b(i) = \min d(X_i,C_j)$, $C_j \neq C_a$. This value represents the dissimilarity of X_i to its neighbor cluster, and the silhouette $s(i)$ is given by:

$$s(i) = \frac{b(i) - a(i)}{\max\{ a(i), b(i)\}} \tag{2}$$

The higher $s(i)$ the better the assignment of X_i to a given cluster. In addition, if $s(i)$ is equal to zero, then it is not clear whether the feature should have been assigned to its current cluster or to a neighboring one [4]. Finally, if C_a is a singleton, then $s(i)$ is not defined and the most neutral choice is to set $s(i) = 0$ [8]. The average of $s(i)$, $i = 1$, $2, ... , M$, can be used as a criterion to assess the quality of a given feature partition. By doing so, the best clustering is achieved when the silhouette value is maximized.

The computation of the simplified silhouette [6], depends only on the achieved partition and not on the adopted clustering algorithm. Thus, this silhouette can be applied to assess partitions (taking into account the number of clusters) obtained by several clustering algorithms. We adopt the well-known k-medoids algorithm to obtain partitions to be evaluated by the simplified silhouette (SS). This algorithm is interrupted as soon as medoids from two consecutive iterations are equal. Roughly speaking, k-medoids is designed to minimize the sum of distances between features and nearest medoids. From the SS criterion viewpoint, good partitions are also obtained when this minimization is suitably performed, as well as when clusters are well separated.

The SS is a numeric criterion that allows estimating the number of clusters automatically. Thus, it can provide a way of circumventing an important limitation of k-medoids, i.e. the number of clusters k must be determined a priori. In this sense, one can perform multiple runs of k-medoids (for different values of k) and then choose the best available partition, which corresponds to the maximum achieved value for the SS. It is also well-known that k-medoids may get stuck at suboptimal solutions for a given k [18]. To alleviate this problem, one can perform multiple runs of k-medoids for a fixed k.

The sampling strategy depicted in Fig. 1 summarizes the method used here to estimate the *correct* number of clusters $k*$ according to the SS, as well as to find the corresponding clusters of features. For each cluster obtained, the corresponding medoid is chosen as the representative feature. By doing so, a subset of $k*$ features is achieved and, thus, the proposed filter is capable of automatically determining the number of selected features. We call this feature selection method as Simplified Silhouette Filter (SSF). Provided that the clustering method (Fig. 1) is applied to vectors from the feature space, the number of clusters, $k*$, here corresponds to the number of selected features from the feature set X.

---

1. Choose the minimum and the maximum number of clusters in a set of values for k, respectively k_{min} and k_{max}, as well as a number of different initial partitions (n_p) for k-medoids.
2. SSV← –Infinity;   / SSV = Simplified Silhouette Value /
3. For each $k∈ \{k_{min}, k_{min}+1,..., k_{max}-1 ,k_{max}\}$ do:
   3.1 Generate n_p random initial partitions of features into k nonempty clusters;
   3.2 Run k-medoids for each initial partition generated in Step 3.1 and compute its corresponding simplified silhouette. Let the best obtained value be BOV;
   3.3 If (BOV > SSV) then { SSV←BOV;
                          $k* ← k$;
                          Hold the corresponding partition for $k*$. }
4. Return SSV and its corresponding feature clusters for $k*$.

Fig. 1. Sampling strategy for k-medoids

4 Empirical Evaluation

Ten datasets were used to assess the performance of the proposed algorithm, but due to space restrictions we will report results obtained in only three representative datasets. These datasets are widely known and available at the UCI Machine Learning Repository [2], namely: Ionosphere, Wisconsin Breast Cancer, and Spambase.

A summary of the main features of the datasets used in our experiments is provided in Table 1. In all experiments, we have set $k_{min} = 2$, $k_{max} = M$-1, and $n_p = 20$ for clustering features. From a practical viewpoint, one can consider that these values somehow determine the size of the search space to be assessed, as well as the computational effort to find the corresponding solution. Therefore, domain knowledge, when available, can be incorporated into this approach in order to set those parameters in scenarios that present limitations of computational resources.

As discussed in Section 2, the performance of Mitra et al.'s algorithm [13] is highly dependent upon a parameter, k_{NN}, chosen by the user and that controls the cardinality of the subset of selected features. The authors [13] claim that it may be useful to control the representation of the data at different levels of details, performing some kind of exploratory data analysis. Despite the good results shown in [13], a number of experiments reported here illustrate that the choice of the k_{NN} value may be hard to be accomplished in practice. To make this point clearer, we have run Mitra et al.'s algorithm [13] by varying k_{NN} within the range of all possible values - $\{1,..., M-1\}$ - for each dataset. The quality of each feature subset has been assessed by the generalization capability of the obtained classifier, which is estimated using a 10-fold cross-validation process. Feature selection has been performed using only the training folds, and classification accuracy has been estimated in the test folds [14]. The same training/test folds were used for both algorithms. Due to space restrictions, we report here only the best and the worse results obtained by the algorithm proposed by Mitra et al.'s [13], but we shall note that such results are particularly useful to assess the difficulty faced by the user at setting the k_{NN} value. In addition, such results favor the illustration of practical scenarios of particular interest, in which either the Mitra et al.'s algorithm [13] or the Simplified Silhouette Filter (SSF) may be preferred.

Table 1. Summary of the datasets used: N(number of instances) and M (number of features)

Dataset	N	M	# classes (distributions - %)
Ionosphere	351	34	2 (35.9 – 64.1)
Wisconsin	683	9	2 (65.0 – 35.0)
Spambase	4,601	57	2 (39.4 – 60.6)

Tables 2-4 summarize the average results obtained in a 10-fold cross-validation process. Variances appear within brackets. The 1st column of each table presents the algorithms used in the experiment. The acronyms SSF, Mitra, and *All* refer to the simplified silhouette filter, Mitra et al.'s algorithm [13] (where B and W stand for the best and worse results, respectively), and finally to the results found by using all features of the dataset. Also, k_{NN}, SSV, M^*, A(%)-KNN, and A(%)-Bayes stand for initial number of k nearest neighbors (Section 2), SS value (Section 3), number of selected features, average correct classification rate for a k-NN classifier, and average correct classification rate for the Naïve Bayes classifier, respectively. These classifiers were chosen to illustrate the performance of the feature selection methods evaluated in this work due to their widespread use in practice. We have employed the k-NN and Naïve Bayes classifiers available in Weka [15]. In the k-NN classifier the number of neighbors used is the number of objects in the training set with distance weighting by 1/distance. Other parameters were used with default values.

Table 2 presents the results achieved for the Ionosphere dataset. The best results for the classifiers k-NN and Naïve Bayes involve different feature subsets. Due to this reason we have left some cells of Table 2 with no values. A statistical t-test allows concluding that, at the $\alpha = 5\%$ level, significant accuracy differences have not been observed between SSF and the best results provided by the Mitra et al.'s algorithm for k-NN, whereas a significant difference has been perceived for Naïve Bayes. The features selected by SSF have shown to be able to induce more accurate classifiers than those constructed from the features selected by Mitra et al.'s algorithm when the number of nearest neighbors of each feature (k_{NN}) is not appropriately chosen by the user – see the results for Mitra (W) in the tables. The number of features selected by SSF is significantly less than the number of clusters selected by Mitra et al.'s algorithm. To summarize, SSF provided competitive results to Mitra et al.'s algorithm, mainly when the cardinalities of the feature subsets are taken into account.

Table 2. Summary of results for the Ionosphere dataset

Algorithm	k_{NN} or SSV	$M*$	A(%)-KNN	A(%)-Bayes
SSF	0.526 (0.23)	2.2 (0.42)	81.76 (4.72)	75.5 (6.22)
Mitra (B)	31 (0)	2 (0)	81.20 (3.84)	-----
Mitra (B)	1 (0)	32 (0)	-----	83.78 (4.17)
Mitra (W)	1 (0)	32 (0)	64.67 (1.58)	-----
Mitra (W)	11 (0)	20.80 (0.63)	-----	74.36 (6.87)
All	-----	34	64.67 (1.58)	83.21 (4.04)

For the Wisconsin dataset (Table 3), the better feature subset found by Mitra et al.'s algorithm - after running it for $k_{NN} \in \{1, 2,..., 8\}$ - has provided significantly more accurate classifiers (t-test, $\alpha=5\%$) than those induced by the features selected by SSF. However, it is worth stressing that, contrarily to the use of Mitra et al.'s algorithm, when employing SSF the user is not required to choose the k_{NN} value. In this sense, the results achieved for this dataset also show that SSF allows performing suitable dimensionality reduction in a more automatic fashion.

Table 3. Summary of results for the Wisconsin dataset

Algorithm	k_{NN} or SSV	$M*$	A(%)-KNN	A(%)-Bayes
SSF	0.446 (0.03)	3.8 (1.23)	81.56 (8.66)	94.88 (1.72)
Mitra (B)	7 (0)	2 (0)	89.59 (4.03)	-----
Mitra (B)	2 (0)	7 (0)	-----	96.48 (1.85)
Mitra (W)	1 (0)	8 (0)	69.70 (1.82)	-----
Mitra (W)	8 (0)	1 (0)	-----	75.99 (3.05)
All	-----	9	69.84 (1.59)	96.34 (1.86)

Using Naïve Bayes as a baseline classifier, the features selected by the Mitra et al.'s algorithm have shown significantly better results (t-test, $\alpha = 5\%$) than those selected by SSF in Spambase (Table 4). These results illustrate an interesting tradeoff between efficacy (classification accuracy) and computational efficiency. To make it possible to empirically assess computational efficiency, both algorithms (SSF and Mitra et al.'s algorithm) were implemented in C, using only the strictly necessary commands. This way, more uniform efficiency comparisons can be performed. The same computer (Pentium 4, 3.0GHz, 1Gb RAM), running only the operational system, was used for all the controlled experiments. The best Naïve Bayes classifier was obtained by Mitra et al.'s algorithm after several runs of feature selection - one for each value of $k_{NN} \in \{1, 2,..., 56\}$. In addition, for each k_{NN} value a cross-validation process needs to be performed in order to verify the classifier accuracy. Such experiments (involving feature selection and classifier assessment) have consumed 43.6 seconds of computing time. After this process, only one feature has been removed. SSF, by its turn, has removed approximately 54 features (on average) and consumed 19.8 seconds to perform feature selection and classifier assessment. Moreover, if the

Pearson correlation coefficient is used for computing similarities among features, then the estimated average classification rate (in 10-fold cross-validation) of the resultant classifier is 72.20%. Thus, if computational efficiency is an important issue (e.g., when classifiers less computationally efficient than Naïve Bayes are used), SSF may be preferred, whereas if computational efficiency is not of paramount relevance an exploratory data analysis may be more interesting. Significant statistical difference has not been observed in the accuracy of the k-NN classifier.

Table 4. Summary of results for the Spambase dataset

Algorithm	k_{NN} or SSV	$M*$	A(%)-KNN	A(%)-Bayes
SSF	0.907 (0.02)	2.7 (0.82)	65.01 (4.01)	58.60 (11.652)
Mitra(B)	52 (0)	5 (0)	65.73 (4.60)	-----
Mitra(B)	1 (0)	56 (0)	-----	80.36 (1.55)
Mitra(W)	56 (0)	1 (0)	60.60 (0.10)	35.98 (12.00)
All	-----	57	61.23 (0,16)	79.72 (1.94)

5 Conclusions

This paper explored a filter method based on the partitioning of the feature space into clusters of similar features. The proposed algorithm, called Simplified Silhouette Filter (SSF), relies on selecting one representative feature from each cluster obtained, thus allowing the elimination of redundant features. We have shown that SSF represents an improvement in relation to the state of art method described by Mitra et al. [13]. In particular, SSF is capable of selecting features in a more automatic fashion. The number of clusters, and consequently the cardinality of the subset of selected features, is automatically estimated by SSF from data. Empirical results in three datasets illustrate the performance of SSF, which in general provided competitive results when compared to Mitra et al.'s algorithm [13]. Also, SSF tends to be at least as computationally efficient as the Mitra et al.'s algorithm, which has overall time complexity estimated as $O(M^3)$ when the value of the parameter k_{NN} is varied in the set $\{1, 2, ...,M\text{-}1\}$. More precisely, the SSF time complexity is estimated as $O(M^3)$ when its parameters k_{min} and k_{max} have their values set to two and M-1, respectively. However, if domain knowledge is available that allows the user to restrict the SSF search space by adjusting k_{max} to $M^{1/2}$, for instance, the SSF time complexity is estimated as $O(M^2 \ln M)$. Finally, if the number of features to be selected is a priori known then both algorithms are $O(M^2)$. Considering the magnitude of the constant terms – neglected by asymptotic complexity analysis – we have reported experimental results obtained in the Spambase that illustrate that SSF can be more efficient than its counterpart algorithm.

Despite the promising results achieved in this work, some issues should be further investigated. For instance, provided that SSF in principle does not necessarily require the use of any particular clustering algorithm, the investigation of the suitability of clustering algorithms different from the one used in our study is a promising future work. Also, non-linear correlation measures, such as those based on the information-theoretical concept of entropy, can be potentially used in our algorithmic framework, substituting the *maximal information compression index* used here. Finally, more experiments (e.g, using data sets formed by multiple, unbalanced classes) will be performed to further confirm the potential of SSF.

References

1. Arabie, P., Hubert, L.J.: An Overview of Combinatorial Data Analysis. In: Arabie, P., Hubert, L.J., DeSoete, G. (eds.) Clustering and Classification. World Scientific, Singapore (1999)
2. Asuncion, A., Newman, D.J.: UCI Machine Learning Repository. University of California, Irvine, http://www.ics.uci.edu/~mlearn/MLRepository.htm
3. Au, W., Chan, K.C.C., Wong, A.K.C., Wang, Y.: Attribute Clustering for Grouping, Selection, and Classification of Gene Expression Data. IEEE/ACM Transactions on Computational Biology and Bioinformatics 2(2), 83–101 (2005)
4. Everitt, B.S., Landau, S., Leese, M.: Cluster Analysis. Arnold Publishers, London (2001)
5. Guyon, I., Elisseeff, A.: An Introduction to Variable and Feature Selection. Journal of Machine Learning Research 3, 1157–1182 (2003)
6. Hruschka, E.R., Campello, R.J.G.B., de Castro, L.N.: Evolving Clusters in Gene-Expression Data. Information Sciences 176(13), 1898–1927 (2006)
7. John, G., Kohavi, R., Pfleger, K.: Irrelevant features and the subset selection problem. In: Proc. of the Eleventh Int. Conf. on Machine Learning. Morgan Kaufmann, San Francisco (1994)
8. Kaufman, L., Rousseeuw, P.J.: Finding Groups in Data – An Introduction to Cluster Analysis. Wiley Series in Probability and Mathematical Statistics (1990)
9. Kohavi, R., John, G.H.: Wrappers for Feature Subset Selection. Artificial Intelligence 97(1-2), 273–324 (1997)
10. Koller, D., Sahami, M.: Toward optimal feature selection. In: Proc. of the 13th Int. Conf. on Machine Learning, pp. 284–292 (1996)
11. Liu, H., Motoda, H.: Feature Selection for Knowledge Discovery and Data Mining. Kluwer Academic Publishers, Dordrecht (1998)
12. Liu, H., Yu, L.: Toward Integrating Feature Selection Algorithms for Classification and Clustering. IEEE Transactions on Knowledge and Data Engineering 17(3), 1–12 (2005)
13. Mitra, P., Murthy, C.A., Pal, S.K.: Unsupervised Feature Selection using Feature Similarity. IEEE Trans. on Pattern Analysis & Machine Intelligence 24(4), 301–312 (2002)
14. Reunanen, J.: Overfitting in Making Comparisons Between Variable Selection Methods. Journal of Machine Learning Research 3, 1371–1382 (2003)
15. Witten, I.H., Frank, E.: Data Mining – Practical Machine Learning Tools and Techniques with Java Implementations. Morgan Kaufmann Publishers, USA (2000)
16. Yang, Y., Pederson, J.: A comparative study on feature selection in text categorization. In: Proc. of the Fourteenth International Conference on Machine Learning (1997)
17. Yu, L., Liu, H.: Efficient Feature Selection via Analysis of Relevance and Redundancy. Journal of Machine Learning Research (5), 1205–1224 (2004)
18. Zhang, T., Ramakrishnan, R., Livny, M.: BIRCH: A New Data Clustering Algorithm and Its Applications. Data Mining and Knowledge Discovery 1(2), 141–182 (1997)

Automatic Clustering Using a Synergy of Genetic Algorithm and Multi-objective Differential Evolution

Debarati Kundu[1], Kaushik Suresh[1], Sayan Ghosh[1], Swagatam Das[1],
Ajith Abraham[2], and Youakim Badr[2]

[1] Department of Electronics and Telecommunication Engineering
Jadavpur University, Kolkata, India
swagatam@etce.jdvu.ac.in
[2] National Institute of Applied Sciences of Lyon, INSA-Lyon, Villeurbanne, France
ajith.abraham@ieee.org, youakim.badr@insa-lyon.fr

Abstract. This paper applies the Differential Evolution (DE) and Genetic Algorithm (GA) to the task of automatic fuzzy clustering in a Multi-objective Optimization (MO) framework. It compares the performance a hybrid of the GA and DE (GADE) algorithms over the fuzzy clustering problem, where two conflicting fuzzy validity indices are simultaneously optimized. The resultant Pareto optimal set of solutions from each algorithm consists of a number of non-dominated solutions, from which the user can choose the most promising ones according to the problem specifications. A real-coded representation of the search variables, accommodating variable number of cluster centers, is used for GADE. The performance of GADE has also been contrasted to that of two most well-known schemes of MO.

1 Introduction

Optimization-based automatic clustering algorithms greatly rely on a cluster validity function (optimization criterion) the optima of which appear as proxies for the unknown "correct classification" in a previously unhandled dataset [1]. Different formulations of the clustering problem vary in the optimization criterion used. Most existing clustering methods, however, attempt to optimize just one such clustering criterion modeled by a single cluster validity index. This often results into considerable discrepancies observable between the solutions produced by different algorithms on the same data. The single-objective clustering method may prove futile (as judged by means of expert's knowledge) in a context where the criterion employed is inappropriate. In situations where the best solution corresponds to a tradeoff between different conflicting objectives, common sense advocates a multi-objective framework for clustering.

Although there has been a plethora of papers reporting several single-objective evolutionary clustering techniques (a comprehensive survey of which can be found in [1, 2]), very few research works have so far been undertaken towards the application of evolutionary multi-objective optimization algorithms (EMOA) for pattern clustering [3, 4]. A state-of-the-art literature survey indicates that DE has already proved itself as a promising candidate in the field of evolutionary multi-objective optimization (EMO) [5 – 8]. Earlier it has also been successfully applied to

E. Corchado et al. (Eds.): HAIS 2009, LNAI 5572, pp. 177–186, 2009.

single-objective partitional clustering [9 – 11]. The work reported in [3] is based on Deb *et al.*'s celebrated NSGA (Non Dominated Sorting genetic Algorithm)-II [12] and the clustering method described in [4] is based on PESA (Pareto Evolution based Selection) II [13], and both the algorithms are multi-objective variants of Genetic Algorithm (GA). However, the multi-objective variants of DE have not been applied to the general data clustering problems till date, to the best of our knowledge. Since DE, by nature, is a real-coded population-based optimization algorithm, we here resort to centroid-based representation scheme for the search variables. A MOO algorithm, in general, ends up with a number of Pareto optimal solutions. Here we consider the Xie-Beni index [14] and the Fuzzy C Means (FCM) measure (J_m) [15] as the objective functions. The performance of GADE has also been contrasted with two best-known EMOA-based clustering methods till date. The first of these is MOCK by Handl and Knowles [4] while the second one is based on NSGA II and was used by Bandyopadhyay *et al.* for pixel clustering in remote sensing satellite image data [3]. Here we report the results for ten representative datasets including the microarray Yeast sporulation data [16].

2 Multi-objective Optimization Using DE

2.1 The MO Problem

In many practical or real life problems, there are many (possibly conflicting) objectives that need to be optimized simultaneously. Under such circumstances there no longer exists a single optimal solution but rather a whole set of possible solutions of equivalent quality. The field of Multi-objective Optimization (MO) [17 – 19] deals with simultaneous optimization of multiple, possibly competing, objective functions.

2.2 The Differential Evolution (DE) Algorithm

DE [20, 21] is a population-based global optimization algorithm that uses a floating-point (real-coded) representation. It uses crossover (binomial in this case) and mutation operations to optimize a given cost function. For want of space, we avoid mentioning the details of the DE algorithm here and refer the reader to the aforementioned literatures.

2.3 The Multi-objective Variant of DE

We have used the Multi-objective DE (MODE) [4]. MODE was proposed by Xue *et al.* [8]. This algorithm uses a variant of the original DE, in which the best individual is adopted to create the offspring. A Pareto-based approach is introduced to implement the selection of the best individual. If a solution is dominated, a set of non-dominated individuals can be identified and the "best" turns out to be any individual (randomly picked) from this set.

3 Multi-objective Clustering Scheme

3.1 Search-Variable Representation and Description of the New algorithm

In the proposed method, for n data points, each d-dimensional, and for a user-specified maximum number of clusters K_{max}, a chromosome is a vector of real numbers of dimension $K_{max} + K_{max} \times d$. The first K_{max} entries are positive floating-point numbers in [0, 1], each of which controls whether the corresponding cluster is to be activated (i.e. to be really used for classifying the data) or not. The remaining entries are reserved for K_{max} cluster centers, each d-dimensional. For example, the i-th vector is represented as:

$$\vec{X}_i(t) = \boxed{\begin{array}{|c|c|c|c|c|c|c|} T_{i,1} & .... & T_{i,K_{max}} & \vec{m}_{i,1} & \vec{m}_{i,2} & ..... & \vec{m}_{i,K_{max}} \end{array}}$$

The j-th cluster center in the i-th chromosome is active or selected for partitioning the associated dataset if $T_{i,j} = 1$. On the other hand, if $T_{i,j} = 0$, the particular j-th cluster is inactive in the i-th vector in DE population. Thus the $T_{i,j}$ s behave like control genes.

IF $T_{i,j} = 1$ THEN the j-th cluster center $\vec{m}_{i,j}$ is ACTIVE

ELSE $\vec{m}_{i,j}$ is INACTIVE.   (1)

Conjunction of GA and DE algorithms:

The Differential Evolution algorithm is applied on the first K_{max} members of the chromosome (as activated by the corresponding control genes), whereas, the control genes form a binary encoded GA population, which are operated by the Genetic operators of Selection, Crossover and Mutation. Binary tournament selection is employed in this case. The different GA operators are not reiterated here due to space limitations.

Simple generational genetic algorithm pseudo code:

1. Choose initial population
2. Evaluate the fitness of each individual in the population
3. Repeat until termination: (time limit or sufficient fitness achieved)
    1. Select best-ranking individuals to reproduce
    2. Breed new generation through crossover and/or mutation (genetic operations) and give birth to offspring
    3. Evaluate the individual fitnesses of the offspring

Replace worst ranked part of population with offspring.

3.2 Selecting the Objective Functions

Conflict among the objective functions is often beneficial since it guides to globally optimal solutions. In this work we choose the Xie-Beni index XB_q and the FCM objective function J_q as the two objectives. The FCM measure J_q may be defined as:

$$J_q = \sum_{j=1}^{n} \sum_{i=1}^{k} u_{ij}^q \cdot d^2(\vec{Z}_j, \vec{m}_i), \ 1 \le q \le \infty \tag{2}$$

where q is the fuzzy exponent, d indicates a distance measure between the j-th pattern vector and i-th cluster centroid, and u_{ij} denotes the membership of j-th pattern in the i-th cluster. The XB index is defined as a function of the ratio of the total variation σ to the minimum separation sep of the clusters. Here σ and sep may be written as:

$$\sigma = \sum_{i=1}^{k} \sum_{p=1}^{n} u_{ip}^2 \cdot d(\vec{m}_i, \vec{Z}_p) \tag{3}$$

$$\text{and} \quad sep(Z) = \min_{i \ne j} \left\{ d^2(\vec{m}_i, \vec{m}_j) \right\} \tag{4}$$

The XB index is then written as:

$$XB_q = \frac{\sigma}{n \times sep(Z)} = \frac{\sum_{i=1}^{k} \sum_{p=1}^{n} u_{ip}^q \cdot d^2(\vec{m}_i, \vec{Z}_p)}{n \times \min_{i \ne j} \left\{ d^2(\vec{Z}_i, \vec{Z}_j) \right\}} \tag{5}$$

Let the set of centers be denoted by $\{\vec{m}_1, \vec{m}_2, ..., \vec{m}_k\}$. The membership value of the j-th pattern in i-th cluster $u_{ij}, i = 1, 2, ...k$ and $j = 1, 2, ..., n$ are computed as:

$$u_{ij} = \frac{1}{\sum_{p=1}^{k} \left(\frac{d(\vec{m}_i, \vec{Z}_j)}{d(\vec{m}_p, \vec{Z}_j)} \right)^{\frac{2}{q-1}}} \tag{6}$$

Note that while computing the u_{ij} s, using equation (12), if $d(\vec{m}_p, \vec{Z}_j)$ is equal to zero for some p, then u_{ij} is set to zero for all $i = 1, 2, ...k$, $i \ne j$, while u_{pj} is set equal to one. Subsequently the centers encoded in a vector are updated using:

$$\vec{m}_p = \frac{\sum_{j=1}^{n} (u_{pj})^q \cdot \vec{Z}_j}{\sum_{j=1}^{n} (u_{pj})^q} \tag{7}$$

3.3 Avoiding Erroneous Vectors

There is a possibility that in our scheme, during computation of the XB or J_q, a division by zero may be encountered. This may occur when one of the selected cluster centers in a DE-vector is outside the boundary of distributions of the data set. To avoid this problem we first check to see if any cluster has fewer than two data points in it. If so, the cluster center positions of this special chromosome are re-initialized by an average computation.

3.4 Selecting the Best Solution from Pareto-Front

For choosing the most interesting solutions from the Pareto front, we apply Tibshirani *et al.* Gap statistic [24], a statistical method to determine the number of clusters in a data set.

3.5 Evaluating the Clustering Quality

In this work, the final clustering quality is evaluated using two external measures. Specifically we choose the Adjusted Rand Index [25] (which is a generalization of the Rand index [26]) and the Sihouette index [27]. Silhouette width reflects the compactness and separation of the clusters. Given a set of data points $Z = \{\vec{Z}_1,....,\vec{Z}_n\}$ and a given clustering solution $C = \{C_1, C_2,..., C_k\}$, the silhouette width $s(\vec{Z}_j)$ for each data $\vec{Z}_j$ belonging to cluster C_i indicates a measure of the confidence of belongingness, and it is defined as:

$$s(\vec{Z}_j) = \frac{b(\vec{Z}_j) - a(\vec{Z}_j)}{\max(a(\vec{Z}_j), b(\vec{Z}_j))}. \tag{8}$$

Here $a(\vec{Z}_j)$ denotes the average distance of data point $\vec{Z}_j$ from the other data points of the cluster to which the data point $\vec{Z}_j$ is assigned (i. e. cluster C_i). On the other hand, $b(\vec{Z}_j)$ represents the minimum of the average distances of data point $\vec{Z}_j$ from the data points belonging to clusters C_r, $r = 1,2,...,k$ and $r \neq i$. The value of $s(\vec{Z}_j)$ lies between -1 and +1. Large values of $s(\vec{Z}_j)$ (near to 1) indicate that the data point $\vec{Z}_j$ is well clustered. Overall silhouette index $s(C)$ of a clustering solution $C = \{C_1, C_2,..., C_k\}$ is defined as the mean silhouette width over all the data points:

$$s(C) = \frac{1}{n}\sum_{j=1}^{n} s(\vec{Z}_j). \tag{9}$$

4 Experimental Results

4.1 Datasets Used

The experimental results showing the effectiveness of multi-objective DE based clustering has been provided for six artificial and four real life datasets. Table 1 presents the details of the datasets. The real-life datasets are iris, wine, breast-cancer [28] and the yeast sporulation data. The sporulation dataset is available from [31].

4.2 Parameters for the Algorithms

GADE has been used with 40 parameter vectors in each generation and each run of each algorithm was continued for 100 generations. The value of scale factor F is a random value between 0.5 and 1. The other parameters for the multi-objective GA (NSGA II) based clustering are fixed as follows: number of generations = 100, population size = 50, crossover probability = 0.8, mutation probability $= \dfrac{1}{Chromosome\_length}$. Please note that GADE and the NSGA II use the same parameter representation scheme. Clustering with MOCK was performed with the source codes available from [32].

4.3 Presentation of Results

The mean Silhouette index values of the best-of-run solutions provided by six contestant algorithms over the 10 datasets have been provided in Table 2. The best entries have been marked in boldface in each row. Table 3 enlists the adjusted rand index values except for Yeast sporulation data as no standard nominal classification is known for this dataset.

Table 1. Details of the datasets used

Dataset	Number of points	Number of clusters	Number of Characteristics
Dataset_1	900	9	2
Dataset _2	76	3	2
Dataset _3	400	4	3
Dataset _4	300	6	2
Dataset _5	500	10	2
Dataset_ 6	810	3	2
Iris	150	3	4
Wine	178	3	13
Breast-Cancer	683	2	9
Yeast Sporulation	474	7	7

Table 2. Mean value of sil index found and standard deviations (in parentheses) by contestant algorithm over 30 independent runs on ten datasets

Dataset	Algorithms Compared					
	GADE		NSGA II		MOCK	
	k	Silhouette Index	k	Silhouette Index	k	Silhouette Index
Dataset_1	**9.12** (1.46)	**0.735312** (0.254134)	9.37 (1.72)	0.669317 (0.0892)	8.52 (2.81)	0.66342 (0.0736)
Dataset_2	**3.36** (0.65)	**0.664993** (0.123610)	3.16 (0.072)	0.654393 (0.00927)	3.33 (1.03)	0.658921 (0.004731)
Dataset_3	4.14 (0.36)	**0.872521** (0.127479)	3.57 (0.51)	0.765691 (0.005686)	3.78 (1.25)	0.768419 (0.006721)
Dataset_4	6.04 (0.25)	0.705079 (0.115610)	6.28 (0.46)	0.827618 (0.02871)	**6.08** (0.51)	**0.832527** (0.007825)
Dataset_5	9.24 (3.89)	**0.771040** (0.042776)	12.43 (0.939)	0.768379 (0.005384)	10.41 (0.80)	0.769342 (0.006208)
Dataset_6	5.19 (0.93)	**0.792000** (0.208000)	4.65 (1.58)	0.642091 (0.002833)	5.16 (0.38)	0.640957 (0.008349)
Iris	2.31 (0.76)	0.429655 (0.331443)	2.16 (1.06)	0.566613 (0.082651)	**3.05** (0.37)	**0.6003725** (0.005129)
Wine	**3.16** (0.46)	**0.582197** (0.00427)	3.88 (0.67)	0.5767342 (0.009415)	3.59 (0.46)	0.576834 (0.000812)
Breast Cancer	**2.08** (0.38)	**0.648297** (0.00734)	2.57 (0.60)	0.6004642 (0.004561)	2.10 (0.53)	0.626719 (0.01094)
Yeast Sporulation	**7.08** (0.12)	**0.641630** (0.212575)	7.22 (0.68)	0.641306 (0.04813)	6.67 (0.857)	0.613567 (0.005738)

Table 3. Mean value of adjusted rand index found and standard deviations (in parentheses) by contestant algorithm over 30 independent runs on ten datasets

Dataset	Algorithms Compared		
	GADE	NSGA2	MOCK
Dataset_1	**0.884288 (0.101020)**	0.802180(0.004782)	0.810934 (0.0059348)
Dataset_2	**0.951535 (0.179265)**	0.9378123(0.006821)	0.946547 (0.004536)
Dataset_3	0.850030 (0.152226)	**0.963841(0.0046719)**	0.878732 (0.0712523)
Dataset_4	0.785995(0.137284)	0.957818 (0.004678)	**0.978761 (0.006734)**
Dataset_5	0.788450 (0.019142)	**0.947641 (0.006646)**	0.9454568 (0.0012043)
Dataset_6	0.692516 (0.168323)	0.881395 (0.056483)	**0.910294 (0.016743)**
Iris	**0.843862 (0.076887)**	0.715898 (0.005739)	0.786574 (0.075763)
Wine	**0.875849 (0.0087642)**	0.828645(0.0074653)	0.864764 (0.0034398)
Breast Cancer	**0.956456 (0.0053)**	0.944236(0.006521)	0.9465731 (0.006748)

4.3.1   Significance and Validation of Microarray Data Clustering Results

In this section the best clustering solution provided by different algorithms on the sporulation data of yeast has been visualized using the cluster profile plot (in parallel coordinates[30]) in MATLAB 7.0.4 version. It is a common way of visualizing high-dimensional geometry. Cluster profile plots (in parallel coordinates) of seven clusters for the best clustering result (provided by GADE) on yeast sporulation data has been shown in Figure 1. The blue polylines indicate the member genes within a cluster while the black polyline indicates the centroid of that gene. The heatmap and fatigo results may be obtained from [33].

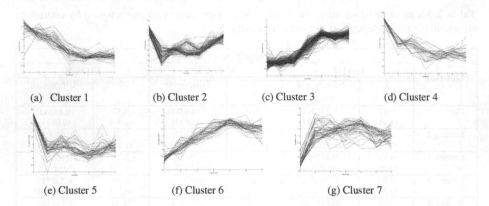

<table>
<tr><td>(a) Cluster 1</td><td>(b) Cluster 2</td><td>(c) Cluster 3</td><td>(d) Cluster 4</td></tr>
</table>

(e) Cluster 5   (f) Cluster 6   (g) Cluster 7

Fig. 1. Cluster profile plots (in parallel coordinates) of seven clusters for the best clustering result (provided by GADE) on yeast sporulation data

5 Conclusions

This paper compared and contrasted the performance of GADE in an automatic clustering framework with two other prominent multi-objective clustering algorithms. The multi-objective GADE-variant used the same variable representation scheme. Tables 2 to 4 indicate that GADE was usually able to produce better final clustering results as compared to MOCK or NSGA II in terms of both adjusted Rand index and Silhouette index when all the algorithms were let run for an equal number of generations. Future research may extend the multi-objective GADE-based clustering schemes to handle discrete chromosome representation schemes that no longer depend on cluster centroids and thus are not biased in any sense towards spherical clusters.

References

[1] Jain, A.K., Murty, M.N., Flynn, P.J.: Data clustering: a review. ACM Computing Surveys 31(3), 264–323 (1999)

[2] Xu, R., Wunsch, D.: Clustering. Series on Computational Intelligence. IEEE Press, Los Alamitos (2008)

[3] Bandyopadhyay, S., Maulik, U., Mukhopadhyay, A.: Multiobjective genetic clustering for pixel classification in remote sensing imagery. IEEE Transactions Geoscience and Remote Sensing (2006)

[4] Handl, J., Knowles, J.: An evolutionary approach to multiobjective clustering. IEEE Transactions on Evolutionary Computation 11(1), 56–76 (2007)

[5] Abbass, H.A., Sarker, R.: The pareto differential evolution algorithm. International Journal on Artificial Intelligence Tools 11(4), 531–552 (2002)

[6] Xue, F., Sanderson, A.C., Graves, R.J.: Pareto-based multi-objective differential evolution. In: Proceedings of the 2003 Congress on Evolutionary Computation (CEC 2003), Canberra, Australia, vol. 2, pp. 862–869. IEEE Press, Los Alamitos (2003)

[7] Robic, T., Filipic, B.: DEMO: Differential Evolution for Multiobjective Optimiza-tion. In: Coello Coello, C.A., Hernández Aguirre, A., Zitzler, E. (eds.) EMO 2005. LNCS, vol. 3410, pp. 520–533. Springer, Heidelberg (2005)

[8] Iorio, A.W., Li, X.: Solving rotated multi-objective optimization problems using differen-tial evolution. In: Webb, G.I., Yu, X. (eds.) AI 2004. LNCS, vol. 3339, pp. 861–872. Springer, Heidelberg (2004)

[9] Paterlinia, S., Krink, T.: Differential evolution and particle swarm optimisation in parti-tional clustering. Computational Statistics & Data Analysis 50(5), 1220–1247 (2006)

[10] Omran, M., Engelbrecht, A.P., Salman, A.: Differential evolution methods for unsuper-vised image classification. In: Proceedings of Seventh Congress on Evolutionary Compu-tation (CEC 2005), IEEE Press, Los Alamitos (2005)

[11] Das, S., Abraham, A., Konar, A.: Automatic clustering using an improved differential evolution algorithm. IEEE Transactions on Systems Man and Cybernetics - Part A 38(1), 1–20 (2008)

[12] Deb, K., Pratap, A., Agarwal, S., Meyarivan, T.: A fast and elitist multiobjective genetic algorithm: NSGA-II. IEEE Transactions on Evolutionary Computation 6(2) (2002)

[13] Corne, D.W., Knowles, J.D., Oates, M.J.: The pareto-envelope based selection algo-rithm for multiobjective optimisation. In: Schoenauer, M., Deb, K., Rudolph, G., Yao, X., Lut-ton, E., Merelo, J.J., Schwefel, H.-P. (eds.) PPSN 2000. LNCS, vol. 1917, pp. 869–878. Springer, Heidelberg (2000)

[14] Xie, X., Beni, G.: Validity measure for fuzzy clustering. IEEE Trans. Pattern Anal. Ma-chine Learning 3, 841–846 (1991)

[15] Bezdek, J.C.: Cluster validity with fuzzy sets. Journal of Cybernetics (3), 58–72 (1974)

[16] Chu, S., et al.: The transcriptional program of sporulation in budding yeast. Science 282, 699–705 (1998)

[17] Sawaragi, Y., Nakayama, H., Tanino, T.: Theory of multiobjective optimization. Mathe-matics in Science and Engineering, vol. 176. Academic Press Inc., Orlando (1985)

[18] Deb, K.: Multi-Objective Optimization using Evolutionary Algorithms. John Wiley & Sons, Chichester (2001)

[19] Coello Coello, C.A., Lamont, G.B., Van Veldhuizen, D.A.: Evolutionary Algorithms for Solving Multi-Objective Problems. Springer, Heidelberg (2007)

[20] Storn, R., Price, K.: Differential evolution – a simple and efficient heuristic for global op-timization over continuous spaces. Journal of Global Optimization 11(4), 341–359 (1997)

[21] Storn, R., Price, K.V., Lampinen, J.: Differential Evolution - A Practical Approach to Global Optimization. Springer, Berlin (2005)

[22] Mattson, C.A., Mullur, A.A., Messac, A.: Smart Pareto filter: Obtaining a minimal repre-sentation of multiobjective design space. Eng. Optim. 36(6), 721–740 (2004)

[23] Branke, J., Deb, K., Dierolf, H., Osswald, M.: Finding knees in multi-objective optimiza-tion. In: Yao, X., Burke, E.K., Lozano, J.A., Smith, J., Merelo-Guervós, J.J., Bullinaria, J.A., Rowe, J.E., Tiňo, P., Kabán, A., Schwefel, H.-P. (eds.) PPSN 2004. LNCS, vol. 3242, pp. 722–731. Springer, Heidelberg (2004)

[24] Tibshirani, R., Walther, G., Hastie, T.: Estimating the number of clusters in a dataset via the Gap statistic. J. Royal Statist. Soc.: SeriesB (Statistical Methodology) 63(2), 411–423 (2001)

[25] Hubert, L., Arabie, P.: Comparing partitions. Journal of Classification, 193–218 (1985)

[26] Rand, W.M.: Objective criteria for the evaluation of clustering methods. Journal of the American Statistical Association 66, 846–850 (1971)

[27] Rousseeuw, P.J.: Silhouettes: A graphical aid to the interpretation and validation of cluster analysis. J. Comput. Appl. Math. 20(1), 53–65 (1987)

[28] Blake, C., Keough, E., Merz, C.J.: UCI repository of machine learning database (1998),
     http://www.ics.uci.edu/~mlearn/MLrepository.html
[29] Theodoridis, S., Koutroumbas, K.: Pattern Recognition, 2nd edn. Elsevier/ Academic
     Press (2003)
[30] Keim, D.A., Kriegel, H.-P.: Visualization techniques for mining large databases: a com-
     parison. IEEE Transactions on Knowledge and Data Engineering 8(6), 923–938 (1996)
[31] http://cmgm.stanford.edu/pbrown/sporulation
[32] http://dbkgroup.org/handl/mock/
[33] http://swagatamdas19.googlepages.com/

Credibility Coefficients
in Hybrid Artificial Intelligence Systems

Roman Podraza

Institute of Computer Science
Warsaw University of Technology
Nowowiejska 15/19, 00-665 Warsaw, Poland
R.Podraza@ii.pw.edu.pl

Abstract. ARES System is an application dedicated to data analysis supported by Rough Set theory. Currently the system is expanded by such approaches as Emerging Patterns and Support Vector Machine. A unique feature of ARES System is applying credibility coefficients to identify improper objects within information systems. The credibility coefficient is a measure, which attempts to assess a degree of typicality of each object in respect to the rest of information system. The paper presents a concept of credibility coefficients in context of hybrid artificial intelligence systems combined on ARES System platform. Ordinal credibility coefficient supports aggregation of number incomparable credibility coefficients based on different approaches.

Keywords: credibility coefficients, information system, classification, Rough Sets, Emerging Patterns, SVM.

1 Introduction

Credibility of data is a necessary condition to draw useful conclusions and acquire valuable knowledge. Data analysis is such a process where credibility of data has to be assessed to identify outliers - the data with questionable reliability. Depending on the decision regarding such data, they can be removed, corrected or left without modification. Outliers can be recognized only if the domain of data is known and understood. Our goal is to discover the incredible data automatically by a system.

The proposed solution is an automatic credibility analysis incorporated into data analysis of information system. The "suspicious data" should be recognized by analyzing available data set only. Credibility coefficients [1] [2] are calculated for all objects of information system. This approach to credibility analysis is universal and can be applied to any information system regardless of its domain. The main goal of introducing the credibility coefficients was to find out exceptions to rules because very often they can be more interesting that the rules themselves.

Credibility coefficients were implemented within ARES System [2], which is a common platform for a number of different data analysis approaches. Its initial functionalities were based on Rough Set theory [3] and then have been expanded by introducing discovering Emerging Patterns (EP) [4] and Support Vector Machine (SVM) methodology [5].

E. Corchado et al. (Eds.): HAIS 2009, LNAI 5572, pp. 187–194, 2009.

The problem of identifying incredible data is common for all data analysis systems. The context of hybrid artificial intelligence systems gives the potential of many approaches in calculating credibility coefficients. Ordinal credibility coefficient [6] is a basic concept of transforming results of many heuristic algorithms of credibility coefficients. The concept is used in designing Multi Credibility Coefficient Method [6], which can be a tool for a hybrid artificial intelligence system.

In the next section an overview of ARES System is sketched. Then follow sections describing credibility coefficients and their theoretical aspects. Subsequently a concept of ordinal credibility coefficient is followed by presentation of Multi Credibility Coefficient Method. Finally conclusions complete the paper.

2 ARES System

The system has been designed to give a full interactive access to process of data analyses involving different approaches. A multi-window graphical user interface enables observing and/or comparing different steps or phases of data analysis. Many independent data sets can be processed and presented simultaneously. Hierarchically organized items accessible through a directory browser represent stages of data analysis. The items can be opened in windows within a workspace view and this is the way to present the final and partial results of data analysis. Each item has its description containing some statistic and explanatory information as execution time used to produce the item, algorithm applied and some other specific data (e.g. number of generated rules for rule set item).

ARES System processes information systems or decision tables with single decision attribute. Editing functionalities enable cutting off information system by removing objects (rows) and/or attributes (columns). There are available capabilities of data discretization by a number of methods.

The initial version of ARES System [1] comprised modules for performing the following tasks from Rough Set domain such as

- Discovering approximations of decision classes
- Determining discernibility matrices
- Finding relative reducts
- Discovering frequents sets
- Mining decision rules

Each of these tasks has a number of algorithms to be chosen. There are some useful tools presenting the results in context of input data – e.g. for a selected rule a set of objects supporting it, or set of objects supporting its antecedent.

A domain of Rough Set theory in ARES System has been supplemented by module for discovering discriminant of information system. The module comprises three algorithms LEM1, LEM2 and AQ [7]. The last algorithm generates rules with disjunctive representation. There is a possibility to check consistency of information system, which is a necessary condition to calculate a discriminant of information system.

ARES System has incorporated the KTDA system [6] [8] based on Emerging Patterns (EP) approach. Two different algorithms of discovering EPs are supported– using maximal frequent itemsets proposed in [4] and using decision tree [9], but with some extensions and improvements. The former one reflects the classical approach and requires stating minimal growth rate and minimal support in the target class, while the latter one uses Fisher's Exact Test used to discover only such EPs which are statistically significant.

EPs enable data classification for which CAEP (Classification by Aggregating Emerging Patterns) algorithm [10] is applied. For this classifier set of EPs discovered by maximal frequent itemsets algorithms gives usually slightly better classification. On the other hand algorithm using decision tree produces more compact and more significant knowledge, which probably can be more interesting for expert trying to update his/her knowledge on the analyzed problem.

Support Vector Machines is yet another methodology being integrated to ARES System. Currently there are attempts to expose results of classification of data done with this approach. Only credibility coefficients calculated for each object from information system are available from ARES System now, but their calculations involve the classification itself.

ARES System provides numerous algorithms to calculate credibility coefficients for objects of information systems. The credibility coefficients are presented in the following section.

3 Characteristics of Credibility Coefficient

Credibility coefficients define relative measures of credibility of particular object within information system. If the evaluation is performed after some modifications in the information system, then in most cases the values of credibility coefficients are different. The domain of credibility coefficient is interval <0; 1>.

Values of credibility coefficients evaluated by various methods are incomparable even for the same object of information system. The same values of credibility coefficients calculated with various methods do not have to represent the same strength of persuasion that the particular object is credible. Values of credibility coefficients reckoned by the same method can be compared to assess if a particular object has the same, lower or higher credibility than another one.

Credibility coefficients can be used solely to introduce an ordering of objects from information system. The value of credibility coefficient itself, without values of other objects, is usually useless. The exception to this rule is situation where interpretation of the value is known and a user would like to exploit it. However in the system of knowledge acquisition, the only purpose of credibility coefficients is to identify outliers and ranking of credibility is sufficient for it.

Several methods of evaluation of credibility coefficients are available [1] [2]. Nevertheless straight combining all of them is impossible, because any their aggregation is nonsense mathematical operation on numbers (resembling an attempt to reckon an average for set of physical values with various units of measure).

So far the research on credibility analysis was mostly concentrated on inventing new heuristic methods for reckoning credibility coefficients. The proposed methods were tested to check how effectively a particular method can identify the outliers. This kind of research requires still much effort to get some intuition, however it seems to be insufficient to draw general conclusions regarding meaning, applications and efficiency of credibility coefficients.

The system of credibility analysis has to find out frequent schemas of data and dependencies between them or statistical relations appearing for analyzed data set (information system). The only sidelight to identify a particular object as outlier is to discover its inferior regularity in respect to other objects or, in other words, its non-typicality. Hence a credibility coefficient is not exactly measure of credibility, but it is a measure of typicality. If an object is less typical, so it does not precisely conform to knowledge discovered in the data set. The object not supporting relationships observed frequently enough should receive lower evaluation of typicality.

Typical data object appears frequently enough to observe some rules or laws satisfied by their attributes. Probability of appearance of typical data is higher than for other data. The probability distribution (in a multidimensional space of domains of objects' attributes) is usually unknown, difficult to be modeled or estimated. However this idea can help in understanding the goal of credibility coefficients. An attempt to assess typicality of object is (or should be) an estimation of this probability. We assume that typical data object is correct and credible one, and appears more frequently than other incorrect ones. Evaluation of credibility coefficients is a process in which we try to assess a probability of appearance of a given object from an information system, treating the information system as a sample. Since the problem is too complex and too difficult to be solved then we have to reduce our expectations. The goal of using credibility coefficients is not to find out a precise value of this probability, but to determine the same ordering of data objects, which could result from the values of probabilities. The hypothetical ideal credibility coefficients should represent probability distribution for all objects from information system. However such a task is impossible for an automatic approach, which is based only on numbers representing particular values of many attributes describing the object.

Currently in ARES System a number of algorithms for calculations of basic credibility coefficients have been implemented. Heuristics of the algorithms were based on the following concepts:

- Approximation of Rough Set classes,
- Statistics of attribute values,
- Hybrid one combining the previous two,
- Frequent Sets,
- Extracted Rules (Rough Set approach),
- Voting Classifier (CAEP - for Emerging Patterns),
- Support Vector Machines.

The first five algorithms for calculations of credibility coefficients belong to the original version of ARES System and exploit concepts of Rough Set theory [1] [2]. Credibility coefficient based on voting classifier was incorporated as a part of KTDA system [6] [8]. It takes into account a vector of weights of votes determined by the classifier. Any voting classifier can be used, like CAEP and SVM classifiers in ARES System. Some experiments with credibility coefficients presented their ability to identify corrupted data "injected" into original data sets; however it is difficult to prove superiority of particular approaches over others. The effectiveness of different credibility coefficients vary depending on their applications (e.g. identifying falsified data, incrementing measures of knowledge quality indicators and/or discovering new and/or "better" rules by removing the most improper data) and the benchmarks (information systems).

4 Ordinal Credibility Coefficient

We proposed ordinal credibility coefficient [6], which is a new kind of credibility co-efficient. Its values are not computed directly from attributes of data objects but they use credibility coefficient's values obtained for the data set from an arbitrary chosen calculating method.

Let us have a non-empty information system IS and a method of calculating credibility coefficients. For a data object $x \in IS$ let $cred(x)$ mean the value of credibility coefficient obtained from a given method for x. A value:

$$\frac{\left\|\{y \in IS : cred(y) \le cred(x)\}\right\|}{|IS|} \tag{1}$$

for $x \in IS$ is named ordinal credibility coefficient for object x belonging to information system IS. It is denoted as $cred_{ORD}(x)$.

The ordinal credibility coefficient based on arbitrary chosen basic credibility coefficient method and reckoned for a particular object belonging to a given information system expresses the relative amount of records with credibility coefficients less or equal to the credibility coefficient for this record. To designate ordinal credibility coefficients this transformation has to be performed for all data objects.

Example
Let us consider a simple information system with eight objects identified by letters from a to h. Let us assume that a credibility coefficient calculating method produced some values, denoted as $cred(x)$, as shown in the second column in Table 1. The columns of the table show respectively sets of objects with values of credibility coefficients less or equal to the given object's, cardinalities of these sets and values of ordinal credibility coefficient.

Table 1. Calculation of ordinal credibility coefficients

x	cred(x)	Y_x = { y: cred(y) ≤ cred(x) }	$\|Y_x\|$	$cred_{ORD}(x)$
a	1.00	{ a, b, c, d, e, f, g, h }	8	8/8 = 1
b	1.00	{ a, b, c, d, e, f, g, h }	8	8/8 = 1
c	0.99	{ c, d, e, f, g, h }	6	6/8 = 0.75
d	0.98	{ d, e, f, g, h }	5	5/8 = 0.625
e	0.80	{ e, f, g, h }	4	4/8 = 0.50
f	0.78	{ f, g, h }	3	3/8 = 0.375
g	0.65	{ g, h }	2	2/8 = 0.25
h	0.65	{ g, h }	2	2/8 = 0.25

Properties
Ordinal credibility coefficients have the following properties.
- The domain of ordinal credibility coefficient values is (0, 1>. The minimal value of ordinal credibility coefficient for an object from information system IS is $1/|IS|$.

- Basic credibility coefficient values and their ordinal counterparts introduce the same ordering (or ranking) for objects of information system. The transformation from basic to ordinal credibility coefficient preserves information on relative typicality of objects.
- Ordinal credibility coefficient has characteristics of all credibility coefficients.
- Ordinal credibility coefficient has its interpretation. For a given credibility coefficient and information system IS, value $cred_{ORD}(x)$, for $x \in IS$, denotes this part of information system IS, which has assessed credibility less or equal to the estimated credibility of object x and value $1\text{-}cred_{ORD}(x)$ shows portion of information system IS, which has estimated credibility grater than credibility of object x.
- If we try to designate ordinal credibility coefficient using it as a basic one (we can call it as ordinal credibility coefficient of the second degree) then the same original input values are produced. This is a consequence derived directly from the definition. Ordinal credibility coefficient of any degree is equivalent to it of the first degree.

The impact of the fourth property (on interpretation of ordinal credibility coefficient) is very significant. This is the common platform, where different credibility coefficients may be processed. For instance ordinal credibility coefficients can be compared. It is possible to aggregate ordinal credibility coefficient's values corresponding to the same object. For example, we have two different ordinal credibility coefficients for object $x \in IS$. They were derived from credibility coefficients computed by method M_1 and M_2, respectively, for information system IS. It is possible to have the average of these two ordinal credibility coefficient values for object $x \in IS$. The obtained average is meaningful. It denotes an average part of information system IS with credibility less or equal to the credibility estimated for object x while considering both computing methods M_1 and M_2.

Finally, thanks to the definition of ordinal credibility coefficient, its value for a single object provides useful information itself without a need of referencing to credibility coefficients for other objects. This is not a case for a basic credibility coefficient, where its value for an individual object is meaningful only in the context of coefficient values for other objects or if the computing method is known and can be interpreted.

5 Multi Credibility Coefficient Method

There was a problem with many independent methods of reckoning credibility coefficients to draw conclusions for a particular object in information system, because it required independent analysis of incomparable results. So process of enriching capabilities of identifying incredible data was accompanied by necessity of human/expert interference. This drawback contradicted the main assumption on automatic identifying improper data. For large information systems and for many credibility analysis approaches the task could be too tedious.

So it was crucial to find a complex measure, which could express a common output for many partial results. This measure should be tunable, to give appropriate flexibility for the whole methodology and for a particular application. The final result was feasible thanks to concept of ordinal credibility coefficient. We proposed a new approach of combining a number of credibility calculation algorithms using the notion of ordinal credibility coefficient to obtain an aggregate outcome.

Let us have a non-empty information system *IS*. Let $M_1, ..., M_N$ denote N arbitrary chosen methods of computing credibility coefficients. Let $cred^{(i)}(x)$, $1 \leq i \leq N$ denote a credibility coefficient for object $x \in IS$ obtained by applying method M_i to information system *IS*. Let $cred^{(i)}_{ORD}(x)$, $1 \leq i \leq N$ denote an ordinal credibility coefficient for object $x \in IS$ derived from all values $cred^{(i)}(y)$, $y \in IS$. The Multi Credibility Coefficient Method evaluates credibility coefficient for object $x \in IS$ as value:

$$\frac{\sum_{i=1}^{N} w_i * cred^{(i)}_{ORD}(x)}{\sum_{i=1}^{N} w_i} \tag{2}$$

Although values computed by Multi Credibility Coefficient Method are results of aggregating ordinal credibility coefficients, these values are not ordinal credibility coefficients themselves. Of course they can be transformed to ordinal coefficients.

In Fig. 1 a scheme for the proposed method is presented. Symbol "T" denotes transformation from basic credibility coefficients to the corresponding ordinal credibility coefficients. As the result of this process we get a weighted average of ordinal credibility coefficient for each object. Multi Credibility Coefficient Method combines its components into a signature, which may be fine-tuned by adjusting the weights.

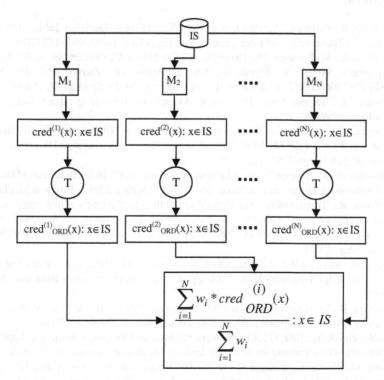

Fig. 1. Symbolic scheme for Multi Credibility Coefficient Method

6 Conclusions

The paper presents idea of credibility coefficients, which were designed to identify incredible objects in information systems. They were implemented in ARES System based on the Rough Set theory and extended by Emerging Patterns approach and Support Vector Machines methodology.

Credibility coefficients are independent heuristic metrics, which were mutually incomparable. The nature of credibility coefficients contradicted applying them directly into the task of identifying improper data in hybrid artificial intelligence environment. It seemed to be the main obstacle in extending ARES System by subsequent data analysis methodologies.

Ordinal credibility coefficients derived from basic credibility ones were the solution to the problem. They have clear and common interpretation. This makes them a platform for combining results of a number of credibility analyses. Multi Credibility Coefficient Method is an example of applying ordinal credibility coefficients to produce a collective outcome reflecting several different credibility analysis. Multi Credibility Coefficient Method uses a set of weights allowing for tuning impact of each applied credibility assessments. The concept has to be verified in further research and practical experiments in hybrid artificial intelligence environment.

References

[1] Podraza, R., Dominik, A.: Problem of Data Reliability in Decision Tables. Int. J. of Information Technology and Intelligent Computing (IT&IC) 1(1), 103–112 (2006)

[2] Podraza, R., Walkiewicz, M., Dominik, A.: Credibility Coefficients in ARES Rough Set Exploration System. In: Ślęzak, D., Yao, J., Peters, J.F., Ziarko, W.P., Hu, X. (eds.) RSFDGrC 2005 Part II. LNCS (LNAI), vol. 3642, pp. 29–38. Springer, Heidelberg (2005)

[3] Pawlak, Z.: Rough Sets. Theoretical Aspects of Reasoning about Data. Kluwer, Dordrecht (1991)

[4] Dong, G., Li, J.: Efficient Mining of Emerging Patterns: Discovering Trends and Differences. In: Proc. of the SIGKDD (5th ACM Int. Conf. on Knowledge Discovery and Data Mining), San Diego, USA, pp. 43–52 (1999)

[5] Schmilovici, A.: Support Vector Machines. In: Maimon, O., Rokach, L. (eds.) Data Mining and Knowledge Discovery Handbook, pp. 257–274. Springer Sci. + Business Media (2005)

[6] Podraza, R., Tomaszewski, K.: Ordinal credibility coefficient - a new approach in the data credibility analysis. In: An, A., Stefanowski, J., Ramanna, S., Butz, C.J., Pedrycz, W., Wang, G. (eds.) RSFDGrC 2007. LNCS (LNAI), vol. 4482, pp. 190–198. Springer, Heidelberg (2007)

[7] Grzymała-Busse, J.W.: Rule Induction. In: Maimon, O., Rokach, L. (eds.) Data Mining and Knowledge Discovery Hand-book, pp. 277–295. Springer Sci. + Business Media Inc. (2005)

[8] Podraza, R., Tomaszewski, K.: KTDA: Emerging Patterns Based Data Analysis System. Annales UMCS, Informatica, AI, Lublin, Poland, vol. 4, 279–290 (2006)

[9] Boulesteix, A.-L., Tutz, G.: A Framework to Discover Emerging Patterns for Application in Microarray Data. Institute for Statistics, Ludwig-Maximilian-Universität, Munich (2003)

[10] Dong, G., Zhang, X., Wong, L., Li, J.: CAEP: Classification by Aggregating Emerging Patterns. In: Proc. of 2nd Int. Conf. on Discovery Science, Tokyo, Japan, pp. 30–42 (1999)

An Evolutionary Algorithm for Missing Values Substitution in Classification Tasks

Jonathan de A. Silva and Eduardo R. Hruschka

University of Sao Paulo (USP), Brazil
{jandrade,erh}@icmc.usp.br

Abstract. This paper proposes a method for substituting missing values that is based on an evolutionary algorithm for clustering. Missing values substitution has been traditionally assessed by some measures of the prediction capability of imputation methods. Although this evaluation is useful, it does not allow inferring the influence of imputed values in the ultimate modeling task (e.g., in classification). In this sense, alternative approaches to the so called prediction capability evaluation are needed. Therefore, we here also discuss the influence of imputed values in the classification task. Preliminary results obtained in a bioinformatics data set illustrate that the proposed imputation algorithm can insert less classification bias than three state of the art algorithms (i.e., KNNimpute, SKNN and IKNN). Finally, we illustrate that better prediction results do not necessarily imply in less classification bias.

Keywords: Missing Values, Clustering, Classification, Bioinformatics.

1 Introduction

The absence of values is common in real-world data sets and it can occur for a number of reasons like, for instance, malfunctioning measurement equipment and refusing of some respondents to answer certain questions in surveys. Such missing data are usually problematic. Therefore, several approaches have been proposed to deal with them - e.g. see [1,2]. A simple approach to deal with missing values involves ignoring instances and/or attributes containing missing values, but the waste of data may be considerable and incomplete data sets may lead to biased statistical analyses. Alternatively, some approaches for data analysis can be tolerant to missing values. Finally, a significant number of data mining methods only work with complete data sets. For these methods, approaches aimed at filling in missing values are particularly relevant.

The task of filling in missing data is often referred to as missing values substitution or imputation and it can be performed in a number of ways like, for instance, by the widely used mean/mode imputation. However, this approach considerably underestimates the population variance and does not take into account the between-attribute relationships, which are usually relevant to the process of missing values replacement. Moreover, data mining methods usually explore relationships between attributes and, thus, it is critical to preserve them,

E. Corchado et al. (Eds.): HAIS 2009, LNAI 5572, pp. 195–202, 2009.

as far as possible, when replacing missing values [1]. In this sense, imputation is aimed at carefully substituting missing values, trying to avoid the insertion of bias in the data set. If imputation is performed in a suitable way, higher quality data becomes available, and the data mining outcomes can be improved.

Recently, a number of algorithms capable of dealing with missing values have been developed, including nearest-neighbors based imputation algorithms like those described in [3,4,5] and that have been shown very useful in bioinformatics applications. Despite the encouraging results achieved by such imputation algorithms, most of the experimental settings reported in the literature only assess their prediction capabilities, obtained from the simulation of missing entries for some attributes whose values are actually known. From this standpoint, artificially generated missing values are substituted and then compared to their corresponding known values. Although this approach is valid and widely adopted, the prediction results are not the most important issue to be analyzed, mainly in classification problems [6]. In reality, the substitution process should generate values that least distorts the original characteristics of the data set for the classification process. In this context, the main contributions of this paper are:

- A description of an imputation algorithm (named EAC-I, from Evolutionary Algorithm for Clustering based Imputation);
- The presentation of preliminary experimental results that suggest that: (i) EAC-I can insert less classification bias than three state of the art algorithms for imputation (i.e., KNNimpute [5], SKNN [4], and IKNN [3]); (ii) better prediction results may not necessarily lead to less classification bias.

The remainder of this paper is organized as follows. The next section describes EAC-I. Section 3 briefly reviews a methodology [6] to estimate the bias inserted by imputation methods in the context of classification problems. Section 4 reports some experimental results obtained in a bioinformatics data set. Finally, Section 5 concludes this paper.

2   EAC-I: Evolutionary Algorithm for Clustering Based Imputation

Our imputation method relies on the assumption that clusters of (partially unknown) data can provide useful information for imputation purposes. In particular, data clusters can be viewed as information granules that summarize the spatial distribution of data. Such information granules can provide a workable estimate to fulfill missing values that least distorts the values that are actually present in the data set. Having this purpose in mind, several clustering algorithms can be adapted for imputation. For instance, the popular k-means would be a spontaneous choice for being simple and scalable. However, it is sensitive to initialization of prototypes and requires that the number of clusters be specified in advance. This can be restrictive in practice, leading us to alternatively consider the Evolutionary Algorithm for Clustering (EAC) [7] - which is capable of automatically estimating the number of clusters from data - as part of our framework. In a nutshell, EAC [7] has been designed to evolve data partitions with variable number of clusters (k) by eliminating, splitting, and merging clusters that are systematically refined by the k-means algorithm. EAC also has been shown, from a statistical perspective, to be more computationally efficient than a traditional approach based on multiple runs of k-means. We refer the reader interested in further details of the EAC to [7]. In the following we will concentrate on the fundamental concepts necessary to understand the principles of the proposed EAC-I.

EAC-I has a simple encoding scheme. In order to explain it, let us consider a data set composed of N instances. A partition is encoded as an integer string (genotype) of N positions. Each string position corresponds to a data set instance, i.e., the i-th position represents the i-th instance of the data set. Thus, each string component has a value over the possible cluster labels $\{1, 2, 3, ..., k\}$.

The underlying idea behind EAC-I is to repeatedly performing a clustering step followed by an imputation step. More precisely, let each instance i of a given data set be described by both a vector of m attribute values $\mathbf{x}^i = [x_1^i, x_2^i, ..., x_m^i]$ and its corresponding class c_i, which can take any value from a set of predefined values $C = \{c_1, c_2, ..., c_k\}$. A data set can be represented by a matrix $\mathbf{X}_D$ formed by a set of vectors $\mathbf{x}^i$ ($i = 1, ..., N$), each one with an associated $c_i \in C$. This matrix is formed by the values of the attributes a_l ($l = 1, ..., m$) for each instance i and its respective class. Thus, x_l^i is the value of the l-th attribute of the i-th instance in $\mathbf{X}_D$. In general, $\mathbf{X}_D$ is formed by both complete instances (without any missing value) and by instances with at least one missing value. Let $\mathbf{X}_C$ be the subset of instances of $\mathbf{X}_D$ that do not have any missing value, and $\mathbf{X}_M$ be the subset of instances of $\mathbf{X}_D$ with at least one missing value, i.e., $\mathbf{X}_D = \mathbf{X}_C \cup \mathbf{X}_M$. In this context, imputation methods fill in the missing values of $\mathbf{X}_M$, originating a filled matrix $\mathbf{X}_F$. Now we can focus on the main steps of EAC-I summarized in Fig. 1. In classification problems, EAC-I can be adapted for supervised imputation, which involves applying it for the instances of each class separately, as done in this work. Finally, some EAC-I steps deserve further attention:

a) In step 1, as well as in Step 2.1 for $t = 1$, the computation of the (Euclidean) dissimilarities between instances takes into account missing values. In brief, only attributes a_l ($l = 1, ..., m$) without missing values are used for computing dissimilarities between two instances (or between an instance and a centroid). Attributes a_o ($o = 1, ..., m$) containing missing values in any instance for which a particular pair wise dissimilarity computation is being performed are omitted in the distance function. To avoid falsely low distances for instances with a lot of missing values, we average each computed dissimilarity by the number of attributes considered in the computation.

b) In Step 2.2, a filled matrix $\mathbf{X}^{(t)}{}_F$ is derived from estimating values for fulfilling missing entries of $\mathbf{X}_M$. We shall refer to such values as estimates (or alternatively, partial imputations) because they are likely to vary for $t = 1, ..., G$. From this standpoint, we consider that values have been actually imputed when $t = G$. In what concerns such partial imputations, they are performed by considering only the instances of the cluster for which a given instance belongs to (at a given t). In particular, a traditional k-NN imputation method is applied by considering only a subset of $\mathbf{X}^{(t)}{}_D$, which is given by instances of a specific cluster. Similarly to KNNImpute [5], the computation of the values to be imputed are based on a weighted (Euclidean) distance function for which the closer the instance the more relevant it is for imputation purposes.

---

1. Initialize a set (population) of randomly generated data partitions (represented by genotypes) using $\mathbf{X}_D$. $\mathbf{X}^{(0)}{}_D \leftarrow \mathbf{X}_D$;
2. For $t = 1, ..., G$ do: // G is the number of generations (iterations) of EAC-I //
   2.1. Apply k-means to each genotype - using $\mathbf{X}^{(t-1)}{}_D$;
   2.2. Assess genotypes (data partitions) and estimate values to be imputed according to the best available genotype; Store the estimated values in $\mathbf{X}^{(t)}{}_F$ for future reference;
   2.3. Select genotypes using the well-known proportional selection;
   2.4. In 50% of the selected genotypes, apply the mutation operator 1, which eliminates some randomly selected clusters, placing its instances into the nearest remaining clusters. In the remaining genotypes, apply the mutation operator 2, which splits some randomly selected clusters, each of which into two new clusters.
   2.5. Replace the old genotypes by those just formed in step 2.4;
   2.6. Update the data matrix: $\mathbf{X}^{(t)}{}_D \leftarrow (\mathbf{X}_C \cup \mathbf{X}^{(t)}{}_F)$;
3. $\mathbf{X}'_D \leftarrow \mathbf{X}^{(G)}{}_D$;

Fig. 1. Main Steps of EAC-I (further details are provided in the text)

3   Bias in Classification

Recall from Section 2 that: $\mathbf{X}_D$ is formed by both complete instances and by instances with at least one missing value, $\mathbf{X}_C$ is the subset of instances of $\mathbf{X}_D$ that do not have any missing value, and $\mathbf{X}_M$ be the subset of instances of $\mathbf{X}_D$ with at least one missing value, i.e., $\mathbf{X}_D = \mathbf{X}_C \cup \mathbf{X}_M$. In this context, imputation methods fill in the missing values of $\mathbf{X}_M$, originating a filled matrix $\mathbf{X}_F$. We assume that the *class value* is known for every instance. In an ideal situation, the imputation method fills in the missing values, originating filled values, without inserting any bias in the dataset. In a more realistic view, imputation methods are aimed at decreasing the amount of inserted bias to acceptable levels, in such a way that a data set $\mathbf{X}'_D = \mathbf{X}_C \cup \mathbf{X}_F$, probably containing more information than $\mathbf{X}_C$, can be used for data mining (e.g., considering issues such as attribute selection, combining multiple models, and so on). From this standpoint, it is particularly important to emphasize that we are assuming that the known values in $\mathbf{X}_M$ may contain important information for the modeling process. This information would be partially (or completely) lost if the instances and/or attributes with missing values were ignored.

Two general approaches have been used in the literature to evaluate the bias inserted by imputations. We shall refer to them as *prediction* and *modeling* approaches. In a prediction approach, missing values are simulated, i.e., some known values are removed and then imputed. For instance, some known values from $\mathbf{X}_C$ could be artificially eliminated, simulating missing entries. In this way, it is possible to evaluate how similar the imputed values are to the real, a priori known values. The underlying assumption behind this approach is that the more similar the imputed value is to the real value, the better the imputation method is. Although the prediction approach is valid and widely adopted, the prediction results are not the most important issue to be analyzed as discussed, for instance, in [6]. In brief, the prediction approach does not allow estimating the inserted bias from a modeling perspective. More precisely, the substitution process must generate values that least distorts the original characteristics of $\mathbf{X}_D$, which can be assumed to be the between-attribute relationships, for the modeling process. These relationships are often explored by classification algorithms.

Several authors (e.g., see [6] for a review) have argued that it is more important to take into account the influence of imputed values in the modeling process (e.g., preserving the relationships between attributes) than to get more accurate predictions. Roughly speaking, although the imputed values are predictions, it is not the accuracy of these predictions that is of most importance when replacing missing values. It is more important that such predictions produce a workable estimate that least distorts the values that are actually present in the dataset. In other words, the main purpose of imputation is not to use the values themselves, but to make available to the modeling tools the information contained in the other variables' values that are present. For all these reasons, we have focused on the inserted biases in terms of classification results, which somehow allow evaluating to what extent the relationships between attributes are being maintained after imputation. Finally, one must acknowledge that in real-world applications the imputed values cannot be compared with any value.

The bias inserted by imputation can be defined as [6] "the magnitude of the change in the between-attribute relationships caused by patterns introduced by an imputation process". The problem is that the relationships between attributes are hardly known a priori (before data mining is performed). Therefore, usually the inserted bias cannot be directly measured, but it can be estimated. In classification problems, the underlying assumption is that between-attribute relationships are induced by a particular classifier. Consequently, the quality of such discovered relationships can be indirectly estimated by classification measures like, for instance, the Average Correct Classification Rate (ACCR). In this sense, we here adopt a methodology to estimate the inserted bias detailed in [6] and addressed in the sequel.

In data mining applications, different classifiers are often assessed for a given data set, in such a way that the best available classifier is then chosen according to some criterion of model quality (e.g., the ACCR). Our underlying assumption is that the best classifier (BC) - in relation to $\mathbf{X}_C$ and to the available classifiers - provides a suitable model for classifying instances after imputations have been performed. Thus, it is important to assess if the imputed values adjust themselves to the BC model. It is a common practice to evaluate classifier performance in a test set. The same concept can be adapted to evaluate imputations, considering $\mathbf{X}_C$ as the training set and $\mathbf{X}_F$ as the test set. Then, inserted bias can be estimated by means of the procedure in Fig. 2. According to this procedure, a positive bias is achieved when the ACCR in $\mathbf{X}_F$ (step 2) is greater than in the cross-validation process in $\mathbf{X}_C$ (step 1). In this case, the imputed values are likely to improve the classifier's ACCR in $\mathbf{X}'_D$. Accordingly, a negative bias is inserted when the imputed values are likely to worsen the classifier's ACCR in $\mathbf{X}'_D$. Finally, no bias is likely inserted when the classifier's accuracies in $\mathbf{X}_F$ and in the cross-validation process in $\mathbf{X}_C$ are equal. Assuming that the imputation process should not introduce artificial patterns into the data, this is the ideal situation. Indeed, these artificial patterns, not present in the known values, may be later discovered during the data mining process in $\mathbf{X}'_D$. Therefore, the inclusion of such artificial patterns, which are simply an artifact of the imputation process, should be avoided. According to our elaboration, not only a negative bias but also a positive bias is not desirable, as both imply that artificial patterns have been likely incorporated into the dataset.

---

1) Evaluate the classifier's ACCR by cross-validation in $\mathbf{X}_C$, obtaining ACCRC;
2) Evaluate the classifier's ACCR in $\mathbf{X}_F$ (here viewed as a test set) considering that $\mathbf{X}_C$ is the training set. In other words, this step involves building the classifier in $\mathbf{X}_C$ and then testing it in the instances of $\mathbf{X}_F$, thus obtaining ACCRF.
3) The bias (b) inserted by the performed imputations is estimated from the difference between the results achieved in steps 2) and 1): $\widehat{b}$ =ACCRF− ACCRC.

Fig. 2. Estimating the inserted bias on classification

4   Illustrative Experimental Results

In our experiments, we have used the popular Yeast data set, which deals with the prediction of the cellular localization sites of proteins, from the UCI Repository. This data set has 1,484 instances described by 8 attributes and categorized into 9 classes. Following Rubin's typology, we have simulated missing values according to the distribution of missingness known as "missing completely at random". In particular, in each simulation we have removed values from the real attributes according to different rates, viz. 10%, 30%, 50%, and 70%. For each of these missing rates, 10 simulations were performed for different quantities of attributes (1, 2, and 3), resulting in 120 data sets with missing values. Since we are using Euclidean distance to compute dissimilarities, only the real attributes were considered for imputation.

Considering EAC-I, we have arbitrarily adopted the following parameters: populations formed by five genotypes, five k-means iterations for each genotype, and $G = 20$. Sensitivity analysis on these parameters will be the subject of future work. For better confidence, we run EAC-I five times in each simulated

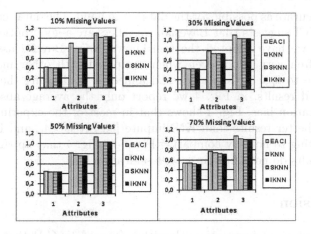

Fig. 3. Prediction results - NRMSE

data set, and the achieved results have been averaged for comparison purposes. The results obtained by EAC-I were compared to those achieved by three state of the art algorithms for imputation - KNNimpute [5], SKNN [4], and IKNN [3] - setting their parameters as suggested by the original references.

For illustration purposes, let us initially consider the results obtained from a prediction point of view (as addressed in Section 3). The accuracy of predictions was evaluated by calculating the error between actual (known) values and the respective imputed values. To do so, we have employed the widely known *Normalized Root Mean Squared Error* (NRMSE). The obtained results are illustrated in Fig. 3. As it can be seen from this figure, in most of the experiments EAC-I has provided slightly worse results than the other algorithms under study.

Let us now concentrate on the inserted bias on classification - by means of the procedure detailed in Section 3. To do so, it would be desirable to employ a

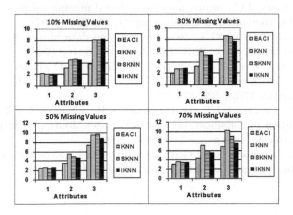

Fig. 4. Absolute values of the inserted bias (%) on classification.

classifier as accurate as possible for the data set in hand. The data set description provided by UCI suggests that Bayesian classifiers are a good option for Yeast. In this sense, we have adopted the well known Naïve Bayes classifier, whose accuracy is often very good, even when compared with other more complex classifiers. Due to space restrictions and aimed at facilitating the illustration of the achieved results, in Fig. 4 we report only the average absolute values of the classification bias. Fig. 4 shows that in most of the experiments EAC-I has provided better results than KNNimpute [5], SKNN [4], and IKNN [3]. It is also interesting to note by comparing Fig. 3 to Fig. 4 that apparently better prediction results do not imply in less classification bias.

5   Conclusion

We have described an imputation algorithm (named EAC-I) that is based on evolving clusters of data. Preliminary experimental results suggest that EAC-I can insert less classification bias than three state of the art algorithms for imputation (i.e., KNNimpute [5], SKNN [4], and IKNN [3]). Also, our experiments indicate that better prediction results may not necessarily lead to less classification bias.

Acknowledgments

The authors acknowledge Capes, CNPq, and FAPESP for their financial support.

References

1. Pyle, D.: Data Preparation for Data Mining (The Morgan Kaufmann Series in Data Management Systems). Morgan Kaufmann, San Francisco (1999)
2. Schafer, J.L.: Analysis of Incomplete Multivariate Data. CRC, Boca Raton (2000)
3. Brás, L.P., Menezes, J.C.: Improving cluster-based missing value estimation of dna microarray data. Biomolecular Engineering 24, 273–282 (2007)
4. Kim, K.Y., Kim, B.J., Yi, G.S.: Reuse of imputed data in microarray analysis increases imputation efficiency. BMC Bioinformatics 5, 160 (2004)
5. Troyanskaya, O.G., Cantor, M., Sherlock, G., Brown, P.O., Hastie, T., Tibshirani, R., Botstein, D., Altman, R.B.: Missing value estimation methods for dna microarrays. Bioinformatics 17(6), 520–525 (2001)
6. Hruschka, E.R., Garcia, A.J.T., Hruschka Jr., E.R., Ebecken, N.F.F.: On the influence of imputation in classification: Practical issues. Journal of Experimental and Theoretical Artificial Intelligence 21, 43–58 (2009)
7. Hruschka, E.R., Campello, R.J.G.B., de Castro, L.N.: Evolving clusters in gene-expression data. Information Science 176(13), 1898–1927 (2006)

A Generic and Extendible Multi-Agent Data Mining Framework

Kamal Ali Albashiri and Frans Coenen

Department of Computer Science, The University of Liverpool,
Ashton Building, Ashton Street, Liverpool L69 3BX, United Kingdom
{ali,frans}@csc.liv.ac.uk

Abstract. A generic and extendible Multi-Agent Data Mining (MADM) framework, EMADS (the Extendible Multi-Agent Data mining System) is described. The central feature of the framework is that it avoids the use of agreed meta-language formats by supporting a system of wrappers. The advantage offered is that the system is easily extendible, so that further data agents and mining agents can simply be added to the system. A demonstration EMADS framework is currently available. The paper includes details of the EMADS architecture and the wrapper principle incorporated into it. A full description and evaluation of the framework's operation is provided by considering two MADM scenarios.

1 Motivation and Goals

Multi-Agent Data Mining (MADM) seeks to harness the general advantages of Multi-Agent Systems (MAS) in the application domain of Data Mining (DM). MAS technology has much to offer DM, particularly in the context of various forms of distributed and cooperative DM. The main issues with MADM are the disparate nature of DM and the wide range of tasks encompassed. Any desired generic MADM framework therefore requires a sophisticated communication mechanism to support it. In the work described here we address the communication requirements of MADM by using a system of mediators and wrappers coupled with an Agent Communication Language (ACL) such as FIPA ACL [8]. We believe this can more readily address the issues concerned with the variety and range of contexts to which a generic MADM can be applicable. The use of wrappers also avoids the need for agreed meta-language formats.

To investigate and evaluate the expected advantages of wrappers and mediators in the context of generic MADM, we have developed and implemented (in JADE) a multi-agent framework, EMADS (the Extendible Multi-Agent Data mining System). The primary goal of the EMADS framework is extendibility; we wish to provide a means for integrating new DM algorithms and data sources in our MADM framework. However, EMADS also seeks to address some of the issues of DM that would benefit from the use of a generic framework. EMADS provides:

- Flexibility in assembling communities of autonomous service providers, including the incorporation of existing applications.

E. Corchado et al. (Eds.): HAIS 2009, LNAI 5572, pp. 203–210, 2009.

- Minimisation of the effort required to create new agents, and to wrap existing applications.
- Support for end users to express DM requests without having detailed knowledge of the individual agents.

The paper's organisation is as follows. A brief review of some related work on MADM is presented in Section 2. The conceptual framework, together with an overview of the wrapper principle, is presented in Section 3 and Section 4. The framework's operation is illustrated in Section 5 using two DM scenarios, and finally some conclusions are presented in Section 6.

2 Related Work

MAS have shown much promise for flexible, fault-tolerant, distributed problem solving. Some MADM frameworks focus on developing complex features for specific DM tasks, without attempting to provide much support for usability or extendibility [10]. The success of peer-to-peer systems and negotiating agents has engendered a demand for more generic, flexible, robust frameworks.

There have been only few such generic MADM systems. An early example was IDM [6], a multi-agent architecture for direct DM to help businesses gather intelligence about their internal commerce agent heuristics and architectures for KDD. In [3] a generic task framework was introduced, but designed to work only with spatial data. The most recent system was introduced in [9] where the authors proposed a multi-agent system to provide a general framework for distributed DM applications. In this system the effort to embed the logic of a specific domain has been minimised and is limited to the customisation of the user. However, although its customisable feature is of a considerable benefit, it still requires users to have very good DM knowledge. The EMADS system which we describe below aims to allow DM algorithms to be embedded in a flexible framework with minimum effort by the user.

3 System Architecture

The EMADS framework has several different modes of operation according to the nature of the participant. Each mode of operation has a corresponding category of *User Agent*. Broadly, the supported categories are:

- **Developers:** Developers are participants, who have full access and may contribute DM algorithms in the form of *Data Mining Agents* (DM Agents).
- **Data Miners:** These are participants, with restricted access to the system, who may pose DM requests through User Agents and *Task Agents* (see below for further details).
- **Data Contributors:** These are participants, again with restricted access, who are prepared to make data available, by launching *Data Agents*, to be used by DM agents.

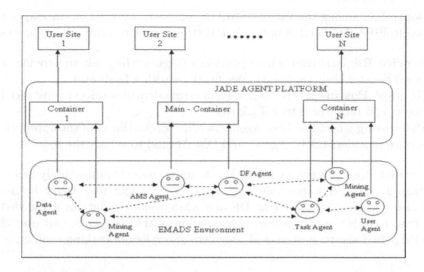

Fig. 1. EMADS Architecture as Implemented in Jade

Conceptually the nature of the requests that may be posted by EMADS users is extensive. In the current demonstration implementation a number of generic requests are supported directed at classification and Association Rule Mining (ARM) scenarios. Two exemplar scenarios are used to illustrate this paper (Section 5).

Fig. 1 presents the EMADS architecture as implemented in JADE (The Java Agent Development Environment) [4]. It shows a sample collection of several application agents and housekeeping agents, organised as a community of peers by a common relationship to each other, that exist in a set of containers. In particular the main container holds the housekeeping agents (an Agent Management System (AMS) agent and a Directory Facilitator (DF) agent). These are specialized server agents responsible for facilitating agents to locate one another.

A user agent runs on the user's local host and is responsible for: (i) accepting user input (request), (ii) launching the appropriate Task Agent to process user requests, and (iii) displaying the results of the (distributed) computation. The user expresses a task to be executed using standard interface dialogue mechanisms by clicking on active areas in the interface and, in some cases, by entering threshold values. Note that the user does not need to specify which agent or agents should be employed to perform the desired task. For instance, if the question "What is the best classifier for my data?" is posed in the user interface, this request will trigger a Task Agent. The Task Agent requests the facilitator to match the action part of the request to capabilities published by other agents. The request is then routed by the Task Agent to the appropriate combination of agents to execute the request. On completion the results are sent back to the user agent for display.

Cooperation among the various EMADS agents is achieved via messages expressed in FIPA ACL and is normally structured around a three-stage process:

1. **Service Registration** where providers (agents who wish to provide *services*) register their capability specifications with a facilitator.
2. **Request Posting** where User Agents (*requesters* of services) construct requests and relay them to a Task Agent,
3. **Processing** where the Task Agent coordinates the efforts of the appropriate service providers (Data Agents and DM Agents) to satisfy the request.

Note that Stage 1 (service registration) is not necessarily immediately followed by stage 2 and 3; it is possible that a services provider may never be used. Note also that the facilitator (the DF and AMS agents) maintains a knowledge base that records the capabilities of the various EMADS agents, and uses this knowledge to assist requesters and providers of services in making contact.

4   System Extendibility

One of the principal objectives of EMADS is to provide an easily extendible MADM framework that can easily accept new data sources and new data mining techniques. The desired extendibility is achieved by a system of wrappers. EMADS wrappers are used to "wrap" data mining artefacts so that they become EMADS agents and can communicate with other EMADS agents. As such EMADS wrappers can be viewed as agents in their own right that are subsumed once they have been integrated with data or tools to become data or data mining agents. The wrappers essentially provide an application interface to EMADS that has to be implemented by the end user; this has been designed to be a fairly trivial operation.

EMADS provides the definition of an abstract parent agent class and every instance agent object (i.e., a program that implements a learning DM algorithm) is then defined as a subclass of this parent class. Through the variables and methods inherited by all agent subclasses, the parent agent class describes a simple and minimal interface that all subclasses have to comply to. As long as an agent conforms to this interface, it can be introduced and used immediately as part of the EMADS system. Two broad categories of wrapper have been defined: (i) data wrappers and (ii) tool wrappers.

4.1   Data Wrappers

Data wrappers are used to "wrap" a data source and consequently create a data agent. A data wrapper holds the location (file path) of a data source, so that it can be accessed by other agents; and meta information about the data. To assist end users in the application of data wrappers a data wrapper GUI is available. Once created, the data agent announces itself to the DF agent as a consequence of which it becomes available to all EMADS users.

4.2   Tool Wrappers

Tool wrappers are used to "wrap" up data mining software systems and thus create a mining agent. Generally the software systems will be data mining tools of various kinds (classifiers, clusters, AR miners, etc.) although they could also be (say) data normalisation/discretization or visualization tools. It is intended that EMADS will incorporate a substantial number of different tool wrappers each defined by the nature of the desired I/O which in turn will be informed by the nature of the generic data mining tasks that it us desirable for EMADS to be able to perform.

5   System Demonstration

The operation of EMADS is described in the following two subsections by considering two demonstration applications (scenarios).

5.1   Meta ARM (Association Rule Mining) Scenario

Meta Mining is defined here as the process of combining individually obtained results of N applications of a DM activity. The motivation behind the scenario is that data relevant to a particular DM application may be owned and maintained by different, geographically dispersed, organizations.

The meta ARM scenario comprises a set of N data agents, N ARM mining agents and a meta ARM agent. Note that each ARM mining agent could have a different ARM algorithm associated with it, although, it is assumed that a common data structure is used to facilitate data interchange. For the scenario described here a set enumeration tree structure called a T-tree [7] was used. Once generated the N *local* T-trees are passed to the Meta ARM agent which creates a *global* T-tree. During the global T-tree generation process the Meta ARM agent interacts with the various ARM agents. There are a number of strategies that can be adopted with respect to when in the process intra agent communication should be made. The authors identified five distinct strategies (Benchmark, Apriori, Brute Force, Hybrid 1 and Hybrid 2). A full description of the algorithms can be found in [1].

5.1.1   Experimentation and Analysis

To evaluate the five Meta ARM algorithms, in the context of the EMADS vision, a number of experiments were conducted designed to analyze the effect of: (i) the number of data agents, (ii) the size of the data agents' datasets in terms of number of records, and (iii) the size of the data agents' datasets in terms of number of attributes. For each of the experiments we measured: (i) processing time, (ii) the overall size of the communications (Kbytes), and (iii) the number of individual communications.

The results shown in Fig. 2 indicate, with respect to Meta ARM, that EMADS offers positive advantages in that all the Meta ARM algorithms were more computationally efficient than the bench mark algorithm (no intra agent cooperation). The results of the analysis also indicated that the Apriori Meta ARM

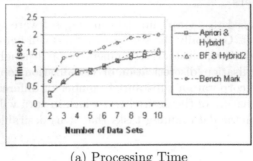

(a) Processing Time

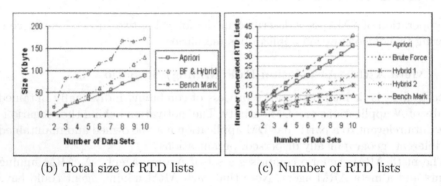

(b) Total size of RTD lists   (c) Number of RTD lists

Fig. 2. Effect of number of data sources

approach coped best with a large number of data sources, while the Brute Force and Hybrid 1 approaches coped best with increased data sizes (in terms of column/rows).

5.2 Classifier Generation Scenario

The Classifier Generation scenario is that of an end user who wishes to obtain a "best" classifier founded on a given, pre-labelled, data set; which can then be applied to further unlabelled data. The assumption is that the given data set is binary valued and that the user requires a single-label, as opposed to a multi-labelled, classifier. The request is made using the individual's user agent which in turn will spawn an appropriate task agent. For this scenario the task agent interacts with mining agents that hold *single labelled classifier* generators that take binary valued data as input. Each of these mining agents generate a classifier, together with an accuracy estimate. Once received the task agent selects the classifier with the best accuracy and returns this to the user agent.

From the literature there are many reported techniques available for generating classifiers. For the scenario reported here the authors used implementations of eight different algorithms[1]. These were placed within an appropriately defined

---

[1] Taken from the LUCS-KDD software repository at
  $http://www.csc.liv.ac.uk/frans/KDD/Software/$

Table 1. Classification Results

Num	Data Set	Classifier Name	Accuracy	Gen. Time (sec)
1	connect4.D129.N67557.C3	RDT	79.76	502.65
2	adult.D97.N48842.C2	IGDT	86.05	86.17
3	letRecog.D106.N20000.C26	RDT	91.79	31.52
4	anneal.D73.N898.C6	FOIL	98.44	5.82
5	breast.D20.N699.C2	IGDT	93.98	1.28
6	dematology.D49.N366.C6	RDT	96.17	11.28
7	heart.D52.N303.C5	RDT	96.02	3.04
8	auto.D137.N205.C7	IGDT	76.47	12.17
9	penDigits.D89.N10992.C10	RDT	99.18	13.77
10	soybean-large.D118.N683.C19	RDT	98.83	13.22

tool wrapper to produce eight (single label binary data classifier generator) DM agents. This was found to be a trivial operation indicating the versatility of the wrapper concept.

5.2.1   Experimentation and Analysis

To evaluate the classification scenario, a sequence of data sets taken from the UCI machine learning data repository [5] were used (pre-processed by data agents so that they were discretized/normalized into a binary valued format). The results are presented in Table 1. Each row in the table represents a particular request and gives the name of the data set, the selected best algorithm as identified from the interaction between agents, the resulting best accuracy and the total EMADS execution time from creation of the initial task agent to the final "best" classifier being returned to the user.

The results demonstrate firstly that EMADS can usefully be adopted to produce a best classifier from a selection of classifiers. Secondly that the operation of EMADS is not significantly hindered by agent communication overheads, although this has some effect. Generation time, in most cases does not seem to be an issue, so further classifier generator mining agents could easily be added. The results also reinforce the often observed phenomenon that there is no single best classifier generator suited to all kinds of data set. Further details of this process can be also found in Albashiri et al. [2].

6   Conclusions

This paper described EMADS, a generic multi-agent framework for DM. The principal advantages offered by the system are that of experience and resource sharing, flexibility and extendibility, protection of privacy and intellectual property rights and information hiding. The framework's operation was illustrated using meta ARM and classification scenarios. Extendibility is demonstrated by showing how wrappers are used to incorporate existing software into EMADS. Experience to date indicates that, given an appropriate wrapper, existing DM software can very easily be packaged to become a DM agent. Flexibility is illustrated using the classification scenario. Information hiding is demonstrated in that users need have no knowledge of how any particular piece of DM software works or the location of the data used.

A good foundation has been established for both DM research and genuine application based DM. The research team is at present working towards increasing the diversity of mining tasks that can be addressed. There are many directions in which the work can (and is being) taken forward. One interesting direction is to build on the wealth of distributed DM research that is currently available and progress this in a MAS context. The research team is also enhancing the system's robustness so as to make it publicly available. It is hoped that once the system is live other interested DM practitioners will be prepared to contribute algorithms and data.

References

1. Albashiri, K., Coenen, F., Sanderson, R., Leng, P.: Frequent Set Meta Mining: Towards Multi-Agent Data Mining. In: Bramer, M., Coenen, F.P., Petridis, M. (eds.), pp. 139–151. Springer, London (2007)
2. Albashiri, K., Coenen, F., Leng. P.: EMADS: An Extendible Multi-Agent Data Miner. In: Bramer, M., Coenen, F.P., Petridis, M. (eds.), pp. 263–276. Springer, London (2008)
3. Baazaoui, H., Faiz, S., Hamed, R., Ghezala, H.: A Framework for data mining based multi-agent: an application to spatial data. In: 3rd World Enformatika Conference, Istanbul (2005)
4. Bellifemine, F., Poggi, A., Rimassi, G.: JADE: A FIPA-Compliant agent framework. In: Proceedings Practical Applications of Intelligent Agents and Multi-Agents, pp. 97–108 (1999), http://sharon.cselt.it/projects/jade
5. Blake, C., Merz, C.: UCI Repository of machine learning databases. Irvine, CA: University of California, Department of Information and Computer Science (1998), http://www.ics.uci.edu/mlearn/MLRepository.html
6. Bose, R., Sugumaran, V.: IDM: An Intelligent Software Agent Based Data Mining Environment. In: Proceedings of IEEE Press, San Diego, CA (1998)
7. Coenen, F., Leng, P., Goulbourne, G.: Tree Structures for Mining Association Rules. Journal of DM and Knowledge Discovery 8(1), 25–51 (2004)
8. Foundation for Intelligent Physical Agents, FIPA 2002 Specification. Geneva, Switzerland (2002), http://www.fipa.org/specifications/index.html
9. Giuseppe, D., Giancarlo, F.: A customisable multi-agent system for distributed data mining. In: Proceedings ACM symposium on applied computing (2007)
10. Klusch, M., Lodi, G.: Agent-based distributed data mining: The KDEC scheme. In: Klusch, M., Bergamaschi, S., Edwards, P., Petta, P. (eds.) Intelligent Information Agents. LNCS, vol. 2586, pp. 104–122. Springer, Heidelberg (2003)

A Modular Distributed Decision Support System with Data Mining Capabilities

Leonardo Gualano and Paul Young

Department of Mechanical and Manufacturing Engineering, Dublin City University,
Glasnevin, Dublin, Ireland
Leonardo.Gualano@dcu.ie, Paul.Young@dcu.ie

Abstract. Although Decision Support Systems (DSSs) to help control strategies have been developed and improved for about two decades, their technology is still limited to end-to-end solutions. This paper proposes a framework for developing a Modular Distributed Decision Support System (MDDSS), capable of handling global knowledge expressed in various forms and accessible by different users with different needs. The interaction of human experts in different parts of the world and intelligent distributed modules will allow the system to deal with increased volume of knowledge (impossible with existing systems) and also to be easily upgradable to future technologies.

Keywords: Decision Support Systems, Distributed Knowledge, Evolutionary Techniques.

1 Introduction

Decision Support Systems (DSS) is a field of research since 1960 [1]. Because DSS is a very wide field of research, there is no uniform definition of such systems. For this reason, much effort has been put in the last decade in defining standards, generalizations and guidelines to develop DSSs which are both efficient and compatible with each other. Various researches have proposed different generic DSS framework, aimed to be valid with any user or organization [2]. In this context, artificial intelligence has found its way into DSS technologies, mainly through the use of Artificial Neural Networks (ANNs), Fuzzy Logic (FL) [3] and Evolutionary Techniques (ETs) [4]. These unconventional techniques have not only been applied directly to DSS fields, but their major contribution resides in the fact that they constitute the basis on which more specific research has been built. In other words, the adaptive characteristics of the above artificial intelligence techniques have been further investigated and new concepts have developed DSS which can undergo different adaptive and/or evolutionary processes in order to be more robust and reliable in changing environments. Different environmental dynamics have been classified and different approaches to DSS evolution/adaptation have therefore been proposed to accommodate them.

ETs have also found their applications in Data Mining, which is a critical aspect of DSSs as it represents the source of the information on which decisions are based. Due to the large amount of multivariate data present in organizational warehouses, however, data mining is usually a difficult task to implement in DSS, so difficult that

E. Corchado et al. (Eds.): HAIS 2009, LNAI 5572, pp. 211–218, 2009.

some researchers have given up automatic data mining and left this task to human action [4, 5]. Other researchers, however, used ETs in order to overcome this limitation, proposing new evolutionary data mining frameworks [6 - 8].

Nowadays, several DSS standards are available and more are being developed to deal with ever larger databases. However, all the current DSSs are limited to specific end-to-end solutions and targeted to single users, groups of users or specific organizations. There is no current development of a DSS which can deal with global knowledge (due to the massive amount of data such system would have to handle) and which can be used simultaneously in an indefinite number of applications (due to the current need of categorizing all the data available). The Modular Distributed Decision Support System (MDDSS) framework proposed in this paper seeks to overcome these limitations, with the case study in Section 3 illustrating how the system might be implemented.

2   The MDDSS Framework

The MDDSS framework proposed is based on the idea of cooperative human and machine intelligence. A set of subsystems (FSs), each of them performing different target tasks, will interact together and will exchange information. Each FS is also constructed from a number of subsystems (modules), each of them implementing a different function.

2.1   Evolutionary Techniques

ETs are algorithms which recursively generate, test and modify populations of functional units (individuals) in order to evolve an acceptable solution to perform a task. Usually, the following steps are involved in an ET:

1.   Create an initial population of individuals.
2.   Test each individual with a fitness function and select the fittest individuals.
3.   By applying some operators (such as crossover and/or mutation operators) to the selected individuals, generate a new population of individuals.
4.   Repeat from step 2 until a termination criterion is satisfied.

For the scope of this paper, it is important to briefly explain how ETs are usually applied to evolve ANNs. This both because ANN technology offers a potential data mining support in the MDDSS framework and because the evolution of the links between the FSs will occur in a very similar way in which the different neurons of ANNs are evolved (See Case Study, Section 3). An ANN is a network of neurons, each of them performing a basic task (applying a transfer function to their inputs). Neurons are connected in different ways and learning algorithms are used to strengthen or weaken their connections in order to create a neurons-connections combination able to perform complex tasks. ANNs are mostly used for system modelling and data categorization and they are more efficient than other techniques when dealing with large number of input/outputs or when the input-output relationships of the implemented functions are not well understood. ETs are mainly applied to ANNs in two different ways:

1. By evolving an efficient network of connections between the neurons.
2. By strengthening/weakening the neurons connections. By evolving new learning algorithms.

A detailed explanation of how ETs are applied to ANNs is given by Xin Yao [9]. The application of ETs improves the ANNs data mining capability and their adaptation to different problems. This technique is therefore proposed for implementation in the MDDSS framework in order to create modules (section 2.2) which will support the data mining tasks. Global knowledge is infinitively vast and it is therefore impossible for it to be included in a database (even if this is distributed).Instead, the approach here is that global knowledge will be continuously investigated by the system. To this purpose, it is worth clarifying that the MDDSS is intended as man-machine cooperation, and so are the evolution and the data mining tasks involved in it. MDDSS is proposed as a system which helps humans in their decisions, not as a system substituting them. The role of human experts will therefore be of paramount importance for the functioning of the MDDSS.

2.2 Modules

Modules are the basic units on which all the MDDSS structure is based. Each module produces some outputs by performing a function only on its inputs. Each module is therefore independent from the rest of the system. Because different users will have different requirements from the system, modules will often need to be customized and this means that each user should be free to program their own modules and add them to the system without affecting the overall structure or behavior of the full MDDSS. Hence, some standards may need to be developed, defining rules for all the modules to safely communicate to each..

Although generic and customized modules will interact in the same system, they can all be classified under three main categories (Fig. 1):

1. Input modules, A, aimed at acquiring knowledge, will have their own way of extrapolating useful information from different sources (Data series, text files…) and will forward the input data, under a standardized form, to a population of Inner Modules. On a first stage of the MDDSS development, only two or three different types of knowledge will be addressed. This will prove the functioning of the proposed framework. However, modules capable of dealing with other types of knowledge can always be added in future.
2. Inner Modules, B, (which can be Intelligent Evolutionary Modules) will receive data from Input modules and will then perform some required tasks on the data (mining, modeling, …).
3. Output modules, C, will transform the knowledge acquired from the Inner modules into standard forms (web reports, data series...) for use in decision making.

Different users will adopt combinations of modules similar to that in Fig. 1 so that knowledge will be stored in a distributed manner (Each FS will have its own memory).

2.3 Functional Structures

An FS will be formed by a customized combination of modules, adapted to the needs of the user. A general framework of an FS is shown in Fig. 1.

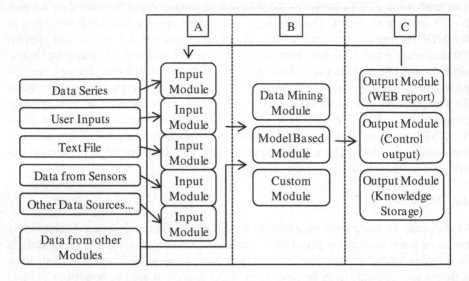

Fig. 1. Proposed framework of Functional Structure for Decision Support

A FS is basically a macro-module in the meaning that it is independent from other FSs in the MDDSS and it only performs operations on its inputs. This means that, in a similar way as modules, there must be a common standard of exchanging knowledge between FSs.

Each FS can perform its own data mining, if needed, through the use of dedicated modules, which will implement both conventional and unconventional techniques to detect from other nearby FS's (where 'nearby' can be meant in terms of geography, discipline, ...) only those data which are relevant to the scope of the FS itself. To this purpose it is worth to remind that data mining modules can also be modules interacting with human experts in order to take advantage of the human data mining abilities. Some central nodes, containing lists of 'nearby' FSs (and other nodes) available in the network and the type of data associated with each of them, can also be used in order for an FS to know which other FS are available and the data they may contain. This will allow more accurate knowledge to be distilled to the different local users.

2.4 MDDSS

The MDDSS will be a network of FSs., where different FSs can be organized to form a network of inter- and extra-organizational links. Because of the evolutionary and data mining capabilities, the links will be continuously changing: Links to obsolete knowledge will be weakened until confined with time to their specific applications or destroyed, links to useful knowledge will be strengthened and new links will be continuously created. The MDDSS will then be able to deal with global knowledge and users will both take advantage of and contribute to global knowledge for decision support. This framework allows the MDDSS to be adaptive and easily upgradable to future technologies by allowing more advanced modules and FSs to be added to existing ones without the need for the full system to be changed.

3 Case Study – DeLisi s.r.l.

DeLisi s.r.l. manufactures metal wire straightening, bending and cutting machines. Considerable overheads are constituted by a recurrent problem of bad castings from their suppliers. DeLisi s.r.l. buys certain quantities of castings from two local companies, offering the same alloy at different prices and different qualities. Bad castings are often shipped back to the supplier for re-casting. There are fewer problems with the higher quality alloy castings, but there is an added transportation cost for these which is not incurred with the low quality alloy.

However the following factors have to be considered:

* Returning the castings creates considerable disturbance in production.
* When it is possible, extra work is performed by DeLisi s.r.l. to enable a bad casting to be rendered usable in order to avoid these delays.
* However, this extra work cannot be performed immediately as the machines must be reconfigured for this, and faulty parts may queue for prolonged periods of time. When this happens, calculating the extra costs incurred becomes a difficult task.
* The delay due to casting issues often delays the readiness of certain products. This plays a negative role on customers waiting for machines, leading to further loss, which has never been fully quantified.

3.1 FS Implementation

DeLisi s.r.l. wants to implement some decision support to help the management strategy with the problem of bad castings, and two different FSs as shown in Fig. 2 are implemented with the following main objectives:

FS1 -   Decide the most convenient purchase (low quality or high quality).
FS2 -   Decide whether it is more convenient to ship back the bad castings or to undertake extra work to recuperate them.

3.2 Evolution of FS1 and Its Links

However, many of the required inputs for FS1 have never been recorded previously and therefore they rely on the experience and instinct of the management team. FS1 can therefore be improved with time as it builds up a record of all the inputs. Once this record is available, some extra modules are added to FS1 and some of the inputs are substituted by more accurate ones. Simultaneously the company also creates a record of the real costs incurred with the two different casting qualities. Because there

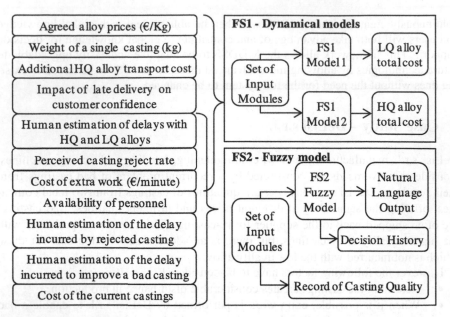

Fig. 2. First implementation of the two FSs at DeLisi s.r.l. The different models implemented in the two FSs are simple mathematical/fuzzy models.

will be some error between the estimated costs from FS1 and the actual costs, FS1 may be further upgraded as follows:

- The two models implemented in FS1 will be substituted by two different ANNs Records of past costs is now available and training sets can be constructed.
- An extra input variable is added, produced by FS2: The percentage of cases bad fusions are present with the different qualities of alloy.
- The lower quality alloy company has agreed to disclose some of its information regarding the production of the alloy. An extra input is therefore added to FS1, produced by an FS located in the low quality alloy company: The technical Characteristics of the low quality alloy.

It is not certain that all the new added inputs will prove valuable for use in FS1, and the ANNs will perform the data mining task by weighting the connections with these inputs in relation to their contribution to the FS1 objectives. After training, in fact, it is found that the weights of the ANNs corresponding to the two of the extra inputs approach zero. This means that the two new extra inputs are insignificant for the results of the models and the links between the FS1 and FS2 and the low alloy company FS, corresponding to the new added variables are weakened after training to an extent which allows these links to be deleted. In fact, the apparent extra information in these sources is implicit in the other inputs already in use for FS1.

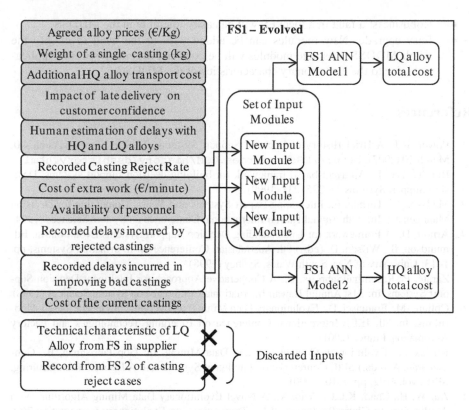

Fig. 3. FS1: New modules, ANN models and new input set including the discarded inputs

3.3 Summary of Case Study

This case study has shown the evolution in time of one FS and how the links between different FSs are evolved. The ANN implementation has been chosen here as its training methods and the way ANNs strengthen/weaken their input connections resemble more complex GA techniques which can also be used for data mining or other purposes in the MDDSS. It has also been shown how the role of human experts is vital for the start of the evolution of the MDDSS.

4   Conclusions and Expected Outcomes

This research will allow the development of a novel framework of a DSS (the MDDSS) which can deal with global knowledge and previously unreachable amount of data. The distributed structure of the developed DSS framework will have the following four main advantages:

- Less local storage space required: No massive databases needed.
- The task of categorizing knowledge [10 - 11] is sensibly reduced

- Robustness: a fault in a local FS will be easily tolerable in the MDDSS..
- Easy upgrade: New modules can be added without the need to modify the whole MDDSS. Obsolete modules will be confined in time to specific applications due to the evolutionary characteristics of the MDDSS.

References

1. Power, D.J.: A Brief History of Decision Support Systems., DSSResources, version 4.0, March 10 (2007), http://DSSResources.COM/history/dsshistory.html
2. Bui, T., Lee, J.: An agent-based framework for building decision support systems. Decision Support Systems 25, 225–237 (1999)
3. Medsker, L., Turban, E.: Integrating Expert Systems and Neural Computing for Decision Management. In: 27th Annual Hawaii International Conference on System Sciences (1994)
4. Arnott, D.: A Framework for Understanding Decision Support Systems Evolution. In: Edmundson, B., Wilson, D. (eds.) 9th Australasian Conference on Information Systems, pp. 1–13. University of New South Wales, Sydney (1998)
5. Zaraté, P., Rosenthal-Sabroux, C.: A Cooperative Approach for Intelligent Decision Support Systems. In: 31st Annual Hawaii International Conference on System Sciences (1998)
6. Collard, M., Francisci, D.: Evolutionary Data Mining: an overview of Genetic-based Algorithms. In: 8th IEEE International Conference on Emerging Technologies and Factory Automation, France (2001)
7. Kusiak, A.: Evolutionary Computation and Data Mining. In: Gopalakrishnan, B., Gunasekaran, A. (eds.) SPIE Conference on Intelligent Systems and Advanced Manufacturing, SPIE, vol. 4192, pp. 1–10 (2000)
8. Au, W.-H., Chan, K.C.C., Yao, X.: A Novel Evolutionary Data Mining Algorithm With Applications to Churn Prediction. IEEE Transactions on Evolutionary Computation 7(6), 532–545 (2003)
9. Yao, X.: Evolving Artificial Neural Networks. IEEE 87(9), 1423–1447 (1999)
10. Saaty, T.L.: Fundamentals of Decision Making and Priority Theory with the Analytic Hierarchy Process. RWS Publ. (2000) ISBN 0-9620317-6-3
11. Lai, Y.J., Liu, T.Y., Hwang, C.L.: TOPSIS for MODM. European Journal of Operational Research 76, 486–500 (1994)

A Fuzzy Quantitative Integrated Metric Model for CMMI Appraisal

Ching-Hsue Cheng[1,*], Jing-Rong Chang[2], Chen-Yi Kuo[1], and Shu-Ying Liao[1]

[1] Department of Information Management, National Yunlin University of Science and Technology, 123, Section 3, University Road, Touliu, Yunlin 640, Taiwan
{chcheng,g8923731,g9023706}@yuntech.edu.tw
[2] Department of Information Management, Chaoyang University of Technology, 168, Jifong East Road, Wufong Township, Taichung County 41349, Taiwan
chrischang@cyut.edu.tw

Abstract. In Capability Maturity Model Integrated (CMMI) systems, the Lead Appraiser evaluates the processes of one company according to the qualitative methods, such as questionnaire, interview, and document. A Fuzzy Quantitative Integrated Metric Model (FQIMM) is proposed in this paper due to the subjective measurement and non-quantitative representations of Lead Appraiser. The FQIMM integrates Quantitative Software Metrics Set, linguistic variables and interval of confidence. FQIMM can help software development companies to evaluate competitiveness by quantitative approach and then know their position more quickly and effectively.

1 Introduction

In recent years, software development companies and departments follow software process methodology (SPM) including ISO9000, Capability Maturity Model (CMM) and Capability Maturity Model Integrated (CMMI) for improving competitiveness and quality. Moreover, the impact of software process improvement on organizations performance and the situations was investigated by many researches, such as Ashrafi [2], Chen, Li and Lee [4], Jiang et al. [9], and Osmundson, et al. [12].

In order to evaluate the maturity of the software development process in a software company, software measurements are needed and used to measure specific attributes of a software product or software process. This study area uses software measures to derive various objectives [6,7]. In CMMI, the Lead Appraisers can know the effects and the performance of institutionalized process in one company according to Specific and Generic Goals, and Generic and Specific Practices defined in Process Area. Lead Appraisers can also evaluate in qualitative description based on questionnaire, interview and document by Appraisal Requirements for CMMI, and Standard CMMI Appraisal Method for Process Improvement (SCAMPI).

---

* Corresponding author.

E. Corchado et al. (Eds.): HAIS 2009, LNAI 5572, pp. 219–226, 2009.

The CMMI evaluation results are usually dependent on the Lead Appraiser's subjective judgment. Therefore, Chen and Huang [3] built Quantitative Software Metrics Set by Appraisal Requirements for CMMI and Function Point Analysis [5] and chose one category of each process area in CMMI Continuous for Quantitative Software Metrics Set. In Quantitative Software Metrics Set, 'benchmark' means the computing unit is calculated by function point, but can't benchmark with certificated company. Benchmark was defined as "evaluating the practices of best-in-class organizations and adapting processes to incorporate the best of these practices [1]". This paper accepts above definitions and improves this process.

However, Quantitative Software Metrics Set only provided benchmarking partly by some function point unit metrics without designing an understanding questionnaire. In order to solve this problem, a new Fuzzy Quantitative Integrated Metric Model (FQIMM) is proposed in this paper due to the subjective measurement and non-quantitative representations of Lead Appraiser. The FQIMM integrates Quantitative Software Metrics Set, linguistic variables, and interval of confidence. FQIMM can help companies to evaluate their competitiveness by quantitative approach and then know their position more quickly and effectively. This study also invites two experts to provide the importance of Generic Practices and Specific Practices for verification, and they serve as hosts of the CMMI group in the Institute for Information Industry in Taiwan. Two software companies are invited to provide the performance value of the practices; Company C1, with more than 25 years history, has passed CMMI Level 2 in 2003 as a benchmarking company. And, Company C2, with more than 35 years history, does not have CMMI Level 2 certification as a self-assessment company.

Following section one, the preliminaries are introduced. The proposed FQIMM is described in section three. We also make some briefly conclusion according the results of case companies in the last section.

2 Preliminaries

In this section, the Capability Maturity Model Integrated and fuzzy set theory adopted in this paper are briefly introduced.

2.1 Capability Maturity Model Integrated (CMMI)

In 1980, the researchers in Software Engineering Institute (SEI) have engaged in the research of software maturity in different software development organizations. In order to evaluate the performance of contractors, Department of Defense (DOD) provided a set guideline of software process management (SPM). Many researchers also developed variety models that described the activities and models of software process called Capability Maturity Model® (CMM) [13,14].

Software Engineering Institute has published several modules of CMM, and there are five kinds CMM so far. Although they are applied to many organizations, it exits a lot of problems when an organization uses more than one model simultaneously. Because of different architectures, contexts, and methods among several models, organizations must spend more resources to finish integrating the different models. In order to solve such problem, Software Engineering Institute has developed a set of model called CMMI (Capability Maturity Model Integrated), which can solve this problem and integrate training and appraisal models. Therefore, CMMI Version 1.02 was published in 2000, and CMMI Verison1.1 in 2001.

There are lots of researches about the improvement and discussion for CMMI. For example, Huang and Han [8] proposed a decision support model hat assists managers in determining the priorities of the CMMI process areas based on the characteristics of the is being developed. Jung and Goldenson [10] think that that higher process maturity is associated with improved project performance and product quality, so they provides a empirical evidence to support this proposition. Their results indicate that organizational size does not influence the relationship, while geographical region is deemed to be an independent variable. In 2008, Lee, et al. [11] develop an ontology-based intelligent decision support agent (OIDSA) to apply to project monitoring and control of CMMI. It is composed of a natural language processing agent, a fuzzy inference agent, and a performance decision support agent.

2.2 Fuzzy Set Theory

Zadeh [15,16] introduced the concepts of fuzzy set, which was a technique designing to cope with imprecise linguistic concepts or fuzzy terms. It allowed users to provide inputs in imprecise terms and received either fuzzy or precise advice.

The definitions and arithmetic operations of fuzzy number described by Zimmermann [19] are introduced below:

[**Definition 1**]. Let X is a universe of discourse corresponding to an object whose current status is fuzzy, and the status value is characterized by a fuzzy set $\tilde{A}$ in X. A membership function (MF) $\mu_{\tilde{A}}(x): X \rightarrow [0,1]$, is called the membership function of $\tilde{A}$.

Fuzzy numbers are usually described by membership functions. A common and very used membership function type is triangular membership functions that can be described as a triplet $\tilde{A} = (a_1, a_2, a_3)$, and the arithmetic operations of fuzzy numbers depends on the arithmetic operations of the interval. Some main operations for fuzzy numbers are described in ref. [16].

A linguistic variable is "a variable whose values are words or sentences in a natural or artificial language" [17,18]. For instance, some matters are characterized by linguistic term in nature, such as 'good', 'medium', and 'bad'. Each linguistic variable may be assigned one or more linguistic values, which are in turn connected to a numeric value through the mechanism of membership functions. Figure 1 shows the membership function of a linguistic variable.

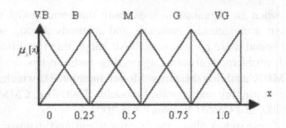

Fig. 1. The membership function of a linguistic variable

3   A Fuzzy Quantitative Integrated Metric Model (FQIMM)

In this section, the proposed model and the stepwise algorithm will be described. The collected data and tentatively results are also included in section 3.2.

3.1   FQIMM Appraisal Framework

In light of Quantitative Software Metrics Set only provides benchmarking partly by some function point unit metrics without an understanding questionnaire, in addition, a questionnaire is integrated by mechanisms. Therefore, a new CMMI appraisal model with complete benchmarking for diagnosing companies is proposed in this study. The Appraisal Framework of FQIMM can be divided into three components:

- Experts' opinions: Ask domain experts to give the weightings of General Practice (GP) or Specific Practice (SP) for the integration computing of FQIMM.
- FQIMM: FQIMM consists of Quantitative Software Metrics Set, Linguistic variable, interval of confidence, and the ranking process, and simplifies Quantitative Software Metrics Set to an understanding questionnaire and removes redundant question items to satisfy face validity; with linguistic variable, decision makers can present imprecise values.
- Recommendations: This paper will provide recommendations for self-assessment companies by the metrics and the ranking results.

3.2   Algorithm of FQIMM

FQIMM includes four process areas, where are Planning Project (PP), Process and Product Quality Assurance (PPQA), Requirements management (RM), and Organizational Process Performance (OPP). For each process area, the new company will benchmark with the certificated company, and provides metric values and ranking results and a total ranking result. The proposed algorithm for FQIMM appraisal and the illustrated data are listed as below:

Step 1.   Build fuzzy weighting matrix $\widetilde{w}$

To get the importance for various general and specific practices and transform them to a fuzzy weighting matrix $\widetilde{w}$.

The questionnaire is designed for collecting opinions of the experts in the CMMI field. Therefore, this study invites two experts to provide the importance of General

Practice and Specific Practice, and they services as managers of the CMMI group in the Institute for Information Industry in Taiwan. They are invited to provide the importance of General Practice and Specific Practice by linguistic variables. From Fig. 1, we transform the linguistic variables and average them to a fuzzy weighting matrix $\tilde{w}$ for Project Planning in Table 1.

Step 2. Build metrics performance matrix $\tilde{M}$

To get the values of the question items and compute the metrics $\tilde{M}$. Company C_1 and Company C_2 provides the interval values to the question items. Then, we summarize as Table 2.

Step 3. Re-organize $\tilde{M}$ to fuzzy numbers matrix $\tilde{A}$

The integrated fuzzy numbers matrix $\tilde{A}$ is composed by metric $\tilde{M}$ according to Quantitative Software Metrics Set's practices. The values of the metrics $\tilde{M}$ for Project Planning are presented in Table 2. The translation of practices matrix $\tilde{A}$ for Project Planning is shown in Table 3.

Table 1. The importance of practices for Project Planning

Activities	Expert		Fuzzy Weight ($\tilde{w}$)
	E1	E2	
SG1 Establish Estimates			
Estimate the Scope of the Project	Very important	Important	(0.625,0.875,1.0)
Establish Estimates of Work Product and Task Attributes	Very important	Very important	(0.75,1.0,1.0)
Define Project Life Cycle	Very important	Very important	(0.75,1.0,1.0)
Determine Estimates of Effort and Cost	Very important	Important	(0.625,0.875,1.0)
SG2 Develop a Project Plan			
Establish the Budget and Schedule	Very important	Important	(0.625,0.875,1.0)
Identify Project Risks	Very important	Important	(0.625,0.875,1.0)
Plan for Data Management	Very important	Important	(0.625,0.875,1.0)
Plan for Project Resources	Very important	Important	(0.625,0.875,1.0)
Plan for Needed Knowledge and Skills	Very important	Important	(0.625,0.875,1.0)
Plan Stakeholder Involvement	Very important	Important	(0.625,0.875,1.0)
Establish the Project Plan	Very important	Very important	(0.75,1.0,1.0)
SG3 Obtain Commitment to the Plan			
Review Plans that Affect the Project	Very important	Important	(0.625,0.875,1.0)
Reconcile Work and Resource Levels	Very important	Important	(0.625,0.875,1.0)
Obtain Plan Commitment	Very important	Very important	(0.75,1.0,1.0)
GG2 Institutionalize a Managed Process			
Establish an Organizational Policy	Very important	Very important	(0.75,1.0,1.0)
Plan the Process	Very important	Very important	(0.75,1.0,1.0)
Provide Resources	Very important	Important	(0.625,0.875,1.0)
Assign Responsibility	Very important	Important	(0.625,0.875,1.0)
Train People	Very important	Important	(0.625,0.875,1.0)
Manage Configurations	Very important	Important	(0.625,0.875,1.0)
Identify and Involve Relevant Stakeholders	Very important	Important	(0.625,0.875,1.0)
Monitor and Control the Process	Very important	Very important	(0.75,1.0,1.0)
Objectively Evaluate Adherence	Very important	Important	(0.625,0.875,1.0)
Review Status with Higher Level Management	Very important	Very important	(0.75,1.0,1.0)
GG 3 Institutionalize a Defined Process			
Establish a Defined Process	Median	Unimportant	(0.125,0.375,0.625)
Collect Improvement Information	Median	Unimportant	(0.125,0.375,0.625)

SG: Specific Goal.   GG: General Goal.

Step 4. Aggregate evaluation $\tilde{R}$

To multiple a fuzzy number matrix $\tilde{A}$ and fuzzy weighting matrix $\tilde{w}$, then get fuzzy aggregative evaluation matrix ($\tilde{R}$), as equation one shows below:

$$\tilde{R} = \tilde{A} \otimes \tilde{w} \tag{1}$$

Step 5. Rank results

The fuzzy aggregative evaluation matrix ($\tilde{R}$) will be ranked by the defuzzified method by the centroid method [19].

Table 2. The metric values of practices for Project Planning

Metric	Definition	Metric values for company	
		C1	C2
PA-PR-UB-1	Process planning integrity (%)	[34, 34]	[20, 20]
PA-PR-UB-2	Audited process components (%)	[71.4,71.4]	[100,100]
PA-PR-BM-1	Estimated workload of modifying process' shortages (%)	[75,80]	[100,100]
PP-SP1.1-UB-1	Description of work items and products	(0.6,0.85,1.0)	(0.45,0.7,0.95)
PP-SP1.1-UB-2	WBS	(0.75,1.0,1.0)	(0.25,0.5,0.75)
PP-SP1.2-BM-1	Technical solutions (%)	[38,100]	100~100
PP-SP1.2-BM-2	Software function size	[900,2400]	[6,6]
PP-SP1.3-UB-1	Life-cycle definition	(0.625,0.875,1)	(0.5,0.75,1.0)
PP-SP1.4-BM-1	Project workload	[720,720]	[3,3]
PP-SP1.4-BM-2	Software cost estimation (million)	[320,320]	[2,2]
PP-SP2.1-BM-1	Project-budget estimation (million)	[57.6,57.6]	[0.36,0.36]
PP-SP2.1-BM-2	Project-schedule estimation (/months)	[46.8,187.2]	[0.3375,0.3375]
PP-SP2.2-UB-1	Project-risk estimation	(0.58,0.83,1.0)	(0.35,0.6,0.85)
PP-SP2.3-UB-1	Project information management	(0.42,0.67,0.89)	(0.39,0.64,0.89)
PP-SP2.4-UB-1	Project needed resource planning	(0.75,1.0,1.0)	(0.35,0.6,0.85)
PP-SP2.5-UB-1	Needed knowledge & skill	(0.45,0.7,0.95)	(0.45,0.7,0.95)
PP-SP2.6-UB-1	Involved personal planning	(0.54,0.79,0.96)	(0.46,0.71,0.96)
PP-SP3.1-UB-1	Audited plans affected project	(0.625,0.875,1)	(0.44,0.69,0.94)
PP-SP3.2-BM-1	Project cost	[-3.1,-3.1]	[-33.3,-33.3]
PP-SP3.2-BM-2	Project schedule (%)	[0.0,0.0]	[0.0,0.0]
PP-SP3.2-BM-3	Key-stakeholder commitments	[0,0]	[0,0]
PP-SP3.3-UB-1	Accepted commitments (%)	[0.0,0.0]	[0.0,0.0]
PP-GP2.1-UB-1	Organization policy	(0.5,0.75,0.94)	(0.19,0.44,0.69)
PP-GP2.2-UB-1	Project planning process	(0.69,0.94,1.0)	(0.38,0.63,0.88)
PP-GP2.3-UB-1	Resource sufficiency	(0.625,0.875,1.0)	(0.29,0.54,0.79)
PP-GP2.4-UB-1	Assigned responsibility adaptation	(0.75,1.0,1.0)	(0.5,0.75,1.0)
PP-GP2.5-BM-1	Train hours	[12,12]	[30,30]
PP-GP2.6-UB-1	Configurations management	(0.69,0.94,1.0)	(0.39,0.64,0.89)
PP-GP2.7-UB-1	Key-stakeholder involvement	(0.54,0.79,1.0)	(0.46,0.71,0.96)
PP-GP2.8-BM-1	Project workload changes (%)	[0.0,0.0]	[0.0,0.0]
PP-GP2.9-UB-1	Project rules	(0.71,0.96,1.0)	(0.29,0.54,0.79)
PP-GP2.10-UB-1	Audited results with high manager	(0.5,0.75,1.0)	(0.5,0.75,1.0)
PP-GP3.1-UB-1	Project planning process definition	(0.58,0.83,1.0)	(0.38,0.63,0.88)
PP-GP3.2-UB-1	Improve information collection	(0.5,0.75,1.0)	(0.44,0.69,0.94)

Table 3. The transferred metric values of practices for Project Planning

Activities	Company	
	C₁	C₂
SG1 Establish Estimates		
Estimate the Scope of the Project	(0.57,0.82,0.95)	(0.44,0.69,0.84)
Establish Estimates of Work Product and Task Attributes	(0.6,0.85,0.95)	(0.45,0.7,0.8)
Define Project Life Cycle	(0.53,0.78,0.94)	(0.5,0.75,0.875)
Determine Estimates of Effort and Cost	(0.6,0.85,0.95)	(0.3,0.55,0.7)
SG2 Develop a Project Plan		
Establish the Budget and Schedule	(0.6,0.85,0.95)	(0.3,0.55,0.7)
Identify Project Risks	(0.52,0.77,0.94)	(0.46,0.71,0.83)
Plan for Data Management	(0.48,0.73,0.91)	(0.47,0.72,0.85)
Plan for Project Resources	(0.56,0.81,0.94)	(0.46,0.71,0.84)
Plan for Needed Knowledge and Skills	(0.48,0.74,0.93)	(0.49,0.74,0.86)
Plan Stakeholder Involvement	(0.51,0.76,0.93)	(0.49,0.74,0.87)
Establish the Project Plan	(0.5,0.75,0.92)	(0.5,0.75,0.83)
SG3 Obtain Commitment to the Plan		
Review Plans that Affect the Project	(0.53,0.78,0.94)	(0.48.0.73.0.86)
Reconcile Work and Resource Levels	(0.63,0.88,0.96)	(0.63,0.88,0.92)
Obtain Plan Commitment	(0.56,0.81,0.94)	(0.56,0.81,0.88)
GG2 Institutionalize a Managed Process		
Establish an Organizational Policy	(0.42,0.67,0.90)	(0.31,0.56,0.73)
Plan the Process	(0.48,0.73,0.92)	(0.38,0.63,0.79)
Provide Resources	(0.46,0.71,0.92)	(0.35,0.60,0.76)
Assign Responsibility	(0.5,0.75,0.92)	(0.42,0.67,0.83)
Train People	(0.33,0.58,0.83)	(0.42,0.67,0.83)
Manage Configurations	(0.55,0.80,0.94)	(0.47,0.72,0.85)
Identify and Involve Relevant Stakeholders	(0.43,0.68,0.92)	(0.40,0.65,0.82)
Monitor and Control the Process	(0.56,0.81,0.94)	(0.56,0.81,0.88)
Objectively Evaluate Adherence	(0.49,0.74,0.92)	(0.35,0.60,0.76)
Review Status with Higher Level Management	(0.42,0.67,0.92)	(0.42,0.67,0.83)
GG 3 Institutionalize a Defined Process		
Establish a Defined Process	(0.44,0.69,0.92)	(0.38,0.63,0.79)
Collect Improvement Information	(0.42,0.67,0.92)	(0.40,0.65,0.81)

SG: Specific Goal.   GG: General Goal.

4 Conclusions

Previous research has focused on investigate the impact of software process improvement on organizations performance and the situations [2,4]. There are few studies on discussing how to make the appraisal quickly and specifically. In this paper, a Fuzzy Quantitative Integrated Metric Model (FQIMM) is proposed to measure the performance of CMMI in quantitative way and to rank the ordering for decision makers. By mean of a practical example, two real companies are illustrated by the proposed method. This paper only measures and appraises these four process areas, because the Quantitative Software Metrics Set only focuses on four of twenty-five process areas. In the future, the proposed model will be validated by some other software companies' data and we will prove it is consistent with decision makers' thoughts. Besides, more simulation results and comparison with other methods will be made in the future.

Acknowledgements

The authors thank the anonymous referees for their constructive and useful comments, and also thank the partially support sponsored by the National Science Council of Taiwan under the Grant NSC97-2410-H-234-018.

References

1. APQC (American Productivity & Quality Center), The benchmarking management guide. Productivity Press, Oregon (1993)
2. Ashrafi, N.: The impact of software process improvement on quality: in theory and practice. Information & Management 40, 677–690 (2003)
3. Chen, C.Y., Huang, S.J.: Using Quantitative Software Metrics to Support CMMI Deployment and Appraisal. Software Industry Service Newsletter 54, 46–57 (2003)
4. Chen, E.Y., Li, H.G., Lee, T.S.: Software process management of top companies in Taiwan: a comparative study. Total Quality Management 13, 701–713 (2003)
5. Garmus, D., Herron, D.: Function Point Analysis: Measurement Practices for Successful Software Projects. Addison-Wesley, Reading (2001)
6. Grady, R.B.: Measuring and Managing Software Maintenance. IEEE Software 4, 35–45 (1987)
7. Grady, R.B.: Work-Product Analysis: The Philosopher's Stone of Software. IEEE Software 7, 26–34 (1990)
8. Huang, S.-J., Han, W.-M.: Selection priority of process areas based on CMMI continuous representation. Information & Management 43, 297–307 (2006)
9. Jiang, J.J., Klein, G., Hwang, H.-G., Huang, J., Hung, S.-Y.: An exploration of the relationship between software development process maturity and project performance. Information & Management 41(3), 279–288 (2004)
10. Jung, H.-W., Goldenson, D.R.: Evaluating the relationship between process improvement and schedule. Information and Software Technology 51, 351–361 (2009)
11. Lee, C.-S., Wang, M.-H., Chen, J.-J.: Ontology-based intelligent decision support agent for CMMI project monitoring and control. International Journal of Approximate Reasoning 48, 62–76 (2008)
12. Osmundson, J.S., Michael, J.B., Machniak, M.J., Grossman, M.A.: Quality management metrics for software development. Information & Management 40, 799–812 (2003)
13. Paulk, M., Curtis, B., Chrissis, M., Weber, C.: Capability maturity model for software, Technical Report CMU/SEI-91-TR-024, Software Engineering Institute, Carnegie Mellon University, Pittsburgh, PA 15213 (1991)
14. Paulk, M., Curtis, B., Chrissis, M., Weber, C., et al.: The capability maturity model for software, version 1.1, Technical Report CMU/SEI-93-TR-024, Software Engineering Institute, Carnegie Mellon University, Pittsburgh, PA 15213 (1993)
15. Zadeh, L.A.: Fuzzy sets. Information and Control 8, 338–353 (1965)
16. Zadeh, L.A.: Fuzzy Logic. IEEE Computer 21, 83–93 (1988)
17. Zadeh, L.A.: The concept of a linguistic variable and its application to approximate reasoning (I). Information Sciences 8, 199–249 (1975)
18. Zadeh, L.A.: The concept of a linguistic variable and its application to approximate reasoning (II). Information Sciences 8, 301–357 (1975)
19. Zimmermann, H.J.: Fuzzy Set Theory and its Applications. Kluwer Academic Publisher, London (1991)

Analyzing Transitive Rules on a Hybrid Concept Discovery System

Yusuf Kavurucu, Pinar Senkul, and Ismail Hakki Toroslu

Middle East Technical University, Computer Engineering
Ankara, Turkey
{yusuf.kavurucu,senkul,toroslu}@ceng.metu.edu.tr
http://www.ceng.metu.edu.tr

Abstract. Multi-relational concept discovery aims to find the relational rules that best describe the target concept. An important challenge that relational knowledge discovery systems face is intractably large search space and there is a trade-off between pruning the search space for fast discovery and generating high quality rules. Combining ILP approach with conventional association rule mining techniques provides effective pruning mechanisms. Due to the nature of Apriori algorithm, the facts that do not have common attributes with the target concept are discarded. This leads to efficient pruning of search space. However, under certain conditions, it fails to generate transitive rules, which is an important drawback when transitive rules are the only way to describe the target concept. In this work, we analyze the effect of incorporating unrelated facts for generating transitive rules in an hybrid relational concept discovery system, namely C^2D, which combines ILP and Apriori.

Keywords: Concept Discovery, ILP, MRDM, Transitive Rule.

1 Introduction

Real life data intensive applications require the use of databases in order to store complex data in relational form. The need for applying data mining and learning tasks on such data has led to the development of multi-relational learning systems that can be directly applied to relational data [1]. Relational upgrades of data mining and concept learning systems generally employ first-order predicate logic as representation language for background knowledge and data structures. The learning systems, which induce logical patterns valid for given background knowledge, have been investigated under a research area, called Inductive Logic Programming (ILP) [2]. In ILP-based concept discovery systems, the aim is to discover a complete and consistent hypothesis for a specific target concept in the light of past experiences.

A challenging problem of relational concept discovery is dealing with intractably large search space. Several systems have been developed employing various search strategies, language pattern limitations and hypothesis evaluation criteria, in order to prune the search space. However, there is a trade-off

E. Corchado et al. (Eds.): HAIS 2009, LNAI 5572, pp. 227–234, 2009.

between pruning the search space and generating high-quality patterns. In this work, we analyze this trade-off problem on discovery of transitive rules. A *transitive rule* includes body predicates that do not have common attributes with the head. Transitive rule generation problem is analyzed in Confidence-based Concept Discovery (C^2D) system [3], [4]. Major features of C^2D are as follows:

1. Instead of strong declarative biases such as input-output modes, the information inside the relational database schema such as argument types and primary-foreign key relationships, a confidence-based pruning mechanism and APRIORI method [5] are used to prune the search space in C^2D.

2. C^2D directly works on relational databases without any requirement of negative instances.

3. Aggregate predicates are defined and incorporated into concept discovery process. In addition, a simple method was developed to handle comparison operators on numeric attributes, which generally accompany aggregate predicates.

The experimental results of C^2D revealed promising performance on the quality of concept discovery in comparison with similar works [3], [4]. However, we encountered the following problem with transitive rule generation:

Concept discovery systems utilizing Apriori technique for specialization may not be able to generate transitive rules if the target concept has common attributes only with some of the background predicates. Since in Apriori, only previously incorporated predicates are used in the rule extension, the rest of the predicates (unrelated relations) can not take part in the hypothesis. This prevents the generation of transitive rules through such predicates. Our previously developed concept discovery system C^2D also suffers from the same problem as well, because it only uses related facts for the construction of most general rules. In this work, in order to remove this drawback, we extended the generalization mechanism of C^2D in such a way that transitive rules through unrelated facts will also be captured.

This paper is organized as follows: Section 2 presents the related work. Section 3 gives an overview of C^2D. Section 4 gives the basic definitions for transitive rule construction. Section 5 presents the experiments to discuss the usage of indirectly related relations in C^2D. Finally, Section 6 includes concluding remarks.

2   Related Work

FOIL, PROGOL, ALEPH and WARMR are some of the well-known ILP-based systems in the literature. FOIL [6] is one of the earliest concept discovery systems. It is a top-down relational ILP system, which uses refinement graph in the search process. In FOIL, negative examples are not explicitly given; they are generated on the basis of CWA.

PROGOL [7] is a top-down relational ILP system, which is based on inverse entailment. A bottom clause is a maximally specific clause, which covers a positive example and is derived using inverse entailment. PROGOL extends clauses by traversing the refinement lattice. ALEPH [8] is similar to PROGOL, whereas it is possible to apply different search strategies and evaluation functions.

Design of algorithms for frequent pattern discovery has become a popular topic in data mining. Almost all algorithms have the same level-wise search technique known as APRIORI algorithm. WARMR [9] is a descriptive ILP system that employs Apriori rule to find frequent queries having the target relation.

The proposed work is similar to ALEPH as both systems produce concept definition from given target. WARMR is another similar work in a sense that, both systems employ Apriori-based searching methods. Unlike ALEPH and WARMR, C^2D does not need mode declarations. It only requires type specifications of the arguments, which already exist together with relational tables. Most of the systems require negative information, whereas C^2D directly works on databases which have only positive data. Similar to FOIL, negative information is implicitly described according to CWA. Finally, it uses a novel confidence-based hypothesis evaluation criterion and search space pruning method.

ALEPH and WARMR can use indirectly related relations and generate transitive rules only with using strict mode declarations. In C^2D, transitive rules are generated without the guidance of mode declarations.

3 C^2D: Confidence-Based Concept Discovery Method

The proposed technique is a part of the concept discovery system named C^2D [3], [4] and is coupled with the rule generation heuristics of the system. Therefore, in this section, we describe the outline and the important features of C^2D.

C^2D is a concept discovery system that uses first-order logic as the concept definition language and generates a set of definite clauses having the target concept in the head. When pure first-order logic is used in learning systems, it may also generate unreasonable clauses such as the ones in which predicates having arguments with actually different types are related. C^2D prevents this by disallowing predicate arguments with different types to be unified. In C^2D, four mechanisms are effective for pruning the search space.

1. Only the related facts of the selected target concept instance are used to find the candidate clauses in the generalization step. Although it is very effective for reducing the search space, it may fail to generate some transitive rules. This mechanism is modified in order to solve the mentioned problem.

2. The second one is a generality ordering on the concept clauses based on θ-subsumption and is defined as follows: A definite clause C θ-subsumes a definite clause C', i.e. at least as general as C', if and only if $\exists \theta$ such that: $head(C) = head(C')$ and $body(C') \supseteq body(C)\theta$.

3. The third one, which is novel in C^2D, is the use of *confidence* as follows: If the confidence of a clause is not higher than the confidence values of the two parent clauses in the Apriori lattice, then it is pruned. By this way, in the solution path, each specialized clause has higher confidence value than its parents.

4. The fourth one is about primary-foreign key relationship between the head and body relations. If such a relationship exists between them, the foreign key argument of the body relation can only have the same variable as the primary key argument of the head predicate in the generalization step.

The algorithm of C^2D has the following main steps:

Generalization: The generalization step is similar to the approach in [10]. After picking the first uncovered positive example, C^2D searches facts related to selected concept instance in the database, including the related facts that belong to the target concept in order for the system to induce recursive rules. By default, the system generalizes the concept instance with all related facts. In order to generate transitive rules through indirectly related facts, "add indirectly related facts" option is added to the system. This option provides the functionality that is described in Section 4.

Refinement of Generalization: C^2D refines the two literal concept descriptions with an Apriori-based specialization operator that searches the definite clause space from general to specific. As in Apriori, the search proceeds level-wise in the hypothesis space and it is mainly composed of two steps: frequent clause set selection from candidate clauses and candidate clause set generation as refinements of the frequent clauses in the previous level. The standard Apriori search lattice is extended in order to capture first-order logical clauses.

The candidate clauses for the next level of the search space are generated in three important steps:

1. Frequent clauses of the previous level are joined to generate the candidate clauses via union operator. In order to apply the union operator to two frequent definite clauses, these clauses must have the same head literal, and bodies must have all but one literal in common.

2. For each frequent union clause, a further specialization step is employed that unifies the existential variables of the same type in the body of the clause.

3. Except for the first level, the candidate clauses that have confidence value not higher than parent's confidence values are eliminated.

Evaluation: For the first uncovered instance of the target concept, the system constructs the search tree consisting of the frequent and confident candidate clauses that induce the current concept instance. Then it eliminates the clauses having less confidence value than the confidence threshold. Finally, the system decides on which clause in the search tree represents a better concept description than other candidates according to f-metric definition.

Covering: After the best clause is selected, concept instances covered by this clause are removed from the concept instances set. The main iteration continues until all concept instances are covered or no more possible candidate clause can be found for the uncovered concept instances.

4   Transitive Rules in C^2D

Discarding unrelated facts of target concept instance is an efficient technique for pruning the search space. But, in some cases, the hypothesis set may contain transitive rules in which unrelated (indirectly related) facts should take part. We firstly give the basic definitions used in this paper.

Definition 1 *(related/unrelated facts). For $t(a_1,...,a_n)$ (the target concept instance - a_i's are constants), the background facts which contain any a_i (with* *the same type according to a_i in target instance) are related facts of $t(a_1, ..., a_n)$. And the background facts which do not contain any of the a_i's as arguments are called unrelated facts of $t(a_1, ..., a_n)$. For the target concept instance $r(a,b)$, the background facts $m(a,c)$ and $n(d,b,e)$ are related facts.*

Definition 2 *(indirectly related facts). For $t(a_1, ..., a_n)$ and the related fact $f(b_1, ..., b_m)$, the background facts which do not contain any a_i but contain at least one b_i (with the same type as b_i in f) are indirectly related facts of $t(a_1, ..., a_n)$. For example, given the target instance $t(a,b)$ and related fact $m(a,c)$, the background instance $n(c,d)$ is an indirectly related fact of $t(a,b)$.*

Definition 3 *(related/unrelated relation). For $t(a_1, ..., a_n)$, a background relation (predicate) p is a related relation to t if p has any instance that is a related fact to $t(a_1, ..., a_n)$. Similarly, a background relation r is an unrelated relation to t if r does not have any instance that is a related fact to $t(a_1, ..., a_n)$.*

Definition 4 *(transitive rule). For the target concept $t(X_1, ..., X_n)$ and the background predicates $b_1(Y_1, ..., Y_m), ..., b_k(Z_1, ..., Z_r)$, a transitive rule is a rule, in which;*

- *target predicate is the head of the rule,*
- *some or all of the background predicates take part in the body of the rule,*
- *in a body predicate, at least one argument is common with other predicates,*
- *at least one body predicate has no common argument with the head predicate.*

For example, for the target predicate t(A,B) and the background predicates f(C,D), p(E,F) and r(G,H), the following is a transitive rule:

   t(A, B) :- f(A, C), p(C, D), r(D, B). (A, B, C, and D are variables)

In the earlier version, C^2D considers only related facts to construct the 2-predicate (one head and one body predicate) generalized rules. Due to inverse resolution in the generalization step, a related relation may take part in the rule body without having any common attribute with the head. Therefore, this earlier approach can generate transitive rules. However, this mechanism falls short for the domains including background predicates that are unrelated relations to all target facts. Then, they can never take part in rule generation. Michalski's trains problem [11] is a typical case for this situation. In this data set, the target relation *eastbound(train)* is only related with *has_car(train, car)* relation. The other background relations are only related with *has_car* relation.

The previous version of C^2D precedes as follows: The *eastbound* relation has 5 records and the system takes the first target instance (*east1*). The target relation has one parameter and its type is *train*. Only (*has_car*) relation is related with *eastbound* and the other background relations are not related. So, it is not possible to join different relations in the specialization phase. Because of this, C^2D finds rules that include only *has_car* relation in the body. This leads to the problem that generated rules are very general and do not reflect the characteristic of the *train* concept.

In order to solve this problem, Generalization step of C^2D, which is described in Section 3, is modified as follows:

(New) Generalization: Let $t(a_1,...,a_n)$ be an uncovered positive example. As the first step, set S that contains all related facts of $t(a_1,...,a_n)$, is generated. As the second step, set S' that contains related facts of each element of set S, is generated. Note that S' contains indirectly related facts of $t(a_1,...,a_n)$. Thirdly, set S is set to be S $\cup$ S'. With the elements of set S, two literal generalizations of t are generated. The second and third steps constitute the modifications for including indirectly related relations.

For example, in the Michalski's train example, the first uncovered target instance is eastbound(east1). Set S for this instance is generated as {has_car(east1, car_11), has_car(east1, car_12), has_car(east1, car_13), has_car(east1, car_14)}, set S' is {closed(car_12), ..., load(car_11, rectangle, 3)}. Therefore, set S is set to be {has_car(east1, car_11), ..., load(car_11, rectangle, 3)}. As a result of this extension, C^2D finds the following transitive rule:

eastbound(A):-has_car(A,B),closed(B). (s=5/5, c=5/7).

5   Experimental Results

An example for using indirectly related facts for transitive rule construction is the *kinship* data set that is a slightly different version of data given in [12]. There are totally 104 records in the data set. In this experiment, a new relation called *elti(A,B)* was defined, which represents the family relation between the wives of two brothers. (The term *elti* is the Turkish word for this family relationship). In the data set, the people in *elti* relation have no brothers. Therefore, *brother* is an unrelated relation for all target facts of *elti*.

If *add indirectly related facts* option is not selected, then C^2D cannot find high quality and semantically correct rules for the target relation *elti*. If this option is selected, then C^2D adds some records of the *brother* relation into the Apriori lattice in the generalization step. Finally, it finds the following rules that can capture the description of *elti*:

This experiment is also conducted on PROGOL and ALEPH systems. However, they could not found any rules under several mode declarations.

elti(A,B) :- husband(C,A), husband(D,B), brother(C,D).
elti(A,B) :- husband(C,A), husband(D,B), brother(D,C).
elti(A,B) :- husband(C,A), wife(B,D), brother(C,D).
elti(A,B) :- husband(C,A), wife(B,D), brother(D,C).
elti(A,B) :- husband(C,B), wife(A,D), brother(C,D).
elti(A,B) :- husband(C,B), wife(A,D), brother(D,C).
elti(A,B) :- wife(A,C), wife(B,D), brother(C,D).
elti(A,B) :- wife(A,C), wife(B,D), brother(D,C).

As discussed in Section 4, transitive rules may be constructed even though there are no unrelated facts in the domain. The related facts used in 2-predicate generalized rules can be combined in the specification phase in such a way that the variable names are unified to form a transitive rule. In the other direction, under the proposed extension, eventhough the background has missing information, it is still possible to discover rules that define the target concept.

As an example, consider the *kinship* data set again. We extend this data set with a new relation called *dunur(A,B)* to represent the family relationship of two persons who are the parents of a married couple (The term *dunur* is the Turkish word for this family relationship). The *dunur* relation has 16 records and has two arguments having *person* type. In the data set, the *dunur* relation is selected as the target relation. In this run, *add indirectly related facts* option is not selected, which means that the indirectly related facts are not included in the generalization phase and the following rules are generated:

dunur(A,B) :- daughter(C,A), husband(D,C), son(D,B).
dunur(A,B) :- daughter(C,B), husband(D,C), son(D,A).
dunur(A,B) :- daughter(C,A), wife(C,D), son(D,B).
dunur(A,B) :- daughter(C,B), wife(C,D), son(D,A).

As seen above, the resulting rule set contains the transitive rules that captures the semantics of *dunur* predicate. However, if some of the records, such as the husband or wife of the parents, are deleted in the data set, then *husband* and *wife* relations are not directly related to the *dunur* relation. (Note that it is a different situation than *elti* example. In this case, due to missing information, some of the related relations appear to be unrelated). When *add indirectly related facts* option is off again, the system cannot generate successful rules as a solution. However, when *add indirectly related facts* option is selected, the above rules can be generated under the missing and reduced background information.

The same experiments are conducted on PROGOL and ALEPH systems. Neither of the systems could find any rules in both of the experiments.

The general overview of the experimental results are given in Table 1.

Table 1. The experimental results

Experiment	Including Unrelated Facts	Without Unrelated Facts
Eastbound Train		
Accuracy	0.7	0
Coverage	1.0	0
Elti		
Accuracy	1.0	0.5
Coverage	1.0	0.5
Dunur		
Accuracy	1.0	1.0
Coverage	1.0	0.75

6   Conclusion

In this work, we analyze transitive rule generation in an ILP-based concept discovery system named C^2D and propose a technique for improving the transitive rule generation. This technique is based on inclusion of unrelated facts into only the most general rule generation step of the concept discovery process. By this way, it is possible to generate rules that captures transitive relations through unrelated facts and to extract transitive rules under missing background information. Such rules are not very common, however, they become of importance for the cases where they are the only rules that describe the target concept.

Similar relational knowledge discovery systems can generate transitive rules, however they use mode declarations which requires high level logic programming and domain knowledge. Under the proposed enhancements, C^2D handles transitive rule generation without mode declarations.

Inclusion of unrelated facts extend the search space by a linear function of number of related facts in the generalization step. Experimentally, it is observed that this extension increases the execution time only slightly, which is approximately between 10% and 20% according to the test data set. We added this capability as a parametric property that can be turned on only for the domains that have unrelated facts in the background information.

References

1. Džeroski, S.: Multi-relational data mining: an introduction. SIGKDD Explorations 5(1), 1–16 (2003)
2. Muggleton, S.: Inductive Logic Programming. In: The MIT Encyclopedia of the Cognitive Sciences (MITECS). MIT Press, Cambridge (1999)
3. Kavurucu, Y., Senkul, P., Toroslu, I.H.: Confidence-based concept discovery in multi-relational data mining. In: International Conference on Data Mining and Applications (ICDMA), Hong Kong, pp. 446–451 (March 2008)
4. Kavurucu, Y., Senkul, P., Toroslu, I.H.: Aggregation in confidence-based concept discovery for multi-relational data mining. In: IADIS European Conference on Data Mining (ECDM), Amsterdam, Netherland, pp. 43–50 (July 2008)
5. Agrawal, R., Mannila, H., Srikant, R., Toivonen, H., Verkamo, A.I.: Fast discovery of association rules. In: Advances in Knowledge Discovery and Data Mining, pp. 307–328. AAAI/MIT Press (1996)
6. Quinlan, J.R.: Learning logical definitions from relations. Mach. Learn. 5(3), 239–266 (1990)
7. Muggleton, S.: Inverse entailment and Progol. New Generation Computing, Special issue on Inductive Logic Programming 13(3-4), 245–286 (1995)
8. Srinivasan, A.: The aleph manual (1999)
9. Dehaspe, L., Raedt, L.D.: Mining association rules in multiple relations. In: Džeroski, S., Lavrač, N. (eds.) ILP 1997. LNCS, vol. 1297, pp. 125–132. Springer, Heidelberg (1997)
10. Toprak, S.D., Senkul, P., Kavurucu, Y., Toroslu, I.H.: A new ILP-based concept discovery method for business intelligence. In: ICDE Workshop on Data Mining and Business Intelligence (April 2007)
11. Michalski, R., Larson, J.: Inductive inference of vl decision rules. In: Workshop on Pattern-Directed Inference Systems, Hawaii, SIGART Newsletter, vol. 63, pp. 33–44. ACM Press, New York (1997)
12. Hinton, G.: UCI machine learning repository kinship data set (1990), http://archive.ics.uci.edu/ml/datasets/Kinship

Survey of Business Intelligence for Energy Markets

Manuel Mejía-Lavalle[1], Ricardo Sosa R.[2], Nemorio González M.[2],
and Liliana Argotte R.[1]

[1] Instituto de Investigaciones Eléctricas, Reforma 113, 62490 Cuernavaca, Morelos, México
[2] Centro Nacional de Control de Energía CFE, Manuelito 32, 01780 México, D.F.
mlavalle@iie.org.mx, ricardo.sosa@cfe.gob.mx

Abstract. Today, there is the need for establishing a strong relationship between Business Intelligence (BI) and Energy Markets (EM). This is crucial because of enormous and increasing data volumes generated and stored day by day in the EM. The data volume turns impossible to obtain clear data understanding through human analysis or with traditional tools. BI can be the solution. In this sense, we present a comprehensive survey related with the BI applications for the EM, in order to show trends and useful methods for tackling down every day EM challenges. We outline how BI approach can effectively support a variety of difficult and challenging EM issues like prediction, pattern recognition, modeling and others. We can observe that hybrid artificial intelligence systems are common in EM. An extensive bibliography is also included.

1 Introduction

Energy Markets are the consequence of the deregulation of electric power market. In these markets, power suppliers and consumers are free to negotiate the terms of their contracts, e.g., energy prices, power quantities and contract durations in several auctions structures. Several issues have to be considered and analyzed before taking actions. There are physical, economical and political restrictions that have to be considered together [1]. Although enormous data volumes related with this process are stored day-by-day, and hour-by-hour, it is impossible (through human analysis or with traditional technology) to obtain knowledge from this data, in order to take wise decisions.

In this complex scenario, BI approach can be an important solution due to its intrinsic way to manage very large data sets in an intelligent and automatic fashion. BI is focused on transform data into knowledge (or intelligence) to improve corporation central process. At the end, BI is a discipline formed with tools emerged from Artificial Intelligence and Database technology, which main purpose is to give people the information or knowledge that they need to do their jobs. It is predicted that, in the near future, BI will become a need of all huge corporation [2].

We survey recent applications of BI for the EM, seeking for the most relevant, important or characteristic state of art works. This survey can help to detect diverse EM problems and its BI solutions, and can be a useful guide for EM decision- makers from countries with deregulated schemas, or from countries next to participate in this global environment. In addition, our survey can be helpful to detect and create new and innovative solution proposals.

E. Corchado et al. (Eds.): HAIS 2009, LNAI 5572, pp. 235–243, 2009.
© Springer-Verlag Berlin Heidelberg 2009

To address these topics, in Section 2, we describe the EM dynamics and the essential problems. Next, in Section 3 we survey works that combine EM problems with BI solutions. Finally, we conclude and discuss future work challenges in Section 4. An extensive reference Section is also included.

2 Energy Markets

The electrical power industry around the world has been continuously deregulated and reconstructed from a centralized monopoly schema to a competitive market structure. This transformation is taking place since the last century.

Today, there is many people and governments that agree in the fact that in most cases the competitive market is a more proper mechanism for energy supply to free the customer's choices, improve overall social welfare and allows markets to operate more efficiently and cost effectively [3].

The Energy Market is a very dynamic and inter-related system. In the open market, power companies and consumers submit their generation or consumption bids and corresponding prices to the market operator [4].

When market prices do not fully reflect environmental and social costs, consumers' choices are distorted. Sending an appropriate market signal to encourage environmentally, rational energy use may be accomplished through a variety of economic facts, from direct taxes to tradable permits, including by reducing consumption and changing its composition in favor of more environmentally friendly goods and services. The government policies will need to take into account particular and national circumstances.

Additional to policies, there are EM technical factors and restrictions too. For instance, unlike other commodities, the electricity cannot be stored and power system stability requires consistent balance between supply and demand. Therefore, besides this factor, there are many physical and technical constraints for the transmission and delivery of energy. Moreover, unpredictable and uncontrollable contingencies in power system increase the complexities. As such, EM exhibits a high volatile nature and is probably the most volatile market.

All these circumstances create a complex multi-disciplinary challenge. Broadly speaking, we can say that EM face problems related to issues that concern prediction, pattern recognition, modeling and other areas, and they can be solve with the BI approach.

3 Business Intelligence and Energy Markets Related Work

Our survey on *EM problems-BI solutions* found papers from very different countries, but almost sharing the same information needs. For instance, according to our survey, we found papers from: USA (17%), Australia (13%), Spain (13%), China (9%), Portugal (7%), Brazil (6%), Greece (4%), Italy (4%), Japan (4%), and others countries with almost 2%, like Germany, Argentina, Canada, Russia, Iraq, UK, Colombia, Taiwan, Belgium, Switzerland, Ireland, Netherlands and Singapore. Related to the BI algorithms, techniques, methods or disciplines applied to solve different problems from the EM scenario, we observed that there is a great diversity and hybrid artificial intelligence systems are common, as Table 1 shows.

Table 1. BI algorithms / disciplines applied to EM problems

Algorithm / Discipline	Reference
Artificial Neural Networks	[4], [5], [6], [7], [8], [9], [10]
Bagging	[11]
Clustering	[12], [13]
Data Warehouse	[9], [14]
Fuzzy logic	[8], [15], [16]
Genetic algorithms	[6], [17]
Induction and Decision Trees, Learning Classifier Systems (C4.5, J48)	[7], [12], [18], [19], [20], [21], [22], [23]
Information Entropy Theory	[5]
Information Gap Theory	[24]
Intelligent Agents and Multi Agents Frameworks	[23], [25], [26], [27], [28], [29], [30], [31], [32]
K-Nearest Neighbor	[22], [27]
Knowledge Discovery in Databases	[33], [34], [35]
Multivariate Adaptive Regression Splines	[36]
Naive Bayes	[37], [38], [39]
Support Vector Machines	[4], [11], [22], [37], [39], [40]
Data Mining, Time Series, Discrete Fourier Transform, Rough Sets	[1], [13], [41], [42], [43], [44], [45], [46]
Wavelet Transforms	[4], [41], [47]

Attending the application type, we grouped our findings in four broad categories: prediction, pattern recognition, modeling and others, as we show in Table 2. In the next sub-sections, we addressed these topics, and described briefly some of these works.

3.1 Prediction

Deregulation in EM brings demand and electricity prices uncertainty, placing higher requirements on forecasting; commonly for estimating load and prices levels more accurately.

Forecasting is essential to support the decision making for an appropriate scheduling and planning in the operation of power systems and for developing bidding strategies or negotiation skills in the EMs. It is being considered as a helpful tool for consumers and producers.

There have been several research projects around the world involving forecasting of EM prices, electricity demand and transient stability; taking into account that lower operation cost and higher reliability of electricity supply is the main operation goal.

There are many research works on electricity market price forecasting. However, most of them are focused on the expected price analysis rather than price spikes forecasting. In Australia, two methods (Support Vector Machines and Naive Bayes) are chosen to be the spike occurrence predictors [39].

Table 2. BI´s EM applications by year

Year	Prediction			Pattern recognition			Modeling			Others	
	Price	Demand	Transient	Operation	Characterization of Customers	Analysis of supply and demand	Simulation with agents	Model behavior	Simulation / Monitoring	Data Quality	Integrated systems
2007	[22], [39]					[8], [16], [48]			[35]		
2006	[36]	[18], [27]		[19], [20]	[10], [44], [46]		[26], [28]	[17], [24]	[34]	[49]	
2005	[37], [38]	[5]				[1], [7]	[29]				[50]
2004	[30]	[33]	[11], [40]	[45]			[23], [30]				[51]
2003		[41]			[43]		[25]	[32]			[14]
2002					[21]			[31]			
2001				[12]		[42]				[47]	
2000					[52]			[6]			

In [40] they propose a novel approach for power system small signal stability analysis and control which combines the state-of-the-art small signal analysis tools, Data Mining, and the synchronized phase measurement techniques.

3.2 Pattern Recognition

During the process of data analysis is looking to find patterns from which to draw a series of rules and/or identify factors that facilitate the understanding of market behavior of electric power, detection of flaws or any other aspect of interest.

In [45] the historical data of distribution feeder faults of Taiwan Power Company was used for validation and this study aims to use Rough Set theory as a Data Mining tool to derive useful patterns and rules for distribution feeder faulty equipment diagnosis and fault location.

In [21] they mention that the electric power industry is an industry that had allowed local monopolies to date in the background of economies of scale, so it was unnecessary for electric power utilities to deeply understand customers. It presents a concept of customer load profile analysis based on Customer Relationship Management (CRM) concept and applications of Data Mining techniques. In [8] the dynamic behavior and non-linear demand in relation to the parameters that determine, as: calendar, climate, economy and energy intensity, increase the complexity of constructing models of analysis of the demand. They use Fuzzy Artificial Neural Networks or Neuro-Fuzzy as a technique.

3.3 Modeling

One area of concern in other countries has been displaying the future behavior of the market for electric power in liberalized markets, for which projects have been developed for simulation and modeling of the behavior of market power by using different techniques.

To study electricity markets behavior and evolution Praca [25] propose a Multi-Agent simulator where agents represent several entities that can be found in electricity markets, such as generators, consumers, market operators and network operators, but also entities that are emerging with the advent of liberalization, such as traders.

The simulator probes the possible effects of market rules and conditions by simulating the strategic behavior of participants. In this paper, a special attention is devoted to the strategic decision processes of seller, buyer and trader agents, in order to gain advantage facing the new emerging competitive market.

One project of model behavior is in [17], it presents an Evolutionary Algorithm to develop cooperative strategies for power buyers in a deregulated electrical power market. Cooperative strategies are evolved through the collaboration of the buyer with other buyers defined by the different group memberships.

The paper explores how buyers can lower their costs by using the algorithm that evolves their group sizes and memberships. The algorithm interfaces with Power World Simulator to include in the technical aspect of a power system network, particularly the effects of the network constraints on the power flow. Simulation tests on an IEEE 14-bus transmission network are conducted and power buyer strategies are observed and analyzed.

The project in [34] describes research into the application of Data Mining algorithms for deriving new control monitoring patterns that can improve the plant's performance.

It proposes the use of a knowledge engineering methodology, named CommonKADS, as a tool for completing the Knowledge Discovery in Databases phases that precede Data Mining. The application of Data Mining classification algorithms resulted in new control monitoring rules, which improve performance without demanding installation of new equipment.

The derived rules can also be used to trace a possible malfunction of a measurement instrument and even more to replace the recording values with those resulting from the Data Mining algorithms.

3.4 Others

There is other alternative solutions to implement the BI approach to electric power markets where the authors have developed other projects such as: building Data Warehouses for OLAP analysis, development of Integrated Systems or systems for Quality Assessment (QA) data. Below we describe some works related with these topics.

The project in [47] presents a priority-based data transmission protocol for sending large volumes of data (Data Cubes) over congested networks, which allows the incremental implementation of OLAP applications. A Wavelet-based lossy compression algorithm that is the core of the transmission protocol is also presented. In [49] QA is addressed, and in [14], [50] and [51] Integrated Systems for EMs are presented.

4 Conclusions and Future Work

We present a comprehensive survey related with the BI solutions for the EM problems, in order to show trends and useful methods for tackling down every day EM challenges. We can observe that hybrid artificial intelligence systems are common and important strategies in EM. We think that it is fundamental for the EM and the computer science communities to know how is the BI´s level of applicability to define present and future research work.

In these sense, this paper is an effort to bring attention from electric power industry, in particular the EM community, about the outmost relevance for applying the BI approach and its methods, because those methods have been designed to overcome the limitations that traditional approaches and methods have.

We highlight the novelty of our paper because, to the best of our knowledge, we didn't find any state-of-the-art BI for EM published work. Furthermore, because we cannot found anything directly written from research institutions of several countries, we claim that there are many opportunities in these countries to work and contribute in this field.

Additionally, we observed a great variety of BI´s methods applied to the EM problems. Broadly speaking, we detect four principal areas of interest: prediction, pattern recognition, modeling and other areas like integrated systems and data quality projects; but we point out that there are still open research topics for EM to explore and apply, like change point detection algorithms, associative rules, feature selection methods and variable inter-relations studios.

All these are material for future work. A final contribution of this paper is an extensive list of bibliographic references to be consulted as a source for researches interested in solving EM problems using the BI approach.

References

1. Bhatnagar, R., Rao, C.: Energy resource management based on data mining and AI. In: Proceedings ACEEE Summer Study on Energy Efficiency in Industry, Cincinnati, vol. 6, pp. 14–23 (2005)
2. McKay, L.: Business intelligence comes out of the back office. CRM magazine (June 2008)
3. Wu, F.F., Varaiya, P.: Coordinated multilateral trades for electric power networks: theory and implementation 1. Electrical Power and Energy Systems 21, 75–102 (1999)
4. Xu, Z., Dong, Z.Y., Liu, W.Q.: Neural network models for electricity market forecasting. In: Proc. of 4th Australian Workshop on Signal Processing and its applications, Australia (1996)
5. Sun, W., Lu, J., He, Y.: Information Entropy Based Neural Network Model for Short-Term Load Forecasting. In: Transmision and Distrib. Conference: Asia and Pacific IEEE/PES, pp. 1–5 (2005)
6. Sheble, G.: Artificial life techniques for market data mining and emulation. Artificial life techniques for market data mining and emulation 3, 1696–1699 (2000)
7. Sánchez-Ubeda, E.F., Muñoz, A., Villar, J.: Minería y visualización de datos del mercado eléctrico español. III Taller Nacional de Minería de Datos y Aprendizaje. Madrid (2005)
8. Andrade, F., Camargo, C., Teive, R.: Aplicação de técnicas de mineração de dados para a previsão de demanda no médio prazo. Encontro Regional Ibero-americano do CIGRÉ. Brasil (2007)

9. Richeldi, M.: A business intelligence solution for energy budget control. In: Proc. of the 3rd Int. Conf. on the Practical Application of Knowledge Discovery and Data Mining, pp. 167–178 (1999)
10. Verdu, V., Garcia, S.: Classification, filtering, and identification of electrical customer load patterns through the use of SOMs. IEEE Transactions on Power Systems, 1672–1682 (2006)
11. Xu, T., He, R.: Appls. of data mining for power system trans. stability prediction. In: Proc. IEEE Int. Conf.Electric Utility Deregulation, Restructuring & Power Tehnologies, pp. 389–392 (2004)
12. Sánchez, E., Peco, J., Raymont, P., Gomez, T.: App. of data mining techniques to identify structural congestion problems under uncertainty. Porto Power Tech. Proceedings 1, 6–14 (2001)
13. Wu, S., Wu, Y., Wang, Y., Ye, Y.: An algorithm for time series data mining based on clustering. Sch. of Computing Science and Technology (2006)
14. Felden, C., Chamoni, P.: Web farming and data warehousing for energy tradefloors. In: Proceedings IEEE/WIC International Conference on Web Intelligence, pp. 642–647 (2003)
15. Widjaja, M., Mielczarski, W.: A Fuzzy-Based Approach to Analyze System Demand in the Australian Electricity Market. In: Power Engineering Society Summer Meeting, pp. 918–923. IEEE, Los Alamitos (1999)
16. Samper, M.E., Vargas, A.: Costos de generación distribuida bajo incertidumbres empleando técnicas Fuzzy. In: Enc. Regional Ibero-americano CIGRÉ Foz do Iguazú, Paraná, pp. 20–24 (2007)
17. Srinivasan, D., Woo, D., Rachmawati, L., Wei Lye, K.: Evolving Cooperative Behavior in a Power Market. In: Genetic and evolutionary computation conference. Proceedings of the 8th annual conference on Genetic and evolutionary computation, pp. 1599–1600 (2006)
18. Lobato, E., Ugedo, A., Rouco, R., Echavarren, F.M.: Decision Trees Applied to Spanish Power Systems Applications. Sch. of Engineering (2006)
19. Voumvoulakis, E.M., Gavoyiannis, A.E., Hatziargyriou, N.: Decision trees for dynamic security assessment and load shedding scheme. In: Power Eng. Soc. General Meeting, Atenas, pp. 7–15 (2006)
20. Li, F., Runger, G.C., Tuv, E.: Change-point detection with supervised learning and feature selection. International Journal of Production Research, 2853–2868 (2006)
21. Kitayama, M., Matsubara, R., Izui, Y.: Data mining for customer load profile analysis. In: Proc. IEEE Power Eng. Society Transmission and Distribution Conference, pp. 654–655 (2002)
22. Troncoso, A., Riquelme, J., Gómez, A., Martínez, J., Riquelme, J.: Electricity Market Price Forecasting Based on Weighted Nearest Neighbors Techniques. IEEE Transactions on Power Systems (2007)
23. Bagnall, A.J.: Autonomous Adaptive Agents in Simulations of the UK Market in Electricity. University of East Anglia (2004)
24. Cheong, M., Sheble, G.: Knowledge extraction and data mining for the competitive electricity auction market. In: Int. Conf. on Prob. Methods Applied to Power Systems, Iowa, pp. 221–229 (2006)
25. Praca, I., Ramos, C., Vale, Z.: A new agent-based framework for the simulation of electricity markets. In: Intelligent Agent Technology, IEEE/WIC Int. Conference, pp. 469–473 (2003)
26. Geoff, J., Cohen, D., Dodier, R., Platt, G., Palmer, D.: A Deployed Multi-Agent Framework for Distributed Energy Applications. In: Int. Conf. on Autonomous Agents, 676–678 (2006)

27. Herrera, F.: Propuesta del Grupo de la Universidad de Sevilla para MINDAT-PLUS. Minería de datos para los usuarios en diferentes áreas de aplicación (2006)
28. Da Silva Lima, W., De Andrade Freitas, E.N.: A multi agent based simulator for Brazilian wholesale electricity energy market. In: First-Order Dynamic Logic. LNCS, pp. 68–77. Springer, Heidelberg (1979)
29. Kok, J.K., Kamphuis, I.G.: PowerMatcher: Multiagent Control in the Electricity Infrastructure. In: International Conference on Automous Agents, pp. 75–82 (2005)
30. Fujii, Y., Okamura, T., Inagaki, K.: Basic analysis of the pricing processes in modeled electricity markets with multi-agent simulation. Restructuring & Power Tech., Tokyo, 257–261 (2004)
31. Karbovskii, I.N., Lukatskii, A.M., Fedorova, G.V., Shapot, D.V.: A small model of market behavior. Automation and Remote Control, Moscow, 654–665 (2002)
32. Debs, A., Hansen, C., Yu-Chi, W., Hirsch, P., Szot, L.: Bidding strategies for GENCO's in the short-term electricity market simulator (STEMS). In: Proceedings of the IASTED International Conference on Intelligent Systems and Control, Salzburg, Austria, pp. 135–139 (2003)
33. Rico, J., Flores, J., Sotomane, C., Calderon, F.: Extracting temporal patterns from time series data bases for prediction of electrical demand. In: Monroy, R., Arroyo-Figueroa, G., Sucar, L.E., Sossa, H. (eds.) MICAI 2004. LNCS (LNAI), vol. 2972, pp. 21–29. Springer, Heidelberg (2004)
34. Athanasopoulou, C., Chatziathanasiou, V., Komminou, M., Petkani, Z.: Applying knowledge engineering and data mining for optimization of control monitoring of power plants. In: Proceeding (521) European Power and Energy Systems (2006)
35. Cruz, R.D., Carrillo, G., Pinzón, S.: Metodología para la monitorización del mercado eléctrico a partir de Inteligencia Competitiva. Encontro Regional Ibero-americano do CIGRÉ (2007)
36. Zeripour, H., Bhattacharya, K., Canizares, C.: Forecasting the hourly Ontario energy price by multivariate adaptive regression splines. In: IEEE Power Eng. Society General Meeting, pp. 7–15 (2006)
37. Zhao, J.H., Li, X., Wong, K.P.: A general method for electricity market price spike analysis. In: IEEE Power Engineering Society General Meeting, pp. 286–293 (2005)
38. Lu, X., Dong, Z.Y., Li, X.: Electricity market price spike forecast with data mining techniques. Electric Power Systems Research, 19–29 (2005)
39. Zhao, J.H., Dong, Z.Y., Li, X., Wong, K.P.: A framework for electricity price spike analysis with advanced data mining methods. IEEE Trans. on Power Systems 22(1), 376–385 (2007)
40. Dongjie, X., Peng, W., Renmu, H., Tao, X.: Small signal stability analysis and control in market environment. In: IEEE International Conference on Electric Utility Deregulation, Restructuring and Power Technologies, vol. 1, pp. 322–325 (2004)
41. Zhenfyou, H., Qingquan, Q., Zhibing, W.: Wavelet Analysis and Electric Power Market E-Commerce. In: Proceedings of the 4th International Conference on Parallel and Distributed Computing, Applications and Technologies, pp. 725–729 (2003)
42. Vucetic, S., Tomsovic, K., Obradovic, Z.: Discovering price-load relationships in California's electricity market. IEEE Transactions on Power Systems, 280–286 (2001)
43. Duarte, F., Rodrigues, F., Figueiredo, V., Vale, Z., Cordeiro, M.: Data Mining Techniques Applied to Electric Energy Consumers Characterization. In: Proceedings of the Seventh IASTED International Conference on Artificial Intelligence and Soft Computing, Oporto, pp. 105–110 (2003)

44. Nizar, A.H., Dong, Z.Y., Zhao, J.H.: Load profiling and data mining techniques in electricity deregulated market. In: IEEE Power Engineering Society General Meeting, vol. 7 (2006)
45. Peng, J.-T., Chien, C.F., Tseng, T.L.: Rough set theory for data mining for fault diagnosis on distribution feeder. In: IEEE Proc. of Generation, transmission and distribution, pp. 689–697 (2004)
46. Ramos, S., Vale, Z., Santana, J., Rodrigues, F.: An application to the consumer-supplier relationship supported by data mining techniques for MV customers. WSEAS Trans.Power Systems, 1350–1357 (2006)
47. Cannataro, M., Curci, W., Giglio, M.: A priority-based transmission protocol for congested networks supporting incremental computations. In: Proceedings International Conference on Information Technology: Coding and Computing, pp. 383–388 (2001)
48. Frezzi, P., Garcés, F., Haubrich, H.J.: Analysis of short-term strategic behavior in power markets. Encontro Regional Ibero-americano do CIGRÉ (2007)
49. Wirth, J.: Information quality. Top 10 Trends in Business Intelligence (2006)
50. Taft, J.: The Intelligent Power Grid (2005)
51. Cherian, S., Ambrosio, R.S.: Towards realizing the GridWise [trademark] vision: Integrating the operations and behavior of dispersed energy devices, consumers, and markets. In: IEEE PES Power Systems Conference and Exposition, pp. 1–6 (2004)
52. Alcocer Morales, Y.E., Chiesara Sanchez, A.F., Galiasso, Y.R.E.: Nuevo mercadeo de la energia para nuevos usuarios. Vision Tecnologica, 121–130 (2000)

Hybrid Multilogistic Regression by Means of Evolutionary Radial Basis Functions: Application to Precision Agriculture[*]

P.A. Gutiérrez[1,**], C. Hervás-Martínez[1], J.C. Fernández[1], and F. López-Granados[2]

[1] Department of Computer Science and Numerical Analysis, University of Cordoba, Spain
Tel.: +34 957 218 349; Fax: +34 957 218 630
pagutierrez@uco.es
[2] Institute of Sustainable Agriculture, CSIC, Córdoba, Spain

Abstract. In this paper, a previously defined hybrid multilogistic regression model is extended and applied to a precision agriculture problem. This model is based on a prediction function which is a combination of the initial covariates of the problem and the hidden neurons of an Artificial Neural Network (ANN). Several statistical and soft computing techniques have been applied for determining these models such as logistic regression, ANNs and Evolutionary Algorithms (EAs). This paper proposes the use of Radial Basis Functions (RBFs) transformations for this model. The estimation of the coefficients of the model is basically carried out in two phases. First, the number of RBFs and the radii and centers' vector are determined by means of an EA. Afterwards, the new RBF nonlinear transformations obtained for the best individual in the last generation are added to the covariate space. Finally, a maximum likelihood optimization method determines the rest of the coefficients of the multilogistic regression model. In order to determine the performance of this approach, it has been applied to a problem of discriminating cover crops in olive orchards affected by its phenological stage using their spectral signatures obtained with a high-resolution field spectroradiometer. The empirical results for this complex real agronomical problem and the corresponding Dunnet statistical test carried out show that the proposed model is very promising in terms of classification accuracy and number of wavelengths used by the classifier.

Keywords: evolutionary neural networks, radial basis functions, multilogistic regression, multiclassification, spectral signatures, cover crops.

1 Introduction

The traditional statistical approach to pattern recognition is a natural application of the Bayesian decision theory. In order to achieve a correct classification, either the

---

[*] This work has been partially subsidized by the TIN2008-06681-C06-03 project of the Spanish Inter-Ministerial Commission of Science and Technology (MICYT), FEDER funds and the P08-TIC-3745 project of the "Junta de Andalucía" (Spain).
[**] Corresponding author.

E. Corchado et al. (Eds.): HAIS 2009, LNAI 5572, pp. 244–251, 2009.

class densities or the posterior class probability distributions are required. This distinction results in two types of techniques: those that model the class densities, such as kernel methods (Gaussian and Gaussian mixture classifiers), and those that model the posterior probabilities directly, such as classification trees, linear and nonparametric Logistic Regression (LR), multilayer neural networks with sigmoidal or product units (global functions), or Radial Basis Functions (RBFs, local functions)... [1] These last classifiers simultaneously estimate the minimum-error Bayesian *a posteriori* probabilities for all classes. However, in spite of the great number of these techniques developed to solve classification problems, there is no optimum methodology or technique to solve specific problems. This point has encouraged the combination or hybridization of different types of classifiers [2]. The LR model has been widely used in statistics for many years and has recently been the object of extensive study in the machine learning community. In general the prediction function of this model is linear. Hastie, Tibshirani and Friedman [3] proposed the use of non-linear functions in the prediction function for solving complex real problems of classification. Based on that proposal, a recently multilogistic methodology [4] hybridizes a linear model and nonlinear Product-Unit Neural Network (PUNN) models for binary and multi-classification, LIPUNN. Following these ideas, we present in this paper an extension and a different aspect of the methodology by designing a multilogistic learning algorithm which combines different elements such as LR, RBF Neural Networks (RBFNNs), an Evolutionary Programming (EP) algorithm and a specific multilogistic algorithm for simplifying the models obtained. The methodology involves augmenting/replacing the input covariate vector with RBFs, which are transformations of the input variables, and then using linear models in this new space of derived covariates. The approach is named Logistic Initial and Radial Basis Function regression (LIRBF). RBFNNs, as an alternative to multilayer perceptrons, have been found to be very helpful to many engineering problems because: (1) they are universal approximators [5]; (2) they have more compact topology than other neural networks; and (3) their learning speed is fast because of their locally tuned neurons.

The basic differences of LIPUNN methodology and the LIRBF presented here are: (1) in this paper we are using RBFs hidden units (which are *local* approximators), while in [4] we used product unit basis functions (which are *global* approximators); ii) the initialization of the population procedure is different as it incorporates a standard k-means process for the initialization of the centers of the Gaussian RBFs; iii) the LR algorithms used in this paper are more advanced and perform an automatic structural simplification of the model (MLogistic [6] and SLogistic [7]); iv) a more exhaustive 10-fold cross-validation with 10 repetitions per fold has been performed.

This methodology is applied to a real problem of discriminating cover crops in olive orchards as affected by its phenological stage using a high-resolution field spectroradiometer. Olive (Olea europaea L.) is the main perennial Spanish crop where soil management in olive orchards is mainly based on intensive and tillage operations. These operations have a great relevancy in terms of increase of atmospheric CO_2, desertification, erosion and land degradation [8]. Due to these negative environmental impacts, the European Union only subsidizes those cropping systems which require the implementation of conservation techniques, which mainly consist of altering as little as possible the natural soil and protecting it with cover crops. Remotely sensed data may offer the ability to efficiently identify and map crops and cropping methods over large areas [9]. These techniques can signify lower costs, faster work and better reliability than ground visits. The accuracy of the thematic map is extremely important since this map could be used to help the administrative follow-up for making the decision on conceding or not the subsidy. To map olive trees and cover crops, it is necessary that suitable differences exist in spectral reflectance among them and bare soil and this is the main objective of this study. The paper is aimed to establish the misclassification percentage of the spectral signature discrimination problem and to validate the classification accuracy of the methodologies presented by using a 10-fold cross-validation procedure. Three models were tested: a) Evolutionary RBFNNs, ERBFNN, b) MultiLogistic using Initial and RBF covariates (without structural simplification, MLIRBF), and c) SimpleLogistic using Initial and RBF covariates (with structural simplification, SLIRBF).

The rest of the paper is structured as follows: The LIRBF model is defined in Section 2. Section 3 presents the different steps of the learning algorithm. Section 4 explains the experimental design applied for the real agronomical problem. Finally, the work is summarized and the conclusions are drawn in Section 5.

2   LIRBF Model

In classification problems, measurements $\mathbf{x}_i$, $i = 1, 2, ..., k$, are taken on a single individual (or object), and the individuals are to be classified into one of J classes on the basis of these measurements. It is assumed that J is finite, and the measurements $\mathbf{x}_i$ are random observations from these classes. Based on the training sample $D = \{(\mathbf{x}_n, \mathbf{y}_n); n = 1, ..., N\}$, where $\mathbf{x}_n = (x_{1n}, ..., x_{kn})$ is the vector of measurements taking values in $\Omega \subset R^k$, and $\mathbf{y}_n$ is the class level of the n-th individual, we wish to find a decision function $C : \Omega \rightarrow \{1, 2, ..., J\}$ for classifying the individuals. A misclassification occurs when the decision rule C assigns an individual of the training sample to a class j when it is actually coming from a class $l \neq j$. To evaluate the performance of the classifiers we define the Correct Classification Rate (CCR) by $CCR = \dfrac{1}{N} \sum_{n=1}^{N} I(C(\mathbf{x}_n) = \mathbf{y}_n)$, where $I(.)$ is the zero-one loss function. We adopt the common technique of representing the class levels using a "1-of-J" encoding vector $\mathbf{y} = \left(y^{(1)}, y^{(2)}, ..., y^{(J)} \right)$, such as $y^{(l)} = 1$ if $\mathbf{x}$ corresponds to an example belonging to class l and $y^{(l)} = 0$ otherwise. Usually it is assumed that the training data are independent and identically distributed samples from an unknown probability distribution.

Suppose that the conditional probability that $\mathbf{x}$ belongs to class l verifies: $p_l(\mathbf{x}) = p\left(y^{(l)} = 1 \middle| \mathbf{x} \right) > 0$, $l = 1, 2, ..., J$, $\mathbf{x} \in \Omega$. Under a multinomial logistic regression, the probability that $\mathbf{x}$ belongs to class l is based in the equation:

$$p_l(\mathbf{x}, \boldsymbol{\beta}_l) = \frac{\exp(f_l(\mathbf{x}, \boldsymbol{\beta}_l))}{\sum\limits_{l=1}^{J} \exp(f_l(\mathbf{x}, \boldsymbol{\beta}_l))}, \quad l = 1, ..., J, \tag{1}$$

where $\boldsymbol{\beta}_l = (\beta_{l,0}, \beta_{l,1}, ..., \beta_{l,m})$, the $x_0 = 1$ value has been added to the input covariates vector $\mathbf{x}$ and the prediction function $f_l(\mathbf{x}, \boldsymbol{\beta}_l) = \boldsymbol{\beta}_l^T \mathbf{x}$ is linear in the covariates. The vector components $\boldsymbol{\beta}_l$ are estimated from the training data set D. If we use the normalization probability axiom, we have that $f_J(\mathbf{x}, \boldsymbol{\theta}_J) = 0$ and, in this way, it is not necessary to estimate the parameters $\boldsymbol{\beta}_J$.

Our Logistic Regression model proposal is based on the combination of the standard Initial covariates and nonlinear RBF transformed covariates (LIRBF). The general expression of the model is given by the following equation:

$$f_l(\mathbf{x}, \boldsymbol{\theta}_l) = \boldsymbol{\alpha}_l^T \mathbf{x} + \boldsymbol{\beta}_l^T \mathbf{B}(\mathbf{x}, \mathbf{W}), \quad l = 1, 2, ..., J - 1, \tag{2}$$

where $\mathbf{x} = (1, x_1, ..., x_k)$ and $\mathbf{B}(\mathbf{x}, \mathbf{W}) = (B_1(\mathbf{x}, \mathbf{w_1}), ..., B_m(\mathbf{x}, \mathbf{w_m}))$, with $B_j(\mathbf{x}, \mathbf{w}_j)$ being a Gaussian RBF:

$$B_j(\mathbf{x}, \mathbf{w}_j) = \exp\left(-\frac{\|\mathbf{x} - \mathbf{c_j}\|^2}{r_j^2}\right). \tag{3}$$

Then, the prediction function of the LIRBF model is:

$$f_l(\mathbf{x}, \boldsymbol{\theta}_l) = \alpha_0^l + \sum_{i=1}^{k} \alpha_i^l x_i + \sum_{j=1}^{m} \beta_j^l \exp\left(-\frac{\|\mathbf{x} - \mathbf{c_j}\|^2}{r_j^2}\right), \quad l = 1, 2, ..., J - 1, \tag{4}$$

Let $\boldsymbol{\theta}_l = (\boldsymbol{\alpha}^l, \boldsymbol{\beta}^l, \mathbf{W})$ be, where $\boldsymbol{\alpha}^l = (\alpha_0^l, \alpha_1^l, ..., \alpha_k^l)$, $\boldsymbol{\beta}^l = \left(\beta_1^l, ..., \beta_m^l\right)$ are the coefficients of the LR model and $\mathbf{W} = (\mathbf{w}_1, \mathbf{w}_2, ..., \mathbf{w}_m)$ are the parameters of the RBFs, where $\mathbf{w}_j = (w_{j0}, w_{j1}, ..., w_{jk})$, $\mathbf{c}_j = (w_{j1}, ..., w_{jk})$ is the centre of the j-th Gaussian RBF and $r_j = w_{j0}$ is the corresponding radius. The probability that $\mathbf{x}$ belongs to class l is given by substituting $f_l(\mathbf{x}, \boldsymbol{\theta}_l)$ of Eq. 1 by the value of Eq. 4.

To perform the Maximum Likelihood (ML) estimation of $\boldsymbol{\theta} = (\boldsymbol{\theta}_1, \boldsymbol{\theta}_2, ..., \boldsymbol{\theta}_{J-1})$, one can minimize the negative log-likelihood function:

$$L(\boldsymbol{\theta}) = \frac{1}{N} \sum_{n=1}^{N} \left[-\sum_{l=1}^{J-1} y_n^{(l)} f_l(\mathbf{x}_n, \boldsymbol{\theta}_l) + \log \sum_{l=1}^{J-1} \exp f_l(\mathbf{x}_n, \boldsymbol{\theta}_l) \right]. \tag{5}$$

The classification rule coincides with the optimal Bayes' rule. In other words, an individual should be assigned to the class which has the maximum probability:

$$C(\mathbf{x}) = \hat{l}, \text{ where } \hat{l} = \arg\max_l p_l(\mathbf{x}, \boldsymbol{\theta}_l^*), \text{ for } l = 1, 2, ..., J$$

The non-linearity of the model with respect to the θ_l parameters and the indefinite character of the associated Hessian matrix do not recommend the use of gradient-based methods to maximize the log-likelihood function. Moreover, the optimal number of basis functions of the model (i.e. the RBFs in the RBFNN) is unknown. Thus, the estimation of the vector $\hat{\theta}$ is carried out by means of a combination of an Evolutionary Programming (EP) algorithm and a standard ML optimization method.

3 Learning the LIRBF Model Coefficients

The process is structured in different steps. The first step obtains an RBFNN using and EP algorithm (designing its structure and obtaining the weights), and the second obtains the LIRBF model. The EP algorithm determines the number m of RBFs in the model, and the corresponding vector W of centres and radii. Once the basis functions have been determined by the EP algorithm, we consider a transformation of the input space by adding the nonlinear transformations given by these RBFs. The model is now linear in these new variables and the initial covariates. The remaining coefficient vector α and β are calculated by the ML optimization method.

To apply evolutionary neural network techniques, we consider a population of RBFNN models. There are no connections between the nodes of a layer and none between the input and output layers either. The activation function of the j-th node in the hidden layer is given by Eq. 3. The activation function of the output node l is given by $g_l(\mathbf{x}, \boldsymbol{\beta}^l, \mathbf{W}) = \beta_0^l + \sum_{j=1}^{m} \beta_j^l B_j(\mathbf{x}, \mathbf{w}_j)$, where β_j^l is the weight of the connection between the hidden node j and the output node l. The transfer function of all output nodes is the identity function. The main objective of the algorithm is to design a RBFNN with the better structure and weights for the classification problem tackled. The search begins with an initial population of RBFNNs, to which a population-update algorithm is applied in each iteration. The algorithm shares many characteristics and properties with other previous algorithms [10, 11]. Individuals are subject to the operations of replication and mutation, but crossover is not used due to its potential disadvantages in evolving ANNs [10]. Although in this step a concrete value for the β vector is obtained, we only consider the estimated weight vector $\hat{\mathbf{W}}$, which builds the RBFs. The parameters of the EP algorithm are the following: the centres w_{ji} and the coefficients β_j are initialized in the $[-2,2]$ interval and the radii in the $[0,1]$ interval. The number of RBF nodes is $m = 3$. The population size is $1,000$ and the maximum number of generations is 50.

In a second step, we consider a transformation of the input space, by including the nonlinear RBFs obtained for the best model in the last generation by the EP algorithm, that is, $z_1 = B_1(\mathbf{x}, \hat{\mathbf{w}}_1)$,...., $z_m = B_m(\mathbf{x}, \hat{\mathbf{w}}_m)$. Then, we minimize the negative log-likelihood function for N observations (Eq. 5). Now, the Hessian matrix associated to this expression using the new variables $x_1, x_2,..., x_k, z_1, z_2,..., z_m$ is semi-definite positive and exist a global optimum. In this final step, the MLogistic and

SLogistic algorithms have been used for obtaining the parameters matrix $\boldsymbol{\theta}_j = (\alpha^l, \beta^l, \mathbf{W})$, where the coefficients $\mathbf{W}$ are given by the EP algorithm. Both algorithms are available in the WEKA machine learning workbench [12].

4 Experiments

The study was conducted in Andalusia, southern Spain, in a location named "Cortijo del Rey", in early spring and early summer. Forty spectral signatures of live cover crop, twenty of dead cover crops, ten of olive trees, and ten of bare soil were taken on spring and summer. Measurements were collected using an ASD Handheld FieldSpec Spectroradiometer. The hyperspectral range was between 400 and 900 nm. Then, these collected hyperspectral measurements were reduced and averaged to represent 25 nm-wide measurements between 575 and 725 nm, as previous experiments determined that these seven measurements are sufficient for the determination of the original spectral curves. The experimental design for the two phenological stages was conducted using a 10-fold cross-validation procedure with 10 repetitions for fold. The experiments were carried out using a software package developed in JAVA by the authors, as an extension of the JCLEC framework (http://jclec.sourceforge.net/) [13].

Table 1. Statistical results (Mean±Standard Deviation) in CCR_G for the different classifiers

	(1)	(2)	(3)	(4)	(5)
Cortijo spring	91.63±9.74	95.25±7.70	95.25±7.70	87.50±11.51	84.50±11.53
Cortijo summer	89.60±14.06	91.00±12.19	91.40±11.81	77.40±14.95	72.00±12.39
	(6)	(7)	(8)	(9)	(10)
Cortijo spring	91.00±8.90	91.25±8.79	86.00±5.11	92.13±7.88	**96.13±7.04**
Cortijo summer	80.00±14.49	79.60±14.77	86.60±15.06	91.40±10.73	**92.80±10.45**

(1) MLogistic; (2) SLogistic; (3) LMT; (4) C4.5; (5) NBTree; (6) ABoost10; (7) ABoost100; (8) ERBFNN; (9) MLIRBF; (10) SLIRBF.

We compare our LIRBF approaches (without structural simplification, MLIRBF, and with structural simplification, SLIRBF) to recent algorithms [14]: LR without characteristic selection (MLogistic); LR with characteristics selection (SLogistic); Logistic Model Trees, LMT [14]; C4.5 trees; Naïve Bayesian Trees (NBTree); boosted C4.5 trees using AdaBoost.M1 with 10 and 100 boosting iterations (ABoost(10) and ABoost(100)) and our Evolutionary RBFNN algorithm (ERBFNN). Table 1 shows the average and standard deviation of the Correct Classification Rates for the generalization set (CCR_G) of the learners generated by the previously mentioned algorithms. A descriptive analysis of the results leads to the following remark: the SLIRBF method obtains the best mean CCR_G ($CCR_G = 96.13\%$ in spring and 92.80% in summer). To ascertain if there are significant differences in mean CCR_G, and, under the hypothesis of normality of the results, we statistically compare all the classifiers. Our decisions on the comparisons will be determined by using confidence intervals for mean differences. We choose the Dunnett [15] procedure to build these intervals; it is a multiple comparison test with a control method, the SLIRBF methodology, to identify which of these algorithms are worse (Win, W or 1), similar (Draw, D or 0) or better (Lose, L or -1) than SLIRBF for each phonological stage. The results can be seen in table 2, the Dunnett test applied for a 95% confidence level. We conclude that, SLIRBF obtains significant better results than MLogistic, C4.5, NBTree, ABoost10, ABoost100, ERBFNN and MLIRBF for $\alpha = 0.05$ in spring. The differences are significant with respect to C4.5, NBTree, ABoost10, ABoost100 and ERBFNN in summer. Computational requirements for training MLIRBF models were nearly insignificant once the ERBFNN models are built. If we need to produce a very accurate thematic map ready to be used for decision-making procedures by administrations, the criteria for selecting the models should be based on the accuracy of the classifications and these more sophisticated and accurate models would be highly recommended.

Table 2. Results of the Dunnet test comparing SLIRBF(10) to MLogistic(1), SLogistic(2), LMT(3), C4.5(4), NBTree(5), ABoost10(6), ABoost100(7), ERBFNN(8) and MLIRBF(9)

	(1)	(2)	(3)	(4)	(5)	(6)	(7)	(8)	(9)	(10)	W	D	L
C. spring	1	0	0	1	1	1	1	1	1	-	7	2	0
C. summer	0	0	0	1	1	1	1	1	0	-	5	4	0

Finally, Table 3 presents the best models obtained in each phenological stage. In spring, the best model uses three RBFs centred in the $\lambda_{625}*$, $\lambda_{650}*$ and $\lambda_{725}*$ wavelengths, where the form of the spectral signatures is non linear. In summer, the discriminant functions are very simple with only two RBFs located around the $\lambda_{575}*$ and $\lambda_{600}*$ wavelengths. Therefore, we have very simple classifiers with high CCR_G values, which allow reducing the cost of future remote sensed images, because a low number of wavelengths will be necessary.

Table 3. Discrimination equations provided by the best SLIRBF models in spring and summer for the classification of the spectral signatures

Cortijo spring. Discriminant equations
$F_1 = -18.83 - 17.64(\lambda_{575}*) - 1.80(\lambda_{675}*) + 9.96(\lambda_{700}*) + 19.43$ (RBF$_1$)
$F_2 = -16.53 - 10.58(\lambda_{575}*) + 4.10((\lambda_{675}*) + 9.02((\lambda_{700}*) - 4.77(\lambda_{725}*) + 4.71(RBF_1) + 21.68(RBF_3)$
$F_3 = 7.44 - 4.83((\lambda_{575}*) + 9.02((\lambda_{700}*) + 4.71$ (RBF$_1$) $- 30.17$ (RBF$_2$)
RBFs
$RBF_1 = (\exp(-0.5(((\lambda_{725}* - 1.18)^2)^{0.5} / (1.28)^2))$
$RBF_2 = (\exp(-0.5*(((\lambda_{625}* + 0.67)^2)^{0.5} / (0.93)^2))$
$RBF_3 = (\exp(-0.5*(((\lambda_{625}* + 0.39)^2 + (\lambda_{650}* + 0.21)^2)^{0.5} / (1.13)^2))$
Cortijo summer. Discriminant equations
$F_1 = 0.68 + 1.53((\lambda_{675}*) + 1.28(\lambda_{725}*)$
$F_2 = -1.19 + 1.53*((\lambda_{675}*) + 5.57$ (RBF$_1$)
$F_3 = -1.44 + 1.53*(\lambda_{675}*) + 4.30$ (RBF$_2$)
RBFs
$RBF_1 = (\exp(-0.5(((\lambda_{600}* + 0.72)^2)^{0.5} / (0.66)^2))$
$RBF_2 = (\exp(-0.5((((\lambda_{575}* - 1.23)^2)^{0.5} / (0.87)^2))$
$F_i = \log\ odd\ p_i,\ \lambda_i* \in (-2,2)$

5 Conclusions

The end goal of this work was to assess the potential of the methodologies presented for discriminating cover crops in two different phenological stages (spring and summer). From the analysis of the results obtained, several conclusions can be drawn. The covariate selection process incorporated in the SLIRBF methodology is necessary. In both phonological states, our methodology obtains mean CCR_G values significantly better or similar than the rest of methodologies. From the observation of the best models, it can be concluded that the models presented are able to reduce substantially the covariates but with a high classification accuracy. Thus, the interpretability of the models is enhanced and the cost of future remote sensed images is reduced. The obtained results would provide information to program the suitable wavelengths of airborne hyperspectral sensors such as Compact Airborne Spectrographic Imager (CASI) for administrative follow-up of agro-environmental measures in olive orchards under conservation agriculture.

References

1. Hastie, T.J., Tibshirani, R.J.: Nonparametric regression and classification. From statistics to neural networks. Nato ASI Series 136, 70–82
2. Major, R.L., Ragsdale, C.T.: Aggregating expert predictions in a networked environment. Computers & Operations Research 28(12), 1231–1244 (2001)
3. Hastie, T.J., Tibshirani, R.J., Friedman, J.: The elements of statistical learning. In: Data mining, inference and prediction. Springer, Heidelberg (2001)
4. Hervás-Martínez, C., Martínez-Estudillo, F., Carbonero, M.: Multilogistic Regression by means of Evolutionary Product-Unit Neural Networks. Neural Networks 21, 951–961 (2008)
5. Park, J., Sandberg, I.W.: Universal approximation using radial basis function networks. Neural Computation 3(2), 246–257 (1991)
6. le Cessie, S., van Houwelingen, J.: Ridge estimators in logistic regression. Applied Statistics 41(1), 191–201 (1992)
7. Friedman, J., Hastie, T., Tibshirani, R.: Additive logistic regression; a statistical view of boosting. The Annals of Statistics 38(2), 337–374 (2000)
8. Schlesinger, W.H.: Carbon sequestration in soils: some cautions amidst optimism. Agric. Ecosyst. Environ. 82, 121–127 (2000)
9. South, S., Qi, J., Lusch, D.P.: Optimal classification methods for mapping agricultural tillage practices. Remote Sensing of Environment 91, 90–97 (2004)
10. Angeline, P.J., Sauders, G.M., Pollack, J.B.: An evolutionary algorithm that constructs recurren neural networks. IEEE Transactions on Neural Nets 5, 54–65 (1994)
11. Martínez, F.J., Hervás-Martínez, C., Gutiérrez, P.A., Martínez-Estudillo, A.C.: Evolutionary product-unit neural networks classifiers. Neurocomputing 72(1-2), 548–561 (2008)
12. Witten, I.H., Frank, E.: Data Mining: Practical Machine Learning Tools and Techniques. In: Data Management Systems, 2nd edn. Morgan Kaufmann, Elsevier (2005)
13. Ventura, S., Romero, C., Zafra, A., Delgado, J.A., Hervas, C.: JCLEC: a Java framework for evolutionary computation. Soft Computing 12(4), 381–392 (2008)
14. Landwehr, N., Hall, M., Frank, E.: Logistic model trees. Machine Learning 59(1-2), 161–205 (2005)
15. Hochberg, Y., Tamhane, A.: Multiple Comparison Procedure. J. Wiley and Sons, Chichester (1987)

Economic Load Dispatch Using a Chemotactic Differential Evolution Algorithm

Arijit Biswas[1], Sambarta Dasgupta[1], Bijaya K. Panigrahi[2], V. Ravikumar Pandi[2], Swagatam Das[2], Ajith Abraham[3], and Youakim Badr[3]

[1] Dept. of Electronics and Telecommunication Engg,
Jadavpur University, Kolkata, India
[2] Department of Electrical Engineering, IIT, Delhi, India
bkpanigrahi@ee.iitd.ac.in, swagatamdas19@yahoo.co.in
[3] National Institute of Applied Sciences of Lyon, INSA-Lyon, Villeurbanne, France
ajith.abraham@ieee.org, youakim.badr@insa-lyon.fr

Abstract. This paper presents a novel stochastic optimization approach to solve constrained economic load dispatch (ELD) problem using Hybrid Bacterial Foraging-Differential Evolution optimization algorithm. In this hybrid approach computational chemotaxis of BFOA, which may also be viewed as a stochastic gradient search, has been coupled with DE type mutation and crossover of the optimization agents. The proposed methodology easily takes care of solving non-convex economic load dispatch problems along with different constraints like transmission losses, dynamic operation constraints (ramp rate limits) and prohibited operating zones. Simulations were performed over various standard test systems with different number of generating units and comparisons are performed with other existing relevant approaches. The findings affirmed the robustness and proficiency of the proposed methodology over other existing techniques.

1 Introduction

Economic load dispatch (ELD) problem [1,2] is a constrained optimization problem in power systems that have the objective of dividing the total power demand among the online participating generators economically while satisfying the various constraints. Over the years, many efforts have been made to solve the problem, incorporating different kinds of constraints or multiple objectives, through various mathematical programming and optimization techniques. The conventional methods include Lambda iteration method [3, 4], base point and participation factors method [3, 4], gradient method [3, 5], etc. Among these methods, lambda iteration is most common one and, owing to its ease of implementation, has been applied through various software packages to solve ELD problems. But for effective implementation of this method, the formulation needs to be continuous. The basic ELD considers the power balance constraint apart from the generating capacity limits. However, a practical ELD must take ramp rate limits, prohibited operating zones, valve point loading effects, and multi fuel options [6] into consideration to provide the completeness for the ELD problem formulation. The resulting ELD is a non-convex optimization problem, which is a challenging one and cannot be solved by the

E. Corchado et al. (Eds.): HAIS 2009, LNAI 5572, pp. 252–260, 2009.
© Springer-Verlag Berlin Heidelberg 2009

traditional methods. An ELD problem with valve point loading has also been solved by dynamic programming (DP) [7, 8]. Though promising results are obtained in small sized power systems while solving it with DP, it unnecessarily raises the length of solution procedure resulting in its vulnerability to solve large size ELD problems in stipulated time frames.

Moreover, evolutionary and behavioral random search algorithms such as Genetic Algorithm (GA) [9 – 11], Particle Swarm Optimization (PSO) [12, 13] etc. have previously been implemented on the ELD problem at hand. In addition, an integrated parallel GA incorporating ideas form simulated annealing (SA) and Tabu search (TS) techniques was also proposed in [14] utilizing generator's output power as the encoded parameter. Yalcinoz has used a real-coded representation technique along with arithmetic genetic operators and elitistic selection to yield a quality solution [15]. GA has been deployed to solve ELD with various modifications over the years. In a similar attempt, a unit independent encoding scheme has also been proposed based on equal incremental cost criterion [16]. In spite of its successful implementation, GA does posses some weaknesses leading to longer computation time and less guaranteed convergence, particularly in case of epistatic objective function containing highly correlated parameters [17, 18].

This paper proposes a new optimization approach, to solve the ELD using a hybrid Bacterial Foraging (BF) [19] –Differential Evolution (DE) [20, 21] algorithm, which is a recently emerged stochastic optimization technique. Passino proposed the Bacterial Foraging optimization technique, where the social foraging behavior of *Escherichia coli* (those living in our intestines) has been studied thoroughly. On the other hand DE is a simple Genetic Algorithm (GA) [22], which implements a differential mutation operator that distinguishes it from traditional GA. In this work the chemotaxis step of bacterial foraging is made adaptive and merged with the DE in order to tackle real world problems in a more elegant way.

2   Problem Description

In a power system, the unit commitment problem has various sub-problems varying from linear programming problems to complex non-linear problems. The concerned problem, i.e., Economic Load Dispatch (ELD) problem is one of the different non-linear programming sub-problems of unit commitment. The ELD problem is about minimizing the fuel cost of generating units for a specific period of operation so as to accomplish optimal generation dispatch among operating units and in return satisfying the system load demand, generator operation constraints with ramp rate limits and prohibited operating zones. The ELD problem with smooth and non-smooth cost functions is considered in this paper.

2.1   ELD Problem Formulation

The objective function corresponding to the production cost can be approximated to be a quadratic function of the active power outputs from the generating units. Symbolically, it is represented as

$$\text{Minimize } F_t^{\cos t} = \sum_{i=1}^{N_G} f_i(P_i) \tag{1}$$

where $f_i(P_i) = a_i P_i^2 + b_i P_i + c_i, \quad i = 1,2,3,...,N_G$ \hfill (2)

is the expression for cost function corresponding to i^{th} generating unit and a_i, b_i and c_i are its cost coefficients. P_i is the real power output (MW) of ith generator corresponding to time period t. N_G is the number of online generating units to be dispatched. This constrained ELD problem is subjected to a variety of constraints depending upon assumptions and practical implications. These include power balance constraints to take into account the energy balance; ramp rate limits to incorporate dynamic nature of ELD problem and prohibited operating zones. These constraints are discussed as under.

1) Power Balance Constraints or Demand Constraints:
   This constraint is based on the principle of equilibrium between total system generation ($\sum_{i=1}^{N_G} P_i$) and total system loads (P_D) and losses (P_L). That is,

$$\sum_{i=1}^{N_G} P_i = P_D + P_L \tag{3}$$

where P_L is obtained using B- coefficients, given by

$$P_L = \sum_{i=1}^{N_G} \sum_{j=1}^{N_G} P_i B_{ij} P_j \tag{4}$$

2) The Generator Constraints: The output power of each generating unit has a lower and upper bound so that it lies in between these bounds. This constraint is represented by a pair of inequality constraints as follows.

$$P_i^{min} \le P_i \le P_i^{max} \tag{5}$$

where, P_i^{min} and P_i^{max} are lower and upper bounds for power outputs of the ith generating unit.

3) The Ramp Rate Limits: One of unpractical assumption that prevailed for simplifying the problem in many of the earlier research is that the adjustments of the power output are instantaneous. However, under practical circumstances ramp rate limit restricts the operating range of all the online units for adjusting the generator operation between two operating periods. The generation may increase or decrease with corresponding upper and downward ramp rate limits. So, units are constrained due to these ramp rate limits as mentioned below.

If power generation increases, $P_i - P_i^{t-1} \le UR_i$ \hfill (6)

If power generation decreases, $P_i^{t-1} - P_i \le DR_i$ \hfill (7)

where P_i^{t-1} is the power generation of unit i at previous hour and UR_i and DR_i are the upper and lower ramp rate limits respectively. The inclusion of ramp rate limits modifies the generator operation constraints (5) as follows.

$$\max(P_i^{min}, UR_i - P_i) \le P_i \le \min(P_i^{max}, P_i^{t-1} - DR_i) \tag{8}$$

4) Prohibited Operating Zone: The generating units may have certain ranges where operation is restricted on the grounds of physical limitations of machine components or instability e.g. due to steam valve or vibration in shaft bearings. Consequently, discontinuities are produced in cost curves corresponding to the prohibited operating zones. So, there is a quest to avoid operation in these zones in order to economize the production. Symbolically, for a generating unit *i*,

$$P_i \leq \breve{P}^{pz} \text{ and } P_i \geq \hat{P}^{pz} \qquad (9)$$

where $\breve{P}^{pz}$ and $\hat{P}^{pz}$ are the lower and upper are limits of a given prohibited zone for generating unit *i*.

2.2 ELD Constraints Handling

The equality and inequality constraints of the ELD problem are considered in the Fitness function (*Jerror*) itself by incorporating a penalty function

$$PF_i = \begin{cases} k_i \left(U_i - U_i^{lim} \right)^2 & \text{if violated} \\ 0 & \text{otherwise} \end{cases} \qquad (10)$$

Where k_i is the constant, called penalty factor for the ith constraint. Now the final solution should not contain any penalty for the constraint violation. Therefore the objective of the problem is the minimization of generation cost and penalty function due to any constraint violation as defined by the following equation

$$J_{error} = F_t^{cost} + \sum_{i=1}^{nc} PF_i \text{ ,where "nc" is the number of constraints.} \qquad (11)$$

3 The Hybrid Algorithm

DE has reportedly outperformed powerful meta-heuristics like genetic algorithm (GA) and particle swarm optimization (PSO) [23]. Practical experiences suggest that DE may occasionally stop proceeding towards the global optima, while the population has not converged to a local optima or any other point. Occasionally even new individuals may enter the population but the algorithm does not progress by finding any better solutions. This situation is usually referred to as *stagnation* [24]. In the present work, we have incorporated an adaptive chemotactic step borrowed from the realm of BFOA into DE. The computational chemotaxis in BFOA serves as a stochastic gradient descent based local search .It was seen to greatly improvise the convergence characteristics of the classical DE. The resulting hybrid algorithm is referred here as the CDE (Chemotactic Differential Evolution).

The CDE (Chemotactic DE) Algorithm:

Initialize parameters S, N_C, N_S, $C(i)(i=1,2...N)$, F, CR.
where,
S: The number of bacteria in the population,
D:Dimension,
N_C: No. of chemotactic steps,
$C (i)$: the size of the step taken in the random direction specified by the tumble.
F: Scale factor for DE type mutation
CR: Crossover Rate.

Set $j = 0, t = 0$;

Chemotaxis loop: $j = j + 1$;

Differential evolution mutation loop: $t = t + 1$;

$\theta(i, j, t)$ denotes the position of the ith bacterium in the jth chemotactic and t th differential evolution loop.

for $i = 1, 2, ..., S$, a chemotactic step is taken for i-th bacterium.

  (a)Chemotaxis loop:
      (i) Value of the objective function $J(i, j, t)$ is computed, where $J(i, j, t)$ symbolizes
          value of objective function at j th chemotaxis cycle for i- th bacterium at t-th DE
          mutation step;
      (ii) $J_{last} = J(i, j, t)$ we store this value of objective function for
          comparison with values of objective function yet to be obtained in future.
      (iii) **Tumble:** generate a random vector $\Delta(i) \in \Re^D$ with each element
          $\Delta_m(i), m = 1, 2, ..., D$ is a random number on [-1, 1].

      (iv) **Move:** $\theta(i, j+1, t) = \omega.\theta(i, j, t) + C(i).(\Delta(i) / \sqrt{\Delta(i).\Delta^T(i)})$;
          Where, ω = inertia factor which is generally equals to 1 but becomes 0.8
                 if the function has an optimal value close to 0.
          $$C(i) = \text{step size for k th bacterium} = 0.1. \frac{J(i, j, t))}{(J(i, j, t) + 1000)}$$
              Step size is made an increasing function of objective function value to
          have a feedback arrangement.
      (v) $J(i, j, t)$ is computed.
      (vi)   **Swim:** We consider here only i-th bacterium is moving and others are not moving.
Now Let $m = 0$;

while $m < N_s$ (no of steps less than max limit).

Let $m = m + 1$;

If $J(i, j, t) < J_{last}$ (if going better)
          $J_{last} = J(i, j, t)$;

  And let, $\theta(i, j+1, t) = \omega.\theta(i, j, t) + C(i).(\Delta(i) / \sqrt{\Delta(i).\Delta^T(i)})$

  Else, $m = N_s$ (end of while loop);

for $i = 1, 2, ..., S$, a differential evolution mutation step is taken for i-th bacterium.

(b) Differential Evolution Mutation Loop:

(i) For each $\theta(i, j+1, t)$ trial solution vector we choose randomly three other distinct

vectors from the current population namely $\theta(l), \theta(m), \theta(n)$ such that

$i \neq l \neq m \neq n$

(ii) $V(i, j+1, t) = \theta(l) + F.(\theta(m) - \theta(n))$;

Where, $V(i, j+1, t)$ is the donor vector corresponding to $\theta(i, j+1, t)$.

(iii) Then the donor and the target vector interchange components probabilistically to
yield a trial vector $U(i, j+1, t)$ following:

$U_p(i, j+1, t) = V_p(i, j+1, t)$ If ($rand_p(0,1) \leq CR$) or $(p = rn(i))$

$\theta_p(i, j+1, t)$ If ($rand_p(0,1) > CR$) or $(p \neq rn(i))$ for p-th

dimension.

where $rand_p(0, 1) \in [0,1]$ is the p-th evaluation of a uniform random number   generator.

$rn(i) \in \{1,2,....,D\}$ is a randomly chosen index which ensures that $U(i, j+1, t)$ gets at

least one component from $V(i, j+1, t)$.

(iv) $J(i, j+1, t)$ is computed for trial vector;

(v) If $J(U(i, j+1, t)) < J(\theta(i, j+1, t))$, $\theta(i, j+1, t+1) = U(i, j+1, t)$;

Original vector is replaced by offspring if value of objective function for it is
smaller.

If $j < N_c$,start another chemotaxis loop.

4 Experiment Results and Discussions

4.1 ELD with Smooth and Non Smooth Cost Function Considering Ramp Rate Limits and Prohibited Operating Zones

The applicability and viability of the aforementioned technique for practical applications has been tested on four different power system cases. The obtained results are compared with the reported results of GA, PSO [12], CPSO [25], PSO-LRS, NPSO and NPSO-LRS [26] and Chaotic Differential Evolution [27, 28] methods. The cases taken for our study comprises of 6, 13, 15 and 40 generator systems. Following subsections deal with the detailed discussion of the obtained results.

4.2 Six-Unit System

The system contains six thermal generating units. The total load demand on the system is 1263 MW. The results are compared with the elitist GA [12], PSO [12], NPSO-LRS [26] and CPSO [25] methods for this test system. Parameters of all the thermal units are reported in [12]. Results obtained using the proposed hybridized Bacterial Foraging algorithm is listed in table 1. It can be evidently seen from table 1 that the technique provided better results compared to other reported evolutionary algorithm techniques. It is also observed that the minimum cost using the proposed approach is less than the reported minimum cost using some of other methods. The standard deviation of the cost is 0.0147 $.

4.3 Thirteen-Unit System

This test system consists of 13 generating units with valve point loading as mentioned in [29]. The expected load demand for this system is 1800MW. Since this is larger system with more nonlinearities, it has more local minima and difficult to obtain the global solution. The best result obtained is reported in Table 2 and compared with other recently reported results. Another reported result for minimum cost in [29] is $ 17994.07.

Table 1. Result for a six-unit system for demand of 1263 MW

Generator Power Output (MW)	BF_DE Hybrid	PSO[12]	GA[12]	NPSO-LRS [34]	CPSO1[33]
P_{G1}	446.7146	447.4970	474.8066	446.96	434.4236
P_{G2}	173.1485	173.3221	178.6363	173.3944	173.4385
P_{G3}	262.7945	263.4745	262.2089	262.3436	274.2247
P_{G4}	143.4884	139.0594	134.2826	139.5120	128.0183
P_{G5}	163.9163	165.4761	151.9039	164.7089	179.7042
P_{G6}	85.3553	87.1280	74.1812	89.0162	85.9082
Total Power Generation (MW)	1275.4	1276.01	1276.03	1275.94	1276.0
Minimum Cost ($/hr)	15444.1564	15450	15459	15450	15447
Ploss (MW)	12.4220	12.9584	13.0217	12.9361	12.9583
Mean Cost ($/hr)	15444.7815	15454	15469	15450.5	15449
Standard Deviation of Cost ($/hr)	0.0147	-	-	-	-

-: Not reported in the referred literature.

Table 2. Result for a 13-unit system for a demand of 1800 MW

Generator Power output (MW)	Bacterial Foraging Differential Evolution Hybrid	Chaotic Differential Evolution [36]
P_{G1}	628.3185306858442	628.3173
P_{G2}	149.59965011285834	149.2407
P_{G3}	222.753309362145	223.1685
P_{G4}	109.86655008717487	109.8540
P_{G5}	109.86327261039418	109.8657
P_{G6}	109.86654988406237	109.8666
P_{G7}	109.86337243612016	109.8211
P_{G8}	109.86654836418003	109.8664
P_{G9}	59.99957824230915	60.000
P_{G10}	39.999657552894476	40.000
P_{G11}	39.997977001623795	40.000
P_{G12}	54.99916355936233	55.000
P_{G13}	54.999507665171905	55.000
Total Power Generation (MW)	1799.99	1800.00
Minimum Cost ($/hr)	17960.3966	17963.9401
Mean Cost ($/hr)	17960.6258	17973.1339
Standard Deviation of Cost ($/hr)	0.1371	1.9735

5 Conclusions

The paper has employed the hybridized bacterial foraging-differential evolution algorithm on the constrained economic load dispatch problem. Practical generator operation is modeled using several non linear characteristics like ramp rate limits, prohibited operating zones. The proposed approach has produced results comparable or better than those generated by other evolutionary algorithms and the solutions obtained have superior solution quality and good convergence characteristics. From this limited comparative study, it can be concluded that the applied algorithm can be effectively used to solve smooth as well as non-smooth constrained ELD problems. In future, efforts will be made to incorporate more realistic constraints to the problem structure and the practical large sized problems would be attempted by the proposed methodology.

References

[1] Choudhary, B.H., Rahman, S.: A Review of recent advances in economic dispatch. IEEE Trans. on Power System 5(4), 1248–1259 (1990)

[2] Happ, H.H.: Optimal power dispatch-a comprehensive survey. IEEE Trans on Power Apparatus and Systems PAS-96, 841–854 (1971)

[3] Wood, A.J., Wollenberg, B.F.: Power Generation, Operation and Control. John Wiley & Sons, New York (1984)

[4] Chen, C.L., Wang, S.C.: Branch and bound scheduling for thermal generating units. IEEE Trans. on Energy Conversion 8(2), 184–189 (1993)

[5] Lee, K.Y., et al.: Fuel cost minimization for both real and reactive power dispatches. IEE Proc. C. Gener. Trsns. & distr. 131(3), 85–93 (1984)

[6] Lin, C.E., Viviani, G.L.: Hierarchical economic dispatch for piecewise quadratic cost functions. IEEE Trans. on Power Apparatus Systems PAS103, 1170–1175 (1984)

[7] Bakirtzis, A., Petridis, V., Kazarlis, S.: Genetic algorithm solution to the economic dispatch problem. Proc. Inst. Elect. Eng. Gen., Transm. Distrib. 141(4), 377–382 (1994)

[8] Lee, F.N., Breipohl, A.M.: Reserve constrained economic dispatch with prohibited operating zones. IEEE Trans. Power Syst. 8, 246–254 (1993)

[9] Sheble, G.B., Brittig, K.: Refined genetic algorithm-economic dispatch example. In: IEEE Paper 94 WM 199-0 PWRS, presented at the IEEE/PES, Winter Meeting (1994)

[10] Walters, D.C., Sheble, G.B.: Genetic algorithm solution of economic dispatch with valve point loading. IEEE Trans. on Power Systems 8(3), 1325–1332 (1993)

[11] Ma, A.A., El-Keib, Smith, R.E.: A genetic algorithm-based approach to economic dispatch of power systems. In: IEEE conference (1994)

[12] Gaing, Z.-L.: Particle swarm optimization to solving the economic dispatch considering the generator constraints. IEEE Trans. on Power Systems 18(3), 1187–1195 (2003)

[13] Park, J.B., Lee, K.S., Shin, J.R., Lee, K.Y.: A Particle Swarm optimization for economic dispatch with non-smooth cost functions. IEEE Transactions on Power Systems 20(1), 34–42 (2005)

[14] Chen, P.H., Chang, H.C.: Large scale economic dispatch by genetic algorithm. IEEE Trans. Power Syst. 10(4), 1919–1926 (1995)

[15] Yalcionoz, T., Altun, H., Uzam, M.: Economic dispatch solution using a genetic algorithm based on arithmetic crossover. In: Proc. IEEE Proto Power Tech. Conf., Proto, Portugal (September 2001)

[16] Fung, C.C., Chow, S.Y., Wong, K.P.: Solving the economic dispatch problem with an integrated parallel genetic algorithm. In: Proc. PowerCon Int. Conf., vol. 3, pp. 1257–1262 (2000)

[17] Fogel, D.B.: Evolutionary Computation: Toward a New Philosophy of Machine Intelligence, 2nd edn. IEEE Press, Piscataway (2000)

[18] Eberhart, R.C., Shi, Y.: Comparison between genetic algorithms and particle swarm optimization. In: Proc. IEEE Int. Conf. Evol. Comput., May 1998, pp. 611–616 (1998)

[19] Passino, K.M.: Biomimicry of bacterial foraging for distributed optimization and control. IEEE. Control Syst. Mag., 52–67 (June 2002)

[20] Price, K., Storn, R., Lampinen, J.: Differential Evolution - A Practical Approach to Global Optimization. Springer, Heidelberg (2005)

[21] Biswas, A., Dasgupta, S., Das, S., Abraham, A.: A Synergy of Differential Evolution and Bacterial Foraging Algorithm for Global Optimization. International Journal on Neural and Mass-Parallel Computing and Information Systems, Neural Network World 17(6), 607–626 (2007)

[22] Goldberg, D.E.: Genetic Algorithms in Search, Optimization and Machine Learning. Addison-Wesley, Reading (1989)

[23] Vesterstrøm, J., Thomson, R.: A Comparative Study of Differential Evolution, Particle Swarm Optimization, and Evolutionary Algorithms on Numerical Benchmark Problems. In: Proc. Sixth Congress on Evolutionary Computation (CEC 2004). IEEE Press, Los Alamitos (2004)

[24] Lampinen, J., Zelinka, I.: On Stagnation of the Differential Evolution Algorithm. In: Ošmera, P. (ed.) Proceedings of MENDEL 2000, 6th International Mendel Conference on Soft Computing, Brno, Czech Republic, June 7–9 (2000)

[25] Jiejin, C., Xiaoqian, M., Lixiang, L., Haipeng, P.: Chaotic particle swarm optimization for economic dispatch considering the generator constraints. Energy conver. and Management 48, 645–653 (2007)

[26] Immanuel Selvakumar, A., Thanushkodi, K.: A new particle swarm optimization Solution to nonconvex economic dispatch problems. IEEE Trans. on power systems 22(1), 42–51 (2007)

[27] Coelho, L.D.S., Mariani, V.C.: Combining of chaotic differential evolution and quadratic programming for economic dispatch optimization with valve point effect. IEEE Trans. on Power systems 21(2), 989–996 (2006)

[28] Coelho, L.D.S., Mariani, V.C.: Erratum Correlation to - Combining of chaotic differential evolution and quadratic programming for economic dispatch optimization with valve point effect. IEEE Trans. on Power systems 21(3), 1465 (2006)

[29] Sinha, N., Chakrabarti, R., Chattopadhyay, P.K.: Evolutionary programming techniques for economic load dispatch. IEEE Trans. Evol. Comput. 7(1), 83–94 (2003)

Cellular Automata Rule Detection Using Circular Asynchronous Evolutionary Search

Anca Gog and Camelia Chira

Babes-Bolyai University, M Kogalniceanu 1, 400084 Cluj-Napoca, Romania
{anca,cchira}@cs.ubbcluj.ro

Abstract. A circular evolutionary model is proposed to produce *Cellular Automata (CA)* rules for the computationally emergent task of density classification. The task refers to determining the initial density most present in the initial cellular state of a one-dimensional cellular automaton within a number of update steps. This is a challenging problem extensively studied due to its simplicity and potential to generate a variety of complex behaviors. The proposed circular evolutionary model aims to facilitate a good exploitation of relevant genetic material while increasing the population diversity. This goal is achieved by integrating a fitness guided population topology with an asynchronous search scheme. Both selection and recombination take place asynchronously enabling a gradual propagation of information from the fittest individuals towards the less fit members of the population. Numerical experiments emphasize a competitive performance of the circular search algorithm compared to other evolutionary models indicating the potential of the proposed model.

1   Introduction

Cellular Automata (CA) are discrete dynamical systems having the ability to generate highly complex behaviour starting from a simple initial configuration and set of update rules [1]. The large interest in CA can be explained by their potential to perform computations and their suitability for studying complex phenomena [2,3].

A key aspect in CA research refers to evolving rules for CA able to perform computational tasks for which global coordination is needed (highlighting an interesting emergent behaviour). The one-dimensional binary-state CA capable of performing computational tasks has been extensively investigated. The most widely studied problem refers to the density classification task - a prototypical distributed computational task for CAs [2,4,5]. The task refers to finding the density most present in the initial cellular state. This is not a trivial task because finding the density of the initial configuration is a global task while CA relies only on local interactions between cells with limited information and communication.

In this paper a new evolutionary approach is proposed to address the density classification problem. The proposed model called *Circular Evolutionary Algorithm (CEA)* is enabled by a new selection scheme based on which pairs of

E. Corchado et al. (Eds.): HAIS 2009, LNAI 5572, pp. 261–268, 2009.

individuals are recombined. A gradual propagation of the fittest individuals' genetic material into the population is facilitated by considering and interpreting both a time dimension and a space dimension for the algorithm. On one hand, selection and recombination take place asynchronously allowing the improvement of individuals during the process of selection and recombination within the same generation. On the other hand, the population has a geometrical structure given by the circular placement of individuals according to their fitness. The population topology allows the definition of a neighborhood notion for individuals, recombination taking place only between individuals belonging to the same neighborhood.

Numerical experiments and comparisons regarding the CA density task indicate a good performance of the proposed CEA model. CEA is able to produce high-quality strategies in a significantly larger number of runs compared to other evolutionary approaches to the density classification problem.

2   The Density Classification Task and Related Work

The case of one-dimensional CA is considered. A one-dimensional lattice of N two-state cells is used for representing such CAs. The state of each cell changes according to a function depending on the current states in the neighborhood. The neighborhood of a cell is given by the cell itself and its r neighbors on both sides of the cell, where r represents the radius of the CA. The initial configuration of cell states (0s and 1s) for the lattice evolves in discrete time steps updating cells simultaneously according to the CA rule.

The objective of the density classification task is to find a binary one-dimensional CA able to classify the density of 1s in the initial configuration. Let ρ_0 denote the fraction (the density) of 1s in the initial configuration. If $\rho_0 > 1/2$ (1 is dominant in the initial configuration) then the CA must reach a fixed-point configuration of 1s otherwise it must reach a fixed-point configuration of 0s within a certain number of time steps.

The density classification task is a challenging problem for which different adaptive algorithms have been proposed and studied in the literature. In a 1978 study of reliable computation in CA, Gacs, Kurdyumov and Levin proposed the well known GKL rule which approximately computes whether $\rho_0 > 1/2$ but gives significant classification errors when the density of 1s in the initial configuration is close to 1/2 [1]. The GKL rule represents the starting point for studying the density classification task and inspired Packard [6] to make the first attempts to use evolutionary computation for finding CA rules.

A genetic programming approach to the density task has been proposed in [7] producing a new rule with a slightly better performance compared to previous results. Ferreira [8] used gene expression programming for the same task and discovered two new rules with even better results.

The potential of genetic algorithms for computational emergence in the density classification task has been extensively investigated by Mitchell et al [4,1,3,9]. The three main strategies discovered are default, block-expanding (a refinement of the default strategy) and particle revealing increasing computational complexity [9]. These rules are among the best known CA rules for the density task being able to facilitate global synchronization and opening the prospect of using evolutionary models to automatically evolve computation for more complex systems.

Juille and Polack [5] used coevolutionary learning for the density classification task reporting the best results for this problem (in 86 out of 100 runs high-quality strategies are produced). Coevolutionary models can also lead however to non-successful search usually associated with the occurrence of Red Queen dynamics of the CA and intial configurations.

In a comparative study of evolutionary and coevolutionary search for the CA density classification problem, Pagie and Mitchell [9] indicate that an evolutionary model produces the particle strategy in only 2 out of 100 runs compared to the very high 86% efficacy obtained by coevolutionary models. The effectiveness of the model is significantly diminished when the ability of the coevolving population to exploit the CA is removed. Furthermore, Pagie and Mitchell indicate that this exploitation can lead to the evolution of CA with lower performance value [9].

The potential of evolutionary models to efficiently approach the problem of detecting CA rules to facilitate a certain global behavior is confirmed by various current research results [3,10,11]. Recent studies indicate that evolutionary algorithms are able to obtain high-quality results for other CA computational tasks (besides the density problem) such as the synchronization task [10].

3   The Circular Evolutionary Model

The proposed circular evolutionary model adopts a fitness guided mate selection strategy. This new type of selection for recombination is possible due to the topology of the population and the asynchronous application of the search operators - the key features of the proposed *Circular Evolutionary Algorithm (CEA)*.

3.1   The Space Dimension

The population size is fixed during all stages of the algorithm and is chosen to be a square number in order to allow a certain topology of the population. Let n^2 be the size of the population (where n is an even number).

Individuals are sorted according to their fitness and are distributed over $\frac{n}{2}$ concentric circles according to the following rule: the fittest individuals are placed on the smallest circle while the less fit individuals are placed on the largest circle. The number of individuals placed on circle $i(i = 0, \frac{n}{2} - 1)$ is $4 * (n - 2 * i - 1)$. This means that the individuals belonging to the concentric circles can be easily manipulated using a two-dimensional grid.

Each individual from the population gets the chance of being improved by involving it in a recombination process. The diversity will be thus increased as genetic material of both very fit individuals and less fit individuals is used in the search process. The selection scheme specifies the second parent involved in each recombination with an emphasis on the exploitation of the search space.

The individuals from the most inner circle - the circle labeled $\frac{n}{2} - 1$ (the fittest individuals of the population) are copied in the next population just as they are. This elitist choice is extremely suitable particularly for algorithms using a relative fitness that is slightly different for each generation. Copying the best individuals in the next generation means that these individuals will be tested again but using a different fitness function and they will survive only if they also have a very high quality in the new generation.

An individual belonging to circle $i(i = 0, \frac{n}{2} - 2)$ selects a mate from circle $i + 1$. This means that each individual from the population is recombined with a better individual but still in the same neighborhood. Although individuals from the most inner circle are not directly involved in recombination, they will be chosen as mates for individuals belonging to circle $\frac{n}{2} - 2$.

3.2   The Time Dimension

A strong feature of the proposed model refers to the asynchronous search scheme. Both selection and recombination take place asynchronously.

Let us denote by (p_1, p_2) the pair of individuals selected for recombination (where p_1 belongs to circle i while p_2 belongs to circle $i + 1$). Let x denote the best offspring resulted after recombination and x^* the best offspring mutated. The first parent p_1 is replaced by the best between x and x^* if the new individual has a better quality. This elitist scheme that allows only better individuals to replace the first parents is counteracted by the fact that all individuals from the population are involved in recombination.

Once all individuals from circle i have been considered for recombination, circle i may contain some improved individuals that will be involved within the same generation in other recombination processes with individuals from circle $i - 1$.

This asynchronous search scheme facilitates the propagation of good genetic material (from the fittest individuals) to the closest fit individuals, which after being improved gradually transfer it further down the circles until good genetic material collected from the entire population reaches the less fit individuals placed on the outer circles.

4   Circular Evolutionary Approach to the Density Classification Task

The proposed CEA model is engaged for evolving rules in the density classification task for one-dimensional binary CAs based on the evolutionary setup proposed by Mitchell et al [1].

A potential solution of the problem is a one-dimensional array of bits of size 2^{2r+1} (where r is the radius in the density task) and represents a rule table for the cellular automaton. The initial population of size n^2 is randomly generated.

The potential solutions are evaluated by means of a real-valued fitness function $f : X \to [0,1]$, where X denotes the search space of the problem. As already stated,$|X| = 2^s$, where $s = 2^{2r+1}$.

The fitness function represents the fraction of correct classification over 100 randomly generated initial configurations. A relative fitness is used as the set of initial configurations is generated anew for each generation of the algorithm. This way, solutions having a high fitness in one generation and which survive in the next generation will be evaluated again using another set of 100 initial configurations. Every set of 100 initial configurations is generated so that their densities are uniformly distributed over $\rho_0 \in [0,1]$.

It is important to emphasize the difference between the fitness of a rule and the performance of a rule. While the fitness is evaluated by using 100 uniformly distributed initial configurations, the performance of a rule is computed as the fraction of correct classifications over 10^4 initial configurations generated such that each cell of an initial configuration has the same probability of being 1 or 0 (unbiased distribution).

For each individual belonging to circle $i(i = 0, \frac{n}{2} - 2)$ a mate is selected from circle $i + 1$. The local selection scheme used for choosing a mate from the circle $i + 1$ is a tournament scheme with a tournament size of $2 * (n - 2 * i - 1)$, where $4 * (n - 2 * i - 1)$ represents the number of individuals that belong to the circle $i(i = 0, \frac{n}{2} - 2)$. The selection for recombination is performed asynchronously starting with the individuals belonging to circle $(\frac{n}{2} - 2)$ and continuing until mates for the individuals belonging to circle 0 are selected.

Once mating pairs are selected, a two-point crossover is used for the experiments reported in this paper. The recombination process starts with the fittest individuals from the population, thus giving them the opportunity to improve their fitness before being recombined with less fit individuals.

The offspring resulted after a recombination is mutated at exactly two randomly chosen positions. A weak mutation is considered for the experiments reported in this paper (the probability of obtaining a different value for the chosen position being equal to the probability of obtaining the very same value).

The replacement of the first parent with the best offspring obtained after recombination and mutation takes place asynchronously. The offspring will replace the first parent only if it has a better fitness.

5   Numerical Experiments

CEA is applied for the most frequently studied version of the density classification problem where the size of the one-dimensional binary-state CA is $N = 149$ and the radius is 3. This means that each cell is connected to 3 neighbors from both sides giving a neighborhood size of 7. The lattice size is chosen to be odd in order to avoid the case $\rho_0 = \frac{1}{2}$. The radius of the CA gives a rule size of $2^{2r+1} = 128$. An exhaustive evaluation of all possible rules would not be possible as there are a total of $2^{128} \simeq 10^{36}$ possible rules.

The CA is iterated until it reaches a fixed-point configuration of 1s or 0s but for no more than $M \simeq 2N$ time steps. The population size, the number of generations and the number of runs considered for the current experiments are all set to 100.

The rule evolved by most of the 100 algorithm runs is the so-called "block expanding" rule. Even if the fitness of this solution is often around 0.98, the performance of this rule is only around 0.65. This rule is based on the idea of quickly relaxing to a fixed-point configuration of 0s unless there is a sufficiently large block (7 or more) of adjacent cells with the value 1, interpreted as a good predictor of $\rho_0 > \frac{1}{2}$. The succesful evolution of a CA according to this rule is depicted in Figure 1. It is a simple rule where the emergence of global behavior from local interactions is not well emphasized. Details about this rule can be found in [4].

Fig. 1. The evolution of a CA for a 0s block-expanding rule (011200000000A57F7FFFFFDFF777FFFFF) with a performance of 0.6485 and a fitness of 0.98. On the left, the density of 1s is less than 1/2 while, on the right, the density of 1s is higher than 1/2.

A more sophisticated rule - called the particle rule [4] - where global coordination is more visible has been obtained in approximately 5% of the runs which is a considerably higher procentage of runs compared to other evolutionary approaches [9]. The particle rule has a performance around 0.76 and this is the rule with the best performance obtained so far for CA, for this particular task and parameters. The evolution of a CA using this rule has been depicted in Figure 2.

The proposed CEA is able to detect rules with a high performance early in the search process (after a relatively small number of generations). Figure 3 shows the fitness and the performance of detected rules over 100 generations.

Compared to current evolutionary approaches [9], the proposed CEA technique reports a better success rate particularly for the difficult-to-obtain particle rule. It should be emphasized that these results are obtained with a very basic CEA parameter setting (the number of generations is only 100 for CEA compared 5000 in [9] and the population size is 100 for CEA compared to 900 in [9]). On one hand, this simple CEA parameter setting is able to the reduce the complexity of the algorithm. On the other hand, it is expected that the promising results already obtained by CEA can be further improved using a more appropiate setup for the number of intial configurations, population size, number of generations and number of runs.

Fig. 2.   The   evolution   of   a   CA   for   a   particle   rule
(07100400100000C752B7D2FF2FFFFFFBF) having the performance 0.7125 and
the fitness 1. On the left, the density of 1s is less than 1/2 while, on the right, the
density of 1s is higher than 1/2.

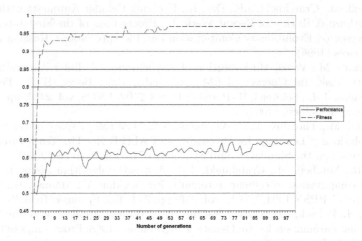

Fig. 3. The fitness and the performance values obtained by CEA over 100 generations

6   Conclusions and Future Work

An evolutionary search model (called *CEA - Circular Evolutionary Algorithm*)
relying on a space dimension as well as a time dimension is proposed. The space
dimension refers to the population topology: individuals are sorted according to
their fitness and distributed over concentric circles. The time dimension refers
to the asynchronous search process involved in proposed CEA.

The proposed CEA model is engaged in CA rule detection. The density classification problem in one dimensional binary state CA is addressed using CEA
with an appropriate representation. CEA has a better success rate in producing
the particle strategy compared to recent evolutionary models for this task. The
experimental results are highly competitive indicating the great potential of the
proposed circular evolutionary model.

Ongoing work focuses on obtaining the numerical reasults for CEA in CA rule detection using a more productive parameter setting (higher number of initial configurations, number of generations, population size and number of algorithm runs). Furthermore, experiments are carried out for the density classification task in asynchronous CA - where a single cell or a subset of cells is updated in one time step. Recent studies also indicate that the asynchronous update in CA represents a more realistic approach to modeling complex systems inspired by nature [2]. Moreover, future work refers to studying and extending the proposed CEA model for addressing other computationally emergent tasks and complex real-world problems (such as those in [10,11]).

References

1. Mitchell, M., Crutchfield, J.P., Das, R.: Evolving Cellular Automata with Genetic Algorithms: A Review of Recent Work. In: Proceedings of the First International Conference on Evolutionary Computation and Its Applications, Russian Academy of Sciences (1996)
2. Tomassini, M., Venzi, M.: Evolution of Asynchronous Cellular Automata for the Density Task. In: Guervós, J.J.M., Adamidis, P.A., Beyer, H.-G., Fernández-Villacañas, J.-L., Schwefel, H.-P. (eds.) PPSN 2002. LNCS, vol. 2439, pp. 934–943. Springer, Heidelberg (2002)
3. Mitchell, M., Thomure, M.D., Williams, N.L.: The role of space in the Success of Coevolutionary Learning. In: Proceedings of ALIFE X - The Tenth International Conference on the Simulation and Synthesis of Living Systems (2006)
4. Das, R., Mitchell, M., Crutchfield, J.P.: A genetic algorithm discovers particle-based computation in cellular automata. In: Davidor, Y., Männer, R., Schwefel, H.-P. (eds.) PPSN 1994. LNCS, vol. 866, pp. 344–353. Springer, Heidelberg (1994)
5. Juille, H., Pollack, J.B.: Coevolving the 'ideal' trainer: Application to the discovery of cellular automata rules. In: Genetic Programming 1998: Proceedings of the Third Annual Conference (1998)
6. Packard, N.H.: Adaptation toward the edge of chaos. In: Dynamic Patterns in Complex Systems, pp. 293–301. World Scientific, Singapore (1988)
7. Koza, J.R.: Genetic Programming: On the Programming of Computers by Means of Natural Selection. MIT Press, Cambridge (1992)
8. Ferreira, C.: Gene Expression Programming: A New Adaptive Algorithm for Solving Problems. Complex Systems 13(2), 87–129 (2001)
9. Pagie, L., Mitchell, M.: A comparison of evolutionary and coevolutionary search. Int. J. Comput. Intell. Appl. 2(1), 53–69 (2002)
10. Mariano, A.S., de Oliveira, G.M.B.: Evolving one-dimensional radius-2 cellular automata rules for the synchronization task. In: AUTOMATA 2008 Theory and Applications of Cellular Automata, pp. 514–526. Luniver Press (2008)
11. Zhao, Y., Billings, S.A.: Identification of the Belousov-Zhabotinsky Reaction using Cellular Automata Models. International Journal of Bifurcation and Chaos 17(5), 1687–1701 (2007)

Evolutionary Non-linear Great Deluge for University Course Timetabling

Dario Landa-Silva and Joe Henry Obit

Automated Scheduling, Optimisation and Planning Research Group
School of Computer Science, The University of Nottingham, UK
{jds,jzh}@cs.nott.ac.uk

Abstract. This paper presents a hybrid evolutionary algorithm to tackle university course timetabling problems. The proposed approach is an extension of a non-linear great deluge algorithm in which evolutionary operators are incorporated. First, we generate a population of feasible solutions using a tailored process that incorporates heuristics for graph colouring and assignment problems. That initialisation process is capable of producing feasible solutions even for the large and most constrained problem instances. Then, the population of feasible timetables is subject to a steady-state evolutionary process that combines mutation and stochastic local search. We conduct experiments to evaluate the performance of the proposed hybrid algorithm and in particular, the contribution of the evolutionary operators. Our results show that the hybrid between non-linear great deluge and evolutionary operators produces very good results on the instances of the university course timetabling problem tackled here.

Keywords: hybrid evolutionary algorithm, non-linear great deluge, course timetabling.

1 Introduction

Finding good quality solutions for timetabling problems is a very challenging task due to the combinatorial and highly constrained nature of these problems [10]. In recent years, several researchers have tackled the course timetabling problem, particulary the set of 11 instances proposed by Socha et al. [14]. Among the algorithms proposed there are: a MAX-MIN ant system by Socha et al. [14]; a tabu search hyper-heuristic strategy by Burke et al. [7]; an evolutionary algorithm, ant colony optimisation, iterated local search, simulated annealing and tabu search by Rossi-Doria et al. [13]; fuzzy multiple heuristic ordering by Asmuni et al. [5]; variable neighbourhood search by Abdullah et al. [1]; iterative improvement with composite neighbourhoods by Abdullah et al. [2,4]; a graph-based hyper-heuristic by Burke et al. [9] and a hybrid evolutionary algorithm by Abdullah et al. [3].

This paper proposes a two-stage hybrid meta-heuristic approach to tackle course timetabling problems. The first stage constructs feasible timetables while the second stage is an improvement process that also operates within the feasible

E. Corchado et al. (Eds.): HAIS 2009, LNAI 5572, pp. 269–276, 2009.

region of the search space. The second stage is a combination of non-linear great deluge [12] with evolutionary operators to improve the quality of timetables by reducing the violation of soft constraints.

The rest of this paper is organised as follows. In Section 2, the subject problem and test instances are described. Then, Section 3 gives details of the proposed hybrid evolutionary algorithm. Results and experiments are presented and discussed in Section 4 while conclusions are given in Section 5. The key contributions of this paper are: an initialisation heuristic that generates feasible timetables everytime and a simple yet effective hybrid evolutionary algorithm that is very competitive with much more elaborate algorithms already presented in the literature.

2   University Course Timetabling

In general, university course timetabling is the process of allocating, subject to predefined constraints, a set of limited timeslots and rooms to courses, in such a way as to achieve as close as possible a set of desirable objectives. In timetabling problems, constraints are commonly divided into *hard* and *soft*. A timetable is said to be feasible if no hard constraints are violated while soft constraints may be violated but we try to minimise such violation in order to increase the quality of the timetable. In this work, we tackle the course timetabling problem defined by Socha et al. [14] where there are: n events $E = \{e_1, e_2, \ldots, e_n\}$, k timeslots $T = \{t_1, t_2, \ldots, t_k\}$, m rooms $R = \{r_1, r_2, \ldots, r_m\}$ and a set S of students. Each room has a limited capacity and a set F of features that might be required by events. Each student must attend a number of events within E. The problem is to assign the n events to the k timeslots and m rooms in such a way that all hard constraints are satisfied and the violation of soft constraints is minimised.

Hard Constraints. There are four in this problem: (1) a student cannot attend two events simultaneously, (2) only one event can be assigned per timeslot in each room, (3) the room capacity must not be exceeded at any time, (4) the room assigned to an event must have the features required by the event.

Soft Constraints. There are three in this problem: (1) students should not have exactly one event timetabled on a day; (2) students should not have to attend more that two consecutive events on a day; (3) students should not have to attend an event in the last timeslot of the day.

The benchmark data sets proposed by Socha et al. [14] are split according to their size into 5 small, 5 medium and 1 large. For the small instances, $n = 100$, $m = 5$, $|S| = 80$, $|F| = 5$. For the medium instances, $n = 400$, $m = 10$, $|S| = 200$, $|F| = 5$. For the large instances, $n = 400$, $m = 10$, $|S| = 400$, $|F| = 10$. For all instances, $k = 45$ (9 hours in each of 5 days). It should be noted that although a timetable with zero penalty exists for each of these problem instances (the data sets were generated starting from such a timetable [14]), so far no heuristic method has found the ideal timetable for the medium and large instances. Hence, these data sets are still very challenging for most heuristic search algorithms.

3   Evolutionary Non-linear Great Deluge Algorithm

3.1   The Hybrid Strategy

We now describe the overall hybrid strategy, an extension of our previous algorithm which maintains a single-solution during the search [12]. Here, we extend that algorithm to a population-based evolutionary approach by incorporating tournament selection, a mutation operator and a replacement strategy.

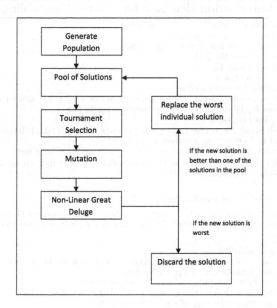

Fig. 1. Evolutionary Non-linear Great Deluge Algorithm

Figure 1 shows the components of this hybrid algorithm. It begins by generating an initial population which becomes the pool of solutions. Then, in each generation the algorithm works as follows. First, tournament selection takes place where 5 individuals are chosen at random and the one with the best fitness is selected (x^t). Then, a mutation operator is applied to x^t while maintaining feasibility obtaining solution x^m. This is followed by applying the non-linear great deluge algorithm to x^m to obtain an improved solution x^i. Then, the worst solution in the pool of solutions, x^w (ties broken at random) is identified and if x^i is better than x^w then x^i replaces x^w. This hybrid algorithm is executed for a pre-determined amount of computation time according to the size of the problem instance. Note that this is a steady-state evolutionary approach that uses non-linear great deluge for intensification and a mutation operator for diversification. The following subsections describe each of the algorithm components is more detail.

3.2   Initialisation Heuristic

The pseudo-code for the initialisation heuristic is shown in Algorithm 1. Two well-known graph colouring heuristics are incorporated. Largest degree (LD) refers to the event with the largest number of conflicting events. Saturation degree (SD) of an event refers to the number of available timeslots to timetable that event without conflicts in the current partial solution.

Algorithm 1. Initialisation Heuristic for Course Timetabling

Input: set of events in the *unscheduled events list E*
Sort *unscheduled events list* using LD
while *(unscheduled events list* is not empty*)* **do**
  Choose event E_j with the LD
  Calculate SD for event E_j
  if *(SD equals zero)* **then**
    Select a timeslot t at random
    From those events already scheduled in timeslot t (if any), move those that conflict with event E_j (if any) to the *rescheduled events list*
    Place E_j into timeslot t
    for *(each event E_i in the rescheduled events list* with SD different to zero*)* **do**
      Find a timeslot t at random to place E_i
      Recalculate SD for all events in the *rescheduled events list*
    Move all events that remain in the *rescheduled events list* (those with SD equal to zero) to the *unscheduled events list*
  else
    Find a timeslot t at random to place E_j
  if *(unscheduled events list* is not empty and $time_U$ has elapsed*)* **then**
    One by one, place events from the *unscheduled events list* into any random selected timeslot
while *(solution not feasible)* **do**
  Select move $move_1$ or move $move_2$ at random and then perform Local Search
  if *(solution not feasible and loop >10)* **then**
    Identify an event E_i that violates hard constraints
    Apply Tabu Search for ts_{max} iterations using move $move_1$ to reschedule E_i (loop is reset to zero at the end of each tabu search)
  loop++;
Output: A feasible solution (timetable)

In the first while loop, the initialisation heuristic attempts to place all events into timeslots while avoiding conflicts. In order to do that, the heuristic uses the saturation degree criterion and a list of rescheduled events to temporarily place conflicting events. The heuristic tries to do this for a given $time_U$ but once that time has elapsed, all remaining unscheduled events are placed into random timeslots. That is, if by the end of the first while loop the solution is not yet feasible, at least the penalty due to hard constraint violations is already very low. In the second while loop, the heuristic uses simple local search and tabu search to achieve feasibility. Two neighbourhood moves $move_1$ and $move_2$ (described below) are used. The local search attempts to improve the solution but it also works as a disturbing operator, hence the reason for the maximum of 10 trials before switching to tabu search. The tabu search uses the move $move_1$ only and is carried out for a fixed number of iterations ts_{max}. In our experiments, this initialisation heuristic always finds a feasible solution for all the problem instances considered.

3.3   Mutation Operator

With a probability equal to 0.5, the mutation operator is applied to the solution selected from the tournament (x^t). The mutation operator selects at random 1 out of 3 types of neighbourhood moves in order to change the solution while maintaining feasibility. These moves are described below.

- $move_1$. Selects one event at random and assigns it to a feasible timeslot and room. Note that in the tabu search step of Algorithm 1, $move_1$ selects only events that violate hard constraints.
- $move_2$. Selects two events at random and swaps their timeslots and rooms while ensuring feasibility is maintained.
- $move_3$. Selects three events at random, then it exchanges the position of the events at random and ensuring feasibility is maintained.

3.4   Non-linear Great Deluge

The standard great deluge algorithm was developed by Dueck [11]. The basic idea behind this approach is to explore neighbour solutions accepting those that are not worse than the current one. Otherwise, the candidate solution is accepted only if its penalty is not above a pre-defined water level which decreases linearly as the search progresses. The water level determines the speed of the search. The higher the decay rate the faster the water level goes down and the faster the algorithm terminates. Burke et al. [6] proposed to initialise the water level equal to the initial cost function. The decay rate at each iteration is constant and they interpreted the parameter as a function of the expected search time and the expected solution quality. To calculate the decay rate B, they first estimate the desired result (solution quality) $f(S')$ and then calculate $B = B_0 - f(S')/$Number of moves. In our previous work [12], we proposed a great deluge approach in which the decay rate of the water level is non-linear and is determined by the following expression:

$$B = B \times (\exp^{-\delta(rnd[min,max])}) + \beta \qquad (1)$$

The parameters in Eq. (1) control the speed and the shape of the water level decay rate. Therefore, the higher the values of min and max the faster the water level decreases. In return, the improvement is quickly achieved but it will suffer from this greediness by trapping itself in local optima. To counterbalance this greediness, floating the water level (relaxation) is necessary. Then, in addition to using a non-linear decay rate for the water level B, we also allow B to go up when its value is about to converge with the penalty cost of the candidate solution. We increase the water level B by a random number within the interval $[B_{min}, B_{max}]$. For all instances types, the interval used was $[2, 4]$. Full details of this strategy to control the water level decay rate in the modified great deluge can be seen in [12].

Then, the non-linear great deluge component in Figure 1 is our previous single-solution approach which produced very good results on the same instances of the course timetabling problem considered here (see [12]). In this paper we show that by incorporating some components from evolutionary algorithms, the resulting hybrid approach produces some better results than those currently reported in the literature.

4   Experiments and Results

We now evaluate the performance of the proposed hybrid algorithm. For each type of dataset, a fixed computation time ($time_{max}$) in seconds was used as the stopping condition: 2600 for small problems, 7200 for medium problems and 10000 for the large problem. This fixed computation time is for the whole process including the construction of the initial population. In the initialisation heuristic (Algorithm 1), we set $time_U$ to 0.5, 5 and 25 seconds for small, medium and large instances respectively and $ts_{max} = 500$ iterations for all problem instances. We executed the proposed hybrid algorithm 10 times for each problem instance.

Table 1 shows the average results obtained by the previous Non-linear Great Deluge, the Evolutionary Non-linear Great Deluge proposed here, and the results from other algorithms reported in the literature. For each dataset, the best results are indicated in bold. The main goal of this comparison is to assess whether extending the non-linear great deluge to a hybrid evolutionary approach helps to produce better solutions for the course timetabling problem.

We can see in Table 1 that the hybrid evolutionary algorithm described here (ENGD) clearly outperforms our previous single-solution algorithm (NGD). It

Table 1. Comparison of results obtained by the approach proposed in this paper against the best known results from the literature on the course timetabling problem. ENGD is the Evolutionary Non-Linear Great Deluge proposed here, ENGD(-m) is ENGD without Mutation, NGD is the Non-Linear Great Deluge in [12], MMAS is the MAX-MIN Ant System in [14], CFHH is the Choice Function Hyper-heuristic in [7], VNS-T is the Hybrid of VNS with Tabu Search in [2], HEA is the Hybrid Evolutionary Algorithm in [3]. In the first column, S1-S5 are small problem instances, M1-M5 are medium problem instances, while L1 is the large problem instance.

	NGD	ENGD(-m)	ENGD	Best Known
S1	3	3	**0**	0 (VNS-T)
S2	4	2	1	0 (VNS-T)
S3	6	2	**0**	0 (CFHH)
S4	6	3	**0**	0 (VNS-T)
S5	**0**	**0**	**0**	0 (MMAS)
M1	140	157	**126**	146 (CFHH)
M2	130	178	**123**	147 (HEA)
M3	189	240	**185**	246 (HEA)
M4	112	152	116	164.5 (MMAS)
M5	141	142	**129**	130 (HEA)
L1	876	995	821	**529** (HEA)

is also evident that the tailored mutation operator makes a significant contribution to the good performance of ENGD as the results obtained by ENGD(-m) are considerably worse. The proposed hybrid evolutionary approach matches the best known solution quality for almost all small problem instances except S2 and improves the best known results for most medium instances except M4. Only on the case of the large problem instance, we see that our algorithm does not match the best known result. It is also important to stress that the proposed initialisation heuristic is an important component of this hybrid algorithm because finding a feasible solution is crucial when tackling course timetabling problems.

Overall, this experimental evidence shows that by combining some key evolutionary components and an effective stochastic local search procedure, we have been able to produce a hybrid evolutionary approach that is still quite simple but much more effective (than the single-solution stochastic local search) in generating best known solutions for a well-known set of difficult course timetabling problem instances. The proposed algorithm seems particularly effective on small and medium problem instances.

5   Conclusions

Solving timetabling problems remains a challenge to many heuristic algorithms. In this paper, we tackled a well-known set of benchmark instances of the university course timetabling problem. Previous to this work, several algorithms ranging from relatively simple iterative neighbourhood search procedures [2] to more elaborate hyper-heuristic approaches [9] have been applied to this problem. We extended our previous approach, a single-solution non-linear great deluge algorithm, towards an evolutionary variant by incorporating some key operators like a population of solutions, tournament selection, a mutation operator and a steady-state replacement strategy. The results from our experiments provide evidence that our hybrid evolutionary algorithm is capable of producing best known solutions for a number of the test instances used here. The tailored mutation operator which uses 3 neighbourhood moves seems to make a substantial contribution to the good performance of the proposed algorithm. Obtaining the best timetables (with penalty equal to zero) for the medium and large instances is still a challenge. Our future work contemplates the investigation of cooperative strategies and information sharing mechanisms to tackle these university course timetabling problems.

References

1. Abdullah, S., Burke, E.K., McCollum, B.: An Investigation of Variable Neighbourhood Search for University Course Timetabling. In: Proceedings of MISTA 2005: The 2nd Multidisciplinary Conference on Scheduling: Theory and Applications, pp. 413–427 (2005)
2. Abdullah, S., Burke, E.K., McCollum, B.: Using a Randomised Iterative Improvement Algorithm with Composite Neighbourhood Structures for University Course Timetabling. In: Proceedings of MIC 2005: The 6th Meta-heuristic International Conference, Vienna, Austria, August 22-26 (2005)

3. Abdullah, S., Burke, E.K., McCollum, B.: A Hybrid Evolutionary Approach to the University Course Timetabling Problem. In: Proceedings of CEC 2007: The 2007 IEEE Congress on Evolutionary Computation, pp. 1764–1768 (2007)
4. Abdullah, S., Burke, E.K., McCollum, B.: Using a Randomised Iterative Improvement Algorithm with Composite Neighborhood Structures for University Course Timetabling. In: Metaheuristics - Progress in Complex Systems Optimization, pp. 153–172. Springer, Heidelberg (2007)
5. Asmuni, H., Burke, E.K., Garibaldi, J.: Fuzzy Multiple Heuristic Ordering for Course Timetabling. In: Proceedings of the 5th United Kingdom Workshop on Computational Intelligence (UKCI 2005), pp. 302–309 (2005)
6. Burke, E.K., Bykov, Y., Newall, J., Petrovic, S.: A Time-predefined Approach to Course Timetabling. Yugoslav Journal of Operations Research (YUJOR) 13(2), 139–151 (2003)
7. Burke, E.K., Kendall, G., Soubeiga, E.: A Tabu-search Hyperheuristic for Timetabling and Rostering. Journal of Heuristics 9, 451–470 (2003)
8. Burke, E.K., Eckersley, A., McCollum, B., Petrovic, S., Qu, R.: Hybrid Variable Neighbourhood Approaches to University Exam Timetabling. Technical Report NOTTCS-TR-2006-2, University of Nottingham, School of Computer Science (2006)
9. Burke, E.K., McCollum, B., Meisels, A., Petrovic, S., Qu, R.: A Graph Based Hyper-Heuristic for Educational Timetabling Problems. European Journal of Operational Research 176, 177–192 (2007)
10. Cooper, T., Kingston, H.: The Complexity of Timetable Construction Problems. In: Burke, E.K., Ross, P. (eds.) PATAT 1995. LNCS, vol. 1153, pp. 283–295. Springer, Heidelberg (1996)
11. Dueck, G.: New Optimization Heuristic: The Great Deluge Algorithm and the Record-to-record Travel. Journal of Computational Physics 104, 86–92 (1993)
12. Landa-Silva, D., Obit, J.H.: Great Deluge with Nonlinear Decay Rate for Solving Course Timetabling Problems. In: Proceedings of the 2008 IEEE Conference on Intelligent Systems (IS 2008), pp. 8.11–8.18. IEEE Press, Los Alamitos (2008)
13. Rossi-Doria, O., Sampels, M., Birattari, M., Chiarandini, M., Dorigo, M., Gambardella, L., Knowles, J., Manfrin, M., Mastrolilli, M., Paechter, B., Paquete, L., Stuetzle, T.: A Comparion of the Performance of Different Metaheuristics on the Timetabling Problem. In: Burke, E.K., De Causmaecker, P. (eds.) PATAT 2002. LNCS, vol. 2740, pp. 333–352. Springer, Heidelberg (2003)
14. Socha, K., Knowles, J., Sampels, M.: A Max-min Ant System for the University Course Timetabling Problem. In: Dorigo, M., Di Caro, G.A., Sampels, M. (eds.) Ant Algorithms 2002. LNCS, vol. 2463, pp. 1–13. Springer, Heidelberg (2002)

Co-operative Co-evolutionary Approach to Multi-objective Optimization

Rafał Dreżewski and Krystian Obrocki

Department of Computer Science
AGH University of Science and Technology, Kraków, Poland
drezew@agh.edu.pl

Abstract. Co-evolutionary algorithms are evolutionary algorithms in which the given individual's fitness value estimation is made on the basis of interactions of this individual with other individuals present in the population. In this paper agent-based versions of co-operative co-evolutionary algorithms are presented and evaluated with the use of standard multi-objective test functions. The results of experiments are used to compare proposed agent-based co-evolutionary algorithms with state-of-the-art multi-objective evolutionary algorithms: SPEA2 and NSGA-II.

1 Introduction

Co-evolutionary algorithms [9] are particular branch of the evolutionary algorithms—robust and effective techniques for finding approximate solutions of global and multimodal optimization problems. Co-evolutionary algorithms allow for solving problems for which it is impossible to formulate explicit fitness function because of their specific property—the fitness of the given individual is estimated on the basis of its interactions with other individuals existing in the population. The form of these interactions—co-operative or competitive—serves as the basic way of classifying co-evolutionary algorithms. Co-evolutionary interactions also promote the population diversity and introduce "arms races" among species (in the case of competitive interactions).

Many real-life decision making and optimization problems are multi-objective in nature. Usually we have to deal with many criteria, and making better the value of one of them usually means that other criteria values are worsening. There are quite many techniques of solving multi-objective problems. One of them is Pareto approach, in which we are interested in the whole set of so called "Pareto optimal" solutions (formal definition of multi-objective optimization problems, Pareto optimality, domination relation, and other basic notions, may be found for example in [2]). Evolutionary algorithms are techniques, which were recently applied with great success to solving multi-objective optimization problems—especially with the use of Pareto approach [2].

One of the problems which may occur during solving multi-objective problems with the use of evolutionary algorithms is the loss of population diversity. It is quite harmful in this case because the Pareto frontier would not be located properly—the algorithm (the population) would only locate selected parts of the frontier and in the case of multimodal multi-objective problems (when many local Pareto frontiers exist [2]) there exists

E. Corchado et al. (Eds.): HAIS 2009, LNAI 5572, pp. 277–284, 2009.

the risk of locating local Pareto frontier instead of a global one. Co-evolution is one of the mechanisms that can be used in order to reduce the negative impact of the loss of population diversity.

The idea of integration of the multi-objective evolutionary algorithm and the co-operative co-evolutionary algorithm was proposed for the first time in [8]. The algorithm was verified with the use of standard multi-objective test problems. The experiments showed that the application of co-operative co-evolution leads to better results when compared to "classical" evolutionary approaches. Because of the principles of functioning of the co-operative co-evolutionary algorithm—multiple populations, which interact only during the fitness estimation—it is quite easy to implement its distributed version. First such attempt was made in distributed co-operative co-evolutionary algorithm (DCCEA) [10].

Agent-based evolutionary algorithms are a result of merging evolutionary computations and multi-agent systems paradigms. In fact two approaches to constructing agent-based evolutionary algorithms are possible. In the first one the multi-agent layer of the system serves as a "manager" for decentralized evolutionary computations. In the second approach individuals are agents, which "live" within the environment, evolve, compete for resources, and make independently all decisions (for example see [5]). Of course, all kinds of hybrid approaches are also possible.

In the case of first approach each agent holds inside its own sub-population of individuals and evolves them. Each agent also manages the computations, in such a way that it tries to minimize the communication delays, search for computational nodes which are not overloaded and migrates to them (with the whole sub-population of individuals), etc.

The paper starts with the presentation of agent-based co-operative co-evolutionary algorithms utilizing multi-agent layer as a "manager" for evolutionary computations. In the next section these algorithms are experimentally verified and compared to two state-of-the-art multi-objective evolutionary algorithms (SPEA2 and NSGA-II) with the use of commonly used multi-objective test problems.

2   Agent-Based Co-operative Co-evolutionary System for Multi-objective Optimization

In this section the agent-based co-operative co-evolutionary system for multi-objective optimization is presented. In the described system agents are used rather as elements that manage the evolutionary computations, not as individuals that evolve themselves (see sec. 1 for the discussion of the possibilities of mixing agent-based systems and evolutionary computations). All versions of the algorithms were implemented with the use of agent-based evolutionary computations framework *jAgE* ([1])—this platform has all mechanisms and elements needed to implement agent-based evolutionary algorithms and it allows for the distributed computations. We will focus here on general system's architecture and implemented algorithms. Three versions of agent-based co-evolutionary algorithms are presented: co-operative co-evolutionary multi-agent algorithm (*CCEA-jAgE*), agent-based co-operative co-evolutionary version of NSGA-II algorithm (*CCNSGA2-jAgE*), and agent-based co-operative co-evolutionary version of SPEA2 algorithm (*CCSPEA2-jAgE*).

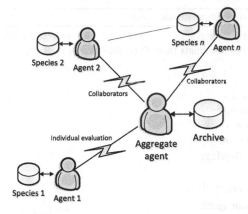

Fig. 1. The architecture of agent-based co-operative co-evolutionary algorithm

In the presented system the co-operative co-evolutionary techniques were adapted to the demands of multi-objective problems and implemented with the use of mechanisms supported by the jAgE platform.

Because in the co-operative co-evolutionary approach the representatives of each species (sub-populations) have to be aggregated (in order to form the complete solution) and also because of the necessity of storing the complete non-dominated solutions, the central computational node (agent-aggregate) was introduced (see fig. 1). Its tasks include forming complete solutions (composed of the representatives of each species) and evaluation of the solutions. It also maintains the set of non-dominated solutions found so far. Each sub-population is responsible only for the selected part of the solution, and evolved by one computational agent.

In co-operative co-evolutionary algorithm computational nodes do not have to communicate very often–communication is needed only during evaluation of the solutions–thus the parallelization of the computations can be realized effectively in the decentralized system, not only on parallel machines.

Co-operative co-evolutionary multi-agent algorithm (CCEA-jAgE) is the agent-based and distributed version of multi-objective co-operative co-evolutionary algorithm based on algorithm proposed in [8].

In the first step of this algorithm each of the computational agents performs the initialization of its sub-population (which is associated with the selected part of the problem—in our case this is one decision variable). Aggregate agent waits for receiving all of the sub-populations. When it receives all sub-populations, it forms complete solutions and computes the contribution of individuals coming from each species (sub-populations) to the whole solution quality. Then the aggregate sends back all sub-populations and puts copies of all non-dominated solutions to the set of non-dominated solutions found so far.

Following step of computational agents is presented in the alg. 1. Actions performed by the aggregate agent in the following steps are presented in alg. 2.

The process of creating complete solutions (aggregating individuals) and computing the contribution of the given individual to the quality of the whole solution is made with the use of standard co-operative co-evolutionary schema. Firstly representatives r_s of all

Algorithm 1. Step of the computational agent

1 receive P^t from aggregate agent ;   /*P^t is the sub-population in time t*/
2 compute the fitness of individuals from P^t on the basis of their contribution to the solution quality;
3 $P^{t+1} \leftarrow \emptyset$;
4 **while** P^{t+1} *is not full* **do**
5   select parents from P^t;
6   generate offspring;
7   apply recombination;
8      $P^{t+1} = P^{t+1}$ + offspring;
9 **end**
10 mutate individuals from P^{t+1};
11 send P^{t+1} to aggregate agent;

Algorithm 2. Step of the aggregate agent

1 **while** *stopping condition is not fulfilled* **do**
2   **for** $a \leftarrow a_1$ to a_n **do**
3   receive P_a^t from agent a;
4   **end**
5   **for** $a \leftarrow a_1$ to a_n **do**
6   P_a^{t+1} = select individuals from $P_a^{t-1} \bigcup P_a^t$;
7   **end**
8      $C^{t+1} \leftarrow$ complete solutions formed from P^{t+1};
9   calculate the contribution of individuals coming from different species to the whole solution quality;
10   **for** $a \leftarrow a_1$ to a_n **do**
11   send P_a^{t+1} to the agent a;
12   **end**
13   update the set of non-dominated solutions A^{t+1} with the use of C^{t+1};
14 **end**

species are chosen, and then for subsequent individuals i_s from subsequent species s the pool c_{pool} of complete solutions is created. For every solution from the pool (which is composed of the given individual i_s and representatives of all other species) the values of all criteria are computed. From the pool one solution is chosen and inserted into the set C of currently generated solutions. The vector of values $F(x)$ of the chosen solution is the measure of contribution of the given individual i_s to the quality of the solution.

As a result of integration of the previously described CCEA-jAgE algorithm and NSGA-II ([3]) the **agent-based co-operative version of NSGA-II (CCNSGA2-jAgE)** was created. CCNSGA2-jAgE is possible to obtain via the proper configuration of the CCEA-jAgE (very similar solution was in fact applied in non-dominated sorting co-operative co-evolutionary genetic algorithm [7]). Thanks to the computed contribution of the given individual to the quality of the complete solution, the fitness computation in agent-based co-evolutionary NSGA-II is realized with the use of non-dominated sorting and crowding distance metric (see [3]). Additionally, the aggregate agent joins the populations of parents and offspring, and chooses (on the basis of elitist selection and within each sub-population separately) individuals which will form the next generation sub-population used for the creation of complete solutions. The applied schema implies that N best (according to non-dominated sorting and crowding distance metric) individuals survive. Other parts of algorithm are realized in the same way as in the case of previously described agent-based co-operative algorithm.

In the case of **agent-based co-operative co-evolutionary version of SPEA2 algorithm (CCSPEA2-jAgE)** some modifications of the algorithms presented previously had to be done. It was caused mainly by the fact that SPEA2 uses additional external set of solutions during the process of evaluating individuals (compare [11]). In the described agent-based co-evolutionary version of SPEA2 algorithm each computational agent has its own, local, external set of solutions (*lA*) used during the fitness estimation. This set is also sent to the aggregate agent, along with the sub-population which is evolved by the given computational agent.

First step of aggregate agent and computational agents is the same as in the case of CCEA-jAgE. Next steps of the algorithm of computational agents begin with the receiving of sub-population P^t and local external set of solutions lA^t from the aggregate agent. On the basis of the contributions of the individuals to the quality of the complete solutions (computed by the aggregate agent), the fitness of individuals is computed. Next the archive lA^{t+1} is updated with the use of mechanisms from SPEA2 ([11]). Parents are selected from lA^{t+1} and children generated with the use of recombination operator are inserted into P^{t+1} (offspring population). Then mutation is applied to the individuals from set P^{t+1} and it is sent to aggregate agent together with the individuals from lA^{t+1}.

In the case of aggregate agent, the changes include receiving and sending additional sets of individuals lA^t. Due to the fact that lA^t is the set of parents, now the step of selecting individuals to the next generation sub-population may be omitted.

3 The Experiments

The system presented in the previous section was experimentally verified with the use of commonly used test problems: DTLZ1, DTLZ2, DTLZ3, DTLZ4, DTLZ5, and DTLZ6 [4]. The system was also applied to multi-objective portfolio optimization problem—results can be found in [6]. The main goal of the experiments was to compare three agent-based co-operative co-evolutionary algorithms with two state-of-the-art multi-objective evolutionary algorithms: NSGA-II and SPEA2.

In all five compared algorithms (CCEA-jAgE, CCNSGA2-jAgE, CCSPEA2-jAgE, NSGA-II and SPEA2) the binary representation was used (32 bits per decision variable). One point crossover and bit inversion was used as genetic operators. Probability of crossover was 0.9. The probability of mutation was $10/L$, where L is the length of the chromosome. Tournament selection with elitism was used in CCEA-jAgE, CCNSGA2-jAgE, NSGA-II algorithms and tournament selection without elitism in the case of CCSPEA2-jAgE and SPEA2. The size of the tournament was 3. The size of the population was set to 50. Maximal size of the set of non-dominated individuals was set to

50. Values presented in the figures are averages from 15 runs of each algorithm against each test problem. Due to space limitations only values of hypervolume metrics ([2]) are presented.

In the figures 2-4 values of hypervolume metric are presented for all six test problems. In the case of DTLZ1 and DTLZ3 (fig. 2a and 3a) problems the best results were obtained with the use of proposed agent-based co-evolutionary algorithms. Also results of slightly slower in this case CCSPEA2-jAgE are better than those of SPEA2 and, the worst in this case, NSGA-II. In the case of DTLZ2 problem (fig. 2b) the best results were obtained by SPEA2 and slightly worse by CCEA-jAgE and NSGA-II. Results generated by CCNSGA2-jAgE and CCSPEA2-jAgE are less satisfying in this case. In the case of problem DTLZ4 (see fig. 3b) all algorithms generated comparable results, with the exception of SPEA2. The quality of the solutions generated for DTLZ5 problem (fig. 4a) is comparable in the case of all algorithms—only in the case of CCSPEA2-jAgE the average value of hypervolume metric is slightly lower than values for other algorithms. In the case of DTLZ6 function (fig. 4b) the Pareto frontier was not properly localized only by CCSPEA2-jAgE. The solutions obtained by other algorithms are

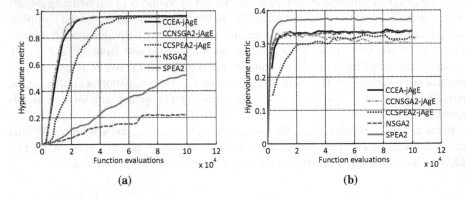

Fig. 2. Average values of hypervolume metric for DTLZ1 (a) and DTLZ2 (b) problems

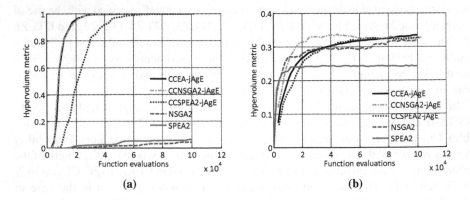

Fig. 3. Average values of hypervolume metric for DTLZ3 (a) and DTLZ4 (b) problems

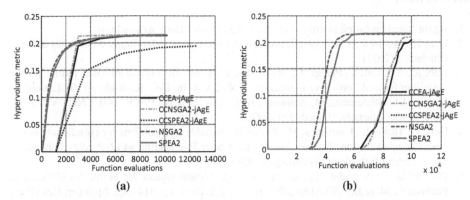

Fig. 4. Average values of hypervolume metric for DTLZ5 (a) and DTLZ6 (b) problems

of comparable quality, but NSGA-II and SPEA2 required about two times less fitness function evaluations to obtain such results.

4 Summary and Conclusions

In this paper agent-based co-operative co-evolutionary algorithm (CCSPEA2-jAgE) was proposed. In such system agents are used generally as the layer which manages the evolutionary computations. Thanks to the properties of co-operative co-evolutionary approach (interaction of individuals only at the stage of fitness evaluation of complete solutions) and properties of multi-agent approach, proposed algorithm was parallelized. The implementation was realized with the use of jAgE agent-based evolutionary framework, which allows for distributed computations. Also, within the same system, agent-based co-operative co-evolutionary versions of SPEA2 and NSGA-II algorithms were implemented.

Three proposed agent-based algorithms were experimentally verified with the use of DTLZ problems and compared to SPEA2 and NSGA-II algorithms. Presented results show that proposed CCSPEA2-jAgE obtained very satisfying results, comparable—and in the case of some problems even better—to those obtained by state-of-the-art SPEA2 and NSGA-II algorithms. Slightly less satisfying were the results obtained by proposed agent-based co-operative versions of SPEA2 and NSGA-II.

Future research will certainly include experiments with other multi-objective problems, not only with test functions but also with some real life problems. It will allow for additional verification of the proposed algorithms and will probably result in some improvements. On the other hand, different approach to agent-based realization of co-operative co-evolution will be further developed—the approach which utilizes agents as individuals living within the environment and independently forming co-operations (complete solutions).

References

1. Agent-based evolution platform (jAgE), http://age.iisg.agh.edu.pl
2. Deb, K.: Multi-Objective Optimization using Evolutionary Algorithms. John Wiley & Sons, Chichester (2001)
3. Deb, K., Agrawal, S., Pratap, A., Meyarivan, T.: A fast elitist non-dominated sorting genetic algorithm for multi-objective optimization: NSGA-II. In: Deb, K., Rudolph, G., Lutton, E., Merelo, J.J., Schoenauer, M., Schwefel, H.-P., Yao, X. (eds.) PPSN 2000. LNCS, vol. 1917, pp. 849–858. Springer, Heidelberg (2000)
4. Deb, K., Thiele, L., Laumanns, M., Zitzler, E.: Scalable test problems for evolutionary multi-objective optimization. Technical report, Computer Engineering and Networks Laboratory, Swiss Federal Institute of Technology (2001)
5. Dreżewski, R.: A model of co-evolution in multi-agent system. In: Mařík, V., Müller, J.P., Pěchouček, M. (eds.) CEEMAS 2003. LNCS, vol. 2691, pp. 314–323. Springer, Heidelberg (2003)
6. Dreżewski, R., Obrocki, K., Siwik, L.: Comparison of multi-agent co-operative co-evolutionary and evolutionary algorithms for multi-objective portfolio optimization. In: Applications of Evolutionary Computing. Springer, Heidelberg (2009)
7. Iorio, A., Li, X.: A cooperative coevolutionary multiobjective algorithm using non-dominated sorting. In: Deb, K., et al. (eds.) GECCO 2004. LNCS, vol. 3102-3103, pp. 537–548. Springer, Heidelberg (2004)
8. Keerativuttitumrong, N., Chaiyaratana, N., Varavithya, V.: Multi-objective co-operative co-evolutionary genetic algorithm. In: Guervós, J.J.M., Adamidis, P.A., Beyer, H.-G., Fernández-Villacañas, J.-L., Schwefel, H.-P. (eds.) PPSN 2002. LNCS, vol. 2439, pp. 288–297. Springer, Heidelberg (2002)
9. Paredis, J.: Coevolutionary algorithms. In: Bäck, T., Fogel, D., Michalewicz, Z. (eds.) Handbook of Evolutionary Computation, 1st supplement. IOP Publishing and Oxford University Press (1998)
10. Tan, K.C., Yang, Y.J., Lee, T.H.: A Distributed Cooperative Coevolutionary Algorithm for Multiobjective Optimization. In: Proceedings of the 2003 Congress on Evolutionary Computation (CEC 2003), pp. 2513–2520. IEEE Press, Los Alamitos (2003)
11. Zitzler, E., Laumanns, M., Thiele, L.: SPEA2: Improving the strength pareto evolutionary algorithm. Technical Report TIK-Report 103, Computer Engineering and Networks Laboratory, Swiss Federal Institute of Technology (2001)

A GA(TS) Hybrid Algorithm for Scheduling in Computational Grids

Fatos Xhafa[1], Juan A. Gonzalez[1], Keshav P. Dahal[2], and Ajith Abraham[3]

[1] Department of Languages and Informatics Systems
Technical University of Catalonia, Barcelona, Spain
fatos@lsi.upc.edu
[2] School of Informatics, University of Bradford, UK
k.p.dahal@Bradford.ac.uk
[3] Center of Excellence for Quantifiable Quality of Service
Norwegian University of Science and Technology, Norway
ajith.abraham@ieee.org

Abstract. The hybridization of heuristics methods aims at exploring the synergies among stand alone heuristics in order to achieve better results for the optimization problem under study. In this paper we present a hybridization of Genetic Algorithms (GAs) and Tabu Search (TS) for scheduling in computational grids. The purpose in this hybridization is to benefit the exploration of the solution space by a population of individuals with the exploitation of solutions through a smart search of the TS. Our GA(TS) hybrid algorithm runs the GA as the main algorithm and calls TS procedure to improve individuals of the population. We evaluated the proposed hybrid algorithm using different Grid scenarios generated by a Grid simulator. The computational results showed that the hybrid algorithm outperforms both the GA and TS for the makespan value but cannot outperform them for the flowtime of the scheduling.

1 Introduction

Meta-heuristics are the *de facto* approach to cope in practice with the computationally hard optimization problems. Meta-heuristics are in fact *hybrid* in their nature since they consist of a high level algorithm that guides the search using other particular methods. For instance, in population based meta-heuristics, such as Genetic Algorithms, the solution space is explored through a population of individuals and there are used methods for generating the initial population, computing the fitness of individuals as well genetic operators to transmit the genetic information from parents to offsprings.

Besides using meta-heuristics as *stand alone* approaches, during the last years, the attention of researchers has shifted to consider another type of high level algorithms, namely hybrid algorithms. These algorithms do not follow any concrete meta-heuristic, but rather they combine other meta-heuristics and/or other methods (e.g. exact methods) yielding thus hybrid meta-heuristics.

The *rationale* behind the hybridization resides in the "no free lunch theorem" [16] stating that "... *all algorithms that search for an extremum of a cost*

E. Corchado et al. (Eds.): HAIS 2009, LNAI 5572, pp. 285–292, 2009.

function perform exactly the same, when averaged over all possible cost functions. In particular, if algorithm A outperforms algorithm B on some cost functions, then loosely speaking there must exist exactly as many other functions where B outperforms A." Based on this theorem, existing algorithms can be used as components for designing new efficient search algorithms and expect improved performance of the newly obtained algorithm for some cost functions.

There are at least two major issues in designing hybrid meta-heuristics: (a) how to choose existing heuristic methods to combine, and (b) how to combine the chosen heuristic methods into new hybrid approaches. Unfortunately, there are no theoretical foundations for these issues. For the former, different classes of search algorithms can be considered for the purposes of hybridization, such as exact methods, simple heuristic methods and meta-heuristics. Moreover, meta-heuristics themselves are classified into local search based methods, population based methods and other classes of nature inspired meta-heuristics. Therefore, in principle, one could combine any methods from the same class or methods from different classes. Regarding the later, there are some attempts for taxonomies of hybrid meta-heuristics [8,5]; in fact, the common approach is to *try out* in smart ways, based on domain knowledge of problem at hand and characteristics of heuristics methods, different hybrid approaches and shed light on the performance of the resulting hybrid approach. Frameworks that facilitate the fast prototyping have been also provided in the meta-heuristics literature [2,4].

In this paper, we present a hybrid algorithm for the problem of scheduling independent tasks in computational grids. A computational grid is a distributed infrastructure of computational resources highly heterogenous, interconnected through heterogenous networks. One key issues in Grids is to design efficient schedulers, which will be used as part of middleware services to provide efficient planning of users' tasks to grid nodes. Recently, heuristic approaches have been presented for the problem [1,7,9,11,10,12], however, proper hybrid approaches are lacking. Our hybrid approach combines Genetic Algorithms (GAs) and Tabu Search (TS) methods. Roughly, our hybrid algorithm runs the GA as the main algorithm and calls TS procedure to improve individuals of the population. Our hybrid algorithms deals with the scheduling problem as a bi-objective optimization problem, in which makespan is considered a primary objective and flowtime a secondary one. Such optimization scheme is usually referred to as hierarchic optimization. The proposed algorithm has been experimentally evaluated and the results are contrasted against both GAs and TS for the problem.

The rest of the paper is organized as follows. In Section 2, we briefly present the scheduling of independent tasks considered as a bi-objective optimization problem in this work. In Section 3, types of hybridizations are presented. The GAs and TS for the problem as well as the hybrid approach are given in Section 4. The experimental study and some computational results are given in Section 5. We conclude in Section 6 with some remarks and indications for future work.

2   Scheduling of Independent Tasks in Computational Grids

Many applications are being developed to be run in computational grids to benefit from the large amount of computational resources in such systems. In simple Grid systems such as enteprise grids or campus grids, the user can use queuing systems such as Condor or Sun Grid Engine; even, manual selection of the appropriate machines for running the application is possible in such grids. In large scale and highly heterogenous grids, however, this tedious task is automatically handled by grid schedulers, which are expected to find planning of users' tasks and applications to most appropriate machines.

One class of grid schedulers are batch schedulers, that is, schedulers that compute a planning of a set of tasks/applications altogether to a set of grid nodes. Meta-heuristic approaches are useful for the design of such schedulers, since they usually provide quality solutions in short times.

In this work we are interested in scheduling of independent tasks to grid resources. The formal definition of the problem is based on the definition of the Expected Time to Compute (ETC) matrix in which $ETC[j][m]$ indicates an estimation of how long will it take to complete task j using resource m. Under the ETC matrix model, the independent scheduling can be defined as follows:

- A number of independent *tasks* to be allocated to grid resources. Each task has to be processed entirely in a single resource and is not preempted (once started, a task runs until completion).
- A number of *machines* candidates to participate in the allocation of tasks.
- The *workload* (in millions of instructions) of each task.
- The *computing capacity* of each machine (in *Mips*).
- The ready times, denoted $ready_m$, indicating when machine m will have finished the previously assigned tasks. At the beginning, usually ready times are considered equal to zero (all machines in the machine set are available for task allocation).
- The *ETC* matrix of size $nb\_tasks \times nb\_machines$, where $ETC[j][m]$ is the value of the expected time to compute of task j in machine m.

The quality of a schedule can be measured using several optimization criteria, such as minimizing the *makespan* (that is, the finishing time of the latest task), the *flowtime* (i.e., the sum of finalization times of all the tasks), the *completion time* of tasks in every machine (closely related to makespan), or maximizing the resource utilization. In this work we consider that the most important criterion is that of minimizing the *makespan*. Additionally, we consider the minimization of the *flowtime* of the grid system as a secondary criterion. These two criteria are formally defined as follows: *makespan*: $\min_{S_i \in Sched}\{\max_{j \in Tasks} F_j\}$ and, *flowtime*: $\min_{S_i \in Sched}\{\sum_{j \in Tasks} F_j\}$, where F_j denotes the time when task j finalizes and $Sched$ is the set of all possible schedules. Notice that by considering the makespan as the main objective to optimize and the flowtime as a secondary goal, we aim at designing a hierarchical algorithm, in which the value for makespan can not be worsened when optimizing the flowtime.

3   Hybridization of Meta-heuristics

As mentioned earlier, the hybridization started as an approach that tries to combine fully or partially two or more algorithms to enhance the performance of stand alone search method for optimization problems. To achieve such goal, the hybridization should be able to embed the best features of the combined algorithms into a new high level algorithm.

Current hybrid models take into account two main aspects: (1) Type of methods to hybridize, and (2) Level of hybridization. The first refers to the type of the methods to be hybridized. Essentially we could consider two cases: (a) meta-heuristics + meta-heuristics and (b) meta-heuristics + specific search method. In the first case the components are meta-heuristics while in the later, a meta-heuristic is combined with another type of search method, which could be an exact algorithm, dynamic programming, constraint programming or other AI techniques. In this work we are considering the first case, being the meta-heuristics the GAs and TS method.

The level of hybridization, on the other hand, refers to the degree of coupling between the meta-heuristics, the execution sequence and the control strategy.

Level of hybridization. *Loosely coupled*: in this case the hybridized meta-heuristics preserve their identity, namely, their flow is fully used in the hybridization. This case is also referred to as *high level of hybridization*. *Strongly coupled*: in this case, the hybridized meta-heuristics inter-change their inner procedures, resulting in a *low level of hybridization*.

Execution sequence. *Sequential* (the meta-heuristics flows are run sequentially) or *Parallel* (the meta-heuristics flows are run in parallel.)

Control strategy. *Coercive*: the main flow is that of one of the meta-heuristics, the other meta-heuristics flow is subordinated to the main flow. *Cooperative*: the meta-heuristics explore the solution space cooperatively (eventually, they can explore different parts of the solution space.)

4   The Proposed GA(TS) Hybrid Approach

For the design of our hybrid approach we consider two well-known meta-heuristics: Genetic Algorithms (GAs) and Tabu Search (TS). Both GAs and TS have been developed for the independent task scheduling in Xhafa et al. [11] and [12] in sequential setting. We have considered the Steady-State GA in this work. The choice of these two meta-heuristics is based on the following observations. First, grid schedulers should be very fast in order to adapt to dynamic nature of computational grids. Therefore, a fast convergence of the main algorithm is preferable in this case, which can be achieved through a good tradeoff between exploration and exploitation of the search. Second, in order to achieve high quality planning in a very short time, it is suggestive to combine the *exploration* of the solution space by a population of individuals with the *exploitation* of neighborhoods of solutions through local search. In such case, GAs and TS are among the best representatives of population based and local search methods, respectively.

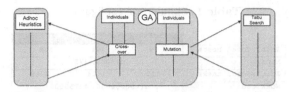

Fig. 1. The hybrid GA(TS) scheme

We are thus considering the case of hybridization of two meta-heuristics running in sequential environment. We have considered a low level hybridization and the coercive control strategy. Roughly, our hybrid algorithm runs the GA as the main algorithm and calls TS to improve individuals of the population.

The hybridization scheme is shown in Figure 1. It should be noted that in the hybridization scheme in Figure 1, instead of replacing the mutation procedure of GAs by the TS procedure, we have added a new function to the GA Population class (namely `apply_TabuSearch`) for applying the TS. This new function could be applied to any individual of the current population, however, this is computationally costly. In our case, given that we want to run the Grid scheduler in short times, the `apply_TabuSearch` is applied with small probability[1]. In fact, this parameter can well be used to tune the convergence of the GA since TS usually provides substantial improvements to individuals.

We shortly present next both the GA and TS meta-heuristics for independent task scheduling in computational grids (refer to [11] and [12] for details.)

4.1 GAs for the Scheduling Problem in Grids

GAs are a population-based approaches where individuals represent possible solutions, which are successively evaluated, selected, crossed, mutated and replaced by simulating the Darwinian evolution found in nature. We have implemented the Steady State version of GAs. In Steady State GAs, a few good individuals of population are selected and crossed. Then, the worst individuals of the population are replaced by the newly generated descendants; the rest of the individuals of the population survive and pass to the next generation. The rest of genetic operators and methods are as follows: *Initialization methods* are MCT and LJFR-SJFR implemented in [14,15]; *Selection operator*: Linear ranking; *Crossover operator*: Cycle Crossover (CX); *Mutation operator*: Mutate Rebalancing. The concrete values for the rest of parameters are given in Section 5.

4.2 Tabu Search for the Scheduling Problem in Grids

Tabu Search (TS) has shown its effectiveness in a broad range of combinatorial optimization problems and distinguishes for its flexibility in exploiting domain/problem knowledge. The main procedures used in TS are summarized next.

---

[1] This is a user input parameter. For the purposes of this work, `apply_TabuSearch` is applied roughly to 30% of individuals.

Table 1. Simulators' configuration

	Small	Medium	Large
Init./Total hosts	32	64	128
Mips		n(1000, 175)	
Init./Total tasks	512	1024	2048
Workload		n(250000000, 43750000)	
Host selection		All	
Task selection		All	
Local policy		SPTF	
Number of runs		30	

The *initial solution* is found using Min-Min method [14]. Regarding *historical memory*, both short and long term memories have been used in TS algorithm. For the *recency* memory, a matrix TL ($nb\_tasks \times nb\_machines$) is used to maintain the tabu status. In addition, a tabu hash table (TH) is maintained in order to further filter the tabu solutions. The neighborhood exploration is done using a steepest descent - mildest ascent method using two types of movements, namely, transfer (moves a task from a machine to another one, appropriately chosen) and swap (two tasks assigned to two different machines are swapped). Further, several *aspiration criteria* are used to remove the tabu status of movements. They are defined using the fitness of solutions as well as information from recency matrix. *Intensification* is implemented using elite solutions while *soft diversification* uses penalties to ETC values, task distribution and task freezing. Finally, *strong diversification* is implemented using large perturbations of solutions.

The concrete values for the rest of parameters are given in Section 5.

5   Experimental Study

We have used a Grid simulator [13] to evaluate our hybrid algorithm.

Simulation environment setting. For the evaluation of the GA(TS) hybrid algorithm, we have used three Grid scenarios: small, medium and large size. They consist, respectively, of 32 hosts/512 machines, 64 hosts/1024 machines, and 128 hosts / 2048 machines. Each scenario is generated from the simulator but the number of tasks and machines are kept constant, that is, for both of them, respectively, the number of initial tasks equals the total number of tasks in the system and and the initial number of machines equals the total number of machines. The objective of these scenarios is to capture characteristics of real grids.

The configuration of simulator follows the parameters given in Table 1. In the table n($\cdot$,$\cdot$) refers to normal distribution; SPTF stands for Shortest Processing Time First local policy. As can be seen from Table 1, the number of hosts are 32, 64 to 128 and the number of jobs are 512, 1024 and 2048, respectively. The computing capacities and task workloads are generated using normal distributions. Moreover, all hosts have been considered available for scheduling and all tasks are scheduled.

The parameter values of the GA and TS algorithms used in the hybrid algorithm are given in Tables 2 and 3.

Table 2. Parameter values of GA

Parameter	Value
evolution steps	$20 \cdot nb\_tasks$
population size	$4 \cdot (\log_2(nb\_tasks) - 1)$
intermediate pop.	$(pop\ size)/3$
cross probab.	1.0
mutation probab.	0.4

Table 3. Parameter values of TS

Parameter	Value
#iterations	$nb\_tasks \cdot nb\_mach$
max. tabu status	$1.5 \cdot nb\_mach$
#repetitions before activating intensific./diversific.	$4 * \ln(nb\_tasks) \cdot \ln(nb\_mach)$
#iterations per intensific./diversific.	$\log_2(nb\_tasks)$
#iterations for aspiration criteria	$max\_tabu/2 - \log_2(max\_tabu)$

Computational results and evaluation. The simulator is run[2] 30 times for each scenario and computational results for makespan and flowtime are averaged. Standard deviation (at 95% confidence interval) is also reported. The results for makespan and flowtime are given in Table 4 and Table 5, resp.

Table 4. Makespan values

	Small	Medium	Large
GA (hierarchic)	2808662.116	2760024.390	2764455.222
	±1,795%	±1,010%	±0,745%
TS (hierarchic)	2805531.301	2752355.018	**2748878.934**
	±1,829%	±1,056%	±0,669%
GA(TS) (hierarchic)	**2805519.428**	**2751989.166**	2812776.300
	±1,829%	±1,058%	±1,176%

Table 5. Flowtime values

	Small	Medium	Large
GA (hierarchic)	**709845463.699**	**1405493291.442**	2811723598.025
	±1,209%	±0,655%	±0,487%
TS (hierarchic)	710189541.278	1408001699.550	2812229021.221
	±1,124%	±0,616%	±0,455%
GA(TS) (hierarchic)	711183944.069	1409127007.870	**2811605453.116**
	±1,174%	±0,604%	±0,465%

As can be seen from Table 4, for makespan value the GA(TS) outperforms both GA and TS for small and medium size grid scenarios but achieves worse value for large size scenario. On the other hand, from Table 5, we can see that GA(TS) performs better than both GA and TS for flowtime value only for large size instances. So, GA(TS) performs better for makespan value, which is considered primary objective in hierarchic version, than for flowtime parameter, which is a secondary objective. In fact, close to (sub-)optimal solutions, makespan and flowtime behave as contradictory objectives and thus under our hierarchic model, the improvements of flowtime are difficult to happen.

6   Conclusions

In this paper we have presented a hybrid GA(TS) algorithm for the problem of independent scheduling in computational grids. The hybridization follows a low level approach in which GA is the main flow and TS is subordinated to it. The

---

[2] AMD Athlon 64 3200+, 2GB RAM.

objective function considered is that of bi-objective in which makespan is primary
objective and flowtime is secondary. The experimental evaluation showed that
GA(TS) outperforms both GA and TS for makespan values of small and medium
size grid scenarios and for flowtime values of large size grid scenarios.

The GA(TS) hybridization scheme is very appropriate for parallel implementation, by running TS method to all individuals of GA population in parallel.

References

1. Abraham, A., Buyya, R., Nath, B.: Nature's heuristics for scheduling jobs on computational grids. In: The 8th IEEE International Conference on Advanced Computing and Communications, India (2000)
2. Alba, E., Almeida, F., Blesa, M., Cotta, C., Díaz, M., Dorta, I., Gabarró, J., León,
   C., Luque, G., Petit, J., Rodríguez, C., Rojas, A., Xhafa, F.: Efficient parallel
   LAN/WAN algorithms for optimization. The Mallba project. Parallel Computing 32(5-6), 415–440 (2006)
3. Braun, T., Siegel, H., Beck, N., Boloni, L., Maheswaran, M., Reuther, A., Robertson, J., Theys, M., Yao, B.: A comparison of eleven static heuristics for mapping
   a class of independent tasks onto heterogeneous distributed computing systems.
   Journal of Parallel and Distributed Computing 61(6), 810–837 (2001)
4. Cahon, S., Melab, N., Talbi, E.: Building with paradisEO reusable parallel and
   distributed evolutionary algorithms. Parallel Computing 30(5-6), 677–697 (2004)
5. Jourdan, L., Basseur, M., Talbi, E.: Hybridizing Exact Method and Metaheuristics:
   A Taxonomy. European Journal of Operational Research (Online, 2008)
6. Lau, H.C., Wan, W.C., Lim, M.K., Halim, S.: A Development Framework for Rapid
   Meta-Heuristics Hybridization. In: Proc. of the 28th Annual International Computer Software and Applications Conference, pp. 362–367 (2004)
7. Ritchie, G., Levine, J.: A fast, effective local search for scheduling independent
   jobs in heterogeneous computing environments. TechRep, Centre for Intelligent
   Systems, University of Edinburgh (2003)
8. Talbi, E.: A Taxonomy of Hybrid Metaheuristics. J. of Heur. 8(5), 541–564 (2002)
9. Xhafa, F.: A Hybrid Evolutionary Heuristic for Job Scheduling in Computational
   Grids, ch. 10. Springer Series: Studies in Comp. Intell., vol. 75 (2007)
10. Xhafa, F., Barolli, L., Durresi, A.: An Experimental Study on Genetic Algorithms
    for Resource Allocation on Grid Systems. JOIN 8(4), 427–443 (2007)
11. Xhafa, F., Carretero, J., Abraham, A.: Genetic Algorithm Based Schedulers for
    Grid Computing Systems. International Journal of Innovative Computing, Infor-
    mation and Control 3(5), 1–19 (2007)
12. Xhafa, F., Carretero, J., Dorronsoro, B., Alba, E.: Tabu Search Algorithm for
    Scheduling Independent Jobs in Computational Grids. Computers and Informatics
    (to appear, 2009)
13. Xhafa, F., Carretero, J., Barolli, L., Durresi, A.: Requirements for an Event-Based
    Simulation Package for Grid Systems. JOIN 8(2), 163–178 (2007)
14. Xhafa, F., Carretero, J., Barolli, L., Durresi, A.: Immediate Mode Scheduling in
    Grid Systems. Int. J. of Web and Grid Services 3(2), 219–236 (2007)
15. Xhafa, F., Barolli, L., Durresi, A.: Batch Mode Schedulers for Grid Systems. In-
    ternational Journal of Web and Grid Services 3(1), 19–37 (2007)
16. Wolpert, D.H., Macready, W.G.: No Free Lunch Theorems for Optimization. IEEE
    Transactions on Evolutionary Computation 1(1), 67–82 (1997)

On the Model–Building Issue of Multi–Objective Estimation of Distribution Algorithms

Luis Martí, Jesús García, Antonio Berlanga, and José M. Molina

GIAA, Dept. of Informatics, Universidad Carlos III de Madrid
Av. Universidad Carlos III 22, Colmenarejo 28270 Madrid, Spain
{lmarti,jgherrer}@inf.uc3m.es, {aberlan,molina}@ia.uc3m.es
http://www.giaa.inf.uc3m.es/

Abstract. It has been claimed that perhaps a paradigm shift is necessary in order to be able to deal with this scalability issue of multi–objective optimization evolutionary algorithms. Estimation of distribution algorithms are viable candidates for such task because of their adaptation and learning abilities and simplified algorithmics. Nevertheless, the extension of EDAs to the multi–objective domain have not provided a significant improvement over MOEAs.

In this paper we analyze the possible causes of this underachievement and propose a set of measures that should be taken in order to overcome the current situation.

1 Introduction

Estimation of distribution algorithms (EDAs) [1] have been hailed as one of the cornerstones of modern evolutionary computation. Like most evolutionary algorithms [2], EDAs are population based optimization algorithms. However, in these algorithms, the step where the evolutionary operators are applied to the population, is substituted by construction of a statistical model of the most promising subset of the population. This model is then sampled to produce new individuals that are merged with the original population following a given substitution policy.

On particular area where EDAs could yield important results is the one pertaining to multi–objective optimization problems (MOPs). In this class of problems the optimizer must find one or more feasible solutions that correspond with the extreme values (either maximum or minimum) of two or more functions subject to a set of constraint. Therefore, an optimizer's solution is a set of equally good, trade–off solutions.

The application of evolutionary computation to MOPs has prompted the creation of what has been called multi–objective optimization evolutionary algorithms (MOEAs) [3]. However, those approaches tend to fail when faced with problems with a relatively large amount of objectives as they require an exponential increase of the resources made available to them (see [4,5] and [3] pp. 414–419]).

E. Corchado et al. (Eds.): HAIS 2009, LNAI 5572, pp. 293–300, 2009.

It has been claimed that perhaps a paradigm shift is necessary in order to be able to deal with this scalability issue. EDAs are viable candidates for such task because of their adaptation and learning abilities and simplified algorithmics. Nevertheless, the extension of EDAs to the multi–objective domain; which has been denominated multi–objective EDAs (MOEDAs) [6], have not provided a significant improvement over MOEAs.

In this paper we analyze the possible causes of this underachievement and propose a set of measures that should be taken in order to overcome the current situation. The rest of this paper first deals with the theoretical background of the discussed matters. We then deal with the model–building issue of MOEDAs and reflect on its impact on those algorithms shortcomings. After that we propose a set of guidelines for constructing MOEDAs capable of better handling many–objective problems.

2   Background

The concept of multi–objective optimization refers to the process of finding one or more feasible solutions of a problem that correspond with the extreme values (either maximum or minimum) of two or more functions subject to a set of constrains:

Definition 1 (Multi–Objective Optimization Problem).

$$\left.\begin{array}{l} \text{minimize } \boldsymbol{F}(\boldsymbol{x}) = \langle f_1(\boldsymbol{x}), \ldots, f_M(\boldsymbol{x}) \rangle \,, \\ \text{subject to } c_1(\boldsymbol{x}), \ldots, c_C(\boldsymbol{x}) \leq 0 \,, \\ \qquad d_1(\boldsymbol{x}), \ldots, d_D(\boldsymbol{x}) = 0 \,, \\ \text{with } \boldsymbol{x} \in \mathcal{D} \,, \end{array}\right\} \tag{1}$$

where $\mathcal{D}$ is known as the decision space. *The functions $f_1(\boldsymbol{x}), \ldots, f_M(\boldsymbol{x})$ are the* objective functions. *The image set, $\mathcal{O}$, product of the projection of $\mathcal{D}$ thru $f_1(\boldsymbol{x}), \ldots, f_M(\boldsymbol{x})$ is called* objective space *($\boldsymbol{F} : \mathcal{D} \to \mathcal{O}$). $c_1(\boldsymbol{x}), \ldots, c_C(\boldsymbol{x}) \leq 0$ and $d_1(\boldsymbol{x}), \ldots, d_D(\boldsymbol{x}) = 0$ express the restrictions imposed to the values of $\boldsymbol{x}$.*

In general terms, this type of problem does not have a unique optimal solution. Instead an algorithm solving the problem defined in (1) should produce a set containing equally good, trade–off, optimal solutions. The optimality of a set of solutions can be defined relying on the so called *Pareto dominance relation*:

Definition 2 (Pareto Dominance Relation). *For the optimization problem specified in (1) and having $\boldsymbol{x}_1, \boldsymbol{x}_2 \in \mathcal{D}$. $\boldsymbol{x}_1$ is said to dominate $\boldsymbol{x}_2$ (expressed as $\boldsymbol{x}_1 \prec \boldsymbol{x}_2$) iff $\forall f_j,\ f_j(\boldsymbol{x}_1) \leq f_j(\boldsymbol{x}_2)$ and $\exists f_i$ such that $f_i(\boldsymbol{x}_1) < f_i(\boldsymbol{x}_2)$.*

The solution of (1) is a subset of $\mathcal{D}$ that contains elements are not dominated by other elements of $\mathcal{D}$.

Definition 3 (Pareto–Optimal Set). *The solution of problem (1) is the set $\mathcal{D}^*$ such that $\mathcal{D}^* \subseteq \mathcal{D}$ and $\forall \boldsymbol{x}_1 \in \mathcal{D}^* \nexists \boldsymbol{x}_2$ that $\boldsymbol{x}_2 \prec \boldsymbol{x}_1$.*

$\mathcal{D}^*$ is known as the *Pareto–optimal set* and its image in objective space is called *Pareto–optimal front*, $\mathcal{O}^*$.

Finding the explicit formulation of $\mathcal{D}^*$ is often impossible. Generally, an algorithm solving (1) yields a discrete local Pareto–optimal set, $\mathcal{P}^*$, that approximates $\mathcal{D}^*$. The image of $\mathcal{P}^*$ in objective space, $\mathcal{PF}^*$, is known as the local Pareto–optimal front.

2.1 Evolutionary Approaches to Multi–Objective Optimization

MOPs have been addressed with a variety of methods. Among them, evolutionary algorithms (EAs) [2] have proven themselves as a valid and competent approach from theoretical and practical points of view. This has led to what has been called multi–objective optimization evolutionary algorithms (MOEAs). Their success is due to the fact that EAs do not make any assumptions about the underlying fitness landscape. Therefore, it is believed they perform consistently well across a wide range of problems.

2.2 Multi–Objective Estimation of Distribution Algorithms

Estimation of distribution algorithms (EDAs) have been claimed as a paradigm shift in the field of evolutionary computation. Like EAs, EDAs are population based optimization algorithms. However in EDAs the step where the evolutionary operators are applied to the population is substituted by construction of a statistical model of the most promising subset of the population. This model is then sampled to produce new individuals that are merged with the original population following a given substitution policy. Because of this model–building feature EDAs have also been called probabilistic–model–building genetic algorithms (PMBGAs) [7].

The introduction of machine learning techniques implies that these new algorithms lose the biological plausibility of its predecessors. In return, they gain the capacity of scalably solving many challenging problems, often significantly outperforming standard EAs and other optimization techniques.

Multi–objective optimization EDAs (MOEDAs) [6] are the extensions of EDAs to the multi–objective domain. Most of MOEDAs consist of a modification of existing EDAs whose fitness assignment strategy is substituted by a previously existing one used by MOEAs.

3 Understanding Model–Building

The cause of MOEDAs underachievement in many–objective problems can be traced back to their model–building algorithms. Not until recently, EDAs practitioners have failed to recognize that machine learning approaches can't be extrapolated as–is to the model–building task. In particular, there are properties shared by most machine–learning approaches that could be preventing MOEDAs from yielding a substantial improvement over MOEAs. They are:

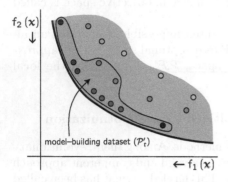
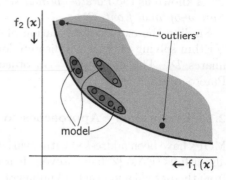

(a) Population subset used for model–building

(b) The resulting model disregards outliers

Fig. 1. The outliers issue of most of current model–building approaches

- the incorrect treatment of data outliers;
- tendency towards the loss of population diversity, and;
- excess of computational effort devoted to finding an optimal model.

These issues can be traced back to the single–objective predecessor of most MOEDAs and their corresponding model–building algorithms. Most model–building schemes used so far by EDAs use off–the–shelf machine learning methods. However, the research community on this area has failed to acknowledge that the model–building problem has particular requirements that those methods do not meet and even have conflicts with. Furthermore, when scaling up in the amount of objectives this situation is frequently aggravated by the implications of the curse of dimensionality.

In the statistical and machine learning areas the data instances that are relatively isolated or diverse from the greater masses of data are known as outliers. Historically, these outliers are handled as not representative, noisy or bogus data (see Figure 1 for a graphical explanation).

On the other hand, in model–building it is known beforehand that all the available data is valid as it represents the best part of the current population. Therefore, no points must be disregarded. Instead, these outliers are essential, as they represent unexplored or recently discovered areas of the current Pareto–optimal front. They should not only be preserved but, perhaps, even reinforced. A model–building algorithm that primes outliers might actually accelerate the search process and alleviate the rate of the exponential dimension–population size dependency.

Another drawback of most MOEDAs, and most EDAs, for that matter, is the lost of diversity lost of population diversity. This fact has been already pointed out and some proposals have been laid out for addressing it [8,9,10]. This loss of diversity could also be traced back to the nature of the model–building algorithm.

The root cause that makes current methods disregard outliers and lose population diversity can be traced to the global error–based learning that take place in those methods. In that type of learning a dataset–wise error is minimized. Because of that, infrequent or poorly represented elements are sacrificed in order to achieve a better overall error.

Similarly, for the model–building problem, there is no need for having the least complex accurate model of the data. In spite of this, most of the current approaches dedicate a sizable effort to finding the optimal model complexity.

4   Building Blocks for Scalable MOEDAs

As commented, a topic that remains not properly dealt with inside the MOEA and MOEDA scope is the scalability of the algorithms or what has been denominated as the *many–objective problems* [11]. For these algorithms their scalability issue is exposed regarding two quantities:

1. the dimension of the decision space, $\mathcal{D}$, that is, the number of variables that take part of the problem, and;
2. the dimension of the objective space, $\mathcal{O}$, or in other words, the number of objective functions to be optimized.

A critical quantity is the dimension of the objective space as it has been experimentally shown to have an exponential relation with the optimal size of the population (see [3] pp. 414–419). This fact implies that, with the increase of the number of objective functions an optimization algorithm needs an exponential amount of resources made available to it.

This question can be reduced to the problem of how to handle a relatively small population in which non–dominated and dominated individuals are not adequately represented rendering useless the Pareto–based approaches.

Therefore, in order to achieve a sizable improvement in the scalability of MOEDAs, it is essential to arm them with:

- an efficient, scalable and robust fitness assignment function;
- an scheme for objective reduction or recombination, and;
- a custom–made model–building algorithm that promotes search and pushes the population towards newly found sub–optimal zones in objective space.

4.1   Improved Fitness Assignment

The issue of finding a better fitness assignment function is a really complex one. As previously discussed, as dimensions grow, there is an exponential explosion in the amount of resources required. This growth implies that mutual comparison and/or sorting processes that take place as part of the fitness assignment become very time consuming. One possible solution is to bypass the exponential relation, but this will probably lead to a situation where there is a high degree of non–domination of individuals and therefore it is impossible to find a correct direction for the search.

There have been a number of works [12,13] that propose the use of performance indicators, in particular the hypervolume indicator [14], as fitness assignment functions. This is a promising line of research as these approaches might overcome the situations where most of the population is non–dominated and a direction of search can not be found . However, the computation of this indicator has been shown to be very computationally complex.

4.2 Objective Reduction

As it was mentioned before, there has been a number of works [15,16,17] directed towards the reduction of the number of objective functions to a minimum and therefore towards the mitigation of the complexity of a problem. Most of these approaches distinguish with of the objectives are of a non–contradictory nature with regard to the rest and, therefore can be safely removed during most of the optimization process or recombined with others.

4.3 Competent Model Builders

Improving the model–building algorithm seems to be a most promising direction for research as, to the best of our knowledge, it has not been properly addressed. So far, MOEDA approaches have mostly used off–the–shelf machine learning methods. However, the task in discussion has different characteristics than those for which those methods were originally meant for.

A previous study [18] compared the behavior of a set of model–building algorithms under equal conditions. That study found that, in high–dimensional problems and under the same experimental conditions, statistically robust algorithms, like Bayesian networks, were outperformed by less robust approaches like k–means algorithm or the randomized leader algorithm.

The cause of this behavior can be attributed to the fact that statistically rigorous methods are not meant specifically for the problem we are dealing with here. These behaviors, although justified in the original field of application of the algorithms, might hinder the performance of the process, both in the accuracy and in the resource consumption senses. Among these behaviors we can find two important ones: the disregarding of outliers and the dedication of an excessive amount of resources to finding the optimal model structure or topology.

The multi–objective optimization neural estimation of distribution algorithm (MONEDA) [19] was devised with the purpose of dealing with the above described issues of model–building and, therefore, address MOEDA scalability. MONEDA uses a modified growing neural gas (GNG) network [20] for model–building (MB–GNG). Thanks to its custom–made model–building algorithm, that solves the above listed requirements, the preservation of elite individuals and the individual replacement scheme MONEDA is capable of scalably solve continuous multi–objective optimization problems with better performance than similar algorithms. In some initial studies [19,21], MONEDA outperformed similar state–of–the–art MOEDAs and MOEAs.

5 Final Remarks and Future Trends

In this work we have dealt with the theoretical aspects of current multi–objective estimation of distribution algorithms (MOEDAs) that prompts their under-achievement when faced with high–dimensional and complex problems. We have discussed the main MOEDA property to which this problem can be attributed, that is, the use of improper model–building algorithms.

Current model builders do not take into account that the model–building problem is of a different nature with regard to "traditional" or "classical" statistical modeling. Taking this into account we have provided a set of guidelines that should be followed in order to build viable alternatives to existing approaches.

Acknowledgement

This work was supported in part by projects CICYT TIN2008–06742–C02–02/TSI, CICYT TEC2008–06732–C02–02/TEC, SINPROB, CAM MADRINET S–0505/TIC/0255 and DPS2008–07029–C02–02.

References

1. Lozano, J.A., Larrañaga, P., Inza, I., Bengoetxea, E. (eds.): Towards a New Evolutionary Computation: Advances on Estimation of Distribution Algorithms. Springer, Heidelberg (2006)
2. Bäck, T.: Evolutionary algorithms in theory and practice: Evolution Strategies, Evolutionary Programming, Genetic Algorithms. Oxford University Press, New York (1996)
3. Deb, K.: Multi-Objective Optimization using Evolutionary Algorithms. John Wiley & Sons, Chichester (2001)
4. Khare, V., Yao, X., Deb, K.: Performance Scaling of Multi-objective Evolutionary Algorithms. In: Fonseca, C.M., Fleming, P.J., Zitzler, E., Deb, K., Thiele, L. (eds.) EMO 2003. LNCS, vol. 2632, pp. 376–390. Springer, Heidelberg (2003)
5. Purshouse, R.C., Fleming, P.J.: On the evolutionary optimization of many conflicting objectives. IEEE Transactions on Evolutionary Computation 11(6), 770–784 (2007)
6. Pelikan, M., Sastry, K., Goldberg, D.E.: Multiobjective estimation of distribution algorithms. In: Pelikan, M., Sastry, K., Cantú-Paz, E. (eds.) Scalable Optimization via Probabilistic Modeling: From Algorithms to Applications. Studies in Computational Intelligence, pp. 223–248. Springer, Heidelberg (2006)
7. Pelikan, M., Goldberg, D.E., Lobo, F.: A survey of optimization by building and using probabilistic models. IlliGAL Report No. 99018, University of Illinois at Urbana-Champaign, Illinois Genetic Algorithms Laboratory, Urbana, IL (1999)
8. Ahn, C.W., Ramakrishna, R.S.: Multiobjective real-coded bayesian optimization algorithm revisited: diversity preservation. In: GECCO 2007: Proceedings of the 9th annual conference on Genetic and evolutionary computation, pp. 593–600. ACM Press, New York (2007)
9. Shapiro, J.: Diversity loss in general estimation of distribution algorithms. In: Parallel Problem Solving from Nature - PPSN IX, pp. 92–101 (2006)

10. Yuan, B., Gallagher, M.: On the importance of diversity maintenance in estimation of distribution algorithms. In: GECCO 2005: Proceedings of the 2005 conference on Genetic and evolutionary computation, pp. 719–726. ACM, New York (2005)
11. Purshouse, R.C.: On the Evolutionary Optimisation of Many Objectives. PhD thesis, Department of Automatic Control and Systems Engineering, The University of Sheffield, Sheffield, UK (September 2003)
12. Zitzler, E., Künzli, S.: Indicator-based Selection in Multiobjective Search. In: Yao, X., Burke, E.K., Lozano, J.A., Smith, J., Merelo-Guervós, J.J., Bullinaria, J.A., Rowe, J.E., Tiño, P., Kabán, A., Schwefel, H.-P. (eds.) PPSN 2004. LNCS, vol. 3242, pp. 832–842. Springer, Heidelberg (2004)
13. Brockhoff, D., Zitzler, E.: Improving hypervolume–based multiobjective evolutionary algorithms by using objective reduction methods. In: Congress on Evolutionary Computation (CEC 2007), pp. 2086–2093. IEEE Press, Los Alamitos (2007)
14. Zitzler, E., Brockhoff, D., Thiele, L.: The hypervolume indicator revisited: On the design of pareto-compliant indicators via weighted integration. In: Obayashi, S., et al. (eds.) EMO 2007. LNCS, vol. 4403, pp. 862–876. Springer, Heidelberg (2007)
15. Deb, K., Saxena, D.K.: On finding Pareto–optimal solutions through dimensionality reduction for certain large–dimensional multi–objective optimization problems. Technical Report 2005011, KanGAL (December 2005)
16. Brockhoff, D., Zitzler, E.: Dimensionality reduction in multiobjective optimization: The minimum objective subset problem. In: Waldmann, K.H., Stocker, U.M. (eds.) Operations Research Proceedings 2006, pp. 423–429. Springer, Heidelberg (2007)
17. Brockhoff, D., Saxena, D.K., Deb, K., Zitzler, E.: On handling a large number of objectives a posteriori and during optimization. In: Knowles, J., Corne, D., Deb, K. (eds.) Multi–Objective Problem Solving from Nature: From Concepts to Applications. Natural Computing Series, pp. 377–403. Springer, Heidelberg (2008)
18. Martí, L., García, J., Berlanga, A., Molina, J.M.: Model-building algorithms for multiobjective EDAs: Directions for improvement. In: Michalewicz, Z. (ed.) 2008 IEEE Conference on Evolutionary Computation (CEC), part of 2008 IEEE World Congress on Computational Intelligence (WCCI 2008), pp. 2848–2855. IEEE Press, Los Alamitos (2008)
19. Mart, L., Garca, J., Berlanga, A., Molina, J.M.: Introducing MONEDA: Scalable multiobjective optimization with a neural estimation of distribution algorithm. In: Thierens, D., Deb, K., Pelikan, M., Beyer, H.G., Doerr, B., Poli, R., Bittari, M. (eds.) GECCO 2008: 10th Annual Conference on Genetic and Evolutionary Computation, pp. 689–696. ACM Press, New York (2008); EMO Track Best Paper Nominee
20. Fritzke, B.: A growing neural gas network learns topologies. In: Tesauro, G., Touretzky, D.S., Leen, T.K. (eds.) Advances in Neural Information Processing Systems, vol. 7, pp. 625–632. MIT Press, Cambridge (1995)
21. Martí, L., García, J., Berlanga, A., Molina, J.M.: On the computational properties of the multi-objective neural estimation of distribution algorithms. In: Pelta, D.A., Krasnogor, N. (eds.) International Workshop on Nature Inspired Cooperative Strategies for Optimization. Studies in Computational Intelligence. Springer, Heidelberg (2008) (in press)

A Hooke-Jeeves Based Memetic Algorithm for Solving Dynamic Optimisation Problems

Irene Moser[1] and Raymond Chiong[2]

[1] Faculty of ICT, Swinburne University of Technology, Melbourne, Australia
imoser@swin.edu.au
[2] School of Computing & Design, Swinburne University of Technology (Sarawak Campus),
Jalan Simpang Tiga, 93350 Kuching, Sarawak, Malaysia
rchiong@swinburne.edu.my

Abstract. Dynamic optimisation problems are difficult to solve because they involve variables that change over time. In this paper, we present a new Hooke-Jeeves based Memetic Algorithm (HJMA) for dynamic function optimisation, and use the Moving Peaks (MP) problem as a test bed for experimentation. The results show that HJMA outperforms all previously published approaches on the three standardised benchmark scenarios of the MP problem. Some observations on the behaviour of the algorithm suggest that the original Hooke-Jeeves algorithm is surprisingly similar to the simple local search employed for this task in previous work.

Keywords: Hooke-Jeeves pattern search, extremal optimisation, dynamic function optimisation, moving peaks problem.

1 Introduction

One of the fundamental issues that makes optimisation problems difficult to solve is the dynamically changing fitness landscapes [1]. In problems with dynamic fitness landscapes, the task of an optimisation algorithm is normally to provide candidate solutions with momentarily optimal objective values for each point in time.

The Moving Peaks (MP) problem is a good example of this kind of problem. It consists of multidimensional landscape with a definable number of peaks, where the height, the width and the position of each peak are altered slightly every time a change in the environment occurs. Created by Branke [2] as a benchmark for dynamic problem solvers, the MP problem has been used by many for the testing of algorithms for dynamic function optimisation.

In previous work [3], an algorithm called Multi-phase Multi-individual Extremal Optimisation (MMEO) was designed. MMEO exhibits great simplicity, but it works extremely well for the MP problem. Thorough analysis [4] has shown that the success story of MMEO is largely due to the local search component in it, and this has motivated us in studying the Hooke-Jeeves (HJ) algorithm, a very simple local search that was proposed by Hooke and Jeeves over 40 years ago [5].

In this paper, we present a new Hooke-Jeeves based Memetic Algorithm (HJMA) for solving the MP problem. HJMA is a hybridisation of HJ pattern search and

E. Corchado et al. (Eds.): HAIS 2009, LNAI 5572, pp. 301–309, 2009.
© Springer-Verlag Berlin Heidelberg 2009

Extremal Optimisation (EO), an optimisation heuristic that was first introduced by Boettcher and Percus [6] in 1999. Based on a very simple principle of mutating a single solution according to a power-law distribution, EO was designed to exclude bad solutions rather than finding good solutions. In other words, EO was not intended to show any convergence behaviour. This characteristic seemed to make EO a very promising choice for dynamic implementations – as no convergence exists, EO is expected to automatically adapt the current working solution according to the feedback received from the objective function. If the objective function returns fitness values that reflect the current search space, the algorithm is expected to be able to adapt to changes regardless of the severity or the frequency of the changes. However, experimental studies [4] revealed that EO alone does not work well for the MP problem, as far as the solution quality is concerned. The need to check for duplicates during the local search phase has an unexpectedly large impact on the solution quality. As such, this study reports on the hybridisation of EO with HJ.

The rest of this paper is organised as follows. Section 2 introduces the background of the MP problem by describing some of the existing solution methods. Following which, section 3 discusses the HJMA in detail. We then present our experimental results in section 4. Finally, section 5 concludes our studies and highlights some potential future work.

2 Background

The MP problem can be formally defined with the following function:

$$F(\vec{x}, t) = \max(B(\vec{x}), \max_{1..m} P(\vec{x}, h_i(t), w_i(t), \vec{p}_i(t))) \tag{1}$$

where $B(\vec{x})$ is the base landscape on which the m peaks move, with each peak P having its height h, width w and location $\vec{p}$. It is necessary to note that the location, width, height, and movement of the peaks are all free parameters. For the purposes of performance comparison, three standardised sets of parameter settings, called Scenario 1, Scenario 2 and Scenario 3 respectively, are defined. Most of the benchmark results have been published predominantly for Scenario 2 with 10 peaks that move at an average distance of 1/10 of the search space in a random direction, mainly due to its appropriateness in terms of difficulty and solvability.

Many solutions for the MP problem have been presented since its inception, and one of the pioneering solutions can be found in the work of its creator, Branke, based on a genetic algorithm (GA) [7]. In his seminal work, Branke used a memory-based multi-population variation of a GA to store and retrieve individuals when a change occurs. However, he found that the approach is less useful for more drastic changes to the landscape. For this reason, he collaborated with others to develop the self-organising scouts algorithm [8]. This extended GA approach is based on a forking mechanism which starts with a single population, and then dividing off subpopulations with a designated search area and size. Comparing to the standard GA, it shows a much better performance. More results of the self-organising scouts approach were subsequently published by Branke and Schmeck in [9], where they also introduced the offline error[1] as a standard metric for performance measure.

Other related works that have applied GA to the MP problem can be found in [10], [11], [12], [13], [14] and [15].

Apart from GA, particle swarm optimisation (PSO) is another popular method that has been used extensively in dynamic optimisation domain. Blackwell [16], who introduced charged particles (hence CPSO) that repel each other and circle around neutral particles of the swarm for better convergence behaviour in dynamic environments, was among the first to study PSO for the MP problem. Afterwards, Blackwell and Branke [17] applied a multi-population version of the same approach as multi-CPSO to the same problem. They also introduced multi-Quantum Swarm Optimisation (multi-QSO), a variation whose charged particles move randomly within a cloud of fixed radius centred around the swarm attractor. All these approaches perform well on the MP problem.

Other PSO-based studies include Parrott and Li [18] who adapted the speciation technique from GA to PSO, Janson and Middendorf [19] who proposed to employ a tree structure where each particle uses the best location found by the individual immediately above it in the tree structure in addition to its own best find, as well as Wang et al. [20] who used Branke's [7] idea of employing 3 populations originally for GA to PSO. However, the best result by PSO comes from Blackwell and Branke [21] who added anti-convergence to the exclusion and quantum/charged particle features they first conceived in [16] and [17] respectively. In [21], Blackwell and Branke reported an offline error of 1.72 from solving Scenario 2.

There are also other types of solutions for the MP problem, such as the differential evolution (DE) [22], the stochastic diffusion search (SDS) [23] inspired by neural networks, and the B-cell algorithm (BCA) [24]. Among these, the DE approach by Mendes and Mohais [22] produced almost equal quality as Blackwell and Branke's PSO [21], with an offline error of 1.75 from solving the same settings.

The good performances of Blackwell and Branke's PSO and Mendes and Mohais' DE have encouraged Lung and Dumitrescu [25] to develop a hybridised algorithm that combines PSO and Crowding DE, called Collaborative Evolutionary-Swarm Optimisation (CESO), in which equal populations of both methods collaborate. Their offline error of 1.38 on Scenario 2 with 10 peaks surpasses those of Blackwell and Branke's as well as Mendes and Mohais'.

While all these approaches are impressive, the best solution in the literature comes in the very simple MMEO algorithm by Moser and Hendtlass [3]. MMEO is a multiphase multi-individual version of EO. As has been established [4], a large proportion of the quality of its hitherto unsurpassed results is contributed by its local search component, which is rather straightforward and deterministic. It outperforms all available approaches to date with an offline error of 0.66 on Scenario 2.

---

[1] At each evaluation, the difference between the maximum height of the landscape and the best-known solution to date is recorded. It is then averaged by the number of evaluations (note that this measure has some vulnerabilities: it is sensitive to the overall height of the landscape, the number of peaks, and the number of evaluations before change).

In spite of the outstanding results, the local search of MMEO still carries redundant steps which cause unnecessary function evaluations. Furthermore, the step lengths used in [3] were chosen intuitively without careful consideration. As such, we believe that there is still room for improvements. In this study, we intend to explore the potential of HJ pattern search, and examine how the abovementioned issues could be overcome through the development of HJMA.

3 A Novel Hybrid Approach

In this section, we present our novel HJMA approach. The basic EO algorithm is used as the basis for performing the global search, while HJ will be incorporated during the local search process.

3.1 Global Search

This task is achieved by an adaptation of the EO algorithm. Unlike other population-based approaches, EO works to improve only a single solution using mutation. This solution consists of multiple components which are assigned individual fitness values. Based on the Bak-Sneppen model [26] of self-organised criticality (SOC), EO eliminates the worst component by exchanging it for another element at each iteration.

The initial solution is always created randomly. Variations are made to this initial solution using a "stepwise" sampling scheme that changes each of the dimensional variables at a time to produce a set of candidates. The sampling scheme produces a predefined number of equally distanced candidates in every dimension (see [27] for details). These candidate solutions are then ordered according to fitness.

This provides a rank k (where 1 is the worst) for each solution. The solution to be adopted can then be chosen with a probability of $k^{-\tau}$ where the only free parameter τ, usually a value between 1.1 and 3.0, is set to infinity. This setting eliminates the possibility for uphill moves which are often beneficial when EO is used as a stand-alone algorithm. In combination with a local search, the use of uphill moves has proved to be less desirable.

The algorithm then adopts one of the candidates as the next solution and proceeds to the local search phase.

3.2 Local Search

After the global search phase, the local search process takes place using the HJ pattern search algorithm. As described in the original paper [5], HJ starts with an exploratory move in which all dimensional variables in turn are changed by a predefined step. As improvements are equally likely in both directions along the dimensional axes, this takes at least twice as many function evaluations as there are dimensions, at most equally many. The pattern move then repeats all changes that were found to be successful in the exploratory move and uses a single function evaluation to evaluate the effect of the combined change.

The implementation of HJ algorithm is formalised as follows:

Hooke-Jeeves Pattern Search Algorithm

1. Obtain initial base point x^t. Determine set of step lengths.

2. Move the base point along every one of the d dimensional axes at a time and evaluate the result. Adopt each new point if improvement on the previous point. This takes at least d, at most $2d$ evaluations. If any of the moves was successful, go to 3. If none was successful, go to 4.

3. Repeat the successful moves in a combined pattern move. If the new point has a better fitness, assume it as the new base point. Return to 2 whichever the outcome.

4. Adjust step length to next smaller step. If there is a smaller step, continue from 2. If not, terminate.

The HJ procedure repeats until no improving change can be made in any dimension. The step size is reduced and the procedure repeated until there are no more step sizes.

For our experiments, the use of exponential decline in step sizes has proved most successful.

$$s_j = s_{j-1} * b^j \tag{2}$$

The sequence described by equation (2) was used for the experiments with HJMA. The initial value s_0 has to be determined on the basis of knowledge about the search space, and was set to $\{8, 9, 10, 11, 12\}$ for our experiments. The power base b was set to the values $\{0.2, 0.3, 0.5\}$, as none of the individual values proved consistently superior to the others. The results presented did not always use the same step length sequences.

3.3 HJMA

The complete HJMA algorithm differs from MMEO [3] only in the local search part. Unlike MMEO, HJMA uses the HJ algorithm, which is used with different step length sequences. Also, the HJMA local search records which direction along each dimensional axis was used for the last improvement and checks this direction first.

In all other respects, the algorithms are identical. HJMA also eradicates duplicates after every exploratory and pattern move, and stores the solution when it cannot be improved further and it is not identified as a duplicate. The HJMA algorithm also comprises a fine-tuning phase where the best solution in memory is improved using a further step on the exponential sequence. The complete algorithm is outlined below:

Hooke-Jeeves based Memetic Algorithm

1. Find new solution by stepwise sampling of the space in each dimension. Evaluate solutions and rank by resulting fitness (quality). Choose new individual using power-law distribution.

2. Use HJ pattern search to optimise the new solution locally. Stop if too close to other solution. Stop when no further improvement is possible.

3. Store new solution if it was not removed as a duplicate in step 2.

4. Check whether the existing individuals can be improved by further local optimisation, i.e. a change has occurred.

5. Fine-tune best individual using HJ but sample closer to current position. Stop when no further improvement is possible.

4 Experiments and Results

In our experiments, we compare HJMA to MMEO [3] and CESO [25] from the literature. Additionally, we also compare it with an improved MMEO where some redundancies in the step lengths have been removed. As in the HJ implementation, the local search in the new MMEO also records the direction in which every dimensional variable is likely to improve.

The experimental results on all three scenarios are summarised in Table 1, averaged over 50 runs with 100 changes to each run. The corresponding standard error is calculated by dividing the standard deviation and the square root of the number of runs.

Table 1. Offline error and standard error for all scenarios

Scenario	CESO [25]	MMEO [3]	new MMEO	HJMA
1	-	0.10 ± 0.01	0.06 ± 0.01	0.06 ± 0.01
2	1.38 ± 0.02	0.66 ± 0.20	0.25 ± 0.08	0.25 ± 0.10
3	-	3.77 ± 1.72	1.43 ± 0.54	1.57 ± 0.73

From Table 1, it is obvious that HJMA clearly outperforms CESO and MMEO. It also shows comparable performance to the newly improved MMEO.

4.1 Varying the Number of Peaks

The results obtained in Table 1 have used a total of 10 peaks for all the scenarios. In general, it is easier to find the global maximum when the number of peaks is small. However, it is easier to score on the offline error when the landscape is elevated (more peaks). To evaluate the performance of HJMA when different number of peaks is present, we test it with experiments on 1, 10, 20, 30, 40, 50 and 100 peaks for comparison. Experimental results obtained for different number of peaks are presented in Table 2.

For CESO, the best result has been obtained with the one peak setup. While HJMA did not perform well with one peak, it obtained better results than CESO and MMEO in all other instances. The improved MMEO shares similar results with HJMA. The reasons behind the similarity in results as well as the exceptionally poor performance in the scenario with the single peak will be the subject of further studies.

Table 2. Offline error and standard error for varying number of peaks

No. peaks	CESO [25]	MMEO [3]	new MMEO	HJMA
1	**1.04 ± 0.00**	11.3 ± 3.56	7.47 ± 1.98	7.08 ± 1.99
10	1.38 ± 0.02	0.66 ± 0.20	**0.25 ± 0.08**	0.25 ± 0.10
20	1.72 ± 0.02	0.90 ± 0.16	0.40 ± 0.11	**0.39 ± 0.10**
30	1.24 ± 0.01	1.06 ± 0.14	0.49 ± 0.10	**0.49 ± 0.09**
40	1.30 ± 0.02	1.18 ± 0.16	**0.56 ± 0.09**	0.56 ± 0.09
50	1.45 ± 0.01	1.23 ± 0.11	0.59 ± 0.10	**0.58 ± 0.09**
100	1.28 ± 0.02	1.38 ± 0.09	**0.66 ± 0.07**	**0.66 ± 0.07**

4.2 Varying the Dimensionality

The dimensionality, tantamount to the complexity of the problem, is expected to have a large impact on the performances of different algorithms. The standard scenarios used to obtain the results in Table 1 have five dimensions. In order to investigate the effect of varying dimensionality in the search space, we test out HJMA and other algorithms on experiments with different dimensionality values. Numerical results on 10, 50 and 100 dimensions are presented in Table 3.

Table 3. Offline error and standard error for varying dimensionality

Dimensions	CESO [25]	MMEO [3]	new MMEO	HJMA
10	2.51 ± 0.04	2.44 ± 0.77	2.25 ± 0.85	**2.17 ± 0.80**
50	6.81 ± 0.07	206.3 ± 35.7	6.22 ± 1.6	**5.79 ± 1.4**
100	24.60 ± 0.25	480.5 ± 70.1	17.8 ± 6.9	**16.5 ± 5.4**

As can be observed from Table 3, CESO reported an average offline error of 2.51 in the 10 dimensions search space. The performance of CESO deteriorated drastically when the dimensionality increases, with average offline errors of 6.81 and 24.60 for 50 dimensions and 100 dimensions respectively. On the other hand, HJMA is able to maintain a fairly competitive performance even when the dimensionality is increased to 100. It is also the first time where HJMA has shown distinguishably better results than the improved MMEO.

5 Conclusion

In this paper, we have proposed a new hybrid algorithm – HJMA – for solving dynamic function optimisation problems. HJMA significantly outperformed the best algorithms for the MP problem currently available in the literature. It has also maintained its outstanding performance in challenging environments, i.e. search spaces with different number of peaks and different dimensionality. The HJ pattern search has been particularly robust compared to other local search algorithms when the dimensionality is high.

In general, there is still room for HJMA to improve, considering the fact that it has been devised within a short period of time. Future work will investigate the portability of HJ with other types of metaheuristics.

References

1. Weise, T., Zapf, M., Chiong, R., Nebro, A.J.: Why is optimization difficult? In: Chiong, R. (ed.) Nature-Inspired Algorithms for Optimisation. Studies in Computational Intelligence, vol. 193, pp. 1–50. Springer, Heidelberg (2009)
2. Branke, J.: The moving peaks benchmark, http://www.aifb.uni-karlsruhe.de/~jbr/MovPeaks/ (viewed 08/11/2008)

3. Moser, I., Hendtlass, T.: A simple and efficient multi-component algorithm for solving dynamic function optimisation problems. In: Proc. IEEE Congress on Evolutionary Computation (CEC 2007), Singapore, pp. 252–259 (2007)
4. Moser, I.: Applying extremal optimisation to dynamic optimisation problems. Ph.D. thesis, Faculty of Information and Communication Technologies, Swinburne University of Technology, Australia (2008)
5. Hooke, R., Jeeves, T.: Direct search solutions of numerical and statistical problems. Journal of the Association for Computing Machinery 8, 212–229 (1961)
6. Boettcher, S., Percus, A.G.: Extremal optimization: methods derived from co-evolution. In: Proc. Genetic and Evolutionary Computation Conference (GECCO 1999), Orlando, Florida, USA, pp. 825–832 (1999)
7. Branke, J.: Memory-enhanced evolutionary algorithms for changing optimization problems. In: Proc. IEEE Congress on Evolutionary Computation (CEC 1999), Washington, DC, USA, pp. 1875–1882 (1999)
8. Branke, J., Kaußler, T., Schmidt, C., Schmeck, H.: A multi-population approach to dynamic optimization problems. In: Parmee, I.C. (ed.) Adaptive Computing in Design and Manufacturing (ACDM 2000), pp. 299–308. Springer, Berlin (2000)
9. Branke, J., Schmeck, H.: Designing evolutionary algorithms for dynamic optimization problems. In: Tsutsui, S., Ghosh, A. (eds.) Theory and Application of Evolutionary Computation: Recent Trends, pp. 239–362. Springer, Berlin (2002)
10. Kramer, G.R., Gallagher, J.C.: Improvements to the *CGA enabling online intrinsic. In: Proc. NASA/DoD Conference on Evolvable Hardware, Chicago, Illinois, USA, pp. 225–231 (2003)
11. Zou, X., Wang, M., Zhou, A., Mckay, B.: Evolutionary optimization based on chaotic sequence in dynamic environments. In: Proc. IEEE International Conference on Networking, Sensing and Control, Taipei, Taiwan, pp. 1364–1369 (2004)
12. Ronnewinkel, C., Martinetz, T.: Explicit speciation with few a priori parameters for dynamic optimization problems. In: GECCO Workshop on Evolutionary Algorithms for Dynamic Optimization Problems, San Francisco, California, USA, pp. 31–34 (2001)
13. Bui, L.T., Branke, J., Abbass, H.A.: Diversity as a selection pressure in dynamic environments. In: Proc. Genetic and Evolutionary Computation Conference (GECCO 2005), Washington, DC, USA, pp. 1557–1558 (2005)
14. Bui, L.T., Branke, J., Abbass, H.A.: Multiobjective optimization for dynamic environments. In: Proc. IEEE Congress on Evolutionary Computation (CEC 2005), Edinburgh, UK, pp. 2349–2356 (2005)
15. Fentress, S.W.: Exaptation as a means of evolving complex solutions. Master's Thesis, School of Informatics, University of Edinburgh, UK (2005)
16. Blackwell, T.M.: Swarms in dynamic environments. In: Proc. Genetic and Evolutionary Computation Conference (GECCO 2003), Chicago, Illinois, USA, pp. 1–12 (2003)
17. Blackwell, T., Branke, J.: Multi-swarm optimization in dynamic environments. In: Raidl, G.R., Cagnoni, S., Branke, J., Corne, D.W., Drechsler, R., Jin, Y., Johnson, C.G., Machado, P., Marchiori, E., Rothlauf, F., Smith, G.D., Squillero, G. (eds.) EvoWorkshops 2004. LNCS, vol. 3005, pp. 489–500. Springer, Heidelberg (2004)
18. Parrott, D., Li, X.: A particle swarm model for tracking multiple peaks in a dynamic environment using speciation. In: Proc. IEEE Congress on Evolutionary Computation (CEC 2004), Portland, Oregon, USA, pp. 105–116 (2004)

19. Janson, S., Middendorf, M.: A hierachical particle swarm optimizer for dynamic optimization problems. In: Raidl, G.R., Cagnoni, S., Branke, J., Corne, D.W., Drechsler, R., Jin, Y., Johnson, C.G., Machado, P., Marchiori, E., Rothlauf, F., Smith, G.D., Squillero, G. (eds.) EvoWorkshops 2004. LNCS, vol. 3005, pp. 513–524. Springer, Heidelberg (2004)

20. Wang, H., Wang, D., Yang, S.: Triggered memory-based swarm optimisation in dynamic environments. In: Giacobini, M., et al. (eds.) EvoWorkshops 2007. LNCS, vol. 4448, pp. 637–646. Springer, Heidelberg (2007)

21. Blackwell, T., Branke, J.: Multi-swarms, exclusion and anti-convergence in dynamic environments. IEEE Transactions on Evolutionary Computation 10(4), 51–58 (2006)

22. Mendes, R., Mohais, A.: DynDE: a differential evolution for dynamic optimization problems. In: Proc. IEEE Congress on Evolutionary Computation (CEC 2005), Edinburgh, UK, pp. 2808–2815 (2005)

23. Meyer, K.D., Nasut, S.J., Bishop, M.: Stochastic diffusion search: partial function evaluation in swarm intelligence dynamic optimization. In: Abraham, A., et al. (eds.) Stigmergic Optimization. Studies in Computational Intelligence, vol. 31, pp. 185–207. Springer, Berlin (2006)

24. Trojanowski, K.: B-cell algorithm as a parallel approach to optimization of moving peaks benchmark tasks. In: Proc. 6th International Conference on Computer Information Systems and Industrial Management Applications (CISIM 2007), Elk, Poland, pp. 143–148 (2007)

25. Lung, R.I., Dumitrescu, D.: A collaborative model for tracking optima in dynamic environments. In: Proc. IEEE Congress on Evolutionary Computation (CEC 2007), Singapore, pp. 564–567 (2007)

26. Bak, P., Sneppen, K.: Punctuated equilibrium and criticality in a simple model of evolution. Physics Review Letters 74, 4083–4086 (1993)

27. Boettcher, S., Percus, A.G.: Nature's way of optimizing. Artificial Intelligence 119(1), 275–286 (2000)

Hybrid Evolutionary Algorithm for Solving Global Optimization Problems

Radha Thangaraj[1], Millie Pant[1], Ajith Abraham[2], and Youakim Badr[2]

[1] Department of Paper Technology, IIT Roorkee, India
t.radha@ieee.org, millifpt@iitr.ernet.in
[2] National Institute of Applied Sciences of Lyon, INSA-Lyon, Villeurbanne, France
ajith.abraham@ieee.org, youakim.badr@insa-lyon.fr

Abstract. Differential Evolution (DE) is a novel evolutionary approach capable of handling non-differentiable, non-linear and multi-modal objective functions. DE has been consistently ranked as one of the best search algorithm for solving global optimization problems in several case studies. This paper presents a simple and modified hybridized Differential Evolution algorithm for solving global optimization problems. The proposed algorithm is a hybrid of Differential Evolution (DE) and Evolutionary Programming (EP). Based on the generation of initial population, three versions are proposed. Besides using the uniform distribution (U-MDE), the Gaussian distribution (G-MDE) and Sobol sequence (S-MDE) are also used for generating the initial population. Empirical results show that the proposed versions are quite competent for solving the considered test functions.

Keywords: Hybrid Algorithm, Differential Evolution, Evolutionary Programming, Global Optimization.

1 Introduction

Evolutionary Algorithms (EAs) are general-purpose stochastic search methods imitating the phenomena of biological evolution. One of the reasons of the success of EAs is their population based strategy which prevents them from getting trapped in a local optimal solution and consequently increases their probability of finding a global optimal solution. Thus, EAs can be viewed as global optimization algorithms. Some frequently used EAs include Evolutionary Programming (EP) [1], Evolution Strategies (ES) [2], Genetic Algorithms (GA) [3], Particle Swarm Optimization [4], and Differential Evolution [5]. These algorithms have been applied successfully to wide range of problems [6] – [9]. Some common features of these algorithms may be given as:

➢ Start with population of points instead of single point
➢ Do not depend on initial guess.
➢ Are able to solve ill-defined or inexplicitly expressed problems as they do not depend on the mathematical properties like continuity or differentiability
➢ Inherits natural parallelism

Despite having several attractive features, these algorithms also have weaknesses and drawbacks like slow convergence, loss of diversity, stagnation of population leading to a

E. Corchado et al. (Eds.): HAIS 2009, LNAI 5572, pp. 310–318, 2009.

suboptimal performance etc. These problems become more prominent in case of multimodal problems having several local and global optima. Several variants of these algorithms have been proposed in the past to improve their performance. In the present study we propose a hybridized version of DE, which is relatively a newer addition to the class of EA.

DE was proposed by Storn and Price [5] in 1995. It soon became a popular tool for solving global optimization problems because of several attractive features like fewer control parameters, ease in programming, efficiency etc. DE has parameters like mutation, crossover and selection for guiding the population towards the optimum solution similar to GAs. However, it's the application of these operators that makes DE different from GA. The main difference between GAs and DE is that; in GAs, mutation is the result of small perturbations to the genes of an individual while in DE mutation is the result of arithmetic combinations of individuals. At the beginning of the evolution process, the mutation operator of DE favors exploration. As evolution progresses, the mutation operator favors exploitation. Hence, DE automatically adapts the mutation increments (i.e. search step) to the best value based on the stage of the evolutionary process. Mutation in DE is therefore not based on a predefined probability density function.

DE has been successfully applied to solve a wide range of real life application problems such as clustering [10], unsupervised image classification [11], digital filter design [12], optimization of non-linear functions [13], global optimization of non-linear chemical engineering processes [14] and multi-objective optimization [15] etc. Also it has reportedly outperformed other optimization techniques [16] – [18].

However like other EA, DE has certain flaws like slow convergence and stagnation of population. Several modified versions of DE are available in literature for improving the performance of basic DE. One class of such algorithms includes hybridized versions where DE is combined with some other algorithm to produce a new algorithm. DE has been hybridized with ACO, Simulated Annealing, PSO, local search methods like Nelder Mead etc. The hybridized versions have also been used successfully for solving practical application problems [19]-[21].

The present study differs from other hybridized algorithms in two ways; firstly we have hybridized DE with EP, which to the best of our knowledge has not been done before and secondly we use different initializing techniques for generation of random numbers like uniformly distributed random numbers, Gaussian distributed random numbers and random numbers generated using quasi random Sobol sequence. The proposed algorithms are named as Modified Differential Evolution having uniform distribution (U-MDE), having Gaussian distribution (G-MDE), having Sobol distribution (S-MDE). The rationale for using different initialization techniques is that the population based search methods generally use computer generated uniformly distributed random numbers. This technique however is not very efficient as the computer generated random numbers may not cover the search domain effectively. In the present work we used Sobol sequence to generate the initial population. It is a quasi random sequence and covers the search space more evenly in comparison to the computer generated random numbers. It has given better results in comparison to the algorithms using uniformly distributed random numbers [22] – [24].

The remaining of the paper is organized as follows: in Sections 2 and 3, we give a brief description of DE and EP algorithms respectively. Section 4, describes the proposed MDE algorithm. Experimental settings and numerical results are given in subsections 5.1 and 5.2 respectively of section 5. The paper concludes with Section 6.

2 Differential Evolution

A general DE variant may be denoted as DE/X/Y/Z, where X denotes the vector to be mutated, Y specifies the number of difference vectors used and Z specifies the cross-over scheme which may be binomial or exponential. Throughout the study we shall consider the mutation strategy DE/rand/1/bin [5]. It is also known as the classical version of DE and is perhaps the most frequently used version of DE. DE works as follows: First, all individuals are initialized with uniformly distributed random num-bers and evaluated using the fitness function provided. Then the following will be executed until maximum number of generation has been reached or an optimum solu-tion is found.

For a D-dimensional search space, each target vector $x_{i,g}$, a mutant vector is generated by

$$v_{i,g+1} = x_{r_1,g} + F*(x_{r_2,g} - x_{r_3,g}) \qquad (1)$$

where $r_1, r_2, r_3 \in \{1,2,....,NP\}$ are randomly chosen integers, must be different from each other and also different from the running index i. F (>0) is a scaling factor which controls the amplification of the differential evolution $(x_{r_2,g} - x_{r_3,g})$. In order to increase the diversity of the perturbed parameter vectors, crossover is introduced. The parent vector is mixed with the mutated vector to produce a trial vector $u_{ji,g+1}$,

$$u_{ji,g+1} = \begin{cases} v_{ji,g+1} & if \ (rand_j \le CR) \ or \ (j = j_{rand}) \\ x_{ji,g} & if \ (rand_j > CR) \ and \ (j \ne j_{rand}) \end{cases} \qquad (2)$$

where j = 1, 2,......., D; $rand_j \in [0,1]$; CR is the crossover constant takes values in the range [0, 1] and $j_{rand} \in (1,2,.....,D)$ is the randomly chosen index.

Selection is the step to choose the vector between the target vector and the trial vector with the aim of creating an individual for the next generation.

3 Evolutionary Programming

Evolutionary programming (EP) originated from the research of L.J. Fogel in 1962 [25] on using simulated evolution to develop artificial intelligence. The concept of self adaptive EP (SAEP) was introduced by Back and Schwefel [26] and Fogel [27] and was shown to be more efficient than the normal EP. The computational steps of SAEP are given below:

Step 1:   Each individual is taken as a pair of real-valued vectors, (x_i , σ_i) for all i=1,...,M. The x_i's give the ith member's object variables and σ_i's the associated strategy parameters.

Step 2:   Evaluate the objective function of each individual.

Step 3:   Mutation: Creates a single offspring (x_i' , σ_i'), from each parent (x_i , σ_i) for all i=1,..., M by

$$\sigma_i'(j) = \sigma_i(j) \exp(\tau N(0,1) + \tau' N_j(0,1))$$
$$x_i'(j) = x_i(j) + \sigma_i'(j) N_j(0,1) \ \text{for all} \ j = 1,.....n.$$

where N(0,1) denotes a random number distributed by Gaussian distribution.

The factors τ and τ' are commonly set to $1/\sqrt{2n}$ and $1/\sqrt{2\sqrt{n}}$ respectively.

Step 4:   Calculate the objective function value of each offspring (x_i', σ_i'), for all i = 1,...,M.

Step 5:   Selection:

Each individual x from the union of parents (x_i, σ_i) and offspring (x_i', σ_i'), is evaluated against q other randomly chosen solutions. For each comparison, a "win" is assigned if x is better than its opponent The M solutions with the greatest number of wins are retained to be parents of the next generation. Parameter q allows tuning selection pressure, typically $q = 10$.

Step 6:   Stop if the stopping criteria is reached otherwise go to step 3.

4 Proposed MDE Algorithm: A Hybridized Version of DE and EP

The proposed DE-PSO as mentioned earlier is a hybrid version of DE and EP. MDE starts like the usual DE algorithm up to the point where the trial vector is generated. If the trial vector is better than the target vector, then it is included in the population otherwise the algorithm enters the EP phase and generates a new candidate solution using EP based mutation. The method is repeated iteratively till the optimum value is reached. The inclusion of EP phase creates a perturbation in the population, which in turn helps in maintaining diversity of the population and producing a good optimal solution. The proposed MDE algorithm initialize with uniform distribution is called as U-MDE, initialize with Gaussian distribution is called as G-MDE and initialize with Sobol sequence is called as S-MDE. The pseudo code of the MDE Algorithm is:

Initialize the population using uniform (/Gaussian/ Sobol sequence) distributed random numbers

For i = 1 to N (Population size) do

Select $r_1, r_2, r_3 \in$ N randomly

// r_1, r_2, r_3 are selected such that $r_1 \neq r_2 \neq r_3$ //

For j = 1 to D (dimension) do

        Select $j_{rand} \in D$

        If (rand () < CR or j = j_{rand})

// rand () denotes a uniformly distributed random number between 0 and 1//

$$U_{ji,g+1} = x_{r_1,g} + F * (x_{r_2,g} - x_{r_3,g})$$

        End if

$$X_{ji,g+1} = \begin{cases} U_{ji,g+1} & if & f(U_{ji,g+1}) < f(X_{ji,g}) \\ TX_{ji} & if & f(TX_{ji}) < f(X_{ji,g}) \\ X_{ji,g} & otherwise \end{cases}$$

Where $TX_{ji} = X_{ji,g} + \sigma_i * N_j(0,1)$

        $\sigma_i = \sigma_i * \exp(\tau_a * N(0,1) + \tau_b * N_j(0,1))$

    End for

End for

5 Experimental Settings and Numerical Results

5.1 Experimental Settings

We considered a set of ten unconstrained benchmark functions namely Rastringin (f_1), Spherical (f_2), Griewank (f_3), Rosenbrock (f_4), Ackley (f_5), Generalized penalized function 1 (f_6), Generalized penalized function 2 (f_7), Levy (f_8), Test2N (f_9) and Circle (f_{10}) to validate the performance of proposed algorithms. The following experimental settings have been considered for the present study:

Experimental settings for proposed MDE algorithms and DE:

For dimension 30: Pop=50, run=30, Max Gne=3000

Experimental settings for proposed MDE and DEPSO [28] (Table 4):

Pop: 30, dim: 30, Max Gne.: 12000 for functions f_1, f_3, f_5 and Max Gne.: 4000 for function f_6

Experimental settings for proposed MDE and BBDE [29] (Table 5):

Pop: 30, dim: 30, Max number of function evaluations: 100000

Experimental settings for proposed MDE and DEPSO [30] (Table 6):

Pop: 20, 40 and 80, dim: 30, Max Gne: 2000

In the above mentioned settings, Pop denotes the population size taken; run denotes the number of times an algorithm is executed; Max Gne denotes the maximum number of generations allowed for each algorithm. All the algorithms were implemented using Dev C++ on a PC compatible with Pentium IV, a 3.2 GHz processor and 2 GB of RAM.

5.2 Numerical Results

The proposed versions are compared with the basic DE and with two other hybrid versions of DE, called BBDE [29] and DEPSO [28], [30]. In Table 1 we give the comparison of the proposed versions with the basic DE in terms of average fitness function value, standard deviation and the number of generations required to satisfy the given stopping criteria. In terms of average fitness function value all the algorithms gave good performance as it is evident from Table 1, although the proposed versions gave a slightly better performance in some of the test cases. If we compare the standard deviation, then also we can observe that all the algorithms converged to the desired objective function value with small value for standard deviation which is less than zero in almost all the test cases. This tendency shows the stability of the algorithms. However when we compare the proposed versions with the basic DE in terms of number of generations, it can be clearly seen that the proposed versions converged much faster in comparison to the basic DE. The performance curves of the proposed algorithms with respect to few selected problems are given in Figure 1. The proposed versions are also compared with three other hybrid versions available in literature. In Table 2, we compare the performance of the proposed algorithms with DEPSO; in Table 3, the comparison is done with BBDE and in Table 4, the comparison is done with DEPSO, another hybridized version of DE. In all the versions taken for comparison, hybridization of DE is done with PSO. The results of other hybridized versions are taken from literature and therefore we have considered the problems which are common to all the algorithms. For example in Table 2, we have only considered test cases f1, f3, f5, f6 because these were the only test cases common to the problems considered for the present study and to the literature. From Tables 2-4, we can easily see that the proposed versions give a much better performance in comparison to other hybridized versions available in literature for almost all the test problems.

Table 1. Average Fitness function Value (AFV), Standard Deviation (STD) and average number of generations for basic DE and the modified versions proposed in the present study

F	DE		U-MDE		G-MDE		S-MDE	
	AFV (STD)	Gen.	AFV (STD)	Gen.	AFV (STD)	Gen.	AFV (STD)	Gen.
f_1	1.22e-05 (8.47e-06)	364	1.09e-05 (5.84e-06)	351	8.00e-07 (4.3e-06)	316	1.17e-05 (3.51e-06)	282
f_2	1.12e-05 (6.07e-06)	173	8.66e-06 (2.17e-06)	167	6.27e-06 (3.38e-06)	165	8.75e-07 (5.42e-06)	135
f_3	9.86e-09 (1.76e-08)	699	1.05e-10 (2.14e-10)	713	7.07e-11 (3.48e09)	140	9.79e-012 (2.75e-09)	112
f_4	0.0953 (0.0479)	3000	0.0259 (0.1437)	3000	0.0044 (0.1723)	3000	0.04857 (1.0967)	3000
f_5	5.39e-05 (1.55e-05)	323	4.90e-05 (1.84e-05)	318	4.66e-06 (1.51e-05)	250	4.56e-05 (1.33e-05)	239
f_6	1.04e-05 (5.93e-06)	214	8.92e-06 (4.97e-06)	214	9.59e-06 (4.37e-06)	116	8.59e-06 (3.37e-06)	123
f_7	-1.15043 (5.94e-06)	232	-1.15043 (7.46e-06)	232	-1.15044 (4.75e-06)	167	-1.15043 (3.29e-06)	162
f_8	-21.2361 (6.00e-06)	223	-21.5023 (5.48e-06)	215	-21.5023 (5.45e-06)	200	-21.5023 (7.53e-06)	223
f_9	-77.2608 (6.49e-06)	195	-78.3323 (4.87e-06)	193	-78.3323 (2.02e-06)	119	-78.3323 (4.84e-06)	191
f_{10}	0.0914 (0.0235)	3000	0.0869 (0.0246)	3000	0.0868 (0.00022)	3000	0.08049 (0.01276)	3000

Table 2. Comparison Results (1): MDE vs DEPSO [28] in terms of average fitness value

F	DEPSO	U-MDE	G-MDE	S-MDE
f_1	24.216 (6.417)	0.0000(0.0000)	0.0000(0.0000)	0.0000(0.0000)
f_3	6.2e-16 (4.1e-16)	0.0000(0.0000)	0.0000(0.0000)	0.0000(0.0000)
f_5	-0.0002 (0.0002)	3.69e-15(0.0000)	3.69e-15(0.0000)	3.69e-15(0.0000)
f_6	3.9e-20 (4.1e-21)	5.51e-13(0.0000)	5.32e-18(0.0000)	4.71e-22(0.0000)

Table 3. Comparison Reslts (2): MDE vs BBDE [29] in terms of average fitness value

F	BBDE	U-MDE	G-MDE	S-MDE
f_1	72.185 (3.018)	1.73e-13(1.36e-13)	1.99e-18(3.66e-18)	0.0000(0.0000)
f_3	0.269e-01 (0.767-02)	2.16e-20(2.65e-20)	5.42e-21(1.62e-20)	5.42e-21(1.62e-20)
f_4	14.295 (0.948)	48.129(25.23)	25.51(0.9108)	25.69(3.91)
f_5	2.1361 (0.159)	2.18e-14(4.03e-15)	1.08e-14(3.55e-15)	7.25e-15(2.31e-15)

Table 4. Comparison Results (3): MDE vs DEPSO [30] in terms of average fitness value

F	Swarm size	DEPSO	U-MDE	G-MDE	S-MDE
f_1	20	0.8656	0.1989	**0.0000**	0.3979
	40	0.009950	2.83e-14	1.73e-19	**0.0000**
	80	3.919E-9	1.11e-13	5.20e-19	**0.0000**
f_3	20	0.009073	1.08e-20	**0.0000**	4.33e-20
	40	0.006930	3.25e-20	1.08e-20	**0.0000**
	80	0.005589	2.16e-20	**0.0000**	**0.0000**
f_4	20	80.8259	43.95	26.64	**25.3198**
	40	66.8730	28.52	24.74	**24.5649**
	80	60.6405	28.7815	24.15	**23.2118**

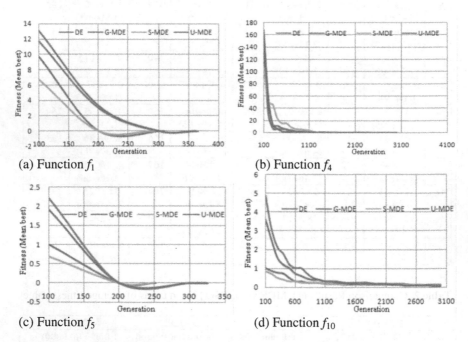

(a) Function f_1 (b) Function f_4

(c) Function f_5 (d) Function f_{10}

Fig. 1. Performance curves of selected benchmark problems

6 Conclusions

In the present study, we propose modified hybridized versions of DE algorithm for solving global optimization problems. The proposed versions used EP type mutation operator stochastically during the selection of trial vector. The inclusion of EP mutation operator helps in increasing the diversity which in turn helps in better exploration of the search space which finally helps in improving the solution quality and the convergence rate of an algorithm. This is evident from the empirical studies done in the present study. Another experiment that we have done is with the initialization of the population. Besides using the computer generated random numbers in U-MDE, we also initialized the population with Gaussian distribution and with a Quasi Random sequence called Sobol sequence. As expected, initialization with a quasi random sequence gave much better results in comparison to the other distributions. This behavior is quite expected because a quasi random sequence covers the search domain more uniformly in comparison to other distributions. We are continuing our work towards the theoretical development of the proposed algorithms and extending them for solving constrained optimization problems.

References

[1] Fogel, L.J., Owens, A.J., Walsh, M.J.: Artificial intelligence through a simulation of evolution. In: Maxfield, M., Callahan, A., Fogel, L.J. (eds.) Biophysics and Cybernetic systems. Proc. of the 2nd Cybernetic Sciences Symposium, pp. 131–155. Spartan Books (1965)

[2] Rechenberg, I.: Evolution Strategy: Optimization of Technical systems by means of biological evolution. Fromman-Holzboog (1973)

[3] Holland, J.H.: Adaptation in Natural and Artificial Systems: An Introductory Analysis with Applications to Biology, Control, and Artificial Intelligence. University of Michigan Press, Ann Arbor

[4] Kennedy, J., Eberhart, R.: Particle Swarm Optimization. In: IEEE International Conference on Neural Networks, Perth, Australia, pp. IV:1942–IV:1948. IEEE Service Center, Piscataway (1995)

[5] Storn, R., Price, K.: Differential Evolution – a simple and efficient adaptive scheme for global optimization over continuous spaces. Technical Report, International Computer Science Institute, Berkley (1995)

[6] Blesa, M.J., Blum, C.: A nature-inspired algorithm for the disjoint paths problem. In: Proc. Of 20th Int. Parallel and Distributed Processing Symposium, pp. 1–8. IEEE press, Los Alamitos (2006)

[7] delValle, Y., Moorthy, G.K.V., Mohagheghi, S., Hernandez, J.-C., Harley, R.G.: Particle Swarm Optimization: Basic Concepts, Variants and Applications in Power Systems. IEEE Trans. On Evolutionary Computation 12(2), 171–195 (2008)

[8] Hsiao, C.-T., Chahine, G., Gumerov, N.: Application of a Hybrid Genetic/Powell Algorithm and a Boundary Element Method to Electrical Impedence Tomograpghy. Journal of Computational Physics 173, 433–453 (2001)

[9] Kannan, S., Slochanal, S.M.R., Pathy, N.P.: Application and Comparison of metaheuristic techniques to generation expansion planning problem. IEEE Trans. on Power Systems 20(1), 466–475 (2005)

[10] Paterlini, S., Krink, T.: High performance clustering with differential evolution. In: Proceedings of the IEEE Congress on Evolutionary Computation, vol. 2, pp. 2004–2011 (2004)

[11] Omran, M., Engelbrecht, A., Salman, A.: Differential evolution methods for unsupervised image classification. In: Proceedings of the IEEE Congress on Evolutionary Computation, vol. 2, pp. 966–973 (2005)

[12] Storn. R.: Differential evolution design for an IIR-filter with requirements for magnitude and group delay. Technical Report TR-95-026, International Computer Science Institute, Berkeley, CA (1995)

[13] Babu, B., Angira, R.: Optimization of non-linear functions using evolutionary computation. In: Proceedings of the 12th ISME International Conference on Mechanical Engineering, India, pp. 153–157 (2001)

[14] Angira, R., Babu, B.: Evolutionary computation for global optimization of non-linear chemical engineering processes. In: Proceedings of International Symposium on Process Systems Engineering and Control, Mumbai, pp. 87–91 (2003)

[15] Abbass, H.: A memetic pareto evolutionary approach to artificial neural networks. In: Stumptner, M., Corbett, D.R., Brooks, M. (eds.) Canadian AI 2001. LNCS, vol. 2256, pp. 1–12. Springer, Heidelberg (2002a)

[16] Vesterstroem, J., Thomsen, R.: A comparative study of differential evolution, particle swarm optimization, and evolutionary algorithms on numerical benchmark problems. Proc. Congr. Evol. Comput. 2, 1980–1987 (2004)

[17] Andre, J., Siarry, P., Dognon, T.: An improvement of the standard genetic algorithm fighting premature convergence in continuous optimization. Advance in Engineering Software 32, 49–60 (2001)

[18] Hrstka, O., Ku°cerová, A.: Improvement of real coded genetic algorithm based on differential operators preventing premature convergence. Advance in Engineering Software 35, 237–246 (2004)

[19] Chiou, J.-P.: Variable scaling hybrid differential evolution for large-scale economic dispatch problems. Electric Power Systems Research 77(3-4), 212–218 (2007)

[20] Wang, F.-S., Su, T.-L., Jang, H.-J.: Hybrid Differential Evolution for Problems of Kinetic Parameter Estimationand Dynamic Optimization of an Ethanol Fermentation Process. Ind. Eng. Chem. Res. 40(13), 2876–2885 (2001)

[21] Luo, C., Yu, B.: Low Dimensional Simplex Evolution—A Hybrid Heuristic for Global Optimization. In: Eighth ACIS International Conference on Software Engineering, Artificial Intelligence, Networking, and Parallel/Distributed Computing, pp. 470–474 (2007)

[22] Kimura, S., Matsumura, K.: Genetic Algorithms using low discrepancy sequences. In: Proc. of GEECO 2005, pp. 1341–1346 (2005)

[23] Nguyen, X.H., Nguyen, Q.U., Mckay, R.I., Tuan, P.M.: Initializing PSO with Randomized Low-Discrepancy Sequences: The Comparative Results. In: Proc. of IEEE Congress on Evolutionary Algorithms, pp. 1985–1992 (2007)

[24] Pant, M., Thangaraj, R., Abraham, A.: Improved Particle Swarm Optimization with Low-discrepancy Sequences. In: Proc. IEEE Cong. on Evolutionary Computation, Hong Kong, pp. 3016–3023 (2008)

[25] Fogel, L.J.: Autonomous Automata. Industrial Research 4, 14–19 (1962)

[26] Bäck, T., Schwefel, H.-P.: An overview of evolutionary algorithms for parameter optimization. Evol. Comput. 1(1), 1–23 (1993)

[27] Fogel, D.B.: Evolutionary Computation: Toward a new Philosophy of Machine Intelligence. IEEE press, Los Alamitos (1995)

[28] Hao, Z.-F., Gua, G.-H., Huang, H.: A Particle Swarm Optimization Algorithm with Differential Evolution. In: Sixth International conference on Machine Learning and Cybernetics, pp. 1031–1035 (2007)

[29] Omran, M.G.H., Engelbrecht, A.P., Salman, A.: Differential Evolution based Particle Swarm Optimization. In: IEEE Swarm Intelligence Symposium (SIS 2007), pp. 112–119 (2007)

[30] Zhang, W.-J., Xie, X.-F.: DEPSO: Hybrid Particle Swarm with Differential Evolution Operator. In: IEEE International Conference on Systems, Man & Cybernetics (SMCC), Washington D C, USA, pp. 3816–3821 (2003)

Fragmentary Synchronization in Chaotic Neural Network and Data Mining

Elena N. Benderskaya[1] and Sofya V. Zhukova[2]

[1] St. Petersburg State Polytechnical University, Faculty of Computer Science,
Russia, 194021, St. Petersburg, Politechnicheskaya 21
bender@sp.ru
[2] St. Petersburg State University, Graduate School of Management,
Russia, 199004, St. Petersburg Volkhovsky Per. 3
sophya.zhukova@gmail.com

Abstract. This paper proposes an improved model of chaotic neural network used to cluster high-dimensional datasets with cross sections in the feature space. A thorough study was designed to elucidate the possible behavior of hundreds interacting chaotic oscillators. New synchronization type - fragmentary synchronization within cluster elements dynamics was found. The paper describes a method for detecting fragmentary synchronization and it's advantages when applied to data mining problem.

Keywords: clustering, cluster analysis, chaotic neural network, chaotic map lattices, fragmentary synchronization.

1 Introduction

General formalization of clustering problem consists in finding out the most rational clusterization K of input data samples X. The division is provided due to some similarity measure between various combinations of the n elements in the dataset ($X = \{x_1, x_2, ..., x_n\}$, $x_i = \{x_{i1}, x_{i2}, ..., x_{ip}\}$). Every element is described by p features (dimension of input space) and can belong simultaneously only to one of m clusters.

The similarity measure depends greatly on mutual disposition of elements in the input dataset. If we have no a priori information about the type of groups (ellipsoidal, ball-shaped, compact, scattered due to some distribution or just chaotically, and this list is endless) then the probability of erroneous measure choice is very high [1, 2].

This is the main reason clusterization to be related to non-formalizable problems. In terms of neural networks it is solved by means of unsupervised learning or learning without a teacher [3], because the system is to learn by itself to extract the solution from input dataset without external aid.

Each clustering technique works under the assumption that input data can be successfully clustered using the concrete similarity measure or their combination.

It can lead to gross mistakes and as a result bring to erroneous solution if this supposition does not fulfill. Moreover it happens to be hard to find express recommendations which method is the most appropriate to cluster a concrete input dataset

E. Corchado et al. (Eds.): HAIS 2009, LNAI 5572, pp. 319–326, 2009.
© Springer-Verlag Berlin Heidelberg 2009

(N-dimensional image) without calling an expert. This in its turn leads us to conclusion that any clustering method allows to obtain only partial solution.

Information about the amount of groups and their topology is frequently unavailable. In this case application of classical algebraic and probabilistic clustering methods does not always provide unique and correct solution. The main idea of these methods is to determine typical representatives of clusters (centers of clusters) in terms of averaging-out [4].

Distributed processing in its pure form is capable to reduce computing complexity but not to improve solution quality. This remark fully agrees with recent applications of self organizing maps (Kohonen's network) and ART neural networks, which provide satisfactory solutions only when clusters separating curves are easily determined in features space or the concept of averaging-out is valid. These artificial neural network models [3] in fact are parallel interpretations of c-means method [1, 2] and thus they possess its drawbacks. Underlying cause lies in the insufficient complexity of the systems. These networks are static. But not only the structure of network (number of elements and links between them) but also its dynamics must meet the requirements of problems' complexity level.

Another promising direction is designing dynamic neural networks. As if in support of the idea numerous investigations in neurophysiology sphere reveal that biological neural networks appear to be nonlinear dynamic systems with chaotic nature of electrochemical signals. Computer science development predetermined great abilities of computer modeling. It became possible to study complex nonlinear dynamics. Great evidence for rich behavior of artificial chaotic systems was accumulated and thus chaos theory came into being [5-7]. Dynamics exponential unpredictability of chaotic systems, their extreme instability generates variety of system's possible states that can help us to describe all the multiformity of our planet. It is assumed to be very advantageous to obtain clustering problem solution using effects produced by chaotic systems interaction. In this paper we try to make next step in the development of universal clustering technique.

2 Oscillatory Clusters and Input Dataset

Emergence of clustering effects turns out to be universal concept in animate nature and in abiocoen. Self-organization occurs in various phenomena such as structures creation, cooperative behavior, etc. Clusters built up from atoms, molecules, neurons are examined in many scientific fields.

Primary results on modeling high dimensional chaotic map lattices were published by K. Kaneko [8]. These works showed up the fact that globally coupled chaotic map lattices exhibit formation of ensembles synchronously oscillating elements. These ensembles were called clusters serving as system's attractors. If there appear to be several clusters then the system is characterized by multistability, when several attractors coexist in the phase space at the same parameters values.

Kaneko's model [8, 9] encompasses a number of identical logistic maps globally coupled with the same strength ε.

In terms of neural networks that means that all synaptic weights w_{ij}, that join element i and element j are equal $w_{ij} = \varepsilon$. Variables change their state in the range [-1, 1] due to special transfer function $f(y(t)) = 1 - \lambda y^2(t)$. The time evolution of the system is given by

$$y_i(t+1) = (1-\varepsilon)f(y_i(t)) + \frac{\varepsilon}{N} \sum_{\substack{j=1 \\ j \neq i}}^{N} f(y_j(t)), \ i, j = \overline{1, N},$$ (1)

where N – number of variables. In [8] was shown that globally coupled chaotic system may occurs several phases: coherent (one cluster), ordered (several big clusters), partially ordered, turbulent (number of clusters coincide with number of variables). However, this abstract clustering phenomenon does not advance us in solving clustering problem.

Leonardo Angelini and his colleagues proposed to apply oscillatory clustering phenomenon to image clustering [10, 12]. The information about input dataset was given to logistic map network by means of inhomogeneous weights assignment

$$W = \{w_{ij}\} = \exp\left(-\frac{d_{ij}^2}{2a}\right), d_{ij} = |x_i - x_j|, \ i, j = \overline{1, N},$$ (2)

where N – number of elements, w_{ij} - strength of link between elements i and j, d_{ij} - Euclidean distance between neurons i and j, a – local scale, depending on k-nearest neighbors. The value of a is fixed as the average distance of k-nearest neighbor pairs of points in the whole system.

Each neuron is responsible for one object in the dataset, but the image itself is not given to inputs, because CNN does not have classical inputs – it is recurrent neural network with one layer of N neurons. Instead, the image (input dataset) predetermines the strength of neurons interactions (similar to Hopfield's network [3]). Evolution of each neuron is governed by

$$y_i(t+1) = \frac{1}{C_i} \sum_{i \neq j}^{N} w_{ij} f(y_i(t)), \ t = 1...T,$$ (3)

$$f(y(t)) = 1 - 2y^2(t)$$ (4)

where $C_i = \sum_{i \neq j} w_{ij}, i, j = \overline{1, N}$, T – time interval, N – number of elements. Neurons state is dependent on the state of all other elements. After transitionary period start to appear synchronous clusters. To reveal them neurons outputs are rounded up to 0 and 1 values and Shannon information is calculated [10, 12] to detect mutual similarity of neurons dynamics.

In the end neurons are joined in clusters several times, because a priori it is unknown the veritable threshold θ that corresponds to real clustering. The value of θ_i controls the resolution at which the data is clustered. Thresholds are chosen with some step in the range of minimum and maximum values in the information matrix. Neural network in which weight coefficients are calculated in compliance with (3) and evolution is given by (4) was called chaotic neural network (CNN). It's name stresses the chaotic functioning dynamics of the system, guaranteed by transfer function (5).

3   CNN Modeling

For a start simple 2D clustering problem illustrated in Fig.1 was solved in terms of Angelini's model. Since we know the answer for a test clustering problem then it is correct to order inputs by their belongings to clusters. It is important to stress that this operation will not change CNN functioning, because initial conditions are set in a random way in the range [-1, 1] as before. This will help us to watch CNN dynamics.

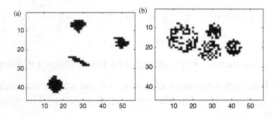

Fig. 1. Clustering problems: (a) - simple clustering problem (134 objects, each described by two coordinates ($p=2$), four clusters are compact and stand apart from each others, clusters comprise correspondingly 49, 36, 30, 19 objects-points); (b) – more complex clustering problem (210 objects arranged with lower density in four clusters close to each other in future space, population is correspondingly 85, 50, 40, 35 points).

In accordance with Angelini's algorithm we received dendrogram on Fig. 2a. Here we can see that decision-making about proper solution can be provided without an expert knowledge, because overwhelming majority of variants (82%) coincide with each other. Fig. 2a shows that when threshold $\theta \in [0.12; 0.69]$ we obtain the only one clusterization, displayed in Fig. 2b that fully agrees with our expectations. The analysis of analogous experimental results indicated that CNN is good enough in solving simple clustering problems as well as all other classical methods.

Next, more complex clustering problem in Fig 1b was solved by means of CNN. Clusterization results showed in Fig. 3 seem to be rather disappointing – only 23% of variants correspond to the expected answer. This makes it impossible to choose the proper clusterization without opinion of an expert, because it is impossible to "guess" automatically that more natural solution can be obtained for $\theta \in [0.48; 0.63]$.

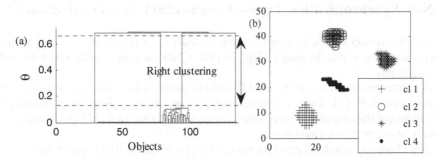

Fig. 2. Clustering results: (a) – dendrogram of solution variants; (b) – the most stable (repeating) clusterization that coincides with real answer and takes 82% of all other variants

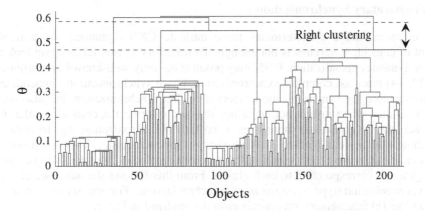

Fig. 3. Dendrogram of clusterizations for problem on Fig. 1b – only 23% of variants coincide with the expected one

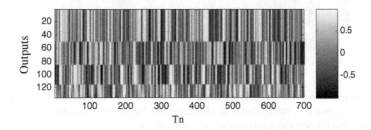

Fig. 4. Chaotic neural network dynamics (statistics of outputs dynamics) while clustering image in Fig. 1a. Color bar to the left shows correspondence between output's absolute value and color in gray scale palette. CNN parameters: $Tp = 1000$, $Tn = 1000$, $k = 20$, random initial conditions.

4   New Synchronization Type – Fragmentary Synchronization

To find the reason of the failure we have to analyze in detail the CNN dynamics. When clustering a simple image (Fig. 1*a*) the CNN outputs evolve synchronously within clusters.

Complete synchronization [8, 13, 14] within clusters takes place in case of simple images, look at Fig. 4. Due to complete agreement within a cluster oscillations and quite different fluctuations of three other clusters the choice of stable clustering result without an expert is not a problem.

CNN clustering results depend on mutual synchronization effects rather than on the processing method used to treat neurons dynamics. To widen the scope of concerned clustering method extensive analysis of possible CNN synchronous regimes has been undertaken.

4.1   Fragmentary Synchronization

More complex image predetermines more intricate CNN dynamics. As a result amount of coincident variants is not enough. To reveal the reason let us again look at outputs dynamics visualization. CNN may produce not only well-known synchronization [13, 14] types as: complete synchronization, phase synchronization, generalized synchronization, lag synchronization, intermittent lag synchronization, but also such synchronization when instant output values in one cluster do not coincide neither by amplitude nor by phase and there is even no fixed synchronization lag. In spite of everything joint mutual synchronization exists within each cluster. This synchronization is characterized by individual oscillation cluster melodies, by some unique "music fragments" corresponding to each cluster. From this follows the name we give to this synchronization type - fragmentary synchronization. For the second problem (look at Fig 1*b*) fragmentary synchronization is visualized in Fig. 5.

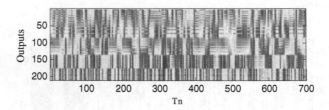

Fig. 5. Chaotic neural network dynamics while clustering image in Fig. 1*b*. Fragmentary synchronization takes place – each cluster is characterized by its own "fragment of a melody". Neurons within one cluster can evolve rather asynchronously.

4.2   Proposed Method to Detect Fragmentary Synchronization

Thorough investigation indicated that unsuccessful results were induced by inefficient method of statistics processing. The use of Shannon's entropy coarse loss of information about transfers consequences. Because in this case real values are replaced by Boolean ones. From results analysis we have inferred the following.

(a) Fragmentary synchronization detection is to be based on the comparison of instant absolute values of outputs but not approximated in this or that way values.

(b) Asynchronous oscillations within one cluster in case of fragmentary synchronization can be nevertheless classified as similar. Let us consider some sequences be more alike than the others and easily related to the same group1. Others may be alike but in a different way and they are joined in group2. To combine group1 and group2 into one we need only one more sequence in the same degree similar to both of groups. In other words joining up not neighboring neurons within one cluster occurs due to similar but not identical dynamics of neurons that lay between them.

In this paper we propose to produce pair comparison of y_i dynamics sequences just as it is done to detect complete synchronization. But besides admissible divergence in instant values of both sequences we introduce bearable percent of moments (counts) where the boundary of admissible divergence may be broken.

Two neurons belong to the same cluster if their dynamics difference is less than ε and this condition is broken less than in p percents in the interval $[Tp+1, Tn]$. In compliance with suggested computational procedure were received experimental results displayed in Fig 6. The bar graph illustrates the appearance frequencies of various clusterizations. Number of analyzed variants is 100. They are received under the condition that threshold ε is changed from 0.05 up to 0.5 with step 0.05, and threshold p is changed from 5% up to 50% with step 5%. To provide a vivid demonstration bar graph represent only variants that appeared more than once. The most frequent clusterization found by means of new method corresponds to the number of four clusters. So the proposed dynamics treatment has obvious advantage over existing processing method, because there appear an: ability to reveal intricate structure of macroscopic attractor represented by means of fragmentary synchronized chaotic sequences and what is more important to find final clusterization without expert assistance because number of identical answers is large enough.

Wide set of clustering experiments with other images showed that fragmentary synchronization detection method can be also used to reveal phase and complete synchronization. Though as it is expected, high solution quality has a considerable cost of great computational complexity.

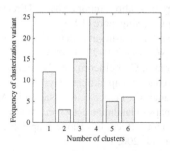

Fig. 6. Clustering results. The most stable variant of clusterization when we have 4 clusters found by improved CNN automatically without a priori information about number and topology of clusters, and the answer coincide with the expected one for problem on Fig. 1b.

5 Conclusion

Increase of system's dimension and nonlinearity degree produces more complex behavior of component parts and the system as a whole and demands complication of analysis techniques. Research results allow to improve clustering quality of chaotic neural network. The main advantage of proposed modifications is the opportunity to solve complex clustering problems without expert assistance in the case if input information about objects is not contradictory (when even the expert cannot provide a decision). The synergism of obtained solution results from multidisciplinary hybridism (neural networks, self-organization, chaos theory, cybernetics) that fully reflects in CNN structure and discovered fragmentary synchronization type. This is one more evidence for the necessity of complex analysis and synthesis. In the paper we received important qualitative results. Future investigations would be devoted to the estimation of quantitative characteristics of the new algorithm, especially computational complexity. And this is sensible only in case of CNN hardware implementation, because both functioning and output processing stages can be provided on the base of array computation and distributed processing.

References

1. Valente de Oliveira, J., Pedrycz, W.: Advances in Fuzzy Clustering and its Applications, p. 454. Wiley, Chichester (2007)
2. Han, J., Kamber, M.: Data Mining. In: Concepts and Techniques. The Morgan Kaufmann Series in Data Management Systems, p. 800. Morgan Kaufmann, San Francisco (2005)
3. Haykin, S.: Neural Networks. In: A Comprehensive Foundation. Prentice Hall PTR, Upper Saddle River (1998)
4. Dimitriadou, E., Weingessel, A., Hornik, K.: Voting-Merging: An Ensemble Method for Clustering. In: Dorffner, G., Bischof, H., Hornik, K. (eds.) ICANN 2001. LNCS, vol. 2130, pp. 217–224. Springer, Heidelberg (2001)
5. Schweitzer, F.: Self-Organization of Complex Structures: From Individual to Collective Dynamics, p. 620. CRC Press, Boca Raton (1997)
6. Mosekilde, E., Maistrenko, Y., Postnov, D.: Chaotic synchronization. World Scientific Series on Nonlinear Science. Series A 42, 440 (2002)
7. Haken, H.: Synergetics. In: Introduction and Advanced Topics, Physics and Astronomy Online Library, p. 758. Springer, Heidelberg (2004)
8. Kaneko, K.: Phenomenology of spatio-temporal chaos. Directions in chaos., pp. 272–353. World Scientific Publishing Co., Singapore (1987)
9. Kaneko, K.: Chaotic but regular posi-nega switch among coded attractors by cluster-size variations. Phys. Rev. Lett. N63(14), 219–223 (1989)
10. Angelini, L., Carlo, F., Marangi, C., Pellicoro, M., Nardullia, M., Stramaglia, S.: Clustering data by inhomogeneous chaotic map lattices. Phys. Rev. Lett. N85, 78–102 (2000)
11. Angelini, L., Carlo, F., Marangi, C., Pellicoro, M., Nardullia, M., Stramaglia, S.: Clustering by inhomogeneous chaotic maps in landmine detection. Phys. Rev. Lett. 86, 89–132 (2001)
12. Angelini, L.: Antiferromagnetic effects in chaotic map lattices with a conservation law. Physics Letters A 307(1), 41–49 (2003)
13. Pikovsky, A., Rosenblum, M., Kurths, J.: Synchronization: A Universal Concept in Nonlinear Sciences. Cambridge University Press, Cambridge (2003)
14. Osipov, G.V., Kurths, J., Zhou, C.: Synchronization in Oscillatory Networks. Springer Series in Synergetics, p. 370. Springer, Heidelberg (2007)

Two-Stage Neural Network Approach to Precise 24-Hour Load Pattern Prediction

Krzysztof Siwek[1] and Stanislaw Osowski[1,2]

[1] Warsaw University of Technology – Dept. of Electrical Engineering
ul. Koszykowa 75, 00-661 Warsaw - Poland
Tel.: +48222347235
[2] Military University of Technology - Dept. of Electronics
ul. Kaliskiego 2, 00-908 Warsaw - Poland
{ksiwek,sto}@iem.pw.edu.pl

Abstract. The paper presents the neural network approach to the precise 24-hour load pattern prediction for the next day in the power system. In this approach we use the ensemble of few neural network predictors working in parallel. The predicted series containing 24 values of the load pattern generated by the neural predictors are combined together using principal component analysis. Few principal components form the input vector for the final stage predictor composed of another neural network. The developed system of prediction was tested on the real data of the Polish Power System. The results have been compared to the appropriate values generated by other methods.

Keywords: load forecasting, neural networks, PCA.

1 Introduction

The prediction of the pattern corresponding to 24 hours ahead load demand is very important for improving the economy of the power system generation and distribution, since enables to deliver the really needed power and in this way reduce the cost of energy. Although many different approaches have been developed in the past [1,3,4,6,7,9,11,13,14] there is still need to improve the accuracy of the predicting systems.

This paper is devoted to the task of 24-hour load pattern forecasting. The authors propose the ensemble [8] of neural predictors working in a parallel way. Each predictor generates the 24-component time series corresponding to 24 hours ahead load demand. These predicted values are concatenated into a longer vector, combining together all predictions. At application of M predictors the length of such vector is $24 \times M$. At this stage we compress this information using the principal component analysis (PCA), reducing the length of such vector to only K most important principal components, at $K << 24 \times M$. The reduced size vectors are used as the input data for the final predictor. This 2-step approach to the power demand forecasting was checked in the numerical experiments concerning the Polish Power System. The results compared to the other neural approaches have shown significant improvement of the accuracy of prediction. For the same data we were able to reduce the total mean squared errors in a significant way.

E. Corchado et al. (Eds.): HAIS 2009, LNAI 5572, pp. 327–335, 2009.

2 The Proposed Methods of Prediction

The key point in our approach is using many independent neural predictors. All of them are fed by the same learning data. Their output signals are grouped together and are subject to PCA analysis. Few first principal components form the input data to the final neural predictor, delivering the finally forecasted values of load pattern for the next 24 hours.

2.1 The Ensemble of Neural Predictors

In the first step of our solution we apply the ensemble of individual neural predictors learned on the same data set. We have used 4 individual predictors: the multilayer perceptron (MLP), Elman recurrent network, Support Vector Machine (SVM) and self-organizing (SO) Kohonen network.

The first three predictors make use of the universal approximation ability of these supervised learned networks. To represent the generally unknown, next day load pattern, we map the past loads of the system into the present forecasted load at dth day and hth hour. Our general supervised model of the load may be described in the following mathematical form [11]

$$\hat{P}(d,h) = f\big(\mathbf{w}, t, s, P(d,h-1), ..., P(d,h-H), P(d-1,h), ..., P(d-D,h-H)\big) \qquad (1)$$

where $\mathbf{w}$ represents the vector of parameters of the network, H and D - the number of past hours and days, respectively, influencing the prediction process, t - the type of the day (workdays, Friday, Saturday or holidays) and s - the season of the year (autumn, winter, spring or summer). The value $\hat{P}(d,h)$ represents the predicted loads and the values $P(d-i,h-j)$ written without hat – the known values of the load from the past. All data samples have been normalized dividing the real load by the mean value of the data base of the Power System of the years taking part in experiments. In this mathematical model we have omitted the temperature because the prediction is concerned with the data corresponding to the territory of the whole country. The temperature is changing a lot in different regions of the country, so it would be difficult to adjust the proper value of the temperature for the particular day. However in the case of forecasting the energy consumption for a small region the inclusion of temperature and gradient of temperature in the model would be beneficial and easy to consider in our model.

The particular form of the applied predictors depends on their structure and way of learning. The MLP network [5] consists of many simple neuron-like processing units of sigmoidal activation function grouped together in layers. The number of hidden layers and neurons of sigmoidal non-linearity are usually subject to adjustment in an experimental way by training different structures and choosing the smallest one, still satisfying the learning accuracy.

Elman network is a two layer recurrent structure of sigmoidal neurons applying the feedback from the output of the hidden layer to the input of the network [5]. This feedback path allows Elman network to learn to recognize and generate temporal patterns, which are of interest in prediction. The learning strategy of Elman network uses the similar principle of minimization of error function defined for learning data.

Support Vector Machine is a very peculiar neural like structure used here for regression task [12]. It contains only one output unit, hence to solve the prediction of the load pattern for 24 hours ahead we have to train 24 separate networks. In training them we use the same input data as for MLP and Elman networks. The learning strategy of SVM network is relied on another philosophy than the other classical neural networks, like MLP, RBF or Elman. Instead of minimizing the error function defined for the learning data it minimizes the weights of the network, while keeping the output signals as close as possible to their destination values. In fact SVM applies the quadratic optimization approach to learning. Excellent presentation of the details of learning strategy of SVM can be found in textbooks of Scholkopf and Smola [12].

To differentiate the types of predictors as much as possible we have additionally applied the self-organizing Kohonen network approach [1,10]. The main task of the self-organizing network is to learn the characteristics of the daily loads (profile vectors) of the system, defined in the way

$$p(d,h) = \frac{P(d,h) - P_m(d)}{\sigma(d)} \tag{2}$$

for each day d and 24 hours h=1, 2, ..., 24, where $P(d,h)$ is the real load of dth day at hth hour, $P_m(d)$ is the mean value of the load of dth day and $\sigma(d)$ is the standard deviation of the load of dth day. The days of the same type belonging to the same seasons of the year have similar profile patterns and form clusters, grouping the similar data. Each cluster is represented by one neuron, acting in the competitive mode. The set of 24-dimensional vector profiles $\hat{\mathbf{p}}(d) = [\hat{p}(d,1), \hat{p}(d,2),..., \hat{p}(d,24)]^T$ for different days of the years taking part in learning process, forms the training data of the network. Once the network is trained, each neuron represents the data closest to its weight vector in the chosen metric space. The prediction of the load for dth day and hth hour may be expressed now in the form

$$\hat{P}(d,h) = \hat{\sigma}(d)\hat{p}(d,h) + \hat{P}_m(d) \tag{3}$$

where the variables with hat mean the predicted values. To make the prediction of the load $\hat{P}(d,h)$ for the particular day and hour we have to know not only the load profile vector $\mathbf{p}(d)$, but also the mean value and standard deviation of load for this day.

The predictions of the mean value and standard deviation for the particular day have been obtained by applying the standard MLP network [11] in a way very similar to the already presented MLP approach. The profile vector prediction for dth day is estimated by averaging the winner vectors of the Kohonen network for this particular day (for example Tuesdays in July) from the past history, i.e.,

$$\hat{\mathbf{p}}(d) = \frac{\sum_{i=1}^{n} k_{di}\mathbf{w}_i}{\sum_{i=1}^{n} k_{di}} \tag{4}$$

where k_{di} is the quantity of appearances of ith neuron among the winners in the past for this particular day type and $\mathbf{w}_i$ is the weight vector of the winner.

2.2 Principal Component Analysis of Data

Consider a data set generated by M predictors arranged in the form of matrix $\mathbf{Z}$ of the size $p \times (24M)$, where p is the number of days under prediction and M the number of predictors. Each predictor output contains 24 predicted power demands for the particular hours of the day. The rows of the matrix $\mathbf{Z}$ are the composition of the concatenated outputs of M predictors for the respective day. The aim of the principal component analysis (PCA) is to map these p high-dimensional vectors into a lower dimensional space [2]. In this way each long vector $\mathbf{z}$ (the rows of the matrix $\mathbf{Z}$) will be represented now by the vector $\mathbf{y}$ of smaller dimension K, containing sufficiently high percentage of the most important part of the original information. This is done by using PCA [2]. On the basis of learning data set we form the PCA matrix $\mathbf{W}$ transforming the 24M-dimensional vector $\mathbf{z}$ into K-dimensional vector $\mathbf{y}$, where $\mathbf{y=Wz}$. This linear transformation forms the low-dimensional vector $\mathbf{y}$, representing the essential part of information concerning the load pattern for the next 24 hours.

2.3 Final Predictor

The set of p low-dimensional vectors $\mathbf{y}$ is used in the second stage as the training data for the final predictor, whose output signals will represent the forecasted 24-component time series. To get high quality of prediction results we have to use the predictor of highest possible accuracy. On the basis of results of the first stage prediction we decided to use two best neural structures: the MLP and SVM. In the case of MLP only one network of 24 linear output neurons is used. Applying SVM we have to train 24 SVM structures (each specializing for the particular hour of the day) fed by the same input data. As the input data for learning final predictor we have used the pairs $(\mathbf{y}_i, \mathbf{d}_i)$ for $i=1, 2, ..., p$. Vectors $\mathbf{y}_i$ result from PCA analysis and $\mathbf{d}_i$ are the known load patterns used in learning the ensemble of predictors in the first step of our approach. Fig. 1 presents the final forecasting system proposed in this paper.

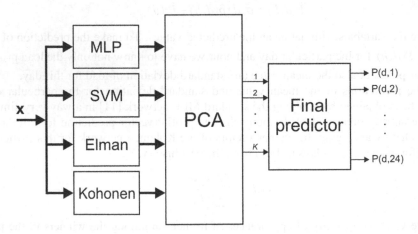

Fig. 1. The diagram of the proposed 2-stage forecasting system

3 Data Base

The numerical experiments have been performed for the data of the Polish Power System (PPS) of three years (over 26280 hours). The same data set has been used for each individual predictor. The first two years (17520 hours) have been applied only in learning stage and the last year, not taking part in learning (8760 hours) has been left for testing all trained predictors. The data samples have been normalized dividing the real load by the mean value of the data base of the Polish Power System of 3 years taking part in the experiments. Fig. 2 presents the hourly load demand (the normalized values) of PPS within the analyzed three years.

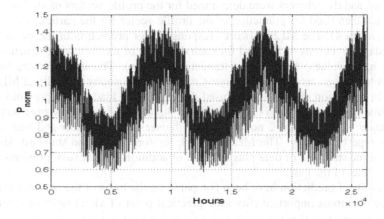

Fig. 2. The hourly change of the normalized power demand in PPS within 3 years

We can see great variation of the load demand, changing from hour to hour. At the total mean equal P_m=16019MW the standard deviation is equal σ=2800MW (the real, not normalized values). This means very high diversity of the load patterns and significant difficulties in their accurate prediction.

4 Numerical Results

The individual predictors have been adapted on the basis of the learning data. The optimal MLP network of the structure 23-20-19-24 was trained using the conjugate gradient algorithm [5]. The input signals of the neural network have been formed by the normalized loads of the nearest past 4 hours of the actual day and 5 hours (the actual hour and 4 nearest past) for 3 previous days (19 components together), as well as the type of the season (two nodes coded in binary way: 00 – spring, 01 - summer, 10 – autumn an 11 – winter) and type of the day (two nodes: 11 – working days, 10 – Saturdays, 01 – Fridays, 00 –holidays). The hidden layer neurons were sigmoidal. The particular numbers of hidden neurons have been determined in the introductory experiments using the validation data set extracted from the learning data (1/5 of the set). Each output linear neuron was responsible for prediction of the load for the particular hour of the day.

The Elman recurrent network structure (23-8-24) applied also 23 input nodes containing the same signals as in MLP, 8 hidden neurons and 24 output neurons. The SVM network of also 23 inputs applies special strategy of prediction since it possesses only one output neuron. We have trained 24 separate SVM networks of Gaussian kernel functions for prediction of 24-point time series (each SVM network responsible for prediction of the load of the particular hour of the day). The modified Platt algorithm [12] was applied in learning all SVM networks, implemented on Matlab platform [15].

In the self-organizing (SO) approach we have applied 100 self-organizing neurons for prediction of the profiles. This number was found after series of introductory experiments. After adapting the Kohonen network the learning data of all days have been tested and the winners were determined for the profile vectors of all days. These winners are then used for prediction of the profile vector for the particular day in the future (relation 4). The MLP networks responsible for prediction of the mean values and standard deviations of the load for each day were of the following structures: 10-6-1 (the mean values) and 14-8-1 (standard deviation). The input vectors for both networks have been arranged by applying the same philosophy as in direct MLP prediction. In prediction of the mean we used the daily mean loads of the previous 3 days of the same week and of 4 days (the actual and 3 previous days) of the previous week. Additionally we have used 2 nodes to code the season of the year and one node to code the type of the day. The MLP network for forecasting the standard deviation used the same structure of data plus the data of additional week (two previous weeks instead of one) of the past.

The results of prediction have been compared on the basis of the committed errors. There are four most important (from the practical point of view) types of errors. Let us denote by $P(h)$ and $\widehat{P}(h)$ the real and predicted load at hth hour, respectively and by n the total number of hours of prediction. We have adopted the following definitions of errors:

- the mean absolute percentage error (MAPE)

$$MAPE = \frac{1}{n}\sum_{h=1}^{n}\frac{\left|P(h)-\widehat{P}(h)\right|}{P(h)}\cdot 100\% \tag{5}$$

- the mean squared error (MSE)

$$MSE = \frac{1}{n}\sum_{h=1}^{n}\left[P(h)-\widehat{P}(h)\right]^{2} \tag{6}$$

- the normalized mean squared error (NMSE)

$$NMSE = \frac{MSE}{\left[mean(P)\right]^{2}} \tag{7}$$

where $mean(P)$ represents the mean value of the load in the time period of prediction

- the maximum percentage error (MAXPE)

$$MAXPE = \max\left\{\frac{\left|P(h)-\widehat{P}(h)\right|}{P(h)}\cdot 100\%\right\} \tag{8}$$

The errors have been calculated separately for the learning and testing data. Here we will present only the testing errors, related to the data not taking part in learning, since this information is the most important from the practical point of view. Taking into account the stochastic nature of the learning algorithms used in training the neural networks we have repeated the learning and testing procedures 20 times, determining their means and standard deviations. Table 1 presents the obtained values of mean and standard deviation of these testing errors of all individual predictors for one year (365 days corresponding to 8760 hours) not taking part in learning.

Table 1. The testing errors of the load forecasting for the Polish Power System using individual predictors

Method	MAPE [%]	MAXPE [%]	MSE [MW²]	NMSE
MLP	2.07±0.14	16.92	(1.75±0.12)e5	(6.82±0.47)e-4
SVM	2.24±0	28.32	2.94e5±0	1.17e-3±0
Elman	2.26±0.08	24.95	(3.14±0.11)e5	(1.22±0.04)e-3
SO	2.37±0.02	18.10	(2.40±0.012)e5	(9.35±0.05)e-4

It is evident that in any respect the most accurate is the MLP predictor. The results generated by all individual predictors have been processed according to the described procedure: first by PCA (24 main principal components selected) and then by second stage prediction using independently MLP and SVM as the final predictor. The mean errors and standard deviations of the final prediction results are gathered in Table 2.

Table 2. The testing errors of the final load forecasting for the Polish Power System

Method	MAPE [%]	MAXPE [%]	MSE [MW²]	NMSE
MLP final predictor	1.48±0.09	14.29	(1.044±0.06)e+5	(4.07±0.35)e-4
SVM final predictor	1.35±0.05	10.74	(9.50±0.05)e+4	(3.70±0.02)e-4

The presented results confirm very high efficiency of the second predicting stage. All error measures have been reduced significantly with respect to the best individual predictor (MLP) The MAPE was reduced by more than 28%, MSE by 46% and MAXPE by 36%.

It is interesting to compare the accuracy of our results with the approaches presented in the other papers. The same data of the Polish Power System have been predicted by the authors of [13]. This paper has given the results only in the form of the normalized mean squared error (NMSE) defined as the real MSE value divided by the square of the mean value of the load. The best resulting NMSE of [13] was equal NMSE=1.8e-3. Our best result corresponding to the same data was equal 3.70e-4.

5 Conclusions

The paper has presented the new approach to forecasting the 24-hour load pattern of the power system. In the proposed solution many different predictors are trained simultaneously and their results combined together using principal component analysis and second stage prediction. In the classical approach the less fortunate predictors are usually discarded and the results of the best one are treated as the final outcome. In the presented approach we analyze all of them and take into account at preparation of the final forecast. In the presented solution we have used four individual predictors although this number may be easily extended without any significant change of the general procedure.

We have tried two different neural structures to integrate the results of individual predictors. One used the MLP neural network of 24 outputs and the second – 24 SVM networks, each responsible for the load of the particular hour of the day. The best results have been obtained at application of the SVM integrator.

The experimental results have shown that the performance of the individual predictors can be improved significantly by the integration of their results. The improvement is observed even at application of different quality predictors. For the data corresponding to the Polish Power System and application of four different predictors we have got 28% relative improvement of MAPE and more than 40% of MSE error over the best individual predictors (the MLP network).

Acknowledgments. This work is supported by Polish Ministry of Science and Higher Education by grant in the years 2008-2010.

References

[1] Cottrell, M., Girard, B., Girard, Y., Muller, C., Rousset, P.: Daily electrical power curve: classification and forecasting using a Kohonen map. In: Sandoval, F., Mira, J. (eds.) IWANN 1995. LNCS, vol. 930, pp. 1107–1113. Springer, Heidelberg (1995)

[2] Diamantras, K., Kung, S.Y.: Principal component neural networks. Wiley, N.Y (1996)

[3] Fidalgo, J.N., Pecas Lopez, J.: Load forecasting performance enhancement when facing anomalous events. IEEE Trans. Power Systems 20, 408–415 (2005)

[4] Gonzalez-Romera, E., Jaramillo-Moran, M.A., Carmona-Fernandez, D.: Monthly electric energy demand forecasting based on trend extraction. IEEE Trans. Power Systems 21, 1946–1953 (2006)

[5] Haykin, S.: Neural networks, a comprehensive foundation. Macmillan, N.Y (2002)

[6] Hippert, H.S., Pedreira, C.E., Souza, R.C.: Neural networks for short-term load forecasting: a review and evaluation. IEEE Trans. on Power Systems 16, 44–55 (2001)

[7] Kandil, N., Wamkeue, R., Saad, M., Georges, S.: An efficient approach for short term load forecasting using artificial neural networks. Electrical Power and Energy Systems 28, 525–530 (2006)

[8] Kuntcheva, L.: Combining pattern classifiers - methods and algorithms. Wiley, New Jersey (2004)

[9] Mandal, P., Senjyu, T., Urasaki, N., Funabashi, T.: A neural network based several hours ahead electric load forecasting using similar days approach. Electrical Power and Energy Systems 28, 367–373 (2006)

[10] Osowski, S., Siwek, K.: The self-organizing neural network approach to load forecasting in power system. In: Int. Joint Conf. on Neural Networks, Washington, pp. 1345–1348 (1999)

[11] Osowski, S., Siwek, K.: Regularization of neural networks for load forecasting in power system. In: IEE Proc. GTD, vol. 149, pp. 340–345 (2002)

[12] Schölkopf, B., Smola, A.: Learning with Kernels. MIT Press, Cambridge (2002)

[13] Sorjamaa, A., Hao, J., Reyhani, N., Li, Y., Lendasse, A.: Methodology for long-term prediction of time series. Neurocomputing 70, 2861–2869 (2007)

[14] Yalcinoz, T., Eminoglu, U.: Short term and medium term power distribution load forecasting by neural networks. Energy Conversion and Management 46, 1393–1405 (2005)

[15] Matlab manual, user's guide, MathWorks, Natick (2002)

Tentative Exploration on Reinforcement Learning Algorithms for Stochastic Rewards

Luis Peña[1], Antonio LaTorre[2], José-María Peña[2], and Sascha Ossowski[1]

[1] Artificial Intelligence Department, Universidad Rey Juan Carlos
{luis.pena,sascha.ossowski}@urjc.es
[2] Computer Architecture Department, Universidad Politécnica de Madrid
{atorre,jmpena}@fi.upm.es

Abstract. This paper addresses a way to generate mixed strategies using reinforcement learning algorithms in domains with stochastic rewards. A new algorithm, based on Q-learning model, called TERSQ is introduced. As a difference from other approaches for stochastic scenarios, TERSQ uses a global exploration rate for all the state/actions in the same run. This exploration rate is selected at the beginning of each round, using a probabilistic distribution, which is updated once the run is finished. In this paper we compare TERSQ with similar approaches that use probability distributions depending on state-action pairs. Two experimental scenarios have been considered. First one deals with the problem of learning the optimal way to combine several evolutionary algorithms used simultaneously by a hybrid approach. In the second one, the objective is to learn the best strategy for a set of competing agents in combat-based videogame.[1]

1 Introduction

Stochastic games (SGs) and Markov decision processes (MDPs) have been studied in the literature as typical agent-based scenarios in which reinforcement learning (RL) has been successfully applied. Although SGs differ from MDPs in the existence of multiple agents performing simultaneous actions (and all their actions determine the next state), there are other multiagent problems, like matrix games (sometimes referred as strategic games), that also consider the existence of multiple actors in the problem. However, SGs and MDPs have one common characteristic: *"state transitions are non-deterministic"*. Transitions depend on the action performed by the agent in MDPs, or the combination of actions of all agents in SGs, but in a stochastic way. This characteristic motivates the use of stochastic-specific learners [1,2,3].

---

[1] The present work has been partially funded by the Spanish Ministry of Science and Innovation through the projects THOMAS-COIN (grant TIN2006-14630-C03-02), PEO-HCDP (grant TIN2007-67148), and also with the Madrid Regional Education Ministry IV PRICT. This project has been carried out in collaboration of CeSViMa supercomputing center.

E. Corchado et al. (Eds.): HAIS 2009, LNAI 5572, pp. 336–343, 2009.

RL methods also use a reward function which evaluates the effect of carrying out one action in a given situation. This expected value is obtained by a reward function on the new state. In many RL problems, reward functions, as well as state transitions, are non-deterministic.

In this paper, a new RL algorithm is presented. This algorithm, named Tentative Exploration by Restricted Stochastic Quota (TERSQ), is compared against two state-of-the-art RL algorithms for stochastic problems. TERSQ optimizes a stochastic quota through different learning executions. This quota parametrizes a binomial decision process that selects whether the algorithm should perform a deterministic or a biased exploration. This quota acts as a variable learning rate in the RL algorithm. In order to evaluate TERSQ performance, two problems (one MDP and a SG) have been considered: (i) adaptative features for hybrid evolutionary algorithms, and (ii) behavior of competing characters in videogames.

The rest of the paper is organized as follows: section 2 reviews the related work on RL algorithms for stochastic games. Section 3 defines TERSQ algorithm. In section 4, the experimental scenarios for both problems are described. Finally, section 5 details the conclusions derived from this study.

2   Related Work

The objective of the agents is to select the best actions (best response) to maximize a reward function (attenuated by a factor of γ). In the context of non-deterministic learning scenarios, the agent should be able to identify mixed strategies which are functions that assign a probability distribution to agent's next action $\rho_i : \mathcal{S} \to PD(\mathcal{A}_i)$.

Despite deterministic RL algorithms (such as Q-learners) are not appropriate to deal with MDPs and SGs, some variants of these algorithms have been successfully applied on these scenarios. PHC and WoLF [1] are extensions to the Q-learner algorithm particularly designed to deal with stochastic scenarios.

In [4], the authors extends WoLF algorithm to incorporate the concept of Infinitesimal Gradient Ascent (IGA) presented by [5] to define the "wining" situations required to update the learning rate in WoLF. GIGA-WoLF [6] is an extension of the latter considering the concept of Generalized IGA [7]. BL-WoLF [2] is an enhanced version of WoLF that provides a bounded-loss where the cost of learning is measured by the losses the learning agent accrues (rather than the number of rounds). Another variant is Hyper-Q [3], in which values of mixed strategies rather than base actions are learned, and in which other agents strategies are estimated from observed actions via Bayesian inference. WPL (Weighted Policy Learner) [8] is a new RL algorithm which does not assumed an agent knew the underlying game structure.

3   TERSQ Algorithm

In this work we introduce the TERSQ algorithm for reinforcement learning environments. The main idea of this algorithm is to use a global stochastic quota,

σ, in order to select the action to be executed. A binomial decision process is performed in such a way that action with best Q-values are selected with a probability of σ while the rest of the actions are stochastically selected with a probability of $1 - \sigma$ according to their Q-value ranking.

Let
$\mathcal{A}$ be the set of possible actions for the state s, and $A_i \in \mathcal{A}$ one action for this state,
α, γ the learning parameters,
$\sigma \in \Gamma = \{0.0, 0.1, \ldots, 1.0\}$ the global quota used to select A_{max},
$\tau(\sigma)$ the average performance of σ

1. Let σ be selected from Γ following the specific criteria of the current phase.
2. Initialize $Q(A_i, s) = 0$.
3. Repeat until the round has finished:
   (a) For each action A_i on each state s, a basic probability $\pi(s, A_i)$ is obtained by a ranking process where actions are sorted according to their Q-values in an increasing order:

$$\{A_i'\} = sort(\{A_i\}) \tag{1}$$

$$\pi(s, \{A_i'\}) = i \times \pi_0 \ / \ \sum \pi(s, \{A_i'\}) = 1 \tag{2}$$

   (b) These probabilities are adjusted by the σ quota as follows,

$$\widehat{\pi(s, A_i)} = \pi(s, A_i) \times (1 - \sigma), \quad A_i \neq A_{max} \tag{3}$$

   and for the A_{max} (action with the best actual Q-value)

$$\widehat{\pi(s, A_{max})} = (\pi(s, A_{max}) \times (1 - \sigma)) + \sigma \tag{4}$$

   (c) Select action A_i with probability $\widehat{\pi(s, A_i)}$.
   (d) Q-values are updated observing reward r and next state s',

$$Q(A_i, s) = (1 - \alpha)Q(A_i, s) + \alpha \left(r + \gamma \max_{A_j} Q(A_j, s') \right)$$

4. Update the $\tau(\sigma)$ according to the evaluation of the round.

The σ value is selected for each round based on three different criteria. From these criteria, three phases can be established: (1) *Tentative Phase* in which the algorithm tries all the possible σ values (from a finite set of values, named Γ) to get an initial estimation of the performance of every possible σ value, (2) σ *Adjustment Phase* where σ values are proportionally chosen according to their average performance $\tau(\sigma)$ (which is updated at the end of each round), and (3) *Optimal σ Phase* where the σ value with highest average performance is selected for the rest of the learning process. The usual Q-learning technique is applied during all the process.

4 Experimental Scenarios

4.1 Hybrid Evolutionary Algorithm

Preliminaries. According to Sinha and Goldberg [9], three are the main reasons for hybridization in Evolutionary Algorithms:

1. An improvement in the performance of the Evolutionary Algorithm (for example, the speed of convergence).
2. An improvement in the quality of the solutions obtained by the Evolutionary Algorithm.
3. To incorporate the Evolutionary Algorithm as a part of a larger system.

One alternative to deal with Hybrid Evolutionary Algorithms is a dynamically adjusted framework, named Multiple Offspring Sampling (MOS). This framework is able to simultaneously handle several evolutionary approaches to produce the new offspring and dynamically adjust the participation of each of these approaches according to their current performance. [10] provides a completed presentation of MOS.

Learning the Hybrid Strategy. When working with hybrid evolutionary algorithms it is hard to guess which is the best way to handle each of the different offspring mechanisms present in the hybrid approach. Static participation ratios could lead to suboptimal results and to a waste of resources (creation of solutions by means of algorithms with a poor performance). A dynamic adjustment of the participation of each of the reproductive techniques that compose the hybrid algorithm can solve most of these problems. However, several executions of the same dynamically adjusted hybrid algorithm can result in different curves of participation. At this point, the question is if it is possible to learn a nearly optimal way to adapt the participation of the different available techniques on the overall search process through different stages of the algorithm execution.

For this purpose, a hybrid evolutionary algorithm with reinforced-learning capabilities has been proposed. It is able to learn the optimal way for adapting participation by using one of the following RL algorithms: TERSQ, described in this contribution, Policy Hill Climbing (PHC) and WoLF [1]. In this algorithm, an action is the creation of an offspring individual by a particular reproductive mechanism. The set of possible states is defined by a discretized participation value for each of the evolutionary techniques. A state transition is performed when the ratio of individuals produced by a technique introduces a change in the discretized participation value.

340   L. Peña et al.

Experimentation. For this experimentation two continuous optimization functions proposed for the CEC'08 Special Session and Competition on Large Scale Global Optimization [11] have been selected. Shifted Rastrigin's function is a multi-modal, shifted and separable function with a huge number of local optima. Schwefel's Problem is an unimodal, shifted and non-separable function. Both are good examples of hard optimization functions were a hybrid evolutionary approach can be successfully exploited to obtain better results than with single algorithms.

	UCUM	BCUM	UCGM	BCGM
Evolutionary Model	GA			
Initializer	Uniform			
Crossover	Uniform	BLX-α	Uniform	BLX-α
Mutator	Uniform		Gaussian	

Previous table presents the set of techniques used by the hybrid evolutionary algorithm. These four reproductive mechanisms are simultaneously used by the hybrid algorithm. Each time an individual is created, this action is recorded and the Q-values updated. The three aforementioned RL algorithms are tested on the two proposed functions. For the TERSQ algorithm, 11 rounds are performed in the Tentative Phase, 50 rounds in the σ Adjustment Phase and 50 rounds in the σ Optimal phase. For the other two algorithms, PHC and WoLF, 111 rounds are executed.

Phase	PHC	WoLF	TERSQ	
Tentative	–	–	1,97E-01	Rastrigin
	–	–	5,87E+00	Schwefel
σ Adjustment	–	–	1,96E-01	Rastrigin
	–	–	5,69E+00	Schwefel
σ Optimal	3,81E-01	2,22E-01	1,76E-01	Rastrigin
	7,87E+00	7,16E+00	5,64E+00	Schwefel

This table presents the results obtained by the three RL algorithms. From these results, we can observe that the average error reported by the TERSQ algorithm is smaller in both functions even in the Tentative Phase. Moreover, the TERSQ algorithm is able to improve its results in both the σ Adjustment and the σ Optimal phases, obtaining average errors 54% and 21% smaller compared to PHC and WoLF, respectively, on the Rastrigin function, and 28% and 21% smaller on the Schwefel Problem.

4.2   Videogame Characters

Environment Description. The second experiment is the combat between two characters to bring enemy's Hit Points counter to 0. Every character has two state counters, Hit Points (HPs) that represents the remaining life for this character and the Exahustion Points (EPs) counter that shows the fatigue level of the character. If HPs reach 0 the character is dead and thus it is defeated. On the other hand, if EPs are below 0 the character cannot do anything but rest until it is recovered.

A character can perform two types of actions: Offensive and Defensive Actions. Offensive Actions take a fixed amount of time to be executed, named Action Points (APs), which represent the time that the action takes to be triggered after it is called. In addition, Offensive Actions consume some EPs when they are triggered. Once an Offensive Action is fired it has a probability of hitting the target and inflicting some damage. The damage of an action can be of three types: (1) HPs damage, (2) EPs damage and (3) Stun damage. The first two damage types represent a direct amount to be substracted to the respective counter of the enemy. The stun damage works in a different way: this type of damage makes the target to cancel his present declared action and makes that the target cannot declare any other action until he gets recovered from the stun.

Table 1. Results for videogames characters scenario

		Wins					
		A-TERSQ	A-PHC	A-WoLF	B-TERSQ	B-PHC	B-WoLF
	A-TERSQ		50,56%	50,44%	44,61%	39,32%	38,38%
	A-PHC	49,44%		50,18%	36,49%	34,91%	33,32%
Vs	A-WoLF	49,56%	49,82%		35,87%	34,57%	33,13%
	B-TERSQ	55,39%	63,51%	64,13%		53,94%	53,02%
	B-PHC	60,68%	65,09%	65,43%	46,06%		49,42%
	B-WoLF	61,62%	66,68%	66,87%	46,98%	50,58%	
	N Wins	55,33%	59,19%	59,45%	41,96%	42,55%	41,34%

(a) Ratio of wins versus other characters

		A-TERSQ	A-PHC	A-WoLF	B-TERSQ	B-PHC	B-WoLF
	A-TERSQ		37,38%	36,52%	52,23%	33,96%	33,05%
	A-PHC	62,62%		49,47%	49,07%	41,62%	40,52%
Vs	A-WoLF	63,48%	50,53%		49,91%	42,93%	41,58%
	B-TERSQ	47,77%	50,93%	50,09%		51,71%	48,71%
	B-PHC	66,04%	58,38%	57,07%	48,29%		48,98%
	B-WoLF	66,95%	59,48%	58,42%	51,29%	51,02%	
	N Wins	61,56%	51,53%	50,47%	50,13%	44,27%	42,61%

(b) Results on fixed σ stage

APs of Defensive Actions represent the time the defense is active when it is declared, consuming the EPs when it finishes. If a character is hit and he has a Defensive Action declared, he has a probability of blocking the attack. If the Defensive Action blocks the attack, the damage taken by the character is reduced by a factor applied to the HPs and EPs damages and, if the Defensive Action specifies it, to the stun damage.

Once the combat begins, every character choices which action he wants to declare and the APs are added to the current instant counter. When an action is triggered, it is resolved depending on its type, and then the character choices another action if he has any remaining EPs. If he has none, the character must rest for a fixed amount of time to recover some EPs.

Experimentation. To evaluate the TERSQ algorithm on this environment, two different character profiles, A and B, have been created. Each profile defines specific HPs, EPs and action characteristics. For each of these profiles, three RL algorithms are used: TERSQ, PHC and WoLF resulting six different characters (each of the two profiles and each of the three RL algorithms). The experiment consists of 10 executions of 200000 combats (rounds). For each combat, two characters are randomly selected from the six available characters. The Q-values and the learning rates are reseted when each execution begins.

Table 1.a) presents the winning percentage for each pair of characters averaged for the 10 executions. These results show a better performance of PHC or WoLF for each of the character profiles.

Table 1.b) shows the winning ratios restricted to the last combats of every experiment, once σ has been selected. These last results emphasize a significative better performance of the two characters controlled by TERSQ algorithm. The two character profiles are different: **B** is worse than **A**, but despite of this B-TERSQ character profile beats nearly half of times against **A**-profiles, outperforming others **B**-profiles.

The figure 1 shows the evolution of the winnig ratio along the combats with the inflexion points that marks the differents phases at 2000 (end of Tentative) and 120000 (end of σ Adjustment).

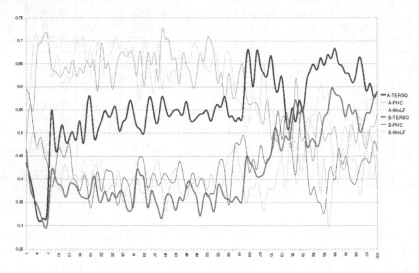

Fig. 1. Ratio evolution across the combats, sampled by 2000

5   Conclusions

In this paper, a new RL algorithm named TERSQ has been presented. This algorithm differs from PHC and WoLF in how the selection probability for each state-action pair is computed. PHC and WoLF maintain a separate matrix for these probabilities. This forces these two algorithms to learn both the attenuated reward value (Q-value) and the selection probability. TERSQ implements a mechanism for directly computing this probability from the Q-values using a global quota, σ. Two different experimental scenarios have been proposed. In the first scenario, the proposed RL algorithm shows better average performance on the two proposed functions. Furthermore, TERSQ is able to improve its own average performance in each of the subsequent phases. In the second scenario, once the optimal σ value has been selected, its results clearly outperform those of the competing algorithm. This behavior can be explained as a more explorative training phase, in which results are not very good, provides better information that can be exploited in the last phase.

References

1. Bowling, M., Veloso, M.: Multiagent learning using a variable learning rate. Artificial Intelligence 136, 215–250 (2002)
2. Conitzer, V., Sandholm, T.: Bl-wolf: A framework for loss-bounded learnability in zero-sum games. In: International Conference on Machine Learning (ICML), pp. 91–98 (2003)
3. Tesauro, G.: Extending q-learning to general adaptive multi-agent systems. In: Advances in Neural Information Processing Systems, vol. 16, p. 2004. MIT Press, Cambridge (2004)
4. Bowling, M., Veloso, M.: Convergence of gradient dynamics with a variable learning rate. In: Proceedings of the Eighteenth International Conference on Machine Learning, pp. 27–34. Morgan Kaufmann, San Francisco (2001)
5. Singh, S., Kearns, M., Mansour, Y.: Nash convergence of gradient dynamics in general-sum games. In: Proceedings of the Sixteenth Conference on Uncertainty in Artificial Intelligence, pp. 541–548. Morgan Kaufmann, San Francisco (2000)
6. Bowling, M.: Convergence and no-regret in multiagent learning. In: Advances in Neural Information Processing Systems, vol. 17, pp. 209–216. MIT Press, Cambridge (2005)
7. Zinkevich, M.: Online convex programming and generalized infinitesimal gradient ascent. In: Proceedings of the Twentieth International Conference on Machine Learning, pp. 928–936 (2003)
8. Abdallah, S., Lesser, V.: A Multiagent Reinforcement Learning Algorithm with Non-linear Dynamics. Journal of Artificial Intelligence Research 33, 521–549 (2008)
9. Sinha, A., Goldberg, D.: A survey of hybrid genetic and evolutionary algorithms. Technical Report 2003004, Illinois Genetic Algorithms Laboratory, IlliGAL (2003)
10. LaTorre, A., Peña, J., González, S., Robles, V., Famili, F.: Breast cancer biomarker selection using multiple offspring sampling. In: Proceedings of the ECML/PKDD 2007 Workshop on Data Mining in Functional Genomics and Proteomics: Current Trends and Future Directions, Warsaw, Poland. Springer, Heidelberg (2007)
11. Tang, K., Yao, X., Suganthan, P., MacNish, C., Chen, Y., Chen, C., Yang, Z.: Benchmark functions for the cec 2008 special session and competition on large scale global optimization. Technical report, Nature Inspired Computation and Applications Laboratory, USTC, China (2007)

Comparative Evaluation of Semi-supervised Geodesic GTM

Raúl Cruz-Barbosa[1,2] and Alfredo Vellido[1]

[1] Universitat Politècnica de Catalunya, 08034, Barcelona, Spain
{rcruz,avellido}@lsi.upc.edu
[2] Universidad Tecnológica de la Mixteca, 69000, Huajuapan, Oaxaca, México

Abstract. In many real problems that ultimately require data classification, not all the class labels are readily available. This concerns the field of semi-supervised learning, in which missing class labels must be inferred from the available ones as well as from the natural cluster structure of the data. This structure can sometimes be quite convoluted. Previous research has shown the advantage, for these cases, of using the geodesic metric in clustering models of the manifold learning family to reveal the underlying true data structure. In this brief paper, we present a novel semi-supervised approach, namely Semi-Supervised Geo-GTM (SS-Geo-GTM). This is an extension of Geo-GTM, a variation on the Generative Topographic Mapping (GTM) manifold learning model for data clustering and visualization that resorts to the geodesic metric. SS-Geo-GTM uses a proximity graph built from Geo-GTM manifold as the basis for a label propagation algorithm that infers missing class labels. Its performance is compared to those of a semi-supervised version of the standard GTM and of the alternative Laplacian Eigenmaps method.

1 Introduction

In many real applications, class labels are either completely or partially unavailable. The first case scenario is that of unsupervised learning, where the most common task to be performed is data clustering. The second case is less frequently considered and becomes a task at the interface between supervised and unsupervised models: semi-supervised learning (SSL,[1]). Disparate approaches to SSL are found in recent academic literature, including (though not limited to) generative models, low-density separation methods, and graph-based techniques.

We are here specifically interested in graph-based methods. A two-stage technique was recently proposed in [2]. In the first stage, data items are clustered using the Emergent Self-Organizing Map (ESOM) [3]. Then, ESOM is considered as a proximity graph and a modified label propagation (LP) is carried out in the second stage. In this paper, we present a semi-supervised approach, inspired by that proposed in [2]. It is based on Geo-GTM [4], which is an extension of the statistically principled Generative Topographic Mapping (GTM) [5]. Geo-GTM prioritizes neighbourhood relationships along a generated manifold embedded in the observed data space. In doing so, the model can account for convoluted

E. Corchado et al. (Eds.): HAIS 2009, LNAI 5572, pp. 344–351, 2009.

foldings in the data structure and still reveal their underlying structure. In the proposed semi-supervised Geo-GTM (SS-Geo-GTM), the prototypes are inserted and linked by the nearest neighbour to the data manifold constructed by Geo-GTM. The resulting graph is considered as a proximity graph for which an *ad hoc* version of LP is proposed. Following the same methodology, we also develop a semi-supervised version for the standard GTM (SS-GTM), which uses the Euclidean metric. We compare the performance of the proposed SS-Geo-GTM with that of SS-GTM, as well as with that of Laplacian Eigenmaps [6] (a popular graph-based semi-supervised method). Several experiments with artificial and real data sets, using different percentages of available class labels and also with the presence of different levels of uninformative noise, show that SS-Geo-GTM overall outperforms both SS-GTM and Laplacian Eigenmaps.

2 Semi-supervised Geo-GTM

2.1 GTM and Geodesic Metric

The standard GTM is a latent variable model of the manifold learning family, aimed to provide simultaneous data clustering and visualization. Unlike many projection methods, GTM defines a mapping from the low-dimensional latent space onto the observed data space. Such mapping is carried through by a set of basis functions generating a constrained mixture density distribution and is defined as a generalized linear regression model

$$y = \phi(u)W, \tag{1}$$

where ϕ are R basis functions $\phi(u) = (\phi_1(u), ..., \phi_R(u))$ (spherically symmetric Gaussians in the standard formulation); u is a point in (a usually 2-dimensional) latent space; and W is a matrix of adaptive weights w_{rd} .

To avoid computational intractability, a regular grid of M points u_m is sampled from the latent space, which acts as visualization space. Each of them has a fixed prior probability $p(u_m) = 1/M$ and is mapped to a data *prototype* $y_m(u_m, W)$ using Eq. 1. These prototypes define the low-dimensional manifold nonlinearly embedded in the data space. A probability distribution for the multivariate data $X = \{x_n\}_{n=1}^N$ and a corresponding log-likelihood can be defined as $L(W, \beta|X) = \sum_{n=1}^N \ln\{\frac{1}{M} \sum_{m=1}^M (\frac{\beta}{2\pi})^{D/2} \exp\{-\beta/2 \|y_m - x_n\|^2\}\}$, where β is the inverse of the noise variance, which accounts for the fact that data points might not strictly lie on the low dimensional embedded manifold generated by the GTM. The EM algorithm can be used to estimate the parameters of the model, which are the adaptive matrix of weights W and β. In the E-step of the EM algorithm, the mapping is inverted and the responsibilities z_{mn} (the posterior probability of cluster m membership for each data item x_n) can be directly computed as

$$z_{mn} = p(u_m|x_n, W, \beta) = \frac{p(x_n|u_m, W, \beta)p(u_m)}{\sum_{m'} p(x_n|u_{m'}, W, \beta)p(u_{m'})}, \tag{2}$$

where $p(x_n|u_m, W, \beta) = \mathcal{N}(y(u_m, W), \beta)$.

Manifold learning methods work on the assumption that multivariate data can be faithfully represented by lower-dimensional manifolds embedded in the data space. Distances between data point or between these and the model pro-toypes are calculated in the standard GTM using the Euclidean metric. Other manifold methods such as ISOMAP [7] and Curvilinear Distance Analysis [8] use the geodesic distance as a basis for generating the data manifold. In doing so, some of the distortions that the use of the Euclidean distance may introduce (undesired manifold curvature effects) can be avoided. The otherwise computa-tionally intractable geodesic metric can be approximated by graph distances [9], so that instead of finding the minimum arc-length between two data items on a manifold, we find the length of the shortest path between them, where such path is built by connecting the closest successive data items. In this paper, this is done using the K-rule, which allows connecting the K-nearest neighbours. A weighted graph is then constructed by using the data and the set of allowed con-nections. The data are the vertices, the allowed connections are the edges, and the edge labels are the Euclidean distances between the corresponding vertices. If the resulting graph is disconnected, some edges are added using a minimum spanning tree procedure in order to connect it. Finally, the distance matrix, d_g, of the weighted undirected graph is obtained by repeatedly applying Dijkstra's algorithm [10], which computes the shortest path between all data samples.

2.2 Geo-GTM

Geo-GTM is an extension of GTM that favors the similarity of points along the manifold, while penalizing the similarity of points that are not contiguous in the manifold, even if close by Euclidean distance. This is achieved by mod-ifying the calculation of the responsibilities in standard GTM in proportion to the discrepancy between the geodesic (approximated by the graph) and the Eu-clidean distances. Following [11], such discrepancy is explicited by defining the exponential distribution

$$\mathcal{E}(d_g|d_e, \alpha) = \frac{1}{\alpha} \exp\left\{ -\frac{d_g(\mathbf{x}_n, \mathbf{y}_m) - d_e(\mathbf{x}_n, \mathbf{y}_m)}{\alpha} \right\}, \tag{3}$$

where $d_e(\mathbf{x}_n, \mathbf{y}_m)$ and $d_g(\mathbf{x}_n, \mathbf{y}_m)$ are the Euclidean and graph distances between data item $\mathbf{x}_n$ and the GTM prototype $\mathbf{y}_m$. Responsibilities become:

$$z_{mn}^{geo} = p(\mathbf{u}_m|\mathbf{x}_n, \mathbf{W}, \beta) = \frac{p'(\mathbf{x}_n|\mathbf{u}_m, \mathbf{W}, \beta)p(\mathbf{u}_m)}{\sum_{m'} p'(\mathbf{x}_n|\mathbf{u}_{m'}, \mathbf{W}, \beta)p(\mathbf{u}_{m'})}, \tag{4}$$

where $p'(\mathbf{x}_n|\mathbf{u}_m, \mathbf{W}, \beta) = \mathcal{N}(\mathbf{y}(\mathbf{u}_m, \mathbf{W}), \beta)\mathcal{E}(d_g(\mathbf{x}_n, \mathbf{y}_m)^2|d_e(\mathbf{x}_n, \mathbf{y}_m)^2, 1)$.

Here d_g and d_e are used as squared distances in order to be consistent with standard GTM [5]. When there is no agreement between the graph approxi-mation of the geodesic distance and the Euclidean distance, the value of the numerator of the fraction within the exponential in (3) increases, pushing the modified responsibility in (4) towards smaller values, i.e., punishing the discrep-ancy between metrics.

2.3   A Semi-supervised Extension of Geo-GTM

The basic idea underlying the proposed semi-supervised approach is that neighbouring points are most likely to share their label and that these labels are best propagated through neighbouring nodes according to proximity. We modify an existing LP algorithm [12] to account for the information provided by the trained Geo-GTM and the result is the proposed SS-Geo-GTM.

Unlike from original LP, where a fully connected graph is constructed using as nodes both labeled and unlabeled data items, here a label vector $\mathbf{L}_m \in [0,1]^k$ is first associated to each Geo-GTM prototype $\mathbf{y}_m$. These label vectors can be considered as nodes in a proximity graph. The weights of the edges are derived from the graph distances d_g between prototypes. For this, the prototypes are inserted and linked to the graph (data manifold constructed by Geo-GTM) through the nearest data point. Only non-empty clusters (corresponding to prototypes with, at least, one data point assigned to them) are retained. The edge weight between nodes m and m' is calculated as

$$w_{mm'} = \exp(-\frac{d_g^2(m, m')}{\sigma^2}), \qquad (5)$$

where the σ parameter behaves as a radius that defines the level of sparseness in the graph for label information. Here, we propose an *ad hoc* criterion for the choice of σ that consists on assigning it the value of what we call the main reference inter-prototype (MRIP) distance. For this, we first calculate the Cumulative Responsibility (CR), which is the sum of responsibilities over all data items in X, for each cluster m, $CR_m = \sum_{n=1}^{N} z_{mn}^{geo}$. The prototypes with highest CR are considered as the most representative in the data set. We then choose MRIP to be the graph distance $d_g(\mathbf{y}_{m1}, \mathbf{y}_{m2})$ between the two non-contiguous prototypes $\mathbf{y}_{m1}, \mathbf{y}_{m2}$ of highest CR.

Following [2], the available label information of $\mathbf{x}_n \in X$ with class attribution $c(\mathbf{x}_n) = C_t \in \{C_1, \ldots, C_k\}$ will be used to fix the label vectors of the prototypes to which they are assigned ($\mathbf{x}_n$ is assigned to $\mathbf{y}_m$ through $\mathbf{u}_m = \arg\max_{\mathbf{u}_i} z_{in}^{geo}$), so that $L_{m,j} = 1$ if $j = t$, and $L_{m,j} = 0$ otherwise. Label vectors of unlabeled nodes can be initialized at random [12]. Unlabeled prototypes will then update their label by propagation according to

$$\mathbf{L}_m^{new} = \frac{\sum_{m'} w_{mm'} \mathbf{L}_{m'}}{\sum_{m'} w_{mm'}} \qquad (6)$$

until no further changes occur in the label updating. Subsequently, unlabeled data items are labeled by assignment to the class more represented on the label vector of the prototype $\mathbf{y}_m$ bearing the highest responsibility for them, according to $c(\mathbf{x}_n) = \arg\max_{C_j \in \{C_1, \ldots, C_k\}} L_{m,j}$. The same method can be used to build a semi-supervised version of the standard GTM model (SS-GTM).

348   R. Cruz-Barbosa and A. Vellido

Table 1. Classification accuracy as an average percentage over one hundred runs (with its corresponding standard deviation). The statistical significance (calculated through a one-way ANOVA test) of the differences between SS-Geo-GTM and SS-GTM is indicated with '*' if $p < 0.01$ and with '**' if $p < 0.05$. Also, $p < 0.01$ was obtained between any SS version and LapEM.

data set	SS-Geo-GTM (% ± std)	SS-GTM (% ± std)	LapEM (% ± std)
*Dalí**	99.54± 2.24	90.71± 7.99	54.57±3.13
*Iris***	88.71 ± 7.88	85.74 ± 8.72	50.39±3.37
*Oil-Flow**	77.43±8.31	36.74±3.29	63.50±12.08

3   Experimental Results and Discussion

Geo-GTM, SS-Geo-GTM, and SS-GTM were initialized following a procedure described in [5]. The latent grid was fixed to a square layout of approximately $(N/2)^{1/2} \times (N/2)^{1/2}$, where N is the number of points in the data set.

Three data sets were selected for the reported experiments: The first one is the artificial 3-D *Dalí* set. It consists of two groups (acting as classes) of 300 data items each that are images of the functions $\mathbf{x}_1 = (t\cos(t), t_2, t\sin(t))$ and $\mathbf{x}_2 = (t\cos(t), t_2, -t\sin(t) + 20)$, where t and t_2 follow $\mathcal{U}(\pi, 3\pi)$ and $\mathcal{U}(0, 10)$, respectively. The second is the well-known *Iris* data, available from the UCI repository. The third is the more complex *Oil-Flow* set, also available online[1], which simulates measurements in an oil pipe corresponding to three possible configurations (classes). It consists of 1,000 items described by 12 attributes.

The central goal of the experiments is the comparison of the performances of SS-Geo-GTM, SS-GTM and the alternative method of Laplacian Eigenmaps (LapEM, [6]) in terms of classification accuracy. We first assume that the choice of the MRIP, described in the previous section, as a value for σ is appropriate. We then evaluate (average accuracy over one hundred runs) the models in the most extreme semi-supervised setting: when the class label is available for only one input item for each class while the rest is unlabeled. The corresponding results are shown in Table 1. SS-Geo-GTM significantly outperforms SS-GTM and LapEM for all data sets and, most notoriously, for the data sets of more convoluted geometry. The differences with SS-GTM are less notorious for the less convoluted *Iris* data set. LapEM yields a very poor behaviour in this setting.

We then proceed to evaluate the performance of the SS-Geo-GTM model in this same setting for a range of different values of σ, both higher and lower than the MRIP, to assess the adequacy of the proposed MRIP choice. We explore the interval $\sigma \in [MRIP - \epsilon, MRIP + \epsilon]$, where $\epsilon > 0$, and measure the performance of SS-Geo-GTM over a hundred runs. These results are shown in Table 2. The models with $\sigma = MRIP$ yield the best results in the range of selected σ values, which confirms the fact that the MRIP value is at least near the optimum value for σ. Consequently, from here on MRIP will be used as the default value for σ.

---

[1] http://research.microsoft.com/~cmbishop/PRML/webdatasets/datasets.htm

Table 2. Average classification accuracy and its std. deviation over 100 runs for different values of σ parameter in the SS-Geo-GTM setting.

Dalí		Iris		Oil-Flow	
$\sigma < MRIP$	% ± std	$\sigma < MRIP$	% ± std	$\sigma < MRIP$	% ± std
5.0	98.06±3.73	0.05	85.72±8.93	0.10	74.74±8.63
10.0	98.46±4.69	0.10	87.24±8.97	0.20	75.03±9.08
15.0	99.19±2.44	0.12	87.37±7.46	0.25	75.24±9.26
20.0	99.37±2.22	0.14	86.94±9.73	0.30	74.38±10.10
25.0	99.48±2.13	0.15	88.20±8.14	0.35	75.74±8.98
MRIP = 31.36	99.54±2.24	**MRIP= 0.21**	88.71±7.88	**MRIP = 0.43**	77.43±8.31
$\sigma > MRIP$		$\sigma > MRIP$		$\sigma > MRIP$	
35.0	98.54±3.96	0.30	88.30±7.46	0.50	75.97±8.51
40.0	98.43±4.54	0.40	88.69±8.93	0.55	74.71±8.56
45.0	97.95±4.77	1.0	88.64±7.63	0.60	74.70±8.80
50.0	96.84±6.55	3.0	88.59±5.32	0.65	73.98±8.77
55.0	95.35±8.01	4.0	83.03±7.29	0.75	72.08±9.88

Table 3. Average classification accuracy and its std. deviation over 100 runs, for all models. A randomly increasing percentage of pre-labeled items per class was chosen in each run. The '$\star$' symbol means that the experiment was not carried out because the corresponding percentage of available labels was less than or equal to one label per class. A super-index '*' indicates that the differences between the corresponding model and SS-Geo-GTM were significant at $p < 0.01$ in the ANOVA test for all percentages of class labels. A super-index '**' indicates that no differences were significant.

% of avail. labels	Classification accuracy (% ± std)					
	Dalí			Iris		
	SS-Geo	SS-GTM *	LapEM *	SS-Geo	SS-GTM **	LapEM *
1	100±0	93.43±5.46	64.91±4.52	$\star$	$\star$	$\star$
2	100±0	96.96±3.41	76.00±5.88	$\star$	$\star$	$\star$
3	100±0	97.74±2.05	79.65±9.29	$\star$	$\star$	$\star$
4	100±0	98.29±1.80	75.24±10.56	90.00±8.11	89.46±5.24	58.10±4.01
5	100±0	98.61±1.32	88.72±8.05	89.96±6.98	89.18±6.48	57.01±4.57
6	100±0	98.66±1.64	95.01±4.95	91.30±7.37	91.66±3.02	63.68±4.48
7	100±0	98.98±0.80	97.68±3.16	90.74±7.62	90.94±3.03	64.22±4.86
8	100±0	99.19±0.82	98.64±2.13	91.91±5.31	91.90±3.03	69.84±5.26
9	100±0	99.30±0.70	98.88±1.87	92.35±4.90	91.88±2.48	70.19±4.97
10	100±0	99.24±0.73	99.39±1.39	93.19±4.36	92.32±2.42	74.87±5.92

% of avail. labels	Oil-Flow (% ± std)		
	SS-Geo	SS-GTM *	LapEM *
1	83.93±5.60	39.96±3.44	76.43±7.55
2	90.08±3.49	55.88±10.95	83.36±5.48
3	91.79±3.07	64.71±7.95	87.56±4.42
4	94.28±2.60	70.69±6.06	89.71±3.51
5	95.14±2.20	74.11±5.05	91.63±3.25
6	95.97±2.01	76.51±4.30	92.63±2.76
7	96.43±1.81	79.10±4.24	93.77±2.36
8	96.65±1.53	80.88±4.27	94.41±2.11
9	97.11±1.66	82.19±3.43	95.18±2.07
10	97.53±1.22	83.91±3.58	95.58±1.53

Will this difference of performance remain when the label availability condition is relaxed? To answer this question, the ratio of randomly selected labeled data is increased from a single one to a 1%, and from there, up to a 10%. The corresponding results are shown in Table 3. SS-Geo-GTM clearly (and again significantly according to an ANOVA test) outperforms SS-GTM for *Dalí* and *Oil-Flow* and, as expected, the performance monotonically improves with the

Table 4. Average classification accuracy and its std. deviation over 100 runs, for all models in the presence of increasing levels of uninformative noise. An increasing percentage of pre-labeled items per class was randomly chosen in each run.

Dataset	noise level	model	Percent of available labels				
			2	4	6	8	10
Dalí	0.1	SS-Geo	100±0	100±0	100±0	100±0	100±0
		SS-GTM	96.29±3.37	98.15±1.97	99.09±1.0	99.31±0.99	99.28±0.89
		LapEM	*75.48±6.56*	*75.73±10.38*	*94.48±4.66*	*98.07±2.02*	*98.50±1.96*
	0.3	SS-Geo	99.83±1.11	100±0	100±0	100±0	100±0
		SS-GTM	95.57±4.0	98.11±1.45	98.56±0.83	98.77±0.75	98.88±0.69
		LapEM	*74.47±5.27*	*75.11±11.11*	*95.55±4.82*	*99.03±1.96*	*99.54±1.12*
	0.5	SS-Geo	99.04±3.16	100±0	100±0	100±0	100±0
		SS-GTM	96.52±3.09	98.05±2.16	98.99±1.40	99.31±1.06	99.39±0.78
		LapEM	*77.67±6.79*	*76.56±10.30*	*95.06±4.53*	*97.49±2.76*	*98.87±1.61*
	1.0	SS-Geo	95.14±5.52	97.75±2.94	98.71±1.98	99.23±0.73	99.28±0.92
		SS-GTM	96.12±3.79	98.36±1.53	98.66±1.21	99.04±0.45	99.06±0.35
		LapEM	*73.86±6.07*	*70.73±10.57*	*92.15±5.34*	*97.23±3.09*	*98.93±1.39*
	2.0	SS-Geo	94.78±3.66	96.45±1.63	96.96±0.67	97.11±0.58	97.19±0.48
		SS-GTM	92.96±3.0	94.28±1.96	94.73±1.75	95.45±1.01	95.36±1.07
		LapEM	*74.02±5.72*	*72.11±11.66*	*90.00±5.91*	*94.54±3.37*	*95.99±1.86*
Oil-Flow	0.01	SS-Geo	88.13±4.05	93.87±2.71	95.63±2.24	96.87±1.45	97.26±1.18
		SS-GTM	55.54±11.94	70.66±5.84	77.14±4.65	80.25±3.58	84.15±3.39
		LapEM	*81.35±5.67*	*88.17±3.41*	*91.80±2.67*	*93.20±2.30*	*94.77±1.70*
	0.03	SS-Geo	88.60±4.06	93.34±2.94	95.46±1.94	96.31±1.64	96.98±1.23
		SS-GTM	55.14±10.71	71.54±6.00	77.26±4.53	81.40±3.63	82.60±3.24
		LapEM	*79.79±7.18*	*90.50±3.72*	*94.00±2.72*	*95.91±1.98*	*96.59±1.13*
	0.05	SS-Geo	90.10±4.38	94.94±2.49	96.34±1.93	97.42±1.69	97.84±1.23
		SS-GTM	53.39±11.81	70.52±7.42	75.79±4.77	81.32±4.52	83.84±4.34
		LapEM	*78.26±7.82*	*92.04±2.81*	*94.86±2.22*	*95.79±1.68*	*96.62±1.37*
	0.1	SS-Geo	60.40±12.81	81.48±8.91	88.95±4.89	91.19±3.59	92.49±2.59
		SS-GTM	49.88±10.11	70.30±8.63	78.20±4.48	82.68±4.50	85.08±4.23
		LapEM	*66.78±11.12*	*87.81±4.79*	*92.50±2.95*	*94.23±2.23*	*95.42±1.78*
	0.2	SS-Geo	59.89±11.38	75.76±6.16	79.50±5.03	83.0±3.78	85.41±2.63
		SS-GTM	44.94±9.92	56.18±10.59	66.01±7.04	72.31±5.55	75.37±4.27
		LapEM	*63.75±7.44*	*77.32±4.55*	*82.22±3.31*	*85.47±2.15*	*86.58±1.84*

increasing percentage of labels. The differences for the latter set, more complex and high-dimensional, are striking. Also, SS-Geo-GTM outperforms LapEM for all data sets. For *Dalí*, SS-Geo-GTM achieves a 100% accuracy even with a 1% of labeled data, while SS-GTM and LapEM do not reach that average accuracy even with a 10%. The *Iris* data set benefits less of the addition of class labels and the performances of SS-Geo-GTM and SS-GTM models are comparable. This confirms that the use of the geodesic metric is likely to improve the results mainly for data sets of convoluted underlying geometry.

In previous research [13], the Geo-GTM model was shown to behave better than standard GTM in the presence of noise, as measured by the test log-likelihood. We now extend these results to the semi-supervised setting to gauge and compare the robustness of the analyzed methods in the presence of noise in some illustrative experiments. For this, Gaussian noise of zero mean and increasing standard deviation was added to: a noise-free version of the *Dalí* set (added noise from $\sigma = 0.1$ to $\sigma = 2.0$) and the most difficult dataset, *Oil-Flow* (added noise from $\sigma = 0.01$ to $\sigma = 0.2$). The noise scale magnitude is in correspondence with the data scale. As in the previous experiment, we also analyze the evolution of the performance of these models as the percentage of available labels for each dataset is increased from 2% to 10%.

These new results are shown in Table 4. In accordance to the results presented in [13], the geodesic variant SS-Geo-GTM consistently outperforms SS-GTM (and LapEM) across data sets and noise levels, with few exceptions. The robustness of the semi-supervised procedure for SS-GTM is surprisingly good, though. For the more complex *Oil-Flow* set, both models deteriorate significantly at high noise levels. Overall, these results indicate that the resilience of the models is mostly due to the inclusion of the geodesic metric and not to the semi-supervised procedure itself. It is worth noting that the results for LapEM only become comparable as the percentage of available labels increases.

Given that the proposed semi-supervised procedure depends on the quality of the graph constructed by Geo-GTM, it might have limitations to deal with very sparse data sets. Future research will be devoted to explore this issue, as well as to gauge the effect of the presence of outliers in the performance of the proposed semi-supervised model.

References

1. Chapelle, O., Schölkopf, B., Zien, A. (eds.): Semi-Supervised Learning. MIT Press, Cambridge (2006)
2. Herrmann, L., Ultsch, A.: Label propagation for semi-supervised learning in self-organizing maps. In: Proceedings of the 6th WSOM 2007 (2007)
3. Ultsch, A.: Maps for the visualization of high-dimensional data spaces. In: Proceedings of WSOM 2003, 225–230 (2003)
4. Cruz-Barbosa, R., Vellido, A.: Unfolding the manifold in Generative Topographic Mapping. In: Corchado, E., Abraham, A., Pedrycz, W. (eds.) HAIS 2008. LNCS (LNAI), vol. 5271, pp. 392–399. Springer, Heidelberg (2008)
5. Bishop, C.M., Svensén, M., Williams, C.K.I.: The Generative Topographic Mapping. Neural Computation 10(1), 215–234 (1998)
6. Belkin, M., Niyogi, P.: Using manifold structure for partially labelled classification. In: Procs. of NIPS, vol. 15. MIT Press, Cambridge (2003)
7. Tenenbaum, J.B., de Silva, V., Langford, J.C.: A global geometric framework for nonlinear dimensionality reduction. Science 290, 2319–2323 (2000)
8. Lee, J.A., Lendasse, A., Verleysen, M.: Curvilinear distance analysis versus isomap. In: Procs. of the ESANN, pp. 185–192 (2002)
9. Bernstein, M., de Silva, V., Langford, J., Tenenbaum, J.: Graph approximations to geodesics on embedded manifolds. Technical report, Stanford U., CA (2000)
10. Dijkstra, E.W.: A note on two problems in connection with graphs. Numerische Mathematik 1, 269–271 (1959)
11. Archambeau, C., Verleysen, M.: Manifold constrained finite gaussian mixtures. In: Cabestany, J., Prieto, A.G., Sandoval, F. (eds.) IWANN 2005. LNCS, vol. 3512, pp. 820–828. Springer, Heidelberg (2005)
12. Zhu, X., Ghahramani, Z.: Learning from labeled and unlabeled data with label propagation. Technical report, CMU-CALD-02-107, Carnegie Mellon University (2002)
13. Cruz-Barbosa, R., Vellido, A.: Geodesic Generative Topographic Mapping. In: Geffner, H., Prada, R., Machado Alexandre, I., David, N. (eds.) IBERAMIA 2008. LNCS (LNAI), vol. 5290, pp. 113–122. Springer, Heidelberg (2008)

Application of Interval Type-2 Fuzzy Logic Systems for Control of the Coiling Entry Temperature in a Hot Strip Mill

Gerardo M. Méndez[1,*], Luis Leduc-Lezama[2], Rafael Colas[3],
Gabriel Murillo-Pérez[2], Jorge Ramírez-Cuellar[2], and José J. López[1]

[1] Department of Electrical and Electronics Engineering, Instituto Tecnológico de Nuevo
León, Cd. Guadalupe, N.L. México, CP 67170
gmm_paper@yahoo.com.mx
[2] Department of Process Engineering, Ternium, San Nicolas de los Garza,
N.L. México, CP 66452
lleduc@ternium.com.mx
[3] Faculty of Mechanical and Electrical Engineering, Autonomous University of Nuevo Leon,
San Nicolas de los Garza, N.L. México, CP 66452
rcolas@mail.uanl.mx

Abstract. An interval type-2 fuzzy logic system is used to setup the cooling
water applied to the strip as it traverses the run out table in order to achieve the
coiler entry temperature target. The interval type-2 fuzzy setup model uses as
inputs the target coiling entry temperature, the target strip thickness, the predicted finish mill exit temperature and the target finishing mill exit speed. The
experimental results of the application of the interval type-2 fuzzy logic system
for coiler entry temperature prediction in a real hot strip mill were carried out
for three different types of coils. They proved the feasibility of the systems developed here for coiler entry temperature prediction. Comparison with an on-
line type-1 fuzzy logic based model shows that the interval type-2 fuzzy logic
system improves performance in coiler entry temperature prediction under the
tested condition.

Keywords: Type-2 fuzzy inference systems, temperature modeling and control,
uncertain rule-based fuzzy logic systems.

1 Introduction

The aim of this work is to present and discuss the implementation of the control function called coiling temperature control (CTC) using interval type-2 (IT2) fuzzy logic
systems (FLS). The IT2 CTC model comprises two principal tasks: head end setup
and feedback. Its purpose is to achieve and maintain target head strip temperature at
the coiler entry pyrometer. The IT2 CTC model controls coiler entry temperature by
applying cooling water to the strip as it traverses the run out table and adjusting the
flow of water for the next bar to compensate for changes in strip gage, strip speed,
finish mill exit temperature, and coiler entry temperature. The amount of water

---

*Corresponding author.

E. Corchado et al. (Eds.): HAIS 2009, LNAI 5572, pp. 352–359, 2009.
© Springer-Verlag Berlin Heidelberg 2009

applied to the strip is calculated from the interval type-2 fuzzy logic system and updated from measured data of each bar that is rolled. The IT2 CTC setup model runs off-line and calculates cooling water requirements and the spray references. Once per bar, the feedback task updates the IT2 CTC coiling temperature model parameters.

IT2 FLS is an emerging technology [1] that accounts for random and systematic components [2] of industrial measurements. Non-linearity of the processes is handled by FLS as identifiers and universal approximators of nonlinear dynamic systems [3]-[6]. Such characteristics give IT2 FLS a great potential to model and control industrial processes.

In coiling temperature control, the inputs of the IT2 FLS setup model, used to predict the coiler entry temperature, are the target coiling temperature, the target strip thickness, the predicted finish mill exit temperature and the target finishing mill exit speed. Currently, the surface temperature is measured using a pyrometer located at the coiler entry side. Scale grows at the strip surface producing a noisy temperature measurement. The measurement is also affected by environment water steam as well as pyrometer location, calibration, resolution and repeatability. Although coiler entry (CLE) temperature prediction (y) is a critical issue in a hot strip mill (HSM) the problem has not been fully addressed by interval type-2 fuzzy logic control systems [7].

2 The Hot Strip Mill Process

In a HSM, as in any other industrial process, keeping the quality requirements such as thickness, finishing temperature and coiler temperature (the latter determines strip mechanical properties) is a major concern. The most critical section of the coil is the head-end. This is due to the uncertainties involved at the head-end of the incoming steel bar, and the varying conditions from bar to bar. Currently, in order to achieve the head-end quality requirements, there are automation systems based on physical modeling, particularly in the reheat furnace, roughing mill (RM), finishing mill (FM) and the run out cooling zone [7], [8].

Fig. 1 depicts a simplified diagram of a HSM, from its initial stage, the reheat furnace entry, to the final stage, the coilers.

The market is becoming more competitive and worldwide and therefore more demanding [9], [10]. It requires a more stringent control of quality parameters and a

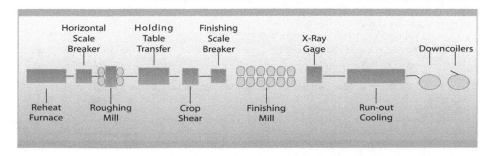

Fig. 1. Schematic view of a typical hot strip mill

more flexible manufacturing, capable of rolling a wider gamma of products in shorter periods. Most commercial systems compensation techniques (P or PI based) only compensate for error under current conditions, therefore, the first of batch coils frequently present out of specification head-ends [11]. In recent years, research on estimation of process variable in a HSM by adaptive neural networks (ANN) and fuzzy logic systems has received particular attention worldwide[12]-[15]. ANN and FLS offer the advantages of reliably representing highly non-linear relations, automatically updating the knowledge they contain and providing fast processing times[10],[16].

The strip temperature at coiler entry, which has a large contribution to the final properties of the product, depends on the flow of cooling water, strip gage, temperature and speed at finishing mill exit zone. On the other hand, temperature measurement is highly uncertain. Scale breaker (SB) entry mean and surface temperatures are used by the finishing mill setup (FSU) model [8] to preset the finishing mill stand screws and to calculate the transfer bar (TB) thread speed, both required to achieve, respectively, the finishing mill exit target head gage and finishing mill exit target head temperature

3 IT2 Design

The IT2 CTC setup model calculates the flow of run out table cooling water required to achieve desired coiling temperature on the head of the piece as a function of target coiling temperature, target strip thickness, predicted finish mill exit temperature, and target finishing mill exit speed. The total required flow is then translated into some number of sprays at particular flow levels. The translation is based on spray flow characteristics and run out table sprays in service.

The feedback process adapts the parameters of the IT2 CTC model. Fig. 2 shows the water cooling sprays at runout table. The adaptation, which permits the model to respond to changing mill and process conditions, is a function of measured coiler entry temperature, measured finish mill exit temperature, measured strip thickness, measured finishing mill exit speed, and the target coiler entry temperature.

Sprays in service, are operator-selected input. Spray locations is part of the configuration data. The (x_1) antecedent-input space was divided into ten fuzzy sets, (x_2) was divided into twenty-five fuzzy sets, (x_3) was divided into five fuzzy sets, and (x_4) was divided into ten fuzzy sets thus, having twelve thousand five hundred rules. The output (consequent, y) is the head-end CLE surface temperature.

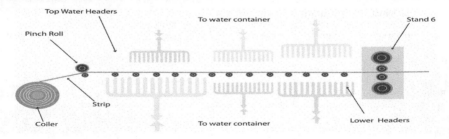

Fig. 2. Schematic view of water-cooling sprays at runout table

Gaussian primary membership-functions with uncertain means were chosen for both, antecedents and consequents. Each rule of the IT2 CTC model was characterized by twelve antecedent membership function parameters (two for left-hand and right-hand bounds of the mean and one for standard deviation, for each of the four antecedent Gaussian membership functions) and two consequent parameters (one for left-hand and one for right-hand end points of the centroid of the consequent IT2 fuzzy set).

The primary membership functions for each input of IT2 CTC model using non-singleton type-1 fuzzy sets was of the form:

$$\mu_{X_k}(x_k) = \exp\left[-\frac{1}{2}\left[\frac{x_k - x'_k}{\sigma_{X_k}}\right]^2\right]. \tag{1}$$

where: $k=1, 2, 3, 4$ (the number of type-1 non-singleton inputs) and $\mu_{Xk}(x_k)$ centered at the measured input $x_k = x'_k$. The standard deviation of the inputs was initially set as $\sigma_{X1} = 40.0$ °C, $\sigma_{X2} = 0.2$ mm, $\sigma_{X3} = 30.0$ °C, and $\sigma_{X4} = 0.43$ m/s. These values were also selected experimentally.

Noisy input-output data pairs of three different coil types with different target gage, target width and steel grade were taken and used as training and validation data, see Table 1, and experiments were carried out for these different coil types. The standard deviation of the noise of the four inputs was initially set as $\sigma_{n1} = 13.0$ °C, $\sigma_{n2} = 0.04$ mm, $\sigma_{n3} = 7.0$ °C, and $\sigma_{n4} = 0.08$ m/s. The Gaussian primary membership function with uncertain means for each antecedent was defined as:

$$\mu_k^l(x_k) = \exp\left[-\frac{1}{2}\left[\frac{x_k - m_{kn}^l}{\sigma_k^l}\right]^2\right]. \tag{2}$$

where $m^l_k \in [m^l_{k1}, m^l_{k2}]$ is the uncertain mean, $k=1, 2, 3, 4$ (the number of antecedents), $l=1, 2,..12500$; $n=1, 2$ (the lower and upper bounds of the uncertain mean) and σ_k^l is the standard deviation.

Using the calculated mean and standard deviation from measurements of all inputs, the values of the antecedent intervals of uncertainty were established. The initial intervals of uncertainty for input (x_1) were selected as shown in Fig. 3.

Table 1. Material type coils

	Target Gage (mm)	Target Width (mm)	Steel grade (SAE/AISI)
Coil A	1.981	1067.0	1006
Coil B	2.006	991.0	1006
Coil C	2.159	952.0	1006

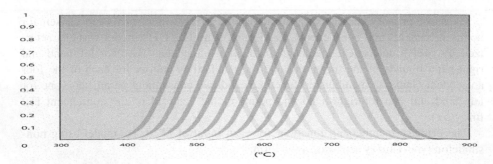

Fig. 3. Membership functions for the antecedent fuzzy sets of (x_1) input

The values of the initial intervals of uncertainty for input (x_2), input (x_3) and input (x_4) are shown in Fig. 4, Fig. 5 and Fig. 6, respectively.

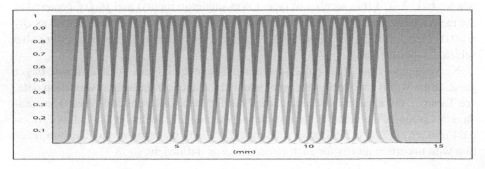

Fig. 4. Membership functions for the antecedent fuzzy sets of (x_2) input

The resulting IT2 CTC model uses type-1 non-singleton fuzzification, maximum t-conorm, product t-norm, product implication, and center-of-sets type-reduction. For parameter optimization, the learning method BP-BP was used.

The IT2 fuzzy rule base consists of a set of IF-THEN rules that represents the model of the system. The IT2 CTC model have four inputs $x_1 \in X_1$, $x_2 \in X_2$, $x_3 \in X_3$ and $x_4 \in X_4$ and one output $y \in Y$, which have a corresponding rule base size of $M = 12500$ rules of the form:

$$R^l : IF \quad x_1 \quad is \quad \tilde{F}_1^l \quad and \quad x_2 \quad is \quad \tilde{F}_2^l, \quad THEN \quad y \quad is \quad \tilde{G}^l . \tag{3}$$

where $l = 1, 2, \ldots 12500$. These rules represent a fuzzy relation between the input space $X_1 \times X_2 \times X_3 \times X_4$ and the output space Y, and it is complete, consistent and continuous [4].

The primary membership function for each consequent is a Gaussian function with uncertain means, as defined in equation (2), while y_l^l and y_r^l are the consequent parameters.

4 Experimental Results

An IT2 CTC model was trained to predict the CLE temperature. Three different sets of data for the three different coil types mentioned and shown in Table 1 were taken from a real-life mill. Experiments were run for each product type set independently. For each input-output data pairs, for each product type set, the twelve thousand five hundred rules were tuned.

The performance evaluation for each of the learning methods is based on the root mean-squared error (RMSE) criteria:

$$RMSE = \sqrt{\frac{1}{n}\sum_{k=1}^{n}\left[\mathbf{Y}(k) - \mathbf{f}_{s2}\left(\mathbf{x}^{(k)}\right)\right]^2} \tag{4}$$

where $\mathbf{Y}(k)$ is the output validation data vector i.e. the actual CLE temperature measurements vector for system evaluation, different to the training data vector but from the same coil type, and $\mathbf{f}_{s2}(\mathbf{x}^{(k)})$ is the temperature vector predicted by the tested IT2 CTC model.

Fig. 7, shows the RMSE of the off-line IT2 CTC setups for type A coils after twenty-five epoch computations. In this Figure, the horizontal axis represents the number of training epochs, while the vertical axis, id the RMSE of the validation obtained with the test set after the corresponding number of training epochs, as outlined elsewhere [17].

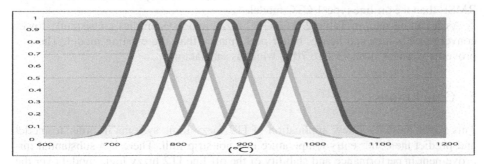

Fig. 5. Membership functions for the antecedent fuzzy sets of (x_3) input

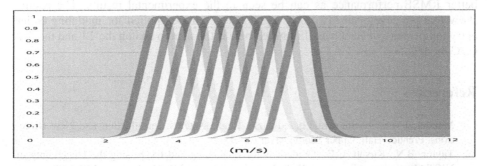

Fig. 6. Membership functions for the antecedent fuzzy sets of (x_4) input

Table 2. Comparison between RMSE after 25 epochs of T1 and IT2 CTC models

	IT2 FLS	T1 FLS	Difference (%)
Coil A – non-singleton	4.4	14.3	70
Coil B – non-singleton	5.2	7.2	28
Coil C – non-singleton	5.6	8.4	33

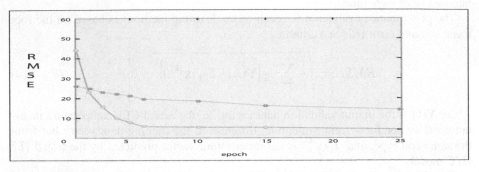

Fig. 7. RMSE for Type A coil: (*) T1 and (o) IT2 CTC models for non-singleton inputs

Fig. 7 also shows the results of the on-line type-1 (T1) based CTC model. This corresponding to the IT2 FLS experiment showed in same figure. By comparison, it can be concluded that the off-line IT2 CTC model have better performance in terms of RMSE than the on-line type-1 CTC model.

As it can be seen in Table 2, IT2 CTC fuzzy logic system has consistently lower convergence values and hence, better performance than the existing model. The improvement ranges from 33% to 70%, which is satisfactory.

5 Conclusions

This paper presents a new application of IT2 fuzzy logic systems in order to model and predict the coiler entry temperature in a hot strip mill. There is a substantial improvement in performance and stability of the off-line IT2 fuzzy logic model over the on-line system based on type-1 fuzzy logic systems. The IT2 CTC model achieves the better RMSE performance as can be seen in the experimental results. The results show that the developed IT2 fuzzy logic system can be applied for modeling coiler entry temperature of steel bars. In future, there is the plan to switch the T1 and the IT2 CTC models.

References

1. Mendel, J.M.: Uncertain rule-based fuzzy Logic systems: Introduction and New Directions. Prentice-Hall, Upper Saddle River (2001)
2. Taylor, B.N., Kuyatt, C.E.: Guidelines for Evaluating and Expressing the Uncertainty of NIST Measurement Results (NIST), Technical Report 1297, Gaitherburg, MD (1994)

3. Wang, L.X.: Fuzzy Systems are Universal Approximators. In: IEEE Conf. on Fuzzy Systems, San Diego, pp. 1163–1170 (1992)
4. Wang, L.X.: Fuzzy Systems as Nonlinear Mapping. Prentice-hall PTR, Upper Saddle River (1997)
5. Wang, L.X.: Fuzzy Systems as Nonlinear Dynamic System Identifiers. In: 31st IEEE Conference on Decision and Control, Tucson, Arizona, pp. 897–902 (1992)
6. Wang, L.X., Mendel, J.M.: Back-propagation Fuzzy Systems as Nonlinear Dynamic System Identifiers. In: Proceedings of the IEEE Conf. on Fuzzy Systems, San Diego, CA, pp. 1409–1418 (1992)
7. Martinetz, T., Protzel, P., Gramchow, O., Sorgel, G.: Neural Network Control for Rolling Mills, ELITE Foundation. In: EUFIT 1994, Achen, Germany, pp. 147–152 (1994)
8. General Electric, Models Users Reference, Manual 1, Roanoke VA (1993)
9. Sato, N., Kamada, N., Naito, S., Fukushima, T., Fujino, M.: Application of Fuzzy Control System to Hot Strip Mill. In: Proceedings of the IEEE International Conference on Industrial Electronics, Control, Instrumentation, San Diego, CA, pp. 1202–1206 (1992)
10. Lee, D.Y., Cho, H.S.: A Neural Network Approach to the Control of the Plate Width in Hot Plate Mills. In: IEEE International Joint Conference on Neural Networks, Washington, DC, vol. 5, pp. 3391–3396 (1999)
11. Bissessur, Y., Martin, E.B., Morris, A.J., Kitson, P.: Fault Detection in Hot Steel Rolling Using Neural Networks and Multivariate Statistics. IEE Proceedings Control Theory and Application (147), 633–640 (2000)
12. Schalng, M., Poppe, T.: Neural Network for Steel Manufacturing. IEEE Expert II, 8–9 (1996)
13. Yao, X., Tieu, A.K., Fang, X.D., Frances, D.: Neural Network Application to Head & Tail Width Control in Hot Strip Mill. In: IEEE International Conference on Neural Networks, Orlando, FL, vol. (1), pp. 433–437 (1995)
14. Kim, Y.S., Yum, B.J., Kim, M.: Robust Design of Artificial Neural Network for Roll Force Prediction in Hot Strip Mill. In: IEEE International Joint Conference on Neural Network, vol. (4), pp. 2800–2804 (2001)
15. Xie, H.B., Jiang, Z.B., Liu, X.H., Wang, G.D., Tieub, A.K.: Prediction of Coiling Temperature on Run-out Table of Hot Strip Mill Using Data Mining. Journal of Materials Processing Technology 177, 121–125 (2006)
16. Watanabe, T., Narazaki, H., Kitamura, A., Takahashi, Y., Hasegawa, H.: A New Mill-setup System for Hot Strip Rolling Mill that Integrates a Process Model and Expertise. In: IEEE International Conference on Computational Cybernetics and Simulation, Orlando, FL, vol. (3), pp. 2818–2822 (1997)
17. Liang, Q., Mendel, J.M.: Interval Type-2 Fuzzy Logic Systems: Theory and Design. Transactions on Fuzzy Systems (8), 535–550 (2000)

A Review on the Application of Hybrid Artificial Intelligence Systems to Optimization Problems in Operations Management

Oscar Ibáñez, Oscar Cordón, Sergio Damas, and Luis Magdalena

European Centre for Soft Computing, Asturias, Spain
{oscar.ibanez,oscar.cordon,sergio.damas,
luis.magdalena}@softcomputing.es

Abstract. The use of hybrid artificial intelligence systems in operations management has grown during the last years given their ability to tackle combinatorial and NP hard problems. Furthermore, operations management problems usually involve imprecision, uncertainty, vagueness, and high-dimensionality. This paper examines recent developments in the field of hybrid artificial intelligence systems for those operations management problems where hybrid approaches are more representative: design engineering, process planning, assembly line balancing, and dynamic scheduling.

Keywords: hybrid artificial intelligence systems, design engineering, process planning, assembly line balancing, dynamic scheduling, operations management.

1 Introduction

Hybrid artificial intelligence systems (HAIS) comprise a research field of computational intelligence focusing on synergistic combinations of multiple approaches to develop the next generation of intelligent systems. Neural computing, machine learning, fuzzy logic, evolutionary algorithms, and agent-based methods, are among the main components of this discipline. Recently, hybrid intelligent systems are becoming popular due to their capabilities in handling many real-world complex problems involving imprecision, uncertainty, vagueness, and high-dimensionality. They provide us with the opportunity to use both expert knowledge and raw data to solve problems in a more interesting and promising way. This multidisciplinary research field is in continuous expansion in the artificial intelligence research community. Examples of three books extensively describing computational intelligence architectures and models as well as their possible fusion have been authored by Kasabov [1], Koza [2] and Cordon et al. [3], and Lin and Lee [4].

The American Production and Inventory Control Society (APICS) defines operations management as "the field of study that focuses on the effective planning, scheduling, use and control of a manufacturing or service organization through the study of concepts from design engineering, industrial engineering, management information systems, quality management, production management, inventory management, accounting, and other functions as they affect the organization".

E. Corchado et al. (Eds.): HAIS 2009, LNAI 5572, pp. 360–367, 2009.

The aim of this contribution is to illustrate recent developments in the field of HAIS applied to operations management optimization problems. We have made a selection of those problems where the presence of HAISs is more significant. However, this review is not exhaustive since there are more operations management problems not included in this state of the art because of the existence of a small number of HAIS solutions.

2 Design Engineering

The design engineer usually leads the project, designing the overall frameworks and the most far reaching parts. The scientific community has extensively studied this design domain during the last decades for the establishment of general purpose and domain-independent scientific rules and methodologies. Finger et al. [5] surveyed the issues of design theory, design methodologies and design models in a review paper. Another survey on the same topic was conducted by Evbuomwan et al. [6] a few years later. Besides summarizing and reviewing the developed design models and methodologies, the said papers investigated the nature and the characteristics of the design process, classified the design models into categories and located possible research opportunities. The following issues regarding the design process become distinguishable: (a) the design knowledge representation (modeling), (b) the search for optimal solutions, and (c) the retrieval of pre-existing design knowledge and the learning of new knowledge.

Several approaches implementing combinations of artificial intelligence techniques are described in the following. Vico et al. [7] consider design synthesis as an optimization problem. Under this perspective an artificial neural network (ANN) is utilized to implement a fitness function for a genetic algorithm (GA) that searches for the optimal solution. Sasaki et al. [8] propose a method to solve fuzzy multiple-objective optimal system design problems with hybridized genetic algorithms (HGAs). This approach enables the design of flexible optimal systems by applying fuzzy goals and fuzzy constraints. Moreover, generalized upper bounding is applied in order to structure the representation of the chromosomes in the genetic algorithm. Wang et al. [9] suggest an interactive evolutionary approach to synthesize component-based preliminary design engineering problems by combining agent-based hierarchical design representation, set-based design generation with fuzzy design trade off strategy, and evolutionary synthesis. The proposed framework facilitates the human–computer interaction to define the fitness function of solutions incorporating both multi-criteria evaluation and constraint satisfaction. Xiong et al. [10] present a synthetic mixed-discrete fuzzy nonlinear programming optimization method that combines the fuzzy formulation with a GA and a traditional gradient-based optimization strategy. This method can find a global compromise solution for fuzzy optimization problems containing mixed-discrete design variables, even when the objective functions are non-convex and/or non-differentiable. Su et al. [11] propose a hybrid approach on the basis of integration of a knowledge base, ANN, GA, and CAD/CAE/CAM in a single environment, which can be implemented in various stages of the design process. The

GA is used to conduct optimization tasks in the context of achieving the optimal combination of design parameters, as well as the optimal architecture of the ANN used in this hybrid system. Tsai et al. [12] suggest that designers may create a new design in shorter time by modifying previous designs. Taking this as a base, they propose an intelligent design retrieval system based on soft computing techniques. Fuzzy relations and fuzzy composition are used for features associations, while a fuzzy neural network is responsible for the composition of object association functions allowing designers to control the similarity of retrieved designs. Saridakis et al. [13] represent the design problem in terms of qualitative and quantitative design parameters and their associative relationships of different formalisms, with a GA to be deployed to find the optimal solution according to a specific optimization criterion. During genetic optimization, the best solutions are recorded and submitted to a neuro-fuzzy process that limits the number of inputs and outputs and solves problem's complexity by substituting existing associative relations with a fuzzy rule system.

The fusion of artificial intelligence techniques has also been reported in a significant number of design applications, such as the design of adaptive car-following indicator [14], the optimization of clamping forces in a machining operation [15], the human–machine workstation design and simulation [16], etc.

3 Process Planning

Process planning takes as input the design characteristics of a product, and gives as output its complete production plan. This plan should determine the machining processes needed, the tools to be used, and the sequencing of operations. If more than one plan is available, then an optimal plan should be selected. It comprises a series of tasks that are heavily dependent on the type of product that is to be processed. It can be more or less elaborate, according to the processing requirements of a particular part. Horvath et al. [17] illustrated some elements of the process that should be determined by a process plan. Process planning is the link between the design and manufacturing phases of a product.

Bowden and Bullington [18] created a hybrid system called GUARDS, based on unsupervised machine learning and evolutionary algorithms (EAs) in order to optimize the control of a manufacturing process. The system learned to select the optimal process plan according to the status of the plant. Horvath et al. [17] described a complete process planning procedure, from the input of part specifications in the form of CAD files, to the optimization of the constructed process plan. They used an object-oriented approach in the form of "features." A "feature" was an object that defined specific operations and contained all of the relative functional, geometrical, and technological data. Knowledge-based reasoning was used for the generation of plans, which were then optimized with the help of a GA. Hayashi et al. [19] introduced an interesting method for the evaluation of future plans in a manufacturing plant with uncertain parameter values. A binary-coded EA was employed for the evaluation task. The solution was represented as a string of all of the plant's parameters, and the objective of the algorithm was defined according to user's preferences.

4 The Assembly Line Balancing Problem

An assembly line is made up of a number of workstations, arranged in series and in parallel, through which the work progresses on a product flows, thus composing a flow-oriented production system. Production items of a single type (single-model) or of several types (mixed-model) visit stations successively, where a subset of tasks of known duration are performed on them.

The assembly line configuration involves determining an optimal assignment of a subset of tasks to each station of the plant fulfilling certain time and precedence restrictions. In short, the goal is to achieve a grouping of tasks that minimises the inefficiency of the line or its total downtime and that respects all the constraints imposed on the tasks and on the stations. Such problem is called assembly line balancing (ALB) [20] and arises in mass manufacturing with a significant regularity both for the first-time installation of the line or when reconfiguring it. It is thus a very complex combinatorial optimisation problem (known to be NP-hard) of great relevance for managers and practitioners.

The first family of "academic" problems modelling this situation was known as Simple Assembly Line Balancing Problems (SALBP) [20,21], and it only considers the assignment of each task to a single station in such a way that all the precedence constraints are satisfied and no station workload time is greater than the line cycle time. Two versions of this problem are known [20]: SALBP-1, involving the assignment of tasks to stations so that the number of stations is minimised for a given production rate; and SALBP-2, aiming to maximize the production rate.

When other considerations are added to those of the SALBP family, the resulting problems are known by the name of General Assembly Line Balancing Problems (GALBP) in the literature [22]. A generic classification scheme for the field of ALB considering many different variants is also provided in a recent paper by Boysen et al. [23].

In [24] the authors reviewed the state of the art of optimization methods for assembly line design. Analyzing this paper and the updated state of the art, we can conclude that there is a high predominance of pure approaches instead of hybrid ones applied to the ALBP. Heuristic and exact methods are the most common. GAs and other metaheuristics like simulated annealing and Ant Colony Optimization are increasingly applied to tackle this problem. Metaheuristics, in general, and GAs specifically designed for ALBPs, seem to be very useful in the case of multiple objective problems, in which they present a set of Pareto optimal solutions.

A few number of hybrid approaches have been developed to deal with this optimization problem. Most of them hybridize a global search, mainly a GA, with one or more local search methods. This kind of hybridization is called memetic algorithm. In that sense, Haq et al. [25] dealt with mixed-model assembly line balancing for n models. The hybridization consists of using the solution from the modified ranked positional method for the initial solution of a classical GA to reduce the search space within the global solution space, thereby reducing search time. A similar hybridization is used in [26], where Ponnambalam et al. developed a multiple objective GA (MOGA). They used 14 simple heuristics to initialise the population and classical genetic operators to evolve solutions. The method aims to maximize the line efficiency as well as the smoothness index. During the execution of the MOGA, a tentative set of Pareto optimal solutions are stored and updated at each generation.

There are some papers that instead of using classical GAs use a Grouping Genetic Algorithm (GGA) [27], which demonstrated their best performance. In contrast, Tseng et al. [28] proposed hybrid evolutionary multiple-objective algorithms to design product assembly systems pondering simultaneously over assembly sequence planning (ASP) and ALB. The results provide a set of objectives and amend Pareto-optimal solutions to benefit decision makers in the assembly plan. In addition, an implemented decision analytic model supports the preference selection from the Pareto-optimal ones.

Besides the memetic approaches, there are other proposals that combine different metaheuristics, tabu search and swarm algorithms ([29-31]) ([29]), GAs and simulated annealing ([32]). For example, Blum et al. [29], considered the time and space constrained simple assembly line balancing problem (TSALBP-1) with the objective of minimizing the number of necessary work stations. For tackling this problem they proposed a Beam-ACO approach, which is an algorithm that results from hybridizing ACO with beam search.

5 The Dynamic Scheduling Problem

Scheduling problems have been widely studied over the last decades. Due to the complexity and the variety of such problems, most works consider static problems in which activities are known in advance and constraints are fixed [33,34]. However, every scheduling problem is subject to unexpected events. In these cases, a new solution is needed in a preferably short time taking these events into account and being as close as possible to the current solution. The main uncertainties encountered in a real manufacturing system are the following: machine breakdowns including uncertain repair times; increased priority of jobs; and changes in due dates and order cancellations.

Whenever an unexpected event happens in a manufacturing plant, a scheduling decision must be made in real time about the possible reordering of jobs. This process is known as "rescheduling." The main objective of rescheduling is "to find immediate solutions to problems resulting from disturbances in the production system" [35].

The problem can be defined as follows: let $A = (1, \dots n)$ be a set of activities, and $R = (1, \dots, r)$ a set of renewable resources. Each resource k is available in a constant amount R_k. Each activity i has a duration p_i and requires a constant amount r_{ik} of the resource k during its execution. Preemption is not allowed. Activities are related by precedence constraints, and resource constraints require that for each period of time and for each resource, the total demand of resource does not exceed the resource capacity. The objective considered here is to find a solution for which the end time of the schedule is minimized.

In the last few years, hybrid dynamic scheduling systems have been employed for solving this complex problem. We will review the most representative examples as follows. Machine learning is one of the approaches that have traditionally been used in manufacturing environments to face uncertainties. Chiu and Yih [36] proposed such a learning-based methodology for dynamic scheduling. They divided the scheduling process in a series of ordered scheduling points. An EA examined which dispatching rules performed better for each of these points, given a set of plant conditions (system status). The chromosome was comprised by a series of genes, each one representing a respective scheduling point and taking as a value one of the available dispatching rules. The performance of the algorithm was simulated under different plant conditions, forming a knowledge base that described the scheduling rules that were preferable in different cases. A binary decision tree was used to describe the gained knowledge. Aytug et al. [37] presented a different machine learning approach for dynamic scheduling, based on classifier systems. In this case, an initial knowledge base was given, and an EA modified it, using results taken from the simulation of the production line. In that way, the system learned to react to certain unexpected events. HAIS based on ANNs, EAs, and an inductive learning algorithm called trace-driven knowledge acquisition (TDKA) [38] was used by Jones et al. [39] to infer knowledge about the scheduling process. A back-propagation ANN selected a number of candidate dispatching rules out of a larger set of available rules. The schedules formed by these dispatching rules were used as the initial population of an EA that evolved an optimal schedule. The results taken from the simulation of the schedule helped TDKA to create a set of rules that formed the knowledge base. Lee et al. [40] proposed a hybrid scheduling framework which consisted of an inductive learning system for job releasing in the plant, and an EA-based system for the dispatching of jobs at the machines. Goldberg's [41] genetics-based machine learning method and an EA-based status selection method have also been employed by Tamaki et al. [42] and Ikkai et al. [43], respectively, to induce scheduling knowledge from manufacturing systems.

6 Concluding Remarks

In recent years, HAIS are widely used in different research fields. They have demonstrated they are useful optimization techniques given their capabilities to tackle complex problems involving imprecision, uncertainty, vagueness, and high-dimensionality. On the other hand, most of the tasks related to operations management can be formulated as optimization problems. Many of them have been successfully tackled by pure approaches, as GAs. However, many others are specially complicated and they demand hybrid approaches. In this contribution we reviewed the capability of HAIS to deal with this sort of problems with a very good performance in four different operations management domains: design engineering, process planning, assembly line balancing, and dynamic scheduling.

Acknowledgements

This work was partially supported by the Spanish Ministerio de Ciencia e Innovación (ref. MICINN IAP-020100-2008-15).

References

1. Kasabov, K.N.: Foundation of Neural Networks, Fuzzy Systems and Knowledge Engineering. MIT Press, Cambridge (1996)
2. Koza, R.J.: Genetic Programming, On the Programming of Computers by Means of Natural Selection. MIT Press, Cambridge (2000)

3. Cordon, O., Herrera, F., Hoffman, F., Magdalena, L.: Genetic Fuzzy Systems: Evolutionary Tuning and Learning of Fuzzy Knowledge Bases. World Scientific Co. Ltd., Singapore (2001)
4. Lin, C.-T., Lee, C.: Neural Fuzzy Systems: A Neuro-Fuzzy Synergism to Intelligent Systems. Prentice Hall, Upper Saddle River (1996)
5. Finger, S., Dixon, J.R.: A review of research in mechanical engineering design. Part I: descriptive, prescriptive, and computer-based models of design processes. Part II: representations, analysis and design for the life cycle. Research in Engineering Design (1), 51–67 (1989)
6. Evbuowman, N.F.O., Sivaloganathan, S., Jebb, A.: A survey of design philosophies, models, methods and systems. Proc. Insts. Mech. Engrs. Part B: Journal of Engineering Manufacture 210(4), 301–320 (1996)
7. Vico, F.J., Veredas, F.J., Bravo, J.M., Almaraz, J.: Automatic design synthesis with artificial intelligence techniques. Artificial Intelligence in Engineering 13(3), 251–256 (1999)
8. Sasaki, M., Gen, M.: Fuzzy multiple objective optimal system design by hybrid genetic algorithm. Applied Soft Computing 3(3), 189–196 (2003)
9. Wang, J., Terpenny, J.: Interactive evolutionary solution synthesis in fuzzy set-based preliminary engineering design. J. of Intelligent Manufacturing 14(2), 153–167 (2003)
10. Xiong, Y., Rao, S.S.: Fuzzy nonlinear programming for mixed discrete design optimization through hybrid genetic algorithm. Fuzzy Sets and Systems 146(2), 167–186 (2004)
11. Su, D., Wakelam, M.: Evolutionary optimization within an intelligent hybrid system for design integration. Artificial Intelligence for Engineering Design, Analysis and Manufacturing 13(5), 351–363 (1999)
12. Tsai, C.Y., Chang, C.A.: Fuzzy neural networks for intelligent design retrieval using associative manufacturing features. J. of Intelligent Manufacturing 14(2), 183–195 (2003)
13. Saridakis, K.M., Dentsoras, A.J.: Evolutionary neuro-fuzzy modelling in parametric design. In: I Conference in Innovative production machines and systems (2005)
14. Lu, P.C.: The application of fuzzy neural network techniques in constructing an adaptive car-following indicator. Artificial Intelligence for Engineering Design, Analysis and Manufacturing 12(3), 231–242 (1998)
15. Hamedi, M.: Intelligent fixture design through a hybrid system of artificial neural network and genetic algorithm. Artificial Intelligence Review 23(3), 295–311 (2005)
16. Zha, X.F.: Soft computing framework for intelligent human–machine system design, simulation and optimization. Soft Computing 7(3), 184–198 (2003)
17. Horvath, M., Markus, A., Vancza, C.: Process planning with genetic algorithms on results of knowledge-based reasoning. Comput. Integr. Manuf. 9(2), 145–166 (1996)
18. Bowden, R., Bullington, S.F.: Development of manufacturing control strategies using unsupervised machine learning. IIE Trans. 28(4), 319–331 (1996)
19. Hayashi, Y., Kim, H., Nava, K.: Scenario creation method by genetic algorithms for evaluating future plans. In: Proc. IEEE Conf. Evol. Comput. Piscataway, NJ, pp. 880–885 (1996)
20. Scholl, A.: Balancing and Sequencing of Assembly Lines, 2nd edn. Physica-Verlag, Heidelberg (1999)
21. Scholl, A., Becker, C.: State-of-the-art exact and heuristic solution procedures for simple assembly line balancing. European Journal of Operational Research 168, 666–693 (2006)
22. Becker, C., Scholl, A.: A survey on problems and methods in generalized assembly line balancing. European Journal of Operational Research 168(3), 694–715 (2006)
23. Boysen, N., Fliedner, M., Scholl, A.: A classification of assembly line balancing problems. European Journal of Operational Research 183, 674–693 (2007)
24. Rekiek, B., Dolgui, A., Delchambre, A., Bratcu, A.: State of art of optimization methods for assembly line design. Annual Reviews in Control 26(2), 163–174 (2002)

25. Haq, A.N., Rengarajan, K., Jayaprakash, J.: A hybrid genetic algorithm approach to mixed-model assembly line balancing. International Journal of advanced manufacturing technology 28(3-4), 337–341 (2006)
26. Ponnambalam, S.G., Aravindan, P., Mogileeswar, G.: Assembly line balancing using multi-objective genetic algorithm. In: Proc of CARS&FOF 1998, Coimbatore, India, pp. 222–230 (1998)
27. Falkenauer, E.: Genetic Algorithms and Grouping Problems. John Wiley & Sons Ltd., Chichester (1997)
28. Tseng, H.E., Chen, M.H., Chang, C.C.: Hybrid evolutionary multi-objective algorithms for integrating assembly sequence planning and assembly line balancing. International Journal of Production Research 46(21), 5951–5977 (2008)
29. Blum, C., Bautista, J., Pereira, J.: An extended Beam-ACO approach to the time and space constrained simple assembly line balancing problem. In: van Hemert, J., Cotta, C. (eds.) EvoCOP 2008. LNCS, vol. 4972, pp. 85–96. Springer, Heidelberg (2008)
30. Suwannarongsri, S., Limnararat, S., Puangdownreong, D.: A new hybrid intelligent method for assembly line balancing. In: IEEE International Conference on Industrial Engineering and Engineering Management, vol. 1-4, pp. 1115–1119 (2007)
31. Tasan, S.Ö., Tunali, S.: Improving the genetic algorithms performance in simple assembly line balancing. In: Gavrilova, M.L., Gervasi, O., Kumar, V., Tan, C.J.K., Taniar, D., Laganá, A., Mun, Y., Choo, H. (eds.) ICCSA 2006. LNCS, vol. 3984, pp. 78–87. Springer, Heidelberg (2006)
32. Yuan, M.H., Li, D.B., Tong, Y.F.: Research on mixed-model assembly line balance with genetic simulated annealing algorithm. In: Proceedings of the 14th international conference on industrial engineering and engineering management, vol. a and b, pp. 71–75 (2007)
33. Allahverdi, A., Ng, C.T., Cheng, T.C.E., Kovalyov, M.Y.: A survey of scheduling problems with setup times or costs. European Journal of Operational Research 187(3), 985–1032 (2008)
34. T'kindt, V., Billaut, J.-C.: Multicriteria Scheduling. European Journal of Operational Research 167(3), 589–591 (2005)
35. Jain, A.K., Elmaraghy, H.A.: Production scheduling/rescheduling in flexible manufacturing systems. Int. J. Prod. Res. 35(1), 281–309 (1997)
36. Chiu, C., Yih, Y.: A learning-based methodology for dynamic scheduling in distributed manufacturing systems. Int. J. Prod. Res. 33(11), 3217–3232 (1995)
37. Aytug, H., Koehler, G.H., Snowdon, J.L.: Genetic learning of dynamic scheduling within a simulation environment. Comput. Oper. Res. 21(8), 909–925 (1994)
38. Yih, Y.: Trace-driven knowledge acquisition (TDKA) for rule-based real-time scheduling systems. J. Intell. Manuf. 1, 217–230 (1990)
39. Jones, A., Rabelo, L., Yih, Y.: A hybrid approach for real-time sequencing and scheduling. Int. J. Comput. Integrated Manuf. 8(2), 145–154 (1995)
40. Lee, C.-Y., Piramuthu, S., Tsai, Y.-K.: Job-shop scheduling with a genetic algorithm and machine learning. Int. J. Prod. Res. 35(4), 1171–1191 (1997)
41. Goldberg, D.E.: Genetic Algorithms in Search, Optimization and Machine Learning. Addison-Wesley, Reading (1989)
42. Tamaki, H., Ochi, M., Araki, M.: Application of genetics-based machine learning to production scheduling. In: Symp. Flexible Automation: ASME, pp. 1221–1224 (1996)
43. Ikkai, Y., Inoue, M., Ohkawa, T., Komoda, N.: A learning method of scheduling knowledge by genetic algorithms. In: IEEE Symp. Emerging Technol. Factory Automation, pp. 641–648. IEEE, Los Alamitos (1995)

A Pool of Experts to Evaluate the Evolution of Biological Processes in SBR Plants

Davide Sottara[1], Gabriele Colombini[1], Luca Luccarini[2], and Paola Mello[1]

[1] DEIS, Faculty of Engineering, University of Bologna Viale Risorgimento 2, 40100 Bologna (BO) Italy
[2] ENEA - ACS PROT IDR - Water Resource Management Section Via Martiri di Monte Sole 4, 40129 Bologna (BO) Italy

Abstract. In order to minimize the costs and maximize its efficiency, a Sequencing Batch Reactor requires continuous monitoring. The sensors it can be reasonably equipped with provide only indirect information on the state of the chemical reactions taking place in the tank, so the data must be analysed and interpreted. At present, no optimum, completely reliable procedure exists: instead, there exist several criteria which can be applied under different conditions. This paper shows that estimating the confidence in the quality of the response of a criterion can increase the robustness of a criterion. Then, interpreting the responses in terms of possibility distributions, the different answers can be merged, thus obtaining a more reliable overall estimate.

1 Introduction

The biological treatment of waste water is aimed at the removal of pollutants (nitrogen compounds and organic matter), which are actually nutrients for certain specific microbial populations. This operation is performed in plants which allow the required biochemical reactions to take place in sequence: in particular, nitrification is required to convert all ammonia to nitrate and denitrification to convert nitrate to gaseous nitrogen, which is then released in the atmosphere. A Sequencing Batch Reactor is a special type of waste water treatment plant having the advantage of being extremely flexible, allowing operators to configure and control several operating conditions without structural modifications. In particular, nitrification and denitrification are carried out in the same tank, during different operating phases taking place sequentially in time. In order to save time and energy, and to increase the overall performance of the treatment system, the duration of the phases should be set according to the actual duration of the reactions and not on a worst-case basis, as is typically done in real plants. In order to monitor in real time the state of the reactions, the plant should be equipped with ammonia (NH_4^+) and nitrate (NO_3^-) probes, but the high costs of such instruments, at several thousands of euros, make the option economically infeasible, especially for medium and small plants. These plants, instead, are usually equipped with much cheaper probes: pH, redox potential (ORP), dissolved oxygen concentration (DO) and temperature (T), which still provide

E. Corchado et al. (Eds.): HAIS 2009, LNAI 5572, pp. 368–375, 2009.

indirect information on the evolution of the important reactions. With an adequate sensor data analysis, their correct completion can be recognized: data have been used in the development of several monitoring and control system, roughly classified according to the main principle they are based on:

- comparison of the current data with past, classified data ([7], [10], [11])
- detection of relevant features in the signals ([5], [9])
- prediction and analysis of the pollutant concentrations ([6])

The main problem of such control systems is robustness and reliability: the "boundary"conditions, such as influent load and sludge microbial population, can't usually be observed and controlled, so it is difficult to find general, invariant criteria for the analysis of the time series acquired from the probes. In statistical terms, the data used for the training of a control module could have been sampled from a different population than the one actually present in the monitored plant, so the analysis may not be reliable. Even if a continual, on-line training procedure could deal with this issue, the problem of acquiring reliable training data remains. In order to evaluate the actual correlation between concentrations and signals, both type of data should be available: however, while there are plenty of sensor data, chemical samples are critical because, if not from very expensive probes, they must be acquired from very time-consuming laboratory analysis. Since this is rarely possible, a more feasible approach is having the output of a module include a confidence measure to express its reliability. This value should be taken into account in applying the consequent control actions: it would be senseless to switch from a phase to the following if the reaction had been estimated complete on the basis of totally unsound premises. In this paper, the problem of reliability is addressed in two ways and applied, as a case study, to the analysis of the denitrification process. In the first part, an existing criterion based on the estimation of the pollutant concentrations using a neural network [6] is improved and extended with uncertainty and confidence estimations. In the second part, the proposed module is used along with other existing ones ([11], [5])to form an ensemble of experts. The opinion of each one is considered a possibilistic distribution $\pi_{DEN}(t)$ of the current reaction state at different times ([12], [3]): the answers are then combined to obtain a more robust estimate.

2   Signal Data Analysis

The data are acquired from a pilot-plant SBR, completely built in the ENEA PROT IDR laboratories, placed side stream to the Trebbo di Reno (Bologna) municipal treatment plant and fed on real sewage drawn after sieve treatment. The plant is equipped with pH, ORP and DO probes, with a sampling rate of 60Hz, connected to a National Instruments Acquisition Board. An average of the data collected every minute is then stored in a MySQL database. The proposed denitrification monitoring system tries to estimate the state of the reaction, i.e. the nitrate concentration $[NO_3^-(t)]$, from the values measured by the signal probes. Given the sequential nature of the data, and the non-linearity of the state-observation relation, there is no analytical model that can be inverted: so, a black-box approach has been adopted, and an Elman neural network has been chosen in particular. An Elman network models an input-state to output non-linear relation $I \times S \mapsto O$ and is more suitable than a simple input-output feed-forward network when the relation to be modelled is stateful ([4]):

$$\mathbf{Y}(t) = f(\mathbf{U}(t), \mathbf{X}(t-1))$$

Having to learn an inverse model, the role of state and observation is reversed: $\mathbf{X}(t) = [NO_3^-(t)]$, $\mathbf{Y}(t) = \mathbf{X}(t)$ and $\mathbf{U}(t) = \{pH(t), ORP(t), DO(t)\}$, where Y,X and U denote output, state and input respectively.

The training set consisted of 6 track studies, performed at different times over 6 months of operations. During a track study, the signals are acquired continuously, but the reaction is also observed by sampling the tank approximately every ten minutes. Even if the average duration of the denitrification process is approximately 30 minutes, the tank has also been sampled after its completion to observe its natural, free evolution. So, each track $j : 1..6$ yields a 3×60 observation matrix O_j and a 1×6 state matrix S_j which have been used to train the network according to the following procedure.

2.1   Model Learning

Data preprocessing. Each time serie has been denoised using a regularization algorithm ([2]), then the data have been normalized using the mean $\mu_\mathbf{x}$ and the variance σ_0 computed on the whole training dataset. In order to construct an analogue time serie for the nitrate concentration, the sample data have been interpolated using sigmoidal functions. In practice, simple $1 - 6 - 1$ feed-forward neural networks have been trained with pairs $< t(i), S_j(i) >_{i:1..6}$. The network size has been chosen according to the simple heuristic that each sample could require one neuron to model its temporal neighbourhood. After the six 1×60 matrices have been computed, the values have been scaled to the $[0, 1]$ interval, since concentrations can't be negative and have a maximum value which is exceeded with negligible probability (for nitrates, it is set at 20 mg/L).

Training. Given the input and output vectors, the network has a 4-H-1 architecture, with sigmoidal activation functions in the hidden layer and a linear function clamped to [0,1] in the output neuron. In order to decide the size of the hidden layer H, several networks have been generated, with H ranging from 5 to 15. Each network has been trained 6 times, using a leave-one-out cross-validation procedure: the average of the 6 mean squared errors (MSE) shows no real improvement for $H > 12$, so this value has been chosen for a final network, trained on the whole available dataset.

2.2   Confidence Estimation

The network has a MSE $< 1e-4$, so the average estimation error is not supposed to be greater than 1%. This, however, is true for inputs $i(t) =< \mathbf{u}(t), \mathbf{x}(t-1) >$

which are similar to the data the network has been trained on. Sigmoidal Neural networks, in fact, may have good interpolation capabilities, but are demanding at extrapolation. The training data define a domain outside of which the output of the network should be considered unreliable. To evaluate the similarity between the training data and the new inputs, a Self Organizing Map [4] has been trained with the same inputs used to train the Elman network. The trained SOM is composed by 24 neurons $n_{k:1..24}$, deployed on a 6×4 rectangular grid. Each neuron has an associated Gaussian function γ_k : the network does not return the index of the winning neuron, but rather the fuzzy similarity degree $\chi(\mathbf{i})$:

$$\chi(\mathbf{i}) = \max_k\{\gamma_k(\mathbf{i})\} = \max_k\{e^{-\frac{||\mathbf{i}-\mathbf{n_k}||^2}{\rho^2}}\} \tag{1}$$

The parameter ρ determines the boundaries of the domain, setting the maximum distance at which a new input is considered sufficiently included and thus acceptable. A value for ρ can be chosen according to the following considerations: the activation of each neuron of the Elman network is given by hyperbolic tangent function $\sigma(\cdot)$, whose value for a vector input $\mathbf{i}$ depends on the weight vector $\mathbf{w_k}$ and the bias b_k according to the expression $\sigma(<\mathbf{w_k}, \mathbf{i} - i_k^c >)$, where $< \cdot, \cdot >$ denotes the scalar product between two vectors and $i_k^c = -\mathbf{w_k}^{-1}b_k$. Given its exponential nature, the input is within the scope of the sigmoid only if $< \mathbf{w_k}, \mathbf{i} - i_k^c) >\lesssim ||\mathbf{w_k}|| \cdot ||\mathbf{i} - i_k^c|| \lesssim 3$. Hence, $||\mathbf{w_k}||^2$ determines the scope of the Elman neurons and $\min_k \frac{9}{||\mathbf{i}-i_k^c||^2}$ is the smallest one. Choosing $\rho^2 \propto ||\mathbf{w_k}||^2$, if the Elman neuron with the narrowest scope is placed on a SOM neuron, the SOM would certify an input $\mathbf{i}$ to be within the domain of the Elman.

The degree $\chi(\mathbf{i})$ measures the confidence in the estimation provided by the network. The lower χ, the more unreliable the output value should be considered. To take confidence into account, the single-valued output of the network $y(\mathbf{i})$ is transformed into a possibility distribution ([12]), i.e. a fuzzy set which states, for every value of the domain, the degree at which it is possible that the value is the correct one. In the proposed transformation, the value y is expanded into the interval $[y_L, y_U] = [\max\{0, y - (1-\chi)\}, \min\{1, y+(1-\chi)\}]$ and the points y_L, y and y_U are used to define a triangular, normal fuzzy set $\pi^*(y)$. Successively, this set is discounted by setting $\pi(y)_{y\in[y_L,y_U]} = \max\{\pi^*(y), 1-\chi\}$. Notice that when $\chi = 1$, y remains unaltered and $\pi(y) = 1$, while $\chi = 0$ sets $\pi(y) = 1 \forall y \in [0,1]$, denoting complete ignorance.

2.3   Network Output Analysis

Given the estimate $\pi(y(t))$ for the nitrate concentration at time t, the problem is to establish whether it can be equal to zero (denoted by Z) or not. Moreover, the equivalence need not be exact: the chemical equilibrium prevents the nitrate from disappearing completely and the probes add measurement noise, but the law actually requires the concentrations to be below a threshold (in Italy, 10 mg/L). Hence, the actual objective is $\pi(y(t) \approx Z)$. The concept of Zero is fuzzy, so it is modelled using a triangular fuzzy set from $\mu_Z(0) = 1$ to $\mu_Z(10) = 0$.

Called $P(t) = \pi(y(t))$, the compatibility between the estimation and Z is:

$$\pi(y(t) \approx Z) = \frac{|P(t) \wedge Z|}{|P(t) \wedge Z| + |P(t) \wedge \neg Z|} \qquad (2)$$

The operators are the usual fuzzy set operators, so $\wedge$ is the min T-norm, $\neg$ is 1-complementation and $|\cdot|$ is the fuzzy cardinality defined by $|S| = \sum_{s \in S} \mu(s)$. Equation 2 expresses then a ratio between the possibility that the current estimate is zero and the possibility that it is zero or not zero.

Examples. Figure 1 shows the output for a cycle used for the training. Figure 1(a) shows the time series acquired during a whole cycle (denitrification followed by nitrification). The upper part of figure 1(b), instead, shows the predicted concentration compared to the one obtained from the laboratory analysis (in the figures, the outputs have been rescaled from $[0, 1]$ to their natural ranges for clarity). Notice that the Elman network has been trained for the denitrification process only (corresponding roughly to the first half of the cycle): in fact, the predictions are accurate for the first part, but not so for the second one. However, the SOM recalls correctly the former data and not the latter, so the confidence interval $[y_L, y_U]$, drawn using vertical lines, denotes the different degrees of reliability as expected. The possibility $\pi(y(t) \approx Z)$, shown in the lower half of figure 1(b), is low during the first stages of the denitrification process, when the nitrate

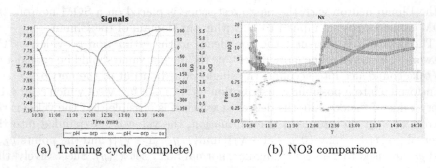

(a) Training cycle (complete)   (b) NO3 comparison

Fig. 1. Validation Results

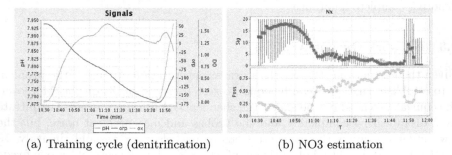

(a) Training cycle (denitrification)   (b) NO3 estimation

Fig. 2. Test Results

concentration is reasonably high, but gets higher as the nitrates are consumed. When the networks no longer recognizes the inputs, it becomes low again as there is no evidence to conclude that the nitrate concentration is still low.

Figure 2 shows the result of the analysis of the signals acquired during a different denitrification cycle, for which no laboratory analysis are available. The possibility estimate, however, agrees strongly with the opinion of a human expert: at the beginning of the reaction, the signals have the expected trend (pH rising and ORP decreasing, see also [10]). The fact that pH is actually lower than the usual adds some uncertainty to the estimate, but the possibility that the nitrate is consumed is low. After some minutes, the data fully become compatible with the training set and nitrate concentration is correctly assumed to be greater than 0. The change in signal trend (pH decreasing and ORP below -150mV) is then associated to a much higher possibility that the nitrates have disappeared. The switch to nitrification conditions (shown by the rise in DO) brings the network back to a state of ignorance, so the possibility lowers.

3 Expert Pooling

The performance of the architecture proposed in section 2 depends strongly on the new inputs being similar to the past ones, but this property cannot be guaranteed by the plant. However, there exist other criteria, with different properties and requirements, that can be exploited to estimate whether the nitrate concentration is near zero. Eventually, the various responses have to be merged: the compact notation π_i will be used to denote the distributions $\pi_{i:1,2,...}(y(t) \approx 0)$ (π_1 is the one in section 2). The problem of knowledge fusion has been addressed in several context with different techniques, such as belief networks (for an example in the waste water domain, see [8]). Possibilistic data fusion, on the other hand, is common in robotics and artificial vision applications (e.g. [1]), but has not been applied extensively to water treatment.

Signal Feature Detection. The algorithm, briefly presented in [5], tries to identify local features in the signals, such as maxima, minima and "knees". It is known from literature ([10]) that the contemporary detection of a maximum in pH and a knee in ORP is a good indicator of the completion of the denitrification process. Since the event detection algorithm returns an exact time t_e and a confidence interval $[t_e^L, t_e^U]$, it is natural to define two triangular possibility distributions $\pi_M(t) = \pi_{max(pH)}(t)$ and $\pi_K(t) = \pi_{knee(ORP)}(t)$. Since the events must be aligned, they are merged using the intersection rule: $\pi_{M \wedge K}(t) = \pi_M(t) \wedge \pi_K(t)$.

The distribution is not normalized, so the maximum possibility is lower as the distance between the individual possibility peaks t_M and t_K increases. Given $\pi_{M \wedge K}(t)$, $\pi_2(t)$, one can set $\pi_2(t) = \max\{\pi_{M \wedge K}(t), \pi_2(t-1)\}$. In general, the information carried by $\pi_{M \wedge K}(t)$ is more valuable since it does not model the possibility that the nitrate concentration is zero at time t, but the possibility that it has reached zero exactly at time t. In fact, $\pi_2(t)$ is computed univocally from $\pi_{M \wedge K}(t)$, but not the opposite.

Sub-phase classification. Another approach ([11]) distinguishes three sub-phases: before ("R"), during ("G") and after ("B") the completion of the denitrification process.

For each time period t, the system tries to recognize the current sub-phase and, combining a SOM neural network with fuzzy logic rules, evaluates the possibility that the individual sub-phases have already begun: given the sequentiality of the three, the end of the first two is marked by the start of the following. The three possibility distributions are defined by the truth values of the fuzzy predicates ackR(t), ackG(t) and ackB(t). Since the process is complete when all three sub-phases have been acknowledged, the intersection degree of ackDen(t) $\equiv$ ackR(t) $\wedge$ ackG(t) $\wedge$ ackB(t) is a candidate distribution $\pi_3(t)$.

Information merging. To combine the answers of the different experts, a merge operator $\cap$ has to be applied, getting $\pi(t) = \cap_{i=1}^3 \pi_i(t)$. A comprehensive review of the existing operators can be found in [3]. Among them, the choice has fallen on the adaptive combination rule:

$$\bigcap_{i=1}^n \pi_i(t) = \frac{\bigwedge_{i=1}^n \pi_i(t)}{\sup_x \bigwedge_{i=1}^n \pi_i(x)} \vee \left(\bigwedge_{i=1}^n \pi_i(t) \wedge \left(1 - \sup_x \bigwedge_{i=1}^n \pi_i(x) \right) \right) \quad (3)$$

When the analysis modules agree, the result is given by the intersection of their individual answers, with a degree of conflict given by $1 - \sup_x \bigwedge_{i=1}^n \pi_i(x)$. The individual modules are robust and their answer usually degrade gracefully towards a cautious, unknown-like answer when they can't recognize the inputs, so the degree of conflict is generally low. A possibility estimate, however, may be erroneous, so conflict is not excluded a priori. In that case, the combination rule returns a distribution which tends to the uniform distribution over the union of the conflicting answers domains as the conflict increases. In equation 3, $\wedge$ and $\vee$ may be any T-norm and co-norm: in practice, $\wedge$ may be the product, but since the distributions are not independent, min is more appropriate.

Figure 3 shows an example of the combined distribution $\pi(t)$. It can be seen that the combined distribution has possibility near to one for $t > 40$, as expected from the observation of the signal probes. In this specific case, the contribution of π_3 has bested the other responses.

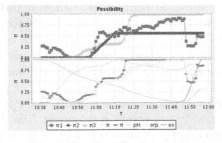

Fig. 3. Merging of Possibility Distributions using $*$ and $\wedge$ T-norms

4   Conclusions

The proposed framework shows a convenient way to combine different existing analysis modules and may be easily extended with additional ones. The use of possibility distributions to take the individual reliability into account has increased the overall robustness of the system. Moreover, expressing the outputs in terms of fuzzy sets will facilitate the integration of such a system with a fuzzy rule engine. Even if the application has been tailored to a specific case, an analogue configuration could be easily applied to the detection of other types of events: in fact, the possibilistic fusion of sensor data is a common application in literature. In the specific case of the SBR plant, the immediate extension would be the analysis of the nitrification process, but other plant configurations will be considered in the future.

References

1. Benferhat, S., Sossai, C.: Reasoning with multiple-source information in a possibilistic logic framework. Information Fusion 7(1), 80–96 (2006)
2. Chartrand, R.: Numerical differentiation of noisy data (submitted, 2007)
3. Dubois, D., Prade, H.: Possibility theory and data fusion in poorly informed environments. Control Engineering Practice 2(5), 811–823 (1994)
4. Haykin, S.: Neural Networks: A Comprehensive Introduction. Prentice-Hall, Englewood Cliffs (1999)
5. Luccarini, L., Bragadin, G.L., Mancini, M., Mello, P., Montali, M., Sottara, D.: Process quality assessment in automatic management of wwtps using formal verification. In: International Symposium on Sanitary and Environmental Engineering-SIDISA 2008 -Proceedings, vol. 1, pp. 152–160. ANDIS (2008)
6. Luca, L., Ettore, P., Alessandro, S., Paolo, R., Selene, G., Sauro, L., Giuseppep, B.: Soft sensors for control of nitrogen and phosphorus removal from wastewaters by neural networks. Wat. Sci. Tech. 45(4-5), 101–107 (2002)
7. Ordóñez, R., Liliana, M.: Multivariate statistical process control and case-based reasoning for situation assessment of sbrs, September 23 (2008)
8. Populaire, S., Thierry, J.B., Mpe, A., Guilikeng, A.: Fusion of expert knowledge with data using belief, August 01 (2002)
9. Rubio, M., Colomer, J., Ruiz, M., Colprim, J., Melndez, J.: Qualitative trends for situation assessment in sbr wastewater treatment process. In: Proceedings of the BESAI 4th ECAI Workshop on Binding Environmental Sciences and Artifical Intelligence, Valencia, Spain (2004)
10. Marsili Libelli, S.: Control of sbr switching by fuzzy pattern recognition. Wat. Res. 40(5), 1095–1107 (2006)
11. Sottara, D., Luccarini, L., Mello, P.: AI techniques for waste water treatment plant control case study: Denitrification in a pilot-scale SBR. In: Apolloni, B., Howlett, R.J., Jain, L. (eds.) KES 2007, Part I. LNCS, vol. 4692, pp. 639–646. Springer, Heidelberg (2007)
12. Zadeh, L.A.: Fuzzy sets as a basis for a theory of possibility. Fuzzy Sets and Systems 100, 9–34 (1999)

A Hybrid Ant-Based Approach to the Economic Triangulation Problem for Input-Output Tables

Camelia-M. Pintea[1], Gloria Cerasela Crisan[2], Camelia Chira[1],
and D. Dumitrescu[1]

[1] Babes-Bolyai University, M. Kogalniceanu 1, 400084 Cluj-Napoca, Romania
{cmpintea,cchira,ddumitr}@cs.ubbcluj.ro
[2] Bacau University, Calea Marasesti 157, 600115 Bacau, Romania, and
Centre Interuniversitaire de Recherche sur les Reseaux d'Entreprise, la Logistique et le
Transport, C.P. 8888, Succursale Centre-ville, Montreal (QC), Canada H3C 3P8
ceraselacrisan@ub.ro

Abstract. The *Triangulation Problem for Input-Output Matrices* has been intensively studied in order to understand the complex series of interactions among the sectors of an economy. The problem refers to finding a simultaneously permutation of rows and columns of a matrix such as the sum of the entries which are above the main diagonal is maximum. This is a linear ordering problem – a well-known NP-hard combinatorial optimization problem. A new hybrid heuristic based on ant algorithms is proposed to efficiently solve the triangulation problem. Starting from a greedy solution, the proposed model hybridizes the *Ant Colony System (ACS)* metaheuristic with an *Insert-Move (IM)* local search mechanism able to refine ant solutions. The proposed *ACS-IM* algorithm is tested with good results on some real-life economic data sets.

Keywords: Linear Ordering Problem, Heuristics, Ant Colony Optimization.

1 Introduction

Arising in many and diverse application areas, combinatorial optimization problems form today a highly investigated domain. Many of these problems are so complex that they cannot be solved within polynomial computation times using available resources. The development of metaheuristics able to efficiently find high-quality near-optimal solutions to such problems with reasonable computational effort is essential.

A special class of combinatorial optimization problems refers to the *linear ordering problems (LOP)* where the objective is to find a permutation with optimum cost of some objects. One such problem is the *Triangulation Problem for Input-Output Matrices* seeking for a permutation of rows and columns in a given matrix of weights W such that the sum of the entries above the main diagonal (of the permuted matrix) is maximized. This problem has been intensively studied by economists [1] in order to understand the interdependencies and complex series of transactions among the sectors of an economy [2]. A variety of models from exact methods to heuristics for solving this NP-hard problem [3] can be found in the literature [3-13].

To the best of our knowledge, this paper presents the first attempt to heuristically approach *LOP* using ant-based models. The proposed model integrates *Ant Colony*

E. Corchado et al. (Eds.): HAIS 2009, LNAI 5572, pp. 376–383, 2009.
© Springer-Verlag Berlin Heidelberg 2009

Optimization (ACO) [14] with local search heuristics to solve *LOP*. The initial solution is generated in a greedy manner and is used as the starting point for an *Ant Colony System (ACS)*. Furthermore, the ant-based model is hybridized with a local search mechanism based on *insert-moves* able to improve the solutions. The proposed hybrid ant-based model is tested for addressing the instances available in the real-world LO-LIB [15] and SGB [16] libraries.

2 The Triangulation Problem for Input-Output Tables

The *Triangulation Problem for Input-Output Matrices* is a *LOP*, a well-known NP-hard combinatorial optimization problem [3]. The economy of a country or region is divided in sectors and an input-output table is built with entries quantifying the transactions between any two sectors. The triangulation of the input-output matrix means the detection of an optimal hierarchy of economic sectors (suppliers should be predominantly arranged before consumers in the matrix).

Using graph theory terms *LOP* can be formulated as the problem of searching for an acyclic tournament having the maximum sum of arc weights in a complete weighted graph [4]. Let $G=(V, A, w)$ be a weighted graph where V represents the set of vertices, A contains the arcs of the graph and the function w refers to the weights associated with arcs. *LOP* aims to maximize the following functional:

$$c_G(\pi) = \sum_{\substack{i \leq j \\ i \neq j}} w(\pi(i), \pi(j)) \tag{1}$$

where π is a permutation of V, $i, j \in V$ and $\leq$ is a total order relation on V.

Existing approaches to solve *LOP* include integer programming methods (branch-and-cut [3], branch-and-bound [5], and interior point method [6]), approximate algorithms [7,8]. Heuristics such as tabu search [9], scatter search [10], iterated local search [11], variable neighborhood search [12] and evolutionary strategies [13] have been successfully employed.

3 The Proposed Hybrid Ant-Based Model

We propose a hybrid model based on the *ACS* metaheuristic [17] for solving the Triangulation Problem. Starting from a greedy initial solution, *ACS* is further hybridized with a particular local search mechanism for solution refinement.

Ant algorithms are designed to solve optimization and distributed control problems by replicating the behavior of social insects to the search space (commonly described as a graph). In the *ACS* model [17], each ant generates a complete tour (associated to a problem solution) by probabilistically choosing the next vertex based on the cost and the pheromone intensity on the connecting arc.

The problem is represented as a complete directed graph with n vertices; the function w assigns real values to arcs defining the static matrix of weights $W = (w_{ij}), 1 \leq i, j \leq n$. Besides the weight, each arc is associated with a pheromone intensity (built up by ants during the search). The pheromone matrix τ is dynamic and has the same dimensions as W.

A problem solution (a permutation of vertices) is a list of n vertices constructed by each artificial ant by moving from one vertex to another. Each ant keeps the list of already visited vertices in order to further avoid them (this is an element of tabu search integrated with the hybrid ant-based model).

3.1 Greedy Solution Initialization

One element of hybridization within the proposed ant-based method for *LOP* refers to the usage of a greedy approach to build the initial solution for each ant.

For a complete solution π (meaning a permutation of length n), the neighborhood $N(\pi)$ contains the permutations that can be obtained from π by left-compounding with a transposition (see equation 2). This means that a neighbor for a permutation is obtained by interchanging two of its elements. Let $\circ$ denote the compounding operator for permutations and $(i\ j)$ refer to a transposition. $N(\pi)$ has $n(n-1)/2$ elements and is given below:

$$N(\pi) = \left\{ \sigma \mid \exists (k\ r): \ \sigma = (\pi(k)\ \pi(r)) \circ \pi \ \ and \ \ 1 \le k < r \le n \right\}. \tag{2}$$

When moving from a permutation π to its neighbor σ the objective function has an added value of:

$$diff(k, r) = (w_{\pi(r)\pi(k)} - w_{\pi(k)\pi(r)}) + \sum_{i=k+1}^{r-1} (w_{\pi(r)\pi(i)} + w_{\pi(i)\pi(k)} - w_{\pi(i)\pi(r)} - w_{\pi(k)\pi(i)}), \tag{3}$$

$1 \le k < r \le n$

The greedy initial local search procedure chooses the best-improvement move – the one with the highest value for $diff(k,r)$ given by (3) – when investigating the neighborhood of a permutation. The proposed ant-based algorithm is initialized with the greedy solution obtained during this initial search.

The pheromone matrix τ is initialized with $\tau_{ij} = \tau_0, 1 \le i, j \le n$, where τ_0 is a small positive constant. The number of artificial ants is denoted by m (a constant value).

3.2 Solution Construction and Pheromone Update

Let us denote by i the current vertex for an ant k. The next vertex j is selected according to the *pseudorandom proportional rule* [14] given by:

$$j = \begin{cases} \arg\max_{l \in N_i^k} \{ \tau_{il} / [w_{il}]^{\beta} \} & if \ q \le q_0 \\ J & otherwise \end{cases} \tag{4}$$

where q is a random variable uniformly distributed in [0, 1], N_i^k refers to the feasible neighborhood of ant k from vertex i, q_0 and β are parameters, and J is a random variable having the following probability distribution [14, 17]:

$$p_{ij}^k = \frac{\tau_{ij} / [w_{ij}]^\beta}{\sum_{l \in N_i^k} \tau_{il} / [w_{il}]^\beta} \quad if \ j \in N_i^k .$$

(5)

Based on the *ACS* model, pheromone update occurs in both the online and offline phases. For the *online* phase, the pheromone is updated during the solution construction, immediately after an ant crosses the arc $(i\ j)$, based on the following updating rule [14]:

$$\tau_{ij} = (1 - \xi)\tau_{ij} + \xi \tau_0 .$$

(6)

where ξ and τ_0 are model parameters.

After all ants have constructed a solution, the *offline* phase implies only the arcs from the *best-so-far* solution and uses the following rule [14]:

$$\tau_{ij} = (1 - \rho)\tau_{ij} + \rho / C_{bs} .$$

(7)

where C_{bs} is the cost of the *best-so-far* solution, and ρ is a parameter.

3.3 Insertion-Based Local Search

The proposed ant-based model for *LOP* is hybridized with a local search mechanism based on insertions aiming to further improve and refine the solution.

The local search mechanism is based on the neighborhood search proposed for *LOP* [12]. *Insert moves (IM)* are used to create a neighborhood of permutations for one solution. An insert move for a permutation π at $\pi(j)$ and i means deleting the element from position j and inserting it between elements $\pi(i-1)$ and $\pi(i)$. This operation results into a permutation σ (obtained from π) for which the objective function (for $i < j$) is the following [12]:

$$c_G(\sigma) = c_G(\pi) + \sum_{k=i}^{j-1} (w_{\pi(j)\pi(k)} - w_{\pi(k),\pi(j)}).$$

(8)

For a complete solution π a neighborhood $N\ (\pi)$ based on the *IM* mechanism is defined as follows:

$$N(\pi) = \{\sigma : \sigma = insert\_move(\pi(j),i), i = 1,2,..., j-1, j+1,...,n\} .$$

(9)

where j is randomly chosen.

$N\ (\pi)$ contains all the permutations that can be obtained from π by applying the insert move mechanism for all positions in the permutation. The current solution π is replaced by a permutation σ belonging to $N\ (\pi)$ if $c_G(\sigma) > c_G(\pi)$.

3.4 Algorithm Description

The proposed model for solving *LOP* starts with a greedy search. *ACS*-based rules are applied and a local search mechanism based on insert moves is engaged. The resulting algorithm is called *Ant Colony System-Insert Move (ACS-IM)* and is outlined below.

```
ACS-IM Algorithm for solving LOP
   procedure GreedyInitialSearch - using (2),(3)
   procedure ACS-IM
        Set parameters, initialize pheromone trails
        while (termination condition not met) do
                ConstructAntsSolutions - using (4),(5),(6)
                Apply Insert Moves - using (8),(9)
                UpdatePheromones - using (7)
        end
```

4 Numerical Results

The proposed *ACS-IM* algorithm is engaged for the 49 problem instances from the real-world *LOLIB* library [15] and the 25 instances from the *SGB* library [16]. *LOLIB* contains input-output tables from European economy sectors while *SGB* includes input-output tables from the United States economy. The parameters considered for *ACS-IM* are: $\beta=2$, $\tau_0=0.1$, $\rho=0.1$, $q_0=0.95$ and the number of ants is equal to the number of vertices.

The results obtained for *LOLIB* are presented in Table 1 (*ACS-IM* with five runs of 50000 iterations). Table 1 shows the reported optimum solution for each instance, the best solution obtained by *ACS-IM*, the number of optimum solutions reported by *ACS-IM* as well as the average deviation of the obtained solution from the optimum one.

The percentage average deviation of the obtained solution from the real optimal solution for all 49 *LOLIB* instances is *0.145* clearly outperforming recent evolutionary models [13] which report an average deviation of *0.714* for the same library.

The results obtained by *ACS-IM* for *LOLIB* are further compared to local search methods [12] in terms of solution quality.

Figure 1 indicates a lower average deviation obtained by *ACS-IM* compared to variable neigborhood search (*VND_ best* and *VND_first*) and local search (*LS_first* and *LS_best*).

Furthermore, instances from the *SGB* library are addressed using *ACS-IM* and the results are compared to variable neighborhood search [12]. Table 2 presents the comparative results (100 runs with 50000 iterations are considered for *ACS-IM*) indicating a similar performance of ACS-IM relative to state-of-the-art techniques – VNS. These are promising preliminary results and we expect their improvement (using other hybridizations) during further development.

There are models combining variable neighborhood search and different strategies for local search able to obtain higher-quality *LOP* solutions [12]. Local search strategies are however limited by the dimension of problem instances. It is expected that the proposed ant-based heuristic for *LOP* is able to produce high quality solutions with reasonable computational efforts even for high-dimension real-world problems.

Table 1. *ACS-IM* results for *LOLIB* [15] obtained for five runs with 50000 iterations

No.	Instances	Optimal	ACS-IM		
			Max	No. Opt.	Avg. Dev
1.	60-stabu3	642050	640424	0	0.0027
2.	60-stabu2	627926	627560	0	0.0027
3.	60-stabu1	422088	421454	0	0.0026
4.	56-tiw56r72	341623	341623	15	0.0011
5.	56-tiw56r67	270497	270497	14	0.0039
6.	56-tiw56r66	256326	256326	25	0.0009
7.	56-tiw56r58	160776	160776	3	0.0016
8.	56-tiw56r54	127390	127390	1	0.0019
9.	56-tiw56n72	462991	462991	3	0.0017
10.	56-tiw56n67	277962	277962	5	0.0045
11.	56-tiw56n66	277593	277593	28	0.0013
12.	56-tiw56n62	217499	217499	7	0.0008
13.	56-tiw56n58	154440	154440	2	0.0015
14.	56-tiw56n54	112767	112757	0	0.0013
15.	50-be75tot	1127387	1127347	0	0.0012
16.	50-be75oi	118159	118158	0	0.00004
17.	50-be75np	790966	790963	0	0.000004
18.	50-be75eec	264940	264638	0	0.0016
19.	44-t75u1xx	63278034	63278034	13	0.0008
20.	44-t75n11xx	113808	113808	77	0.0004
21.	44-t75k11xx	124887	124887	63	0.0002
22.	44-t75i1xx	72664466	72664466	13	0.0005
23.	44-t75e11xx	3095130	3095130	6	0.0021
24.	44-t75d11xx	688601	688601	15	0.0059
25.	44-t74d11xx	673346	673346	15	0.0024
26.	44-t70x11xx	343471236	343471236	65	0.0002
27.	44-t70w11xx	267807180	267807180	47	0.0004
28.	44-t70u11xx	27296800	27296800	4	0.0007
29.	44-t70n11xx	63944	63944	11	0.0010
30.	44-t70l11xx	28108	28108	6	0.0024
31.	44-t70k11xx	69796200	69796200	18	0.0005
32.	44-t70i11xx	28267738	28267738	3	0.0015
33.	44-t70f11xx	413948	413948	14	0.0016
34.	44-t70d11xx	450774	450774	36	0.0013
35.	44-t70d11xn	438235	438235	14	0.0005
36.	44-t70b11xx	623411	623411	47	0.0003
37.	44-t69r11xx	865650	865650	2	0.0041
38.	44-t65w11xx	166052789	166052789	2	0.0014
39.	44-t65n11xx	38814	38814	22	0.0016
40.	44-t65l11xx	18359	18359	81	0.0003
41.	44-t65i11xx	16389651	16389651	5	0.0019
42.	44-t65f11xx	254568	254515	0	0.0014
43.	44-t65d11xx	283971	283969	0	0.0018
44.	44-t65b11xx	411733	411733	2	0.0009
45.	44-t59n11xx	25225	25225	57	0.0019
46.	44-t59i11xx	9182291	9182291	55	0.0006
47.	44-t59f11xx	140678	140678	35	0.0002
48.	44-t59d11xx	163219	163219	42	0.0003
49.	44-t59b11xx	245750	245750	17	0.0008

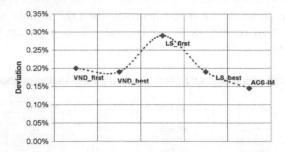

Fig. 1. Average deviation for *LOLIB* instances: *ACS-IM* compared to local search [12]

Table 2. Numerical results for *SGB* instances obtained by *ACS-IM* compared to *Variable Neighborhood Search (VNS)* [12]

Instance	VNS	ACS-IM	
		Max.	Avg.Dev.
sgb75.1	6144679	6144646	0.0004
sgb75.2	6100491	6100377	0.0003
sgb75.3	6165775	6165474	0.0005
sgb75.4	6154958	6154958	0.0005
sgb75.5	6141070	6141070	0.0006
sgb75.6	6144055	6143861	0.0005
sgb75.7	6142899	6142831	0.0008
sgb75.8	6154094	6154062	0.0006
sgb75.9	6135459	6135459	0.0004
sgb75.10	6149271	6149220	0.0006
sgb75.11	6151750	6151722	0.0005
sgb75.12	6150469	6150394	0.0009
sgb75.13	6156935	6156696	0.0005
sgb75.14	6149693	6149440	0.0009
sgb75.15	6150331	6150329	0.0004
sgb75.16	6164959	6164890	0.0006
sgb75.17	6163483	6163481	0.0005
sgb75.18	6063548	6062926	0.0004
sgb75.19	6150967	6150955	0.0002
sgb75.20	6152224	6152223	0.0013
sgb75.21	6159081	6159081	0.0005
sgb75.22	6127019	6127014	0.0002
sgb75.23	6136362	6135885	0.0005
sgb75.24	6168513	6166247	0.0008
sgb75.25	6150026	6149899	0.0004

5   Conclusions and Future Work

The *Triangulation Problem for Input-Output Matrices* is solved with good results using a hybrid ant-based model. The proposed technique integrates an *ACS* model based on a greedy solution initialization and an *Insert-Move* mechanism used for local search. Numerical results obtained on some real-world economical data are encouraging for the potential of nature-inspired metaheuristics in solving linear ordering problems. *ACS-IM* is able to obtain higher-quality solutions to *LOP* compared to evolutionary techniques [13] and results comparable with state-of-the-and neighborhood search methods [12].

Future work focuses on the investigation of other (more efficient) local search mechanisms to be considered for hybridization with the proposed ant-based model in order to improve the quality of *LOP* solutions. Additionally, it is planned to engage the proposed hybrid ant-based model for other available libraries of *LOP* instances to potentially show the benefits of using ant-based models when addressing higher-dimensional problems.

References

1. Pryor, F.L.: Economic evolution and structure: The impact of complexity on the US economic system. Cambridge University Press, Cambridge (1996)
2. Chenery, H.B., Watanabe, T.: International comparisons of the structure of productions. Econometrica 26(4), 487–521 (1958)
3. Reinelt, G.: The linear ordering problem: algorithms and applications. In: Research and Exposition in Mathematics, vol. 8, Heldermann Verlag, Berlin (1985)
4. Chanas, S., Kobylanski, P.: A new heuristic algorithm solving the linear ordering problem. Computational Optimization and Applications 6, 191–205 (1996)
5. Kaas, R.: A branch and bound algorithm for the acyclic subgraph problem. European Journal of Operational Research 8, 355–362 (1981)
6. Mitchell, J.E., Borchers, B.: Solving linear ordering problem with a combined interior point/simplex cutting plane algorithm. In: Frenk, H., Roos, K., Terlaky, T., Zhang, S. (eds.) High Performance Optimization, pp. 349–366. Kluwer Academic Publishers, Dordrecht (2000)
7. Mishra, S., Sikdar, K.: On approximability of linear ordering and related NP-optimization problems in graphs. Discrete Applied Mathematics 136(2-3), 249–269 (2004)
8. Hassin, R., Rubinstein, S.: Approximations algorithms for maximum linear arrangement. In: Halldórsson, M.M. (ed.) SWAT 2000. LNCS, vol. 1851, pp. 231–236. Springer, Heidelberg (2000)
9. Glover, F., Laguna, M.: Tabu Search. Kluwer Academic Publisher, Boston (1997)
10. Glover, F.: A template for Scatter Search and Path Relinking. In: Hao, J.-K., Lutton, E., Ronald, E., Schoenauer, M., Snyers, D. (eds.) AE 1997. LNCS, vol. 1363, pp. 13–54. Springer, Heidelberg (1998)
11. Lourenco, H.R., Martin, O., Stützle, T.: Iterated Local Search. In: Glover, F., Kochenberger, G. (eds.) Handbook of Metaheuristics, International Series in Operations Research & Management Science, vol. 57, pp. 321–353. Kluwer Academic Publishers, Dordrecht (2002)
12. Garcia, C., Perez, D., Campos, V., Marti, R.: Variable neighborhood search for the linear ordering problem. Computers and Operations Research 33, 3549–3565 (2006)
13. Snásel, V., Kromer, P., Platos, J.: Evolutionary approaches to linear ordering problem in DEXA Workshops, pp. 566–570. IEEE Computer Society, Los Alamitos (2008)
14. Dorigo, M., Stützle, T.: Ant Colony Optimization. MIT Press, Cambridge (2004)
15. LOLIB site, http://www.iwr.uni-heidelberg.de/groups/comopt/software/LOLIB/
16. Knuth, D.E.: The Stanford GraphBase: A Platform for Combinatorial Computing. Addison Wesley, New York (1993)
17. Dorigo, M., Gambardella, L.M.: Ant Colony System: A cooperative learning approach in the traveling salesman problem. IEEE Transactions on Evolutionary Computation 1(1), 53–66 (1997)

A Thermodynamical Model Study for an Energy Saving Algorithm

Enrique de la Cal[1], José Ramón Villar[1], and Javier Sedano[2]

[1] University of Oviedo, Computer Science department, Edificio Departamental 1,
Campus de Viesques s/n, 33204 Gijón, Spain
{villarjose,delacal}@uniovi.es
[2] University of Burgos, Electromechanic department,
Escuela Politécnica Superior, Campus Vena, B2
09006 Burgos, Spain
jsedano@ubu.es

Abstract. A local Spanish company that produces electric heaters needs an energy saving device to be integrated with the heaters. It was proven that a hybrid artificial intelligent systems (HAIS) could afford the energy saving reasonably, even though some improvements must be introduced. One of the critical elements in the process of designing an energy saving system is the thermodynamical modeling of the house to be controlled. This work presents a study of different first order techniques, some taken from the literature and other new proposals, for the prediction of the thermal dynamics in a house. Finally it is concluded that a first order prediction system is not a valid prediction model for such an energy saving system.

Keywords: Fuzzy systems, Hybrid Artificial Intelligence Systems, Real World Applications, Thermodynamical Modlling, Electric Energy saving.

1 Motivation

In Spain, a local company has developed a new dry electric heaters production line and a device for electric energy saving (EES) is to be included in the catalogue. In previous works, the development of such a device has been analyzed, and a multi agent hybrid fuzzy system has been proposed [11] [12] [10]. There are two stages in defining the proposed energy distribution algorithm that goes in the CCU: the design stage and the run stage. In the design stage a wide range fuzzy controller is trained and validated out of the hardware device, while in the run stage the whole algorithm is executed in the embedded hardware (CCU)

The first preliminary EES was proposed in [11]. In that proposal, a thermodynamical model (from now on TM) of the house to be controlled was not required, because the fuzzy energy distribution controller (FC) was defined directly by a team of experts.

In our next work, [12], a new system design was presented (see Figure 1) and two improvements were introduced: an optimization step to improve the Experts FC (MOSA) and a TM based on the estimation of thermal parameters by the Simulated Annealing algorithm (SA).

E. Corchado et al. (Eds.): HAIS 2009, LNAI 5572, pp. 384–390, 2009.

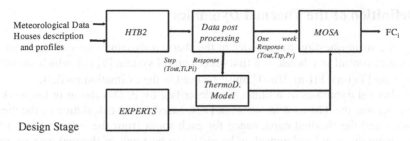

Fig. 1. The Design stage procedure. A fuzzy controller is trained for each configuration pair of climate zone and building topology.

Although in [12] a TM was used and the FC was proven to make a suitable distribution of the available energy with this temperature model, we realised that the TM (based on SA) must be studied in depth. In the present work, techniques for indoor temperature forecasting based on thermodynamical theory [2], SA, GAs and ANNs [7], [9], are to be used and compared to be included in the EES proposed.

The work is organized as follows. The Design stage of the ESS proposed is detailed in Section 2. Section 3 presents the new different TMs. A comparison of the proposed TM is carried out in Section 4. Finally, conclusions and future work are presented in Section 5.

2 The Design Stage

The Design stage includes several steps: the HTB2 simulation, the data post-processing, the identification of the TM and the generation of the best suite FC. The HTB2 simulation software [6] in following HTB2, is a well known tool suitable to analyze the dynamics of heating systems and the energy consumption in buildings using the concentrated parameter model [4] [1]. Also a simulation for determining the thermodynamics of each room in the building is needed. This simulation is the *step response of the house,* where a change in the temperature set point is analyzed.

The post-processing step has two tasks: one consists on preparing the training data for the TM (the step response simulated with HTB2 during one day); and the second task is to prepare a dataset obtained from one week of HTB2 simulation; this will represent the training and validation datasets to be used in the Thermo D. and MOSA steps. A TM for the house is required, because the optimization algorithm (MOSA) needs to calculate an approximation of temperature in a room when a FC is evaluated.

Finally, the MOSA step is the 10 k fold cross validation multi-objective simulated annealing for training the fuzzy controller. The MOSA is the multi-objective adaptation of the simulated annealing presented in [13]. Each individual is the consequent part of the fuzzy controller, and the fitness function is the Distribution algorithm execution. The two objectives to reach are minimizing the energy consumption and keep the house in comfort minimizing the difference between the TM output and temperature set point. The initial individual has been obtained from expert knowledge, and the antecedents and rules are kept the same for all individuals.

3 Definition of the Thermal Dynamics

Firstly, the main question to consider in the thermal dynamics modelling is that the temperature control in a house is a first order control system [9] [2], which means that no past data [Ti(t-n), P(t-n), ∀n>1] must be used in the estimation models.

The thermal dynamics of a house was described by P. Davidsson in his work [2]. This model was the TM used in our work [12], based on the calculation of the thermal resistance and the thermal capacitance for each room from the step response of the house. Using the simulated annealing heuristic the best pair of thermal parameters for each room was found. Here, it's formalized the SA+Davidsson combined proposal presented in our work [12] and presented a set of typical solutions taken from the literature [7] [9].

Identifying the parameters of the Davidsson proposal with the SA

The Davidsson' proposal is based on the consideration that all the thermodynamical characteristics of a room are described by two constants: the thermal resistance, R, which captures the heat losses to the environment, and the thermal capacitance, C, which captures the inertia when heating up/cooling down the entities in the room. (in our experiments we use the sample time of 1 min.) The temperature, T_{xi}, in room x at time i is described by equation (1):

$$T_{xi} = \frac{1}{1 + \frac{1}{R_x C_x}} \cdot \left(T_{x(i-1)} + \frac{P_i + \frac{T_{outi}}{R_x}}{C_x} \right) \tag{1}$$

where P_i is the heating power, T_{outi} the outdoor temperature, and $T_{x(i-1)}$ is the temperature in room x 1 min ago.

It must be considered that the Davisson' work did several simplifications that we didn't: constant outdoor temperature is assumed (10°C), radiation from the sun is negligible (i.e., weather cloudy), [2]. Thus, it's needed to estimate the suitable pair of thermal parameters R and C for each room. Here the Simulated Annealing is used, one of the simplest and fastest optimization algorithm.

Identifying the parameters of an exponential thermal function with the Simulated Annealing (SA)

We have stated that the Davisson'model has a too fast response of the temperature evolution to the influence of the power, despite of having identified the factors R and C ad-hoc for each room. Thus, we think that the exponential model of equation (2) would be more adequate than the Davidsson' alternative.

$$T_{xi} = a_x + b_x \cdot T_{xi-1} + c_x \cdot P_{xi-1} + e^{d_x \cdot T_{xi-1} + e_x \cdot P_{xi-1}} \tag{2}$$

where P_{xi-1} is the heating power in room x 1 min ago, and $T_{x(i-1)}$ is the temperature in room x 1 min ago.

The parameters a_x, b_x, c_x, d_x and e_x are estimated running the SA algorithm for each room x. The fitness was the mse of the multi-step prediction. The training-testing scheme used is 5x2CV.

Identifying a TM with a Genetic Algorithm

A simple generational GA [3] was proposed to learn the parameters of the exponential function (2) too. The individuals are vectors of five real numbers (a_x to e_x parameters) and fitness is the MSE of the multi-step prediction. The configuration of the GA is: an initial population of 100 individuals, crossover probability of 70%, mutation probability of 1% and tournament selection of size 4. The training-testing scheme used was 5x2CV.

Identifying a TM with a fast-forward Artificial Neural Networks

The use of ANN to define prediction models for building variables as indoor temperature or relative humidity, has grown in last years [9] [7]. Our proposal is based on a two-layer network of fast-forward type, with tan-sigmoid transfer function in the hidden layer and a linear transfer function in the output layer. This is a useful structure for function approximation (or regression) problems. As an initial guess, we use the inputs <Tindoor$_t$, Tout$_t$, Power$_t$>, thirty neurons in the hidden layer. The network should have one output neuron since there is only one target, the indoor temperature. We will use the Levenberg-Marquardt [5] [8] algorithm for training.

4 Experiments and Commented Results

This work deals with the comparison of different approaches of the thermodynamical modelling of the indoor temperature in a house with the one proposed based on the Davisson' model and SA [12]. The climate zone chosen for the experiment is the city of Avila, an E1[1] city, and the building topology is a three bedrooms condominium house, with orientation North-South, composed of seven heated spaces.

The environmental data such as the outdoor temperature or the solar radiation have been gathered from statistical data of the city of Avila in the winter of the year 2007. Other files, as the occupancy, the temperature set point profile for each room in the house, small power devices profile, etc. have been designed attending to realistic profiles. Finally, files like ventilation file have been generated attending to the regulations.

The HTB2 simulation produces two datasets: the step response of the house for one day (training dataset) and the evolution of the heating power required to keep the comfort level (21°C for all the rooms) and temperature in the house during one week (testing dataset). The training and testing error of the selected thermal prediction techniques applied over the seven simulated rooms are shown in table 1.

---

[1] The Spanish regulations define 5 climate zones, from the less severe –A3– to the most severe –E1–.

Table 1. Training and testing mean/standard deviation of the percentual error of the thermal models

	David.	SA+David	SA+Expo	GA+Expo	ANN
Train	247,6/158,0	27,0/7,2	16,2/8,9	5,4/0,6	26,2/14,5
Test	130,1/59,8	58,2/18,3	23,9/5,3	28,7/5,4	38,6/28,4

The original Davidsson thermal model [2] with parameters R and C for a typical small room (R = 0.1, C = 3000) were applied to all the rooms. It can be observed that the testing error is lower than the training error, but it's still too high (130,11%). Thus, we think that the Davidsson's formula parameters can be optimized and a meta-searching algorithm SA was used to learn the Davidsson' parameters. However, the results of the SA was better than the direct Davidsson' model (58,23%) but not good enough to be used in the Fuzzy Controller Optimization Algorithm (MOSA). The remaining proposals overpass the Davidsson' proposals but the testing errors are too high again. In spite of the results of the table 1, we consider important to analyze the distribution of the error between the testing ticks (one tick per minute during seven days).

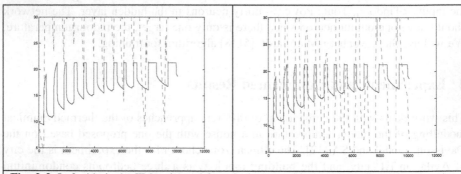

Fig. 2-3. Left side is the TM estimated for a bedroom with the original Davisson function with R = 0.1, C = 3000; Right Side is the TM estimated for a bedroom with the Davisson function parameters optimized with SA.The Solid line is the reference temperature and dotted line is the estimated temperature.

In Figures 2 to 6 we can see the comparison of the reference temperature evolution (solid line) with the predicted temperature (dotted line) for a bedroom. We can state that the Davidsson's models confirm the high error of Table 1. Perhaps the SA+David proposal (Figure 3) follows better the crests than original function (Figure 2), but both models are not suitable temperature models. The exponential models (Figures 4) are not good enough, because they follow the crests with a variable amplitude lag. Finally the solution based on ANN is presented in Figure 5, this proposal fits with high precision the last crests but it's not a valid model for the FC learning step (MOSA) because of its high medium error.

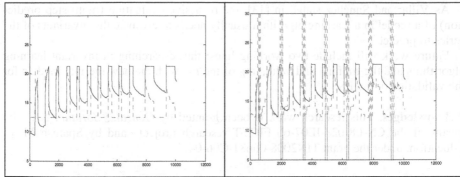

Fig. 4-5. Left side is the TM estimated for a bedroom with the exponential function parameters optimized with SA Left side is the TM estimated for a bedroom with the exponential function parameters optimized with GA. The Solid line is the reference temperature and dotted line is the estimated temperature.

In short, all the first order solutions presented here have an unstable behaviour due to the recursive evaluation so they can be used in a real context. This fact was published in [13], where it said that a recursive evaluation (multi-step prediction) in a learning algorithm usually may not converge to an appropriate model, unless we include some terms that depend on estimates of certain properties of the model (so called 'invariants' of the chaotic series).

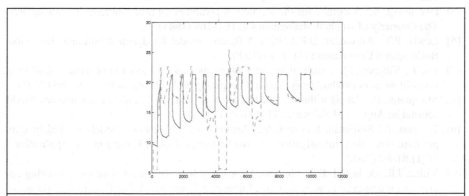

Fig. 6. TM estimated for a bedroom with an ANN. The Solid line is the reference temperature and dotted line is the estimated temperature.

5 Conclusions and Future work

This work compares the thermal prediction proposal presented in preliminary works based on the SA algorithm against with first order prediction AI techniques. Although the results of the new AI technique improve the preliminary work results, the multistep prediction evaluation used in all of them give very unstable prediction of the temperature.

As Villar and Sanchez stated in [13], the recursive evaluation (multi-step prediction) of a model in a learning algorithm usually needs to estimate the invariants of the series to predict.

Future work will include two working lines: the researching in invariant learning algorithms and the training and validating of the FC with the new thermal models for the validation of the ESS.

Acknowledges. This research work is been granted by *Gonzalez Soriano, S.A.* –by means of the CN-08-028-IE07-60 FICYT research project– and by Spanish M. of Education, under the grant TIN2008-06681-C06-04.

References

[1] Bojic, M., Despotovic, M., Malesevic, J., Sokovic, D.: Evaluation of the impact of internal partitions on energy conservation for residential buildings in serbia. Building and Environment (42), 1644–1653 (2007)

[2] Davidsson, P., Boman, M.: Distributed monitoring and control of office buildings by embedded agents. Information Sciences 171, 293–307 (2005)

[3] Holland, J.H.: Adaptation in Natural and Artificial Systems, vol. xx. University of Michigan Press, Ann Arbor (1975)

[4] Koroneos, C., Kottas, G.: Energy consumption modeling analysis and environmental impact assessment of model house in thessaloniki—Greece. Building and Environment 42, 122–138 (2007)

[5] Levenberg, K.: A method for the solution of certain non-linear problems in least squares. The Quarterly of Applied Mathematics 2, 164–168 (1944)

[6] Lewis, P.T., Alexander, D.K.: Htb2: A flexible model for dynamic building simulation. Building and Environment (1), 7–16 (1990)

[7] Lu, T., Viljanen, M.: Artificial neural network models for indoor temperature prediction: investigations in two buildings. Neural Computing & Applications 16(1), 81–89 (2007)

[8] Marquardt, D.: An algorithm for least-squares estimation of nonlinear parameters. SIAM Journal on Applied Mathematics 11, 431–441 (1963)

[9] Thomas, B., Soleimani-Mohseni, M.: Artificial neural network models for indoor temperature prediction: investigations in two buildings. Neural Computing & Applications, 16(1):81–89 (2007)

[10] Villar, J.R., de la Cal, E.A., Sedano, J.: A fuzzy logic based efficient energy saving approach for domestic heating systems. Integrated Computer-Aided Engineering (in press, 2008)

[11] Villar, J.R., de la Cal, E.A., Sedano, J.: Energy saving by means of fuzzy systems. In: Yin, H., Tino, P., Corchado, E., Byrne, W., Yao, X. (eds.) IDEAL 2007. LNCS, vol. 4881, pp. 155–161. Springer, Heidelberg (2007)

[12] Villar, J.R., de la Cal, E.A., Sedano, J.: Minimizing energy consumption in heating systems under uncertainty. In: Corchado, E., Abraham, A., Pedrycz, W. (eds.) HAIS 2008. LNCS, vol. 5271, pp. 583–590. Springer, Heidelberg (2008)

[13] Villar, J.R., Sánchez, L.: Obtaining transparent models of chaotic systems with multiobjective simulated annealing algorithms. Information Sciences 178(4), 952–970 (2008)

A Fuzzy Approach of the Kohonen's Maps Applied to the Analysis of Biomedical Signals

Andrilene Maciel[1], Luis Coradine[1], Roberta Vieira[1], and Manoel Lima[2],

[1] Computer Institute – Federal University of Alagoas
Postal Code 57.072-970 – Tabuleiro dos Martins – Maceió, AL – Brazil
andrilenef@yahoo.com.br, lccoral@gmail.com , rv2l@hotmail.com
[2] Informatics Center – Federal University of Pernambuco
Postal Code 15.064 – 91.501-970 - Recife - PE - Brasil
mel@cin.ufpe.br

Abstract. Self-organizing maps have been used successfully in pattern classification problems related to many areas of knowledge and also applied as a tool for statistical multivariate data analysis. Data classification via self-organizing maps deals specifically with relations between objects, meaning that there are limitations to define class limits when an object belonging to a particular class "migrates" to another one. To address this issue, a solution involving self-organizing maps and fuzzy logic is proposed with the objective of generating a neighborhood between these classes. The developed system receives the network output and automatically generates self-organizing maps. This unified vision of the model is used in the analyzing biomedical signals in diabetic patients for monitoring blood glucose stage. Early diagnosis and glucose signals monitoring can prevent or delay the initiation and development of clinical complications related to diabetes.

Keywords: Self-Organizing Maps, Fuzzy Logic, Diabetes Mellitus, Neural Network.

1 Introduction

Diabetes mellitus affects over 171 million individuals worldwide, and this number could reach a total of 366 million by 2030 [1]. Diabetes is a disorder associated with the metabolism process characterized by the inability of the pancreas to secrete sufficient insulin. When glucose levels remain high for an extended period of time (hyperglycemia) where the patient is at risk of having neuropathy, nephropathy, and other long-term vascular complications [2]. Recently, studies employing self-organizing maps (SOM) and neural networks as classifiers are becoming increasingly present, and these approaches is been applied as alternative tools in the context of multivariate data analysis - including data mining with satisfactory levels of performance [4] and [9]. Recent research addresses the use of Kohonen maps to identify the components of visual clustering [9]. As a rule, Kohonen maps represent the relationship between an object (for example, measure of glucose) related to a particular patient. The SOM networks do not evaluate objects of that class which are on the threshold set by the

E. Corchado et al. (Eds.): HAIS 2009, LNAI 5572, pp. 391–400, 2009.

class, or when an object, that is within the normal range of migrating to a new phase that identifies that this measure can be part of a new class with abnormality characteristics. Recent studies have approach fuzzy logic in analyzing biomedical signals. In Grant [14], a new approach to diabetic control is presented: Fuzzy logic and insulin pump technology. In Lascioa [15], A fuzzy-based methodology for analyzing diabetic neuropathy. In Man [16], a Meal simulation model involving the glucose insulin system is presented. In Dua [17], a Model is used - based blood glucose control for type 1 diabetes via parametric programming. In Owens [18], a run-to-run control of blood glucose concentrations for people with type 1 diabetes mellitus is adopted.

This paper proposes a hybrid model using Kohonen SOM [7] to classify the types of signals from diabetes in accordance with the stage of glucose and fuzzy technical logic to generate the neighborhoods between the classes. The developed system receives the output of the SOM network and automatically generates the unified fuzzy vision of this structure. In this section, the motivation to carry out the work and the main features of the proposal were exposed. Section 2 presents the methodology used in the analysis involving the classification of the Kohonen map and fuzzy logic, while section 3 shows a preliminary analysis and data division applied in processing and analyzing biomedical signals. Finally, section 4 discusses the obtained results while section 5 contains the conclusions and perspectives.

2 Methods

2.1 Self-Organizing Maps (SOM)

The principal goal of SOM is to transform an incoming pattern of arbitrary dimension into one or two-dimensional discrete map. There are three processes involved in SOM: *competition*, *cooperation* and *adaptation*. The modeling by SOM network can be summarized as follows: when an input pattern $\mathbf{x}$ is presented to the network, the SOM algorithm looks for the most similar unit to $\mathbf{x}$. During the training phase, the network increases the similarity of the chosen node and its neighbors to the pattern $\mathbf{x}$. The SOM network uses an algorithm of competitive learning, where the nodes of the output layer compete among themselves to become active among those that generate the highest output. For each input pattern, only one output or node (within a group of neurons) becomes active [9]. The synaptic weight vector for each neuron in the network has the same dimension as the input space. Taking $\mathbf{x} = [x_1, x_2,..., x_n]^T$, let the weight vector of neuron j be denoted by:

$$\mathbf{w}_j = \left[w_{j1}, w_{j2}, w_{j3}, ..., w_{jn}\right]^T for \ j = 1, 2, ..., m, \tag{1}$$

where m represents the total number of neurons in the network. Now, we use the index $i(\mathbf{x})$ to identify the neuron that best matches the input vector $\mathbf{x}$, where, by applying the condition to Euclidian distance, determine $i(\mathbf{x})$:

$$i(\mathbf{x}) = \arg min_j \|\mathbf{x} - \mathbf{w}_i\| , j = 1, 2, ..., m. \tag{2}$$

The winning neuron, identified by $i(\mathbf{x})$, maps an area of input for a discrete array of neurons. The answer comes from updating the vector weights associated with the winning neuron according to the topological neighborhood. By minimizing the distance between the vectors **x** and **w** based on maximizing the inner product of $\mathbf{w}_i^T\mathbf{x}$ [9], the winning neuron is the one who has the smallest distance from the data **x** (which represents best the data). In the cooperation process, the inner product   of $\mathbf{w}_i^T\mathbf{x}$ requires that the neurons next to it also have its synaptic weight vectors adjusted in direction of the data vector. Thus, the winning neuron should excite the neurons that belong to its neighborhood. This observation leads us to make the topological neighborhood around the winning neuron i decaying smoothly with the lateral distance. Let $N_{j,i}$ denote the neighborhood centered on winning neuron i, and encompassing a set of excited neuron, a typical one of which is denoted by j. Let $d_{i,j}$ represent the lateral distance between the winning neuron and the excited neuron, then we may assume that the topological neighborhood is a unimodal function of lateral distance, here represented by:

$$N_{j,i\,(\mathbf{x})} = \exp\left(\frac{-d_{j,i}^2}{2\sigma^2}\right) \tag{3}$$

In order to maintain this cooperation, it is necessary that the topological neighborhood depend on the lateral distance in the output space rather than on some distance measure in the original input space. This lateral distance can be defined as:

$$d_{j,i} = \|\mathbf{r}_j - \mathbf{r}_i\|, \tag{4}$$

where the discrete vector $\mathbf{r}_j$ defines the position of the exciting neuron and $\mathbf{r}_i$ defines the discrete position of winning, both of which are measured in the discrete output space. In the adaptive process, the weight vector of the most activated neuron, with distance $\mathbf{r}_j$, is updated towards the input vector **x**. The process of adapting the synaptic weight vectors of the neuron j at the time k is:

$$\mathbf{w}_j(k+1) = \mathbf{w}_j(k) + \eta\,(k)N_{j,i(\mathbf{v})}(k)\big[\mathbf{x} - w_j(k)\big], \quad \text{for } j=1,2,\ldots m, \tag{5}$$

where $_{(k)}$ defines the learning rate, $N_{j,i(\mathbf{x})}$ defines the adaptation degree of the neuron in relation to winning. The learning rate $\eta\,(k)$ must change gradually with time, in order to allow an adequate convergence of the SOM network [9]. Normally, $\eta\,(k) \rightarrow 0$ when $^k k \rightarrow \infty$, being this requirement met by choosing an exponential decay for $\eta\,(k)$[9][19]:

$$\eta\,(k) = \eta_0 \exp - \frac{k}{\tau^2}, k = 0,1,2\ldots,m, \tag{6}$$

where τ_2 is another time constant of the SOM's algorithm.

2.2 Fuzzy Interpretation of the Topographical Map

A crisp set in a discourse universe X can be defined by a binary pertinence function, where it assumes the value of 0 if $x \notin A$ and 1 if $x \in A$, in the form:

$$\mu_A(x) = \begin{cases} 1, & \text{if } x \in A \\ 0, & \text{if } x \notin A \end{cases} \tag{7}$$

Then, the function associates each element of a degree of relevance $\mu_A(x)$, between 0 e 1, such that $x \in A$. In general, all the variables of a process assume a unique value of its reference universe of a given situation. In this way, there is a linguistic set variable (U, X, T_U) where U is a variable defined under a reference set X. The set defined by $T_U = [A_1, A_2, \ldots]$, finite or infinite, contains normalized fuzzy sets of used to obtain a characterization of U . Where A_i is associated with its linguistic values {"low", "medium", and "high"} characterized as fuzzy subsets whose functions of relevance are represented by $T_U = \{low, medium, high\}$. A Mamdani System has fuzzy inputs and a fuzzy output, assuming a mapping from an input space into an output space [20]. The design of fuzzy controllers is commonly a time- consuming activity involving knowledge acquisition, definition of the controller structure, definition of rules, and other controller parameters. A Fuzzy Approach of the Lee [22] at present as one of the important issue in fuzzy logic systems is how to reduce the total number of involved rules and their corresponding computation requirements. In a standard fuzzy system, the number of rules increases exponentially as the number of variable increases [23]. Suppose there are n input variables and m membership functions for each variable, then it needs m^n rules to construct a complete fuzzy controller. As n increases, the rule base will quickly overload the memory and make the fuzzy controller difficult to implement. These hierarchical fuzzy systems consist of a number of low-dimensional fuzzy systems in a hierarchical form [20],[22] and [23]. The hierarchical fuzzy systems have an advantage due to the fact that the total number of rules increases only linearly with the number of input variables [23]. For the hierarchical fuzzy system, with $4(n = 4)$ input variables: V_1- breakfast (07:00 to 09:00), V_2 - lunch (11:00 to 13:00), V_3- pre-dinner (15:00 to 17:00), V_4 dinner (18:00 to 19:00) and output variable is $V_5 V_5$ - stage of glucose composed of $3(m = 3)$ membership functions, then each low-dimensional fuzzy system. The number of rules in the standard fuzzy system is $3^4 = 81$ possible. Using the approach of Lee [22] from the Limpid Fuzzy-Hierarchical System, the number of fuzzy rules grows to reduce the basis of rules. In the following, an example given is as shown in table 1. The control rule base of the conventional single layer of the fuzzy logic system is given in table 1.

Table 1. Map of rules reduced for random access composed of: Stage of Glucose (SG), Minimum blood glucose (M), Normal blood glucose (N) and Amended blood glucose (A).

SG	V1			V2			V3			V4		
	M	N	A	M	N	A	M	N	A	M	N	A
M	M	A	N	A	M	A	A	N	A	N	M	N
N	N	A	M	A	M	A	A	M	A	N	M	A
A	M	M	N	M	N	A	A	N	A	M	A	M

From table 1, the four input variables are V_1, V_2, V_3, V_4 these are first layer fuzzy logic unit and V_5 and the second layer fuzzy logic unit. In Table 1, the fuzzy linguistic output is obtained by random access so as to generate a basis of the mapping rule base. Defining each variable with three linguistic terms such that M is the minimum blood glucose, N is normal blood glucose and A the Amended blood glucose. And the fuzzy rules of table 1 are as follows: is M, and V_2 is M, and V_3 is M, and V_4 is M, then V_5 is M, respectively. According to Table 2, it is seen that the relationship between $(V_1, V_2, V_3, V_4) = V$ and V_5 is equivalent to the relationship between the mapping variables (A, B, C, D, E, F, G, H) and V.

Table 2. (A) The sorting processes, (B) The relationship of the involved column of mapping variables

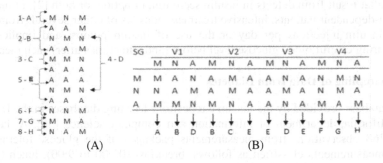

(A)   (B)

So, table 3 can be built, which is the fuzzy logic unit of V_5. And the rules are as follows: **If** V_1 **is A, and** V_2 **is B, and** V_3 **is E, and** V_4 **is F, then** SG **is** M, respectively.

Table 3. The rules reduced for SG

SG	V (input)			
(output)	V1	V2	V3	V4
M	A	B	E	F
N	B	C	D	G
A	D	E	E	H

In defuzzifier, the linguistic variable output value output is inferred by the fuzzy rules that will be translated into a discrete value in order to obtain one that best represents the fuzzy values inserted in the linguistic variable output. Thus, using the central method center of the area as a criterion for defuzzifier which is aimed at calculating the center of the composed area which represents the fuzzy output term (μ_{OUT}).

$$u^* = \frac{\sum_{j=1}^{N} u_i \, \mu_{OUT}(u_i)}{\sum_{j=1}^{N} \mu_{OUT}(u_i)} \qquad (8)$$

Where (μ_{OUT}) the area of a function of relevance (for example: minimum, normal and amended) is modified by the result of fuzzy inference, and   is the position of the centroid of the function of individual relevance, other methods of defuzzifier are seen in more details in [10-13] and [19-21].

3 Case Study

3.1 Biomedical Signals - Patients with Diabetes

Diabetes, apart from being a mounting problem worldwide, it is the fourth or fifth leading cause of death in most developed countries and there is a substantial evidence that it is epidemic in many developing and newly industrialized nations [1]. Diabetes mellitus is a group of metabolic diseases characterized by high blood sugar (glucose) levels that result from defects in insulin secretion, or action, or both [2]. In the case of insulin-dependent patients, intensive treatment consists of either several (from three to four) insulin injections per day or the use of insulin pumps, and implies a self-monitoring careful blood glucose level before (and sometimes after) each meal [2].

3.2   Sampling of Data from Diabetes

The database provided by the machine repository learning databases in the University of California, department of information and computer science [30]. A data frame with 768 observations from measurements packages of the glucose tolerance tests from measurements classified as follows: breakfast (07:00 to 9:00), lunch (11:00 to 13:00), pre-dinner (15:00 to 17:00) and dinner (18:00 to 19:00). The data was coded with labels in the range of [1 to 100]. For example, the signal identified by the code 1, is the measurement of glucose in the range of breakfast (07:00 to 09:00) and so forth which can be identified on the self-organizing map.

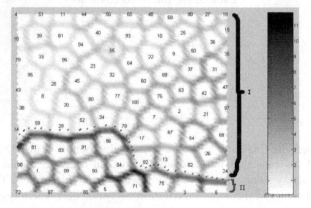

Fig. 1. The contextual map of the signals of glucose

The contextual map shows the presence of two major classes (I and II). The main body of those classes focuses on smaller subclasses according to the specification of each signal glycemic. By example, in class I, representing the division's clearer signals with similar characteristics forming small subclass individualized. The classification held by the statement of Kohonen shown in figure 1, allows the characterization of similarity of glucose signals in accordance with the timetable set out in the glucose tolerance test, divided into two categories I and II. Within the classes (I and II) smaller subclasses according to the specification of each signal glycemic are focused on. For example, in class I, representing the divisions of clearer signals with similar characteristics forming small subclass individualized, the darker rooms represent the dissimilarity of the signals. The projection of two major classes (I and II) and its 98 sub classes as outlined in figure 1, represents a diversification of glucose measurements, and could hamper the analysis of the glucose stage pattern. It also allows the addition of variables to the fuzzy controller, a significant fact for increasing the number of rules adopted by the Mamdani method [20].

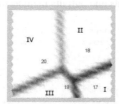

Fig. 2. Contextual Map by patient

Considering the contextual map in figure 2, the best results were obtained from a bi-dimensional grade (25 x 25), reducing the stage of seasons in the first phase and increasing the map refinement in the second. In refinement phases the seasons were fixed (neighborhood = 1) [9] and [19]. As a criterion for minimizing the process of composition of rules, we chose to analyze the patient's profile, seeking to see fluctuations in glucose measurements obtained from the four classes shown in figure 5 according to the intervals of the signals: class I (signal 17 (breakfast (07:00 to 09:00)), class II - signal 18 (lunch (11:00 to 13:00), class III - signal 19 (Pre-dinner (15:00 to 17:00)) and class IV - signal 20 (Post-dinner (18 to 19:00)), using a contextual map 25x25 in size, hexagonal neighborhood, standardized data, 30 times the first phase of $3 \rightarrow 1$ for a thicker mapping of the statement and its refinement used out 15 seasons with fixed neighborhood of $1 \rightarrow 1$, the quantization error and QE = 0.000 ET = 0.80 (irrelevant to the fuzzifier) were observed, more details on the errors can be seen in [7]. Meanwhile, the number of columns of the matrix represents the inputs of the controller. An alternative way to reduce the number of input variables in the fuzzy controller is related to the process of daily measured represented by the range of glucose according to the patient. The algorithm was developed in MATLAB using an adapted version of SOM toolboxes and Fuzzy Logic of MATLAB version 7.0 [21].

4 Implementations and Results

The data matrix A represents a set of controller input data defined by the linguistic variables $U_i = [\, u_1, u_2, u_3, u_4 \,]$ where i represents the daily monitoring of blood glucose rate (BGR) (item 3.2). For example, the linguistic variable entry can be represented by breakfast (07:00 to 9:00) and its linguistic terms $T_{breakfast} = \{\, Minimum\,, Normal\; e\; Amended\}$, during the interval [0-300]. The classification of the ranges of glucose control can be seen in more details in Mazzaferri [2]. For a measure of blood glucose, a "Minimum blood glucose" with a value equal to 1.0 of relevance, "Normal blood glucose", with a value of relevance equal to 0.0 and "Amended blood glucose" with a value equal to 0.0 of relevance in succession to each measuring system are considered. Thus, any entry or output variable measured will always be associated with a variable of linguistic values with a value of relevance between 0 and 1.

Table 4. Inputs and outputs membership functions.

Linguistic variables Inputs (measurements of glucose)	Trapezoidal Minimum	Triangular Normal	Trapeizodal Amended
Breakfast (07:00 to 09:00)	-3.5 -2.7 39.2 69.4	51.2 72.6 98.8	71 107 303.6 304
Lunch (11:00 to 13:00)	-3.5 -2.7 39.2 69.4	51.2 72.6 98.8	71 107 303.6 304
pre-dinner (15:00 to 17:00)	-3.5 -2.7 39.2 69.4	51.2 72.6 98.8	71 107 303.6 304
Dinner (18:00 to 19:00)	-3.5 -2.7 39.2 69.4	51.2 72.6 98.8	71 107 303.6 304
Outputs	Hypoglycemia	Normal	/ Amended
Stage of glucose	-3.5 -2.7 39.2 69.4	51.2 72.6 98.8	71 107 303.6 304

A finite number of rules that can be set after the characterization of the number of inputs and outputs needful to the fuzzy system, the choice of language predicates (number, distribution and shape of the functions of relevance) to each of these variables are used for forming the basis of rules. Meanwhile, using the rules of inferences or the Limpid fuzzy-hierarchical system [22] associations operating in a manner based on assumptions or conditions, which generate a certain result. The construction of the table [10-13] may be useful in verifying certain details of the operation of the controller, mainly to eliminate redundant predicates and check the consistency of rules. A table of fuzzy rules becomes similar to fuzzy relationship, from the cartesian product and its respective components with all ordered pairs, where the inputs into the matrix of fuzzy relationship on values of relevance are in the interval [0.1]. The rule map is a fuzzy implication matrix, in which inputs imply in certain outputs, or similarly, outputs associated with certain inputs. The membership functions and rules are design tools that give opportunity to model a control surface and controller properties. During the process of manipulating data and training the SOM network, the original data was submitted to the normalized process. For display of signals, the data were not normalized after training the SOM network.

The pattern of data is adopted from the output of the controller crisp fuzzy. It is observed that the variability of glucose signal is around 217.6 to 227.2 (glucose amended), typical case of a patient with a history of diabetes mellitus.

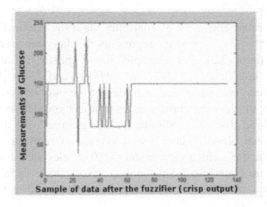

Fig. 3. Crisp output of fuzzy controller from the stage blood glucose by type of patient

5 Conclusions and Perspectives

The integration of SOM network models based on fuzzy rules shows that the nonlinear projection of data provided by self organizing maps does not allow the identification of class intervals or any measures that could represent a minimum threshold. The implementation of fuzzy controllers based on rules presents some practical advantages easily understood and its purpose and result can be easily interpreted. Also, all control functions associated with the rules can be individually tested and could also increase the ease of maintenance because the simplicity of the rules allows usage by less trained personnel. Various topics related to this work may be cited as a suggestion for future research: the use of genetic algorithm for drawing up and implementing the basic control rules based on evolutionary programming, adopting a parametric approach using the Takagi-Sugeno method, and also adopting a topographically generative mapping model. Furthermore, this model could be applied to other types of field. The methodology proposed in this article can be applied to identify other types of diseases from measurements of glucose in visible signals of glycemia.

References

1. World Health Organization. Diabetes Action Now. Switzerland (2004)
2. Mazzaferri, E.L.: Endocrinology. A Review of Clinical Endocrinology. New York - USA (1978)
3. Fayyad, U.M., et al.: Advances In Knowledge Discovery and Data Mining. AAAI Press / The MIT Press, California (1996)
4. Jain, A.K., Dubes, R.C.: Algorithms for Clustering Data. Prentice-Hall, New Jersey (1988)
5. Rencher, A.C.: Methods of Multivariate Analysis, 2nd edn. Wiley Interscience, Canada (2002)
6. Jain, A.K., Murty, M.N., Flynn, P.J.: Data Clustering: A Review. ACM Computing Surveys 31(3), 264–323 (1999)
7. Kohonen, T.: Self-Organizing Maps. Springer, Berlin (1997)
8. Haykin, S.: Neural Networks: a comprehensive foundation, 2nd edn. Prentice Hall, New York (2001)
9. Hussain, M., Eakins, J.P.: Component-based visual clustering using the self-organizing map. Neural Networks, 260–273 (2007)
10. Zadeh, L.A.: Fuzzy Sets. Information and Control 8, 338–353 (1965)
11. Zadeh, L.A.: A fuzzy-algorithm approach to the definition of complex or imprecise concepts. Int. J. Man Machines Studies 8, 249–295 (1976)
12. Gomide, F., Pedrycz, W.: An Introduction to Fuzzy Sets: Analysis and Design. MIT Press, Cambridge (1998)
13. Zadeh, L.A., et al.: Fuzzy Sets, Fuzzy Logic, and Fuzzy Systems. World Scientific, Singapore (1996)
14. Grant, P.: A new approach to diabetic control: Fuzzy logic and insulin pump technology. Medical Engineering Physics, 824–827 (2006)
15. Di Lascioa, L.: A fuzzy-based methodology for the analysis of diabetic neuropathy. Fuzzy and Systems, 203–228 (2007)

16. Dalla Man, C., Rizza, R.A., Cobelli, C.: Meal simulation model of the glucose insulin system. IEEE - Transactions on Biomedical Engineering 54(10), 1740–1749 (2007)
17. Dua, P., Doyle, F., Pistikopoulos, E.: Model - based blood glucose control for type 1 diabetes via parametric programming. IEEE - Transactions on Biomedical Engineering 53(8), 1478–1491 (2006)
18. Owens, C.: Run-to-run control of blood glucose concentrations for people with type 1 diabetes mellitus. IEEE Transactions on Biomedical Engineering 53(6), 990–996 (2006)
19. Mendel, J.M.: Fuzzy Logic Systems for Engineering: A tutorial, pp. 345–377. IEEE, Los Alamitos (1995)
20. Mamdani, E.H., Assilian, S.: An experiment in linguistic synthesis with a fuzzy logic controller. International Journal of Man-Machine Studies 7(1), 1–13 (1975)
21. Silvanandam, S.N., Sumatri, S., Deepa, S.N.: Introduction to Fuzzy Logic Using Matlab. Springer, Berlin (2007)
22. Lee, M.-L., Chung, H.-Y., Yu, F.-M.: Modeling of hierarchical fuzzy systems. Fuzzy Sets and Systems 138, 343–361 (2003)
23. Stufflebeam, J., Prasad, N.R.: Hierarchical fuzzy control. In: IEEE – International Fuzzy Systems Conference Proceedings, pp. 498–503 (1999)
24. Newman, D.J., Hettich, S., Blake, C.L., Merz, C.J.: UCI Repository of machine learning databases. University of California, Department of Information and Computer Science, Irvine (1998), http://www.ics.uci.edu/~mlearn/MLRepository.html
25. Ham, F.M.: Principles of neurocomputing for science and engineering. McGraw-Hill, New York (2001)

Unearth the Hidden Supportive Information for an Intelligent Medical Diagnostic System

Sam Chao and Fai Wong

Faculty of Science and Technology, Unversity of Macau
Av. Padre Tomás Pereira S.J., Taipa, Macao
{lidiasc,derekfw}@umac.mo

Abstract. This paper presents an intelligent diagnostic supporting system – i^+DiaKAW (Intelligent and Interactive Diagnostic Knowledge Acquisition Workbench), which automatically extracts useful knowledge from massive medical data to support real medical diagnosis. In which, our two novel pre-processing algorithms MIDCA (Multivariate Interdependent Discretization for Continuous-valued Attributes) and LUIFS (Latent Utility of Irrelevant Feature Selection) for continuous feature discretization (CFD) and feature selection (FS) respectively, assist in accelerating the diagnostic accuracy by taking the attributes' supportive relevance into consideration during the data preparation process. Such strategy minimizes the information lost and maximizes the intelligence and accuracy of the system. The empirical results on several real-life datasets from UCI repository demonstrate the goodness of our diagnostic system.

Keywords: Discretization and Feature Selection, Data Pre-processing, Latent Supportive Relevance, Intelligent Diagnostic System, Medical Data Mining.

1 Introduction

Data pre-processing is an important and inevitable step that presents the equal challenge to a data mining process. The results of medical care are life or death and this rule is applied to everybody [1]. Therefore the quality of data highly affects the intelligence and learning performance of a system, where feature selection (FS) and continuous feature discretization (CFD) are treated as the dominant issues, and have been the active and fruitful fields of research for decades in data mining [2]. Nevertheless, many FS and CFD methods focus on univariate or high relevant correlations only. This may sometimes loose the significant useful hidden information for the learning task. Especially in medical domain, a single symptom seems useless regarding the diagnosis, may be potentially important when combined with other symptoms. An attribute that is completely useless by itself can provide a significant performance improvement when taken with others; while two attributes that are useless by themselves can be useful together [3], [4].

In this paper, a practical intelligent diagnostic system – i^+DiaKAW (Intelligent and Interactive Diagnostic Knowledge Acquisition Workbench) is presented to automatically extract useful knowledge from massive medical data to support real medical

E. Corchado et al. (Eds.): HAIS 2009, LNAI 5572, pp. 401–408, 2009.

diagnosis. Meanwhile, it accelerates the performance in terms of intelligence and accuracy by employing two novel pre-processing methods: MIDCA (Multivariate Interdependent Discretization for Continuous Attributes) and LUIFS (Latent Utility of Irrelevant Feature Selection). These two methods take the attributes' supportive correlations into consideration, so that minimizes the information lost and maximizes the diagnostic performance. In the next section, the nature of medical data is analyzed as well as its problematic issues. Our diagnostic system is briefly introduced in section 3. While two innovative algorithms MIDCA and LUIFS that focus on discovering the latent usefulness of irrelevant attributes, are described in section 4 and 5 respectively. Section 6 demonstrates the experimental results on several real-life datasets from UCI repository. Finally, we summarize our methods and present the future directions of our research.

2 Nature of Medical Data

As before-mentioned, a single symptom seems useless regarding the diagnosis may be potentially important by providing supportive information to other symptoms. For instance, when learning the medical data for diagnosing *cardiovascular* disease, suppose a patient's database contains attributes such as *age, gender, height, weight, blood pressure, pulse rate, ECG result* and *chest pain*, etc. During FS process, most often attribute *age* or *height* alone will be treated as the least important attribute and discarded accordingly. However, in fact these two attributes together with attribute *weight* may express a potential significance: whether a patient is *overweight*? Hence, none of them could be ignored in the pre-processing step. On the other hand, although attribute *blood pressure* may be treated as important regarding classifying a *cardiovascular* disease, while together with a seemingly useless attribute *age*, they may reveal more specific sense: whether a patient is *hypertensive*? Obviously, the compound features *overweight* and *hypertensive* have more diagnostic power than the individual attributes *weight* and *blood pressure* respectively. It is also proven that if a person is *overweight* and/or *hypertensive* may have more probabilities to suffer from a cardiovascular disease [5].

Moreover, when processing CFD, the discretized intervals should make sense to human expert [6], [7]. As we know that a person's blood pressure is increasing as one is growing up. Therefore it is improper to generate a cutting point such as 140mmHg and 90mmHg for systolic pressure and diastolic pressure, respectively. Since the standard for diagnosing hypertension is a little bit different from young people (regular is 120-130mmHg/80mmHg) to the old people (regular is 140mmHg/90mmHg) [8]. If the blood pressure of a person aged 20 is 139mmHg/89mmHg, one might be considered as a potential hypertensive. In contrast, if a person aged 65 has the same blood pressure measurement, one is definitely considered as normotensive. Obviously, to discretize the continuous-valued attribute *blood pressure*, it must take at least the attribute *age* into consideration, while discretizing other continuous-valued attribute may not take *age* into consideration. The only solution to address the above mentioned problems is to take the usage of seemingly irrelevant attributes into consideration rather than purely the multivariate strategies.

3 *i*⁺DiaKAW

i⁺DiaKAW is a multi-purpose, potentially valuable system that aims at handling the practical problems in real world by extracting useful knowledge from massive medical data automatically and intelligently. The major philosophy behind *i*⁺DiaKAW is in multi-threads: (1) constructing a specialized medical knowledge base and being embedded in other intelligent information system like expert system; (2) assisting physicians or medical experts in diagnosing new case objectively and reliably by providing practical diagnostic rules that acquired from historical medical data; (3) self-diagnosing a person's physical condition according to a set of available symptoms or test results, and meanwhile the helpful therapy advices and beneficial medical knowledge are probably provided. Fig. 1 briefly outlines a general architecture, where its inputs are various forms of raw data or a variety of patient's symptoms, etc; and its outputs are the valued knowledge in terms of medical rules, therapy advices or diagnostic results.

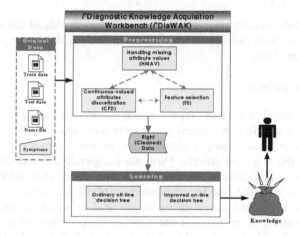

Fig. 1. Overview architecture of *i*⁺DiaKAW system

i⁺DiaKAW comprises three essential fundamental components: (1) data pre-processing module, which is the primary key factor of a successful data mining system; (2) learning module that offers to the users both on-line and off-line learning strategies in terms of a variety of learning algorithms; and (3) supporting module that provides a variety of functions to manipulate data files. All these features enable *i*⁺DiaKAW to minimize the human biases and mistakes, while the generated reliable diagnostic knowledge benefits not only the patients, but also the physicians. In which, data pre-processing module is in the key focus of this paper, while the latter two modules are out of the scope of the current paper.

4 MIDCA

MIDCA is interested mainly in discovering the best interdependent attribute relative to a continuous-valued attribute being discretized. It incorporates the symmetric relief [9], [10] and information theory [11], [12], to look for a best correlated attribute to the continuous-valued attribute as its interdependent attribute to carry out the multivariate discretization. We believe that in an attribute space A, for each continuous-valued attribute a_i, there exists at least one a_j (where $a_i, a_j \in A$ and $i \neq j$), so that a_j is the most correlated with a_i, or vice versa since the correlation is measured symmetrically. For the purpose of finding out such a best interdependent attribute a_j for each a_i, their interdependent weight W_{mvi} is measured in equation (1), where symmetric measures are employed to reduce the bias. The symmetric concept in our method is applied to two correlated attributes, where the measure is calculated twice that treated each attribute in turn as class and then the results are averaged. The detailed formulae for symmetric information gain and relief can be referred to the corresponding papers.

$$ W_{mvi}\left(a_i, a_j\right) = \left[\frac{SymGain(a_i, a_j)}{\sqrt{\sum_{M \neq i}^{A} SymGain(a_i, a_M)^2}} + \frac{SymRelief(a_i, a_j)}{\sqrt{\sum_{M \neq i}^{A} SymRelief(a_i, a_M)^2}} \right] / 2 . \qquad (1) $$

The best interdependent attribute against a continuous-valued attribute being discretized, is the one with the highest interdependent weight W_{mvi} amongst all candidate interdependent attributes.

4.1 MIDCA Algorithm

As a best interdependent attribute regarding the continuous-valued attribute has been discovered, the discretization process carries out subsequently. We adopt the most efficient supervised discretization algorithm – minimal entropy partitioning with MDLP [13] as the stopping criteria. Now we compendiously present our MIDCA algorithm as a five-step process: (1) sort the numeric values of an attribute in ascending order; (2) discover the best interdependent attribute according to the measure of W_{mvi}; (3) for each interval, calculate its conditional information gain measure and select the best cutting point; (4) evaluate whether stopping the calculation according to MDLP method; (5) repeat step (3) if the test failed; else, order the best cutting points to generate a discretized data set.

4.2 Intelligence of MIDCA

MIDCA generates multi-interval discretization, but ensures at least binary discretization. This feature is important, since if a continuous-valued attribute generates null cutting point means that the attribute is useless, hence increases the classification uncertainty. We discovered that the figures of most continuous-valued attributes in medical domain have their special meanings. They may express the degrees or seriousness of a specific illness. For instance, *blood pressure* may indicate the level of hypertension; higher *heart rate* may represent the existence of cardiovascular disease; while *plasma glucose* is an index for diabetes and so on, hence their discretization cannot be ignored. Yet important, MIDCA creates one interdependent attribute for each continuous-valued attribute rather than using one for all continuous-valued attributes. It carries out the discretization with respect to the perfect interdependent attribute for the attribute being discretized in addition to the class attribute. Such novel method is able to concern each continuous-valued attribute differently and humanly, while handles it with maximum intelligence. This is also the main factor for improving the final classification accuracy accordingly.

5 LUIFS

LUIFS claims that an irrelevant attribute in medical domain is the one that provides neither explicit information nor supportive or implicit information. LUIFS takes the inter-correlation between the irrelevant attribute and the other attributes into consideration to measure the latent importance of an irrelevant attribute, thus is beneficial to the diagnostic results. LUIFS generates an optimal feature subset in two phases: (1) Relevant Attributes Seeking (RAS): for each attribute in a dataset, work out its relevant weight W_{rel} regarding the target class, selects the ones whose weights are greater than a pre-defined threshold ϖ into the optimal feature subset; (2) Latent Supportive of Irrelevant Attribute (LSIA) Discovery: for each unselected (false irrelevant) attribute that filtered from phase (1), determine its supportive importance by performing a multivariate interdependent measure I_{mim} with other attributes. Then a false irrelevant attribute becomes relevant if its I_{mim} meets the minimum requirements; otherwise it is really a true irrelevant attribute and discarded accordingly.

5.1 RAS Process

In this phase, each attribute is calculated its relevant weight W_{rel} respect to the target class only. Information gain theory [11], [12] is applied as the measurement, which may find out the most informative (important) attributes in an attribute space A. Attributes are then sorted in descending order, from the most important one to the least useful one. Meanwhile, a pre-defined threshold ϖ is introduced to distinguish the weightiness of an attribute. The value of a threshold either too high or too low may cause the attributes insufficient or surplus. Therefore it is defined as a mean value excluding the ones with the maximum and minimum gain measures, in order to eliminate as much bias as possible. Then, an attribute will be selected into an optimal feature subset if its W_{rel} is greater than the threshold, i.e. $W_{rel} > \varpi$. This phase requires only linear time in the number of given features M, i.e. $O(M)$.

5.2 LSIA Discovery

This phase acts as a key role in LUIFS, its objective is to uncover the usefulness of the latent or supportive relevant attributes. This phase is targeted at those false irrelevant attributes that filtered out from RAS process, looking for their latent utilities in supporting other relevant attributes. To determine whether a false irrelevant attribute is potentially important or not, we evaluate the interdependence measure I_{mim} between it and another attribute regarding the class attribute. Combinative relief theory is employed instead of the original form, which is able to discover the interdependences between attributes. Equation (2) illustrates the measure,

$$I_{mim}(a_{irr}, a_i) = P(different\ value\ of\ \langle a_{irr}, a_i \rangle \mid different\ class)$$
$$-P(differnt\ value\ of\ \langle a_{irr}, a_i \rangle \mid same\ class) \tag{2}$$

where P is a probability function for calculating how well the values of $<a_{irr}, a_i>$ distinguish among the instances from the same and the different classes. It measures the level of hidden supportive importance for a false irrelevant attribute a_{irr} to another attribute a_i, hence higher the weighting, more information it will provide, such that better the diagnostic result.

LSIA discovery phase is more complicated that requires nonlinear time in performance. In the worst case, suppose a dataset $D = A \cup C$, where the feature space $A=\{a_1, a_2, ..., a_M\}$ and class $C=\{c_1, c_2, ..., c_k\}$. If there is only one important attribute a_i discovered in RAS phase and added into the optimal feature subset $A_{opt}=\{a_i\}$, that is, there are $(M-1)$ false irrelevant attributes ignored and unselected. For each false irrelevant attribute a_{irr}, where $irr=(1,2,..,i-1,i+1,...,M)$, calculate its interdependent weight I_{mim} with another attribute. Again in the worst case, if a_{irr} could not encounter an attribute that makes it becoming useful, then the process should be repeated for $(M-1)$ times. Whereas the algorithm is symmetric, i.e. $I_{mim}(a_{irr}, a_j) = I_{mim}(a_j, a_{irr})$, so the total time should be in half respect to the number of false irrelevant attributes, which equals to $(M-1)/2$. Therefore, the complexity of LSIA discovery phase is $(M-1)*[(M-1)/2]$ for the worst case, i.e. $O(M^2)$. Finally, the total complexity of LUIFS algorithm is same as LSIA discovery procedure as $O(M^2)$.

6 Experimental Results

To reveal the intelligence of i^+DiaKAW system, as well as the effectiveness of our preprocessing methods, experiments are performed on twelve real life datasets from UCI repository [14]. The solid evidences demonstrate our belief: by uncovering potential attributes relevance during pre-processing step (either FS or CFD) and taking them into the classification task, the diagnostic accuracy can have significant improved. Table 1 gives the empirical result under two classification algorithms ID3 [15] and Naïve Bayes [16]. Besides, algorithms *Discretize* and *FFS* (combined as *PProcess* that downloaded from UCI repository and used as pre-processing prior classification) are involved into the comparison regarding MIDCA+LUIFS (named as *ML*), while diagnosis without pre-processing is also indicated as *Origin*.

The evaluation result in Table 1 exhibits a noticeable clue that our preprocessing methods promote the classification accuracy for major datasets. Furthermore, from the average results computed in the last row of the above table, the reductions of error rates for *ML* over *Origin* and *PProcess* are 34.9% and 23.9%, 32.2% and 30.6% for ID3 and Naïve Bayes algorithms respectively. This again exhibits significantly the superiority and intelligence of our methodology. However, our methodology slightly downgrades the performance on two datasets *Diabetes* and *Iris* amongst all, which contain only numeric attributes. Since the method needs an additional discretization step prior either CFD or FS process. This may increase the uncertainty, hence increases the error rate accordingly. Nevertheless, our methodology treats each attribute differently and humanly, they do take effect to most cases, and benefit the subsequent classification and diagnostic process with the maximum intelligence.

Table 1. Classification results in error rate of ID3 and Naive Bayes algorithms without/with different preprocessing methods

Dataset	ID3 in Error Rate (%)			Naïve Bayes in Error Rate (%)		
	Origin	PProcess	ML	Origin	PProcess	ML
Cleve	24.11	22.08	15.48	13.42	20.47	12.69
Hepatitis	19.42	20.71	9.71	13.4	16.42	10.32
Hypothyroid	1.36	0.76	0.25	1.87	0.76	0.62
Heart	25.93	18.89	14.81	12.88	21.85	11.27
Sick-euthyroid	3.65	2.75	0.03	11.33	2.47	0.04
Auto	19.52	20.1	14.32	34.5	27.45	25.66
Breast	5.43	4.87	4.29	3.93	3.58	3.01
Diabetes	29.44	21.63	28.92	17.71	21.24	19.1
Iris	6	4.67	6	3.78	5.33	4.28
Crx	17.96	14.49	13.47	19.22	13.91	10.99
Australian	19.78	16.74	13.91	19.4	14.2	11.6
Horse-colic	21.67	18.67	5.33	17.44	17.39	5
Average	**16.19**	**13.86**	**10.54**	**14.07**	**13.76**	**9.55**

7 Conclusions

This paper presents a practical intelligent medical diagnostic system – i^+DiaKAW with data mining technology, which remarkably assists in revolutionizing the traditional medical services from passive, subjective and human-oriented to active, objective and facts-oriented. i^+DiaKAW equips with a novel concept for improving its performance in terms of intelligence and accuracy by uncovering the potential attributes relevance during data preprocessing. Furthermore, our two preprocessing methods MIDCA and LUIFS have been proven efficient, effective and with human intelligence as a pre-processing tool for i^+DiaKAW system.

Our next principal direction for the future research will be focused on the optimization of two preprocessing methods. A feasible solution should be investigated to eliminate the uncertainties and to reduce the complexities as much as possible, in order to adapt to the future development trend in data mining. Last but not least, it would be valuable for i^+DiaKAW system to involve diverse algorithms either of pre-processing or classification, e.g. Focus [17], SVM [18], and ANNs [19] etc., to strengthen and extend its functionality, workability and completeness.

References

1. Cios, K.J., Moore, G.W.: Uniqueness of Medical Data Mining. Journal of Artificial Intelligence in Medicine 26(1), 1–24 (2002)
2. Yu, L., Liu, H.: Efficient Feature Selection via Analysis of Relevance and Redundancy. Journal of Machine Learning Research 5, 1205–1224 (2004)
3. Guyon, I., Elisseeff, A.: An Introduction to Variable and Feature Selection. Journal of Machine Learning Research 3, 1157–1182 (2003)
4. Caruana, R., Sa, V.R.: Benefiting from the Variables that Variable Selection Discards. Journal of Machine Learning Research 3, 1245–1264 (2003)

5. Jia, L., Xu, Y.: Guan Xing Bing De Zhen Duan Yu Zhi Liao. Jun Shi Yi Xue Ke Xue Chu Ban She (2001)
6. Bay, S.D.: Multivariate Discretization of Continuous Variables for Set Mining. In: Sixth ACM SIGKDD International Conference on Knowledge Discovery and Data Mining, pp. 315–319 (2000)
7. Bay, S.D.: Multivariate Discretization for Set Mining. Knowledge and Information Systems 3(4), 491–512 (2001)
8. Gu, W.Q.: Xin Xue Guan Ji Bing Jian Bie Zhen Duan Xue. Xue Yuan Chu Ban She (2006)
9. Kira, K., Rendell, L.: A Practical Approach to Feature Selection. In: International Conference on Machine Learning, pp. 249–256. Morgan Kaufmann, Aberdeen (1992a)
10. Kira, K., Rendell, L.: The Feature Selection Problem: Traditional Methods and New Algorithm. In: AAAI 1992, San Jose, CA (1992b)
11. Mitchell, T.M.: Machine Learning. McGraw-Hill Companies, Inc., New York (1997)
12. Zhu, X.L.: Fundamentals of Applied Information Theory. Tsinghua University Press (2000)
13. Fayyad, U.M., Irani, K.B.: Multi-interval Discretization of Continuous-valued Attributes for Classification Learning. In: Thirteenth International Joint Conference on Artificial Intelligence, pp. 1022–1027 (1993)
14. Blake, C.L., Merz, C.J.: UCI Repository of Machine Learning Databases. Irvine, CA: University of California, Department of Information and Computer Science (1998)
15. Quinlan, J.R.: Induction of Decision Trees. Machine Learning 1(1), 81–106 (1986)
16. Langley, P., Iba, W., Thompsom, K.: An Analysis of Bayesian Classifiers. In: Tenth National Conference on Artificial Intelligence, pp. 223–228. AAAI Press and MIT Press (1992)
17. Almualim, H., Dietterich, T.G.: Learning with Many Irrelevant Features. In: Ninth National Conference on Artifical Intelligence, pp. 547–552. AAAI Press/The MIT Press, Anaheim, California (1992)
18. Scholkopf, B., Smola, A., Williamson, R.C., Bartlett, P.L.: New Support Vector Algorithms. Neural Computation 12, 1207–1245 (2000)
19. Kasabov, N.K.: Foundations of Neural Networks, Fuzzy Systems, and Knowledge Engineering. The MIT Press, Cambridge (1998)

Incremental Kernel Machines for Protein Remote Homology Detection

Lionel Morgado[1] and Carlos Pereira[1,2]

[1] CISUC - Center for Informatics and Systems of the University of Coimbra
Polo II - Universidade de Coimbra, 3030-290 Coimbra, Portugal
lionel@student.dei.uc.pt, cpereira@dei.uc.pt
[2] ISEC - Instituto Superior de Engenharia de Coimbra
Quinta da Nora, 3030-199 Coimbra, Portugal
cpereira@isec.pt

Abstract. Protein membership prediction is a fundamental task to retrieve information for unknown or unidentified sequences. When support vector machines (SVMs) are associated with the right kernels, this machine learning technique can build state-of-the-art classifiers. However, traditional implementations work in a batch fashion, limiting the application to very large and high dimensional data sets, typical in biology. Incremental SVMs introduce an alternative to batch algorithms, and a good candidate to solve these problems. In this work several experiments are conducted to evaluate the performance of the incremental SVM on remote homology detection using a benchmark data set. The main advantages are shown, opening the possibility to further improve the algorithm in order to achieve even better classifiers.

Keywords: Kernel machines, incremental learning, protein classification.

1 Introduction

A traditional issue in bioinformatics is the classification of protein sequences into functional and structural groups based on sequence similarity. Despite being relatively easy to recognize homologues with high levels of similarity, remote homology detection is a much harder task. Approaches used for remote homology detection can be divided into three main groups: pairwise sequence comparison methods, generative models and discriminative classifiers. The most successful methods for remote homology detection are the discriminative, that combine SVMs [5] with special kernels [19, 20, 21, 22, 23]. The SVM is a powerful machine learning technique that combines high accuracy with good generalization, achieving state-of-the-art results. However, traditional SVM batch implementations present some limitations when faced with the high dimensional and large number of examples available in biology. Incremental SVMs can potentially bring the solutions to these issues, by means of their ability to add new information to an existing, already trained model.

In this work, some experiments are performed with a benchmark data set from SCOP [6] previously used on remote homology detection in order to evaluate the performance of an incremental SVM against the batch algorithm and PSI-BLAST.

E. Corchado et al. (Eds.): HAIS 2009, LNAI 5572, pp. 409–416, 2009.
© Springer-Verlag Berlin Heidelberg 2009

An overview on incremental SVMs is presented in Section 2. Section 3 presents the description of the spectrum, mismatch and profile kernels which have been used, and Section 4 presents the experiments and results analysis. Final conclusions and reference to future work are given in the last Section.

2 Incremental Kernel Machines

Nowadays, the advances in technology allow collecting enormous amounts of data and joining it in very large data sets. Computational biology is one of such fields where the millions of available examples can also be characterized by very high and variable dimensionality. However, because traditional SVMs use all data in a batch train, both the models and algorithms complexity are overlapping the computational capacities available, limiting the application to this field. Since less information usually implies simpler models and lower memory requirements, reducing the number of train instances and the dimensionality of the data have been both explored approaches. By common sense, the easiest way to decrease the processing burden is to train only over one smaller set with randomly chosen examples. However the probability of excluding important information with this methodology is very high. A larger training set also represents an advantage, since the extra information can contribute to create more accurate models. Therefore, it is important to analyse every individual example, at least briefly. On the other hand, our knowledge in proteomics and genomics is constantly changing, taking repositories to suffer considerable modifications in relatively short periods of time, that demand frequent time consuming actualisations of the discriminative models. Considering these facts, a SVM that builds models step by step in an incremental/decremental fashion using a smaller number of instances each time should be a reality in computational biology.

The first incremental method proposed takes under consideration that the SVM solution only depends on the support vectors, therefore retraining a model consecutively in new blocks of data and the support vectors obtained from previous training sessions will yield the same result as training with all available points at once, because the support vectors are preserved along the process [7]. The exact formulation of incremental SVM learning was presented some years later [8], and brought the possibility to decrement or "unlearn" a model. The algorithm was extended to leave-one-out procedures, and adapted in a way to minimize the computational cost of recalculating a new solution when regularization parameter C and kernel parameters are changed [9]. Nevertheless, this algorithm presents some limitations associated to the use of all the already seen examples to get the final exact solution. An alternative that tries to solve this matter is SimpleSVM [10]. The SimpleSVM algorithm extends Poggio's principles to the soft-margin case and combines it with block training to keep optimality over unconstrained Lagrangian multipliers. SimpleSVM has a good performance on data sets with few support vectors, however for large scale problems, Sequential Minimal Optimisation (SMO) is preferred [11]. SMO breaks the optimization problem down into two-dimensional sub-problems that may be solved analytically, eliminating the need for a numerical optimization algorithm such as conjugate gradient methods, this way shortening the processing time and the computational burden.

It is precisely from SMO that LASVM is derived [12]. This algorithm is an online kernel classifier based on the soft-margin SVM, that incrementally builds a discriminative model by adding or removing support vectors iteratively, using two different points each turn. New support vectors come from a direction search called PROCESS that involves at least one non support vector from the current kernel expansion, while REPROCESS can eliminate support vectors by changing to zero the weight coefficients of one or both the points analysed. In order to incrementally build the final discriminative model, each iteration demands storing a set of all the potential support vectors, Lagrange coefficients of the kernel expansion and the partial derivatives. A significant difference that arises when comparing LASVM to SimpleSVM is that the former doesn't seek the precise solution of the QP problem in each step but instead an approximation that improves the dual function. So, a finishing step similar to a simplified SMO may be necessary to improve performance on noisy data sets.

In fact, real-life problems are dynamic/online rather than static/batch, because information is prone to change. Some work has been developed around incremental/online classification [13, 14, 15] and regression problems [16, 17, 18], but a lot of research is still needed, in particular for biological data analysis.

3 Kernels for Proteins

Several kernels have been proposed for protein classification [19, 20, 21, 22, 23, 27]. The kernel function aims emphasizing important biological information while converting variable length strings that represent amino acids or nucleotides, into numeric fixed size feature vectors. This mapping is mandatory in the sense that the learning machine demands feature vectors with a fixed number of attributes and largely affect the final accuracy and complexity of the learning machine.

3.1 The Spectrum Kernel

The spectrum kernel [20] is a string kernel type that acts over an input space composed of all finite sequences of characters from an alphabet A with l elements, and maps it to a feature space with l^k dimensions that represent all the possible k-length contiguous subsequences that may be contained in a protein.

The feature map for sequence x is given by:

$$\Phi_k(x) = (\phi_\alpha(x))_{\alpha \in A^k}, \tag{1}$$

where $\phi_\alpha(x)$ contains the number of times subsequence α occurs in x.

Taking into account the definition of kernel, the k-spectrum kernel comes from the dot product:

$$K_k(x, y) = \langle \Phi_k(x), \Phi_k(y) \rangle \tag{2}$$

3.2 The Mismatch Kernel

The mismatch kernel [21] is an extension of the spectrum kernel. It measures sequence similarity based on shared occurrences of fixed-length patterns in the data, allowing mutations between them.

A k-length subsequence α of aminoacids can be described in a (k, m)-neighborhood $N_{(k,m)}(\alpha)$ defined by all the k-length subsequences β that differ from the original α by at most m mismatches.

The entry space uses the feature map:

$$\Phi_{(k,m)}(\alpha) = \left(\phi_\beta(\alpha) \right)_{\alpha \in A^k}, \tag{3}$$

where $\phi_\beta(\alpha)$ contains the number of occurrences and where β belongs to $N_{(k,m)}(\alpha)$.

The mismatch kernel is given by:

$$K_{(k,m)}(x, y) = \langle \Phi_{(k,m)}(x), \Phi_{(k,m)}(y) \rangle, \tag{4}$$

and is equivalent to the spectrum kernel when no mismatches are allowed $(m = 0)$.

3.3 The Profile Kernel

The profile kernel [23] doesn't take as input the protein itself but rather profiles $P(x)$ of a sequence x. Profiles are statistically estimated from close homologues stored in a large sequence database, and can be defined as:

$$P(x) = \{p_i(a), a \in A\}_{i=1}^N, \tag{5}$$

with p_i being the emission probability of aminoacid a in position i and $\sum_{a \in A} p_i(a) = 1$ for every position i. Similarly to the mismatch kernel, mutations are considered. A significant difference is that here the probability of a mutation to occur is measured and only some cases are allowed, considering a score dependent on the position of the substring in the protein chain and a given threshold.

4 Experiments

Remote homology detection was used to evaluate the performance, structure complexity and processing time of incremental SVM algorithms comparatively to batch implementations. The following algorithms were applied: LIBSVM [24] (version 2.85) as the batch SVM, the incremental algorithm LASVM and PSI-BLAST, the most used method by the scientific community.

Due to the very high dimensionality of the feature space generated using string kernels, these were pre-computed in order to avoid computation problems. This methodology also allows planning computation in a way to avoid calculus redundancy. For the implementation, it was necessary to adapt LASVM to accept this kind of data as input.

A 2.4 GHz Intel Core 2 Quad CPU desktop computer with 4 GB RAM was used. PSI-BLAST was executed under Microsoft Windows XP, LIBSVM models were trained under the same operating system running a MATLAB interface, and LASVM was executed under gOS.

The profiles for the profile kernel were obtained with PSI-BLAST using 2 search rounds.

4.1 Data Set Description

The algorithms were tested with a SCOP benchmark data set previously used on remote homology detection [22]. The data set has 7329 domains and was divided according to 54 families. Remote homology detection is simulated by considering all domains for each family as positive test examples and sequences outside the family but belonging to the same superfamily as positive train examples. Negative examples are from outside the positive sequences fold, and were randomly divided into train and test sets in the same ratio as the positive examples.

To evaluate the quality of the created classifiers receiver operating characteristics (ROC) was used. A ROC curve consists in the plot of the true positives rate as a function of true negatives rate at varying decision thresholds, and expresses the ability of a model to correctly rank examples and separate distinct classes. The area under a ROC curve (AUC), also known as ROC score, is the most used performance measure extracted from ROC. A good model has AUC=1, a random classifier is expressed by an AUC $\approx 0, 5$ and the worst case comes when AUC=0.

4.2 Results

The ROC scores (AUC) for the batch SVM, LASVM and PSI-BLAST, are given in Table 1. The kernel notation indicates the length of the subsequences taken under consideration and the number of mismatches allowed (for mismatch kernel) or the threshold value (for profile kernel).

As expected, the SVM with the profile kernel is the one that achieves better results, followed by the mismatch and spectrum kernel. Profile and mismatch kernels create models with even better performance than PSI-BLAST, showing its ability to evidence important biological information based on amino acid sequences alone. This quality is not an exclusive property of the batch algorithms, since LASVM exhibits an identical behaviour, creating models with equal or even superior results for some protein families. It was also verified that processing time is similar when training new models from the beginning with all data points.

The ability of the incremental algorithm to achieve an inferior number of support vectors than LIBSVM (as seen in Figure1), when describing the discriminative decision hyperplane, reveals an important contribution to complexity reduction, making this methodology suitable for large scale problems.

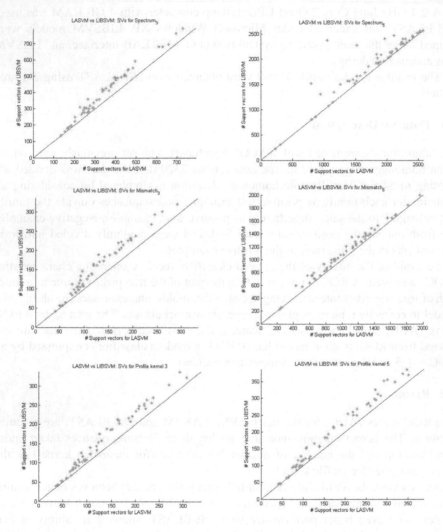

Fig. 1. Support vectors for LASVM plotted against LIBSVM for each family model. Results obtained for spectrum (3), spectrum (5), mismatch (3, 1), mismatch (5, 1), profile (3, 7.5) and profile (5, 7.5) in top left, top right, center left, center right, bottom left and bottom right, respectively.

Table 1. Mean ROC values for all experiments

Algorithm	Kernel	ROC (mean)
PSI-BLAST	-	0.8183
LIBSVM	Spectrum(3)	0.788
	Spectrum(5)	0.720
	Mismatch(3,1)	0.856
	Mismatch(5,1)	0.866
	Profile(3,7.5)	0.89
	Profile(5,7.5)	0.92
LASVM	Spectrum(3)	0.788
	Spectrum(5)	0.699
	Mismatch(3,1)	0.855
	Mismatch(5,1)	0.865
	Profile(3,7.5)	0.89
	Profile(5,7.5)	0.92

5   Conclusions and Future Work

This work proposes incremental SVM algorithms for protein remote homology detection. The presented results show that the incremental formulation, namely LASVM, achieves state-of-the-art results for this kind of task, bringing some advantages over the batch SVM, which by itself can get superior results to the widely accepted PSI-BLAST. The incremental SMO based SVM showed proficiency to generate discriminative models as good as or even better than batch LIBSVM, keeping, for the most families a reduced number of support vectors.

These good results and the potential of the approach encourage the application of the incremental algorithm with different kernels for online classification tasks, and in particular to large biological data sets.

Acknowledgments. This work was supported by FCT – Fundação para a Ciência e a Tecnologia, under Project BIOINK – PTDC/EIA/71770/2006.

References

1. Altschul, S.F., Gish, W., Miller, W., Myers, E.W., Lipman, D.J.: A basic local alignment search tool. J. Mol. Biol. 215, 403–410 (1990)
2. Smith, T.F., Waterman, M.S.: Identification of common molecular sub- Sequences. J. Mol. Biol. 147, 195–197 (1981)
3. Krogh, A., Brown, M., Mian, I., Sjolander, K., Haussler, D.: Hidden markov models in computational biology: Applications to protein modeling. J. Mol. Biol. 235, 1501–1531 (1994)
4. Altschul, S.F., Madden, T.L., Schaffer, A.A., Zhang, J., Zhang, Z., Miller, W., Lipman, D.J.: Gapped BLAST and PSI-BLAST: A new generation of protein database search programs. Nucleic Acids Research 25, 3389–3402 (1997)
5. Vapnik, V.: Statistical Learning Theory. Wiley, New York (1998)

6. Murzin, A.G., Brenner, S.E., Hubbard, T., Chothia, C.: SCOP: A structural classification of proteins database for the investigation of sequences and structure. J. Mol. Biol. 247, 536–540 (1995)
7. Syed, N.A., Liu, H., Sung, K.K.: Incremental Learning with Support Vector Machines (1999)
8. Cauwenberghs, G., Poggio, T.: Incremental and Decremental Support Vector Machine Learning. Advances in Neural Information Processing Systems, vol. 13. MIT Press, Cambridge (2001)
9. Diehl, C.P., Cawenberghs, G.: SVM Incremental Learning, Adaptation and Optimization. In: Proceedings of the International Joint Conference on Neural Networks (2003)
10. Vishwanathan, S.V.N., Smola, A.J., Murty, M.N.: SimpleSVM. In: Proceedings of the Twentieth International Conference on Machine Learning, Washington DC (2003)
11. Platt, J.C.: Sequential Minimal Optimization: A Fast Algorithm for Training Support Vector Machines. Advances in Kernel Methods – Support Vector Learning. MIT Press, Cambridge (1999)
12. Bordes, A., Ertekin, S., Weston, J., Bottou, L.: Fast Kernel Classifiers with Online and Active Learning. Journal of Machine Learning Research (2005)
13. Tax, D.M.J., Laskov, P.: Online SVM Learning: From Classification to Data Description and Back. Neural Networks for Signal Processing (2003)
14. Rüping, S.: Incremental Learning with Support Vector Machines. In: Proceedings of the 2001 IEEE International Conference on Data Mining (2001)
15. Laskov, P., Gehl, C., Krüger, S., Müller, K.: Incremental Support Vector Learning: Analysis, Implementaton and Applications. The Journal of Machine Learning Research 7, 1909–1936 (2006)
16. Martin, M.: On-line Support Vector Machine Regression. In: Proceedings of the 13th European Conference on Machine Learning (2002)
17. Ma, J., Theiler, J., Perkins, S.: Accurate On-line Support Vector Regression. Neural Computation 15, 2683–2703 (2003)
18. Parrella, F.: Online Support Vector Regression – A thesis presented for the degree of Information Science. Department of Information Science, University of Genoa, Italy (2007)
19. Jaakkola, T., Diekhans, M., Haussler, D.: Using the Fisher Kernel Method to Detect Remote Protein Homologies. In: Proceedings of the Seventh International Conference on Intelligent Systems for Molecular Biology (1999)
20. Leslie, C., Eskin, E., Noble, W.: The spectrum kernel: a string kernel for SVM protein classification. In: Pacific Symposium on Biocomputing, vol. 7, pp. 566–575 (2002)
21. Leslie, C., Eskin, E., Weston, J., Noble, W.: Mismatch string kernels for SVM protein classification. Adv. Neural Inf. Process. Syst. 15, 1441–1448 (2002)
22. Weston, J., Leslie, C., Zhou, D., Elisseeff, A., Noble, W.S.: Semi-Supervised Protein Classification using Cluster Kernels. In: NIPS, vol. 17 (2003)
23. Kuang, R., Ie, E., Wang, K., Siddiqi, M., Freund, Y., Leslie, C.: Profile-based string kernels for remote homology detection and motif extraction. In: 3rd International IEEE Computer Society Computational Systems Bioinformatics Conference, Stanford, CA, pp. 152–160. IEEE Computer Society Press, Los Alamitos (2004)
24. Chang, C.C., Lin, C.J.: LIBSVM: a Library for Support Vector Machines (2004), http://www.csie.ntu.edu.tw/~cjlin/libsvm
25. Fawcett, T.: An introduction to ROC analysis. Pattern Recognition Letters 27, 861–874 (2006)
26. Bradley, A.P.: The use of the area under the ROC curve in the evaluation of machine learning algorithms. Pattern recognition 30(7), 1145–1159 (1997)
27. Busuttil, S., Abela, J., Pace, G.J.: Support Vector Machines with Profile-Based Kernels for Remote Protein Homology Detection. Genome Informatics 15(2), 191–200 (2004)

Use of Classification Algorithms in Noise Detection and Elimination

André L. B. Miranda[1], Luís Paulo F. Garcia[1], André C. P. L. F. Carvalho[1],
and Ana C. Lorena[2]

[1] Instituto de Ciências Matemáticas e Computação – Universidade de São Paulo USP
Caixa Postal 668 – 13560-970 – São Carlos – SP – Brazil
{andrelbm,lpgarcia}@grad.icmc.usp.br, andre@icmc.usp.br
[2] Centro de Matemática, Computação e Cognição – Universidade Federal do ABC UFABC
09090-400 – Santo André – SP – Brazil
ana.lorena@ufabc.edu.br

Abstract. Data sets in Bioinformatics usually present a high level of noise. Various processes involved in biological data collection and preparation may be responsible for the introduction of this noise, such as the imprecision inherent to laboratory experiments generating these data. Using noisy data in the induction of classifiers through Machine Learning techniques may harm the classifiers prediction performance. Therefore, the predictions of these classifiers may be used for guiding noise detection and removal. This work compares three approaches for the elimination of noisy data from Bioinformatics data sets using Machine Learning classifiers: the first is based in the removal of the detected noisy examples, the second tries to reclassify these data and the third technique, named hybrid, unifies the previous approaches.

Keywords: Noise, Machine Learning, Gene Expression and Classification.

1 Introduction

Noisy examples are characterized by differentiating from the remaining data in a data set, not following the common patterns present in these data.

Several Machine Learning (ML) techniques are able to induce classifiers from data sets composed of known examples, which can then be used to make predictions for new data. For such, a training set is initially extracted from original data set for the training of the ML techniques. Since the performance of an induced classifier is directly influenced by the quality of its training set, the presence of noise in the training data may affect its performance, decreasing its accuracy, while increasing its complexity and training time [1].

For these reasons, removing noisy examples from training sets may increase data reliability and quality. In this work, we propose three techniques for noise detection and removal. [2]

For such, initially, in a pre-processing stage, four classifiers based four different ML techniques are induced using the original data sets. The combined predictions of these classifiers are used to guide the noise detection process. Next, three strategies

E. Corchado et al. (Eds.): HAIS 2009, LNAI 5572, pp. 417–424, 2009.

are proposed and evaluated to identify and deal with noisy data: (1) removal; (2) re-classification; (3) hybrid. The new data sets generated in each of the previous cases are then provided to the four ML techniques for the induction of new classifiers.

This paper is organized as follows: Section 2 introduces ML main concepts and the classifiers employed in this work. Section 3 presents the noise handling techniques proposed in this work and discusses the importance of noise detection and removal from a data set. Section 4 presents the data sets used in the experiments performed in order to evaluate the pre-processing technique. Section 5 presents the results achieved in these experiments. Section 6 presents the conclusions from this work.

2 Machine Learning Classifiers

Machine Learning (ML) is a research area within Artificial Intelligence (AI) and Statistics concerned with the automatic acquisition of knowledge from data sets, studying techniques able to improve their performance from experience [3]. One of the main areas of ML is data classification, where classifiers are induced using a training set to assign new, previously unseen examples to their correct class.

The classification techniques employed in this work follows different learning paradigms, presenting distinct bias. This choice was made such that different predictors could be ensured, improving their combination for noise detection. The following ML algorithms were chosen: Support Vector Machines, Artificial Neural Networks, CART and k-nearest neighbour.

Support Vector Machines (SVMs) are based on the Statistical Learning theory. They split data from two classes with a maximal margin hyperplane, which can then be used in the classification of new data [4]. For non-linear classification, data are first mapped to a higher dimension space, becoming linearly separable.

Artificial Neural Networks (ANNs) are composed of simple processing units, simulating the biological neurons, which are named nodes of the ANN [5]. The ANN training consists of adjusting weights of connections between the artificial nodes, aiming to approximate the outputs of the ANN to the desired labels for the training data. The ANN employed in this work has one hidden layer with ten nodes.

CART is a Decision Tree (DT) induction algorithm. A DT is composed of leaf nodes, representing the classes, and by decision nodes, representing tests applied to the values data attributes [6]. The classification of new data is performed by traversing the tree according to the results of the tests until a leaf node is reached.

k-Nearest Neighbour (kNN) is an instance-based technique where the classification function is approximated locally in order to obtain predictions for new data. The k-Nearest Neighbours, in which k is an integer, define the classification of a new example [3]. The value of k used in this work was 1, 3 and 5.

3 Noise

When dealing with Bioinformatics data sets, as those employed in this work, it is necessary to take into account the existence of noisy data in attribute values, including the classes. They correspond to data that suffered negative interference during collection or handling (due to poor calibration in instruments, for example) and do not follow any distribution. The presence of noise in data sets can harm the performance of a classifier, besides increasing its complexity taken for its induction [7]; [8].

Two different combinations of the classifiers results are considered. The first is majority voting, in which an example is noisy if the majority of the predictions made by the combined classifiers is not in accordance to its recorded label. The second considers a consensus of the predictions, by which an example is considered noisy if all classifiers make erroneous classifications for it.

Each learning paradigm must have the same influence in the predictions. Therefore, although kNN is run for different k values (1, 3 and 5), only the most voted output of the three kNN classifiers is considered for comparison against the outputs of SVM, ANN and CART. If there are ties in the classes suggested by the kNN classifiers, the class of lower value is chosen by default.

Since there are four classifiers being combined for noise detection, a data item is considered noisy by consensus only if it is misclassified by all four classifiers, while in majority voting combination a data item is noisy if at least two classifiers make incorrect predictions for it.

After the noise is identified, it is necessary to address what to do with it. This work investigates three approaches for the generation of noise-free data sets. The first approach removes data identified as possible noise from the training data set. The second approach, instead of removing noisy examples, reclassifies them and attributes new labels to these data. The new class is the most predicted by the noise-detection classifiers. The third approach is a hybrid of the previous techniques. If the data item is identified as possible noise, kNN decides whether it should be removed or relabelled.

4   Data Sets

Gene expression data sets were employed in this work, because they usually contain a high amount of noise. The nature of laboratory experiments in Biology is affected by the occurrence of several types of errors. Common examples are contaminations of laboratory samples or errors in the calibration of equipments.

Table 1 summarizes the main characteristics of the data sets used. It shows, for each data set, the number of examples (data items), the number of attributes, the number of classes and their reference number.

Table 1. Data sets main characteristics

Data set	# Examples	# Attributes	# Classes	References
Colon16	62	16	2	[10]
Expgen	207	79	5	[11]
Golub64	72	64	2	[12]
Leukemia	327	271	7	[13]
Lung	197	1000	4	[14]

5 Experiments

The presence of noise in data sets tends to harm the predictive accuracy of ML classifiers. The experiments reported in this section aimed to detect and eliminate noisy data from ML training data sets. It is expected that this procedure will lead to new training data sets that shall reduce the complexity of the induced models, reducing their induction times and also improving their overall accuracy performance.

As described in Section 3, four ML classifiers were combined for the identification of noisy data and for its posterior removal or relabelling. The original and noise-clean data sets were then employed in the induction of classifiers, whose performances were compared to verify whether the pre-processing step was beneficial or not.

The ML algorithms were implemented with the R programming language [R-Devel 2008].

5.1 Methodology

The experiments were divided in two parts: pre-processing and classification. The first step, pre-processing, is where noisy data are detected and removed/relabelled with the use of ML classifiers. In the classification step, the original and pre-processed data sets are used in the induction of classifiers.

To obtain better estimates of the classifiers performance, the 10-fold cross-validation methodology was employed in this work, generating 10 triples of train-validation-test folds.

For each training partition, the four classifiers described in Section 2 are induced and evaluated using the validation sets. New training data sets are generated after noisy removal/relabelling is performed. New classifiers are then be trained using the new training data sets, and their accuracies are again evaluated using the validation folds. If the new performance recorded is better than that obtained previously, the pre-processing cycle is repeated. Pre-processing stops when a performance degradation occurs.

Therefore, noisy data will be continuously eliminated until there is no further improvement in the prediction performance in the validation data sets, when pre-processing is concluded. This process aims to find a quasi-optimum number of examples removed/altered, while also avoiding the elimination of excessive data items, which would harm the data sets representativeness.

The classification step is done for the same four ML techniques used in noise detection. All pre-processed data sets according to the three different approaches from Section 3 are submitted to the ML techniques, besides the original data sets. The mean prediction errors of each classifier before and after noise pre-processing are presented for each data set from Section 4, allowing to compare their performance.

5.2 Results

The number of possible noisy examples identified by pre-processing and their percentage for each data set are presented in Table 2. It can be observed that, for all data sets, consensus voting was too much conservative, identifying noisy data only for data set *colon16* using the noise removal technique. Majority voting was conservative in some data sets, especially *golub64*, nevertheless it identified a high amount of noise in data set *colon16*.

It can also be noticed from Table 2 that the consensus combination for the hybrid technique is not shown. This is justified by the fact that the use of consensus with the hybrid technique results in the same data set than that obtained when using consensus combination with the removal technique. If a noise is identified by consensus, all classifiers made wrong predictions for it. This includes kNN, which would then decide to remove the noisy item, as performed by the removal approach.

Table 2. Number and percentage of noisy instances identified

	Original	Relabelling		Removal		Hybrid
		Majority	Consensus	Majority	Consensus	Majority
colon16	62	7 (11,3%)	0 (0,0%)	8 (11,1%)	4 (5,5%)	7 (11,3%)
expgen	207	6 (2, 9%)	0 (0,0%)	6 (7,7%)	0 (0.0%)	2(0,96%)
golub64	72	1 (1,4%)	0 (0,0%)	1 (1,4%)	0 (0.0%)	1 (1,4%)
leukemia	327	7 (2,1%)	0 (0,0%)	16 (4,9%)	0 (0.0%)	7 (2,1%)
lung	197	13 (6,6%)	0 (0,0%)	14 (7,1%)	0 (0,0%)	9(4,56%)

The average and standard deviation of the error rates obtained by SVM, CART, ANN and kNN (for k=1, 3 and 5) in each data set from Section 4 may be found in tables 3 to 7, respectively. To ease comparisons, the pre-processed data sets for which the classifiers error performance were better compared to those induced using original data are highlighted in italics, while the best error performances for each data set are highlighted in bold (despite being for original or pre-processed data).

We tried 3 values for k: 1, 3 and 5. Since the results were similar, we show only the results for 3-NN.

In data set *colon16* relabelling was most effective than removal and hybrid techniques, since the best results for each classifier belongs to this particular strategy. The good performance of majority voting must also be highlighted.

Table 3. Average error rates of classifiers for data set *colon16*

colon16	Original	Relabelling	Removal		Hybrid
		Majority	Majority	Consensus	Majority
SVM	11,19 +- 7,78	*0 +- 0*	*0+-0*	4,76+-7,69	*1,66 +- 5,27*
CART	24,04 +- 20,43	*6,42 +- 11,44*	9,00+-12,37	13,42+-10,66	*6,42 +- 11,44*
ANN	12,85 +- 13,05	*1,66 +- 5,27*	*1,66+-5,27*	10,42+-9,12	*5,00 +- 8,05*
3-NN	14,28 +- 11,55	*1,42 +- 4,51*	*1,66+-5,27*	8,42+-8,98	*3,09 +- 6,54*

Tables 4 to 7 do not present consensus voting results, since no data sets alterations were obtained with this combination approach for these cases. The results with consensus voting may be considered the same obtained for the original data sets. For now on, the nomenclature pre-processed data will refer to majority voting with either removal, relabelling or hybrid techniques.

Results in Table 4, except from one case, suggest improvements in the performance of the classifiers for pre-processed data. The exception previously mentioned is

Table 4. Average error rates of classifiers for data set *expgen*

expgen	Original	Relabelling	Removal	Hybrid
SVM	6,53 +- 4,78	*4,69 +- 3,67*	***2,33+-4,32***	*4,69 +- 4,16*
CART	10,93 +- 7,96	*10,75 +- 8,63*	***8,35+-7,02***	11,01 +- 7,14
ANN	5,17 +- 3,82	*3,32 +- 3,21*	***2,35+-4,34***	*2,39 +- 2,53*
3-NN	7,15 +- 3,81	*4,64 +- 4,15*	***2,38+-4,36***	*3,28 +- 3,06*

Table 5. Average error rates of classifiers for data set *golub64*

golub64	Original	Relabelling	Removal	Hybrid
SVM	1,25 +- 3,95	***0 +- 0***	***0+-0***	***0 +- 0***
CART	19,28 +- 9,87	***11,19 +- 13,38***	***11,19+-13,38***	***11,19 +- 13,38***
ANN	2,5 +- 5,27	3,09 +- 6,54	*1,42+--4,51*	***0 +- 0***
3-NN	**5,59 +- 7,31**	**5,59 +- 7,31**	**5,59 +- 7,31**	**5,59 +- 7,31**

the average error of CART for the hybrid technique on *expgen* data set. In Table 4 it is also possible to notice that removal technique clearly overcome the remaining approaches, while relabelling and hybrid strategies had similar results.

In Table 5, referent to data set *golub64*, the results of SVM and k-NN for k=3 stand out. Their error rates were the same for all pre-processed data, 0% and *5,59 +- 7,3*, respectively. Besides, for ANN, relabelling pre-processing had a negative impact, showing a higher error rate than that obtained for the original data set. This increase was from 2,5 +- 5,27 for original data to 3,09 +- 6,54 for pre-processed data.

As in previous data sets, in Table 6 there was a prominence of the noise removal technique, although all approaches showed improved results over original data. Noise removal technique had the best results for all classifiers. The similarity between relabelling and hybrid techniques must also be noticed, their performance was the same in three of the four tested models.

Table 6. Average error rates of classifiers for data set leukaemia

leukemia	Original	Relabelling	Removal	Hybrid
SVM	7,26 +- 5,36	*3,92 +- 4,13*	***3,23+-4,12***	*3,92 +- 4,13*
CART	19,27 +- 5,57	*14,98 +- 4,88*	***13,83+-6,15***	*14,98 +- 4,88*
ANN	7,38 +- 3,64	*5,08 +- 4,86*	***4,44+-3,56***	*5,10 +- 3,69*
3-NN	6,98 +- 3,95	*5,11 +- 3,06*	***3,21+-2,10***	*5,11 +- 3,06*

In data set *lung* noise in only one case pre-processing did not improve the classifiers performance. This occurred for CART when submitted to the hybrid preprocessed data. In this case, despite noise elimination, the performance was worst than that obtained for original data. In Table 7 no noise pre-processing technique stood out. Each of the three approaches showed improved results for tested classifiers.

Table 7. Average error rates of classifiers for data set *lung*

lung	Original	Relabelling	Removal	Hybrid
SVM	4,53 +- 4,32	*3,98 +- 5,01*	*0,50+-1,58*	*1,97 +- 2,55*
CART	10,14 +- 7,06	*5,98 +- 5,07*	*5,50+-7,85*	11,62 +- 7,72
ANN	18,35 +- 9,69	*8,03 +- 7,89*	*10,5+-10,32*	*9,56 +-10,23*
3-NN	6,52 +- 4,10	*4,53 +- 3,69*	*2,19+-2,83*	***2,00 +-3,49***

A statistical test for multiple comparisons was performed in order to verify whether the results of the ML classifiers on pre-processed data could indeed be improved when compared to those obtained for original data. Using the Friedman statistical test with the Bonferroni-Dunn post-test at 95% of confidence level [9], the following conclusions could be drawn regarding each pre-processing technique:

- Relabelling: improved the results on data sets *colon16* and *lung*;
- Removal (with majority combination): improved the results for data sets *colon16*, *expgen*, leukemia and *lung*;
- Hybrid: improved the results for data set *lung*.

Therefore, the good results of the pre-processing techniques could be confirmed statistically, with a prominence for data set *lung*, where all pre-processing techniques were beneficial. The removal approach also stood out, being able to improve results in four out of the five data sets employed in the experiments.

6 Conclusion

This work showed the influence of noisy data in the performance of ML classifiers and proposed three techniques for noise pre-processing.

For noise detection, a combination of the classifiers was performed. According to a majority voting or consensus voting, these classifiers decide whether a given data item may be considered noisy or not. Hereafter, the noise identified were either removed or relabelled, with techniques: (1) removal of noisy examples; (2) relabelling of these examples; (3) a hybrid approach, which combines the previous strategies.

Consensus voting was too conservative, finding noisy data only in one data set using the noise-removal technique. Majority voting, on the other hand, identified low levels of noise. Despite that, the obtained results were in general improved for all data sets and classification techniques.

Analysing the noise elimination strategies, in general the noise removal technique was more effective than techniques relabelling and hybrid. However, other parameters must be considered when choosing a particular technique for a given data set, as the characteristics of the problems and the data under consideration.

The results obtained reinforce that the presence of noisy data affects the performance of ML classifiers and how a pre-processing step for noise elimination may benefit data quality and reliability.

Acknowledgments

The authors would like to thank the Brazilian research agencies CAPES.CNPq and FAPESP for their support.

References

1. Zhu, X., Wu, X.: Class noise vs. Attribute noise: A quantitative study of their impacts. Artificial Intelligence Review 22(3), 177–210 (2004)
2. Van Hulse, J.D., Khoshgoftaar, T.M., Huang, H.: The pairwise attibrute noise detection algorithm. Knowl. Inf. Syst. 11(2), 171–190 (2007)
3. Mitchell, T.: Machine Learning. McGraw-Hill, New York (1997)
4. Noble, W.S.: Kernel Methods in Computational Biology. In: Support vector machines applications in computational biology, vol. 3, pp. 71–92. MIT Press, Cambridge (2004)
5. Haykin, S.: Neural Network – A Compreensive foundation, 2nd edn. Prentice-Hall, New Jersey (1999)
6. Breiman, L., Friedman, F., Olshen, R.A., Stone, C.J.: Classification and regression trees. Wadsworth (1984)
7. Verbaeten, S., Assche, A.V.: Ensemble Methods for noise elimination in Classification problems. In: Windeatt, T., Roli, F. (eds.) MCS 2003. LNCS, vol. 2709, pp. 317–325. Springer, Heidelberg (2003)
8. Hodge, V., Austin, J.: A survey of outlier detection methodologies. Artificial Intelligence Review 22, 85–126 (2004)
9. Demsar, J.: Statistical Comparisons of Classifiers over Multiple Data Sets. The Journal of Machine Learning Research 7, 1–30 (2006)
10. Mack, D.H., Tom, E.Y., Mahadev, M., Dong, H., Mittman, M., Dee, S., Levine, A.J., Gingeras, T.R., Lockhart, D.J.: Biology of Tumors. In: Mihich, K., Croce, C. (eds.), pp. 123–131. Plenum, New York (1998)
11. Brown, M., Grundy, W., Lin, D., Christianini, N., Sugnet, C., Haussler, D.: Support Vector Machines Classication of Microarray Gene Expression Data, Technical Report UCSC-CRL 99-09, Department of Computer Science, University California Santa Cruz, Santa Cruz, CA (1999)
12. Dudoit, S., Fridlyand, J., Speed, T.P.: Comparison of discrimination methods for the classication of tumors using gene expression data. Technical Report 576, Department of Statistics, UC Berkeley (2000)
13. Yeoh, E.J., Ross, M.E., Shurtle, S.A., Williams, W.K., Patel, D., Mahfouz, R., Behm, F.G., Raimondi, S.C., Relling, M.V., Patel, A., Cheng, C., Campana, D., Wilkins, D., Zhou, X., Li, J., Liu, H., Pui, C.H., Evans, W.E., Naeve, C., Wong, L., Downing, J.R.: Classification, subtype discovery, and prediction of outcome in pediatric acute lymphoblastic leukemia by gene expression profiling. Cancer Cell 1(2), 133–143 (2002)
14. Monti, S., Tamayo, P., Mesirov, J., Golub, T.: Consensus clustering: A resampling based method for class discovery and visualization of gene expression microarray data. Machine Learning 52(1-2), 91–118 (2003)

SGNG Protein Classifier by Matching 3D Structures

Georgina Mirceva, Andrea Kulakov, and Danco Davcev

Faculty of Electrical Engineering and Information Technologies, Skopje, Macedonia
{georgina,kulak,etfdav}@feit.ukim.edu.mk

Abstract. In this paper, a novel 3D structure-based approach is presented for fast and accurate classification of protein molecules. We have used our voxel and ray based descriptors for feature extraction of protein structures. By using these descriptors, in this paper we propose a novel approach for classifying protein molecules, named Supervised Growing Neural Gas (SGNG). It combines the Growing Neural Gas (GNG) as a hidden layer, and Radial Basis Function (RBF) as an output layer. GNG and its supervised version SGNG have not yet been applied for protein retrieval and classification. Our approach was evaluated according to the SCOP method. The results show that our approach achieves more than 83,5% by using the voxel descriptor and 98,4% classification accuracy by using the ray descriptor, while it is simpler and faster than the SCOP method. We provide some experimental results.

1 Introduction

Functionally important sites of proteins are potentially conserved to specific 3D structural folds. To understand the structure-to-function relationship, life sciences researchers and biologists need to retrieve similar structures from protein databases and classify them into the same protein fold. Therefore, the 3D representation of a residue sequence and the way this sequence folds in the 3D space are very important. The protein structures are stored in the world-wide repository Protein Data Bank (PDB) [1], which is the primary repository for experimentally determined protein structures. With the technology innovation the number of protein structures increases every day, so, retrieving structurally similar proteins using current structural alignment algorithms may take hours or even days. Therefore, improving the efficiency of protein structure retrieval and classification becomes an important research issue.

The SCOP (Structural Classification of Proteins) protein database, describes the structural and evolutionary relationships between proteins [2]. Since the existing tools for the comparing secondary structure elements cannot guarantee 100 percent success in the identification of protein structures, SCOP uses experts' experience to carry out this task. Evolutionary relationship of the proteins is presented as they are classified in hierarchical manner. The main levels of the hierarchy are "Family" (based on the proteins' evolutionary relationships), "Superfamily" (based on some common structural characteristics), and "Fold"

E. Corchado et al. (Eds.): HAIS 2009, LNAI 5572, pp. 425–432, 2009.

(based on secondary structure elements). The deepest level at the hierarchy is the domain, so if we can predict the domain, we know all upper levels. There are four main structural classes of proteins: a, b, a/b, and a+b.

Also CATH and FSSP/DALI methods provide classification of protein molecules. All these sophisticated methods for protein classification are very time consuming, especially SCOP where most of the phases are made manually. That's why the number of released proteins in PDB, but still not classified by SCOP, increases every day, as it is shown on Fig. 1. Therefore, a need for fast and accurate methods for protein classification is obvious.

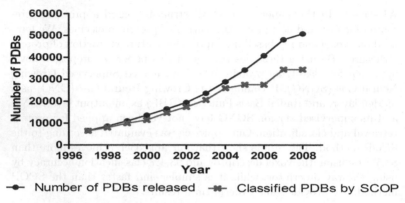

Fig. 1. Number of proteins released in PDB versus number of proteins classified by SCOP

There are many classification algorithms that can be used for protein classification as Naive Bayesian classifier, nearest neighbour classifier, decision trees, neural networks, Support vector machines and so on.

In [3], nine different protein classification methods are used: the profile-HMM, support vector machines (SVMs) with four different kernel functions, SVM-pair wise, SVM-Fisher, decision trees and boosted decision trees.

Nearest neighbour classifier used in [4] can also be applied, but in this way, the unclassified protein is compared with all proteins in the database that can last too long. This can be avoided by using Selforganizing Maps (SOMs) or similar classification methods or neural networks like Growing Neural Gas - GNG. First, we would like to state the properties shared by all the models of such neural networks. The network structure is a graph consisting of a set of nodes and a set of edges connecting the nodes. Each node has an associated position (reference vector) in the input space. Adaptation, during learning, is done by moving the position of the nearest (winning) node and its topological neighbours in the graph toward the input signal. We can differentiate Static SOMs and Growing SOMs. The Static SOMs have a predefined structure which is chosen a priori and does not change during the parameter adaptation. Growing SOMs, however, have no predefined structure, this is generated by successive additions (and possibly deletions) of nodes and edges, thus can be better adapted to the distribution of the training data. In [5] some neural networks of this type are presented. Some of them use Competitive Hebbian Learning (CHL) [6] to build up a topology during selforganization. The principle of CHL is simply to create an edge between the winning and the second winning node at each adaptation step.

The Growing Neural Gas (GNG) model [7] does not impose any explicit constraints on the graph topology. The graph is generated and updated by CHL. The topology of a GNG network reflects the topology of the input data distribution.

In this paper we present a system for classification of protein structures. First, we extract some features of the protein structures, by applying our voxel and ray based descriptors given in ([8], [9]). After proper positioning of the structures, the Spherical Trace Transform is applied to them to produce descriptor vectors, which are rotation invariant. Additionally, some properties of the primary and secondary structure of the protein are taken, forming better integrated descriptor. Also, adapted ray based descriptor applied on the interpolated backbone of the protein was extracted.

There are many algorithms used for protein classification as Naive Bayesian classifier, nearest neighbour classifier, decision trees, support vector machines, and so on. In this paper, we have used Supervised Growing Neural Gas (SGNG) [10] on the previously extracted descriptors. The evaluation of our classification algorithm is made according to the SCOP method.

Our approach is given in section 2; section 3 gives some experimental results. Section 4 concludes the paper and gives some future work directions.

2   Our Approach

In this paper, we present an accurate and fast system that allows the users to classify protein structures. The information about protein structure is stored in PDB files which contain information about primary, secondary and tertiary structure of the proteins. First, we extract some properties of the protein structure, thus forming our voxel and ray based descriptors. Then, we use SGNG in order to classify each newly protein in corresponding protein domain in the SCOP hierarchy.

2.1   Voxel Based Descriptor

We will use our voxel based descriptor presented in ([8], [9]). First, we extract the features of the tertiary structure of the protein by using the voxel based algorithm for 3D object retrieval [11]. Since the exact 3D position of each atom and its radius are known, it may be represented by a sphere. The surface of each sphere is triangulated. In this way, a sphere consists of a small set of vertices and a set of connections between the vertices. After triangulation, we perform Voxelization. Voxelization transforms the continuous 3D space, into the discrete 3D voxel space. First, we divide the continuous 3D space into voxels. Depending on positions of the polygons of a 3D mesh model, to each voxel a value is attributed equal to the fraction of the total surface area of the mesh which is inside the voxel. The information contained in a voxel grid is processed further to obtain both correlated information and more compact representation of voxel attributes as a feature. We applied the 3D Discrete Fourier Transform (3D-DFT) to obtain a spectral domain feature vector which provides rotation invariance of the descriptors. Since the symmetry is present among obtained coefficients, the feature vector is formed from all non-symmetrical coefficients. This vector presents geometrical properties of the protein.

Additionally, some attributes of the primary and secondary structure of the protein molecules are extracted, forming attribute-based descriptor vectors as in [12]. By incorporating the features of primary and secondary structure of the protein, we get better integrated descriptor.

2.2 Ray Based Descriptor

Calpha atoms form the backbone of the protein molecule. There are some residues that hang up on the Calpha atoms, but are not important in the classification. The analyses showed that by taking into account only the Calpha atoms of the protein and extracting some suitable descriptor, we can get better results. According to this, in our ray descriptor presented in [9], we approximated the backbone of the protein with fixed number of points, which are equidistant along the backbone, in order to represent all proteins with descriptors with same length, although they have distinct number of Calpha atoms. Finally, the elements of the descriptor were calculated as distances from approximated points to the centre of mass.

2.3 Supervised Growing Neural Gas (SGNG)

Supervised Growing Neural Gas (SGNG) is described by Bernd Fritzke as a new algorithm for constructing RBF networks. It combines the Growing Neural Gas (GNG) [7] as a hidden layer, and RBF [13] as an output layer (which contains separate output node for each class). The SGNG network topology is shown on Fig. 2.

Assume we want to classify n-dimensional vectors into M classes. The output layer will have one output node for each class, so, the output node with the greatest value is the only one considered in each response from the RBF network. Training is done by presenting pairs of input and expected output vectors. As in GNG, we start with two randomly positioned nodes connected by an edge. The edge has no weight, since it is not part of the actual RBF network, but represents the fact that two nodes are neighbours. The neighbour information is maintained in the same manner as in GNG, by application of Competitive Hebbian learning (CHL). The adaptation of the hidden nodes is also performed as in GNG, the winner node s is moved some fraction of the distance to the input, and the neighbours of s are moved an even smaller fraction of their distance to the input.

Let the input vector be denoted by x, desired output vector denoted by d, the output vector of the output layer is denoted as y. Let σ_j be the standard

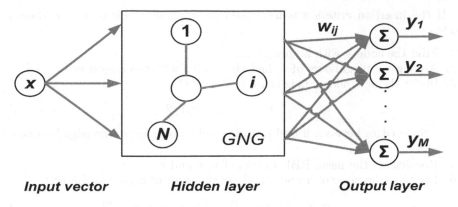

Fig. 2. Supervised Growing Neural Gas

deviation, or width, of the RBF node j. Let z_i be the output from the hidden node i, and w_{ij} the weight from hidden node i to output node j. Let η be the learning rate.

First, we create two randomly positioned hidden nodes, connected by an edge with $age = 0$ and set their errors to 0.

Generate an input vector x conforming to some distribution, with corresponding desired output vector d. Locate the two nodes s and t nearest to x, with reference vectors w_s and w_t. Evaluate the net using the input vector x. Adjust the output layer weights by applying the delta rule (1).

$$w_{ij} = w_{ij} + \eta(d_i - y_i)z_j \qquad (1)$$

The winning node s must update its local error, so we add the distance between actual output and target.

$$error_s = error_s + dist(d, y) \qquad (2)$$

Move s and its neighbours towards x by fractions e_w and e_n of the distance.

$$w_s = w_s + e_w(x - w_s) \qquad (3)$$

$$w_i = w_i + e_n(x - w_i), \forall i \in Neighbour(s) \qquad (4)$$

For each node j that was just moved, set the width of the RBF to the mean of the distances between j and all the neighbours of j.

Increment the age of all edges from node s to its topological neighbours. If s and t are connected by an edge, then reset the age of that edge. If they are not connected, then create an edge with $age = 0$ between them. If there are any edges with an age larger than $agemax$, then remove them. If, after this, there are nodes with no edges, then remove them. Recalculate the mean RBF widths of the affected nodes.

If the insertion criterion is met, then insertion of a new node r is done as follows:

1. Find the node u with largest error.
2. Among the neighbours of u, find the node v with the largest error.
3. Insert the new node r between u and v by (5):

$$w_r = (w_u + w_v)/2 \tag{5}$$

4. Create edges between u and r, and v and r and remove the edge between u and v.
5. Recalculate the mean RBF widths of u, v and r.
6. Decrease the error of u and v and set the error of node r as follows:

$$error_u = error_u/2 \qquad error_v = error_v/2 \tag{6}$$

$$error_r = (error_u + error_v)/2 \tag{7}$$

7. Initialise the weights from node r to all nodes in the output layer. The newly inserted node r receives the interpolated position of its "parents" and the parents errors are reduced by 50%, see (5), (6) and (7). The purpose is to prevent the next insertion to occur in the same place. If more nodes need to be inserted in this region, the errors will reflect this. Decrease all error variables of all nodes j by a factor $errordecay$. This gives recently measured errors greater influence than older ones.

$$error_j = error_j * (1 - errordecay) \tag{8}$$

If the stopping criterion is not met, then repeat the process.

By observing the local accumulated squared errors, we can identify nodes that exist in regions of input space where many misclassifications occur. It is logical to assume that nodes with large local accumulated errors received patterns from different classes since if they did not, the delta rule would have successfully modified the output layer weights and the local accumulated errors would not have grown so high. With this argument in mind, it would seem sensible to use the local error information when deciding possible locations of new nodes.

The criteria for inserting new nodes could be at a fixed number of iterations. A better method would be to observe the mean squared error per pattern or some independent validation set. If the squared error stops decreasing, can be interpreted as the node movements not being able to adjust sufficiently to lower the error, this in turn means that for the current network size this is as good as it can and that it is time to insert a new node. In this implementation we have used the first insertion criteria.

The stop criteria could be defined as a maximum size that the network may reach, but it requires knowledge about the distribution that we might not have. Another method is by defining a maximum error allowed, and train until some threshold is reached. Yet, another approach would be to have an upper limit on the allowed number of misclassifications (since the number of misclassifications converges towards lower values much faster than the error). In this implementation we have used the last stop criteria.

3   Experimental Results

We have implemented a system for protein classification. We have used part of the SCOP 1.73 database. We will examine the classification accuracy by using the voxel and ray based descriptors. We have trained two SGNG networks (for the voxel and ray based descriptor) with 1147 proteins from 30 SCOP domains. The classification accuracy was tested with 127 proteins from 30 SCOP domains. In Table 1 we provide some details about the classification results.

Table 1. Experimental results of our SGNG classifier

Descriptor	Descriptor length	Number of GNG nodes	Accuracy (%)	Classification time for all test proteins (*sec*)
Voxel based	450	163	83.5	3
Ray based	64	78	98.4	0.5

As it can be seen, by using the ray based descriptor we get better classification accuracy. This is due to the fact that proteins that belong to same domain have very close ray based descriptors, and they are far enough from the descriptors of the proteins from other domains. Also time taken for classification is smaller due to smaller number of nodes in the classifier and lower dimensionality of the descriptor. SGNG is faster than k nearest neighbor classifier because in this approach we compare the descriptor of the query protein only with 78/163 (GNG has 78 nodes by using the ray descriptor, and 163 nodes by using the voxel descriptor) vectors, against 1147 in the nearest neighbors case.

4   Conclusion

We have presented a system for protein molecules classification by using information about their primary, secondary and tertiary structure. We have applied the voxel based method for generating geometry descriptor. Additionally, some properties of the primary and secondary structure of the protein were taken, thus forming better integrated descriptor. We have also adapted the ray based descriptor and applied to the interpolated backbone of the protein molecule, forming another descriptor. Supervised Growing Neural Gas was used as classifier of the protein molecules. The SCOP database was used to evaluate the classification accuracy of the system. The results showed that this approach achieves more than 83,5% classification accuracy by using the voxel descriptor and 98,4% classification accuracy by using the ray descriptor, while it is simpler and faster (takes seconds) than the SCOP method (hours, days).

Our future work will be concentrated on increasing the efficiency of the algorithm by investigating new 3D descriptors. We can also incorporate additionally features of the protein structures in the descriptors. Our SGNG classifier can be used for hierarchical classification, by training separate SGNG network for each parent in the SCOP hierarchy.

References

1. Berman, H.M., Westbrook, J., Feng, Z., Gilliland, G., Bhat, T.N., Weissig, H., Shindyalov, I.N., Bourne, P.E.: The Protein Data Bank. Nucleic Acids Research 28, 235–242 (2000)
2. Murzin, A.G., Brenner, S.E., Hubbard, T., Chothia, C.: Scop: A Structural Classification of Proteins Database for the Investigation of Sequences and Structures. J. Mol. Biol. 247, 536–540 (1995)
3. Pooja Khati: Comparative analysis of protein classification methods. Master Thesis. University of Nebraska, Lincoln (2004)
4. Ankerst, M., Kastenmuller, G., Kriegel, H.P., Seidl, T.: Nearest Neighbor Classification in 3D Protein Databases. In: Proc. Seventh Int'l Conf. Intelligent Systems for Molecular Biology (ISMB 1999) (1999)
5. Prudent, Y., Ennaji, A.: A K Nearest Classifier design. Electronic Letters on Computer Vision and Image Analysis 5(2), 58–71 (2005)
6. Martinetz, T.: Competitive Hebbian learning rule forms perfectly topology preserving maps. In: Proc. ICANN 1993, pp. 427–434. Springer, Heidelberg (1993)
7. Fritzke, B.: A growing neural gas network learns topologies. Advances in Neural Information Processing Systems 7, 625–632 (1995)
8. Kalajdziski, S., Mirceva, G., Trivodaliev, K., Davcev, D.: Protein Classification by Matching 3D Structures. In: Proceeding of IEEE Frontiers in the Convergence of Bioscience and Information Technologies, Jeju Island, Korea, pp. 147–152 (2007)
9. Mirceva, G., Kalajdziski, S., Trivodaliev, K., Davcev, D.: Comparative analysis of three efficient approaches for retrieving protein 3D structures. In: IEEE CD Proc. of 4-th Cairo International Biomedical Engineering Conference (CIBEC 2008), Cairo, Egypt (2008),
   ftp://www.bioinf.nileu.edu.eg/pub/CIBEC08/CIBEC08/BF/
   CIBEC'08_BF_29.pdf
10. Holmström, J.: Growing Neural Gas, Experiments with GNG, GNG with Utility and Supervised GNG. Master Thesis. Uppsala University, Uppsala, Sweden (2002)
11. Vranic, D.V.: 3D Model Retrieval. Ph.D. Thesis. University of Leipzig (2004)
12. Daras, P., Zarpalas, D., Axenopoulos, A., Tzovaras, D., Strintzis, M.G.: Three-Dimensional Shape-Structure Comparison Method for Protein Classification. IEEE/ACM Transactions on computational biology and bioinformatics 3(3), 193–207 (2006)
13. Chen, S., Cowan, C.F., Grant, P.M.: Orthogonal least squares learning algorithms for radial basis function networks. IEEE Trans. Neural Networks 2, 302–309 (1991)

Memetic Pareto Differential Evolution for Designing Artificial Neural Networks in Multiclassification Problems Using Cross-Entropy Versus Sensitivity

Juan Carlos Fernández[1], César Hervás[1], Francisco José Martínez[2],
Pedro Antonio Gutiérrez[1], and Manuel Cruz[1]

[1] Department of Computer Science and Numerical Analysis of the University of Cordoba,
Campus de Rabanales, 14071, Cordoba, Spain
fernandezcaballero@gmail.com, {chervas,i02gupep,i42crram}@uco.es
[2] Department of Management and Quantitative Methods, ETEA, Escritor Castilla Aguayo 4,
14005, Cordoba, Spain
fjmestud@etea.com

Abstract. This work proposes a Multiobjective Differential Evolution algorithm based on dominance Pareto concept for multiclassification problems using multilayer perceptron neural network models. The algorithm include a local search procedure and optimizes two conflicting objectives of multiclassifiers, a high correct classification rate and a high classification rate for each class, of which the latter is not usually optimized in classification. Once the Pareto front is built, we use two automatic selection methodologies of individuals: the best model with respect to accuracy and the best model with respect to sensitivity (extremes in the Pareto front). These strategies are applied to solve six classification benchmark problems obtained from the UCI repository. The models obtained show a high accuracy and a high classification rate for each class.

Keywords: Accuracy, Differential Evolution, Local Search, Multiclassification, Multiobjective, Neural Networks, Pareto, Sensitivity.

1 Introduction

Pattern classification occurs when an object needs to be assigned into a predefined class based on a number of observed attributes related to that object. Different methods for pattern classification [1] are shown in the literature, but in recent years Artificial Neural Networks (ANNs) have been an important tool for it [2].

Training Artificial Neural Networks by Evolutionary Pareto-based algorithms [3] is known as Multiobjective Evolutionary Artificial Neural Networks (MOEANNs), and it has been used in recent years to solve classification tasks, having some of its main exponents in H. Abbass [4] and Y. Jin [3].

In this paper we present a Memetic Pareto Differential Evolution (MPDE) algorithm, which is, a MultiObjective Evolutionary Algorithm (MOEA) [5] based on Differential Evolution (DE) [6] and on the Pareto dominance concept for solving multiclass classification problems. MPDE is improved with a local search algorithm, specifically with the *improved Resilient Backpropagation* (*iRprop+*) [7].

E. Corchado et al. (Eds.): HAIS 2009, LNAI 5572, pp. 433–441, 2009.

Many techniques to improve the overall generalization capability for the classifier designed have been proposed, but a few maintain the classification capability in all classes (correctly classified rate per class), something that, in some datasets, is essential to ensure the benefits of a classifier against another. The objective pursued when using MOEAs in classifications with ANNs is mainly designing classifiers with the biggest possible accuracy and with a small structural complexity [3], [8]. Our proposal aims to achieve a high classification rate in the testing dataset with a good classification for each class. There are multi-objective works for classification that optimize the Accuraccy and the Sensitivity or Specificity, but only work with two classes or compare one of the classes with the rest.

The rest of the paper is organized as follows: in section 2 the accuracy and sensitivity measures are proposed and their properties are briefly discussed. Section 3 presents a brief overview about DE in Multiobjective Evolutionary Neural Networks. Section 4 describes the MPDE algorithm. Section 5 shows the experimental design, and finally the conclusions are drawn in Section 6.

2   Accuracy Versus Sensitivity

Accuracy cannot capture all the different behavioral aspects found in two different classifiers [9] so, in this section, we present two measures to evaluate a classifier:

- Accuracy $C = (1/N)\sum_{j=1}^{Q} n_{jj}$, that represents the number of times that the patterns are correctly predicted by a classifier with Q classes and N training or testing patterns, and where n_{jj} is the number of patterns from class $j\text{-}th$ that are correctly classified.

- Minimum Sensitivity $S = \min\{S_i; i = 1,...,Q\}$ given by the minimum of the classification rate per class, where $S_i = n_{ii}/\sum_{j=1}^{Q} n_{ij}$, being $\sum_{i,j=1}^{Q} n_{ij} = N$. Henceforth when we talk about sensitivity, we refer to the minimal sensitivity of all classes.

Assuming that all misclassifications are equally costly and there is no profit for a correct classification, we understand that a good classifier should obtain a high accuracy level as well as an acceptable level for each class, the two-dimensional measure (S,C) is considered in this work for this reason.

Let us consider a Q –class classification problem. Let C and S be respectively the accuracy and the sensitivity associated with a classifier g , then $S \leq C \leq 1-(1-S)p^*$, where p^* is the minimum of the estimated prior probabilities. Therefore, each classifier will be represented as a point in the triangular region in Fig. 1 part B. Simultaneously minimize the $Q(Q-1)$ misclassification rates given by the off-diagonal elements of the confusion matrix has a main shortcoming, the dimension of the Pareto optimal front grows at the rate of the square of the number of classes, making the resolution of the problem extremely complex.

The feasible region within the (S,C) space is reduced considerably as we approach to the $(1,1)$ point; not taking the Pareto front obtained by multiobjective techniques a great diversity in terms of number of elements. It should be noted that for a fixed value of Accuracy C, a classifier will be better when it corresponds to a point nearer to the diagonal of the square. In general, accuracy and sensitivity could be cooperative, but as we approach the $(1,1)$ point or optimum, the objectives become competitive and an improvement in one objective tends to involve a decrease in the other one, which justifies the use of a MOEA.

3 Differential Evolution in Multiobjective Evolutionary Artificial Neural Networks

A particular and simple yet powerful Evolutionary Algorithm (EA) that has been used for multiobjective optimization on ANNs is the Differential Evolution (DE) algorithm proposed by Price and Storn [6]. The main idea in DE with respect to EAs is to use vector differences in the creation of new candidate solutions $C[i]$ as one of the i elements in a population of size N. All applications of DE are distinguished by the strategy used to create and insert new individuals in the population and by the self-adaptation of the parameters of crossover and mutation [10].

DE is used in the literature for multiobjective optimization and applications, and to a lesser extent, for the design of ANNs in classification. DE works well when the objective function has features such as nonlinearity, high dimensionality, the existence of multiple local optimal, undifferentiated or noise. For these reasons and because the article by Abbass [4] has been widely cited and used we have done an improved version of their algorithm.

To the best of our knowledge, sensitivity is nowhere used for improving the capability of generalization, quality and comparison between classifiers. Abbass [4] was one of the first authors in apply DE in Multiobjective Problems with ANNs and, in several works, he employs DE with/within local search procedures to create new individuals and to keep only the nondominated ones as the basis for the next generation, but the objectives to optimize are the accuracy and the net complexity. Ning [11] uses a Modified Differential Evolution algorithm introducing the reorganization of Evolution Strategies during the mutation and optimizing the weights of the feed-forward multilayer neural network, but only uses the mean square error as objective function.

4 The Memetic Pareto Multiobjective Evolutionary Differential Evolution Algorithm (MPMEDE)

4.1 Base Classifier Framework and Objective Functions

We consider standard feed forward MLP neural networks with one input layer with k inputs variables of the problem, one hidden layer with m maximum sigmoidal basis functions which depends on the problem, and one linear output layer with J outputs, one for each class in the problem. In this way, the functional model considered is the following: $f_l(\mathbf{x}, \boldsymbol{\theta}_l) = \beta_0^l + \sum_{j=1}^{M} \beta_j^l \sigma_j(\mathbf{x}, \mathbf{w}_j)$, $l = 1, 2, ..., J$, where $\boldsymbol{\theta} = (\boldsymbol{\theta}_1, ..., \boldsymbol{\theta}_J)^T$ is the transpose matrix containing all the neural net weights, $\boldsymbol{\theta}_l = (\beta_0^l, \beta_1^l, ..., \beta_M^l, \mathbf{w}_1, ..., \mathbf{w}_M)$ is the vector of weights of the l output node, $\mathbf{w}_j = (w_{1j}, ..., w_{Kj})$ is the vector of weights of the connections between input layer and the j hidden node, $\mathbf{x}$ is the input pattern and s the sigmoidal basis function.

We interpret the outputs of neurons on the output layer from a probability point of view, which considers the softmax activation function given by the following expression: $g_l(\mathbf{x}, \boldsymbol{\theta}_l) = \dfrac{\exp f_l(\mathbf{x}, \boldsymbol{\theta}_l)}{\sum_{l=1}^{J} \exp f_l(\mathbf{x}, \boldsymbol{\theta}_l)}$, $l = 1, 2, ..., J$, where $g_l(\mathbf{x}, \boldsymbol{\theta}_l)$ represents the probability of pattern $\mathbf{x}$ belonging to class j. Taking this consideration into account, it can be seen that the class predicted by the neuron net corresponds to the neuron on the output layer whose output value is the greatest.

In this multiobjective context we consider two multiobjective functions to maximize, where the first function is cross-entropy error and is given by the following expression for J classes: $l(\boldsymbol{\theta}) = -\dfrac{1}{N} \sum_{n=1}^{N} \sum_{l=1}^{J} y_n^{(l)} \log g_l(x_n, \boldsymbol{\theta}_l)$, where $\boldsymbol{\theta} = (\boldsymbol{\theta}_1, ..., \boldsymbol{\theta}_J)$. The advantage of using the error function $l(\boldsymbol{\theta})$ instead of $(1-C)$ is that it is a continuous function, then small changes in network parameters produce small changes in the fitness function, which allows improve the convergence. Then, the first fitness measure to maximize is a strictly decreasing transformation of the entropy error $l(\boldsymbol{\theta})$ given by $A(g) = \dfrac{1}{1+l(\boldsymbol{\theta})}$, where g is a sigmoidal basis function neural network model represented by the multivaluated function $g(\mathbf{x}, \boldsymbol{\theta}) = (g_1(\mathbf{x}, \boldsymbol{\theta}_1), ..., g_J(\mathbf{x}, \boldsymbol{\theta}_J))$. The second objective to maximize is the sensitivity $S(g)$ of the classifier as the minimum value of the sensitivities for each class. Both $A(g)$ and $S(g)$ fitness functions, are necessary for the evaluation of the individuals in Algorithm 1 (see step 3).

4.2 MPDE Algorithm

In Algorithm 1 we describe our Memetic Pareto Differential Evolution (MPDE) algorithm. The approach evolves architectures and connection weights simultaneously, each individual being a fully specified ANN. The ANNs are represented using an object-oriented approach and the algorithm deals directly with the ANN phenotype. The fundamental characteristics are the following:

- The maximum number of non-dominated solutions in each generation was set to $(populationSize / 2)$. If it is exceeded, a nearest neighbor distance function [12] is adopted by preventing a agglomerative structure of the Pareto front (step 9-13).
- *Crossover* operator is proposed from step 17, where three parents have been previously selected randomly; being the child a perturbation of the main parent. First, with some probability P_c for each hidden neuron, h, if the probability is met, the

---

Algorithm 1. Memetic Multiobjective Differential Evolution (MPDE)

1: Create a random initial population of potential solutions.
2: **Repeat**
3:      Evaluate the individuals in the population and label those who are non-dominated.
4:      **If** the number of non-dominated individuals is less than 3 **then**
5:          **Repeat**
6:              Find a non-dominated solution among those who are not labeled.
7:              Label the solution as non-dominated.
8:          **Until** the number of non-dominated individuals is greater than or equal to 3.
9:      **Else If** number of non–dominated solutions is greater than ($populationSize$ / 2) **then**
10:          **Repeat**
11:              Calculate the distance of each individual with its nearest neighbor.
12:              Delete the individual with smaller distance.
13:          **Until** the number of non-dominated individuals is equal to ($populationSize$ / 2) .
14:      Delete all dominated solutions from the population.
15:      **Repeat**
16:          Select at random an individual as the main parent $a1$ and two individuals, $a2$, $a3$ as supporting parents.
17:          **Crossover:** with a crossover probability P_c for each hidden neuron, do
18:          $$\rho_h^{child} \leftarrow \begin{cases} 1 & if\left(\rho_h^{a1} + N\left(0,1\right)\left(\rho_h^{a2} - \rho_h^{a3}\right)\right) \geq 0.5 \\ 0 & otherwise \end{cases}$$
19:          $$w_{ih}^{child} \neg w_{ih}^{a1} + N\left(0,1\right)\left(w_{ih}^{a2} - w_{ih}^{a3}\right)$$
20:          otherwise
21:          $$w_{ih}^{child} \neg w_{ih}^{a1}$$
22:          $$r_h^{child} \neg r_h^{a1}$$
23:          and with crossover probability P_c for each output neuron, do
24:          $$w_{ho}^{child} \neg w_{ho}^{a1} + N\left(0,1\right)\left(w_{ho}^{a2} - w_{ho}^{a3}\right)$$
25:          otherwise
26:          $$w_{ho}^{child} \neg w_{ho}^{a1}$$
27:          **If** the child is equal to the main parent **then**
28:              A random link is perturbed by adding a Gaussian distribution $N\left(0,1\right)$.
29:          **Mutation:** with a mutation probability P_m for each neuron do
30:          $$\rho_h^{child} \leftarrow \begin{cases} 1 & if \rho_h^{child} = 0 \\ 0 & otherwise \end{cases}$$
31:          A child has been created. Store the best child so far. NumCreated $\neg$ NumCreated + 1
32:          **If** the candidate dominates the parent **then**
33:              Apply $iRprop+$ local search to the child.
34:              Add the candidate to the population.
35:          **Else If** there is no dominance relation between main parent and child **then**
                Add the candidate to the population.
36:          **Else If** NumCreated = 100 (here the main parent dominates to the child) **then**
37:              Add the best of these 100 children to the population.
38:              NumCreated $\neg$ 0
39:          **Else**   The candidate is discarded and to go to step 15 (No child is added).
40:      **Until** the population size is N.
41: **Until** termination conditions are satisfied, go to 2 above.

---

neuron selected in the child will be maintained ($r = 1$) or deleted ($r = 0$), depending of the value of the expression that is shown in step 18. In the first case the weight w_{ih} between the *i-th* input variable and the *h-th* hidden node will be modified by the expression proposed in step 19. If the crossover probability is not met then the structure of the main parent is inherited by the child (steps 21-22). Third, a similar weight modification is reached with a P_c probability for each output neuron, o , in the output layer (steps 23-26).

- *Mutation* operator consists on adding or deleting neurons in the hidden layer depending on a P_m probability for each them. Taking into account the maximum number of hidden neurons that may exist in an individual in a specific problem, the probability will be used as many times as number of neuron has the classifier. If the neuron exists, is deleted, but if it does not exist, then it is created and the weights are established randomly, see step 29.
- Local search, steps 32-34, has been carried out based in the adaptation of a version of the *Resilient Backpropagation (Rprop)*, the improved Rprop or *iRprop+* [7]. The adaptation is made using a backtracking strategy to the *softmax* activation function and to the cross-entropy error function, modifying the gradient vector. The local search is applied only to the child that dominates to the main parent, after the crossover and mutation have been applied, decreasing in this way the computational cost. Other works perform this operation for each child created before checking if the child dominates or not to the main parent.
- There are significant differences with the Abbass' algorithm proposed in [4]. First, in the crossover we used a P_c probability for each neuron and not for each layer as Abbass does, being our algorithm less aggressive with the changes in the ANNs. The mutator probability also is used in independent way for each neuron and not for the hidden layer; because we believe that the changes proposed by Abbass produce such drastic changes in the ANNs, in which their generalization capability can be reduced significantly. Third, the way in which individuals are added to the population, Abbass adds to the population only those children who dominate the main parent and this decision may leave the algorithm running between the steps 15-40 for a long time, because when the number of generations increase is more difficult to improve the main parent. In our case, the way individuals are added in steps 31-39 is more relaxed, so children that dominates or not to the main parent can be added. In this way the computational time is reduced.

5 Experiments

For the experimental design we consider 6 datasets taken from the UCI repository [16], Autos, Balance, Breast-Cancer, Newthyroid, Pima and HeartStatlog, with 6, 3, 2, 3, 2 and 2 classes respectively. The design was conducted using a stratified holdout procedure with 30 runs, where 75% of the patterns were randomly selected for the training set and the remaining 25% for the test set. The population size is established to $N_p = 100$. The crossover probability is established to 0.8 and the mutation probability to 0.1. For *iRprop+* the adopted parameters are $h^- = 0.5$, $h^+ = 1.2$, $D_0 = 0.0125$ (the initial value of the D_{ij}), $D_{max} = 50$, $D_{min} = 0$ and *Epochs* = 5.

Once the Pareto front is built in each run we use two automatic selection methodologies of individuals: First, the extreme values in training are chosen, that is, the best individual on Entropy, EI, and the best individual on Sensitivity, SI (see Fig 1 A). Once this is done, we get the values of Accuracy C and Sensitivity S in testing of EI, $EI_{testing} = (C_{EI\_testing}, S_{EI\_testing})$ and SI, $SI_{testing} = (C_{SI\_testing}, S_{SI\_testing})$. This is repeated for each run and the average and standard deviation from the EI and SI individuals are estimated obtaining $\overline{EI}_{testing} = (\overline{C}_{EI\_testing}, \overline{S}_{EI\_testing})$ and $\overline{SI}_{testing} = (\overline{C}_{SI\_testing}, \overline{S}_{SI\_testing})$. Therefore, the first expression $\overline{EI}_{testing}$ is the average obtained taking into account the Entropy as primary objective when we choose an individual from the first Pareto front, and the second $\overline{SI}_{testing}$ taking into account the Sensitivity, getting two automatic methodologies called MPDE-E and MPDE-S respectively.

We compare our algorithm with a modified and memetic version of NSGA2 (for details see [13]), which we also have used for designing ANNs models in the same framework shown in this work, using *iRprop+* and a mutation operator, although other implementations can be found in the framework Paradiseo-MOEO [14] . Also, we compare with the SVM methodology from the SMO algorithm with the defaults values that provides Weka [15].

In Table 1 we present the values of the average and the standard deviation for C and S obtained for the best models in each run over the testing set. We can observe that in Balance and Breast-Cancer, MPDE-S obtains the best values in S, and very close to those modified NSGA2 in C. In Autos, the best result in C is obtained by MPDE-E but the best value in S is achieved by MNSGA2-S. In Newthyroid MPDE obtains the best values in S and C, and in Pima and HeartStatlog, MPDE-S obtains the best values in S and very similar to those obtain by MNSGA2-E in C.

Table 1. Statistical results for MPDE and the modified NSGA2 version, MNSGA2, in testing. In **bold** the best result and in *Italic* the second best result.

Dataset	Algorithm	C(%)	S(%)	Dataset	Algorithm	C(%)	S(%)
	MPDE-E	**68.79±5.59**	28.75±21.40		MPDE-E	91.43±1.01	54.36±26.25
	MNSGA2-E	*66.67±4.07*	*39.64±14.92*		MNSGA2-E	**94.01±1.52**	42.66±17.00
Autos	MPDE-S	64.15±5.63	12.26±20.54	Balance	MPDE-S	91.41±1.53	**87.42±4.32**
	MNSGA2-S	66.04±4.78	**42.28±10.98**		MNSGA2-S	*92.47±2.16*	*83.72±8.19*
	SVM	67.92	0.00		SVM	88.46	0.00
	MPDE-E	67.27±2.71	38.09±11.59		MPDE-E	*96.66±2.02*	*81.42±10.74*
	MNSGA2-E	**69.34±2.30**	28.88±9.09		MNSGA2-E	95.12±2.30	74.81±10.07
BreastC	MPDE-S	65.39±3.40	*57.04±7.01*	Newthy	MPDE-S	**96.66±1.84**	**81.64±9.76**
	MNSGA2-S	63.99±3.10	*53.08±6.57*		MNSGA2-S	95.55±2.15	75.07±10.66
	SVM	64.79	23.81		SVM	88.89	55.56
	MPDE-E	*78.59±1.59*	61.94±4.10		MPDE-E	76.17±1.41	61.11±2.20
	MNSGA2-E	**78.99±1.80**	60.44±2.59		MNSGA2-E	**78.28±1.75**	61.88±2.08
Pima	MPDE-S	77.11±2.20	**73.12±2.98**	HeartStlg	MPDE-S	*76.27±1.57*	**63.66±2.37**
	MNSGA2-S	76.96±2.08	*72.68±3.06*		MNSGA2-S	77.5±1.73	*62.66±2.38*
	SVM	78.13	50.75		SVM	76.47	60.00

In Fig. 1 we can see the results obtained by MPDE for Balance dataset in the (S,C) space in one specific run, which presents the best individual on Entropy in training. Observe (Fig 1. A) that the (S,C) values do not form Pareto fronts in testing (Fig 1. B), and the individuals which in the training graphics were in the first Pareto front, can now be located within the (S,C) space in a worst region, since there is no direct relation between training Entropy and testing Accuracy C.

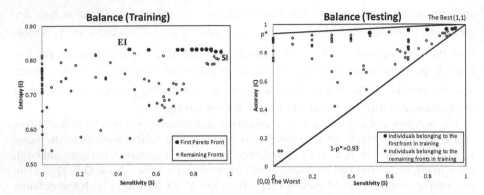

Fig. 1. A) Pareto front in training. B) Feasible region in the (S,C) space for testing.

6   Conclusions

The methodology uses a MOEA which tries to boost Accuracy and Sensitivity as conflicting objectives. A memetic version of DE with the *iRprop+* algorithm as local optimizer, designs the ANNs architecture finding the adequate number of neurons in the hidden layer and the optimal number of connections along with the corresponding weights. The features of the Pareto optimal front allowed us to consider two automatic selection methodologies of individuals: the best model in accuracy and the best model in sensitivity (extremes in the Pareto front). Through optimizing both measures, as is shown in the results, it is possible to obtain classifiers that combine a high classification level with a very good classification rate for each class. In our opinion, the perspective and the memetic DE approach reveal a new point of view for dealing with multi-classification problems.

Acknowledgements

This work has been partially subsidized by TIN 2008-06681-C06-03 project of the Spanish Ministerial Commission of Science and Technology (MICYT), FEDER funds and the P08-TIC-3745 project of the "Junta de Andalucía" (Spain). The research done by J.C. Fernández and P.A. Gutiérrez has been financed respectively by the FPI (grant reference BES-2006-12543) and FPU (grant reference AP-2006- 01746) Predoctoral Programs (Spanish Ministry of Education and Science).

References

1. Duda, R.O., Hart, P.E., Stork, D.G.: Pattern Classification, 2nd edn. Wiley Interscience, New York (2000)
2. Zhang, G.P.: Neural Networks for Classification: A Survey. IEEE Transactions on Systems, Man, and Cybernetics - Part C: Applications and Reviews 30, 451–462 (2000)
3. Jin, Y., Sendhoff, B.: Pareto-Based Multiobjective Machine Learning: An Overview and Case Studies. IEEE Transaction on Systems, Man and Cybernetics, Part. C: Applications and reviews 38, 397–415 (2008)
4. Abbass, H.: An Evolutionary Artificial Neural Networks Approach for Breast Cancer Diagnosis. Artificial Intelligence in Medicine 25, 265–281 (2002)
5. Coello, C.A., Lamont, G.B., Veldhuizen, D.A.V.: Evolutionary Algorithms for Solving Multi-Objective Problems. Springer, Heidelberg (2007)
6. Storn, R., Price, K.: Differential Evolution. A fast and efficient heuristic for global optimization over continuous spaces. Journal of Global Optimization 11, 341–359 (1997)
7. Igel, C., Hüsken, M.: Empirical evaluation of the improved Rprop learning algorithms. Neurocomputing 50, 105–123 (2003)
8. Braga, A.P., Takahashi, R.H.C., Costa, M.A., Teixeira, R.A.: Multi-objective Algorithms for Neural Networks Learning. Studies in Computational Intelligence 16 (2006)
9. Provost, F., Fawcett, T.: Robust classification system for imprecise environments. In: Proccedings of the Fithteenth National Conference on Artificial Intelligence, pp. 706–713 (1998)
10. Zielinski, K., Laur, R.: Variants of Differential Evolution for Multi-Objective Optimization. In: Proceedings of the 2007 IEEE Symposium on Computational Intelligence in Multicriteria Decision Making (MCDM 2007), pp. 91–98 (2007)
11. Ning, G., Zhou, Y.: A Modified Differential Evolution Algorithm for Optimization Neural Network
12. Abbass, H.A., Sarker, R., Newton, C.: PDE: a Pareto-frontier differential evolution approach formulti-objective optimization problems. In: Proceedings of the 2001 Congress on Evolutionary Computation, Seoul, South Korea, vol. 2 (2001)
13. Fernández, J.C., Gutiérrez, P.A., Hervás, C., Martínez, F.J.: Memetic Pareto Evolutionary Artificial Neural Networks for the determination of growth limits of Listeria Monocytogenes. In: Hybrid Intelligent Systems Conference, HIS 2008, pp. 631–636. IEEE, Barcelona (2008)
14. Liefooghe, A., Basseur, M., Jourdan, L., Talbi, E.: ParadisEO-MOEO: A Framework for Evolutionary Multi-objective Optimization. In: Obayashi, S., Deb, K., Poloni, C., Hiroyasu, T., Murata, T. (eds.) EMO 2007. LNCS, vol. 4403, pp. 386–400. Springer, Heidelberg (2007)
15. Witten, I.H., Frank, E.: Data Mining: Practical machine learning tools and techniques, 2nd edn. Morgan Kaufmann, San Francisco (2005)

Pareto-Based Multi-output Model Type Selection

Dirk Gorissen[1], Ivo Couckuyt[1], Karel Crombecq[2], and Tom Dhaene[1]

[1] Ghent University - IBBT, Department of Information Technology (INTEC), Gaston
Crommenlaan 8, Bus 201, 9050 Ghent, Belgium
[2] University of Antwerp, Dept. of Computer Science, Middelheimlaan 1, 2020 Antwerp,
Belgium

Abstract. In engineering design the use of approximation models (= surrogate
models) has become standard practice for design space exploration, sensitivity
analysis, visualization and optimization. Popular surrogate model types include
neural networks, support vector machines, Kriging models, and splines. An engineering simulation typically involves multiple response variables that must be approximated. With many approximation methods available, the question of which
method to use for which response consistently arises among engineers and domain experts. Traditionally, the different responses are modeled separately by independent models, possibly involving a comparison among model types. Instead,
this paper proposes a multi-objective approach can benefit the domain expert
since it enables automatic model *type* selection for each output on the fly without resorting to multiple runs. In effect the optimal model complexity and model
type for each output is determined automatically. In addition a multi-objective approach gives information about output correlation and facilitates the generation
of diverse ensembles. The merit of this approach is illustrated with a modeling
problem from aerospace.

1 Introduction

Regardless of the rapid advances in High Performance Computing (HPC) and multi-
core architectures, it is rarely feasible to explore a design space using high fidelity
computer simulations. As a result, data based surrogate models (otherwise known as
metamodels or response surface models) have become a standard technique to reduce
this computational burden and enable routine tasks such as visualization, design space
exploration, prototyping, sensitivity analysis, and optimization.

This paper is concerned with efficiently and automatically generating accurate *global*
surrogates (valid over the complete design space) using a minimal number of computationally expensive simulations (as opposed to Surrogate Based Optimization (SBO)).
Optimization of the simulation output is not the main goal, rather we are concerned with
optimization of the surrogate model parameters (hyperparameter optimization). Remark
also that since data is expensive to obtain, it is impossible to use traditional, one-shot
experimental designs. Data points must be selected iteratively, there where the information gain will be the greatest (active learning). An important consequence hereof is that
the task of finding the best approximation (= an optimization problem over the model
hyperparameters) becomes a dynamic problem instead of a static one, since the optimal
model parameters will change as the amount and distribution of data points changes.

E. Corchado et al. (Eds.): HAIS 2009, LNAI 5572, pp. 442–449, 2009.

In engineering design simulators are typically modeled on a per-output basis [1]. Each output is modeled independently using separate models (though possibly sharing the same data). Instead of this single objective approach, the authors propose to model the system directly using multi-objective algorithms. This benefits the practitioner by giving information about output correlation, facilitating the generation of diverse ensembles (from the Pareto-optimal set), and enabling the automatic selection of the best model type for each output without having to resort to multiple runs. This paper will discuss these issues and apply them to an aerospace modeling problem.

2   Modeling Multiple Outputs

It is not uncommon that a simulation engine has multiple outputs that all need to be modeled (e.g., [2]). Also many Finite Element (FE) packages generate multiple performance values for free. The direct approach is to model each output independently with separate models (possibly sharing the same data). This, however, leaves no room for trade-offs nor gives any information about the correlation between different outputs. Instead of performing multiple modeling runs (doing a separate hyperparameter optimization for each output) both outputs can be modeled directly if models with multiple outputs are used in conjunction with a multi-objective optimization routine. The resulting Pareto front then gives information about the accuracy trade-off between the outputs and allows the practitioner to choose the model most suited to the particular context. In addition, the final Pareto front enables the generation of diverse ensembles, where the ensemble members consist of the (partial) Pareto-optimal set (see references in [3]). This way all the information in the front can be used. Rainfall runoff modeling and model calibration in hydrology [4] are examples where this multi-objective approach is popular. Models are generated for different output flow components and/or derivative measures and these are then combined into a weighted ensemble or fuzzy committee.

In particular a multi-objective approach enables integration with the automatic surrogate model type selection algorithm described in [5]. This enables automatic selection of the best model type (Kriging, neural networks, support vector machines (SVM), ...) for each output automatically, without having to resort to multiple runs. This is the topic of this paper.

3   Related Work

There is a growing body of research available on multi-objective hyperparameter optimization and model selection strategies. An extensive and excellent overview of the work in this area is given by Jin et. al. in [3]. By far the majority of the cited work uses multi-objective techniques to improve the training of learning methods. Typically an accuracy criterion (such as the validation error) is used together with some regularization parameter or model complexity measure in order to produce more parsimonious models [6]. In engineering design this is closely related to the *"The 5 percent problem"* [7], which arises since single criteria are inadequate at objectively gaging the quality of an approximation model [8]. Another topic that has been the subject of extensive research is that of multi-objective surrogate based optimization (MOSBO). Two examples are

ParEGO [9], and the surrogate based variants of NSGA-II [10]. Though the research into MOSBO is still very young, an excellent overview of current research is already available in [11].

When applying surrogate modeling methods, a recurring question from domain experts is *"Which approximation method is best for my data?"*. Or, as [11] states: *"Little is known about which types of model accord best with particular features of a landscape and, in any case, very little may be known to guide this choice"*. Thus an algorithm to automatically solve this problem is very useful [12]. This is also noticed by [10] who compare different surrogate models for approximating each objective during optimization. The primary concern of a domain expert is obtaining an accurate replacement metamodel for their problem as fast as possible and with minimal overhead. Model selection, model parameter optimization, sampling strategy, etc. are of lesser or no interest to them. Little work has been done to tackle this problem. A solution based on Particle Swarm Optimization (PSO) has been proposed for classification in [13] and a GA based solution for regression in [5]. With this paper we extend the GA based solution so the benefits of automatic model type selection can be carried over to the multi-output case.

4   Core Algorithm

The approach of this paper is built around an island model GA and is illustrated in figure 1. The general idea is to evolve different model types cooperatively in a population and let them compete to approximate the data. Models that produce more accurate fits will have a higher chance of propagating to the next generation. The island model is used since it is the most natural way of incorporating multiple model types into the evolutionary process without mixing them too fast. It also naturally allows for hybrid solutions as will be discussed later. Different sub-populations, called *demes*, exist (initialized differently) and sporadic migration can occur between islands allowing for the exchange of genetic material between species and inter-species competition for resources. Selection and recombination are restricted per deme, such that each sub-population may evolve towards different locally optimal regions of the search space.

The implementation is based on the Matlab GADS toolbox, which itself is based on NSGA-II. Recall from section 1 that data points are selected iteratively (active learning). The general control flow is as follows (see [5] and the Matlab documentation for implementation details): After the initial Design Of Experiments (DOE) has been calculated, an initial sub-population M_i is created for each model type T_i ($i = 1,..,n$ and $M = \bigcup_{i=1}^{n} M_i$). The exact creation algorithm is different for each model type so that model specific knowledge can be exploited. Subsequently, each deme is allowed to evolve according to an elitist GA. Parents are selected using tournament selection and offspring undergo either crossover (with probability p_c) or mutation (with probability $1 - p_c$). The models M_i are implemented as Matlab objects (with full polymorphism) thus each model type can choose its own representation and mutation/crossover implementations. While mutation is straightforward, the crossover operator is more involved since migration causes model types to mix. This raises the question of how to meaningfully recombine two models of different type (e.g., a SVM with a rational function). The solution is to generate a hybrid model by combining both into an ensemble. Once

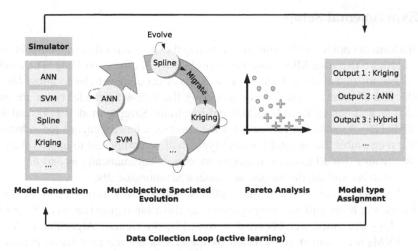

Fig. 1. Automatic multi-objective model generation

every deme has gone through a generation, migration between individuals is allowed to occur at migration interval m_i, with migration fraction m_f and migration direction m_d (a ring topology is used). Finally, an extinction prevention algorithm is used to ensure no model type can disappear completely from the population.

The fitness function calculates the quality of the model fit, according to some generalization estimator ξ (e.g., cross validation). Note that ξ itself can also constitute multiple criteria (see [7] for a discussion on this). Once the GA has terminated, control passes back to the main global surrogate modeling algorithm. At that point M contains the best set of models that can be constructed for the given data. If M contains at least one model that satisfies the user defined targets for ξ the main loop terminates. If not, a new set of maximally informative sample points is selected based on several criteria (quality of the models, non-linearity of the response, etc.) and scheduled for evaluation. Once new simulations become available the GA is resumed.

Note that sample evaluation and model construction/hyperparameter optimization run in parallel to allow an optimal use of computational resources. Some initial results with this algorithm can be found in [7].

5   Langley Glide-Back Booster (LGBB)

For this paper we consider a modeling problem from aerodynamics. NASA's Langley Research Center is developing a small launch vehicle (SLV) [2] that can be used for rapid deployment of small payloads to low earth orbit at significantly lower launch costs, improved reliability and maintainability. In particular, NASA is interested in the aerodynamic characteristics (lift, drag, pitch, side-force, yaw, roll) as a function of three inputs (Mach number, angle of attack, and side slip angle). Simulations are performed with a Cart3D flow solver with a running time of 5-20 hours on a high end workstation. A fixed data set of 780 adaptively chosen points was generated for metamodeling purposes. Thus the active learning loop is switched off.

6 Experimental Setup

The platform of choice for the implementation and experiments described in this paper is the Matlab SUrrogate MOdeling Toolbox (SUMO Toolbox) v6.1 [1]. The toolbox is available for download at http://www.sumo.intec.ugent.be to allow for full reproduction of these results. For the modeling the following model types are used: Artificial Neural Networks (ANN), rational functions, Kriging models [14], and RBF LS-SVMs [15]. As stated in subsection 4 the result of a heterogeneous recombination will be an ensemble. So in total 5 model types will be competing to approximate the data. Remember that all model parameters are chosen automatically as part of the GA. No user input is required, the models are generated automatically.

The ANN models are based on the Matlab Neural Network Toolbox and are trained with Levenberg Marquard backpropagation with Bayesian regularization (300 epochs). The topology and initial weights are determined by the Genetic Algorithm (GA). For the LS-SVMs the c and σ are selected by the GA as are the correlation parameters for Kriging (with a linear regression and Gaussian correlation function). The rational functions are based on a custom implementation, the free parameters being the orders of the two polynomials, the weights of each parameter, and which parameters occur in the denominator. A full explanation of the genetic operators used for each model would consume too much space. All the code is available as part of the SUMO Toolbox and details can be found in the associated technical reports.

The population size of each deme type is set to 15 and the GA is run for 50 generations. The migration interval m_i is set to 5, the migration fraction m_f to 0.1 and the migration direction is *forward* (copies of the m_f best individuals from island i replace the worst individuals in island $i+1$). The fitness function is defined as the Bayesian Estimation Error Quotient ($BEEQ(y, \tilde{y}) = \frac{\sum_{i=1}^{n} |y_i - \tilde{y}_i|}{\sum_{i=1}^{n} |y_i - \bar{y}|}$) calculated on a 20% validation set where the minimal distance between validation points is maximized. The variables $y, \tilde{y}$ and $\bar{y}$ represent the true, predicted and mean true values respectively.

7 Results

For ease of visualization all outputs were modeled in pairs. Figure 2 shows the resulting search trace for three representative combinations (the others were omitted to to save space). The algorithm has been proven to be quite robust in in its model selection ([5,7]) thus we can safely infer from these results.

Let us first regard the $lift - drag$ trace. We notice a number of things. First all the models are roughly on the main $x = y$ diagonal. This means that a model that performs well on $lift$ performs equally well on $drag$. This implies a very strong correlation between the behavior of both outputs, which can actually be expected given the physical relationship between aerodynamic lift and drag. Only the rational models do not consistently follow this trend but this has probably more to do with their implementation as will be discussed later. Since all models are on the main diagonal the model type selection problem has become a single objective problem. It turns out that the ANN models (or actually an ensemble of ANN models if you look closely) are the best to

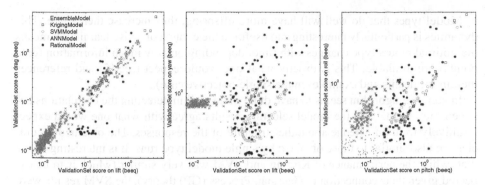

Fig. 2. Heterogeneous Pareto traces: $lift - drag$ (left), $lift - yaw$ (center), $pitch - roll$ (right)

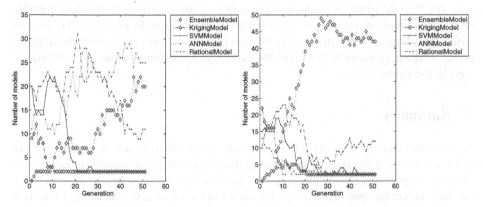

Fig. 3. Evolution of the population make-up $lift - yaw$ (left) and $pitch - yaw$ (right)

use in this case, performing much better than the other model types. Kriging and SVM perform about the same with SVM doing slightly better.

The situation in the second plot ($lift - yaw$) is different. There is a clear trade-off between both outputs. The ANN models still perform best on $lift$ but the yaw output is fitted most accurately by rational models and their ensembles. The difference between SVM models and Kriging models is now more marked, with the SVM models performing much better on the yaw output. Remark also that the distribution of the rational models is much more spread-out and chaotic than that of the other model types (which are relatively clustered). This could actually be seen for all output combinations. The reason is due to the way the order selection of the rational functions is implemented. The order selection contains too much randomness and not enough exploitation of the search space, leading to erratic results. This is currently being remedied.

The $pitch - roll$ plot has the same general structure, though the $pitch$ output turns out much harder to approximate than $lift$. It is interesting to see though that ensembles turn out to be significantly superior to standalone models. It turns out that most of these ensembles turn out to be combinations of the best performing model types.

The results presented so far are of course static results. In reality the best performing model type changes as the evolution progresses. This can be seen from figure 3 which shows the relative share of each model type of the total population for two cases.

Model types that do well will have more offspring, thus increase their share. This dynamics is particularly interesting and useful in the context of active learning. It means the optimal model type changes with time, depending on how much information (data points) are available. This is exactly what one would expect (see [5] and references therein for details and examples in the single objective case).

In sum the algorithm seems to have done quite well in detecting the correlations between the outputs, and the model selection results agree with what one would expect intuitively from knowledge about the structure of the responses. The obtained accuracies are also similar to those obtained by single model type runs. It is interesting to see that, while the performance of Kriging and SVM was very similar (which is to be expected given their connection to Gaussian Process (GP) theory), the SVM results were consistently better and more robust to changes in their hyperparameters. The exact reasons for this are being investigated. It was also quite surprising to see the ensemble models to so well. In virtually all experiments the final Pareto front consisted of only of ensembles. Depending on the model selection process they are predominantly homogeneous (containing multiple models of the same type) or heterogeneous (thus 'filling in' gaps in the Pareto front).

8 Summary

The efficient identification of the best approximation method to use for a given problem is a consistently recurring question among domain experts. Particularly so in the case of multi-output problems. In this paper the authors have presented a new approach that can help tackle this problem through the use of dynamic (if active learning is enabled) multi-objective optimization with speciated evolution. This allows multiple outputs to be modeled together, giving information about the trade-off in the hyperparameter space and identification of the most suitable model type (which can be a hybrid). It further facilitates the generation of diverse ensembles (from the final Pareto front). At the same time the computational cost is roughly the same as performing multiple, independent single model type runs (though this cost is still outweighed by the simulation cost). In addition, a multi-objective evolutionary approach gives more information, naturally permits hybrid solutions (ensembles), and naturally allows the best solution to change depending on how much information (data) is available. Note that the same general approach can also be used to deal with multiple performance/accuracy criteria instead of multiple outputs.

Acknowledgments

The authors would like to thank NASA and Robert Gramacy from the University of California, Santa Cruz for providing the data and information on the LGBB example. Ivo Couckuyt is funded by the Institute for the Promotion of Innovation through Science and Technology in Flanders (IWT-Vlaanderen).

References

1. Gorissen, D., De Tommasi, L., Crombecq, K., Dhaene, T.: Sequential modeling of a low noise amplifier with neural networks and active learning. Neural Computing and Applications (accepted, 2008)
2. Pamadi, B.N., Covell, P.F., Tartabini, P.V., Murphy, K.J.: Aerodynamic characteristics and glide-back performance of langley glide-back booster. In: Proceedings of 22nd Applied Aerodynamics Conference and Exhibit, Providence, Rhode Island (2004)
3. Jin, Y., Sendhoff, B.: Pareto-based multiobjective machine learning: An overview and case studies. IEEE Transactions on Systems, Man, and Cybernetics, Part C: Applications and Reviews 38(3), 397–415 (2008)
4. Fenicia, F., Solomatine, D.P., Savenije, H.H.G., Matgen, P.: Soft combination of local models in a multi-objective framework. Hydrology and Earth System Sciences Discussions 4(1), 91–123 (2007)
5. Gorissen, D., De Tommasi, L., Croon, J., Dhaene, T.: Automatic model type selection with heterogeneous evolution: An application to rf circuit block modeling. In: Proceedings of the IEEE Congress on Evolutionary Computation, WCCI 2008, Hong Kong (2008)
6. Fieldsend, J.E.: Multi-objective supervised learning. In: Knowles, J., Corne, D., Deb, K. (eds.) Multiobjective Problem Solving from Nature From Concepts to Applications. Natural Computing Series. LNCS. Springer, Heidelberg (2008)
7. Gorissen, D., Couckuyt, I., Dhaene, T.: Multiobjective global surrogate modeling. Technical Report TR-08-08, University of Antwerp, Middelheimlaan 1, 2020 Antwerp, Belgium (2008)
8. Li, X.R., Zhao, Z.: Evaluation of estimation algorithms part I: incomprehensive measures of performance. IEEE Transactions on Aerospace and Electronic Systems 42(4), 1340–1358 (2006)
9. Knowles, J.: Parego: A hybrid algorithm with on-line landscape approximation for expensive multiobjective optimization problems. IEEE Transactions on Evolutionary Computation 10(1), 50–66 (2006)
10. Voutchkov, I., Keane, A.: Multiobjective Optimization using Surrogates. In: Parmee, I. (ed.) Proceedings of the Seventh International Conference on Adaptive Computing in Design and Manufacture 2006, Bristol, UK, pp. 167–175 (2006)
11. Knowles, J., Nakayama, H.: Meta-modeling in multiobjective optimization. In: Multiobjective Optimization - Interactive and Evolutionary Approaches. LNCS. Springer, Heidelberg (in press, 2008)
12. Keys, A.C., Rees, L.P., Greenwood, A.G.: Performance measures for selection of metamodels to be used in simulation optimization. Decision Sciences 33, 31–58 (2007)
13. Escalante, H.J., Gomez, M.M., Sucar, L.E.: Psms for neural networks on the ijcnn 2007 agnostic vs prior knowledge challenge. In: IJCNN, pp. 678–683 (2007)
14. Lophaven, S.N., Nielsen, H.B., Søndergaard, J.: Aspects of the matlab toolbox DACE. Technical report, Informatics and Mathematical Modelling, Technical University of Denmark, DTU, Richard Petersens Plads, Building 321, DK-2800 Kgs. Lyngby (2002)
15. Suykens, J., Gestel, T.V., Brabanter, J.D., Moor, B.D., Vandewalle, J.: Least Squares Support Vector Machines. World Scientific Publishing Co., Pte, Ltd., Singapore (2002)

A Comparison of Multi-objective Grammar-Guided Genetic Programming Methods to Multiple Instance Learning

Amelia Zafra and Sebastián Ventura

Department of Computer Science and Numerical Analysis, University of Cordoba

Abstract. This paper develops a first comparative study of multi-objective algorithms in Multiple Instance Learning (MIL) applications. These algorithms use grammar-guided genetic programming, a robust classification paradigm which is able to generate understandable rules that are adapted to work with the MIL framework. The algorithms obtained are based on the most widely used and compared multi-objective evolutionary algorithms. Thus, we design and implement SPG3P-MI based on the Strength Pareto Evolutionary Algorithm, NSG3P-MI based on the Non-dominated Sorting Genetic Algorithm and MOGLG3P-MI based on the Multi-objective genetic local search. These approaches are tested with different MIL applications and compared to a previous single-objective grammar-guided genetic programming proposal. The results demonstrate the excellent performance of multi-objective approaches in achieving accurate models and their ability to generate comprehensive rules in the knowledgable discovery process.

1 Introduction

Multiple Instance Learning (MIL) introduced by Dietterich et al. [1] consists of generating a classifier that will correctly classify unseen patterns. The main characteristic of this learning is that the patterns are bags of instances where each bag can contain different numbers of instances. There exists information about the bags, a bag receives a special label, but the labels of instances are unknown. According to the standard learning hypothesis proposed by Dietterich et al. [1] a bag is positive if and only if at least one of its instances is positive and it is negative if none of its instances produce a positive result. The key challenge in MIL is to cope with the ambiguity of not knowing which of the instances in a positive bag are actually the positive examples and which are not. In this sense, this learning problem can be regarded as a special kind of supervised learning problem where the labeling information is incomplete. This learning framework is receiving growing attention in the machine learning community because numerous real-world tasks can be very naturally represented as multiple instance problems. Among these tasks we can cite text categorization [2], content-based image retrieval [3], image annotation [4], drug activity prediction [5,6], web index page recommendation [7], stock selection [5] and computer security [8].

E. Corchado et al. (Eds.): HAIS 2009, LNAI 5572, pp. 450–458, 2009.

The problem of evaluating the quality of a classifier, whether in MIL perspective or in traditional supervised learning, is naturally posed as a multi-objective problem with several contradictory objectives. If we try to optimize one of them, the others are reduced. All previously used proposals to solve this problem from a MIL perspective do not take into account the multi-objective problem and only obtain one optimal solution combining the different objectives to obtain a high quality classifier. However, this approach is unsatisfactory due to the nature of optimality conditions for multiple objectives. It is well-known that in the presence of multiple and conflicting objectives, the resulting optimization problem gives rise to a set of optimal solutions, instead of just one optimal solution. Multiple optimal solutions exist because no single solution can be a substitute for multiple conflicting objectives and it is shown that algorithms which consider the set of optimal solutions obtain better general results.

In this paper, a first comparative study of the most widely analyzed, compared and tested approaches under various problems and criteria which generate Pareto Optimal Front (POF) is elaborated. We design and implement classic multi-objective evolutionary algorithms using Grammar Guided Genetic Programming (G3P) and adapt them to handle multi-instance problems. Our proposals are the Strength Pareto Grammar-Guided Genetic Programming for MIL (SPG3P-MI) based on the Strength Pareto Evolutionary Algorithm (SPEA2)[9], the Non-dominated Sorting Grammar-Guided Genetic Programming for MIL (NSG3P-MI) based on the Non-dominated Sorting Genetic Algorithm (NSGA2) [10] and Multi-objective genetic local search with Grammar-Guided Genetic Programming for MIL (MOGLSG3P-MI) based on Multi-objective genetic local search (MOGLS)[11]. These algorithms represent classification rules in IF-THEN form which make it possible to determine if a bag is positive or negative and the quality of each classifier is evaluated according to two conflicting quality indexes, sensitivity and specificity. Computational experiments show that multi-objective techniques are robust algorithms which achieve better results than G3P-MI [12] other previously used technique based on G3P and a single-objective. Moreover, multi-objective proposals obtain classifiers which contain simple rules which add comprehensibility and simplicity in the knowledge discovery process.

The paper is organized as follows. In Section 2, a description of the approaches proposed is presented. In Section 3, experiments are conducted. Finally, conclusions and some possible lines for future research are discussed in Section 4.

2   Using Multi-objective G3P for Classification Rule Generation

In our approach, we use an extension of traditional GP systems, called grammar-guided genetic programming (G3P) [13]. G3P facilitates the efficient automatic discovery of empirical laws providing a more systematic way to handle typing using a context-free grammar which establishes a formal definition of the syntactical restrictions. The motivation to include this paradigm is that it retains a significant position due to its flexible variable length solution representation and the low error rates that achieves both in obtaining classification rules, and in other tasks related to prediction, such as feature selection and the generation of discriminant functions. On the other hand, the main motivation to include multi-objective strategies in our proposals is due to the measurements to evaluate a classifier are conflictive, so if the value of any of them is maximized, the value of the others can be significantly reduced. Thus, it is very interesting to obtain the POF and introduce preference information to analyze which of them could be the best to classify new examples.

Multi-objective techniques for evolutionary computation have been widely used on classification topics where significant advances in results have been achieved [14]. If we evaluate its use in Genetic Programming (GP), we can find that it provides better solutions than those obtained using standard GP and lower computational cost [15,16].

In this section we specify different aspects which have been taken into account in the design of the these proposals, such as individual representation, genetic operators and fitness function. The main evolutionary process is not described because it is based on the well-known SPEA2 [9], NSGA2 [10] and MOGLS [11].

2.1   Individual Representation

In our systems, as G3P-MI [12], individuals express the information in the form of IF-THEN classification rules. These rules determine if a bag should be considered positive (that is, if it is a pattern of the concept we want to represent) or negative (if it is not).

If $(cond_B(\text{bag}))$ **then**
   the bag is an instance of the concept.
Else
   the bag is not an instance of the concept.
End-If

where $cond_B$ is a condition that is applied to the bag. Following the Dietterich hypothesis, $cond_B$ can be expressed as:

$$cond_B(bag) = \bigvee_{\forall instance \in bag} cond_I(instance) \tag{1}$$

where $\vee$ is the disjunction operator, and $cond_I$ is a condition that is applied over every instance contained in a given bag. Figure 1 shows the grammar used to represent the condition of the rules.

2.2   Genetic Operators

The process of generating new individuals in a given generation of the evolutionary algorithm is carried out by two operators, crossover and mutator. Depending on the philosophy of the algorithm one or both will be used. In this section, we briefly describe their functioning.

$\langle$S$\rangle \rightarrow \langle$cond$_I\rangle$
$\langle$cond$_I\rangle \rightarrow \langle$cmp$\rangle$| **OR** $\langle$cmp$\rangle$ $\langle$cond$_I\rangle$| **AND** $\langle$cmp$\rangle$ $\langle$cond$_I\rangle$
$\langle$cmp$\rangle \rightarrow \langle$op-num$\rangle$ $\langle$variable$\rangle$ $\langle$value$\rangle$| $\langle$op-cat$\rangle$ $\langle$variable$\rangle$
$\langle$op-cat$\rangle \rightarrow$ **EQ** | **NOT EQ**
$\langle$op-num$\rangle \rightarrow$ **GT** | **GE** | **LT** | **LE**
$\langle$term-name$\rangle \rightarrow$ *Any valid term in dataset*
$\langle$term-freq$\rangle \rightarrow$ *Any integer value*

Fig. 1. Grammar used for representing individuals' genotypes

Crossover Operator. This operator chooses a non-terminal symbol randomly with uniform probability from among the available non-terminal symbols in the grammar and two sub-trees (one from each parent) whose roots are equal or compatible to the symbol selected are swapped. To reduce bloating, if any of the two offspring is too large, they will be replaced by one of their parents.

Mutation Operator. This operator selects with uniform probability the node in the tree where the mutation is to take place. The grammar is used to derive a new subtree which replaces the subtree underneath that node. If the new offspring is too large, it will be eliminated to avoid having invalid individuals.

2.3 Fitness Function

The fitness function is a measure of the effectiveness of the classifier. There are several measures to evaluate different components of the classifier and determine the quality of each rule. We consider two widely accepted parameters for characterizing models in classification problems: sensitivity (Se) and specificity (Sp). Sensitivity is the proportion of cases correctly identified as meeting a certain condition and specificity is the proportion of cases correctly identified as not meeting a certain condition. Both are specified as follows:

$$sensitivity = \frac{t_p}{t_p + f_n}, \quad \begin{cases} t_p & \text{number of positive bags correctly identified.} \\ f_n & \text{number of negative bags not correctly identified.} \end{cases}$$

$$specificity = \frac{t_n}{t_n + f_p}, \quad \begin{cases} t_n & \text{number of negative bags correctly identified.} \\ f_p & \text{number of positive bags not correctly identified.} \end{cases}$$

We look for rules that maximize both *Sensitivity* and *Specificity* at the same time. Nevertheless, there exists a well-known trade-off between these two parameters because they evaluate different and conflicting characteristics in the classification process. Sensitivity alone does not tell us how well the test predicts other classes (that is, the negative cases) and specificity alone does not clarify how well the test recognizes positive cases. It is necessary to optimize both the sensitivity of the test to the class and its specificity to the other class to obtain a high quality classifier.

3   Experimental and Results

A brief description of the application domains used for comparing along with a description of the experimental methodology are presented in the next section. Then, the results and a discussion about the experimentation are detailed.

3.1   Problem Domains Used and Experimental Setting

The datasets used in the experiments represent two well-known applications in MIL, *drug activity prediction* which consists of determining whether a drug molecule will bind strongly to a target protein [1] and *content-based image retrieval* which consists of identifying the intended target object(s) in images [2] Detailed information about these datasets is summarized in Table 1. All datasets are partitioned using 10-fold stratified cross validation [17] on all data sets. Folds are constructed on bags, so that every instance in a given bag appears in the same fold. The partitions of each data set are available at *http:www.uco.es/grupos/ ayrna/mil.*

Table 1. General Information about Data Sets

DATASET	BAGS Positive	BAGS Negative	BAGS Total	ATTRIBUTES	INSTANCES	AVERAGE BAG SIZE
Musk1	47	45	92	166	476	5.17
Musk2	39	63	102	166	6598	64.69
Mutagenesis-Atoms	125	63	188	10	1618	8.61
Mutagenesis-Bonds	125	63	188	16	3995	21.25
Mutagenesis-Chains	125	63	188	24	5349	28.45
Elephant	100	100	200	230	1391	6.96
Tiger	100	100	200	230	1220	6.10
Fox	100	100	200	230	1320	6.60

The algorithms designed have been implemented in the JCLEC software [18]. All experiments are repeated with 10 different seeds and the average results are reported in the results table in the next section.

3.2   Comparison of Multi-objective Strategies

In this section, we compare the different multi-objective techniques implemented, MOGLSG3P-MI, NSG3P-MI and SPG3P-MI. In a first section a quantitative comparison of the performance of different multi-objective algorithms is carried out. Then, the different multi-objective techniques are compared with the accuracy, sensitivity and specificity results of G3P-MI, a previous single-objective G3P algorithm [12].

Analysis of the quality of Multi-Objective strategies. The outcome in the multi-objective algorithms used is an approximation of the Pareto-optimal front (POF). An analysis of the quality of these approximation sets is evaluated to compare the different multi-objective techniques. Many performance measures which evaluate different characteristics have been proposed. Some of the most popular performance measurements as spacing, hypervolume and coverage of sets [19] are analyzed in this work and their average results on the different data sets studied are shown in Table 2. The spacing [19] metric describes the spread of non-dominated set. According to the results showed the non-dominated front of NSG3P-MI has all solutions more equally spaced than the other algorithms. The hypervolume indicator [19] is defined as the area of coverage of non-dominated set with respect to the objective space. The results show that the non-dominated solutions of NSG3P-MI cover more area than the other techniques. Finally, coverage of two sets [19] is evaluated. This metric can be termed relative coverage comparison of two sets. The results show that NSG3P-MI obtains the highest values when it is compared with the other techniques, then by definition the outcomes of NSG3P-MI dominate the outcomes of the other algorithms. Taking into account all the results obtained in the different metrics, NSG3P-MI achieves a better approximation of POF than the other techniques.

Table 2. Analysis of quality of POFs considering average values for all data sets studied

ALGORITHM	HYPERVOLUME (HV)	SPACING (S)	TWO SET COVERAGE (CS)	
MOGLSG3P-MI	0.844516	0.016428	CS(MOGLSG3P-MI,NSG3P-MI)	0.357052
			CS(MOGLSG3P-MI,SPG3P-MI)	0.430090
NSG3P-MI	0.890730	0.007682	CS(NSG3P-MI,MOGLSG3P-MI)	0.722344
			CS(NSG3P-MI,SPG3P-MI)	0.776600
SPG3P-MI	0.872553	0.012290	CS(SPG3P-MI,MOGLSG3P-MI)	0.508293
			CS(SPG3P-MI,NSG3P-MI)	0.235222

Comparison Multi-Objective strategies with a Single-Objective previous version. We compare the results of accuracy, sensitivity and specificity of different multi-objective techniques implemented with the results of a previous single-objective G3P algorithm [12]. The average results of accuracy, sensitivity and specificity for each data set are reported in Table 3. The Friedman test [20] is used to compare the different algorithms. The Friedman test is a nonparametric test that compares the average ranks of the algorithms. These ranks let us know which algorithm obtains the best results considering all data sets. In this way, the algorithm with the value closest to 1 indicates the best algorithm in most data sets. The ranking values for each measurement are also shown in Table 3.

The Friedman test results are shown in Table 4. This test indicates that these are significantly differences both in accuracy and specificity measurements and there is no significant difference for sensitivity measurement. A post-hoc test was used, the Bonferroni-Dunn test [20], to find significant differences occurring between algorithms. Figure 2(a) shows the application of this test on accuracy. This graph represents a bar chart, whose values are proportional to the mean

Table 3. Experimental Results

Algorithm	MOGLSG3P-MI			NSG3P-MI			SPG3P-MI			G3P-MI		
	Acc	*Se*	*Sp*	*Acc*	*Se*	*Sp*	*Acc*	*Se*	*Sp*	*Acc*	*Se*	*Sp*
Elephant	0.8900	0.8700	0.9100	0.9400	0.9400	0.9400	0.9250	0.9400	0.9100	0.8800	0.9300	0.8300
Tiger	0.8850	0.9400	0.8300	0.9350	0.9200	0.9500	0.9200	0.9200	0.9200	0.8700	0.9400	0.8000
Fox	0.7600	0.7800	0.7400	0.7800	0.8600	0.7000	0.8350	0.8900	0.7800	0.7050	0.7900	0.6200
MutAtoms	0.8421	0.9385	0.6333	0.9158	0.9462	0.8500	0.8790	0.9308	0.7667	0.8526	0.8462	0.8167
MutBonds	0.8421	0.9077	0.7000	0.8737	0.9231	0.7667	0.8684	0.9308	0.7333	0.8210	0.8462	0.7833
MutChains	0.8737	0.9462	0.7167	0.9211	0.9462	0.8667	0.9053	0.9000	0.9167	0.8105	0.9231	0.7333
Musk1	0.9778	0.9600	1.0000	1.0000	1.0000	1.0000	0.9667	0.9800	0.9500	0.9445	1.0000	0.9000
Musk2	0.9400	0.9500	0.9333	0.9301	0.9607	0.9095	0.9400	0.9750	0.9167	0.8800	1.0000	0.9000
RANKING	2.8125	3.0000	2.7500	1.3750	2.0000	1.8125	1.9375	2.3750	2.1875	3.8750	2.6250	3.2500

Table 4. Results of the Friedman Test (p=0.1)

	Valor Friedman	Valor $\chi_2(1-\alpha=0.1)$	Conclusion
Acc	17.1375	4.642	Reject null hypothesis
Se	2.5500	4.642	Accept null hypothesis
Sp	5.7375	4.642	Reject null hypothesis

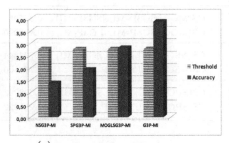

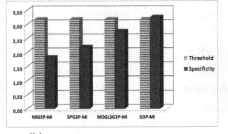

(a) Bonferroni-Dunn for Accuracy   (b) Bonferroni-Dunn for Specificity

Fig. 2. Bonferroni Dunn Test ($p < 0.1$)

rank obtained from each algorithm. This test sets a *Threshold* (represented with one of the grated bars); those values that exceed this bar are algorithms with significantly worse results than the control algorithm (associated in this case with NSG3P-MI because it is the lowest rank value). The threshold in this case is fitted to 2.7486 (with, $1 - \alpha = 0.1$). Observing this figure, the algorithms that exceed the threshold determined by Bonferroni are MOG3P-MI and G3P-MI, therefore they could be considered worse proposals.

With respect to the specificity measurement, Figure 2(b) shows the application of Bonferroni-Dunn post-hoc test on it. The threshold in this case is 3.1861 (with, $1 - \alpha = 0.1$). Observing this figure, the algorithm that exceeds the threshold determined by Bonferroni is G3P-MI, again NSG3P-MI is the best proposal.

We can conclude that statistically there are hardly any differences between multi-objective proposals, except for the accuracy values of the MOGLS algorithm. On the other hand, the differences are more noticeable for the single-objective G3P algorithm that obtains worse results than the rest of the techniques for all measurements and for accuracy and specificity obtains significant differences statistically. Moreover, a better trade-off between the different measurements can be seen in the multi-objective techniques.

4 Conclusions and Future Works

This paper has done a first comparative study of multi-objective evolutionary algorithms on MIL. To do so, the renowned algorithms, SPEA2, NSGA2 and MOGLS have been adapted to work with a G3P paradigm and to handle a MIL scenario. The comparison between the different multi-objective techniques and a previous single-objective G3P algorithm (G3P-MI) has shown that all multi-objective proposals obtain more accurate models. The Friedman test determine that NSG3P-MI is the best proposal with respect to the rest of the algorithms for all measurements considered. Statistically, it can be concluded that there are significant differences between the algorithms with respect to accuracy and specificity values. For these values, a post-test is carried out and this the Bonferroni-Dunn test concludes that G3P-MI is considered to be statistically worse algorithm for both measurements.

This is only a preliminary study and there are still some forthcoming considerations. Thus, it would be interesting to make a more detailed study which evaluate the performance of multi-objectives proposals. Moreover, it would be interesting to do a thorough investigation involving the most representative MIL algorithm in the rest of paradigms used in MIL.

References

1. Dietterich, T.G., Lathrop, R.H., Lozano-Perez, T.: Solving the multiple instance problem with axis-parallel rectangles. Artifical Intelligence 89(1-2), 31–71 (1997)
2. Andrews, S., Tsochantaridis, I., Hofmann, T.: Support vector machines for multiple-instance learning. In: NIPS 2002: Proceedings of Neural Information Processing System, Vancouver, Canada, pp. 561–568 (2002)
3. Pao, H.T., Chuang, S.C., Xu, Y.Y., Fu, H.: An EM based multiple instance learning method for image classification. Expert Systems with Applications 35(3), 1468–1472 (2008)
4. Yang, C., Dong, M., Fotouhi, F.: Region based image annotation through multiple-instance learning. In: Multimedia 2005: Proceedings of the 13th Annual ACM International Conference on Multimedia, New York, USA, pp. 435–438 (2005)
5. Maron, O., Lozano-Pérez, T.: A framework for multiple-instance learning. In: NIPS 1997: Proceedings of Neural Information Processing System 10, Denver, Colorado, USA, pp. 570–576 (1997)
6. Zhou, Z.H., Zhang, M.L.: Solving multi-instance problems with classifier ensemble based on constructive clustering. Knowledge and Information Systems 11(2), 155–170 (2007)

7. Zafra, A., Ventura, S., Romero, C., Herrera-Viedma, E.: Multiple instance learning with genetic programming for web mining. In: Sandoval, F., Prieto, A.G., Cabestany, J., Graña, M. (eds.) IWANN 2007. LNCS, vol. 4507, pp. 919–927. Springer, Heidelberg (2007)
8. Ruffo, G.: Learning single and multiple instance decision tree for computer security applications. PhD thesis, Department of Computer Science. University of Turin, Torino, Italy (2000)
9. Zitzler, E., Laumanns, M., Thiele, L.: SPEA2: Improving the Strength Pareto Evolutionary Algorithm. Technical Report 103, Gloriastrasse 35 (2001)
10. Deb, K., Agrawal, S., Pratap, A., Meyarivan, T.: A fast elitist non-dominated sorting genetic algorithm for multi-objective optimisation: NSGA-II. In: Deb, K., Rudolph, G., Lutton, E., Merelo, J.J., Schoenauer, M., Schwefel, H.-P., Yao, X. (eds.) PPSN 2000. LNCS, vol. 1917, pp. 849–858. Springer, Heidelberg (2000)
11. Jaszkiewicz, A., Kominek, P.: Genetic local search with distance preserving recombination operator for a vehicle routing problem. European Journal of Operational Research 151(2), 352–364 (2003)
12. Zafra, A., Ventura, S.: G3P-MI: A genetic programming algorithm for multiple instance learning. In: Information Science. Elsevier, Amsterdam (submitted)
13. Whigham, P.A.: Grammatically-based genetic programming. In: Proceedings of the Workshop on Genetic Programming: From Theory to Real-World Applications, Tahoe City, California, USA, pp. 33–41 (1995)
14. Shukla, P.K., Deb, K.: On finding multiple pareto-optimal solutions using classical and evolutionary generating methods. European Journal of Operational Research 181(3), 1630–1652 (2007)
15. Parrott, D., Xiaodong, L., Ciesielski, V.: Multi-objective techniques in genetic programming for evolving classifiers. In: IEEE Congress on Evolutionary Computation, vol. 2, pp. 1141–1148 (September 2005)
16. Mugambi, E.M., Hunter, A.: Multi-objective genetic programming optimization of decision trees for classifying medical data. In: KES 2003: Knowledge-Based Intelligent Information and Engineering Systems, pp. 293–299 (2003)
17. Wiens, T.S., Dale, B.C., Boyce, M.S., Kershaw, P.G.: Three way k-fold cross-validation of resource selection functions. Ecological Modelling 212(3-4), 244–255 (2008)
18. Ventura, S., Romero, C., Zafra, A., Delgado, J.A., Hervás, C.: JCLEC: A java framework for evolutionary computation soft computing. Soft Computing 12(4), 381–392 (2008)
19. Coello, C.A., Lamont, G.B., Veldhuizen, D.A.V.: Evolutionary Algorithms for Solving Multi-Objective Problems. In: Genetic and Evolutionary Computation, 2nd edn. Springer, New York (2007)
20. Demšar, J.: Statistical comparisons of classifiers over multiple data sets. Journal of Machine Learning Research 7, 1–30 (2006)

On the Formalization of an Argumentation System for Software Agents

Andres Munoz and Juan A. Botia

Department of Information and Communications Engineering
Computer Science Faculty, University of Murcia
Campus de Espinardo, 30100 Murcia, Spain
amunoz@um.es, juanbot@um.es

Abstract. Argumentation techniques for multi-agent systems (MAS) coordination are relatively common nowadays. But most frameworks are theoretical approaches to the problem. ASBO is an Argumentation System Based on Ontologies. It follows an engineering oriented approach to materialize a software tool which allows working with argumentation in MAS. But ASBO has also a formal model in the background. This paper introduces such formal model, as a way to identify and unambiguously define the core elements that argumentation systems should include.

1 Introduction

MAS engineering has proved its validity in a number of different domains, as for example network administration [6]. One particularity in MAS software is that no agent holds a complete vision of the whole problem faced (i.e. data and control are distributed). In such systems, agent's beliefs about the problem compound the personal view each agent has about the part of the problem it is in charge of. Such beliefs may be overlapped or even incomplete.

In these scenarios, conflicts may arise (e.g. some agents believing the access to a resource to be granted, whereas others asserting it to be forbidden) [9]. Two kinds of conflicts are considered in ASBO: semantic-independent conflicts, so-called *contradictions*, and semantic-dependent ones, or *differences*. Contradictions may appear regardless how the domain is modeled (i.e. a fact and its negation). Differences stick to the domain considered (e.g. classifying an object as square and as rectangular in a unique shape domain). To properly manage the second type of conflicts, the domain is modeled in ASBO by creating a formal and explicit model, based on OWL [2,5]. Due to the addition of meta-information on this knowledge, agents can support reasoning operations, as for example deductive processes or consistency checking. Thus, the conflict detection process, both semantic-dependent and independent, can be included in these operations in a natural manner.

A common approach to coordinate agents to autonomously solve conflicts is by reaching agreements about the status of those conflicts. The alternative followed by ASBO for that is employing a *persuasion* dialog between the implicated agents [3,11]. Such kind of dialog consists of an exchange of opinions among agents that are

E. Corchado et al. (Eds.): HAIS 2009, LNAI 5572, pp. 459–467, 2009.

for/against an issue, with the aim of clarifying which one is the most acceptable. Argumentation [4] is considered as a promising materialization of persuasion dialogs in multi-agent systems. In this way, a negotiation protocol is defined via argumentation [10], that leads to a persuasion dialog in which an agent tries to convince others about a specific proposal. But in this case, not only are the proposals exchanged. Furthermore, *arguments* (i.e. premises and rules used to derive those proposals) are also communicated. It allows agents to resolve conflicts more efficiently than just exchanging proposals, as proved elsewhere [7]. Hence, different attacks (i.e. conflictive points of view on proposals, premises and rules) can be defined over arguments. Finally, an acceptability status on each argument is determined depending on the course of the dialog development. The basic idea is that when an argument is *accepted* in the system, it wins the knowledge conflict concerning it, and therefore the proposal derived from the argument is also accepted. Contrarily, a *defeated* argument means that its derived proposal is not accepted and it loses any possible conflict.

This paper is devoted to formalize all the elements mentioned which are related with the persuasion dialog framework used for conflict resolution. Section 2 presents the formal model. Section 3 illustrates how it works with an example. Section 4 outlines most important conclusions and future works.

2   Formalization of the ASBO Argumentation System

The ASBO system paves the way to the construction of agents which are capable of solving conflicts for autonomous coordination by using argumentation. ASBO agents are structured in a layered manner, from an architectural point of view. Figure 1 shows a representation of a couple of ASBO agents based on a block diagram. The top layer includes all tasks related to argumentation. The middle layer constitutes the formal description model. It contains a common representation (i.e. OWL ontologies) for the domain-specific knowledge on which arguments are built. The bottom layer encapsulates a particular implementation for a specific agent platform. Details about the middle and bottom layers are given elsewhere [8]. The rest of the section is devoted to the persuasion framework in the argumentation layer, by giving a formal description of the communication language, the interaction protocol, the context of the dialog, the effect rules and the termination and outcome conditions.

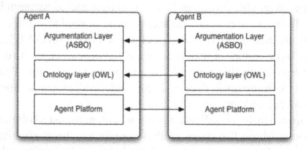

Fig. 1. Two ASBO agents showing the layered disposition of elements

2.1 The Communication Language

ASBO is restricted to the world of software agents. Its main goal is to provide them with an effective mechanism to enable coordination by using dialogs based on deductive mechanisms on personal beliefs. Thus, the ASBO communication language for a persuasion framework, denoted by $ASBO_{cm}$, is defined as the set of performatives which may be exchanged among agents.

Table 1 summarizes all the performatives defined to enable argumentation-based dialogs. They are defined with (1) the form of the utterance, (2) the semantic of the utterance, (3) the intention of the emitter and (4) the effect on the receiver. Beliefs exchanged among agents are denoted with φ and arguments with S, defined as $S = (\varphi, \Phi)$, where φ is the conclusion obtained from the argument, and Φ is the support set which holds that conclusion. In turn, Φ is composed of propositions called *premises*, denoted here by ϕ.

2.2 The Interaction Protocol

The rules of the conversation (i.e. how to start, follow and end an argumentation dialog) are defined in ASBO by using an AUML diagram [1] (see figure 2). Let us denote this protocol with P_{ASBO}. P_{ASBO} starts when the proponent *claims* a proposition φ. The opponent may respond *conceding* that proposition if it agrees, or asking for reasons through the performative *Why*. Then, the proponent *retracts* the proposition φ if no argument can be built for it, or it reacts with a valid argument S (performative *Since*). Now, the opponent may *accept* the argument, or *concede some* premises in S *together opposing to* any other proposal in the argument ($Concede$-but-$Claim$), or directly *oppose to* the conclusion or a premise in S by giving another argument T. The concept of *opposition* to a proposal is represented by the negation symbol $\neg$. The rest of the protocol can easily be understood from figure 2. Moreover, a $[No$-$Response\ p]$ could be sent at any time during the exchange of performatives, although it has not been included in the figure for simplicity. This performative has no responses and it will only be used when an agent can not oppose to a performative p neither p can be accepted.

Table 1. Performatives available in ASBO persuasion dialogs

Utterance	Lliteral meaning	Intention of the emitter	Effect in the receiver
Claim φ	Assertion of proposition φ	To impose φ to other agents	The speaker is associated to φ
Why φ	Challenge of proposition φ, looking for reasons	To obtain arguments for φ	The proposal φ must be justified with arguments
φ Since S	Disclosure of the argument S that supports φ	To prove φ as a valid derivation	The speaker is associated to premises ϕ_i in $S, i = 1..n$
Concede φ	Assumption of proposition φ	To announce that the speaker agrees with φ	The speaker is associated to φ
Concede-but-Claim Ω, v	Assumption of the proposition set $\Omega = \{\omega_1, \ldots, \omega_m\}$, but asserting v at the same time	To announce that the speaker agrees with the propositions in Ω, however it tries to impose v to other agents	The speaker is associated to $\Omega \cup \{v\}$
Retract φ	Rejection of proposition φ	To withdraw φ because a valid argument for it is not found	The speaker withdraws from the proposition φ
Accept S	Acceptation of argument S	To update the speaker's knowledge according to the conclusions in S	The speaker accepts the proposal derived in S
No-Response p	Withdrawal from answering a performative p	To announce that the speaker has no more valid responses to p	The speaker can not continue responding a performative p

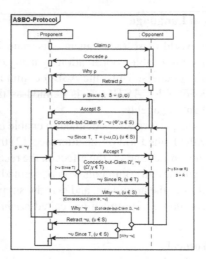

Fig. 2. AUML diagram of the ASBO persuasion protocol

Observe that the protocol in figure 2 offers several options to answer a performative. These options are called the *answer result set* (ARS) of a performative. For example, suppose that an agent receives the message $[Claim\ \varphi]^{1}$. By applying P_{ASBO} to it, the agent obtains the following answer result set: $\{[Why\ \varphi], [Concede\ \varphi], [No\text{-}Response(Claim\ \varphi)]\}$. Notice that the ARS for a performative is not sufficient to generate a suitable response. Thus, an additional piece of information is needed for that: the context of the conversation.

2.3   The Context and the State of a Conversation

The context of a conversation is the set of information with is helpful for an agent in order to take a decision about which possible performative in an ARS is used as an answer. This includes the sequence of messages emitted until that moment, the beliefs the agent holds and the proposals emitted by the other party. In ASBO, a context is defined as follows:

Definition 1 (Context). *Let us consider P, O as two agents which are using P_{ASBO} to exchange messages. Let Δ_P^t be the set of beliefs, intentions and desires P holds in a particular moment t. Let $H_{P,O}^t$ be the set of messages emitted by P to O in t, and let $\Pi_{P,O}^t$ be the set of proposals that O has communicated to P in t. Then, a context for the agent P in a moment t with respect to agent O is defined as the triple $(\Delta_P^t, H_{P,O}^t, \Pi_{P,O}^t)$, denoted as $C_{P,O}^t$.*

Notice that $H_{P,O}^t$ contains the history of all messages sent by P to O until t. This history is used together with P_{ASBO} as follows: when the agent needs to know which utterances may be answered, it uses P_{ASBO} to obtain the corresponding ARS. After

---

[1] And that the conversation is in the appropriate state for considering such message as correct.

this, only those messages compound by a performative and a message body which do not cause cycles can be emitted. Precisely, messages m which cause cycles are these which for a particular message $m' \in H_{P,O}^t$, the performative and message body is the same in m and m' (i.e. either proposals φ in m and φ' in m' or arguments S in m and S' in m' are equivalent). Then, the remaining answers in the ARS are evaluated using Δ_P^t and $\Pi_{P,O}^t$ to decide which is the optimal answer (details are not given here).

Once the context of a conversation for a party is defined, it is possible to define a strategy function which, starting from the ARS of a performative, it generates an adequate response.

Function 1 (Strategy function $\mathcal{P}$). *We define the strategy function in ASBO as*

$$\mathcal{P} : ARS \times C \longrightarrow ASBO_{cm},$$

where ARS is the set of all possible answer result sets and C is the set of all possible contexts. Then, for two particular agents P, O involved in an exchange of performatives in an instant of time t, $\mathcal{P}$ is such that $\forall ars \in ARS$, $\mathcal{P}(ars, C_{P,O}^t) \notin H_{P,O}^t$.

Notice that if, according to the context, no valid responses can be selected in a specific ars because they are equivalent to those in $H_{P,O}^t$, then $\mathcal{P}(ars, C_{P,O}^t) = [No\text{-}Response\ p_t]$, where p_t is the performative being processed in the instant t.

Apart from the context, agents also need to maintain the *state* of the conversation. In this manner, they know which party has the turn of speaking and which performative is currently being processed. In ASBO, it is defined as follows.

Definition 2 (State). *Let P, O be two agents exchanging performatives in a moment t. Let $\mathcal{T}$ be the agent which is to send a performative in t, and let cp be the current performative being processed in t by $\mathcal{T}$. Then, the state of that exchange of performatives in t is the triple $(\mathcal{T}, cp, H_{P,O}^t \cup H_{O,P}^t)$, denoted as $\mathcal{S}^t$.*

2.4 Effect Rules

As $ASBO_{cm}$ performatives are communicative acts, the state $\mathcal{S}$ of a conversation must be updated according to effects of these acts. This process may be performed through two different directions. Suppose $\mathcal{S}^t = (P, p_t, H_{P,O}^t \cup H_{O,P}^t)$. Then, some performatives will update $\mathcal{S}$ in $t+1$ as $\mathcal{S}^{t+1} = (O, p_{t+1}, H_{P,O}^{t+1} \cup H_{O,P}^{t+1})$, i.e. the agent P emitting a performative in t is now the receiver of the next one in $t+1$. Some other performatives, namely *surrendered* (*Concede, Retract, Accept* and *No-Response*), update $\mathcal{S}$ in $t+1$ as $\mathcal{S}^{t+1} = (P, p_{t'}, H_{P,O}^{t+1} \cup H_{O,P}^{t+1})$, where $t' < t$ and $p_{t'} \in H_{O,P}^t$, i.e. P must again utter a new performative as response to a previous message sent by O. This situation occurs when the emitter agent has reached the end of an argumentation line in t, and therefore it is still the responsible for continuing the exchange of performatives. In order to define these exploration mechanisms, ASBO agents need a *backward* function.

Function 2 (Backward function $\mathcal{B}$). *We define a backward function in ASBO as*

$$\mathcal{B} : ASBO_{cm} \times S \longrightarrow ASBO_{cm},$$

where S is the set of all possible states. Then, for a particular perfomative p_{t+n} and the state of the exchange of performatives between agents E and R in $t + n$, $\mathcal{S}^{t+n} = (E, p_{t+n}, H_{E,R}^{t+n} \cup H_{R,E}^{t+n})$, $\mathcal{B}(p_{t+n}, \mathcal{S}^{t+n}) = p_t$, if the following conditions hold:

1. *Let p_t, p_{t+1} and p_{t+n}, $n > 1$, be performatives exchanged among two agents E and R until the instant $t + n$, where $p_t, p_{t+n} \in H_{R,E}^{t+n}$, and $p_{t+1} \in H_{E,R}^{t+n}$*
2. *$p_{t+1} = \mathcal{P}(P_{ASBO}(p_t), C_{E,R}^t)$, and $p_{t+n} = \mathcal{P}(P_{ASBO}(p_{t+1}), C_{R,E}^{t+n-1})$*

Then, if E emits a surrendered performative in $\mathcal{S}^{t+n+1}$ as an answer to p_{t+n}, it must apply $\mathcal{B}(p_{t+n}, \mathcal{S}^{t+n})$, and then $\mathcal{S}^{t+n+1} = (E, \mathcal{B}(p_{t+n}), H_{E,R}^{t+n+1} \cup H_{R,E}^{t+n+1})$. A case of use of $\mathcal{B}$ is given in section 3. Let us see now the effect rules.

Definition 3 (Effect rules). *Let E, R be the emitter and receiver agents of a performative p_{t+1}, respectively. Let $\mathcal{S}^t = (E, p_t, H_{E,R}^t \cup H_{R,E}^t)$ be the state of the exchange of performatives between these agents in the instant t. Let $C_{E,R}^t$ and $C_{R,E}^t$ be the context of agent E and R in t, respectively, such as $p_{t+1} = \mathcal{P}(P_{ASBO}(p_t), C_{E,R}^t)$. Then, the set of effect rules for p_{t+1} is defined as follows:*

ER1. if $p_{t+1} = [Claim\ \varphi]$, then $C_{E,R}^{t+1} = (\Delta_E^t, H_{E,R}^t \cup \{p_{t+1}\}, \Pi_{E,R}^t)$, $C_{R,E}^{t+1} = (\Delta_R^t, H_{R,E}^t, \Pi_{R,E}^t \cup\{\varphi\})$, and $\mathcal{S}^{t+1} = (R, p_{t+1}, H_{E,R}^t \cup H_{R,E}^t \cup \{p_{t+1}\})$

ER2. if $p_{t+1} = [Why\ \varphi]$, then $C_{E,R}^{t+1} = (\Delta_E^t, H_{E,R}^t \cup \{p_{t+1}\}, \Pi_{E,R}^t)$, $C_{R,E}^{t+1} = C_{R,E}^t$, and $\mathcal{S}^{t+1} = (R, p_{t+1}, H_{E,R}^t \cup H_{R,E}^t \cup \{p_{t+1}\})$

ER3. if $p_{t+1} = [\varphi\ Since\ S]$, $S = \{\varphi, \Phi\}$, then $C_{E,R}^{t+1} = (\Delta_E^t, H_{E,R}^t \cup \{p_{t+1}\}, \Pi_{E,R}^t)$, $C_{R,E}^{t+1} = (\Delta_R^t, H_{R,E}^t, \Pi_{R,E}^t \cup \Phi)$, and $\mathcal{S}^{t+1} = (R, p_{t+1}, H_{E,R}^t \cup H_{R,E}^t \cup \{p_{t+1}\})$

ER4. if $p_{t+1} = [Concede\ \varphi]$, then $C_{E,R}^{t+1} = (\Delta_E^t, H_{E,R}^t \cup \{p_{t+1}\}, \Pi_{E,R}^t)$, $C_{R,E}^{t+1} = (\Delta_R^t, H_{R,E}^t, \Pi_{R,E}^t \cup \{\varphi\})$, and $\mathcal{S}^{t+1} = (E, \mathcal{B}(p_t), H_{E,R}^t \cup H_{R,E}^t \cup \{p_{t+1}\})$

ER5. if $p_{t+1} = [Concede\text{-}but\text{-}Claim\ \Omega, v]$, then $C_{E,R}^{t+1} = (\Delta_E^t, H_{E,R}^t \cup\{p_{t+1}\}, \Pi_{E,R}^t)$, $C_{R,E}^{t+1} = (\Delta_R^t, H_{R,E}^t, \Pi_{R,E}^t \cup \Omega \cup \{v\})$, and $\mathcal{S}^{t+1} = (R, p_{t+1}, H_{E,R}^t \cup H_{R,E}^t \cup \{p_{t+1}\})$

ER6. if $p_{t+1} = [Retract\ \varphi]$, then $C_{E,R}^{t+1} = (\Delta_E^t - \{\varphi\}, H_{E,R}^t \cup \{p_{t+1}\}, \Pi_{E,R}^t)$, $C_{R,E}^{t+1} = (\Delta_R^t, H_{R,E}^t, \Pi_{R,E}^t - \{\varphi\})$, and $\mathcal{S}^{t+1} = (E, \mathcal{B}(p_t), H_{E,R}^t \cup H_{R,E}^t \cup \{p_{t+1}\})$

ER7. if $p_{t+1} = [Accept\ S]$, $S = \{\varphi, \Phi\}$, then $C_{E,R}^{t+1} = (\Delta_E^t \cup \{\varphi\}, H_{E,R}^t \cup \{p_{t+1}\}, \Pi_{E,R}^t)$, $C_{R,E}^{t+1} = (\Delta_R^t, H_{R,E}^t, \Pi_{R,E}^t \cup \{\varphi\})$, and $\mathcal{S}^{t+1} = (E, \mathcal{B}(p_t), H_{E,R}^t \cup H_{R,E}^t \cup \{p_{t+1}\})$

ER8. if $p_{t+1} = [No\text{-}Response\ p_t]$, then $C_{E,R}^{t+1} = (\Delta_E^t, H_{E,R}^t \cup \{p_{t+1}\}, \Pi_{E,R}^t)$, $C_{R,E}^{t+1} = C_{R,E}^t$, and $\mathcal{S}^{t+1} = (E, \mathcal{B}(p_t), H_{E,R}^t \cup H_{R,E}^t \cup \{p_{t+1}\})$

2.5 Termination and Outcome Conditions

There are three circumstances in ASBO that produce the end of a dialog discussing the proposition φ. In the first one, the opponent emits $[Concede\ \varphi]$ after accepting whichever argument that supports φ, since it can not find any new opposition to that argument. In the second one, the proponent utters $[Retract\ \varphi]$ when all the arguments supporting φ have been invalidated. Finally, the third condition occurs when whichever party emits $[No\text{-}Response\ p_t]$, such as $p_t = [Claim\ \varphi]$ or $p_t = [Why\ \varphi]$. This last condition means that the agent withdraws without accepting or rejecting the initial claim φ. Moreover, following these termination conditions, the outcome of the dialog can also be defined. If the first condition holds, the proponent wins and the opponent accepts φ. If the second condition holds, then the opponent wins and the proponent must reject φ. In the third case, none of the agents wins and φ is undecided with respect to them. The termination and outcome conditions are formally expressed next.

Definition 4 (Termination and outcome conditions). *Let S^t be the state of a dialog between a proponent P and an opponent O in the instant t. Let $p_1 \in H^t_{P,O} \cup H^t_{O,P}$ be the message $[Claim\ \varphi]$. Then, the dialog terminates when:*

1. $S^t = (O, cp, H^t_{P,O} \cup H^t_{O,P})$, *and* $p_{t+1} = [Concede\ \varphi]$, *or*
2. $S^t = (P, cp, H^t_{P,O} \cup H^t_{O,P})$, *and* $p_{t+1} = [Retract\ \varphi]$, *or*
3. $S^t = (T, p_{t'}, H^t_{P,O} \cup H^t_{O,P}), t' \leq 2$, *and* $p_{t+1} = [No\text{-}Response\ p_{t'}]$

In the first case, P wins the dialog and φ is accepted. In the second case, O is the winner and φ is not accepted. Otherwise, φ is undecided.

3   An Example

In order to illustrate all the ideas explained so far in this section, an example of a complete ASBO persuasion dialog is given next. Let us consider an agent A_{Ex} with arguments $S=(\varphi, \Phi_S)$ and $R=(\varphi, \Phi_R)$. Now, let B_{Ex} be an agent with an argument $T = (\psi, \Phi_T)$, where ψ defeats a premise in Φ_S (i.e. T undercuts S). Both agents start an ASBO persuasion dialog, being A_{Ex} the proponent by claiming φ, and B_{Ex} the opponent to that claim. At the initial moment, $\Delta^0_{A_{Ex}} = \{\{\varphi\} \cup \Phi_S \cup \Phi_R\}$ and $\Delta^0_{B_{Ex}} = \{\{\psi\} \cup \Phi_T\}$. Then, the dialog is developed as shown next:

- $t=0,\ S^0=(A_{Ex}, -, \emptyset),\ C^0_{A_{Ex}}=(\Delta^0_{A_{Ex}}, \emptyset, \emptyset)$, and $C^0_{B_{Ex}}=(\Delta^0_{B_{Ex}}, \emptyset, \emptyset)$. Then, p_1 $=[Claim\ \varphi]$
- $t=1, S^1=(B_{Ex}, p_1, \{p_1\}), C^1_{A_{Ex}}=(\Delta^0_{A_{Ex}}, \{p_1\}, \emptyset)$, and $C^1_{B_{Ex}}=(\Delta^0_{B_{Ex}}, \emptyset, \{\varphi\})$. Then, $\mathcal{P}(P_{ASBO}(p_1), C^1_{B_{Ex}}) = p_2 = [Why\ \varphi]$
- $t=2, S^2=(A_{Ex}, p_2, \{p_1, p_2\}), C^2_{A_{Ex}}=C^1_{A_{Ex}}$, and $C^2_{B_{Ex}}=(\Delta^1_{B_{Ex}}, \{p_2\}, \{\varphi\})$. Then, $\mathcal{P}(P_{ASBO}(p_2), C^2_{A_{Ex}}) = p_3 = [\varphi\ Since\ S]$

Here, A_{Ex} might also have uttered $[\varphi\ Since\ R]$. For the purpose of this example, however, suppose that $[\varphi\ Since\ S]$ is the most appropriate alternative according to A_{Ex}'s rationality.

- $t=3,\ S^3 = (B_{Ex}, p_3, \{p_1, p_2, p_3\}),\ C^3_{A_{Ex}} = (\Delta^2_{A_{Ex}}, \{p_1, p_3\}, \emptyset)$, and $C^3_{B_{Ex}} = (\Delta^2_{B_{Ex}}, \{p_2\}, \{\varphi \cup \Phi_S\})$. Then, $\mathcal{P}(P_{ASBO}(p_3), C^3_{B_{Ex}}) = p_4 = [\psi\ Since\ T]$

Note that argument T defeats argument S since ψ undercuts a premise in Φ_S.

- $t=4$, $\mathcal{S}^4 = (A_{Ex}, p_4, \{p_1, p_2, p_3, p_4\})$, $C_{A_{Ex}}^4 = (\Delta_{A_{Ex}}^3, \{p_1, p_3\}, \{\psi \cup \Phi_T\})$, and $C_{B_{Ex}}^4$ $= (\Delta_{B_{Ex}}^3, \{p_2, p_4\}, \{\varphi \cup \Phi_S\})$. Then, $\mathcal{P}(P_{ASBO}(p_4), C_{A_{Ex}}^4) = p_5 = [Accept\ T]$

Now, p_5 is a surrendered performative answering to p_4. By applying $\mathcal{B}$ to p_4, the performative to process in $\mathcal{S}^5$ by A_{Ex} is p_2 again. This result is due to p_4 is answering to p_3, and in turn, p_3 is a response to p_2.

- $t=5$, $\mathcal{S}^5 = (A_{Ex}, p_2, \{p_1, p_2, p_3, p_4, p_5\})$, $C_{A_{Ex}}^5 = (\Delta_{A_{Ex}}^4 \cup \{\psi\}, \{p_1, p_3, p_5\}, \{\psi \cup \Phi_T\})$, and $C_{B_{Ex}}^5 = (\Delta_{B_{Ex}}^4, \{p_2, p_4\}, \{\varphi \cup \Phi_S \cup \psi\})$. Then, $\mathcal{P}(P_{ASBO}(p_2), C_{A_{Ex}}^5) = p_6 = [\varphi\ Since\ R]$

A_{Ex} opens a new argumentation path after accepting argument T. Hence, it gives a new argument for the proposition φ, answering to the performative $[Why\ \varphi]$ again.

- $t=6$, $\mathcal{S}^6 = (B_{Ex}, p_6, \{p_1, p_2, p_3, p_4, p_5, p_6\})$, $C_{A_{Ex}}^6 = (\Delta_{A_{Ex}}^5, \{p_1, p_3, p_5, p_6\}, \{\psi \cup \Phi_T\})$, and $C_{B_{Ex}}^6 = (\Delta_{B_{Ex}}^5, \{p_2, p_4\}, \{\varphi \cup \Phi_S \cup \psi \cup \Phi_R\})$. Then, $\mathcal{P}(P_{ASBO}(p_6), C_{B_{Ex}}^6) = p_7 = [Accept\ R]$

B_{Ex} emits a surrendered performative in p_7. It responds to performative p_6, which in turn is an answer to p_2. Finally, p_2 is answering to p_1. As a result, $\mathcal{B}(p_6, \mathcal{S}^6) = p_1$ is the peformative to be processed by B_{Ex} in $\mathcal{S}^7$.

- $t=7$, $\mathcal{S}^7 = (B_{Ex}, p_1, \{p_1, p_2, p_3, p_4, p_5, p_6, p_7\})$, $C_{A_{Ex}}^7 = (\Delta_{A_{Ex}}^6, \{p_1, p_3, p_5, p_6\}, \{\psi \cup \Phi_T \cup \varphi\})$, and $C_{B_{Ex}}^7 = (\Delta_{B_{Ex}}^6 \cup \varphi, \{p_2, p_4, p_7\}, \{\varphi, \Phi_S, \psi, \Phi_R\})$. Then, $\mathcal{P}(P_{ASBO}(p_1), C_{B_{Ex}}^7) = p_8 = [Concede\ \varphi]$

Eventually, the dialog has finished since B_{Ex} has conceded the initial proposition φ. The dialog sequence is $[\lfloor Claim\ \varphi \rfloor, \lceil Why\ \varphi \rceil, \lfloor \varphi\ Since\ S \rfloor, \lceil \psi\ Since\ T \rceil, \lfloor Accept\ T \rfloor, \lfloor \varphi\ Since\ R \rfloor, \lceil Accept\ R \rceil, \lceil Concede\ \varphi \rceil]$, where $\lfloor p \rfloor$ is a performative uttered by A_{Ex} and $\lceil p \rceil$ is a performative uttered by B_{Ex}. Consequently, A_{Ex} has won the dialog, and φ is finally accepted by B_{Ex}.

4   Conclusions

Argumentation has demonstrated to be a useful approach when resolving conflicts during the exchange of proposals in a MAS. This is possible due to agents can use their rationality to build and attack arguments that justify those proposals. However, most argumentative approaches in MAS are theoretical. We have developed an Argumentation System Based on Ontologies, ASBO, with an engineering oriented approach. In this manner, ASBO is offered as a software tool to perform argumentation in conventional MAS. This paper has presented the formal background for the persuasion dialog framework used in ASBO.

One future step in ASBO is to adopt a BDI agency paradigm. In this manner, an agent could select the optimal argument during the persuasion dialog according to its intentions and desires.

Acknowledgments. This work has been supported by the Fundación Séneca grant "Programa de Ayuda a los grupos de excelencia 04552/GERM/06 " and thanks to the

Spanish Ministerio de Ciencia e Innovación (MICINN) under the grant AP2006-4154 in frames of the FPU Program and the TIN2005-08501-C03-01 Program.

References

1. Bauer, B., Müller, J.P., Odell, J.: Agent uml: A formalism for specifying multiagent software systems. In: Ciancarini, P., Wooldridge, M.J. (eds.) AOSE 2000. LNCS, vol. 1957, pp. 91–104. Springer, Heidelberg (2001)
2. Berners-Lee, T., Hendler, J., Lassila, O.: The Semantic Web. Scientific American (2001)
3. Brewka, G.: Dynamic argument systems: A formal model of argumentation processes based on situation calculus. JLC: Journal of Logic and Computation 11 (2001)
4. Carbogim, D.V., Robertson, D., Lee, J.: Argument-based applications to knowledge engineering. Knowledge Engineering Review 15(2), 119–149 (2000)
5. Dean, M., Connoll, D., van Harmelen, F., Hendler, J., Horrocks, I., McGuinness, D.L., Patel-Schneider, P.F., Andrea Stein, L.: Web ontology language (OWL). Technical report, W3C (2004)
6. dos Santos, N., ao Miguel Varej, F., de Lira Tavares, O.: Multi-agent systems and network management - a positive experience on unix environments. In: Garijo, F.J., Riquelme, J.-C., Toro, M. (eds.) IBERAMIA 2002. LNCS, vol. 2527, pp. 616–624. Springer, Heidelberg (2002)
7. Karunatillake, N., Jennings, N.R.: Is it worth arguing? In: Rahwan, I., Moraïtis, P., Reed, C. (eds.) ArgMAS 2004. LNCS (LNAI), vol. 3366, pp. 134–250. Springer, Heidelberg (2005)
8. Munoz, A., Botia, J.A.: ASBO: Argumentation system based on ontologies. In: Klusch, M., Pěchouček, M., Polleres, A. (eds.) CIA 2008. LNCS (LNAI), vol. 5180, pp. 191–205. Springer, Heidelberg (2008)
9. Munoz, A., Botia, J.A., Garcia, F.J., Martinez, G., Gomez Skarmeta, A.F.: Solving conflicts in agent-based ubiquitous computing systems: a proposal based on argumentation. In: Agent-Based Ubiquitous Computing (ABUC) Workshop, International Conference on Autonomous Agents and Multiagent Systems (AAMAS 2007), Honolulu, Hawaii (May 2007)
10. Parsons, S., Sierra, C., Jennings, N.R.: Agents that reason and negotiate by arguing. Journal of Logic and Computation 1998(3), 261–292 (1998)
11. Prakken, H.: Coherence and flexibility in dialogue games for argumentation. Journal of Logic and Computation 15(6), 1009–1040 (2005)

A Dialogue-Game Approach for Norm-Based MAS Coordination

S. Heras, N. Criado, E. Argente, and V. Julián

Departamento de Sistemas informáticos y Computación
Universidad Politécnica de Valencia
Camino de Vera s/n, 46022 Valencia, Spain
{sheras,ncriado,eargente,vinglada}@dsic.upv.es

Abstract. Open societies are situated in dynamic environments and are formed by heterogeneous autonomous agents. For ensuring social order, norms have been employed as coordination mechanisms. However, the dynamical features of open systems may cause that norms loose their validity and need to be adapted. Therefore, this paper proposes a new dialogue game protocol for modelling the interactions produced between agents that must reach an agreement on the use of norms. An application example has been presented for showing the performance of the protocol and its usefulness as a mechanism for managing the solving process of a coordination problem through norms.

1 Motivation

Nowadays, Multi-agent Systems (MAS) research on addressing the challenges related to the development of open distributed systems is receiving an increasing interest. In these systems, entities might be unknown and none assumption about their behaviour can be done. Thus, mechanisms for coordinating their behaviours and ensuring social order are essential. Works on Normative Theory have been applied into the MAS area as a mechanism for facing up with undesirable and unexpected behaviours [5].

Two different approximations have been considered as alternatives to the definition of norms: (i) off-line design and (ii) automatic emergence. The latter is more suitable for open systems, in which structural, functional and environmental changes might occur [10]. Therefore, dynamical situations may cause that the norms that regulate an organization lose their validity or should be adapted. In this second case, techniques for reaching an agreement among agents on the employment of norms are needed.

This research is aimed at providing a mechanism for managing norm emergence in open environments. THOMAS framework [3], a development platform for Virtual Organisations (VO) [2], has been selected as a suitable environment to test this proposal. In THOMAS, the coordination among heterogeneous agents is achieved by means of norms. Thus, a normative language for formalising constraints on agent behaviours has been developed [1]. The main idea that inspired this platform was to give support a better integration between the standardised service-oriented computing technologies and the MAS paradigm. Therefore, every operation that can be performed in the THOMAS framework is described and offered by means of Web Services standards. The management of services in THOMAS is carried out by the Service Facilitator (SF) component, which offers an extended yellow pages service. It

E. Corchado et al. (Eds.): HAIS 2009, LNAI 5572, pp. 468–475, 2009.

is in charge of registering and publicising services and also provides discovering and composition services. A description of the THOMAS platform can be found at[1].

Our approach is to apply dialogue games as an argumentation technique to model the interaction among agents that must reach an agreement about the normative context. Dialogue games are interactions between players where each one moves by advancing locutions in accordance to some rules [6]. In MAS, they have been used to specify communication protocols [7] and to evaluate reputation [8]. The application of argumentation techniques to norms definition, however, is a new area of research. This paper is structured as follows: Section 2 describes the proposed dialogue game protocol; Section 3 shows an example application of the proposed protocol for solving a coordination problem; and Section 4 gathers conclusions of this research.

2 Dialogue Game Protocol Specification

In this section, we explain the main features of the proposed dialogue game protocol. First, the classical structure of the *Argument from Expert Opinion* scheme is shown. Then, the social structure that allows agents to evaluate arguments of other agents is described. Based on it, the dialogue game protocol is specified. Finally, how agents can pose arguments and rebut attacks is also detailed.

2.1 Argument from Expert Opinion

We have adapted a general dialogue game, the *Argumentation Scheme Dialogue (ASD)* [9] to formalise the interaction among agents that argue about a normative context. This game extends traditional dialogue games to allow agents to reason with stereotyped patterns of common reasoning called argumentation schemes. We have instantiated this game with a specific scheme, the *Argument from Expert Opinion* [11] that shows the way in which people evaluate the opinions of experts (i.e. agents with knowledge about the norms to use). The structure of the scheme is the following:

- *Major Premise:* Source *E* is an expert in field *F* containing proposition *A*.
- *Minor Premise:* *E* asserts that proposition *A* (in field *F*) is true (false).
- *Conclusion:* Proposition *A* may plausibly be taken to be true (false).

Moreover, this scheme has also a set of *critical questions*, which represent the possible attacks that can be made to rebut the conclusion drawn from the scheme:

1. *Expertise:* How credible is *E* as an expert source?
2. *Field:* Is *E* an expert in the field *F* that *A* is in?
3. *Opinion:* What did *E* assert that implies *A*?
4. *Trustworthiness:* Is *E* personally reliable as a source?
5. *Consistency:* Is *A* consistent with what other experts assert?
6. *Backup Evidence:* Is *E*'s assertion based on evidence?

2.2 Social Structure

Following the normative emergence approach, in this paper we propose the use of a service of normative assistance to solve coordination problems in VOs. This service is

---

[1] http://www.fipa.org/docs/THOMASarchitecture.pdf

provided by a set of *Normative Assistant (NA)* agents that recommend the appropriate modifications to fit the operation of a VO to the current situation by adapting its normative context. The normative assistance services are published by the THOMAS SF and thus, they can be publicly requested by all agents of the organization.

Therefore, whenever an agent of the system wants to solve a coordination problem, it assumes the role of *initiator* and requests the SF a list of providers of the *normative assistance service*. Then, among these providers, it may select a subset of NAs with agents that it personally considers as 'friends' (known agents, if any). This friendship relation comes from past recommendation processes where the initiator was involved. The experience-based friendship relations of all agents of the system can be represented by a social network abstraction. The network topology would be implicitly defined through the confidence relations that an agent has with its friends and thus decentralised. In this network, nodes would represent agents and links would be friendship relations between them (labelled with confidence values).

Therefore, since the first recommendation dialogue where an agent was engaged in as *initiator*, this agent keeps a list of all the agents that participated in the dialogue and its final assessment of its confidence about the recommendations received from other agents. This *confidence degree* $c_{ij} \in [-1, 1]$ is updated at the end of each recommendation process by using a discrete value $u_j \in \{-1: inappropriate, 0: useless or 1: useful\}$ that stands for the final usefulness of each recommendation received by the initiator. In addition, the initiator also informs the NAs that participated in the dialogue of the usefulness value of their recommendations. With this value, each NA can update its *expertise degree* e_j as norm recommenders. These degrees are computed by using equations 1 and 2:

$$c_{ij} = \frac{\sum_{k=1}^{K} u_{j(k)}}{K} \qquad (1) \qquad\qquad e_j = \frac{\sum_{i=1}^{I} c_{ij}}{\deg^+(a_j)} \qquad (2)$$

where $u_{j(k)}$ is the usefulness degree of the recommendation k received by the initiator from its friend a_j, K is the total number of recommendation dialogues performed by the initiator with its friend a_j and $\deg^+(a_j)$ is the number of agents that have a_j as friend (i.e. its *in-degree* in the network).

We also assume that each NA agent stores the set of norms that were effective to solve a coordination problem in the VOs where it has participated in a *norm database*. In addition, for every set of norms, it also stores the attributes that characterised the type of problem addressed. Therefore, the knowledge about the solving process of the different coordination problems is distributed across the network and each NA agent only has a partial view of it based on its own experience. Moreover, if a NA is asked for a set of norms to deal with a coordination problem that it has never been faced with, the agent can also propagate the query to its neighbours in the network by using its friendship relationships (i.e. such NA agents that have eventually provided this NA agent with recommendations in the past). When all NAs have made their proposals, the initiator has been presented with several recommendations about sets of norms. Then, it selects the best proposal by using the proposed dialogue game protocol.

Finally, note that each agent is assumed to have its own reasoning mechanism for evaluating preferences, matching them with its norm database and proposing recommendations. In addition, agents must also know a set of inference rules and the scheme from expert opinion to be able to create arguments. The definition of the individual reasoning mechanisms of agents (e.g. how they manage arguments and evaluate the usefulness degree of recommendations) is out of the scope of this paper. We have mainly focused here on formalising the dialogue-based interaction protocol. Following, the specification of this interaction protocol is detailed.

2.3 Protocol Specification

The dialogue game protocol is defined by a set of commencement rules, locutions, commitment rules, dialogue rules and termination rules. For space restrictions, we only show the dialogue rules of the protocol. The complete specification is provided in [4]. The process consists in a set of parallel dialogues between the initiator and the NAs (who do not speak directly between them). In each step of the dialogue, either the initiator makes a move or the NA answers it by posing the permissible locutions following dialogue rules. Note that, although the information provided by a NA is not directly accessible by another, the initiator acts as a mediator and is able to use this information when speaking with other NAs. The game can end at any step, when the initiator decides to select a recommendation. Next, the dialogue rules are specified:

1. The initiator's request opens the dialogue and each NA can answer it by a request for more information about the properties that characterise the problem, a normative set proposal or a rejection to provide the assistance service.
2. NAs can ask for more information while the initiator accedes to provide it. Finally, the NA makes a proposal (*proponent* role) or rejects to provide the service.
3. Each normative set proposal can be followed by a request for an argument supporting the recommendation, or by its acceptance or rejection.
4. If a recommendation is challenged, the proponent must show its argument for recommending such normative set or withdraw the recommendation.
5. The initiator can reply to the argument of a NA by accepting the recommendation, by rejecting it or by posing a critical attack associated with the *Argument from Expert Opinion* explained in section 2.2. Possible attacks in our context are: questioning the degree of expertise of the NA or demonstrating that the recommendation of the NA is not consistent with other NA recommendations with equal or greater degree of confidence or expertise. Note that we assume (a) every NA can be considered an expert to some extent (*Field question*); (b) that NAs are rational and always propose the recommendation that, using their reasoning mechanisms, mostly fits the features of the current problem (*Opinion question*) and (c) that NAs are honest and their recommendations and arguments are based on their experience (*Backup evidence question*).
6. Trustworthiness and Consistency attacks can also be challenged by the NA. Then, the initiator must provide an argument supporting the attack.
7. NAs can rebut attacks or else, withdraw their recommendations.
8. Finally, the initiator can accept the argument of a NA and choose its recommendation, preliminary accept its argument but pose another attack or reject the argument and hence, the recommendation (ending the dialogue with this NA).

2.4   Decision-Making Process of NAs

The basic decision policy that follows every NA is to do its best to convince the initiator that its normative set recommendation is the most appropriate one to solve the coordination problem. There are two different types of arguments that NAs can use to persuade the initiator: arguments for justifying a recommendation and arguments for rebutting an attack. An argument of the former type consists in a set of common attributes among the problem characterisation and the proposed solution. Regarding the rebutting arguments, they consists on a partial ordering relation between confidence or expertise degrees. The simplest case is to rebut an *expertise attack*, since the NA can only show its expertise degree (note that expertise degrees are private and this attack is thought to provide the initiator with this information).

In the case of a *trustworthiness attack*, the initiator i attacks the recommendation of the proponent p because it has received a different recommendation from other NA n with a higher confidence degree for the initiator. Then, p can rebut the attack if the conditions expressed on equation 3 hold:

$$Argument\ AR = (\ c, <, c)$$
$$partial\ ordering\ relation\ \delta = \{<\}$$
$$attacks \subseteq AR \times AR \tag{3}$$

Case (a)	Case (b)
$(c_{pn}, <, c_{in})$ attacks $(c_{ip}, <, c_{in})$ iff	$(c_{in}, <, c_{pk})$ attacks $(c_{ip}, <, c_{in})$ iff
$c_{pn} < c_{ip}$	$0 < c_{ip} < c_{in} < c_{pk}$

In case (a), p can rebut the attack if it personally knows n and its confidence degree in this NA c_{pn} is lower than the confidence degree c_{ip} of the initiator in p. In case (b), p can rebut the attack if the recommendation was propagated to the NA k and the confidence degree c_{pk} of p in this NA is higher than the confidence degree c_{ip} that the initiator has in p and also the confidence degree c_{in} that the initiator has in the NA n.

In the case of a *consistency attack*, the initiator i attacks the recommendation of the proponent p because it has received a different recommendation from other NA n with a higher expertise degree for the initiator. Then, p can rebut the attack if the conditions expressed on equation 4 hold:

$$Argument\ AR = (\ e, <, e)$$
$$partial\ ordering\ relation\ \delta = \{<\}$$
$$attacks \subseteq AR \times AR \tag{4}$$
$$(e_n, <, e_k)\ attacks\ (e_p, <, e_n)\ iff$$
$$e_p < e_n < e_k$$

In this case, p can rebut the attack if the recommendation was propagated to the NA k and the expertise degree e_k of this NA k is higher than the expertise degree e_p of p and also the expertise degree e_n of the NA n.

3   Normative Context Definition

In order to show the operation of our dialogue game protocol, we have applied it for solving a coordination problem. More concretely, the addressed problem consists in selecting the most suitable norms for a new organization that an agent wants to create.

3.1   Problem Formalization

In our approach, coordination is achieved by means of norms that regulate the activities of agents. Therefore, a coordination problem is formalised as a structure of the type $<problemID, \gamma>$ where $\gamma=\{\ \gamma_0,\ \gamma_1,\ ...,\gamma_k\ \}$ is a set of attributes or properties that characterise the problem; $\gamma_i \in D_i,$ being D_i the domain associated to the property i.

In the application example, the addressed problem consists on the definition of the normative context for a new VO according to its desirable features. This problem is a particular case of a coordination problem. Thus, a problem of normative context definition is defined as $<normDefinition,\gamma>$ where $\gamma=\{\ s,\ m,\ v,\ c,\ f\ \}$ and

- $s=<who,\ how>$ is a property that determines who can change the structural components of the system (i.e. roles, units and norms); where $who \in \{none,\ supervisor,\ all\}$ and $how \in \{increase,\ decrease,\ modify\}$.
- $m \in N$ is the property that specifies the cardinality of the VO members.
- $v \in \{none\ ,supervisor,\ members,\ all\}$ is the visibility property that establishes the access rights to the information of the VO.
- $c \in \{none,\ supervisor,\ all\}$ is the property that determines if the expulsion service can be used as a control mechanism.
- $f=<who,\ how>$ is the property that defines who and how the functionalities, i.e. services, of the VO can be changed; where $who \in \{none,\ supervisor,\ all\}$ and $how \in \{increase,\ decrease,\ modify\}$.

3.2   Application Example

The social network of normative assistants can be implemented in THOMAS as a VO (named *NormativeAssistance*). It is a group of agents formed by all agents that can provide normative assistance (*Assistants*) to other agents. Therefore, agents in charge of the maintenance of the normative context that regulates a specific group can request an advice for norms effective in some scenario. All the communication and functionalities specified by the protocol are carried out as services.

As an example, Figure 1 contains an overview of the services performed by a set of agents which are playing the proposed dialogue game. Let us consider the case that an agent (initiator) has requested the service for creating a new VO in THOMAS. In this situation, the agent needs to define the normative context that will regulate its VO. However, this agent does not know a priory the suitable set of norms for its VO. Thus, it can ask for a NormativeAssistance service to the SF, who will apply its techniques for service searching and composition to discover this recommendation service. If it exists, the initiator agent will obtain the list of service providers from the SF.

As previously mentioned, the *initiator* selects the NAs from the provider list according to its previous experiences. In this example, the *initiator* requests the *NormativeAssistance* service to NA_a and NA_b (Figure 1 steps 1a and 1b). Therefore, the *initiator* carries out two different dialogues concurrently (labelled as a and b). The NA_a does not have the requested information in its norm database and thus, it propagates the request to its friend NA_c (step 2a). NA_c queries its norm database and proposes the normative set NS_c (step 3a and its propagation in step 4a). Moreover, the NA_b proposes NS_b as a solution to the coordination problem characterized by γ. Since the *initiator* does no have enough information for choosing between NS_c and NS_b, it

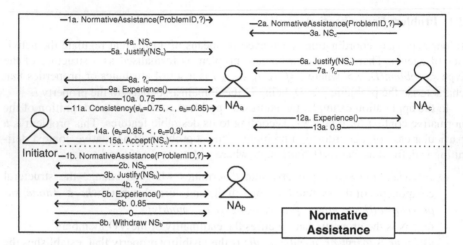

Fig. 1. Example of an execution of the proposed dialogue game

requests a justification of the recommended normative sets (steps 5a, its propagation 6a and step 3b). After receiving the justification of the proposed recommendation (steps 7a, its propagation 8a and step 4b) the *initiator* still does not have enough information for taking a decision. Thus, it poses an expertise attack (steps 9a and 5b) to NA_a and NA_b, which answer by informing about their expertise degree in steps 10a and 6b, respectively. Since the expertise of NA_b is higher than NA_a, the *initiator* poses a consistency attack to NA_a (step 11a). As a consequence, NA_a poses an expertise attack to NA_c (step 12a) to know its expertise degree (provided in step 13a). Hence, NA_a is able to rebut the confidence attack by showing that the expertise level of NS_c is higher than NS_b (step 14a). Then, the initiator poses a consistency attack to NA_b which cannot rebut it and hence, withdraws its recommendation (steps 7b and 8b, respectively). Finally, the initiator accepts the normative set NS_c as the best solution for the current coordination problem.

After this recommendation dialogue, the *initiator* applies the normative set NS_c in its VO and evaluates its effectiveness. The definition of the recommendation evaluation process carried out by agents is out of the scope of this paper. Therefore, the initiator is able to update its confidence degrees of the agents that engaged in the dialogue (NA_a, NA_b and NA_c). This last agent is added to its friend list. In addition, the *initiator* sends feedback information about the utility of the recommendation to all NAs, which update their expertise degrees.

4 Conclusion

This paper addresses the problem consisting in the dynamical definition of norms for agent societies. In open societies, which are situated in a dynamic environment and are formed by heterogeneous autonomous agents, the existence of mechanisms for adapting and modifying norms is essential. Therefore, our approach proposes the employment of dialogue games to model the interactions produced between agents that must reach an agreement on the use of norms. This work takes as basis well known proposals on dialogue games. In addition, the THOMAS framework has been used as an infrastructure for the dialogues among agents.

We have adapted the ASD dialogue game to fit our objective of supporting norm emergence in an open and social context. Finally, an application example has been presented in order to show both the performance of the dialogue game protocol and its usefulness as a mechanism for managing the solving process of a coordination problem through norms. As future work we plan to apply these approach to more complex scenarios in which a more elaborated process is required for solving these coordination problems.

Acknowledgements. This work is supported by TIN2006-14630-C03-01 projects of the Spanish government, FEDER funds and CONSOLIDER-INGENIO 2010 under grant CSD2007-00022, FPU grant AP-2007-01256 awarded to N.Criado.

References

1. Argente, E., Criado, N., Julián, V., Botti, V.: Norms for Agent Service Controlling. In: EUMAS 2008, pp. 1–15 (2008)
2. Boella, G., Hulstijn, J., van der Torre, L.: Virtual organizations as normative multiagent systems. HICSS IEEE Computer Society (2005)
3. Carrascosa, C., Giret, A., Julian, V., Rebollo, M., Argente, E., Botti, V.: Service Oriented Multi-agent Systems: An open architecture. In: AAMAS 2009 (in press, 2009)
4. Heras, S., Navarro, M., Botti, V., Julian, V.: Applying Dialogue Games to Manage Recommendation in Social Networks. In: ArgMAS 2009 (in press, 2009)
5. López y López, F., Luck, M., d'Inverno, M.: Constraining autonomy through norms. In: AAMAS 2002, pp. 674–681 (2002)
6. McBurney, P., Parsons, S.: Dialogue Games in Multi-Agent Systems. Informal Logic. Special Issue on Applications of Argumentation in Computer Science 22(3), 257–274 (2002)
7. McBurney, P., Parsons, S.: Games that Agents Play: A Formal Framework for dialogues between Autonomous Agents. Journal of Logic, Language and Information 12(2), 315–334 (2002)
8. O'Donovan, J., Smyth, B.: Trust in recommender systems. In: 10th International Conference on Intelligent User Interfaces, pp. 167–174 (2005)
9. Reed, C.: Walton. D.: Argumentation Schemes in Dialogue. In: Dissensus and the Search for Common Ground, OSSA 2007, vol. CD-ROM, pp. 1–11 (2007)
10. Verhagen, H.: Norm Autonomous Agents. PhD thesis, Dept. Computer Science, Stockholm University (2000)
11. Walton, D.: Appeal to Expert Opinion. Penn State Press (1997)

Incorporating a Temporal Bounded Execution to the CBR Methodology

M. Navarro, S. Heras, and V. Julián

Departamento de Sistemas Informáticos y Computación
Universidad Politécnica de Valencia
Camino de Vera s/n, 46022 Valencia, Spain
{mnavarro,sheras,vinglada}@dsic.upv.es

Abstract. In real-time Multi-Agent Systems, Real-Time Agents merge intelligent deliberative techniques with real-time reactive actions in a distributed environment. CBR has been successfully applied in Multi-Agent Systems as deliberative mechanism for agents. However, in the case of Real-Time Multi-Agent Systems the temporal restrictions of their Real-Time Agents make their deliberation process to be temporally bounded. Therefore, this paper presents a guide to temporally bound the CBR to adapt it to be used as deliberative mechanism for Real-Time Agents.

1 Introduction

The need for developing software solutions applied to complex, non-dynamic and frequently, non-completely specified environments, has contributed to the confluence of two important research areas, the Artificial Intelligence (AI) and the Real-Time System (RTS). Inside the Artificial Intelligence framework, Multi-Agent Systems paradigm (MAS) represents an appropriate approach for solving inherently distributed problems. The work presented in this paper is planned over the existent relationship between MAS and RTS. This work covers the problem of *Real-Time Agents (RTAs)*, which merge intelligent deliberative techniques with real-time reactive actions in a distributed environment.

A *Real-Time Artificial Intelligence System* (RTAIS) is a system that must accomplish critical processes under a dynamic environment with temporal restrictions by using AI techniques. Here, *anytime algorithms* [4] and *approximate processing* [7] are the most promising algorithms. One line of research in RTAI has been to build applications or architectures that embody real-time concerns in many components [7], such as Guardian [9], Phoenix [10], CIRCA/ SA-CIRCA [8] [17] and ARTIS [6] [2]. An appropriated agent for real-time environments must accomplish its goals, responsibilities and tasks with the added difficulty of temporal restrictions. Thus, a RTA can be defined as an agent with temporal restrictions in, at least, one of its responsibilities. The RTA may have its interactions bounded, and this modification will affect all the communication processes in the MAS where the RTA is located.

The main problem in the architecture of an RTA concerns the deliberation process. This process commonly uses AI techniques as problem-solving methods to compute more intelligent actions. However, the temporal restrictions of RTAs give rise to the

E. Corchado et al. (Eds.): HAIS 2009, LNAI 5572, pp. 476–483, 2009.

necessity of providing techniques that allow their response times to be bounded. These techniques are based on RTAIS techniques [7]. In addition, in a RTA an efficient integration of high-level, deliberative planning processes within reactive processes is necessary. These complex deliberative processes, which allow the agent to reason, to adapt and learn, are unbounded and it is difficult to integrate them in real-time systems. However, their main drawback lies in finding a mechanism that permits their efficient and temporal bounded execution.

In view of the successful applications reported in the literature (see Section 3), we propose a model where RTAs use a CBR method as deliberative mechanism to take decisions. However, the execution of this CBR method must be bounded in order to observe the RTA temporal restrictions. Thus, this paper presents a guide to temporally bound the CBR cycle to adapt it to be used as a deliberative mechanism for RTAs. The paper is structure as follow: Section 2 introduces the concept of Real-Time Agent; Section 3 reviews related successful applications of the CBR method in MAS; Section 4 presents a guide with the principal facts to take into account to temporally bound the CBR cycle and finally, some conclusions are shown in Section 5.

2 Real-Time Agent

A *Real-Time Agent* (RTA) is an agent composed of a series of tasks, some of which have temporal constraints [11]. In these agents, it is also necessary to take into account the temporal correctness, which is expressed by means of a set of temporal restrictions that are imposed by the environment. The RTA must, therefore, ensure the fulfilment of such temporal restrictions. By extension, a Real-Time Multi-Agent System (RTMAS) is a multi-agent system with at least one RTA [11]. Systems of this type require the inclusion of temporal representation in the communication process, management of a unique global time, and the use of real-time communication [23].

It is well-known that a typical real-time system is made up of a set of tasks characterized by a deadline, a period, a worst-case execution time and an assigned priority. These restrictions in the system functionality affect the features of an agent that needs to be modelled as a real-time system. The main problem is that if its tasks are not temporally bounded properly, is not possible to guarantee the fulfilment of the tasks before a deadline is expired and to schedule a plan with these tasks.

The reasoning process of the RTA must be temporally bounded to allow it to perform the tasks for deciding the strategy to reach its objectives. In this way, the RTA will be able to determine whether it has enough time to deliberate and to take into account the temporal cost of its cognitive task when it plans the execution of new tasks. Next sections review CBR applications to MAS and propose a temporally bounded CBR method as deliberative mechanism for the cognitive task of the agent.

3 CBR as Deliberative Mechanism for Agents

In the AI research, the combination of several AI techniques to cope with specific functionalities in hybrid systems has a long history of successful applications. A CBR system provides agent-based systems with the ability to reason and learn autonomously from the experience of agents. These systems propose solutions for solving a current problem by reusing or adapting other solutions that were applied in similar previous problems. With this aim, the system has a case-base that stores its knowledge about past problems together with the solution applied in each case. The most common architecture of a CBR system consists of four phases [1]: the first one is the *Retrieval* phase, where the most similar cases are retrieved from the case-base; then, in the *Reuse* phase, those cases are reused to try to solve the new problem at hand; after that, in the *Revise* phase the solution achieved is revised and adapted to fit the current problem and; finally, in the *Retain* phase, the new case is stored in the case-base and hence, the system learns from new experiences.

The integration of CBR systems and MAS has been studied following many approaches. Therefore, the literature of this area reports research on systems that integrate a CBR engine as a part of the system itself [12], other MAS that provide some or each of their agents with CBR capabilities or even, the development of BDI agents following a CBR methodology [3]. This section is focused on the review of the second approach, CBR applied to MAS, since it fits the scope of our paper.

Since the 90's, the synergies between MAS and CBR are many, although the approaches differ. One early approach was the development of multi-agent CBR systems, which are MAS with cooperative agents characterized by the distribution of their case-bases and/or certain phases of the CBR cycle between them [13] [16] [20]. The main effort in this research area is focused on the policies that agents follow to manage the CBR cycle.

The application of CBR to manage argumentation in MAS is a different and more recent approach that has produced important contributions both in the areas of AI and argumentation theory. In this field, important works are those of Soh and Tsatsoulis, who designed a case-based negotiation model for reflective agents [22], of Pancho Tolchinsky et al., who extended the architecture of the decision support MAS for the organ donation *CARREL+* with a new case-based selection model called *ProCLAIM* [24] and the *Argumentation Based Multi-Agent Learning (AMAL)* framework [19] proposed by Ontañon and Plaza, which features a set of agents that try to solve a classification problem by aggregating their expert knowledge.

Furthermore, an area where the integration between CBR and agent techniques has produced a huge amount of successful applications is the robot navigation domain. An important contribution was proposed by Marling in a case-based model for managing her *ROBOCATS* system [15], playing in the *Robocup* league. Also, Fox developed the *RUPART* system [5], that features a hybrid planner for a mobile robot that delivers mail in real-time. Other application of CBR to manage autonomous navigation tasks was proposed by Likhachev et. al. in a system for the automatic selection and modification of assemblage parameters [14]. Finally, an important research that applies CBR to MAS with mobile robots was developed by Ros et. al. to model team playing behaviour in the robot soccer domain by using CBR [21].

The cited above are outstanding examples of systems that join research efforts and results of both CBR and MAS, but they are only a sample reported in the literature of this prolific area. However, few of these systems cope with the problem of applying CBR as deliberative engine for agents in MAS with real-time constraints. In this case, the case-based reasoning cycle must observe temporal restrictions. The next section tackles this challenge and provides solutions to deal with it.

4   Temporal-Bounded CBR

CBR systems are highly dependent on their application domain. Therefore, designing a general CBR model that might be suitable for any type of real-time domain (hard or soft) is, to date, unattainable. In real-time environments, the CBR phases must be temporal bounded to ensure that solutions are produced on time. In this section, we present some guidelines with the minimum requirements to be taken into account to implement a CBR method in real-time environments.

The design decision about the data structure of the case-base and the different algorithms that implement each phase are important factors to determine the execution time of the CBR cycle. The number of cases in the case-base is another parameter that affects the temporal cost of the retrieval and retain phases. Thus, a maximum number of cases must be defined by the designer. Note that, usually, the temporal cost of the algorithms that implement these phases depend on this number.

For instance, let us assume that the designer chooses a *hash table* as data structure for the case-base. This table is a data structure that associates keys to concrete values. Search is the main operation that it supports in an efficient way: it allows the access to elements (e.g. phone and address) by using a *hash function* to transform a generated key (e.g. owner name or account) to a *hash number* that is used to locate the desired value. The average time to make searches in hash tables is constant and defined as $O(n)$ in the worst case. Therefore, if the cases are stored as entries in a hash table, the maximum time to look for a case depends on the number of cases in the table (i.e. $O(\#cases)$). Similarly, if the case-base is structured as an *auto-balanced binary tree* the search time in the case-base in the worst case would be $O(log\ n)$.

In this research, we propose a modification of the classic CBR cycle to adapt it to be applied in real-time domains. Figure 1 shows a graphical representation of our approach. First, we group the four reasoning phases that implement the *cognitive task* of the real-time agent in two stages: the *learning stage*, which consists of the revise and retain phases and the *deliberative stage*, which includes the retrieve and reuse phases. Both phases will have scheduled their own execution times. In this way, the designer can choose between assigning more time to the deliberative stage (and hence, to design more 'intelligent' agents) or else, keeping more time for the learning stage (and thus, to design agents that are more sensible to updates).

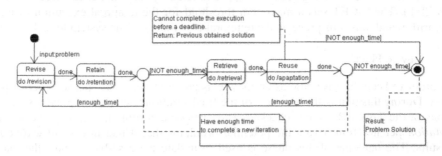

Fig. 1. Temporal-Bounded CBR cycle

Following, the operation of our Time Bounded CBR cycle (TB-CBR) is explained. Firstly, the main difference that can be observed between the classical CBR cycle and the TB-CBR cycle is the starting phase. Our real-time application domain and the restricted size of the case-base gives rise to the need of keeping the case-base as up-dated as possible. Commonly, recent changes in the case-base will affect the potential solution that the CBR cycle is able to provide for a current problem. Therefore, the TB-CBR cycle starts by the learning stage, checking if there are previous cases waiting for being revised and possible stored in the case-base. In our model, the solutions provided at the end of the deliberative stage will be stored in a solution list while a feedback about their utility is received. When each new CBR cycle begins, this list is acceded and while there is enough time, the learning stage of those cases whose solution feedback has been recently received is executed. In case the list is empty, this process is omitted.

After that, the deliberative stage is executed. Thus, the retrieval algorithm is used to search the case-base and retrieve a case that is similar to the current case (i.e. the one that characterises the problem to solve). Each time a similar case is found, it is sent to the reuse phase where it is transformed to a suitable solution for the current problem by using a reuse algorithm. Therefore, at the end of each iteration of the deliberative stage, the TB-CBR method is able to provide a solution for the problem at hand, although this solution can be improved in next iterations as long as the deliberative stage has enough time to perform them.

Hence, the temporal cost of executing the *cognitive task* is greater than or equal to the sum of the execution times of the learning and deliberative stages (as shown in equation 1):

$$T_{cognitive} \geq T_{learn} + T_{deliberative}$$
$$T_{learning} \geq (t_{revise} + t_{retain}) * n \qquad (1)$$
$$T_{deliberative} \geq (t_{retrieve} + t_{reuse}) * m$$

where n and m are the number of iterations of the learning and deliberative stages respectively. The requirements needed to temporally bound each phase of the TB-CBR cycle are explained below. In order to bound the temporal cost of the algorithms that implement these phases and to ensure an adequate temporal control of them, the execution time of these algorithms is approximated by its worst-case execution time (WCET). The WCET sets a maximum threshold for the temporal execution of each algorithm and thus, this prevents the temporal constraints of the system to be broken.

4.1 Revise Phase

In order to keep the case-base as up to date as possible, the revise phase is performed first. During this phase, the accuracy of the final solutions obtained in previous executions of the TB-CBR cycle is checked. The revision algorithm $f_{revision}$ only checks one *solution* per iteration, fixing the potential problems that it had in case of erroneous results. The outcome of this phase is used to update the case-base. Thus, the maximum temporal cost of this phase is bounded by the WCET of the revision algorithm:

$$t_{revise} = WCET(f_{revision}(solution)) \qquad (2)$$

Note that, in order to guarantee a known maximum execution time, this checking must be performed automatically by the computer without human interference. This WCET does not depend on the number of stored solutions or the number of cases in the case-base and again, must be determined by the designer of the algorithm.

4.2 Retain Phase

In this phase it is decided whether a checked *solution* must be added as a new case in the *case-base*. Here, keeping the maximum size of the case-base is crucial, since the temporal cost of most retention algorithms $f_{retention}$ depends on this size. If there is a case in the case-base that is similar enough to the current case, this case (its problem description and solution) is updated if necessary. On the contrary, if there is not a case that represents the problem solved, a new case is created and added to the case-base. In order to keep the maximum size of the case-base, this could entail removing an old case from it. This decision should be taken by the retention algorithm. Nevertheless, the maximum temporal cost that the retain phase needs to execute one iteration is the retention algorithm WCET.

$$T_{retain} = WCET(f_{retention}(solution, case - base)) \qquad (3)$$

4.3 Retrieve Phase

In the retrieve phase, the retrieval algorithm $f_{retrieval}$ is executed to find a case that is similar to the current problem (i.e. *currentcase*) in the *case-base*. Since WCET depends on the structure of the case-base and its number of cases, the designer must calculate this WCET and use this time to estimate the necessary time to execute an iteration of the retrieval algorithm.

$$t_{retrieve} = WCET(f_{retrieval}(currentcase, case - base)) \qquad (4)$$

Each execution of the retrieval algorithm will provide a unique case similar to the current problem (if it exists in the case-base). This result is used as input for the reuse phase. However, in next iterations of the deliberative stage more similar cases can be retrieved with the intention to provide a more accurate solution for the problem.

4.4 Reuse Phase

In this phase, the cases obtained from the retrieve phase are adapted to use them as a potential solution for the current problem. These cases are stored in a list of selected cases (i.e. *cases_list*). Each time the reuse phase is launched, the adaptation algorithm $f_{adaptation}$ searches this list and produces a solution by adapting a single case or a set of cases to fit the context of the current problem to solve. Therefore, the execution time of this algorithm depends on the number of cases that the algorithm is working with.

$$t_{reuse} = \begin{cases} WCET(f_{adaptation}(first\_case)) \\ f_{adaptation}(cases\_list) \end{cases} \qquad (5)$$

To guarantee that the RTA assigns enough time to perform the cognitive task and provides at least one solution, the designer must know the WCET to execute one iteration of the adaptation algorithm (i.e. $f_{adaptation}(first\_case)$). In order to control the execution time of the adaptation algorithm in subsequent iterations (i.e. $f_{adaptation}(cases\_list)$), the RTA must be able to stop the execution of the algorithm in case that it realises that the assigned time to complete the deliberative stage will be overcome. Then, the RTA provides the best solution among the solutions completed in previous iterations. This solution is stored in a list of solutions for being verified in the learning stage.

5 Conclusions

The main contribution of this article is to set some guidelines to develop a CBR method for real-time systems. In this type of systems, timing requirements must be previously known in order to guarantee the correctness of the system. However, since the execution time of each CBR method depends on the specific algorithms that have been used to implement the CBR cycle, a general estimation cannot be made. Therefore, this work provides the designer of the system with a new approach for the CBR cycle, called TB-CBR, which eases the process of bounding each phase of the CBR method. This method implements the CBR cycle in two stages: the deliberative stage, whose execution is mandatory; and the learning stage, whose execution can be optional depending on the temporal requirements.

This TB-CBR proposal has been implemented and tested in an example that consists in a system that manages the mail in a department plant by using mobile robots [18]. In this example[1], robots deliberate to know whether they have enough time to deliver mail. This deliberation is implemented by using a TB-CBR method.

Acknowledgements. This work was partially supported by CONSOLIDER-INGENIO 2010 under grant CSD2007-00022 and by the Spanish government and GVA funds under TIN2006-14630-C0301 and PROMETEO/2008/051 projects.

References

1. Aamodt, A., Plaza, E.: Case-based reasoning; Foundational issues, methodological variations, and system approaches. AI Communications 7(1), 39–59 (1994)
2. Carrascosa, C., Terrasa, A., García-Fornes, A., Espinosa, A., Botti, V.: A Meta-Reasoning Model for Hard Real-Time Agents. In: Marín, R., Onaindía, E., Bugarín, A., Santos, J. (eds.) CAEPIA 2005. LNCS, vol. 4177, pp. 42–51. Springer, Heidelberg (2006)
3. Corchado, J.M., Pellicer, A.: Development of CBR-BDI Agents. International Journal of Computer Science and Applications 2(1), 25–32 (2005)
4. Dean, T., Boddy, M.: An analysis of time-dependent planning. In: Proc. of the 7th National Conference on Artificial Intelligence, pp. 49–54 (1988)
5. Fox, S.E., Anderson-Sprecher, P.: Robot Navigation: Using Integrated Retrieval of Behaviors and Routes. In: Proc. of FLAIRS Conference, pp. 346–351 (2006)

[1] Documentation available at http://www.dsic.upv.es/users/ia/sma/tools/jart

6. Garcia-Fornes, A.: ARTIS: Un modelo y una arquitectura para sistemas de tiempo real inteligentes. Ph.D. Dissertation, DSIC, U. Politecnica Valencia (1996)
7. Garvey, A., Lesser, V.: A survey of research in deliberative Real-Time Artificial Intelligence. The Journal of Real-Time Systems 6, 317–347 (1994)
8. Goldman, R.P., Musliner, D.J., Krebsbach, K.D.: Managing Online Self-Adaptation in Real-Time Environments. In: Proc. of 2nd Int. Workshop on Self Adaptive Software (2001)
9. Hayes-Roth, B., Washington, R., Ash, D., Collinot, A., Vina, A., Seiver, A.: Guardian: A prototype intensive-care monitoring agent. Artificial Intelligence in Medicine 4, 165–185 (1992)
10. Howe, A.E., Hart, D.M., Cohen, P.R.: Addressing real-time constraints in the design of autonomous agents. The Journal of Real-Time Systems 2, 81–97 (1990)
11. Julián, V.J., Botti, V.J.: Developing real-time multi-agent systems. ICAE 11(2), 150–165 (2004)
12. Karacapilidis, N., Papadias, D.: Computer supported argumentation and collaborative decision-making: the HERMES system. Information Systems 26(4), 259–277 (2001)
13. Leake, D.B., Sooriamurthi, R.: When Two Case Bases Are Better Than One: Exploiting Multiple Case Bases. In: Proceedings of ICCBR (2001)
14. Likhachev, M., Kaess, M., Arkin, R.C.: Learning Behavioral Parameterization Using Spatio-Temporal Case-Based Reasoning. In: Proc. of ICRA, vol. 2, pp. 1282–1289 (2002)
15. Marling, C., Tomko, M., Gillen, M., Alexander, D., Chelberg, D.: Case-based reasoning for planning and world modeling in the robocup small size league. In: Workshop on Issues in Designing Physical Agents for Dynamic Real-Time Environments, IJCAI (2003)
16. McGinty, L., Smyth, B.: Collaborative Case-Based Reasoning: Applications in Personalised Route Planning. In: Aha, D.W., Watson, I. (eds.) ICCBR 2001. LNCS, vol. 2080, pp. 362–376. Springer, Heidelberg (2001)
17. Musliner, D.J., Hendler, J.A., Agrawala, A.K., Durfee, E.H., Strosnider, J.K., Paul, C.J.: The Challenge of Real-Time in AI. IEEE Computer, 58–66 (January 1995)
18. Navarro, M., Heras, S., Julián, V.: Ensuring Time in Real-Time Commitments. In: Geffner, H., Prada, R., Machado Alexandre, I., David, N. (eds.) IBERAMIA 2008. LNCS, vol. 5290, pp. 183–192. Springer, Heidelberg (2008)
19. Ontañon, S., Plaza, E.: Learning and Joint Deliberation through Argumentation in Multi-Agent Systems. In: Proc. of AAMAS (2007)
20. Plaza, E., Arcos, J.L., Martín, F.: Cooperative Case-Based Reasoning. In: Weiss, G. (ed.) ECAI 1996 Workshops. LNCS (LNAI), vol. 1221, pp. 180–201. Springer, Heidelberg (1997)
21. Ros, R., López de Màntaras, R., Arcos, J.L., Veloso, M.: Team Playing Behavior in Robot Soccer: A Case-Based Reasoning Approach. In: Weber, R.O., Richter, M.M. (eds.) ICCBR 2007. LNCS, vol. 4626, pp. 46–60. Springer, Heidelberg (2007)
22. Soh, L.-K., Tsatsoulis, C.: A Real-Time Negotiation Model and a Multi-Agent Sensor Network Implementation. AAMAS 11(3), 215–271 (2005)
23. Soler, J., Julian, V.J., García-Fornes, A., Botti, V.: Real-Time Extensions in Multi-agent Communication. In: Conejo, R., Urretavizcaya, M., Pérez-de-la-Cruz, J.-L. (eds.) CAEPIA/TTIA 2003. LNCS (LNAI), vol. 3040, pp. 468–477. Springer, Heidelberg (2004)
24. Tolchinsky, P., Modgil, S., Cortés, U., Sànchez-Marrè, M.: CBR and Argument Schemes for Collaborative Decision Making. In: Proc. COMMA, vol. 144, pp. 71–82 (2006)

Towards Providing Social Knowledge by Event Tracing in Multiagent Systems*

Luis Búrdalo, Andrés Terrasa, Ana García-Fornes, and Agustín Espinosa

Departamento de Sistemas Informáticos y Computación
Universidad Politécnica de Valencia
co/ de Vera SN, 46021, Valencia, Spain
{lburdalo, aterrasa, agarcia, aespinos}@dsic.upv.es
http://www.dsic.upv.es/

Abstract. Social knowledge is one of the key aspects of MAS in order to face complex problems in dynamical environments. However, it is usually incorporated without specific support on behalf of the platform and that does not let agents take all of the advantage of this social knowledge. At present time, the authors of this paper are working in a general tracing system, which could be used by agents in the system to trace other agents' activity and that could be used as an alternative way for agents to perceive their environment. This paper presents first results of this work, consisting of the requirements which should be taken into account when designing such a tracing system.

Keywords: agents, multiagent systems, social knowledge, tracing systems.

1 Introduction

These days, the use and importance of multiagent systems (MAS) has increased because their flexible behavior is very useful to deal with complex problems in dynamic and distributed environments. This is not only due to agents individual features (like autonomy, reactivity or reasoning power), but also to their capability to communicate, cooperate and coordinate with other agents in the MAS in order to fulfil their objectives.

The necessary knowledge to support this social behavior is referred to by Mařik et al in [15] as *social knowledge*. This social knowledge plays an important role in increasing the efficiency in highly decentralized MAS. Social abstractions such as teams, norms, social commitments or trust are the key to face complex situations using MAS; however, these social abstractions are mostly incorporated to the MAS at user level; this is, from the multiagent application itself, without specific support from the multiagent platform, by means of messages

---

* This work is partially supported by projects PROMETEO/2008/051, CSD2007-022 and TIN2008-04446, which is co-funded by the Spanish government and FEDER funds.

E. Corchado et al. (Eds.): HAIS 2009, LNAI 5572, pp. 484–491, 2009.

among agents or blackboard systems. This weak integration of high level social abstractions, also mentioned by Bordini et al in [3], prevents agents in the MAS from exactly knowing what is happening in their environment, since they depend on other agents actively informing them about what they are doing. This dependance on other agents sets out two major problems. First, it can lead to excesive overhead in some of the agents. And second, it is also difficult to trust the information provided directly by other agents using messages in open MAS.

An alternative solution to provide social knowledge could be an event tracing system, integrated within the multiagent platform, which could be used by agents in the system to perceive their environment without having to actively notify each change to the rest of the agents which could be interested in what they do. Such a tracing system, integrated within the multiagent platform and providing a trustworthy event set which were capable to reflect not only communication among agents, but also agents' perceptions, etc, could be used as a way to provide social knowledge to the MAS. Also, such a tracing system would be more trustworthy than agent messages, since the information would not be proportioned by agents, but by the multiagent platform itself. Agents could trust the trace system as much as they can trust the multiagent platform.

Applications which extract information from the system by processing event streams at run time are already considered in the field of event driven architectures [14] and the idea of an standard tracing system available for processes in a system already existed in the field of operating systems (and at present it is contemplated by the POSIX standard[11]). These concepts can be applied to th field of MAS, where event tracing is still considered a facility to help MAS developers in the verification and validation processes.

This paper presents the requirements of such a general, platform-integrated tracing system applied to MAS. These requirements should be taken into account in order to develop a general abstract trace model for MAS, which could be finally incorporated to a real multiagent platform. The rest of the paper presents is structured as folows: Section 2 comments existing work by other authors in the field of tracing MAS. Section 3 presents a set of requirements which should be taken into account in order to design a general tracing system which could be used to improve agents sociability. Finally, section 4 comments this work's main conclusions and future work which is still to be carried out in order to incorporate such a tracing system to a MAS.

2   Event Tracing in Multiagent Systems

One of the most popular tracing facilities for MAS is the Sniffer Agent provided by JADE[1], which keeps tracking of all of the messages sent or received by agents in the system and allows the user to examine their content. JADE also provides an Introspector Agent, which can be used to examine the life cycle of any agent in the system, its behaviors and the messages it has sent or received.

JADEX[18] provides a Conversation Center, which allows a user to send messages directly to any agent while it is executing and to receive answers to those messages from a user-friendly interface. It also provides a DF Browser to track services offered by any agent in the platform at run time and a BDI Tracer which can be used to visualize the internal processes of an agent while it is executing and show causal dependencies among agents' beliefs, goals and plans.

The JACK[20] multiagent platform supports monitoring communication between agents by means of Agent Interaction Diagrams. It also provides a Design Tracing Tool, to view internal details of JACK applications during execution, and a Plan Tracing Tool, to trace the execution of plans and the events that handle them. JACK also provides debugging tools that work at a lower level of abstraction in order to debug the MAS in a more exhaustive way: Audit Logging, Generic Debugging/Agent Debugging.

Other examples of tracing facilities provided by platforms are ZEUS's[10] Society Viewer and Agent Viewer, which display organisational inter-relationships among agents and their messages and agent's internal state. Also, JASON[4,5] provides a Mind Inspector tool to examine agents' internal state.

Apart from those tools provided by multiagent platforms themselves, there are many tracing facilities provided by third party developers. This is the case of Java Sniffer[21], developed by Rockwell Automation based on JADE's Sniffer Agent. Another third party tool based on JADE's Sniffer Agent is ACLAnalyser[9], which intercepts messages interchanged by agents during the execution of the application and stores them in a relational database, which can be lately inspected to detect social pathologies in the MAS. These results can be combined with data mining techniques to help in the multiagent system debugging process [8]. MAMSY, the management tool presented in [19] lets the system administrator monitorize and manage a MAS running over the Magentix multiagent platform[2]. MAMSY provides graphical tools to interact with the MAS and visualize its internal state at run time. In [16], the authors describe an advanced visualisation tools suite for MAS developed with ZEUS, although the authors also claim these tools could be used with CommonKADS.

Tracking messages has also been used in [17], which comments an ampliation of the Prometheus methodology and the related design tool to help the designer to detect protocol violations by tracing conversations among agents in the system and to detect plan selection inconsistencies. Lam et al present in [13] an iterative method based on tracing multiagent applications to help the user understanding the way those applications internally work. They also present a Tracer Tool which implements the described Tracing Method. The Tracer Tool can be applied to any agent system implementation, regardless of agent or system architecture, providing it is able to interface with Java's logging API (directly or via a CORBA interface). Results obtained with this method were presented in [12]. Bose et al present in [7] a combination of this Tracer Tool with a Temporal Trace Language (TTL) Checker presented in [6]. This TTL Checker enables the automated verification of complex dynamic properties against execution traces.

As it can be appreciated, tracing facilities in MAS are usually conceived as debugging tools to help in the validation and verification processes. It is also usual to use these tracing tools as a help for those users which have to understand how the MAS works. Thus, generated events are destinated to be understood by a human observer who would probably use them to debug or to validate the MAS and tracing facilities are mostly human-oriented in order to let MAS users work in a more efficient and also comfortable way. Some multiagent platforms provide their own tracing facilities, although there is also important work carried out by third party developers. However, even those tracing facilities which were not designed by platform developer teams are usually designed for a specific multiagent platform. There is not a standard, general tracing mechanism which let agents and other entities in the system trace each other as they execute like the one provided by POSIX for processes.

3   Tracing System Requirements

From the viewpoint of the tracing process, a MAS can be considered to be formed by set of tracing entities, or components that are susceptible of generating and/or receiving tracing *events*. The tracing system needs to consider, at least, the following list of components inside the MAS as tracing entities: agents, organizational units (or any type of agent aggregation supported by the multiagent platform) and the multiagent platform itself (and its components).

Unlike existing work on tracing MAS, previously mentioned in Section 2, a tracing system which could be used as a knowledge provider must not be human-oriented, but entity-oriented, so that these tracing entities are able to receive events and process them or incorporate them to their reasoning process at run time in order to take advantage from that.

In order to generate trace events, the source code of tracing entities needs to be *instrumented* to include the code which actually produces such events. Attending to where this instrumentation code is placed, trace events can be classified as *platform events* or *application events*.

Platform events are instrumented within the source code of the platform (either in its "core" or in any of its supporting agents). These events represent the generic, application-independent information that the platform designer intends to provide to agents. On the other hand, application events are instrumented within the code of the application agents. These events represent customized run-time information defined by the application designer in order to support specific needs of the application agents.

The rest of the section presents a set of requirements which should be taken into account when developing such a tracing system. These requirements have been classified in three main groups: functional, efficiency and security requirements.

3.1   Functional Requirements

Tracing roles. Any tracing entity in the MAS must be able to play two different roles in the tracing process: *event source* (*ES*) and *event receiver* (*ER*). From the viewpoint of tracing entities, these two tracing roles are dynamic and not exclusive, in the sense that each tracing entity can start and stop playing any of them (or both) at any time, according to its own needs. The relation between ES and ER entities is many to many: it must be possible for events generated by an ES entity to be received by many ER entities, as well as it must also be possible for an ER entity to receive events from multiple ES entities simultaneously.

Chronologically ordered event delivery requierement. Events generated in the system must be delivered to ER entities in chronological order or, at least, include information related to the time when they were produced to allow ER entities to process them in chronological order.

Dynamic definition of event types. Trace events can be classified in event types attending to the information which is generated and attached to them when they are generated. In order to let the event processing be more flexible and efficient, it must be possible for tracing entities to dynamically define new event types at run time. This must be applied to both platform and application event types.

Publication of event types. At any time, ER entities must be able to know which ES entities are producing events and of which types. So, as a consequence of event types being dynamic, the tracing system should keep and up-to-date list of such traceable event types (and ES entities) and to make this list available to all tracing entities in the MAS.

On-line and off-line tracing. In order to let entities work with both historical and run time information, both on-line and off-line tracing should be supported. In on-line tracing, events are delivered to ER entities as they are traced by the tracing system (with a potential delay due to the internal processing of events by the tracing system). In contrast, in off-line tracing, events generated by ES entities are not delivered to running entities, but stored in a log file. Both tracing modes must not be exclusive, meaning that it must be possible for the events generated by any ES entity to be delivered to some ER entities while also being stored in some log files. However, the tracing system does not need to support concurrent access to the events stored in a log file.

3.2   Efficiency Requirements

In any computing system, tracing can be a very expensive process in terms of computational resources. In the case of MAS, the fact that they are by nature highly decentralized systems, both in number of running entities (agents) and hosts, can make their tracing even more expensive. In this context, the tracing process must be optimized in order to minimize the overhead it produces to the system, since a very sophisticated but excessively costly tracing system can become completely useless in practice. The following list introduces a minimal set of efficiency design guidelines that should be considered when designing a tracing system for MAS, in order to make this system realizable and useful. The first two requirements focus on the potential overload of the tracing system while the last one allows entities to set their own limits in the resources devoted to the tracing process.

Selective event delivery. Each ER entity should be able to express which event types it wants to receive, and the tracing system should only deliver events which belong to such types to the entity. Furthermore, each ER entity should be able to change dynamically which events it wants to receive, since entities may need different tracing information at different times during their execution.

Selective event tracing. The tracing system should not spend resources in tracing events which belong to event types that currently no entity wants to receive.

Resource limit control. Each ER entity should be able to limit the maximum amount of its resources to be allocated to receive events, both in on-line and off-line tracing modes. In on-line tracing, if there is some memory data object where events are delivered to until they are retrieved by the corresponding ER entity, then this entity should be able to define the maximum amount of memory devoted to such data object. In off-line tracing, the ER entity that sets the tracing up to the corresponding log file should be able to define the maximum size of the file.

3.3 Security Requirements

Tracing in an open MAS has obvious security issues, since many of the events registered by the tracing system may contain sensitive information that can be used by agents to take advantage from, or even to damage, the MAS. This scenario enforces the necessity of applying some security policy over the events that can be delivered to entities, specially if they are application entities. This policy can be materialized in many different ways, but in essence, it has to allow for the definition of security rules in the MAS that limit the availability of events to the *right* ER entities. The following list of requirements express a minimum set of restrictions by which it is possible to incorporate such security rules to the tracing system.

Authorization to ER entities. Each ES entity in the system must be able to decide which ER entities can receive the events that it generates. This can be accomplished by means of an authorization mechanism, provided by the tracing system, which can be used by ES entities to restrict the event types that are available to each ER entity. Such authorization rules must be dynamic, so that ES entities are able to modify the list of authorized ER entities corresponding to each event type at run time.

Supervisor entities. Situations where an entity must be able to access to other ES entity's events in order to fulfil its objectives, even though the ES entity does not agree with that, are very common in MAS. This can happen, for instance, in normative environments where an agent has to watch the other in order to verify that norms are not being violated and to apply the corresponding sanctions in case they are. The tracing system must also provide mechanisms to let an ER receive events generated by an ES without its authorization under some circumstances.

Delegation of authorizations. If an ER entity is currently authorized to be delivered events corresponding to certain event types, then this entity can delegate this authorization to other ER entities in the system; then, each of them can do so with other entities (potentially forming an *authorization tree*). At any node in the tree, the corresponding entity can add or remove delegations dynamically. If a delegation is removed, all the potential subsequent delegations (subtree) are also removed.

Platform entities authorization. By definition, the tracing system must be granted the authorization for all event types defined in the MAS, both at the platform and application levels. This is required for the tracing system to be able to keep track of any event being generated in the MAS, independently of the privacy rules defined by each ES entity.

4   Conclusions and Future Work

Social knowledge is one of the most important features that make MAS appropiate to deal with complex problems in dynamic and distributed environments. The key to this is the capacity of agents to communicate and coordinate with other agents in the MAS in order to get their objectives. This capacity, though based on high level social concepts such as social commitments, trust, norms or reputation, is usually incorporated to the MAS at user level, using messages or blackboard systems, without support from the multiagent platform. This can produce too much overhead, reducing the scalability of the MAS. Also, it has to be taken into account that sometimes it is difficult to trust information from other agents, specially in open MAS.

A general event tracing system, which agents in the MAS could use to trace other agents in their environment, could be used as a more appropiate and trustworthy social knowledge provider. This paper presents the first step towards defining such a tracing system, which is the identification of its requirements. This paper has identified requirements in different aspects: functionallity, efficiency and security.

Some of the presented requirements set important problems out. Some of these problems are more obvious. For example, the problem of delivering events in chronological order in a distributed MAS. However, others are less evident. For instance, the problem of determining which ES entity is the owner of each trace event, since the instrumented code that produces an event is not always within the source code of the entity which originated it. Just as an example, consider events could as property of those entities which source code has been instrumented to produce them. In this case, all platform events would belong to the multiagent platform, while agents in the MAS would only be owners of application events. It could be more understandable and easier to incorporate considering that events belong to the ES entity which originated them.

Future work will include the design of a general abstract model for MAS which contemplated all of the requirements exposed above and which, after that, could be implemented and incorporated to a multiagent platform.

References

1. U. Agreement. Jade administrator's guide. sharon.cselt.it
2. Alberola, J., Mulet, L., Such, J., Gacía-Fornes, A., Espinosa, A., Botti, V.: Operating system aware multiagent platform design. In: Fifth European Workshop On Multi-Agent Systems (EUMAS 2007), pp. 658–667 (2007)
3. Bordini, R., Dastani, M., Winikoff, M.: Current issues in multi-agent systems development (invited paper). In: O'Hare, G.M.P., Ricci, A., O'Grady, M.J., Dikenelli, O. (eds.) ESAW 2006. LNCS, vol. 4457, pp. 38–61. Springer, Heidelberg (2007)
4. Bordini, R., Jason, J.H.: A java-based interpreter for an extended version of agents-peak, p. 31 (March 2007)
5. Bordini, R., Hubner, J., Vieira, R.: Jason and the golden fleece of agent-oriented programming. In: Bordini, R.H., Dagtani, M., Dix, J., El Fallah Seghrovehni, A. (eds.) Multiagent Systems Artificial Societies And Simulated Organizations, vol. 15, 296 p. (January 2005)
6. Bosse, T., Jonker, C.M., van der Meij, L., Sharpanskykh, A., Treur, J.: Specification and verification of dynamics in cognitive agent models. In: Proceedings of the IEEE/WIC/ACM international conference on Intelligent Agent Technology, December 18-22, pp. 247–254 (2006), doi:10.1109/IAT.2006.112
7. Bosse, T., Lam, D., Barber, K.: Tools for analyzing intelligent agent systems. Web Intelligence and Agent Systems (January 2008)
8. Botia, J., Hernansaez, J., Gomez-Skarmeta, A.: On the application of clustering techniques to support debugging large-scale multi-agent systems. Springer, Heidelberg (2007)
9. Botia, J., Hernansaez, J., Skarmeta, F.: Towards an approach for debugging mas through the analysis of acl messages. In: Lindemann, G., Denzinger, J., Timm, I.J., Unland, R. (eds.) MATES 2004. LNCS, vol. 3187, pp. 301–312. Springer, Heidelberg (2004)
10. Collis, J., Ndumu, D., Nwana, H., Lee, L.: The zeus agent building tool-kit. BT Technology Journal (January 1998)
11. IEEE. 1003.1, 2004 EDITION IEEE Standard for Information Technology Portable Operating System Interface (POSIX) (2004)
12. Lam, D., Barber, K.: Debugging agent behavior in an implemented agent system.. Workshop on Programming Multi-Agent Systems at the Third.. (January 2004)
13. Lam, D., Barber, K.: Comprehending agent software. Proceedings of the fourth international joint conference on.. (January 2005)
14. Luckham, D.: The power of events. Addison-Wesley, Reading (2002)
15. Mafik, V., Pechoucek, M.: Social knowledge in multi-agent systems. Systems (January 2004)
16. Ndumu, D., Nwana, H., Lee, L., Collis, J.: Visualising and debugging distributed multi-agent systems. Proceedings of the third annual conference on Autonomous.. (January 1999)
17. Padgham, L., Winikoff, M., Poutakidis, D.: Adding debugging support to the prometheus methodology. Engineering Applications of Artificial Intelligence (January 2005)
18. Pokahr, A., Braubach, L.: Jadex tool guide, p. 66 (September 2008)
19. Sanchez-Anguix, V., Espinosa, A., Hernández, L., García-Fornes, A.: Mamsy: A management tool for multi-agent systems. In: 7th International Conference on Practical Applications of Agents and Multi-Agent Systems (2009)
20. Software, A.O.: Jack tm tracing manual, p. 85 (May 2008)
21. Tichy, P., Slechta, P.: Java sniffer 2.7 user manual (2006)

A Solution CBR Agent-Based to Classify SOAP Message within SOA Environments

Cristian Pinzón, Belén Pérez, Angélica González,
Ana de Luís y, and J.A. Román

University of Salamanca, Plaza de la Merced s/n, 37008, Salamanca, Spain
{cristian_ivanp,lancho,angelica,adeluis,zjarg}@usal.es

Abstract. This paper presents the core component of a solution based on agent technology specifically adapted for the classification of SOA messages. These messages can carry out attacks that target the applications providing Web Services. An advanced mechanism of classification designed in two phases incorporates a CBR-Agent type for classifying the incoming SOAP messages as legal or malicious. Its main feature involves the use of decision trees, fuzzy logic rules and neural networks for filtering attacks.

Keywords: SOAP message, XML security, multi-agent systems, case-based reasoning.

1 Introduction

The communication among services based on Service Oriented Architecture Web Services (SOA) is carried out by XML-based messages, called SOAP messages. This message exchange process is one of the key elements required in SOA environments for system integration [1]. The SOAP message payload often consists of sensitive information, which is sent through insecure channels such as HTTP connections. If a malicious user playing the role of a middleman intercepts a message between sender and recipient, it can result in a series of malicious tasks carried out over the captured message. A number of technologies and solutions have been proposed for addressing the secure exchange of SOAP message. Some WS standards such as WS-Security [2], WS-Policy [3], among others, continually strive to provide real security. Within academia some solutions in the research & development phase focusing on web service security in greater detail are [1], [4], [5]. However, both the WS-Security Standards and the given solutions still do not provide full security, leaving gaps that can be exploited by any malicious user.

This paper presents the core component of a strong solution based on a multi-agent architecture for tackling the security issue of the Web Service. This core is embedded in a CBR-BDI [6] deliberative agent based on the BDI (Belief, Desire, Intention) [7] model specifically adapted for preventing many attacks over web services. Our study applies a solution in two phases that include novel case-based reasoning (CBR) [8] classification mechanisms. The first phase incorporates decision tree and fuzzy logic rules [9] while the second phase incorporates neural networks capable of making short term predictions [10]. The idea of a CBR mechanism is to exploit the experience

E. Corchado et al. (Eds.): HAIS 2009, LNAI 5572, pp. 492–499, 2009.

gained from similar problems in the past and to adapt a successful solution to the current problem. The CBR-BDI agent explained in this work uses the CBR concept to gain autonomy and improve its problem-solving capabilities. The approach presented in this paper is entirely new and offers a different way to confront the security problem in SOA environments.

The rest of the paper is structured as follows: section 2 presents the problem that has prompted most of this research. Section 3 focuses on the structure of the classifier agent which facilitates classification of SOAP message, and section 4 provides a detailed explanation of the classification model integrated within the classifier agent. Finally, section 5 presents the conclusions obtained by the research.

2 Web Service Security Problem Description

A web service is a software module designed to support interaction between heterogeneous groups within a network. In order to obtain interoperability between platforms, communication between web servers is carried out via an exchange of messages. These messages, referred to as SOAP messages, are based on standard XML (eXtensible Markup Language) and are primarily exchanged using HTTP (Hyper Text Transfer Protocol) [11].

Security is one of the greatest concerns within web service implementations. Attacks usually occur when the SOAP message either comes from a malicious user or is intercepted during its transmission by a malicious node that introduces different kinds of attacks.

The following list contains descriptions of different types of attacks, compiled from those noted in [4], [5], [12].

- Oversize Payload: When it is executed, it reduces or eliminates the availability of a web service while the CPU, memory or bandwidth are being tied up by a massive message dispatch with a large payload.
- Coercive Parsing: Just like a message written with XML, an XML parser can analyze a complex format and lead to a denial of service attack because the memory and processing resources are being used up.
- Injection XML: This is based on the ability to modify the structure of an XML document when an unfiltered user entry goes directly to the XML stream or the message is captured and modified during its transmission.
- Parameter Tampering: A malicious user employs web service entries to manually or automatically (dictionaries attack) execute different types of tests and produce an unexpected response from the server.
- SOAP header attack: Some SOAP message headers are overwritten while they are passing through different nodes before arriving at their destination. It is possible to modify certain fields with malicious code.
- Replay Attack: Sent messages are completely valid, but they are sent en masse over a small time frame in order to overload the web service.

Standards such as WS-Security [2] and WS-Policy [3], among others, have set the standard for solutions to security breaches. One solution proposed by [1] takes information from the actual message structure and adds a new header named *SOAP Account* that contains information on the message structure. One solution based on the

XML firewall [4] was proposed to protect web services in more detail. By applying a syntactic analysis, a validation mechanism, and filtering policies, it is possible to identify attacks in individual or group messages. An adaptive framework for the prevention and detection of intrusions was presented in [5]. Based on a hybrid focus that combines agents, data mining and fuzzy logic, it is supposed to filter attacks that are either already known or new. The solution as presented is an incipient idea still being developed and implemented. Finally, another solution proposed the use of Honeypots [13] as a highly flexible security tool. The focus incorporates 3 components: data extraction based on honeyd, tedpdum data analysis, and extraction from attack signatures. Its main inconvenience is that it depends too much on the ability of the head of security to define when a signature is or is not a type of attack. Even when the techniques mentioned claim to prevent attacks on web services, few provide statistics on the rates of detection, false positives, false negatives and any negative effects on application performance.

The following sections detail the internal model of the CBR-BDI agent, as well as the classification process for SOAP message for identifying malicious messages.

3 Classifier Agent Internal Structure

Agents are characterized by their autonomy; which gives them the ability to work independently and in real-time environments [14]. Because of this and their other capacities, agents are being integrated into security approaches such as IDS [15]. However, the use of agents in these systems focuses on the retrieval of information in distributed environments, which only takes advantage of their mobility capacity.

The classification agent presented in this study interacts with other agents within the architecture. These agents carry out tasks related to capturing messages, syntactic analysis, administration, and user interaction. As opposed to the tasks for these agents, the classification agent executes a classification of SOAP messages in two phases that we will subsequently define in greater detail.

In our research, the agents are based on a BDI model in which beliefs are used as cognitive aptitudes, desires as motivational aptitudes, and intentions as deliberative aptitudes in the agents [7]. However, in order to focus on the problem of the SOAP message attack, it was necessary to provide the agents with a greater capacity for learning and adaption, as well as a greater level of autonomy than a pure BDI model currently possesses. This is possible by providing the classifier agents with a CBR mechanism [8], which allows them to "reason" on their own and adapt to changes in the patterns of attacks. When working with this type of system, the key concept is that of "case". A case is defined as a previous experience and is composed of three elements: a description of the problem that depicts the initial problem; a solution that describes the sequence of actions performed in order to solve the problem; and the final state, which describes the state that has been achieved once the solution is applied. To introduce a CBR engine into a BDI agent, we represent CBR system cases using BDI and implement a CBR cycle which consists of four steps: retrieve, reuse, revise and retain [16].

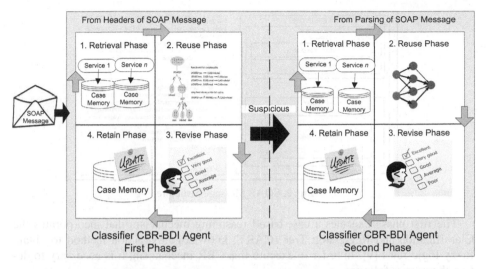

Fig. 1. Classifier CBR-BDI agents in each phase of the mechanism of classification

As previously mentioned, the classifier CBR-BDI agent is the core of the multi-agent architecture and is geared towards classifying SOAP messages for detecting attacks on web services. Figure 1 shows the classifier CBR-BDI agents in each phase of the mechanism of classification.

4   Mechanism for the Classification of SOAP Message Attack

The CBR-BDI classifier agent presented in section 3 incorporates a case-based reasoning mechanism that allows it to classify SOAP messages. The mechanism incorporated into the agent approaches the idea of classification from the perspective of anomaly-based detection. This mechanism requires the use of a database with which it can generate models such as the solution of a new problem based on past experience. In the specific case of SOAP messages, it manages a case memory for each service offered by the architecture, which permits it to handle each incoming message based on the particular characteristics of each web service. Each new SOAP message sent to the architecture is classified as a new case study object. The advantage that the CBR systems provide spans from automatic learning to the ability to adapt and approach new changes that appear in the patterns of attack.

Focusing on the problem that is of interest to us, we will represent a typical SOAP message which consists of a type of wrapping that contains an optional heading and a mandatory body of text with a useful message load, as depicted in figure 2.

Based on the structure of the SOAP messages and the transport protocol used, we can obtain a series of descriptive fields to consider: IPSource, SizeMessage, Time TravelMessage, NumberHopRouting, LengthSoapAction, NumberHeaderBlocks, NumberElementsBody, NestingDepthElements, NumberXMLTagRepeatedBody, NumberLeafNodesBody. Based on this information, we can present a two-part strategy for executing the classification process:

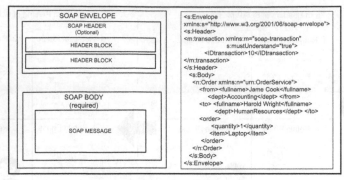

```
SOAP ENVELOPE                        <s:Envelope
   SOAP HEADER                       xmlns:s="http://www.w3.org/2001/06/soap-envelope">
    (Optional)                       <s:Header>
                                     <m:transaction xmlns:m="soap-transaction"
   HEADER BLOCK                                    s:mustUnderstand="true">
                                         <IDtransaction>10</IDtransaction>
   HEADER BLOCK                      </m:transaction>
                                     </s:Header>
                                       <s:Body>
                                         <n:Order xmlns:n="urn:OrderService">
   SOAP BODY                              <from><fullname>Jame Cook</fullname> </from>
   (required)                                <dept>Accounting</dept> </from>
                                           <to> <fullname>Harold Wright</fullname>
                                                <dept>HumanResources</dept> </to>
                                         <order>
   SOAP MESSAGE                             <quantity>1</quantity>
                                           <item>Laptop</item>
                                         </order>
                                       </n:Order>
                                       </s:Body>
                                     </s:Envelope>
```

Fig. 2. SOAP Message Structure

The first phase executes a case-based reasoning mechanism that incorporates the Classification and Regression Tree (CART) knowledge extraction method to obtain fuzzy logic rules [9]. In order to execute this CBR mechanism, it is necessary to define the case as follows:

Table 1. Case Description – CBR (First Phase)

IP Source	Size Message	Time Travel Message	Number Hop Routing	Length SoapAction
String	Int	Int	Int	Int

In each phase of the CBR cycle, certain well-defined tasks are executed. Using CART, we can generate decision-making rules based on the IP Source field extracted from the transport protocol. In order to optimize the system, the reasoning cycle is only executed when it does not have the decision-making rules obtained from previous iterations for an IP from the same class and with similar bytes. In this case, it would only be necessary to recover the rules from the memory of rules. In the opposite case, the CBR cycle is executed in its entirety, as explained below:

- Recovery Phase: Cases will be filtered according to the original IP address so that the cases whose IP is from the same class and with a matching IP are selected. This way, we try to account for the network that the attacks are sent from.
- Reuse: Once the IP filtering process has taken place, we extract the fuzzy logic rules in order to classify the SOAP messages. Knowledge extraction is carried out by applying decision trees specifically with CART, which is then applied to the IPSource, SizeMessage, TimeTravelMessage, NumberHopRouting, Length-SoapAction variables. The final classification that is stored would include three groups: "Malicious", "Legal" and "Suspicious". For each of the decision tree rules, we set up a final classification of the defined groups and the percentage of erroneously classified attacks.
- Revision: Based on the set of previously established rules, we can classify the new SOAP message in such a way that if the final group is "suspicious" or the percentage of errors for the rules is greater than a pre-determined limit, we proceed to the classification phase. Once the classification has been achieved in this

first phase, or if there has been a classification error indicated by an expert, we apply a new decision tree to obtain the rules for the knowledge obtained.
• Retain: In this phase, the decision rules derived from previous phases are stored and associated with the new case based on the IP class and the similar bytes.

Once a solution to the first phase of classification has been obtained, we will determine the need for executing the second phase of the process. This phase is carried out in cases where the message was classified as "Suspicious" or when the percentage of error classification exceeds the defined limit. The second phase involves a much more complex process with an exhaustive syntactic analysis of the SOAP message and the execution of a new CBR cycle. During the execution of the second phase, certain control policies have been established that continuously validate the process overload. The values of the control policies is established by defined variables such as: TimeParsing, CPUTimeParsing, MemoryCostParsing, ValueNestingDepth

Finally, in order to complete the second phase of classification, a CBR mechanism is carried out using a multi-layer perceptron (MLP) in the Reuse phase. This CBR mechanism requires the memory of cases to be specifically defined for each service. The definition of the case is detailed as follows:

Table 2. Case Description – CBR (Second Phase of the Mechanism of Classification)

NumberEle-mentsBody	NumberHeader Blocks	ValueNesting Depth	NumberLeft Nodes	TagXML RepeatedBody
Int	Int	Int	Int	Int

The CBR mechanism executes the following tasks:
• Retrieve: If there happens to be a neural network that is trained for the web service identified in the message, it will be retrieved to perform the classification. If none exists, all of the stored cases corresponding to the web service will be retrieved.
• Reuse: If it has not already received previous training, the neural network will be trained beginning with the retrieved cases. The initial information corresponds to the fields described in table 2, which are transformed into values between 0 and 1 inclusive. Upon completing the training, the new message is classified as either "Malicious" or "Legal".
The MLP uses a Sigmoidal function with a range of possible values at the interval [0,1]. It is used to detect if the request is classified as an attack or not. The value 0 represents a legal message (non attack) and 1 a malicious message (attack). The Sigmoidal activation function is the activation function most used for classifications between two groups.

$$f(x) = \frac{1}{1 + e^{-ax}} \tag{1}$$

The number of neurons in the output layer for the Multilayer Perceptrons is 1, and is responsible for deciding whether or not there is an attack.
• Revise: If the estimated value does not exceed a certain threshold near zero, the message is assumed to be valid; contrarily, it is up to an expert to validate it.

- Retain: The results from the previous stage are stored in the event that the classification of the SOAP message has been either successful or indicated as such by an expert. Finally, in the event that the number of cases related to the network has increased by a percentage, a retraining of the network will be carried out.

5 Conclusions

This research has presented the nucleus of a novel solution that focuses on the protection of web services. The focus incorporates case-based reasoning methods, decision trees, fuzzy logic rules, neural networks, and intelligent agent technology that allows us to approach the problem of web security from a perspective based on learning, adaptability and flexibility.

The solution was designed to be carried out in two phases. In the first phase, a CBR mechanism incorporates decision trees; fuzzy logic rules generate a preliminary robust solution regarding the condition of the message, without sacrificing application performance. If the obtained solution is classified as suspicious, we then proceed to the second phase of the process. This phase does involve a more complex process, with a greater need for resources, and where a second CBR mechanism embeds within a neural network to generate a final result. The proposed solution will continue in the investigation and development for its application in various environments where its performance can be evaluated and real results obtained.

Acknowledgments. This development has been partially supported by the Spanish Ministry of Science project TIN2006-14630-C03-03.

References

1. Rahaman, M.A., Schaad, A., Rits, M.: Towards secure SOAP message exchange in a SOA. In: 3rd workshop on Secure web services, pp. 77–84. ACM, New York (2006)
2. Organization for the Advancement of Structured Information Standards (OASIS): Web Services Security: SOAP Message Security 1.1 (WS-Security (2004), http://docs.oasis-open.org/wss/v1.1/wss-v1.1-spec-os-SOAP MessageSecurity.pdf
3. Bajaj, S., et al.: Web Services Policy Framework, WSPolicy (2004), http://specs.xmlsoap.org/ws/2004/09/policy/ws-policy.pdf
4. Loh, Y., Yau, W., Wong, C., Ho, W.: Design and Implementation of an XML Firewall. Computational Intelligence and Security 2, 1147–1150 (2006)
5. Yee, G., Shin, H., Rao, G.S.V.R.K.: An Adaptive Intrusion Detection and Prevention (ID/IP) Framework for Web Services. In: International Conference on Convergence Information Technology, pp. 528–534. IEEE Computer Society, Washington (2007)
6. Laza, R., Pavon, R., Corchado, J.M.: A Reasoning Model for CBR_BDI Agents Using an Adaptable Fuzzy Inference System. In: Conejo, R., Urretavizcaya, M., Pérez-de-la-Cruz, J.-L. (eds.) CAEPIA/TTIA 2003. LNCS, vol. 3040, pp. 96–106. Springer, Heidelberg (2004)
7. Rao, A., Georgeff, M.: Modeling Rational Agents within a BDI-Architecture. In: KR, pp. 473–484 (1991)

8. Aamodt, A., Plaza, E.: Case-based reasoning: foundational issues, methodological variations, and system approaches. AI Commun. 7, 39–59 (1994)
9. Bittencourt, H., Clarke, R.: Use of classification and regression trees (CART) to classify remotely-sensed digital images. In: Geoscience and Remote Sensing Symposium, IEEE International, vol. 6, pp. 3751–3753 (2003)
10. Shun, J., Malki, H.: Network Intrusion Detection System Using Neural Networks. In: Fourth International Conference on Natural Computation, vol. 5, pp. 242–246 (2008)
11. Snell, J., Tidwell, D., Kulchenko, P.: Programming Web Services with SOAP. O'Reilly, Sebastopol (2001)
12. Jensen, M., Gruschka, N., Herkenhoner, R., Luttenberger, N.: SOA and Web Services: New Technologies, New Standards - New Attacks. In: Fifth European Conference on Web Services-ECOWS 2007, pp. 35–44 (2007)
13. Dagdee, N., Thakar, U.: Intrusion Attack Pattern Analysis and Signature Extraction for Web Services Using Honeypots. In: First International Conference Emerging Trends in Engineering and Technology, pp. 1232–1237 (2008)
14. Carrascosa, C., Bajo, J., Julian, V., Corchado, J.M., Botti, V.: Hybrid multi-agent architecture as a real-time problem-solving model. Expert Syst. Appl. 34, 2–17 (2008)
15. Abraham, A., Jain, R., Thomas, J., Han, S.Y.: D-SCIDS: distributed soft computing intrusion detection system. J. Netw. Comput. Appl. 30, 81–98 (2007)
16. Corchado, J.M., Bajo, J., Abraham, A.: GerAmi: Improving Healthcare Delivery in Geriatric Residences. Intelligent Systems, IEEE 23, 19–25 (2008)

RecMas: A Multiagent System Socioconfiguration Recommendations Tool

Luis F. Castillo[1], Manuel G. Bedia[2], and Ana L. Uribe[1]

[1] Grupo investigación Ingeniería del Software, Universidad Autónoma de Manizales, Colombia
Antigua Estación del Ferrocarril
lfcastil@autonoma.edu.co
[2] Departamento Informática Universidad de Zaragoza
C/ María de Luna, s/n, 50018, Zaragoza, Spain
mgbedia@unizar.es

Abstract. This paper presents a multiagent recommendation system (RecMAS) able to coordinate the interactions between a user agent (AgUser) and a set of commercial agents (AgComs) providing a useful service for monitoring changes in the AgUser's beliefs and decisions based on two parameters: (i) the strength of its own beliefs and (ii) the strength of the AgComs' suggestions. The system was used to test several commercial activities in a shopping centre where the AgComs (AgComs) provided information to an AgUser operating in a wireless device (PDA, mobile phone, etc.) used by a client. The AgUser received messages adapted for conditions of particular offers of interest to the client. Using a theoretical model and a set of simulation experiments, commercial strategies in relation with the socio-dynamics of the system were obtained. This paper concludes with a presentation of a prototype in a real shopping centre.

Keywords: SocioConfiguration, Multiagent systems, Agent-based social simulation.

1 Introduction

Models of artificial societies from different perspectives are useful in a large number of applications. Currently, there exist several different mathematical models that try to explain what types of relations are established in complex social systems [1]. Traditionally, theoretical models used to analyze complex social systems come from the field of social sciences, using qualitative techniques [2] but at present, new models from a quantitative perspective have started to be proposed [3]; in particular, models based in the theories of complexity [4] and emergent phenomena [5]. Nevertheless, although these models include very complex characteristics related with different domains, generally they do not take into consideration complex internal states. In this paper, techniques of the area of the multiagent systems are used for the design of our model. Agent-based systems and multiagent systems [6,11] gather very interesting techniques in order to develop tools which can help us to describe (quantitatively) processes of change of beliefs and social adaptation. In the proposed model there are N BDI-agents, one of which, AgUser, has capacity to buy products adapted to the

E. Corchado et al. (Eds.): HAIS 2009, LNAI 5572, pp. 500–509, 2009.
© Springer-Verlag Berlin Heidelberg 2009

desires of the use it represents in virtual shopping centre stores. The *N–1* remaining agents, AgCom, propose changes in the beliefs of the AgUser involving purchase of new products. The AgUser's immediate beliefs depend on its position in the shopping area (i.e., based on the social interactions); they also depend on how it adapts to new beliefs and knowledge provided by the AgComs.

The paper is divided in the following way: (i) section 2 provides a mathematical model of the system describing the type of interactions between the agents; (ii) section 3 describes how beliefs change due to social interaction and the relation between the importance of a belief and the social impact needed to change it; (iii) section 4 it analyzes different types of social phenomena that emerge in the system and different results obtained from the simulation of the system's behaviour. Cross-points in the processes of belief changes are determined potentially optimizing the agents' commercial strategies; (iv) section 5 demonstrates the implementation process and, in conclusion, section 6 describes the results of a real trial in a shopping centre.

2 Multiagent Recommendation Model (RecMAS)

Assuming a 2-dimensional virtual shopping centre with an area *A* with a community of *N* agents where N – 1 are AgComs which recommend the acquisition of different products, and an AgUser which has the capacity to buy them. In order to have a spatial description of the agents' system, it is assumed that the space is divided in z sectors $A* = A/z$ referred to as *Nj*, the number of agents in the sector *j*. The number of the AgUser's beliefs, *m*, is indicated below and the AgUser's belief concerning a problem is represented by the parameter αi.

$$A = \{ \alpha_i \}_{i=1,.....m} \qquad \alpha_i \in \{ -r, ..., -1,0,1, ..., r \} \qquad (1)$$

αi can change its value in accordance with its interaction with other agents or new external information, from –r to r, where it is assumed that the negative values represent a pre*disposition to reject a proposed task*. Positive values represent a *predisposition for action*, and values in the intermediate region of the interval, a *neutral position waiting for new information*.

2.1 The AgUser's Circulation and the Dynamics of Belief Change

This section considers the equations for the dynamic of: (i) the AgUser 's circulation through the shopping centre and (ii) changes in attitudes as a consequence of interaction with the AgCom. It is assumed that: the AgUser starts his visit to the virtual shopping centre in an area *A* in a sector we will refer to as k_0 (see figure 1a); it engages in purchases in sector *ks* **with** the distance between both positions (which coincides with the diagonal of the virtual environment) referred to as *D(A)*; the AgUser can move to any of the sectors around him (8, in the best cases, see figure 1b); and, always follows the shortest route from *k* ,where it starts, to the destination *ks*. It's route is represented in figure 1a $\phi(k0,\ ks)$, which joins the entrance sector with the shopping sector $\varphi(k,\ ks)$, the route the AgUser would follow from another sector k to ks. Supposing that AgUser is in the position ki and is acting on belief αi; then the equation of the AgUser's circulation (referred to as $\sigma(ki,\ kj)$), from its actual position ki to one of the adjacent positions is as follows:

$$k_j \in V_{ki} \quad \wedge \quad k_i \to V_{ki} = \{\ k_i^1, k_i^2, k_i^3, \ldots, k_i^8\} \tag{2}$$

Consequently:

- if the number of AgComs in ki with belief αi *which coincides* with the AgUser's belief is larger than the number of AgComs that share another belief αj with $j \neq i$, the belief αi is reinforced and the AgUser chooses the direct path to sector ks, represented by,

$$\varphi(k, k_s) = k_s - k \tag{3}$$

- If the number of AgComs in ki that defend a belief αj different from αi is larger than that of those that share αi, the AgUser moves randomly to one of the nearest sectors $kj \in V_{ki}$ in order to attain more beliefs. The formula that summarizes both of the AgUser's circulation strategies is shown below.

$$\sigma(k_i, k_j)\Big|_{k_j \in V_k} = \begin{cases} k_j \ / \ k_j \in k_s - k\ , \ N_{\alpha_i}^{k_i} \geq \sum_{j=1}^{m} \cdots^{k} \\ \mathrm{random}(k_i, k_j) \ , \ N_{\alpha_i}^{k_i} < \sum_{j=1}^{m} N_{\alpha_j}^{n_i} \end{cases} \tag{4}$$

Next, changes in the interaction strength of the agents in the AgCom environment are considered based on the AgUser's route. The previous section introduced two parameters: (i) su that measured the extent to which AgUser is inclined to maintain his belief αi, and (ii) sc that measured the influence the environment had over the AgCom's willingness to modify a belief αj. The parameter $\eta = (sc\ /\ su) \in (0,\ 1)$ measured the relation between both terms, (i) if $\eta = 0$, the AgUser did not alter his beliefs as a result of the interaction with the AgCom, and (ii) if $\eta = 1$, the AgUser gives the same importance to his own beliefs as to those he receives from the environment. Under such conditions it is expected that the influence of the environment on the AgUser who initially had a belief αi, to adopt a belief αj will be defined. α_i^* is referred to as the value for AgUser belief αi, and α_i^n the value the agents AgCom attributes to that belief, where $n = 1,...,N^k$. The influence needed to maintain the belief αi is determined by

$$i_k(\alpha_i) = s_u + \alpha_i^* \sum_{n=1}^{N^k} s_c \cdot \alpha_i^n = s_u(1 + \alpha_i^* \sum_{n=1}^{N^k} \eta \cdot \alpha_i^n) \tag{5}$$

The results obtained and the initial conclusions on how to improve AgCom strategies for attracting greater numbers of clients follow:

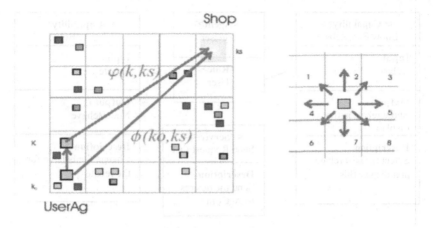

Fig. 1. (a) AgUser's movement in the shopping centre. (b) AgUser's possible movements.

3   RecMAS: Prototype and Implementation

The recommendation system in large shopping malls RecMAS is a computational tool based in MAS that attempts to convince clients to buy certain products. Each client, represented by an AgUser in a mobile device or PDA that he takes with him, interacts through a GPRS or wireless (Bluetooth or WiFi) connection with other agents in the shopping centre.

Each store has n representatives (AgCom) dispersed along the shopping centre with a scope limited to the sector k where they are located. From the point of view of implementation:

- The Gaia methodology [8] was used to model the multiagent system. It is comprised of role definition, protocols, services model, agents model and familiarity diagram. The Gaia methodology enables facile description of the agent system as an organization but has the inconvenience of not enabling a detailed level in the design stage. To correct this deficiency the modelling language Auml [9] was used. Auml is an extension of Uml (Unified Modeling Language) developed specially for agent (see figure 2 AgUser Class Diagram with AUML)
- Each agent has a deliberative architecture of the BDI sort implemented with a combination of its own platform and 3APL language [7]. 3APL language provides programming constructs for implementing agents' beliefs, goals, basic capabilities (such as belief updates, external actions, or communication actions) and a set of practical reasoning rules through which agents' goals can be updated or revised. The 3APL programs are executed on the 3APL platform. Each 3APL program is executed by means of an interpreter that deliberates on the cognitive attitudes of that agent.

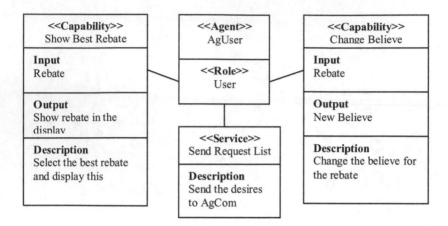

<<Capability>> Show Best Rebate	<<Agent>> AgUser	<<Capability>> Change Believe
Input Rebate	<<Role>> User	**Input** Rebate
Output Show rebate in the display		**Output** New Believe
Description Select the best rebate and display this	<<Service>> Send Request List **Description** Send the desires to AgCom	**Description** Change the believe for the rebate

Fig. 2. AUML AgUser Class Diagram

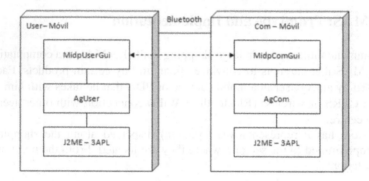

Fig. 3. Deployment Diagram for mobile device application

- As the AgUser's beliefs change after interaction with the AgComs in their coverage area, a procedure is defined for search of nearby agents using a Bluetooth device. Figure 3 show s the deployment diagram application for mobile devices.
- Some devices have experienced complications in sending/receiving processes when connecting via Bluetooth. In particular, Nokia 6620 and Nokia 6680 have worked successfully while the standard Bluetooth protocol JSR-82 has experienced inconsistent behavior. The reason appears to be due to the multiplicity of j2me versions with different restrictions on hardware.

4   AgentComs: Trading Knowledge through Links

The following section seeks to optimize the interaction strategy of the AgCom, i.e., to calculate which percentage of AgComs move away from the diagonal or approach the beginning and end sectors, without decreasing the success level of the gradual strategy while attempting to attract the AgUser when it deviates from the diagonal route.

4.1 Complex Optimization Problem

From an analytical point of view complex *optimization problem* techniques and metrics should be used enabling characterization of a temporal AgCom distribution in two directions (u, v). This measures the AgComs' movement along the diagonal $D(A)$ or anti-diagonal in the initial coordinates system (x, y) (See figure 4).

Movement inter-strategy: First, the progressive movement from a gradual strategy to environment type 2 and type 1 (Image 5A) are also measured. The equations that determine such dynamics are determined by the following equations

$$\varphi_u^{(1)} = \frac{e^{\frac{-v^2}{2(\beta_0^2)}} \, e^{\frac{-u^2 \, (1+\gamma t)^2}{2(\alpha_0)^2}}}{2\pi\beta_0} \cdot \frac{}{\alpha_0}(1+\gamma t) \; ; \varphi_u^{(2)} = \frac{e^{\frac{-v^2}{2(\beta_0^2)}} \, e^{\frac{-(u-\frac{D(A)}{2})^2 \, (1+\gamma t)^2}{2(\alpha_0)^2}}}{2\pi\beta_0} \cdot \frac{}{\alpha_0}(1+\gamma t) \quad (6)$$

Where $^{(2)}$ reflects the case in which the distribution towards the shopping sector with coordinates ($D(A)/2,0$) is studied; α y β being adjustable parameters:

$$\alpha = \frac{\alpha_0}{(1+\gamma t)}, \beta = \beta_0 \qquad (7)$$

Spreading: Secondly, certain metrics are defined to characterize the level of AgCom distribution along the diagonal (in this case,

$$\varphi_v = \frac{e^{\frac{-u^2}{2(\alpha_0^2)}}}{2\pi\alpha_0} \frac{e^{\frac{-v^2}{2(\beta_0-\gamma t)^2}}}{\beta_0+\gamma t} \qquad (8)$$

Optimization of these two functions requires consideration of a multiobjective function where optimization of the distribution of the number of agents is in the directions u and v from the diagonal. The problem is very complex and inefficient; therefore an optimization model will be proposed based in real time adjustment of a

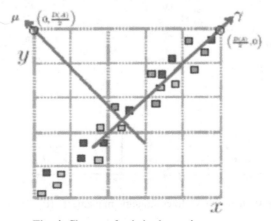

Fig. 4. Change of axis in the environment

number of the AgCom positions: substituting calculus and computational cost through a communication based adaptation plan.

4.2 Optimization Using Wireless Communications

This section introduces a configuration model where rather than searching for a statistically successful static strategy, as in the previous example, the AgCom will follow simple coordination rules to adapt to AgUsers who are not following the diagonal. The proposed model propose tries to find a dynamic adaptation to its environment in order to estimate if the AgUser is moving away from the diagonal route slightly or notably "as it goes". It is assumed that each AgUser can only communicate with other AgComs within their immediate environment. If the message sequence is constant, the distribution switches to the end sector (see figure 5(A)).

$$f_i(m + 1) \propto f_{i-1}(m) + \gamma(u - \tfrac{D(A)}{2})$$ (9)

If the message sequence is intermittent "the agent is close to the diagonal" and the distribution opens the range of values (see figure 5(B)).

$$f_i(m + 1) \propto f_{i-1}(m) + \gamma(v - \tfrac{D(A)}{2})$$ (10)

These simple rules force the AgComs to direct themselves towards (A), the shopping sector, or (B) to increase the range of their positions around the diagonal. The complexity of the communication relations based in these rules is not relevant (linear with regards to the number N of agents). The mobile devices are characterised by having less processing resources compared with computers. For this reason it was important to verify that the communication and search algorithms were not so complex that implementation would be impossible. A multithread routine was defined for

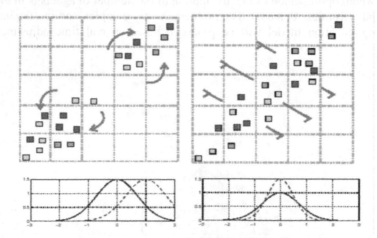

Fig. 5. Adaptation strategies: (A) Inter-strategy movement, (B) "Spreading" of a gradual distribution

the communication process that enabled interaction at a second level while information was presented to the user. One of the limitations of a Bluetooth Piconet is that it only supports up to 8 connected devices. In order to solve this restriction an JBAN library [10] was used enabling connection and control of an unlimited number N of devices in the network.

5 Final Results: Strategies for the Adaptation in Real Time and the Dynamic Attraction of Clients

One of the aspects taken into consideration for the evaluation of results was to gradually increase the area of scope. Initially 8 subregions in the shopping centre were defined. All of them shared the same vertex (entrance of the user) and as the subregion was extended, it contained the previous region. The size ratio between one subregion and the next is based on 40 meters, thus, the smallest is that size and the largest is 8 times bigger (the enlargement is mainly horizontal, because this shopping centre is not square). Different valuations were undertaken (30 users, three times each). In each case the belief they had initially when they entered the shopping centre and the one they bought the product were evaluated. The evaluated strategies where the: (i) "Lying-in-wait" strategy (yellow in the graph), (ii) Gradual strategy (crimson in the graph) and (iii) Adaptive strategy (blue in the graph). These results were as follows:

1. "Lying-in-wait" strategy: For smaller environments the level of success is higher that in larger environments: on average the AgUser coversmore space and has more interactions with other AgComs.
2. "Gradual" strategy: The level of success is higher than in the previous strategy for all the environments and decreases with the size in a trend similar to the previous case. Nevertheless, what is interesting about the result is that in relative terms the slope of the decrease of success falls more quickly.

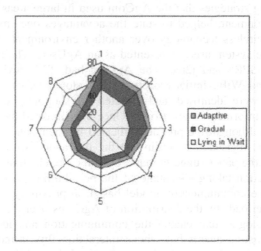

Fig. 6. Graphical representation of the level of the strategies' success: (A) "Lying-in-wait" strategy, (B) Gradual strategy (C) Adaptive strategy

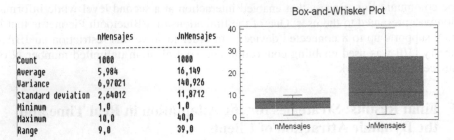

Fig. 7. (i) nMensajes in the optimizing problem and JnMensajes by TradeAgent protocol

3. "Adaptive" strategy: The most interesting result according to the gathered information is that the relative degree in which the success of the strategy decreases is much smaller than in the previous examples. In other words, the "adaptive" strategy is the only one that has acceptable results when the dimensions of the space considered is of a surface of 300 m2 or higher.

1000 simulations were conducted and the data obtained is shown in the figure that follows. The messages in the exchange protocol between AgComs are represented by JnMensajes and nMensajes when the protocol was measured by optimizing of complex problem.

6   Conclusions and Future Work

A model of relations between an AgUser and AgComs inside a multiagent society was developed. This model was designed to represent the internal strength of an AgUser's beliefs and its resistance to belief change processes affecting the products it was predisposed to buying. Next, different environments were proposed representing different advertising strategies that the AgCom used in large areas. The tests, carried out in a real environment, helped to state the advantages one environment enjoyed with the help of wireless technology over another environment. A wireless system was provides to the system user, represented as an AgUser. The mobile device supported Bluetooth, GPRS and j2me. The AgCom had PDA devices that supported Bluetooth, j2me and WiFi. Initially each AgUser stored its user profile, where its initial preferences were identified, and initiated the shopping centre route, visiting stores while being informed of new offers. Once they finished their route, they went to a sector store and bought the product indicated by their beliefs. The evolution of AgUser beliefs was registered according to the interaction with the AgComs. The concluding part of the paper suggests that a dynamic search strategy be sought seeking AgUser interaction taking advantage of the possibilities provided by the use of mobile devices. The communication model based on proximity interaction relations avoids network overloads by the distribution of AgComs over the surface (compared with GPRS technologies that enable the communication and identification of the whole group of existing agents but whose interaction flux is more complex). The result has been very satisfactory: the local scope and the simple rules of interaction have enabled discovery of a strategy that evolves parallel to the AgUser's movement.

The AgCom changed from occupying a static location that responded to a general strategy, to auto-organization based on the messages they receive from their nearest neighbours, anticipating the movements of the AgUser in relation to the diagonal.

Acknowledgements

Dr. Guillermo Alfonso Calvo Mahé of the Universidad Autònoma de Manizales in the Republic of Colombia, assisted in the linguistic review of this article.

References

1. Durfee, E.H.: Designing Organizations for Computational Agents. In: Prietula, M., Carley, K., Gasser, L. (eds.) Simulating Organizations: Computational Models of Institutions and Groups. AAAI Press and MIT Press, Menlo Park, CA (1998)
2. Gilbert, N., Doran, J. (eds.): Simulating Societies: The computer simulation of Social processes. University College, Londos (1994)
3. Helbing, D.: Quantitatuve Sociodynamics, Stochastic Methods and Models of Social Interaction Processes. Kluwer Academic, Dordrecht (1995)
4. Weidlich, W.: Physics and social science - The approach of synergetics. Phys. Rep. 204, 1–163 (1991)
5. Weidlich, W.: Synergetic modelling concepts for sociodynamics with application to collective political opinion formation. J. Math. Sociol. 18, 267–291 (1994)
6. Epstein, J.M., Axtell, R.: Growing Artificial Societies: Social Science from the Bottom Up. MIT Press/Brookings Institution Press (1996)
7. Koch, F., John-Jules, M., Frank, D., Iyad, R.: Programming Deliberative Agents for Mobile Services: the 3APL-M Platform. In: Bordini, R.H., Dastani, M., Dix, J., El Fallah Seghrouchni, A. (eds.) PROMAS 2005. LNCS, vol. 3862, pp. 222–235. Springer, Heidelberg (2006)
8. Wooldridge, M., Jennings, N.R., Kinny, D.: The Gaia Methodology for Agent-Oriented Analysis and Design. Journal of Autonomous Agents and Multi-Agent Systems 15 (2000)
9. Berhard Bauer, J.P.M., Odell, J.: Agent Uml: A Formalism For Specifying Multiagent Interaction. In: Luck, M., Padgham, L. (eds.) Agent-Oriented Software Engineering VIII. LNCS, vol. 4951, pp. 91–103. Springer, Heidelberg (2008)
10. JBAN Library, https://jban.dev.java.net/
11. Corchado, J.M., Gonzalez-Bedia, M., De Paz, Y., Bajo, J., Bajo, J., De Paz, J.F.: Replanning mechanism for deliberative agents in dynamic changing environments. Computational Intelligence, Pág 24(2), 77–107 (2008)

Combining Multiple Classifiers with Dynamic Weighted Voting

R.M. Valdovinos[1] and J.S. Sánchez[2]

[1] Grupo de Cómputo Aplicado, Centro Universitario UAEM Valle de Chalco
Av. Hermenegildo Galeana no. 3, 56615 Valle de Chalco, México
li_rmvr@hotmail.com
[2] Dept. Llenguatges i Sistemes Informàtics, Universitat Jaume I
Av. Sos Baynat s/n, E-12071 Castelló de la Plana, Spain
sanchez@uji.es

Abstract. When a multiple classifier system is employed, one of the most popular methods to accomplish the classifier fusion is the simple majority voting. However, when the performance of the ensemble members is not uniform, the efficiency of this type of voting generally results affected negatively. In this paper, new functions for dynamic weighting in classifier fusion are introduced. Experimental results demonstrate the advantages of these novel strategies over the simple voting scheme.

1 Introduction

A multiple classifier system (MCS) is a set of individual classifiers whose decisions are combined when classifying new patterns. There are many different reasons for combining multiple classifiers to solve a given learning problem. First, MCSs try to exploit the local different behavior of the individual classifiers to improve the accuracy of the overall system. Second, in some cases MCS might not be better than the single best classifier but can diminish or eliminate the risk of picking an inadequate single classifier. Another reason for using MCS arises from the limited representational capability of learning algorithms. It is possible that the classifier space considered for the problem does not contain the optimal classifier.

Let $\mathcal{D} = \{D_1, \ldots, D_L\}$ be a set of L base classifiers. Each classifier D_i ($i = 1, \ldots, L$) assigns an input feature vector $\mathbf{x} \in \Re^n$ to one of the possible C problem classes. The output of an MCS is an L-dimensional vector $[D_1(\mathbf{x}), \ldots, D_L(\mathbf{x})]^{\mathrm{T}}$ containing the decisions of each of the L individual classifiers.

In general, an ensemble is built in two steps: training multiple individual classifiers and then combining their predictions. According to the style of training the base classifiers, current ensemble algorithms can be roughly categorized into two groups: those where base classifiers must be trained sequentially, and algorithms where base classifiers could be trained in parallel. The most popular example of the first category is AdaBoost [7], which sequentially generates a series of individual classifiers where the training instances that are wrongly predicted by a component will be more important in the training of its subsequent classifier. Other methods of this group are Arc-x4 [3] and LogitBoost [8]. The representative of the second category is Bagging [2], which

E. Corchado et al. (Eds.): HAIS 2009, LNAI 5572, pp. 510–516, 2009.

uses bootstrap sampling to generate multiple training sets from the original set and then trains a classifier from each generated training set. Other examples include Wagging [1] and p-Bagging [1].

Regarding to the second step, that is, combining the individual predictions, there appear two main strategies: *selection* and *fusion*. In classifier selection, each individual classifier is supposed to be an expert in a part of the feature space and correspondingly, we select only one classifier to label the input vector $\mathbf{x} \in \Re^n$. In classifier fusion, each component is supposed to have knowledge of the whole feature space and thus, all individual classifiers are taken into account to decide the label of the input vector. Focusing on the fusion strategy, the combination can be made in many different ways. The simplest one employs the majority rule in a plain voting system. More elaborated schemes use weighted voting rules, in which each individual component can be associated with a different weight [15]. Then, the final decision can be made by majority, average, minority, medium, product of votes, or using some other more complex methods [10].

In the present work, several new methods for weighting the individual components of an MCS are proposed. Then the effectiveness of the new approaches is empirically tested over a number of real-problem data sets. All these methods correspond to the so-called dynamic weighting and basically consist of using the distances to the input pattern $\mathbf{x}$ in each individual classifier.

Raudys [12, 13] performs an exhaustive theoretical analysis of trainable fusion rules on large and small sample size cases to determine in which situations neural network ensembles can improve or degrade classification results. Our paper deals with similar ideas, but the focus is on a nearest neighbor classifier and intends to evaluate the performance of a number of weighting strategies.

2   Classifier Fusion

Classifier fusion assumes that all classifiers in the set $\mathcal{D}$ are competitive, instead of complementary. For this reason, each component takes part in the decision of classifying an input vector $\mathbf{x}$. In the simple voting (by majority), the final decision is taken according to the number of votes given by the individual classifiers to each of the C classes, thus assigning $\mathbf{x}$ to the class that has obtained a majority of votes. When working with data sets that contain more than two classes, ties among classes are solved using several criteria. For instance, to take the decision randomly, or to implement an additional classifier whose ultimate goal is to bias the decision towards a certain class.

An important issue that has strongly called the attention of many researchers is the error rate associated to the simple voting method and to the individual components of an MCS. Hansen and Salomon [9] show that if each of the L classifiers being combined has an error rate less than 50%, it may be expected that the accuracy of the ensemble improves when more components are added to the system. However, this assumption not always can be fulfilled. Matan [11] asserts that in some cases, the simple voting might perform even worse than any of the members of the MCS. The employment of some weighting method has been proposed as a way to partially overcome these difficulties.

A weighted voting method has the potential to make the MCS more robust to the choice of the number of individual classifiers. Two general approaches to weighting can be remarked: *dynamic weighting* and *static weighting* of classifiers. In the dynamic strategy, the weights assigned to the individual classifiers of the MCS can change for each input vector in the operational phase. In the static approach, the weights are computed for each classifier in the training phase, and they do not change during the classification of the input patterns.

3 Dynamic Weighted Voting Schemes

Some of the weighting functions introduced in this section are taken from the literature and are here conveniently adapted for classifier fusion in an MCS, while others are now proposed for the first time.

3.1 Dudani's Weighting Function and the Inverse Distance Weight

A voting rule for the k-NN (Nearest Neighbor) rule [4] in which the votes of different neighbors are weighted by a function of their distance to the input pattern was first proposed by Dudani [6]. A neighbor with smaller distance is weighted more heavily than one with a greater distance: the nearest neighbor gets a weight of 1, the furthest neighbor a weight of 0, and the other weights are scaled linearly to the interval in between. From this, the Dudani's weight can be computed as:

$$w_j = \begin{cases} \frac{d_k - d_j}{d_k - d_1} & \text{if} \quad d_k \neq d_1 \\ 1 & \text{if} \quad d_k = d_1 \end{cases} \tag{1}$$

where d_j denotes the distance of the j'th nearest neighbor, d_1 is the distance of the nearest neighbor, and d_k indicates the distance of the furthest (k'th) neighbor.

In order to utilize this weighting function in the classifier fusion, the value of k (i.e., the number of neighbors in Dudani's rule) can be here replaced by the number of classifiers L that constitute the MCS. Moreover, the L distances of $\mathbf{x}$ to its nearest neighbor in each individual classifier have to be sorted in increasing order ($d_1, d_2, \ldots, d_L$). Thus, the original Dudani's weight (Eq. 1) can be now rewritten as follows:

$$w(D_j) = \begin{cases} \frac{d_L - d_j}{d_L - d_1} & \text{if} \quad d_L \neq d_1 \\ 1 & \text{if} \quad d_L = d_1 \end{cases} \tag{2}$$

where d_1 denotes the shortest of the L distances of $\mathbf{x}$ to the nearest neighbor, and correspondingly d_L is the longest of those distances.

Dudani further proposed the *inverse distance weight* [6], which can be expressed as follows:

$$w(D_j) = \frac{1}{d_j} \quad \text{if} \quad d_j \neq 0 \tag{3}$$

3.2 Shepard's Weighting Functions

Another weighting function proposed here is based on Shepard's work [14], who argues for a universal perceptual law which states that the relevance of a previous stimulus for the generalization to a new stimulus is an exponentially decreasing function of its distance in psychological space. This produces the function of Eq. 4, where α and β are constants and determine the slope and the power of the exponential decay function.

$$w(D_j) = e^{-\alpha d_j^\beta} \tag{4}$$

A modification to Shepard's function consists of using a different value of α for each input pattern. Firstly, the L distances of **x** to its nearest neighbor in each individual classifier have to be sorted in decreasing order. Then, the value of α for each input pattern is computed according to $\alpha = L - j + 1$. By this, the higher the distance given by a classifier, the higher the value of α and thereby, the lower the weight assigned to such a classifier.

3.3 Average Distance Weight

Finally, we propose another weighting function, which corresponds to the *average distance weight*. In summary, the aim of this new dynamic weighting procedure is to reward (by assigning the highest weight) the individual classifier with the nearest neighbor to the input pattern.

$$w(D_j) = \frac{\sum_{i=1}^{L} d_i}{d_j} \tag{5}$$

The rationale behind this weight is that the classifier with the nearest neighbor to **x** probably corresponds to that with the highest accuracy in the classification of the given input pattern.

4 Experiments and Results

The results here reported correspond to the experiments over ten real-problem data sets taken from the UCI Machine Learning Database Repository (http://www.ics.uci.edu/~mlearn). Table 1 summarizes the main characteristics of each data set: number of classes, attributes, and patterns. For each data set, the 5-fold cross-validation method was employed to estimate the classification error. On the other hand, it has to be noted that in the present work, all the base classifiers correspond to the 1-NN rule.

The experiments basically consist of computing the classification accuracy when using different voting schemes in an MCS. The weight functions proposed in the present paper (the average distance weight, the Shepard's and modified Shepard's functions, the inverse distance weight, and Dudani's weight) are compared to the simple majority voting. In the experiments here carried out, we have set $\alpha = \beta = 1.0$ for the computation of the original Shepard's weight function (Eq. 4).

The MCSs have been integrated by using four resampling methods: *random selection with no replacement, bagging, boosting,* and *Arc-x4.* Only the result of the best technique on each database has been presented in Table 2. Analogously, for each database,

Table 1. Summary of the databases

	No. Classes	No. Features	No. Patterns
Cancer	2	9	685
Heart	2	13	272
Liver	2	6	347
Pima	2	8	770
Iris	3	4	150
Vehicle	4	18	848
Wine	3	13	180
German	2	24	1002
Phoneme	2	5	5406
Waveform	3	21	5001

related to the number of subsamples to induce the individual classifiers, that is, the number of classifiers in the system, we have experimented with 3, 5, 7, 9, and 15, and the best results have been finally included in Table 2. The 1-NN classification accuracy for each original training set (with no combination) is also reported as the baseline classifier. Note that values in bold type indicate the highest accuracy for each database.

Since the accuracies are very different for the distinct data sets, using these results across the data sets will be inadequate. Instead we calculate ranks for the methods [5]. For each data set, the method with the best accuracy receives rank 1, and the worst receives rank 7. If there is a tie, the ranks are shared. For each method, there are 10 rank values, one for each set. Thus the overall rank of a method is the averaged rank of this method across the 10 data sets. The smaller the rank, the better the method. These ranks are shown in Table 2.

From the results given in Table 2, it is clear that in all databases the employment of an MCS leads to better performance than the individual 1-NN classifier. This is confirmed by the overall rank of the 1-NN, which clearly corresponds to the worst value.

Table 2. Average accuracies and ranks

	Individual 1-NN	Simple voting	Average distance	Shepard's function	Modified Shepard's	Inverse distance	Dudani's weight
Cancer	95.62	**96.35**	96.20	**96.35**	**96.35**	96.20	95.89
Heart	58.15	62.96	**64.81**	61.11	61.85	**64.81**	58.52
Liver	65.22	65.80	**66.09**	65.80	65.80	64.93	60.87
Pima	65.88	**72.81**	72.68	68.37	68.24	71.90	67.58
Iris	96.00	**98.00**	97.33	97.33	96.67	97.33	96.67
Vehicle	64.24	62.34	64.48	**65.56**	65.19	64.48	64.24
Wine	72.35	75.88	**77.65**	73.53	74.12	**77.65**	75.95
German	65.21	70.21	**70.81**	68.11	66.91	**70.81**	67.34
Phoneme	76.08	75.93	76.51	75.97	**76.56**	76.51	76.02
Waveform	77.96	83.20	83.20	83.06	78.20	**83.54**	83.22
Overall Rank	6.35	3.50	2.50	4.00	3.75	2.80	5.10

Second, the application of some weight function generally outperforms the combination of classifiers by means of the simple majority voting. In fact, we can find some weighting scheme with higher (or equal) classification accuracy than that of the simple voting on 8 out of 10 databases.

When comparing the different dynamic weighting schemes, the average distance and the inverse distance weights seem to generally behave better than any other function: each one reaches the highest accuracy on 4 out of 10 databases and correspondingly, the lowest (best) overall ranks (2.50 and 2.80, respectively). On the other hand, when these two methods do not obtain the best results, their accuracies are still very close to that of the winner. For instance, in the case of Phoneme domain, while the modified Shepard's function obtains the highest accuracy rate (76.56%), the classification performance of both the average distance and the inverse distance weights are 76.51%. Similarly, for Cancer database, differences in accuracy with respect to the "best" weighting schemes (Shepard's and modified Shepard's functions) are not significant (only 0.15%). Finally, it is remarkable that the weighting method with the highest (worst) overall rank was found to be Dudani's weight, followed by Shepard's.

5 Concluding Remarks

When an MCS is employed in a classifier fusion scenario, one has to implement some procedure for combining the individual decisions of the base classifiers. Although the plain majority voting rule constitutes a very appealing method due to its conceptual and implementational simplicity, its efficiency can become too poor when the performance of the ensemble members is not uniform. Under this practical situation, more complex voting techniques, mainly in the direction of assigning different weights to each base classifier, should be applied to derive the final decision of the MCS.

In this paper, new methods for dynamic weighting in a MCS have been introduced. More specifically, several weighting functions present in the literature have been adapted to be used in a voting system for classifier fusion. In particular, we have explored the reformulated Dudani's distance, the inverse distance, the average distance, and also the Shepard's function and a modification to it based on a rank of distances.

Experimental results have shown the benefits of using dynamic weighting strategies over the simple majority voting scheme. The average distance and the inverse distance have appeared as the best weighting functions in terms of highest classification accuracy; each one has exhibited the best performance on 4 out of 10 databases, and consequently the best ranks. Results also corroborate that in general, an MCS clearly outperforms the individual 1-NN classifier.

Future work is addressed to investigate other weighting functions applied to classifier fusion. Within this context, the use of data complexity measures could be of interest to conveniently adjust the classifier weights. On the other hand, the results here reported should be viewed as a first step towards a more complete understanding of the behavior of the weighted voting procedures and consequently, it is still necessary to perform a more exhaustive analysis of the dynamic and static weighting strategies over a larger number of synthetic and real databases.

516   R.M. Valdovinos and J.S. Sánchez

Acknowledgements

This work has been partially supported by the Mexican PROMEP/103.5/08/3016, the Spanish CICYT and the Spanish Ministry of Science and Education under grants DPI2006–15542 and CSD2007–00018.

References

1. Bauer, E., Kohavi, R.: An empirical comparison of voting classification algorithms: bagging, boosting, and variants. Machine Learning 36, 105–139 (1999)
2. Breiman, L.: Bagging predictors. Machine Learning 24, 123–140 (1996)
3. Breiman, L.: Arcing classifiers. Annals of Statistics 26, 801–823 (1998)
4. Dasarathy, B.V.: Nearest Neighbor Norms: NN Pattern Classification Techniques. IEEE Computer Society Press, Los Alamos (1991)
5. Demšar, J.: Statistical comparison of classifiers over multiple data sets. Journal of Machine Learning Research 7, 1–30 (2006)
6. Dudani, S.A.: The distance weighted k-nearest neighbor rule. IEEE Trans. on System, Man, and Cybernetics 6, 325–327 (1976)
7. Freund, Y., Schapire, R.E.: Experiments with a new boosting algorithm. In: Proc. 13th Intl. Conf. on Machine Learning, pp. 148–156 (1996)
8. Friedman, J., Hastie, T., Tibshirani, R.: Additive logistic regression: a statistical view of boosting, Technical report, Stanford University (1998)
9. Hansen, L.K., Salomon, P.: Neural network ensembles. IEEE Trans. on Pattern Analysis and Machine Intelligence 12, 993–1001 (1990)
10. Kuncheva, L.I., Bezdek, J.C., Duin, R.P.W.: Decision templates for multiple classifier fusion. Pattern Recognition 34, 299–314 (2001)
11. Matan, O.: On voting ensembles of classifiers. In: Proc. 13th Natl. Conf. on Artificial Intelligence, pp. 84–88 (1996)
12. Raudys, S.: Trainable fusion rules. I. Large sample size case. Neural Networks 19, 1506–1516 (2006)
13. Raudys, S.: Trainable fusion rules. II. Small sample-size effects. Neural Networks 19, 1517–1527 (2006)
14. Shepard, R.N.: Toward a universal law of generalization for psychological science. Science 237, 1317–1323 (1987)
15. Woods, K., Kegelmeyer Jr., W.P., Bowyer, K.: Combination of multiple classifiers using local accuracy estimates. IEEE Trans. on Pattern Analysis and Machine Intelligence 19, 405–410 (1997)

Fusion of Topology Preserving Neural Networks*

C. Saavedra[1], R. Salas[2], H. Allende[1], and C. Moraga[3,4]

[1]Universidad Técnica Federico Santa María, Dept. de Informática, Valparaíso, Chile
[2]Universidad de Valparaíso, Dept. Ingeniería Biomédica, Valparaíso, Chile
[3]European Centre for Soft Computing, 33600 Mieres, Spain
[4]Dortmund University of Technology, 44221 Dortmund, Germany

Abstract. In this paper ensembles of self organizing NNs through fusion are introduced. In these ensembles not the output signals of the base learners are combined, but their architectures are properly merged. Merging algorithms for fusion and boosting-fusion-based ensembles of SOMs, GSOMs and NG networks are presented and positively evaluated on benchmarks from the UCI database.

1 Introduction

Every day large amounts of information are stored, represented as data, to extract relevant information for making decisions. However, the vast complexity of the nature of the data makes it virtually impossible to analyze directly this information due to its high-dimensionality.

Neural maps constitute an important paradigm to address these problems. The main idea of these networks is to project data from some input space onto a position in some output space preserving the neighborhood by means of vector quantizers. As a representative of this idea we can mention Vector Quantization Artificial Neural Networks [2] (VQ-ANN), Self Organizing Maps (SOM) [10], Growing Self Organizing Maps (GSOM)[3] and Neural Gas [11]. Readers are expected to be familiar with these models. The given references may provide additional information.

Performance of these networks may be improved by using ensembles, as will be explained in a later section. In [1] four different ensemble techniques have been applied to topology preserving maps. The methods used for the combination of the base learners are Bagging, Georgakis Fusion, Superposition and Superposition with Re-labelling. The authors show how the ensemble methods can improve the accuracy with respect to the initial models, however with constrained visualization capability. Superposition techniques preserve the visualization although with a rather rigid topology. In this paper, ensembles based on fusion and boosting at the level of architecture will be used, which preserve visualization and flexibility. Preliminary results in this direction were presented at the IWANN Conference 2007 [14].

---

* This work was supported by the Research Grant Fondecyt 1070220, Chile. The work of C. Moraga was partially supported by the Foundation for the Advancement of Soft Computing, Mieres, Spain.

E. Corchado et al. (Eds.): HAIS 2009, LNAI 5572, pp. 517–524, 2009.
© Springer-Verlag Berlin Heidelberg 2009

The remainder or this paper is organized as follows. The next section briefly introduces Machine Learning Ensembles. In the third section, Fusion Algorithms for the Ensemble of selected Neural Networks are stated. Simulation results on synthetic and real data sets are provided in the fourth section. Conclusions and further work are given in the last section.

2   Machine Learning Ensembles

An ensemble consists of a set of base models whose individual decisions are combined to produce results about a phenomenon under study [8], [13]. The reason why a large number of researchers are interested in machine ensembles is that collective decisions are often more accurate than the results of individual machines. There are studies [5] that show the advantages in using an ensemble instead a single machine learning. In a set of machines the individual errors may be compensated and the result is better due the diversity and heterogeneity induced in the set of base learners, reducing the variance, improving the quality and robustness of the results. Research on ensembles has expanded rapidly, often appearing in the literature under different names. See   [6], [8], [12], [13], [15].

There are three types of machine learning combinations: machine selection, machine aggregation and machine fusion. In machine selection, each model is trained to become an expert in some local area of the total feature space, and the output is selected according to the performance of the base models. Thus, when a pattern is presented to the machines the selection of the best among the individuals results is obtained as output. The machines results may also be combined using machine aggregation. This type of combination uses an aggregation technique to obtain a final result from the individual machines decisions. The most well known examples of aggregation techniques are the average, the median, and the majority of votes. In machine fusion all the learners are trained with samples of the entire feature space, the combination process involves merging the individual machine designs to obtain a single (stronger) expert of superior performance. A representation of the fusion approach, using artificial neural networks, is shown in figure 1.

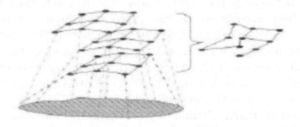

Fig. 1. Machine Learning combination through fusion

3   Ensemble of Vector Quantization Neural Networks Using Fusion

Currently fusion approaches are based on information from the data or in the structure of the models. The method proposed in this paper uses information from both the input data and information obtained by the base models. We call *Fusion-SOM* to the model for merging SOM networks, *Fusion-GSOM* to the fuse GSOM networks; finally the model to fuse neural gas networks is called *Fusion-NG*. Moreover the model *Fusion-NG* has the ability to create connections between the nodes that are similar to each other on the resulting neural network. This is a feature that the base model neural gas does not possess. These ensembles are generated using T self-organized networks which are trained independently and then merged to obtain a final network capable of obtaining better results than a single network.

The construction of one of these models is divided into two parts. In the first part T base structures $\mathcal{M}_t, t = 1 \ldots T$, are created to be trained with bootstrap samples [9] from the input data $D = \{\mathbf{x}_1, \ldots, \mathbf{x}_n\}$. Once the T models are created and trained, the merge process is implemented, which corresponds to the part of the construction of the model. After this stage, the final model is $\mathcal{M}^*$, the result of integrating the former neural networks.

After obtaining the bootstrap samples, all the base models are trained. This training is done independently, each network with one sub-sample collected earlier. Once all networks have been trained, $C_r^{(t)} = \mathcal{C}(\mathbf{w}_r^{(t)})$ can be considered as the input vector that belongs to the Voronoi Polygon of the r-th neuron $\mathbf{w}_r^{(t)}$ in the network $\mathcal{M}_t$. This means that, $C_r^{(t)}$ is the region of the space where entry neuron r is the winner neuron.

To analyze information from the input data and base networks trained previously, for each vector $\mathbf{w}_r^{(t)} \in \mathcal{M}_t, r = 1..M_t$, and for each neural network $t = 1..T$, the *co-association vectors* are defined as follows:

$$\nu_r^{(t)} = \nu(\mathbf{w}_r^{(t)}) = [\nu_r^{(t)}(1), ..., \nu_r^{(t)}(n)] \tag{1}$$

From n, the cardinality of the the data set $\mathcal{D}$, each component of the co-association vector takes the value 1 if the data $\mathbf{x}_i$ belongs to the Voronoi Polygon $C_r^{(t)}$ of the neuron (prototype) $\mathbf{w}_r^{(t)}$ and 0 otherwise.

$$\nu_r^{(t)}(i) = \begin{cases} 1 & \text{if } \mathbf{x}_i \in C_r^t \\ 0 & other\ case \end{cases} \tag{2}$$

Thanks to the co-association vector, neurons with low usage rate in the network may be identified, because their Voronoi Polygons are virtually empty and do not provide relevant information to the final model. That is why at this stage a process of pruning "weak" neurons is generated, where all nodes $\mathbf{w}_r^{(t)}$ that satisfy the inequality $\sum_{i=1}^n \nu_r^{(t)}(i) < \theta_u$ are removed, and θ_u is a threshold that defines the minimum use that a neuron must have. $(0 < \theta_u < n)$.

All the prototypes $\mathbf{w}_r^{(t)} \in \mathcal{M}_t, r = 1..M_t$, of each neural network $t = 1..T$ that were not eliminated in the previous process are grouped into sets $W_k, k = 1..K$

to form new nodes belonging to the neural network $\mathcal{M}^*$. To measure whether two nodes $\mathbf{w}_r$ and $\mathbf{w}_q$ are similar or not, a *dissimilarity measure* as a function of the co-association vectors is defined.

$$ds(\nu_r, \nu_q) = \frac{\sum_{l=1}^n \text{XOR}(\nu_r(l), \nu_q(l))}{\sum_{j=1}^n \text{OR}(\nu_r(j), \nu_q(j))} \tag{3}$$

With this measure the incidence matrix $\mathcal{I} = \{\mathcal{I}_{rq}\}$ of size $M_T \times M_T$ can be formed, where $M_T = \sum_{t=1}^T M_t$, whose elements $\mathcal{I}_{rq}$ take the value 1 if $ds(\nu_r, \nu_q) < \theta_f$ and 0 otherwise, where θ_f is the fusion threshold. $(0 < \theta_f < 1)$. As the idea is to bring together all similar neurons to form K new nodes in the new model a set W_1 was created in which the unit belonging to the first row of the incidence matrix, $\mathbf{w}_{(1)}$, with all the neurons $\mathbf{w}_{(i)}$ compliant $\mathcal{I}_{1i} = 1, i = 1..M_T$ was added. Let s be a counter of the new sets. Then, when making a new unit $\mathbf{w}_{(j)}$ of the matrix, it should be verified whether the neuron belongs to a group created earlier, this being the case, all its similar nodes, i.e. $\mathcal{I}_{ji} = 1, i > j$ must be looked for, and included in all sets W_l where the neuron appears, that is, where $\mathbf{w}_{(j)} \in W_l$. On the other hand, if the neuron does not belong to any set, the counter s is increased by one and the set W_s with the prototype $\mathbf{w}_{(j)}$ and with all the neurons that meet $\mathcal{I}_{ji} = 1, i = j + 1..M_T$ is created.

The position of the new node created $\overline{\mathbf{w}}_k, k = 1..K$ corresponds to the centroid of the set $W_k, k = 1..K$ This is calculated as:

$$\overline{\mathbf{w}}_k = \frac{1}{|W_k|} \sum_{\mathbf{w}_r \in W_k} \mathbf{w}_r \qquad k = 1..K \tag{4}$$

The lattice (neighborhood) in the final model, or the creation of this, is obtained as follows : A connection between neurons $\overline{\mathbf{w}}_k$ and $\overline{\mathbf{w}}_q$ is created, if the dissimilarity between the sets W_k and W_q is lower than the connection threshold θ_c, $(0 < \theta_c \leq 1)$, i.e.

$$\min_{\mathbf{w}_r \in W_k, \mathbf{w}_q \in W_q} ds(\nu_r, \nu_q) < \theta_c \tag{5}$$

Finally, the model created with the initial networks $\mathcal{M}^* = \mathcal{M}(\mathcal{W}^*, \mathcal{N}^*)$ consists of a set of fusion neurons $\mathcal{W}^* = \{\overline{\mathbf{w}}_k, k = 1..K\}$ and all connections $\mathcal{N}^*$ that have been established according to equation (5).

4   Networks Fusion Using Boosting

This model, called *Boost-Fusion*, consists in generating an artificial neural networks ensemble using the Boosting algorithm [7] proposed by H. Drucker. Initially, all examples have the same weights. At each iteration a machine learning model is built, called the base model, using some learning method and taking into account the distribution of weights. Then the weight of each example is adjusted depending on errors obtained by the base model. The final result is obtained merging the base models. The base models can be SOM, GSOM or NG networks.

To obtain diversity in the base models, these models are trained with sampling sets of the original data of the original data obtaining according to the probability distribution given by the weights assigned to the data. Initially the weights of the input data are equal to $h_i = \frac{1}{n}$, for all $i = 1, .., n$.

Once trained the model, the base model performance is evaluated to update the weights of the input data. The error for every input data is given by the equation (6):

$$L_i = \frac{|x_i - m_{c(x_i)}|^2}{SP^2} \tag{6}$$

where $SP = \text{Sup}|x_i - m_{c(x_i)}|$ corresponds to the greatest error obtained from an input data.

The machine confidence β is given by the equation (7).

$$\beta = \frac{L}{1 - L} \quad \text{where,} \quad L = \sum_{i=1}^{n} L_i h_i \tag{7}$$

L corresponds to the errors obtained by the model multiplied by the weights of the input data.

After the evaluation of the model, if the average of data errors L is less than 0.5, i.e., the machine is sufficiently reliable, then the machine is added to the current ensemble. Finally the weights are updated according to the confidence β (7) and the errors obtained L_i (6). The adaptation rule is the following:

$$h_i(t + 1) = h_i(t)\beta^{1-L_i} \quad \forall i = 1, .., n \tag{8}$$

With these weights update a new sample of the original data may be obtained to train the next base model.

5   Experimental Results

In this section the capabilities of our proposed models (*Fusion-SOM, Fusion-NG, BoostFusion-SOM, BoostFusion-NG* and *Fusion-GSOM*) are shown and compared with the classic models in literature (SOM, NG and GSOM) in both Synthetic and Real Data sets. The latter data sets were obtained from a benchmark site [4].

As a first experiment the "Fermat's Spiral" is used, whose equation is $r = \theta^{1/2}$. Figure 2 shows the topology approximation results. Note that all the models are able to learn the topology, but due to their rigid architectures the SOM and GSOM networks show some neurons that are not modeling data and links in blank space, this means that two neurons are shown as neighbors when they are not. The model *Fusion-SOM* effectively locates the neurons and the neighborhood among neurons. Finally, the model *Fusion-GSOM* has a better topology representation than the GSOM model locating effectively the neurons in the final

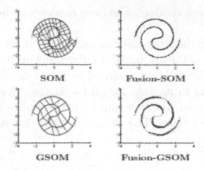

Fig. 2. Simulation results for Fermat's Spiral

model, but the topology representation is worse than in *Fusion-SOM* due the distance between neurons. If these neurons should appear connected in the final model, the connection threshold θ_c would have to be increased. This would however begin to increase the connections among all neurons. Recall that a GSOM is a growing network that tries to obtain an optimal number of neurons, meanwhile a SOM will normally be oversized at the initialization. This explains why the fusion of SOMs may achieve better results than the fusion of GSOMs.

The fusion algorithms were also tested with the following benchmarks data sets: "Wisconsin Cancer" and "Wine" obtained from the UCI Machine Learning Repository [4]. The data set Wisconsin Cancer is composed of 569 instances and 30 attributes meanwhile the Wine data set is composed of 178 instances and 13 attributes.

To compare the performance of the different algorithms the Mean Square Quantization Error $MSQE$ (9) and the Percentage of right Classification PC (10) were computed. Both are defined below.

$$MSQE = \frac{1}{|\mathcal{D}|} \sum_{\mathbf{x}_i \in \mathcal{D}} \|\mathbf{x}_i - \mathbf{w}_{c(\mathbf{x}_i)}\|^2 \tag{9}$$

Where $|\mathcal{D}|$ denotes the cardinality of the training set $\mathcal{D} = \{\mathbf{x}_1, ..., \mathbf{x}_n\}$, and $\mathbf{w}_{c(\mathbf{x}_i)}$ is the BMU neuron for the data $\mathbf{x}_i$ as used in the SOM

$$PC = \frac{1}{|\mathcal{D}|} \sum_{\mathbf{x}_i \in \mathcal{D}} \mathcal{I}(\mathbf{x}_i - \mathbf{w}_{c(\mathbf{x}_i)}) \tag{10}$$

The parameters used in all models for experiments with real data are the following: Base models = 5, $\theta_f = 0.7$, $\theta_c = 0.9$, and $\theta_u = 1$. Moreover for the Wisconsin cancer 36 neurons and for Wine 16 are used. (n_e equals 36 an 16, respectively).

The results for the "Wisconsin Cancer" database are shown in table 1. It may be appreciated that the models based on Boosting (*BoostFusion-SOM* and *BoostFusion-NG*) present the lowest errors for the training process, for the test models the best models are Fusion-SOM and Fusion-NG. Two other networks

Table 1. Results for the "Wisconsin Cancer" DB for SOM, *Fusion-SOM, BoostFusion-SOM*, NG, *Fusion-NG* and *BoostFusion-NG* models. Average over 20 runs.

Model	n_e	$MSQE$ Train	$MSQE$ Test	PC Train	PC Test
SOM	36	289.030	122.955	0.957	0.975
SOM	60	250.540	116.568	**0.962**	0.973
FSOM	61	258.582	**116.400**	0.960	**0.988**
BFSOM	59	**230.806**	117.140	0.957	0.964
NG	36	264.436	126.444	0.956	0.965
NG	60	219.337	117.311	0.959	0.972
FNG	60	222.432	**113.434**	0.961	0.968
BFNG	58	**218.933**	115.866	**0.962**	**0.976**

Table 2. Results for the "Wine" Database for SOM, *Fusion-SOM, BoostFusion-SOM*, NG, *Fusion-NG* and *BoostFusion-NG* models. Average over 20 runs.

Model	n_e	$MSQE$ Train	$MSQE$ Test	PC Train	PC Test
SOM	16	82.404	82.117	0.966	0.941
SOM	25	69.707	77.122	0.967	0.947
FSOM	23	68.530	73.125	**0.976**	0.953
BFSOM	24	**60.804**	**66.115**	0.970	**0.961**
NG	16	80.032	82.848	0.950	0.937
NG	25	62.532	76.205	0.967	0.947
FNG	24	61.377	73.342	**0.972**	0.938
BFNG	25	**58.571**	**72.285**	**0.972**	**0.951**

(a SOM and a NG) with the same number of neurons that the Fusion models were trained as references to compare the performances.

In the second real experiment the "Wine" database was used. The results are shown in table 2. It is easy to appreciate that *Fusion-SOM* and Fusion NG improved the performance of a single base network in error, but the best models are those that rely on the Boosting algorithm. Two other networks (a SOM and a NG) with the same number of neurons that the Fusion models were trained as references to compare the performances. In this case the results are very close between the Fusion models and the base models, but the models based on Boosting have a better performance.

In both real experiments, when the base models used are NG, the best Percentage of right Classification PC was obtained by *BoostFusion* models. In the models based on SOM networks however, the results for the "Wisconsin Cancer" data set are very similar for both *Fusion-SOM* and a SOM network trained with almost the same number of neurons.

6   Conclusions and Further Works

In this paper two techniques were presented to create ensembles of vector quantization neural networks. From these techniques the models *Fusion-SOM, Fusion-GSOM, Fusion-NG, BoostFusion-SOM* and *BoostFusion-NG* were obtained. These models were developed primarily to improve the performance of an individual neural network and to improve the topological representation of data.

These models are based on information from the input data and from the base models trained, this is an advantage with respect to the models in the literature, which are based only in one of two methods: input data or base models.

The strengths of the models presented in this paper are the ability to adjust automatically the number of neurons and connections between them, necessary for optimal representation of the input data, which is very useful when there is no a priori knowledge about the number of neurons needed for a particular database. In addition, the models do not have a predefined architecture, as in the case of a SOM network. The architecture is obtained after fusing the base models, so the final models are able to recognize the distribution of data and accommodate the samples in a more efficient way. Finally, when the base models used do not provide a visual representation of the neighborhood, as is the case of the NG network, the ensemble is capable of generating connections among similar neurons showing explicitly the relationships among units of the model to improve the representation and visualization capabilities of a single neural network.

It was empirically demonstrated that all the discussed models are able to learn and preserve the topology of input data, locating neurons to represent data and the connections to represent the relationship among them improving the representation of topological data obtained by a single network. This improvement is due mainly to their flexible architecture. It should however be mentioned that there is a need for quantitative criteria to determine the quality of a topological representation of a base model in order to be able to make sound comparisons among different models.

References

1. Baruque, B., Corchado, E., Yin, H.: Visom ensembles for visualization and classification. In: Sandoval, F., Prieto, A.G., Cabestany, J., Graña, M. (eds.) IWANN 2007. LNCS, vol. 4507, pp. 235–243. Springer, Heidelberg (2007)
2. Bauer, H.-U., Herrmann, M., Villmann, T.: Neural maps and topographic vector quantization. Neural Networks 12, 659–676 (1999)
3. Bauer, H.-U., Villmann, T.: Growing a hypercubical output space in a self-organizing feature map. IEEE Trans. on Neural Networks 8(2), 218–226 (1997)
4. Blake, C.L., Merz, C.J.: UCI repository of machine learning databases (1998)
5. Brown, G.: Diversity in Neural Networks Ensembles. PhD thesis, School of Computer Science, University of Birminghan (2003)
6. Dasarathy, B.V., Sheela, B.V.: A composite classifier system design: Concepts and methodology. Proceedings of the IEEE 67, 708–713 (1979)
7. Drucker, H.: Improving regressors using boosting techniques. In: Fourteenth International Conference on Machine Learning, pp. 107–115 (1997)
8. Dzeroski, S., Zenko, B.: Is combining classifiers better than selecting the best one? In: International Conference on Machine Learning (ICML), pp. 123–130 (2000)
9. Efron, B.: Bootstrap methods: another look at the jacknife. The Annals of Statistics 7, 1–26 (1979)
10. Kohonen, T.: Self-Organizing Maps. Springer Series in Information Sciences, Third Extended Edition, vol. 30. Springer, Heidelberg (2001)
11. Martinetz, T., Schulten, K.: A "neural gas" network learns topologies. Artificial Neural Networks, 397–402 (1991)
12. Moraga, C.: Multi-nets: new trends in neural networks. Int. Jr. of Systemics, Cybernetics and Informatics 3(3), 11–18 (2008)
13. Polikar, R.: Ensemble based systems in decision making. IEEE Circuits and Systems Magazine, 21–45 (2006)
14. Saavedra, C., Salas, R., Moreno, S., Allende, H.: Fusion of self organizing maps. In: Sandoval, F., Prieto, A.G., Cabestany, J., Graña, M. (eds.) IWANN 2007. LNCS, vol. 4507, pp. 227–234. Springer, Heidelberg (2007)
15. Sharkey, A.J.C.: Combining Artificial Neural Nets. Ensemble and Modular Multi-Net Systems. Springer, Heidelberg (1998)

Adaptive Splitting and Selection Method of Classifier Ensemble Building

Konrad Jackowski and Michal Wozniak

Chair of Systems and Computer Networks, Wroclaw University of Technology,
Wybrzeze Wyspianskiego 27, 50-370 Wroclaw, Poland
Konrad.Jackowski@pwr.wroc.pl

Abstract. The paper presents a novel machine learning method which allows obtaining compound classifier. Its idea bases on splitting feature space into separate regions and choosing the best classifier from available set of recognizers for each region. Splitting and selection take place simultaneously as a part of an optimization process. Evolutionary algorithm is used to find out the optimal solution. The quality of the proposed method is evaluated via computer experiments.

1 Introduction and Related Works

In many review articles multiple classifier systems have been mentioned as one of the most promising in the pattern recognition [10]. In this research the main efforts are concentrated on combining knowledge of the set of elementary classifiers.

There is a number of important issues while building multiple classifier systems. Firstly, how to select classifiers in a pool. Combining similar classifiers should not contribute much to the system being constructed apart from increasing the computational complexity. So it seems interesting to select members of committee with possibly different components. Some authors propose different types of diversity measures [13] which allow to minimize the possibility of a coincidental failure [16].

Another important issue is the choice of a collective decision making method. Some more advanced approaches base on weighing voting [17] while other make fusion of discriminating functions, the main form of which are the *posterior* probability estimators, referring to the probabilistic model of a pattern recognition task [1, 3, 11].

The methodology of choice for classifiers is defined in researches by Rastrigin and Erenstein [20] and is used until today almost without changes. Some proposals from this trend assume a local specialization of particular classifiers and are searching for locally optimal solutions [7, 8]. Clustering and selection algorithm (CS) suggested by Kuncheva [14] is an interesting proposal. In this research the AdaSS algorithm (*Adaptive Splitting and Selection*) is presented. Its principles are as follows:

1. Firstly, a fusion of two steps (clustering and defining field classifiers) into one integrated process is the key issue.
2. Area classifiers will be compound classifiers.

Searching for optimal parameters of the model, including division of space and content of the area classifier committee, is treated as a complex optimizing task, for which the goal is to minimize the error of the whole system.

E. Corchado et al. (Eds.): HAIS 2009, LNAI 5572, pp. 525–532, 2009.

The proposed algorithm is suppose to gain the best results (i.e. improve the quality of the recognition) when it works on the set of week elementary classifiers. Fusion of classifiers which gain good results can lead to extending computation time accompanied with a negligible quality improvement. Therefore, we suggest to use the algorithm when small learning data is available (especially for recognition problems of high dimensionality).

It is also recommended to apply the AdaSS algorithm when elementary classifiers in hand show strict local competences (i.e. gain much better results in some particular regions in comparison to other areas of features space).

2 Model of Compound Classifier

The aim of the pattern recognition task is to classify the object to one of the predefined categories from a set $\mathbf{M}$, on the basis of observation of its features x [4]. To simplify the problem we assume that all of its components are continuous variables. A recognition algorithm

$$\Psi : X \to \mathbf{M} \tag{1}$$

is trained using a learning set (DS), which consists of elements representing feature values x and a corresponding class $j \in \mathbf{M}$

$$DS = \{(x_1, j_1), (x_2, j_2), \dots, (x_N, j_N)\}. \tag{2}$$

Let us assume that we have a set of K simple classifiers:

$$\Pi^\Psi = \{\Psi_1, \Psi_2, \dots, \Psi_K\}. \tag{3}$$

The term *simple* is used here only in order to distinguish it from the compound classifier whose construction is presented in this research. This adjective does not reflect the computational complexity of the model, or any features of algorithms, and in particular the assessment of their usefulness in solving the classifying problems in question. The only required condition is the realization of mentioned mapping (1) corresponding to the discussed problem.

We will assume a model, in which the process of defining competence areas will be realized within a compound classifier training process. The starting point will then be a partitioning of the feature space into constituents [18]:

$$X = \bigcup_{h=1}^{H} \hat{X}_h \qquad \forall k,l \in \{1, \dots, H\} \quad \text{and} \quad k \neq l \qquad \hat{X}_k \cap \hat{X}_l = \varnothing \tag{4}$$

where $\hat{X}_h$ is the hth constituent of X, and H is arbitrary chosen number of constituents. Since the aim of the partitioning is to utilize the competences of elements of Π^Ψ, the corresponding area classifier is defined to each field $\hat{\Psi}_h(x)$, set up of these elements from the pool which lead to reaching the best result in the field.

It leads to the following formula

$$\Psi(x)=i \iff \hat{\Psi}_h(x)=i \quad and \quad x\in \hat{X}_h . \tag{5}$$

Compound classifier $\Psi(x)$ defines the number of class i for an object, given by the area classifier $\hat{\Psi}_h(x)$ connected with field $\hat{X}_h$ where x belongs to.

Now let us move to the thesis which is the basis of the presented AdaSS model.

There exists such a feature space partitioning into constituents, and such a selection of the area classifiers (assigned to respective constituents), that the recognition system, making the decision according to formula (5), could reach the accuracy not worse than the best classifier from Π^{Ψ}.

3 Learning AdaSS Algorithm

Let component $\hat{X}_h$ be represented by the centroid [9] C_h that is a floating point number vector consistent with the feature space dimension. All centroids, arranged into columns, create a set of centroids in the following form:

$$C = \{C_1, C_2 ..., C_H\}. \tag{6}$$

Let us define the function returning the number of the field where the object belongs to, on the basis of its features' values and a given set of centroids:

$$A(x,C) = \underset{h=1}{\overset{H}{\arg\min}}\, d(x,C_h), \tag{7}$$

where d is the Euclidean distance measure.

In order to indicate which elementary classifier chosen from Π^{Ψ} is selected let us define

$$\Lambda_h = [\lambda_{1,h}, \lambda_{2,h}, ..., \lambda_{K,h}]^T . \tag{8}$$

Each element of this vector corresponds with a particular classifier from Π^{Ψ} and can take the value of 1 or 0, which indicates if it is chosen to the committee or not.

Let Λ mean set of all vectors Λ_h. The decision of such a classifier is taken collectively by its all components, according to the majority voting rule, which is presented by the following formula

$$\hat{\Psi}_h(x) = \sum_{h=1}^{H} \delta(A(x,C),h) * \underset{j\in M}{\arg\max} \sum_{k=1}^{K} \Lambda_h(k) * \delta(\Psi_k(x), j) \tag{9}$$

where δ is Kronecker's delta.

The aim is to find out such a compound classifier model, including feature space partitioning as well as the selection of field classifiers, which will assure the lowest frequency of misclassification. We will use the following criterion:

$$\hat{P}_e(\Psi) = \hat{P}_e(C, \Lambda) = \frac{1}{N} \sum_{n=1}^{N} L \left(\sum_{h=1}^{H} \delta(A(x, C), h) * \underset{j \in M}{\text{argmax}} \sum_{k=1}^{K} \Lambda_h(k) * \delta(\Psi_k(x), j), j_n \right) \quad (10)$$

where $L(i, j)$ denotes loss function that returns 1 if wrong decision is made (i.e. $i \neq j$) and 0 if the decision is correct).

Optimization will be held by manipulation of the C_h and Λ_h .vectors.

In order to solve the optimization task, one of a variety of optimization algorithms widely used in theory and practice can be used. In this work it has been decided to engage the evolutionary algorithm [19].

The chromosome, which is denoted by *Chr,* is a model of compound classifier parameters and is a structure consisting of two components. The first of them embodies a set of centroids *C* and represents feature space division. A number of centroids is one of input parameters of the algorithm. The other one consists of a definition of the committees for all areas in the form of matrix Λ. Any procedure that is to be performed on the chromosome will take into account the fact that its both parts have quite a different character, especially that the first one is the matrix of real numbers and the other one is the matrix of logical values. No information exchange will take place between the parts of the chromosomes processed by genetic operators.

Each chromosome $Chr = \{C, \Lambda\}$ corresponds to a given realization of the compound classifier, the quality of which can be obtained by computing the frequencies of misclassifications (10). According to a convention used in evolutionary computation, the quality is denoted by a fitness function of the chromosome:

$$\Phi(Chr) = 1 - P_e(Chr). \quad (11)$$

Phases of AdaSS algorithm is consistent with traditional evolutionary schema with elite promotion [19].

Two main groups of the parameters can be singled out in the algorithm. The first of them consists of parameters that determine a model of the entire recognition system i.e. the space division and committees compositions and their meanings were given in the previous section. The other one gathers those that steer optimization process. Further information about their function comes along with a description of the algorithm that is presented in Tab.1.

It is worth to note that the mutation operator alters a member being processed by adding some random changes to its chromosome. Both components of the chromosome are processed separately. Each of them can be altered with certain probability, that is changing along with optimization progress according to the following schema:

$$P_{mut\,C} = \frac{current\,cycle}{N_{cycle}} \quad \text{and} \quad P_{mut\,\Lambda} = 1 - P_{mut\,C} \quad (12)$$

where $P_{mut\,C}$ is the mutation probability of the centroid vector,

$P_{mut\,\Lambda}$ is the mutation probability of the committee vector.

According to the schema, in the early phase of the optimization, a special emphasis is put on searching possible partitioning of the feature space. In the course of learning progress, attention is shifted onto looking for optimal committees.

Mutation of a centroid involves adding a vector of numbers randomly generated according to a normal density distribution (with the mean equal to 0 and the standard deviation set to Δ_{mut}). The area committee part is altered by negation of a number of randomly drawn components of Λ_h.

The crossover operator generates one offspring member on the basis of two parents. Vectors Λ_h are derived from parents' chromosome according to two-point crossover rule. A set of the child's centroids can be obtained in two alternative ways depending on the parameter value of *Cross_over_model*

- mixing of parents' centroids with accordance to two-point rule
- calculating the average of parents' centroids

The main purpose of the overtraining assessment procedure is to break up the optimization process once it becomes likely that further learning can cause the loss of generalization ability in favor of a too overfitting model to learning set samples. A validation set is used in order to calculate fitness of the individuals in the same way as for regular population assessment. The procedure breaks the process if deterioration of the result is observed in the course of *V_Limit* subsequent learning cycles.

Table. 1. AdaSS parameters

Parameter		Remarks	
Parameters of the system model			
H	number of constituents		
$N_{Committee}$	upper limit of committee member quantity	$[1, K]$	
Parameters of the algorithm			
N_{Cycle}	upper limit of number of cycle of optimization procedure		
$N_{population}$	Population quantity		
α	Elite fraction factor	$[0, 1]$	α
β	mutation fraction factor	$[0, 1]$	$+$ β
χ	crossover fraction factor	$[0, 1]$	$+$
Δ_{mut}	centroid mutation range factor		
COmodel	crossover model	two points cross over	
V_Limit	upper limit of iteration with failing quality	$[1, N_{Cycle}]$	

4 Experimental Investigation

The set of experiments has been carried out using two benchmark databases and 3 generated datasets. The main aims of tests were to verify the AdaSS algorithm performance as a tool for looking for optimal feature space division and select the best area classifiers.

We used 2 databases from UCI Machine Learning Repository [2]:

- *Balance* – 2 class recognition problem, the database consists of 625 instances and is described by 4 attributes.
- *Phoneme* - 2 class recognition problem, the database consists of 5404 instances and is described by 5 attributes,
- and 3 datasets (Banana, Cone Torus, and Higleyman) generated using PRTools toolbox in Matlab environment [6].

The set of five elementary classifiers consisting of undertrained neural networks (denoted as C1, C2, C3, C4, and C5) has been used in both experiments for the purpose of ensuring diversity of simple classifiers that allows their local competences to be exploited. The details of used neural nets are as follow:

- sigmoidal transfer function,
- back propagation learning algorithm,
- 5 neurons in hidden layer,
- number of neurons in last layer equals number of classes of given experiment.

Qualities of the mentioned classifiers were compared with classifier used majority voting of all simple recognizers (MV) and classifiers obtained via described AdaSS learning method for $N_{Committee} = 1$ (AdaSS-1) and $N_{Committee} = 5$ (AdaSS-5).

Additionally the results of Oracle classifier was presented. It is an abstract fusion model, where if at least one of the classifiers produces the correct class label, then the committee of classifiers produces the correct class label too [17].

The setup of experiments was as follow:

1. There are some features of the AdaSS algorithm that can cause some variation in the final results, namely: initial population of individuals are generated randomly, therefore optimization can lead to a different solutions. Performance of learning process is strictly determined by the number of parameters. As there are no simple guidelines how to choose its values with regard to the recognition problem being under analysis, a set of experiments had to be carried out in order to find out the best configuration of the parameters.
2. Classifiers' errors were estimated using the ten-fold cross-validation method [12].
3. All experiments were carried out in Matlab environment using PRtools toolbox [6] and own software. The results of experiments were shown in Fig. 1.
4. Additionally we carried out experiments in order to establish dependencies between number of constituents (*H*) and the quality of proposed method.

First of all, it has to be realized that the presented results have a limited scope and therefore any conclusion drawn on their basis should be treated distantly. Nonetheless, the results did not surprise and meet our intuition. The quality of the compound classifiers outperform the best elementary classifier working on the entire feature space more than 7% up to 10 %. The analysis of the results guarantees us that the full committee classifiers that work on the entire space do not always provide a better result in comparison with the best of its component (as it took place in experiment 2). That property has been widely discussed and is known in literature [5, 15, 21].

The dependencies presented in Fig.2 did not surprise us but we supposed that the number of constituents shouldn't be so big, because it could lead to overtraining. This situation we observed only for *Phoneme* database, and this persuaded us to carry out more exact experiments which will establish this dependencies precisely.

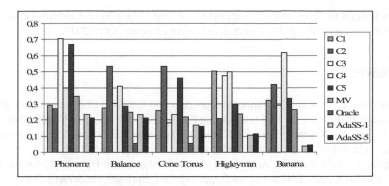

Fig. 1. Frequencies of misclassification for different datasets

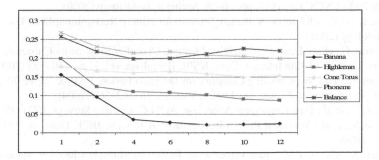

Fig. 2. Frequencies of misclassification versus number of constituents for AdaSS-5 classifier

It seems to be an interesting issue to conduct experiments revealing a relation between the quality of elementary classifiers and the system based on them. It has to be noticed that in the experiments we dealt with undertrained simple classifiers.

5 Conclusions

In the paper a novel model of compound classifier was presented. A proposition of the AdaSS learning algorithm was proposed for the classifier. The classifier fuses decision of area classifiers. Each area classifier makes decision on a particular constituents of the feature space. Contrary to the known algorithms as, e.g. the CS algorithm, space division and selecting classifiers are processed simultaneously in an adaptive manner. The quality of the proposed approach was established in a set of experiments, whose results are promising for future works.

Further research can be focused on:
1. establishing analytical estimation of the quality of the proposed method,
2. exploiting different methods of classifier fusion,
3. detailed analysis of relation between values of all parameters and obtained results,
4. exploiting other optimization algorithms.

Acknowledgment. This work is supported by The Polish State Committee for Scientific Research under the grant which is realizing in years 2006-2009.

References

1. Alexandre, L.A., et al.: Combining Independent and Unbiased Classifiers Using Weighted Average. In: Proc. of the 15th Int. Conf.on Pattern Recognition, pp. 495–498 (2000)
2. Asuncion, A., Newman, D.J.: UCI Machine Learning Repository Irvine. University of California, School of Information and Computer Science, Irvine (2007)
3. Biggio, B., et al.: Bayesian Analysis of Linear Combiners. In: Haindl, M., Kittler, J., Roli, F. (eds.) MCS 2007. LNCS, vol. 4472, pp. 292–301. Springer, Heidelberg (2007)
4. Duda, R.O., et al.: Pattern Classification. Wiley Interscience, Hoboken (2001)
5. Duin, R., et al.: Experiments with Classifier Combining Rules. In: Kittler, J., Roli, F. (eds.) MCS 2000. LNCS, vol. 1857, pp. 16–29. Springer, Heidelberg (2000)
6. Duin, R., et al.: PRTools4, A Matlab Toolbox for Pattern Recognition, Delft University of Technology (2004)
7. Giacinto, G., et al.: Design of Effective Multiple Classifier Systems by Clustering of Classifiers. In: Proc. of 15th Internat. Conf. ICPR, Barcelona, vol. 2, pp. 160–163 (2000)
8. Goebel, K., Yan, W.: Choosing Classifier for Decision Fusion. In: Proc. of the 7th Internat. Conf. on Information Fusion, Stockholm, Sweden, pp. 562–568 (2004)
9. Jain, A.K., et al.: Data Clustering: A Review. ACM Computing Surveys 31(3) (1995)
10. Jain, A.K., et al.: Statistical Pattern Recognition: A Review. IEEE Trans. on PAMI 22(1), 4–37 (2000)
11. Kittler, J., Alkoot, F.M.: Sum versus Vote Fusion in Multiple Classifier Systems. IEEE Trans. on PAMI 20, 226–239 (2003)
12. Kohavi, R.: A study of cross-validation and bootstrap for accuracy estimation and model selection. In: Proc. of the 14th Int. Joint Conf. on AI, San Mateo, pp. 1137–1143 (1995)
13. Krzanowski, W., Partrige, D.: Software Diversity: Practical Statistics for its Measurement and Exploatation, Department of Computer Science, University of Exeter (1996)
14. Kuncheva, L.I.: Cluster-and-selection method for classifier combination. In: Proc. of the 4th Int. Conf. on Knowledge-Based Intel. Eng.Sys. and Allied Technol., Brighton, pp. 185–188 (2000)
15. Kuncheva, L.I., et al.: Decision templates for multiple classifier fusion: an experimental comparision. Pattern Recognition 34, 299–314 (2001)
16. Kuncheva, L.I., et al.: Ten measures of diversity in classifier ensembles: Limits for two classifiers. In: Proc. of the IEE Work. on Intell. Sensor Proc., Birmingham, pp. 10/1–10/6 (2001)
17. Kuncheva, L.I.: Combining pattern classifiers: Methods and algorithms. Wiley Interscience, New Jersey (2004)
18. Kuratowski, K., Mostowski, A.: Set theory. North-Holland Pub. Co., Amsterdam (1996)
19. Michalewicz, z.: Genetics Algorithms + Data Structures = Evolutions Programs. Springer, Berlin (1996)
20. Rastrigin, L.A., et al.: Method of Collective Recognition. Energoizdat, Moscow (1981)
21. Tumer, K., Ghosh, J.: Analysis of Decision Boundaries in Linearly Combined Neural Classifiers. Pattern Recognition 29, 341–348 (1996)

Probability Error in Global Optimal Hierarchical Classifier with Intuitionistic Fuzzy Observations

Robert Burduk

Chair of Systems and Computer Networks, Wroclaw University of Technology,
Wybrzeze Wyspianskiego 27, 50-370 Wroclaw, Poland
robert.burduk@pwr.wroc.pl

Abstract. The paper considers the problem of classification error in pattern recognition. This model of classification is primarily based on the Bayes rule and secondarily on the notion of intuitionistic fuzzy sets. A probability of mis-classifications is derived for a classifier under the assumption that the features are class-conditionally statistically independent, and we have intuitionistic fuzzy information on object features instead of exact information. Additionally, we consider the global optimal hierarchical classifier.

Keywords: hierarchical classifier, error probability, intuitionistic fuzzy set.

1 Introduction

The classification error is the ultimate measure of the performance of a classifier. Competing classifiers can also be evaluated based on their error probabilities. Several studies have previously described the Bayes probability of error for a single-stage classifier [1], [4], [9] and for a hierarchical classifier [8], [10].

Since Zadeh introduced fuzzy sets in 1965 [16], many new approaches and theories treating imprecision and uncertainty have been proposed [7], [11]. In 1986, Atanassov [2] introduced the concept of an intuitionistic fuzzy set. This idea, which is a natural generalization of a standard fuzzy set, seems to be useful in modelling many real life situations, like logic programming [3], decision making problems [14] etc.

In this paper, we consider the problem of classification for the case in which the observations of the features are represented by the intuitionistic fuzzy sets as well as for the cases in which the features are class-conditionally statistically independent and a Bayes rule is used. We consider the global optimal strategy of multistage recognition task. The contents of the work are as follows. Section 2 introduces the necessary background and describes the Bayes hierarchical classifier. In section 3 the introduction to intuitionistic fuzzy sets is presented. In section 4 we presented the difference between the probability of misclassification for the intuitionistic fuzzy and crisp data in Bayes hierarchical classifier.

2 Bayes Hierarchical Classifier

In the paper [10] the Bayesian hierarchical classifier is presented. The synthesis of a multistage classifier is a complex problem. It involves specifying of the following components:

E. Corchado et al. (Eds.): HAIS 2009, LNAI 5572, pp. 533–540, 2009.

- the decision logic, i.e. hierarchical ordering of classes,
- the feature used at each stage of decision,
- the decision rules (strategy) for performing the classification.

This paper is devoted only to the last problem. This means that we will only consider the presentation of decision algorithms, assuming that both the tree structure and the feature used at each non-terminal node have been specified.

The procedure in the Bayesian hierarchical classifier consists of the following sequences of operations presented in Fig. 1. At the first stage, some specific features x_0 are measured. They are chosen from among all accessible features x, which describe the pattern that will be classified. These data constitute a basis for making a decision i_1. This decision, being the result of recognition at the first stage, defines a certain subset in the set of all classes and simultaneously indicates features x_{i_1} (from among x) which should be measured in order to make a decision at the next stage.

Now at the second stage, features x_{i_1} are measured, which together with i_1 are a basis for making the next decision i_2. This decision, – like i_1 – indicates features x_{i_2} that are necessary to make the next decision (at the third stage as in the previous stage) that in turn defines a certain subset of classes, not in the set of all classes, but in the subset indicated by the decision i_2, and so on. The whole procedure ends at the N-th stage, where the decision made i_N indicates a single class, which is the final result of multistage recognition.

Thus multistage recognition means a successive narrowing of the set of potential classes from stage to stage, down to a single class, simultaneously indicating features at every stage that should be measured to make the next decision in a more precise manner.

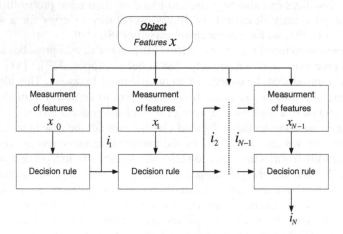

Fig. 1. Block diagram of the hierarchical classifier

2.1 Decision Problem Statement

Let us consider a pattern recognition problem, in which the number of classes equals M. Let us assume that the classes are organised in a $(N+1)$ horizontal decision tree. Let us number all the nodes of the constructed decision-tree with consecutive numbers of $0,1,2,...,$ reserving 0 for the root-node, and let us assign numbers of classes from the $M = \{1,2,...,M\}$ set to terminal nodes so that each one of them can be labelled with the number of the class connected with that node. This allows us to introduce the following notation:

- $M(n)$ – the set of nodes, which distance from the root is n, $n = 0,1,2,...,N$. In particular $M(0) = \{0\}$, $M(N) = M$,

- $\overline{M} = \bigcup_{n=0}^{N-1} M(n)$ – the set of interior nodes (non terminal),

- $M_i \subseteq M(N)$ – the set of class labels attainable from the i-th node ($i \in \overline{M}$),

- M^i – the set of nodes of immediate descendant node i ($i \in \overline{M}$),

- m_i – node of direct predecessor of the i-th node ($i \neq 0$),

- $s(i)$ – the set of nodes on the path from the root-node to the i-th node, $i \neq 0$.

The example of above notation for the two stage binary classifier: $M(0) = \{0\}$, $M(1) = \{5,6\}$, $\overline{M} = \{0,5,6\}$, $M_0 = \{1,2,3,4\}$, $M_5 = \{1,2\}$, $M_6 = \{3,4\}$, $M^0 = \{5,6\}$, $M^5 = \{1,2\}$, $m_5 = \{0\}$, $m_1 = \{5\}$, $s(1) = \{0,5,1\}$, $i_0 = \{0\}$, $i_1 = \{5,6\}$.

We will continue to adopt the probabilistic model of the recognition problem, i.e. we will assume that the class label of the pattern being recognised $j_N \in M(N)$ and its observed features x are realizations of a couple of random variables J_N and X. The complete probabilistic information denotes the knowledge of a priori probabilities of classes:

$$p(j_N) = P(J_N = j_N), \quad j_N \in M(N) \tag{1}$$

and class-conditional probability density functions:

$$f_{j_N}(x) = f(x / j_N), \quad x \in X, \quad j_N \in M(N). \tag{2}$$

Let $x_i \in X_i \subseteq R^{d_i}$, $d_i \leq d$, $i \in M$ denote vector of features used at the i-th node, which have been selected from the vector x. Our aim is now to calculate the so-called multistage recognition strategy $\pi_N = \{\Psi_i\}_{i \in \overline{M}}$, that is the set of recognition algorithms in the form:

$$\Psi_i : X_i \rightarrow M^i, \quad i \in \overline{M}. \tag{3}$$

Formula (3) is a decision rule (recognition algorithm) used at the i-th node that maps observation subspace to the set of immediate descendant nodes of the i-th node. Analogically, decision rule (3) partitions observation subspace X_i into disjoint decision regions $D_{x_i}^k, k \in M^i$, such that observation x_i is allocated to the node k if $k_i \in D_{x_i}^k$, namely: $D_{x_i}^k = \{x_i \in X_i : \Psi_i(x_i) = k\}, k \in M^i, \quad i \in \overline{M}$.

Our aim is to minimise the expected risk function (expected loss function $L(I_N, J_N)$) denoted by:

$$R^*(\pi_N) = \min_{\pi_N} R(\pi_N) = \min_{\pi_N} E[L(I_N, J_N)] \tag{4}$$

where π_N is the strategy of the decision tree classifier. The π_N is the set of classifying rules used at the particular node $\pi_N = \{\Psi_i\}_{i \in \overline{M}}$.

Globally optimal strategy π_N^*. This strategy minimises the mean probability of misclassification on the whole multistage recognition process and leads to an optimal global decision strategy, whose recognition algorithm at the n-th stage is the following:

$$\Psi_{i_n}^*(x_{i_n}) = i_{n+1} \quad if \quad i_{n+1} = \arg\max_{k \in M^{i_n}} Pc(k) p(k) f_k(x_{i_n}) \tag{5}$$

where $Pc(k)$ is the empirical probability of correct classification at the next stages if at the n-th stage decision i_{n+1} is made.

3   Basic Notions of Intuitionistic Fuzzy Events

As opposed to a fuzzy sets in $X = x$, given by $A' = \{< x, \mu_A(x) >: x \in X\}$ where $\mu_A : X \to [0,1]$ is the membership function of the fuzzy sets A', an intuitionistic fuzzy set $A \in X$ is given by $A = \{< x, \mu_A(x), v_A(x) >: x \in X\}$ where $\mu_A : X \to [0,1]$ and $v_A : X \to [0,1]$ with the condition $0 \le \mu_A(x) + v_A(x) \le 1$ $\forall x \in X$ and the numbers $\mu_A(x), v_A(x) \in [0,1]$, denote the degree of membership and non-membership of x to A respectively.

The difference $\pi_A(x) = 1 - \mu_A(x) - v_A(x)$ is called an *intuitionistics index* and the number $\pi_A(x) \in [0,1]$ is treated as a measure of a hesitancy bounded with the appreciation of the degree of the membership or non-membership of an element x to the set A.

In [16] for the first time it was jointed the concept of a fuzzy event ant the probability. The probability of fuzzy event in Zadeh's form is given by:

$$P(A') = \int_{\mathfrak{R}^d} \mu_{A'}(x) f(x) dx. \tag{6}$$

The probability $P(A')$ of a fuzzy event A' defined by (12) represents a crisp number in the interval $[0,1]$.

In [6] the probability of an intuitionistics fuzzy event A is proposed as a crisp number from the interval $[0,1]$:

$$P(A) = \int_{\mathfrak{R}^d} \frac{\mu_A(x) + 1 - v_A(x)}{2} f(x) dx. \tag{7}$$

In the paper [6] there was shown that formula (6) satisfies all classical properties of probability in the theory of Kolmogorov. In our consideration we will use the simple notation for the probability of an intuitionistics fuzzy event A :

$$P(A) = \int_{\Re^d} \tau_A(x) f(x) dx, \qquad (8)$$

where $\tau_A(x) = \frac{\mu_A(x)+1-v_A(x)}{2}$.

Let us consider an intuitionistics fuzzy information. The intuitionistics fuzzy information A_k from $x_k \in \Re^d$, $k=1,...,d$ is a set of intuitionistics fuzzy events $A_k = \{A_k^1, A_k^2,..., A_k^{n_k}\}$ characterised by membership and non-membership functions:

$$A_k = \{< \mu_{A_k^1}(x_k), v_{A_k^1}(x_k) >,,...,< \mu_{A_k^{n_k}}(x_k), v_{A_k^{n_k}}(x_k) >\}. \qquad (9)$$

The value of index n_k defines the possible number of intuitionistics fuzzy events for x_k. In addition, assume that for each observation subspace x_k the set of all available intuitionistics fuzzy observations (9) satisfies the orthogonality constraint:

$$\sum_{l=1}^{n_k} \frac{\mu_{A_k^l}(x_k)+1-v_{A_k^l}(x_k)}{2} = 1 \qquad \forall x \in X. \qquad (10)$$

When we use the probability of the intuitionistics fuzzy event represented by (8) and (10) arises, it is clear that we get $\sum_{l=1}^{n_k} P(A^l) = 1$.

4 Global Optimal Strategy

The decision algorithms for the zero-one loss function in the case the global optimal strategy of multistage recognition are as follows:

$$\Psi_{i_n}^*(A_{i_n}) = i_{n+1} \quad if \qquad (11)$$

$$i_{n+1} = \arg \max_{k \in M^{i_n}} \sum_{j_N \in M_k} p(j_N) q_{IF}^*(j_N / k, j_N) \int_{\Re^d} \tau_{A_{i_n}}(x_{i_n}) f_{j_N}(x_{i_n}) dx_{i_n}$$

for $i_n \in M(n), n = 0,1,2,..., N-1$, where $q_{IF}^*(j_N / i_{n+1}, j_N)$ denotes the probability of accurate classification of the object of the class j_N in further stages using π_N^* strategy rules on the condition that on the n-th stage the i_{n+1} decision has been made. The A_{i_n} denotes the observed intuitionistics fuzzy value of object feature in i_n node.

The probability of error $Pe(\pi_N^*)$ for crisp data and for globally optimal strategy π_N^* is given by [10]:

$$Pe(\pi_N^*) = 1 - \sum_{j_N \in M(N)} p(j_N) \prod_{i_k \in s(j_N)-\{0\}} q(i_k / m_{i_k}, i_k) \tag{12}$$

Similarly, if (10) holds the probability of error $Pe_{IF}(\pi_N^*)$ for intuitionistic fuzzy data and for globally optimal strategy π_N^* is following:

$$Pe_{IF}(\pi_N^*) = 1 - \sum_{j_N \in M(N)} p(j_N) \prod_{i_k \in s(j_N)-\{0\}} q_{IF}(i_k / m_{i_k}, i_k) \tag{13}$$

When we use intuitionistic fuzzy information on object features instead of exact information, we deteriorate the classification accuracy. The upper boundary of the difference between the probability of misclassification for the intuitionistic fuzzy $Pe_{IF}(\pi_N^*)$ and crisp data $Pe(\pi_N^*)$ for the globally optimal strategy of multistage recognition π_N^* is the following:

$$Pe_{IF}(\pi_N^*) - Pe(\pi_N^*) \le \sum_{j_N \in M(N)} p(j_N) \sum_{i_k \in s(j_N)-\{0\}} \varepsilon_{m_{i_k}} \tag{14}$$

where

$$\varepsilon_i = \sum_{A_i \in X_i} \left| \int_{\Re^i} \tau_{A_i}(x_i) \max_{k \in M^i} \{ f_k(x_i) \} dx_i - \max_{k \in M^i} \left\{ \int_{\Re^i} \tau_{A_i}(x_i) f_k(x_i) dx_i \right\} \right|.$$

5   Illustrative Example

Let us consider the two-stage binary classifier. Four classes have identical a priori probabilities that equal 0.25. We use 3-dimensional data $x = [x^{(1)}, x^{(2)}, x^{(3)}]$ where class-conditional probability density functions are normally distributed. For performing the classification at the root-node 0 the first coordinate was used, and components $x^{(2)}$ and $x^{(3)}$ were used at the nodes 5 and 6 respectively. In the data covariance matrices are equal for every class $\sum_{j_2} = 4I$, $j_2 \in M(2)$, and the expected values are the following: $\mu_1 = [0, 0, 0]$, $\mu_2 = [0, 4, 0]$, $\mu_3 = [3, 0, 1]$, $\mu_4 = [3, 0, 8]$. In experiments, the following sets of intuitionistic fuzzy numbers were used: case A: $A = \{A^1, A^2, ..., A^{24}\}$, where

$$\mu_{A^1}(x) = \begin{cases} (x+9)^2 & \text{for} \quad x \in [-9,-8], \\ (x+7)^2 & \text{for} \quad x \in [-8,-7], \\ 0 & \text{for} \quad \text{otherwise,} \end{cases} \qquad \nu_{A^1}(x) = \begin{cases} (x+8)^2 & \text{for} \quad x \in [-9,-7], \\ 0 & \text{for} \quad \text{otherwise,} \end{cases}$$

$$\mu_{A^{24}}(x) = \begin{cases} (x-14)^2 & \text{for} \quad x \in [14,15], \\ (x-16)^2 & \text{for} \quad x \in [15,16], \\ 0 & \text{for} \quad \text{otherwise,} \end{cases} \qquad \nu_{A^{24}}(x) = \begin{cases} (x-15)^2 & \text{for} \quad x \in [14,16], \\ 0 & \text{for} \quad \text{otherwise,} \end{cases}$$

case B: $B = \{B^1, B^2, ..., B^{24}\}$, where

$$\mu_{B^1}(x) = \begin{cases} 4(x+9)^2 & \text{for} \quad x \in [-9, -8.5], \\ 4(x+8)^2 & \text{for} \quad x \in [-8.5, -8], \\ 0 & \text{for} \quad \text{otherwise,} \end{cases} \quad \nu_{B^1}(x) = \begin{cases} 4(x+8.5)^2 & \text{for} \quad x \in [-9, -8], \\ 0 & \text{for} \quad \text{otherwise,} \end{cases}$$

$$\mu_{B^{49}}(x) = \begin{cases} 4(x-15)^2 & \text{for} \quad x \in [15, 15.5], \\ 4(x-16)^2 & \text{for} \quad x \in [15.5, 16], \\ 0 & \text{for} \quad \text{otherwise,} \end{cases} \quad \nu_{B^{49}}(x) = \begin{cases} 4(x-15.5)^2 & \text{for} \quad x \in [15, 16], \\ 0 & \text{for} \quad \text{otherwise,} \end{cases}$$

Table1 shows the difference between the probability of misclassification for intuitionistic fuzzy and non fuzzy data in the globally optimal strategies of multistage classification $Pe(\Psi_F^*) - Pe(\Psi^*)$ calculated form (14). Tab. 1 shows also the difference between the probability of misclassification for individual node of decision tree ε_i. The values for $\varepsilon_i = 5$ are calculated for the translation the class-conditional probability density functions by the value k. The $\varepsilon_i = 5$ denote one of the interior nodes of decision tree meant as 5. These results are calculated for full probabilistic information.

Table 1. The difference between the probability of misclassification for global optimal strategy $Pe(\Psi_F^*) - Pe(\Psi^*)$ and for individual node of decision tree ε_i

Case	$Pe_{IF}(\pi_N^*) - Pe(\pi_N^*)$	$\varepsilon_i = 5$ $\mu_1(x-k)$, $\mu_1(x-k)$, $k =$					
		0	0.1	0.125	0.15	0.25	0.5
A	0.024	0.017	0.015	0.014	0.011	0.010	0.017
B	0.013	0.004	0.003	0.002	0.003	0.004	0.004

The received results show deterioration the quality of classification when we use intuitionistic fuzzy information on object features instead of exact information in Bayes hierarchical classifier. We have to notice that the difference in the misclassification for fuzzy and crisp data does not depend only on the intuitionistic fuzzy observations. The position of the class-conditional probability density in relation to the observed intuitionistic fuzzy features is the essential influence.

6 Conclusion

In the present paper we have concentrated on the Bayes optimal classifier. Assuming a full probabilistic information we have presented the difference between the probability of misclassification for intuitionistics fuzzy and crisp data. Illustrative example shoves that the position of the class-conditional probability density in relation to the observed intuitionistics fuzzy features is the essential influence for the difference $Pe(\Psi_{IF}^*) - Pe(\Psi^*)$. In paper [5] the difference between the probability of misclassification for fuzzy data is presented. Further research should concern medical diagnostic with hierarchical classifier [15]. In medical diagnostic intuitionistic observations are possible [12], [13].

Acknowledgements. This work is supported by The Polish State Committee for Scientific Research under grant for the years 2006–2009.

References

1. Antos, A., Devroye, L., Gyorfi, L.: Lower bounds for Bayes error estimation. IEEE Trans. Pattern Analysis and Machine Intelligence 21, 643–645 (1999)
2. Atanassov, K.: Intuitionistic fuzzy sets. Fuzzy Sets and Systems 20, 87–96 (1986)
3. Atanassov, K., Georgeiv, C.: Intuitionistic fuzzy prolog. Fuzzy Sets and Systems 53, 121–128 (1993)
4. Avi-Itzhak, H., Diep, T.: Arbitrarily tight upper and lower bounds on the bayesian probability of error. IEEE Trans. Pattern Analysis and Machine Intelligence 18, 89–91 (1996)
5. Burduk, R.: Probability of misclassification in Bayesian hierarchical classifier. Advances in Soft Computing 31, 341–348 (2005)
6. Gerstenkorn, T., Mańko, J.: Probability of fuzzy intuitionistic sets. BUSEFAL 45, 128–136 (1990)
7. Goguen, J.: L-fuzzy sets. Journal of Mathematical Analysis and Applications 18(1), 145–174 (1967)
8. Kulkarni, A.: On the mean accuracy of hierarchical classifiers. IEEE Transactions on Computers 27, 771–776 (1978)
9. Kuncheva, L.I.: Combining pattern classifier: Methods and Algorithms. John Wiley, New York (2004)
10. Kurzyński, M.: On the multistage Bayes classifier. Pattern Recognition 21, 355–365 (1988)
11. Pawlak, Z.: Rough sets and fuzzy sets. Fuzzy Sets and Systems 17, 99–102 (1985)
12. Supriya, K.D., Ranjit, B., Akhil, R.R.: An application of intuitionistic fuzzy sets in medical diagnosis. Fuzzy Sets and Systems 117(2), 209–213 (2001)
13. Szmidt, E., Kacprzyk, J.: An intuitionistic fuzzy set based approach to intelligent data analysis: an application to medical diagnosis. Studies In Fuzziness And Soft Computing, 57–70 (2003)
14. Szmidt, E., Kacprzyk, J.: A consensus-reaching process under intuitionistic fuzzy preference relations. International Journal of Intelligent Systems 18(7), 837–852 (2003)
15. Wozniak, M.: Two-stage classifier for diagnosis of hypertension type. In: Maglaveras, N., Chouvarda, I., Koutkias, V., Brause, R. (eds.) ISBMDA 2006. LNCS (LNBI), vol. 4345, pp. 433–440. Springer, Heidelberg (2006)
16. Zadeh, L.A.: Probability measures of fuzzy events. Journal of Mathematical Analysis and Applications 23, 421–427 (1968)

Some Remarks on Chosen Methods of Classifier Fusion Based on Weighted Voting

Michal Wozniak and Konrad Jackowski

Chair of Systems and Computer Networks, Wroclaw University of Technology
Wybrzeze Wyspianskiego 27, 50-370 Wroclaw, Poland
{Michal.Wozniak,Konrad.Jackowski}@pwr.wroc.pl

Abstract. *Multiple Classifier Systems* are nowadays one of the most promising directions in pattern recognition. There are many methods of decision making by the ensemble of classifiers. The most popular are methods that have their origin in voting method, where the decision of the common classifier is a combination of individual classifiers' outputs. This work presents comparative analysis of some classifier fusion methods based on weighted voting of classifiers' responses and combination of classifiers' discriminant functions. We discus which of presented methods could produce classifier better than Oracle one. Some results of computer experiments carried out on benchmark and computer generated data which confirmed our studies are presented also.

1 Introduction

The main goal of the recognition systems [3] is to classify presented objects to one of defined classes. There is much current research into developing ever more efficient and accurate recognition algorithms. Multiple classifier systems are currently the focus of intense research. The subject matter has been known for over 15 years [23]. Some works in this field were published as early as the '60 of the XX century [2], when it was shown that the common decision of independent classifiers is optimal, when chosen weights are inversely proportional to errors made by the classifiers. In many review articles this trend has been mentioned as one of the most promising in the field of the pattern recognition [12]. In the beginning in literature one could find only majority vote, but in later works more advanced methods of finding a common solution to the classifier group problem were proposed. Estimation accuracy of the classifier committee is one of fundamental importance. Known conclusions, derived on analytic way, concern particular case of the majority vote [8] when classifier committee is formed on the basis of independent classifiers. Unfortunately this case has only theoretical character and is not useful in practice. The weighted voting is taken into consideration [7, 9, 16, 19], but a problem of establishing weights for mentioned voting procedure is not simple. Many of authors have proposed treating the block voting as a kind of classifier [10] but the general question is "does the fuser need to be trained?" [5]. An alternative way of common classifier construction is combination of discriminant functions of available classifiers in order to obtain set of common discriminant functions, e.g. via linear combination.

E. Corchado et al. (Eds.): HAIS 2009, LNAI 5572, pp. 541–548, 2009.
© Springer-Verlag Berlin Heidelberg 2009

Paper presents comparative analysis of some methods of classifier fusion based on weighted voting of classifiers' responses and combination of classifiers' discriminant functions. We discus which of presented methods could produce classifier better than Oracle one.

2 Is It Possible to Train Fuser Better Than Oracle?

Oracle is an abstract fusion model, where if at last one of the classifiers recognizes object correctly, then the committee of classifiers points at correct class too. Oracle is usually used in comparative experiments to show limit of classifier committee quality[19]. But the question is if Oracle is really such a limit. Let us consider if it is possible to obtain compound classifier which could achieve higher accuracy than Oracle one. Let us consider some methods of classifier fusion on the basis of classifiers' response and discriminant function they used then we will consider which of presented fusion methods could produce better classifier than Oracle.

2.1 Classifier Fusion Based on Classifier Response

Let us assume that we have n classifiers $\Psi^{(1)}, \Psi^{(2)}, ..., \Psi^{(n)}$. Each of them decides if object belongs to class $i \in M = \{1, ..., M\}$. For making common decision by the group of classifiers we use following common classifier $\overline{\Psi}$:

$$\overline{\Psi} = \arg\max_{j \in M} \sum_{l=1}^{n} \delta\left(j, \Psi^{(l)}\right) w^{(l)} \Psi^{(l)}, \tag{1}$$

where $w^{(l)}$ is the weight of l-th classifier and

$$\delta(j,i) = \begin{cases} 0 & \text{if } i \neq j \\ 1 & \text{if } i = j \end{cases}. \tag{2}$$

Let us note that $w^{(l)}$ plays key-role of the quality of classifier $\Psi^{(l)}$. There are many researches how to set the weights, e.g. in [8, 16] authors proposed to learn the fuser. Let us consider 3 possibilities of weight values set up.

1. Weights assigned to the classifier

This is traditional model of weighted presented in (1). In [16] author showed that an accuracy of classifier $\overline{\Psi}$ is maximized by assigning weights if $w^{(l)} \propto P_{a,l}/1 - P_{a,l}$, where $P_{a,l}$ denotes probability of accuracy of l-th classifier. Unfortunately it is not sufficient for guarantying the smallest classification error. The *prior* probability for each class has to be taken into account also. In real decision problems the values of the *prior* probabilities are usually unknown.

2. Weights assigned to the classifier and the class

Weights $w^{(l)}(i)$ are assigned to each classifier and each classes. For given classifier assigned weights for different classes could be different. Let us note that such model could be used to fusion of heterogeneous classifiers (i.e. which uses different model of recognition task, but their set of class labels are the subsets of M).

3. Weights depended on features values and assigned to the classifier and the class

Weights $w^{(l)}(i, x)$ are assigned to each classifier, each classes, and additionally they are depended on values of feature vector x.

Let us note that for aforementioned models it is not possible to obtain classifier which is better than Oracle classifier because each decision rule making decision according (1) could point to correct class if one classifier produces correct class label at least. Of course the best proposition is mentioned above Oracle classifier. The only model based (partial) on class label which could achieve better results than Oracle is classifier which produces decision on the basis of class labels given by $\Psi^{(1)}, \Psi^{(2)}, ..., \Psi^{(n)}$ and feature vector values. That model was considered in many papers like [11, 17, 18, 21].

2.2 Classifier Fusion Based on Values of Classifiers' Discrimination Function

This classification algorithm is formed by the procedures of classifier fusions on the basis of their discriminating function, the main form of which are *posterior* probability estimators, referring to the probabilistic model of a pattern recognition task [3]. The aggregating methods, which do not require learning, performing fusion with the help of simple operators such as maximum or average. However, they can be used in clearly defined conditions, as it has been presented in a research paper by Duin [5], which limits their practical applications. Weighting methods are an alternative and the selection of weights has a similar importance as it is in case of weighted majority voting. The advantages of this approach include an effective counteraction against the occurrence of elementary classifier overtraining.

Each classifier makes decision based on the value of discriminant function. Let $F^{(l)}(i, x)$ means such a function assigned to class i for given value of x, which is used by the l-th classifier $\Psi^{(l)}$. A common classifier $\hat{\Psi}(x)$ looks as follows

$$\hat{\Psi}(x) = i \quad if \quad \hat{F}(i, x) = \max_{k \in M} \hat{F}(k, x), \tag{3}$$

where

$$\hat{F}(i, x) = \sum_{l=1}^{n} w^{(l)} F^{(l)}(i, x) \text{ and } \sum_{i=1}^{n} w^{(l)} = 1. \tag{4}$$

Varied interpretation of the discriminant function could be given. It could be the *posterior* probability for the classifiers based on Bayes decision theory or outputs of neural network. In general the value of such function means support given for distinguished class.

1. Weights depended on classifier
This is traditional approach where weights are connected with classifier and each discriminant function of given l-th classifier is weighted by the same value $w^{(l)}$. The estimation of probability error of such classifier could be found in [22].

2. Weights depended on classifier and class number
Weight $w^{(l)}(i)$ is assigned to the l-th classifier and the i-th class. For given classifier assigned weights for different classes could be different. Similarly like in 2.1.2 such model could be used to fusion of heterogeneous classifiers.

3. Weights depended on classifier and feature vector
Weight $w^{(l)}(x)$ is assigned to a given classifier l-th and for given x has the same value for each dicriminant functions used by l-th classifier.

4. Weights depended on classifier, class number, and feature vector
Wright $w^{(l)}(i,x)$ is assigned to the l-th classifier but for given x its value could be different for different discriminant function assigned to each classes.

Now let us consider which (and why) one of mentioned-above method could produce classifier better than Oracle. Firstly, let us note that for methods presented in points 1 and 3 such classifier cold not achieve better quality than Oracle because it means that is possible such a combination which produce correct decision if each classifier produces wrong one. For mentioned cases if classifiers point at the wrong decision that meant that value of their discriminant functions assigned to the correct class is lower than assigned to the wrong ones. Let us consider case 3 which is more general than case 1. For such cases where weights' values are independent of class number, fuser produces wrong decision if

$$\sum_{l=1}^{n} w^{(l)}(x) F^{(l)}(correct\_class, x) < \sum_{l=1}^{n} w^{(l)}(x) F^{(l)}(wrong\_class, x). \tag{5}$$

It is not possible to set such values of weights for which presented relation is the inverse one. It is not possible via linear combination, but it is possible in the cases where weights are depended additionally on class label.

2.3 Example of Classifier Fusion Based on Weights Depended on Classifier and Class Number

Let us consider an example presented in Fig.1. It illustrated two class recognition (distinction between "dots" and "squares") problem for which three classifiers are available. The discriminant function of mentioned classifiers are depicted in the first three graphs. We propose a fuser which uses the following common discriminant functions depicted in the forth graphs. For class "dot" (depicted by solid line)

$$\hat{F}(dot, x) = \frac{1}{2} F^{(1)}(dot, x) + \frac{1}{4} F^{(2)}(dot, x) + \frac{1}{4} F^{(3)}(dot, x) \tag{6}$$

and for class "square" (depicted by dashed line)

$$\hat{F}(square, x) = \frac{1}{4} F^{(1)}(square, x) + \frac{1}{4} F^{(2)}(square, x) + \frac{1}{2} F^{(3)}(square, x). \tag{7}$$

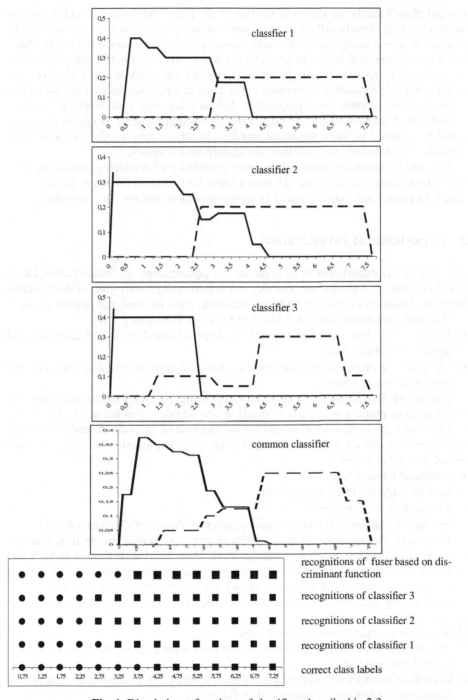

Fig. 1. Discriminant functions of classifiers described in 2.3

Correct class's labels are shown in the last graph. Let us note that for such set of examples the best classifier (the first one) does not recognizes 2 elements correctly. The number of wrong recognitions of oracle classifier is the same, but if we use the fuser based on discriminant functions given by (6) and (7) then its quality (for given set of examples) is better than Oracle, because it does not make any mistake. For such example we used proposition formulated in the point 2, where weights do not depend on x, but we obtain pretty good proposition. Let us thing over case where proposition formulated in point 4 should be used. Please note that weights independent from x could be assigned to linear separated problem, in other cases we should use weights depended on classifier, class number, and feature vector values.

Of course presented example shows only possibility of producing combining classifiers better than oracle but we still do not know how to learn the fuser. Therefore we decide to carried out some computer experiments which confirm our propositions.

3 Experimental Investigation

The aim of the experiment is to compare the performance of fuser of class labels based on weights depended on classifier and feature values with fuser of discriminant functions based on weights depended on classifier, class number, and feature values.

We used 2 databases from UCI Machine Learning Repository [1]:

- *Balance* – 2 class recognition problem, database consists of 625 instances and described by 4 attributes.
- *Phoneme* - 2 class recognition problem, database consists of 5404 instances and described by 5 attributes,

The set of five elementary classifiers consisting of undertrained networks (for which training process were early stopped) or trained ones (denoted as S1, S2, S3, S4, and S5) has been used in experiments for the purpose of ensuring diversity of simple classifiers that allows their local competences to be exploited. The details of used neural nets are as follow:

- sigmoidal transfer function,
- back propagation learning algorithm,
- 5 neurons in hidden layer,
- number of neurons in last layer equals number of classes of given experiment.

They were used to construct the committees proposed in previous section. The first one (based on the class numbers given by single classifiers) was denoted as FCN and FDF stood for the second one (based on the discriminant functions). The fusion block was realized as a neural net with one hidden layer which consisted of 5 neurons.

Additionally the qualities of mentioned classifiers were compared with Oracle classifier (Oracle).

Other set-up of experiments were as follow:

1. All experiments were carried out in Matlab environment using the PRtools toolbox [6] and own software.
2. Classifiers' errors were estimated using the ten fold cross validation method [13]. Results of experiments are presented in Table 1.

Table 1. Frequencies of misclassification of selected classifiers and learning sets

	PHENOMENE		BALANCE		BANANA		HIGLEYMAN	
	train	under-train	train	under-train	train	under-train	train	under-train
S1	29,33%	29,50%	12,02%	59,62%	17,50%	50,00%	49,67%	50,50%
S2	23,94%	55,44%	57,21%	53,85%	14,50%	61,33%	10,83%	51,83%
S3	24,83%	29,44%	42,79%	88,46%	4,67%	34,33%	17,33%	50,67%
S4	26,78%	70,67%	13,94%	50,00%	11,83%	26,33%	13,83%	50,00%
S5	26,78%	34,17%	13,46%	59,13%	12,83%	50,50%	12,17%	66,67%
Oracle	8,50%	0,00%	7,21%	1,44%	2,67%	0,00%	1,00%	0,00%
FDF	21,83%	24,06%	8,17%	22,12%	2,17%	2,33%	4,33%	4,83%
FCN	24,72%	29,33%	11,06%	43,27%	3,67%	26,17%	8,83%	30,83%

3.1 Experimental Results Evaluation

Firstly, one has to note that we are aware of the fact that the scope of computer experiments were limited. Therefore, making general conclusions based on them is very risky.

In the case of the presented experiment:

1. Proposed committees with well trained fuser always gave better results than individual classifiers.
2. FDF classifier gave always better results than FCN. Additionally we have to notice that FDF was definitely better than the best individual classifier, but quality of FCN is only slightly better or similar to the quality of the best classifier.
3. We confirmed our research presented in section 2 and we observed that for *banana* dataset and undertrained classifier we obtain fuser better than Oracle.
4. We observed interesting feature that for undertrained classifier it was possible to obtain ideal committee (classification error of Oracle was equal 0). It could be explained by the high diversity of classifier and this observation proved that works connected with diversity measure propositions are very encouraging and important for the multiple classifier systems projects.

4   Final Remarks

Some methods of classifier fusion were discussed in this paper and two of them were evaluated via computer experiments on benchmark and computer generated databases.

Obtained results justify the use of weighted combination and they are similar as published in [4, 14, 20]. Unfortunately, as it was stated, it is not possible to determine weight values in the analytical way. However, it is hoped that in practical situations the weights can be appropriately set, either with the aid of a suitable expert, or else by applying the data training methods proposed earlier in the paper.

Acknowledgments. This work is supported by The Polish State Committee for Scientific Research under the grant which is realizing in years 2006-2009.

References

1. Asuncion, A., Newman, D.J.: UCI Machine Learning Repository. University of California, School of Information and Computer Science, Irvine (2007), http://www.ics.uci.edu/~mlearn/MLRepository.html
2. Chow, C.K.: Statistical independence and threshold functions. IEEE Trans. on Electronic Computers EC-16, 66–68 (1965)
3. Duda, R.O., et al.: Pattern Classification. Wiley Interscience, Hoboken (2001)
4. Duin, R.P.W., Tax, D.M.J.: Experiments with Classifier Combining Rules. In: Kittler, J., Roli, F. (eds.) MCS 2000. LNCS, vol. 1857, pp. 16–29. Springer, Heidelberg (2000)
5. Duin, R.P.W.: The Combining Classifier: to Train or Not to Train? In: Proc. of the ICPR 2002, Quebec City (2002)
6. Duin, R.P.W., et al.: PRTools4, A Matlab Toolbox for Pattern Recognition, Delft University of Technology (2004)
7. Fumera, G., Roli, F.: A Theoretical and Experimental Analysis of Linear Combiners for Multiple Classifier Systems. IEEE Trans.on PAMI 27(6), 942–956 (2005)
8. Hansen, L.K., Salamon, P.: Neural Networks Ensembles. IEEE Trans. on PAMI 12(10), 993–1001 (1990)
9. Hashem, S.: Optimal linear combinations of neural networks. Neural Networks 10(4), 599–614 (1997)
10. Inoue, H., Narihisa, H.: Optimizing a Multiple Classifier Systems. In: Ishizuka, M., Sattar, A. (eds.) PRICAI 2002. LNCS, vol. 2417, pp. 285–294. Springer, Heidelberg (2002)
11. Jozefczyk, J.: Determination of optimal recognition algorithm in two-level systems. Pattern recognition Letters 4, 413–420 (1986)
12. Jain, A.K., Duin, P.W., Mao, J.: Statistical Pattern Recognition: A Review. IEEE Trans. on PAMI 22(1), 4–37 (2000)
13. Kohavi, R.: A study of cross-validation and bootstrap for accuracy estimation and model selection. In: Proc. of the 14th Int.Joint Conf. on Artificial Intell., San Mateo, pp. 1137–1143 (1995)
14. Kuncheva, L.I., Bezdek, J.C., Duin, R.P.W.: Decision templates for multiple classifier fusion: an experimental comparison. Pattern Recognition 34, 299–314 (2001)
15. Kuncheva, L.I., Whitaker, C.J., Shipp, C.A., Duin, R.P.W.: Limits on the Majority Vote Accuracy in Classier Fusion. Pattern Analysis and Applications 6, 22–31 (2003)
16. Kuncheva, L.I.: Combining pattern classifiers: Methods and algorithms. Wiley, Chichester (2004)
17. Raudys, S.: Trainable fusion rules. I. Large sample size case. Neural Networks 19, 1506–1516 (2006)
18. Raudys, S.: Trainable fusion rules. II. Small sample-size effects. Neural Networks 19, 1517–1527 (2006)
19. Tumer, K., Ghosh, J.: Analysis of Decision Boundaries in Linearly Combined Neural Classifiers. Pattern Recognition 29, 341–348 (1996)
20. Van Erp, M., Vuurpijl, L.G., Schomaker, L.R.B.: An overview and comparison of voting methods for pattern recognition. In: Proc. of IWFHR.8, Canada, pp. 195–200 (2002)
21. Kuncheva, L.: Using degree of consensus in two-level fuzzy pattern recognition. European Journal of Operational Research 80, 365–370 (1995)
22. Wozniak, M.: Experiments on linear combiners. In: Pietka, E., Kawa, J. (eds.) Information technologies in biomedicine, pp. 445–452. Springer, Heidelberg (2008)
23. Xu, L., Krzyzak, A., Suen, C.Y.: Methods of Combining Multiple Classifiers and Their Applications to Handwriting Recognition. IEEE Trans. on SMC (3), 418–435 (1992)

A Hybrid Bumble Bees Mating Optimization - GRASP Algorithm for Clustering

Yannis Marinakis[1], Magdalene Marinaki[2], and Nikolaos Matsatsinis[1]

[1] Decision Support Systems Laboratory, Department of Production Engineering and
Management, Technical University of Crete, 73100 Chania, Greece
marinakis@ergasya.tuc.gr, nikos@ergasya.tuc.gr
[2] Industrial Systems Control Laboratory, Department of Production Engineering and
Management, Technical University of Crete, 73100 Chania, Greece
magda@dssl.tuc.gr

Abstract. A new hybrid algorithm for clustering, which is based on the
concepts of the Bumble Bees Mating Optimization (BBMO) and Greedy
Randomized Adaptive Search Procedure (GRASP), is presented in this
paper. The proposed algorithm is a two phase algorithm which combines
a new algorithm called Bumble Bees Mating Optimization algorithm for
the solution of the feature selection problem and a GRASP algorithm
for the solution of the clustering problem. The performance of the algorithm is compared with other popular metaheuristic and nature inspired
methods using datasets from the UCI Machine Learning Repository. The
high performance of the proposed algorithm is achieved as the algorithm
gives very good results and in some instances the percentage of the correct clustered samples is very high and is larger than 98%.

Keywords: Bumble Bees Mating Optimization, Greedy Randomized
Adaptive Search Procedure, Clustering Analysis.

1 Introduction

During the last decade nature inspired approaches, like Particle Swarm Optimization [4], Ant Colony Optimization [2], bees inspired algorithms [1] have
become increasingly popular. The bees inspired algorithms are divided, mainly,
in two categories according to their behavior in the nature, the foraging behavior
and the mating behavior.

In this paper, a new algorithm that simulates the mating behavior of the Bumble bees, the **Bumble Bees Mating Optimization (BBMO)** is presented.
This algorithm is a population-based swarm intelligence algorithm that simulates the mating behavior that a swarm of bumble bees perform. The algorithm
is used in combination with a Greedy Randomized Adaptive Search Procedure
(GRASP) [3] for the solution of the clustering problem ([6,7]). For the mathematical description of the clustering problem see [5]. More precisely, the proposed
BBMO-GRASP algorithm uses the BBMO for the feature selection phase of the
clustering algorithm while for the clustering phase the GRASP algorithm is applied. In order to assess the efficacy of the proposed algorithm, this methodology

E. Corchado et al. (Eds.): HAIS 2009, LNAI 5572, pp. 549–556, 2009.

is evaluated on datasets from the UCI Machine Learning Repository. Also, the method is compared with the results of a number of other metaheuristic algorithms for clustering analysis that use, mainly, hybridization techniques. The rest of this paper is organized as follows: In the next section the proposed Hybrid BBMO-GRASP algorithm is presented and analyzed in detail. In section 4, the analytical computational results for the datasets used in this study are presented while in the last section conclusions and future research are given.

2   The Proposed Hybrid BBMO-GRASP for Clustering

2.1   Bumble Bees Behavior

Bumble bees are social insects that form colonies consisting of the queen, many workers (females) and the drones (males). Queens are the only members of the nest to survive from one season to the next, as they spend the winter months hibernating in a protected underground overwintering chamber. Upon emerging from hibernation, a queen collects pollen and nectar from flowers and searches for a suitable nest site and when she finds such a place, she prepares wax pots to store food and wax cells into which eggs are laid ([8,9,10,11]).

The bumble bee queen can lay fertilized or unfertilized eggs. The fertilized eggs have chromosomes from the queen and a male or males she mated with the previous year and they develop into workers while the unfertilized eggs contain chromosomes from the queen alone and they develop into males. After the emergence of the first workers, the queen no longer forages as the workers take over the responsibilities of collecting food (foragers) and the queen remains in the nest laying eggs and tending to her young. Some workers, also, remain in the nest and help raise the brood (household workers). Males do not contribute in collecting food or helping rear young as the sole purpose of the males are to mate with the queens. Bumble bee workers are able to lay haploid eggs when the queen's ability to suppress the workers' reproduction diminishes. These eggs are developed into viable male bumble bees ([8,9,10,11]).

A few days after the males leave the nest, new queens will emerge. After new queens and males have gone, the colony begins to deteriorate. The founder queen stops laying eggs and grows weak from old age while the remaining workers continue to forage for food but only for themselves. Away from the colony, the new queens and males live off nectar and pollen and spend the night on flowers or in holes. The queens are eventually mated (often more than once), the sperm from the mating is stored in spermatheca and she searches for a suitable location for diapause. Three different mating behaviors exist in bumble bees. The first mating behavior is where a male perches on a tall structure and waits for queens to fly by and he will pursue them for mating once one queen is spotted. The second mating behavior is when males create a scent trail, marking their flight path with pheromones and, thus, queens of the same species will be attracted to the pheromones and follow the scent trail. The third mating behavior is where males wait at the entrance of a bumble bee nest for queens to leave ([8,9,10,11]).

2.2 BBMO for the Feature Selection Problem

In the BBMO algorithm, there are three kind of bumble bees in the colony, the queen bee, the worker bees and the male bees (drones). Initially, a number of bees are selected randomly. In the feature selection problem, the solutions should have values equal to 0 or to 1, where 0 denotes that the feature is not activated and 1 denotes that the feature is activated. Afterwards, the fitness of each bee is calculated using the GRASP algorithm for clustering (see section 2.3) and the best bee is selected as the queen. All the other bees in the initialization phase of the algorithm are the drones. The queen selects the males that are used for mating by using the second mating behavior where it is assumed in the algorithm that the fittest males let larger amount of pheromone in their flight paths and, thus, the queen selects the most promising paths. Each time the queen successfully mates with a drone, the genotype of the drone is stored in her spermatheca until the maximum number of matings have been reached.

After the mating, the queen finds a place to hibernate and in the next year (iteration) finds a place to create the hive and to begin to lay eggs. There are three kinds of bees that a queen lays: new queens, workers and drones. The first two kinds of bees are created by crossover of the genotype of the queen and the genotype of the drones using a specific crossover operator. In this crossover operator, the points are selected randomly from the selected drones and from the queen. Thus, initially a crossover operator number is selected (Cr_1) that controls the fraction of the parameters that are selected for the drones and the queen. The Cr_1 value is compared with the output of a random number generator, $rand_i(0,1)$. If the random number is less or equal to the Cr_1 the corresponding value is inherited from the queen, otherwise it is selected, randomly, from the solutions of one of drones' genotypes that are stored in spermatheca. Thus, if the solution of the brood is denoted by $b_i(t)$ (t is the iteration number), the solution of the queen is denoted by $q_i(t)$ and the solution of the drone by $d_i(t)$:

$$b_i(t) = \begin{cases} q_i(t), & \text{if } rand_i(0,1) \leq Cr_1 \\ d_i(t), & \text{otherwise.} \end{cases} \quad (1)$$

The fittest of the broods are the candidate for becoming new queens while the rest are the workers. Initially, the new queens are fed from the queen and, afterwards, from the workers and the queen. This is achieved by using a local search phase where each new queen selects from the workers and the queen who is going to feed her by using the following equation:

$$nq_{ij} = nq_{ij} + (b_{max} - \frac{(b_{max} - b_{min}) * lsi}{lsi_{max}}) * (nq_{ij} - q_j) +$$

$$(2)$$

$$\frac{1}{M} * \sum_{k=1}^{M} (b_{min} - \frac{(b_{min} - b_{max}) * lsi}{lsi_{max}}) * (nq_{ij} - w_{kj})$$

where nq_{ij} is the solution of the new queen i, q_j is the the solution of the old queen, w_{kj} is the solution of the worker, M is the number of the workers that each queen selects for feeding her and it is different for each queen, j is the dimension of the problem (number of features), b_{max}, b_{min} are two parameters with values in the interval $(0, 1)$ that control if the new queen is fed from the old queen, from the workers or from both of them, lsi is the current local search iteration and lsi_{max} is the maximum number of local search iterations. Initially, the new queens are fed more from the old queen and as the local search iterations increase, then only the workers feed the new queen. The appropriate choice of the values of b_{max} and b_{min} controls the feeding process, i.e. in order to have the feeding process described previously a large value for b_{max} and a value almost equal to zero for b_{min} are necessary. Afterwards, the new queens leave from the hive. The drones are produced by mutate the queen genotype or by mutate the fittest workers' genotype using a random mutation operator. In this mutation operator, the changes in the genotype of the queen or the workers are performed randomly. The drones, then, leave from the hive and they are looking for new queens for mating. In the next generation, the best fertilized queens survive and all the other members of the population die.

It should be noted that as the proposed Bumble Bees Mating Optimization (BBMO) algorithm is inspired from the mating behavior of the bumble bees, it has a number of differences compared to another nature inspired algorithm that is based on the mating behavior of honey bees, the Honey Bees Mating Optimization (HBMO) algorithm [1,5]. More precisely:

- In the BBMO the workers are different solutions while in the HBMO they are local search phases.
- In the BBMO after the mating of the queen three kinds of bumble bees are produced, the new queens and the workers (by using a crossover operator) and the drones (by using a mutation operator). On the other hand, in the HBMO after the mating of the queen two kinds of honey bees are produced, the queen and the drones (both of them by using a crossover operator).
- In the BBMO the fittest of the broods produced by the crossover operator are the new queens and all the others are the workers while in the HBMO the fittest of the broods is the new queen and all the others are the drones.
- In the BBMO the drones are produced by mutation of the queen or by mutation of the fittest workers. In the HBMO the drones are all the bees produced by the crossover operator except of the queen.
- The feeding procedure in the BBMO is as described previously using the equation 2 while in the HBMO the feeding procedure is local search phases that are applied independently in each brood.

2.3   GRASP for the Clustering Problem

As it was mentioned earlier in the clustering phase of the proposed algorithm a **Greedy Randomized Adaptive Search Procedure (GRASP)** ([3]) is used which is an iterative two phase search algorithm (a **construction phase** and a **local search phase**). An initial solution (i.e. an initial clustering of the samples in the clusters) is constructed step by step and, then, this solution is exposed for improvement in the local search phase of the algorithm. The algorithm can be used for unknown or known a priori number of clusters. If the number of clusters is known a priori, then a number of samples equal to the number of clusters are selected randomly as the initial clusters. In this case, as the iterations of GRASP increase the number of clusters does not change. In each iteration, different samples (equal to the number of clusters) are selected as initial clusters. Afterwards, the best promising candidate samples are selected to create the RCL (the Restricted Candidate List - RCL is the list that is used for the selection of the next element that will be chosen to be inserted to the current solution). After modifications in the RCL, a solution is obtained which is improved using a local search strategy ([5]). If the number of clusters is unknown then, initially a number of samples are selected randomly as the initial clusters. Now, as the iterations of GRASP increase, the number of clusters changes but cannot become less than two. In each iteration a different number of clusters can be found. The creation of the initial solutions and the local search phase work as in the previous case. The only difference compared to the previous case concerns the use of the validity measure in order to choose the best solution ([5]).

3   Computational Results

The performance of the proposed methodology is tested on 9 benchmark instances taken from the UCI Machine Learning Repository. The datasets from the UCI Machine Learning Repository were chosen to include a wide range of domains and their characteristics are given in Table 1. In one case (Breast Cancer Wisconsin) the data set appears with different size of samples because in this data set there is a number of missing values. The problem of missing values was faced with two different ways. In the first way, we put in the missing places the mean values of all existing samples in the corresponding feature (this corresponds to BCW1 in Table 1). In the second way, we did not use the sample that has a missing value in a feature (this corresponds to BCW2 in Table 1). For each data set, Table 1 reports the total number of features and the number of categorical features in parentheses. The parameter settings for BBMO-GRASP algorithm were selected after thorough empirical testing and they are: The number of queens is set equal to 10, the number of workers is set equal to 40, the number of males is set equal to 50, the number of generations is set equal to 100, the lsi_{max} is set equal to 50, the b_{max} is set equal to 0.99, the b_{min} is set equal to 0.001, the size of RCL is set equal to 50 and the number of GRASP's iterations is equal to 100. The algorithm was implemented in Fortran 90 and was compiled using the Lahey f95 compiler on a Centrino Mobile Intel Pentium M 750 at 1.86 GHz, running Suse Linux 9.1.

The results of the proposed algorithm are given in Table 2. After the selection of the final parameters, 10 different runs with the selected parameters

Table 1. Data Sets Characteristics

Data Sets	Observations	Features	Clusters
Australian Credit (AC)	690	14(8)	2
Breast Cancer Wisconsin 1 (BCW1)	699	9	2
Breast Cancer Wisconsin 2 (BCW2)	683	9	2
Heart Disease (HD)	270	13(7)	2
Hepatitis 1 (Hep1)	155	19 (13)	2
Ionosphere (Ion)	351	34	2
Spambase(spam)	4601	57	2
Iris	150	4	3
Wine	178	13	3

were performed for each of the datasets. In Table 2, the results of the best run (selected features and correct clustered samples) of the proposed algorithm, the average results (correct clustered samples) of the 10 runs of the algorithm and the variance of the correct clustered samples are presented. As it can be seen, the proposed algorithm has achieved a very good performance concerning the clustered samples. Also, taking into account the average and the variance, it should be noted that there are no significant differences in the obtained results.

A comparison with other metaheuristic approaches for the solution of the clustering problem is presented in Table 2. In this Table, besides the proposed algorithm, ten other algorithms are used for the solution of the feature subset selection problem and for the clustering problem. Most of these methods are hybrid methods that use in the clustering phase of the algorithm the GRASP algorithm and in the feature selection phase they use other methods, like the Multi Swarm Constriction Particle Swarm Optimization (MSCPSO), a Memetic algorithm, a Honey Bees Mating Optimization (HBMO) algorithm, a Particle Swarm Optimization (PSO) algorithm, an Ant Colony Optimization (ACO) algorithm, a genetic algorithm and a tabu search algorithm. Also, two algorithms that use ACO and PSO in both phases and an algorithm that uses a PSO algorithm in the feature selection phase and an ACO algorithm in the clustering phase are applied. The parameters and the implementation details of all of the algorithms presented in the comparisons are analyzed in the papers [5]. From this table, it can be observed that the Hybrid BBMO-GRASP algorithm performs better (has the largest number of correct clustered samples) than the other ten algorithms in all instances. It should be mentioned that in some instances the differences in the results between the Hybrid BBMO-GRASP algorithm and the other ten algorithms are very significant. Mainly, for the two data sets that have the largest number of features compared to the other data sets, i.e. in the Ionosphere data set the percentage of correct clustered samples for the Hybrid BBMO-GRASP algorithm is 91.45% while for all the other methods the percentage varies between 70.65% to 88.88% and in the Spambase data set the percentage of correct clustered samples for the Hybrid BBMO-GRASP algorithm is 91.61% while for all the other methods the percentage varies between 82.80% to 90.17%. It should, also, be noted that a hybridization algorithm performs always better than a no hybridized algorithm for the problem and for the samples tested in this study. These results prove the significance of the solution of the feature selection

Table 2. Results of the algorithms

Instance	BBMO-GRASP				Memetic-GRASP		HBMO-GRASP	
	Selected Features	Correct Clustered Best	Average	Variance	Selected Features	Correct Clustered	Selected Features	Correct Clustered
BCW2	5	670(98.09%)	668.4(97.86%)	3.37	5	664(97.21%)	5	664(97.21%)
Hep1	5	144(92.90%)	142.6(92%)	0.93	9	139(89.67%)	5	140(90.32%)
AC	7	617(89.42%)	615.1(89.14%)	2.32	8	604(87.53%)	8	604(87.53%)
BCW1	5	684(97.85%)	682.4(97.62%)	2.04	8	677(96.85%)	5	677(96.85%)
Ion	5	321(91.45%)	318.7(90.79%)	2.67	5	305(86.89%)	8	309 (88.03%)
spam	25	4215(91.61%)	4212(91.54%)	4.22	32	4019(87.35%)	31	4028 (87.54%)
HD	5	251(92.96%)	249.1(92.25%)	2.1	9	236(87.41%)	8	237(87.77%)
Iris	3	147(98.00%)	146.7(97.8%)	0.23	3	146(97.33%)	3	146(97.33%)
Wine	6	177(99.43%)	176.4(99.10%)	0.48	7	176(98.87%)	7	176(98.87%)

Instance	PSO-GRASP		ACO-GRASP		Genetic-GRASP		Tabu-GRASP	
	Selected Features	Correct Clustered	Selected Features	Correct Clustered	Selected Features	Correct Clustered	Selected Features	Correct Clustered
BCW2	5	662(96.92%)	5	662(96.92%)	5	662(96.92%)	6	661(96.77%)
Hep1	7	135(87.09%)	9	134(86.45%)	9	134(86.45%)	10	132(85.16%)
AC	8	604(87.53%)	8	603(87.39%)	8	602(87.24%)	9	599(86.81%)
BCW1	5	676(96.70%)	5	676(96.70%)	5	676(96.70%)	8	674(96.42%)
Ion	11	300(85.47%)	2	291(82.90%)	17	266(75.78%)	4	263(74.92%)
spam	51	4009(87.13%)	56	3993(86.78%)	56	3938(85.59%)	34	3810(82.80%)
HD	9	232(85.92%)	9	232(85.92%)	7	231(85.55%)	9	227(84.07%)
Iris	3	145(96.67%)	3	145(96.67%)	4	145(96.67%)	3	145(96.67%)
Wine	7	176(98.87%)	8	176(98.87%)	7	175(98.31%)	7	174(97.75%)

Instance	MSCPSO-GRASP		ACO		PSO		PSO-ACO	
	Selected Features	Correct Clustered	Selected Features	Correct Clustered	Selected Features	Correct Clustered	Selected Features	Correct Clustered
BCW2	5	667(97.65%)	5	662(96.92%)	5	662(96.92%)	5	664(97.21%)
Hep1	5	142(91.65%)	9	133(85.80%)	10	132(85.16%)	6	139(89.67%)
AC	7	610(88.40%)	8	601(87.10%)	8	602(87.24%)	8	604(87.53%)
BCW1	5	681(97.42%)	8	674(96.42%)	8	674(96.42%)	5	677(96.85%)
Ion	5	312(88.88%)	16	258(73.50%)	12	261(74.35%)	7	302(86.03%)
spam	28	4149(90.17%)	41	3967(86.22%)	37	3960(86.06%)	39	4012(87.19%)
HD	6	243(90.00%)	9	227(84.07%)	9	227(84.07%)	9	235(87.03%)
Iris	3	147(98.00%)	3	145(96.67%)	3	145(96.67%)	3	146(97.33%)
Wine	6	177(99.43%)	7	174(97.75%)	7	174(97.75%)	7	176(98.87%)

problem in the clustering algorithm as when more sophisticated methods for the solution of this problem were used the performance of the clustering algorithm was improved. The significance of the solution of the feature selection problem using the BBMO algorithm is, also, demonstrated by the fact that with this algorithm the best solution was found by using less features than the other algorithms used in the comparisons. More precisely, in the most difficult instance, the Spambase instance, the proposed algorithm needed 25 features in order to find the optimal solution, while the other nine algorithms (in the k-means the feature selection problem was not solved) the algorithms needed between 28 - 56 features to find their best solution. It should, also, be mentioned that the algorithm was tested with two options: with known and unknown number of clusters. When the number of clusters was unknown and, thus, in each iteration of the algorithm different initial values of clusters were selected, the algorithm always converged to the optimal number of clusters and with the same results as in the case where the number of clusters was known.

4   Conclusions and Future Research

In this paper, a new metaheuristic algorithm, the Hybrid BBMO-GRASP, is proposed for solving the Clustering Problem. This algorithm is a two phase

algorithm which combines a new Bumble Bees Mating Optimization algorithm
for the solution of the feature selection problem and a Greedy Randomized
Adaptive Search Procedure for the solution of the clustering problem. A number
of other metaheuristic algorithms for the solution of the problem were also used
for comparison purposes. The performance of the proposed algorithm was tested
using various benchmark datasets from UCI Machine Learning Repository. The
significance of the solution of the clustering problem by the proposed algorithm
is demonstrated by the fact that the percentage of the correct clustered samples
is very high and in some instances is larger than 98%. Future research is intended
to be focused on using different algorithms both to the feature selection phase
and to the clustering algorithm phase, like k-means.

References

1. Abbass, H.A.: A monogenous MBO approach to satisfiability. In: Proceeding of the
   International Conference on Computational Intelligence for Modelling, Control and
   Automation, CIMCA 2001, Las Vegas, NV, USA (2001)
2. Dorigo, M., Stutzle, T.: Ant Colony Optimization. A Bradford Book. MIT Press,
   Cambridge (2004)
3. Feo, T.A., Resende, M.G.C.: Greedy randomized adaptive search procedure. Journal of Global Optimization 6, 109–133 (1995)
4. Kennedy, J., Eberhart, R.: Particle swarm optimization. In: Proceedings of IEEE
   International Conference on Neural Networks, vol. 4, pp. 1942–1948 (1995)
5. Marinakis, Y., Marinaki, M., Matsatsinis, N.: A hybrid clustering algorithm based
   on Honey Bees Mating Optimization and Greedy Randomized Adaptive Search
   Procedure. In: Maniezzo, V., Battiti, R., Watson, J.-P. (eds.) LION 2008. LNCS,
   vol. 5313, pp. 138–152. Springer, Heidelberg (2008)
6. Rokach, L., Maimon, O.: Clustering methods. In: Maimon, O., Rokach, L. (eds.)
   Data Mining and Knowledge Discovery Handbook. Springer, New York (2005)
7. Xu, R., Wunsch II, D.: Survey of clustering algorithms. IEEE Transactions on
   Neural Networks 16(3), 645–678 (2005)
8. http://www.bumblebee.org/
9. http://www.everythingabout.net/articles/biology/animals/arthropods/
   insects/bees/bumble_bee/
10. http://bumbleboosters.unl.edu/biology.shtml
11. http://www.colostate.edu/Depts/Entomology/courses/en570/
    papers_1998/walter.htm

A First Study on the Use of Coevolutionary Algorithms for Instance and Feature Selection

Joaquín Derrac[1], Salvador García[2], and Francisco Herrera[1]

[1] University of Granada, Department of Computer Science and Artificial Intelligence,
18071 Granada, Spain
jderrac@correo.ugr.es, herrera@decsai.ugr.es
[2] University of Jaén, Department of Computer Science, 23071 Jaén, Spain
sglopez@ujaen.es

Abstract. Cooperative Coevolution is a technique in the area of Evolutionary Computation. It has been applied to many combinatorial problems with great success. This contribution proposes a Cooperative Coevolution model for simultaneous performing some data reduction processes in classification with nearest neighbours methods through feature and instance selection.

In order to check its performance, we have compared the proposal with other evolutionary approaches for performing data reduction. Results have been analyzed and contrasted by using non-parametric statistical tests, finally showing that the proposed model outperforms the non-cooperative evolutionary techniques.

1 Introduction

One main process in data mining is the one known as data reduction [18]. In classification, it aims to reduce the size of the training set mainly to increase the efficiency of the training phase (by removing redundant instances) and even to reduce the classification error rate (by removing noisy instances).

Instance Selection (IS) and Feature Selection (FS) are two of the most known data reduction techniques in data mining. Both are really effective not only to reduce the size of the train set, but also to filtrate and clean noisy data, thus helping classifiers to improve its accuracy [12,13].

Evolutionary Algorithms (EAs)[5] are general purpose search algorithms that use principles inspired by nature to evolve solutions to problems. EAs have been successfully used in data mining problems[8,9]. Their capacity of tackling IS and FS as combinatorial problems is specially useful [2,14].

Coevolution is a specialized trend of EAs. It tries to simultaneously manage two or more populations (also called species), to evolve them and to allow interactions among individuals of any population. The goal is to improve results achieved from each population separately. The Coevolution model has shown some interesting characteristics in the last years [22]. Also, it has been successfully applied in other problems, like function optimization [11,21].

E. Corchado et al. (Eds.): HAIS 2009, LNAI 5572, pp. 557–564, 2009.
© Springer-Verlag Berlin Heidelberg 2009

Our proposal combines Evolutionary IS and FS with Coevolution techniques, in order to improve the effectiveness of Evolutionary IS and FS applied to nearest neighbours classifiers in terms of accuracy. We have named our proposed model CoCHC (Cooperative Coevolution model using CHC algorithm). A wide range of classification data sets will be used to compare it with other non-coevolutionary models, in order to highlight the benefits of the use of Coevolution.

The rest of this contribution is organized as follow: Section 2 reviews the preliminary theoretical study. Section 3 explains the Cooperative Coevolutive model proposed. Section 4 describes the experimental framework used and presents the analysis of results. Finally, in Section 5, we point out the conclusions achieved.

2   Background and Related Work

This section shows the main topics of the background in which our contribution is based. Section 2.1 describes some evolutionary techniques applied to IS and FS problems. Section 2.2 shows the EAs in which our model is based. Finally, Section 2.3 highlights the main characteristics of Cooperative Coevolution.

2.1   Evolutionary Instance and Feature Selection

EAs have proved to be good mechanisms for data reduction in data mining. They have been widely used to tackle the FS and IS problems.

The FS problem can be defined as a search process of P features from an initial set of M variables, with $P <= M$. It aims to eliminate irrelevant and/or redundant features and to obtain a simpler classification system. Also, this reduction can improve the accuracy of the model in classification [13].

The IS problem can also be defined as a search process, where a reduced set S of instances is selected from the training set. By choosing the most suitable points in the data set as instances for the training data, the classification process can gets greatly increased both its efficiency and accuracy [12]

In [14] is proposed a hybridization of a genetic algorithm with local search operators for FS. In [2], a complete study of the use of EAs in IS is done, highlighting four EAs to complete this task: Generational Genetic Algorithm (GGA) [10], Steady-State Genetic Algorithm (SGA) [20], CHC Adaptive Search Algorithm(CHC) [6] and Population-Based Incremental Learning (PBIL) [1]. They concluded that EAs outperform classical algorithms both in reduction rates and classification accuracy. They also concluded that CHC is the most appropriate EA to make this task, according to the algorithms they compared.

Beyond these applications, it is important to point out that both techniques can be applied simultaneously. Despite the most natural way to combine these techniques is to use one first (i.e IS), to get its results and to apply them to the second technique (i.e FS), some authors have already tried to get some profit from the joint use of both approaches [7].

2.2   CHC Algorithm

As it is exposed in the previous section, CHC is a good example of EA which can be used in IS and FS. We have studied its main characteristics to select it as the baseline EA which will guide the search process of our model (it will be explained in Section 3). During each generation, the CHC algorithm [6] develops the following steps:

1. It uses a parent population of size R to generate an intermediate population of R individuals, which are randomly paired and used to generate R potential offspring.
2. Then, a survival competition is held where the best R chromosomes from the parent and offspring populations are selected to form the next generation.

CHC also implements HUX recombination operator. HUX exchanges half of the bits that differ between parents, where the bit position to be exchanged is randomly determined. It also employs a method of incest prevention: Before applying HUX to two parents, the Hamming distance between them is measured. Only those parents who differ from each other by some number of bits (mating threshold) are mated. If no offspring is inserted into the new population then the threshold is reduced.

No mutation is applied during the recombination phase. Instead, when the search stops making progress the population is reinitialized to introduce new diversity. The chromosome representing the best solution found is used as a template to re-seed the population, randomly changing 35% of the bits in the template chromosome to form each of the other chromosomes in the population.

We have selected CHC because it has been widely studied, being now a well-known algorithm on evolutionary computation. Furthermore, previous studies like [2] support the fact that it can perform well on data reduction problems.

2.3   Cooperative Coevolution

In the context of evolutionary computation, cooperative coevolution can be defined as the co-existence of some interacting populations, evolving simultaneously. Each population evolves individuals representing a component of the final solution. Thus, a full candidate solution is formed by joining an individual chosen from each population [17].

In the underlying evolutionary search procedure, a special fitness function is used. To evaluate an individual, there must be selected one member from the other population (collaborators). The merge of all collaborators will produce a full solution, which can be evaluated by the fitness function.

There are some different proposals of the process of choosing the collaborators. One way, is to evaluate an individual against every single collaborator in the other population [15]. Although it would be a best way to select the collaborators, it will consume a very high number of evaluations in the computation of the fitness function. To reduce this number, there are other choices as the use of just a random individual or the use of the best individual of last generation [16]. The model proposed in the next section will use this last scheme.

3   Cooperative Coevolutive Model Based on Instance and Feature Selection Using CHC

CoCHC employs three populations which simultaneously coexist. They cooperate to get the best possible solution through the evolutionary search procedure. Each population is focused on one reduction data task:

- The first population performs an instance selection.
- The second population performs a feature selection.
- The third population performs both instance and feature selections.

Algorithm 1 shows a basic pseudocode of the model proposed.

Algorithm 1. CoCHC algorithm basic structure

---

1 Generate ISPopulation,FSPopulation and IFSPopulation Randomly;
2 Select initial bestISArray, bestFSArray and bestIFSArray;
3 Evaluate all populations in the multiclassifier;
4 Select bestISArray, bestFSArray and bestIFSArray from each population;
5 **while** *evaluations < max_evaluations* **do**
6   Select best classifier in last generation;
7   Generate simple classifier output from best individuals of the last generation;
8   Do a CHC Generation on every population;
9   Update bestISArray, bestFSArray and bestIFSArray if a better global solution has been found;
10 **end**
    Output: bestISArray, bestFSArray and bestIFSArray

Instruction 1 generates the initial random populations. Instruction 2 evaluates all chromosomes by using simple classifiers (see *Multiclassifier structure* paragraph below), and selects the best individual of each population. Instruction 3 evaluates all chromosomes by using the complete multiclassifier, and instruction 4 selects the new best individual of each population.

In instruction 5 the evolutionary process starts. Instruction 6 selects the best classifier of the last generation (the best simple classifier in accuracy). This will help to break ties in the fitness function evaluation. In Instruction 7, the outputs of the simple classifiers from the best individuals of the last generation are saved.

Instruction 8 performs a CHC generation on each population (see Section 2.2). Instruction 9 updates the best global solution if a better solution (concerning one chromosome from each population)have been found.

When a fixed number of evaluations run out, the evolutionary process is finished. Then, best global solution founded is returned.

At this point, we have to describe three important issues to completely describe CoCHC: The specification of the representation of the chromosomes, the structure of the multiclassifier defined by a full solution and the definition of the fitness function.

Representation: Let us assume a data set with N instances and M attributes. Each chromosome consists of a determinate number of genes, which can represent either an instance or a feature. A binary representation is used, thus each gene has two possible states: 1, if the corresponding feature/instance is included in the training set represented by the chromosome, or 0 if not. The concrete representation and size of the chromosome depend of the population which it belongs:
- IS Population: Each gene represents a instance (chromosome size: N).
- FS Population: Each gene represents a feature (chromosome size: M).
- IFS Population: The first N genes of the chromosome represent instances. Remaining genes represents features (chromosome size: $N+M$).

Multiclassifier structure: To evaluate an individual, one member of each of the other populations must be selected. The merge of all collaborators will produce a full solution, which can be evaluated by the fitness function.

Let U be a chromosome of IS Population, let V be a chromosome of FS Population and let W be a chromosome of IFS Population. The multiclassifier structure is defined by three simple classifiers, based on the 1-NN rule. The first classifier only uses the instances defined by U. The second classifier only uses the features defined by V. Finally, the third classifier uses the instances defined by first N genes of W, and the features defined by the last M genes of W.

To get the output of the multiclassifier, the output of the three simple classifiers have to be computed. Then, a final output must be calculated for each instance, by using a majority vote (ties are broken by using the output of the *best classifier* defined by the global model).

Fitness function: Let G be a chromosome of one population. To compute its fitness their collaborators must be found (they are the best individuals from last generation of the other populations). Let F be a binary string composed of G and its collaborators. We define the next fitness function:

$$Fitness(G) = \alpha \cdot \beta \cdot clasRate(F) + (1-\alpha) \cdot IRed(G) + (1-\beta) \cdot FRed(G) \quad (1)$$

Where $clasRate(F)$ is the percentage of correctly classified objects from the training set by the multiclassifier defined by F, and $IRed(G)$ and $FRed(G)$ are the percentage of reduction achieved on instances and features respectively on the baseline classifier defined by G (reduction rates can be computed on the baseline classifier because it is independent for each classifier, thus is not needed to measure the multiclassifier reduction rate). Finally, α and β are parameters valued between $[0,1]$.

The objective of CoCHC is to maximize the fitness function 1, i.e., to maximize the accuracy and reduction rates of the multiclassifier defined by their best chromosomes.

Before finish this section, it is important to point out that the outputs computed at Instruction 7 make possible that only one simple classifier is needed to be built in every call to fitness function on CHC process, (instead of the three originally required by the fitness function). Thus, the CoCHC model efficiency is greatly increased.

4   Experimental Framework and Results

This section describes the methodology followed in the experimental study conducted in this contribution. Data sets used, parameters of our model and the algorithms used in the comparisons are explained.

4.1   Experimental Framework

To check the performance of CoCHC algorithm, we have used 18 data sets taken from the UCI Machine Learning Database Repository [19]. Table 1 shows their main characteristics. For each data set, it is shown the number of examples, attributes and classes of the problem described.

Table 1. UCI Data sets used in our experiments

Data set	Examples	Attributes	Classes	Data set	Examples	Attributes	Classes
Aut	205	25	6	Housevotes	435	16	2
Bal	625	4	3	Iris	150	4	3
Bands	539	19	2	Mammogr	961	5	2
Bupa	345	6	2	Pima	768	8	2
Car	1728	6	4	Sonar	208	60	2
Cleveland	303	13	5	Tic-tac-toe	958	9	2
Dermat	366	34	6	Vehicle	846	18	4
German	1000	20	2	Wisconsin	699	9	2
Glass	214	9	7	Zoo	101	16	7

The data sets considered are partitioned by using the ten fold cross-validation (10-fcv) procedure. The parameters of CoCHC are: Population size = 50 (for each population), Number of evaluations = 10000, $\alpha = 0.6$, $\beta = 0.98$. The alpha parameter value was taken from the value used on the experiments of [2] (0.5), but slightly increased because the simultaneous use of a FS component. The beta parameter value is near to 1 because in the FS component our model has to remove irrelevant attributes without provoke sudden changes which could decrease the overall accuracy.

Our proposal will be compared with three evolutionary algorithms based on the CHC model, for performing IS, FS and simultaneous IS-FS, respectively. The first one will be denoted by IS-CHC, the second one, FS-CHC; and the last one IFS-CHC. Nearest neighbour rule [3] (1-NN) is also used as a baseline algorithm.

The parameters used for each EA involved in the experimental study are the same as the used by our approach.

4.2   Results

Table 2 shows the average results obtained in test data in terms of accuracy. It also shows the reduction rate achieved in training data. The best results achieved in accuracy for each data set are remarked in bold. Observing Table 2, we can make the following analysis:

– CoCHC achieves the best average result on accuracy.

Table 2. Accuracy obtained in test data

Algorithm	CoCHC		IS-CHC		FS-CHC		IFS-CHC		1-NN
Data set	%Acc.	%Red.	%Acc.	%Red.	%Acc.	%Red.	%Acc.	%Red.	%Acc.
Automobile	79.66	88.72	70.42	91.27	**80.18**	68.00	70.53	98.92	77.43
Bal	88.16	89.28	89.29	98.74	79.04	0.00	**89.43**	98.56	79.04
bands	71.81	82.47	70.14	97.30	71.07	49.47	68.28	99.66	**74.04**
Bupa	68.96	81.96	61.94	96.14	61.93	38.33	**68.98**	99.24	61.08
Car	89.18	70.88	86.69	98.37	**89.58**	18.33	88.66	98.27	85.65
Cleveland	56.76	84.52	**57.81**	97.47	50.83	46.15	57.10	99.46	53.14
Dermatology	95.37	85.60	**97.55**	96.36	95.11	54.71	96.44	99.18	95.35
German	**72.20**	82.80	71.70	98.69	69.30	38.50	71.90	99.89	70.50
Glass	69.99	78.08	69.11	93.14	71.37	43.33	67.06	97.83	**73.61**
Housevotes	**95.14**	89.46	93.32	98.24	94.47	65.00	93.54	99.87	92.16
Iris	**95.33**	81.89	**95.33**	95.56	**95.33**	45.00	94.67	98.11	93.33
Mammographic	**82.00**	97.39	80.23	99.17	72.94	56.00	81.59	99.90	74.72
Pima	72.01	87.73	**76.07**	98.50	68.62	50.00	74.11	99.75	70.33
Sonar	85.55	86.13	76.83	93.75	**86.45**	58.50	79.24	99.76	85.55
Tic-tac-toe	**83.81**	73.29	73.69	97.91	82.78	22.22	75.89	98.91	73.07
Vehicle	**71.99**	82.15	63.36	96.44	71.52	45.56	68.09	99.10	70.10
Wisconsin	**96.28**	91.62	96.27	99.32	95.14	47.78	95.28	99.78	95.57
Zoo	95.58	86.56	**97.00**	86.24	94.75	55.63	87.97	95.76	92.81
Average	**81.66**	84.47	79.26	96.25	79.47	44.58	79.37	99.00	78.75

Table 3. Results of Wilcoxon Signed-Ranks Test

$\alpha = 0.1$	IS-CHC	FS-CHC	IFS-CHC	1-NN
CoCHC	+(.059)	+(.012)	+(.018)	+(.004)

- CoCHC outperforms all the remaining algorithms in 6 of 18 data sets.
- The loss in reduction rate achieved by CoCHC (compared with IS-CHC and IFS-CHC) is not too critical. It increases the average accuracy in 2% with respect to both them and keeps a good reduction rate.

In addition to Table 2, we have performed a two-tailed Wilcoxon Signed-Ranks Test [4], to statistically analyse the results obtained in the experiment. Table 3 shows the p-values obtained by Wilcoxon test.

The results offered by the test indicate us that the proposed model outperforms FS-CHC and IFS-CHC with a level of significance $\alpha = 0.05$, and it is better than IS-CHC considering a level of significance $\alpha = 0.1$.

5 Concluding Remarks

The purpose of this contribution is to present a cooperative coevolutionary model developed to tackle data reduction tasks to improve the classification based on the nearest neighbours technique. The proposal combines processes of evolutionary instance selection and feature selection techniques.

The results show that the use of cooperative coevolution in data reduction based on feature and instance selection can obtain promising results to optimize the performance of nearest neighbour classification.

Acknowledgement. This work was supported by TIN2008-06681-C06-01.

References

1. Baluja, S.: Population-based incremental learning: A method for integrating genetic search based function optimization and competitive learning. Technical report, Pittsburgh, PA, USA (1994)
2. Cano, J.R., Herrera, F., Lozano, M.: Using evolutionary algorithms as instance selection for data reduction in KDD: An experimental study. IEEE Transactions on Evolutionary Computation 7, 561–575 (2003)
3. Cover, T.M., Hart, P.E.: Nearest neighbor pattern classification. IEEE Transactions on Information Theory 13, 21–27 (1967)
4. Demšar, J.: Statistical comparisons of classifiers over multiple data sets. Journal of Machine Learning Research 7, 1–30 (2006)
5. Eiben, A.E., Smith, J.E.: Introduction to Evolutionary Computing. Springer, Heidelberg (2003)
6. Eshelman, L.J.: The CHC adaptative search algorithm: How to safe search when engaging in nontraditional genetic recombination. In: Foundations of Genetic Algorithms, pp. 265–283 (1990)
7. Fragoudis, D., Meretakis, D., Likothanassis, S.: Integrating feature and instance selection for text classification. In: 8th ACM SIGKDD international conference on KDD, pp. 501–506 (2002)
8. Freitas, A.A.: Data Mining and Knowledge Discovery with Evolutionary Algorithms. Springer, New York (2002)
9. Ghosh, A., Jain, L.C.: Evolutionary Computation in Data Mining. Springer, Berlin (2005)
10. Goldberg, D.E.: Genetic Algorithms in Search, Optimization, and Machine Learning. Addison-Wesley, Reading (1989)
11. Au, C., Leung, H.: Guided Mutations in Cooperative Coevolutionary Algorithms for Function Optimization. In: IEEE International Conference on Tools with Artificial Intelligence, pp. 407–414 (2007)
12. Liu, H., Motoda, H.: Instance Selection and Construction for Data Mining. The Springer International Series in Engineering and Computer Science (2001)
13. Liu, H., Motoda, H.: Computational Methods of Feature Selection. Chapman & Hall/Crc Data Mining and Knowledge Discovery Series (2007)
14. Oh, I., Lee, J., Moon, B.: Hybrid Genetic Algorithms for Feature Selection. IEEE Transactions on Pattern Analysis and Machine Intelligence 26, 1424–1437 (2004)
15. Panait, L., Wiegand, R.P., Luke, S.: Improving coevolutionary search for optimal multiagent behaviors. In: International Joint Conferences on Artificial Intelligence, pp. 653–658 (2003)
16. Panait, L., Luke, S., Harrison, J.F.: Archive-Based Cooperative Cooevolutionary Algorithms. In: Genetic and Evolutionary Computation Conference, pp. 345–352 (2006)
17. Potter, M.A., De Jong, K.A.: Cooperative coevolution: an architecture for evolving coadapted subcomponents. Evolutionary Computation 8, 1–29 (2000)
18. Pyle, D.: Data Preparation for Data Mining. The Morgan Kaufmann Series in DMS (1999)
19. Newman, D.J., Hettich, S., Merz, C.B.: UCI repository of ML databases (1998)
20. Whitley, D.: The GENITOR Algorithm and selective preasure: Why Rank Based Allocation of Reproductive Trials is Best. Genetic Algorithms, 116–121 (1989)
21. Jansen, T., Wiegand, R.P.: The Cooperative Coevolutionary (1+1) EA. Evolutionary Computation 12, 405–434 (2004)
22. Wolpert, D., Macready, W.: Coevolutionary Free Lunches. IEEE Transactions on Evolutionary Computation 9, 721–735 (2005)

Unsupervised Feature Selection in High Dimensional Spaces and Uncertainty

José R. Villar[1,*], María R. Suárez[1], Javier Sedano[2], and Felipe Mateos[3]

[1] Computer Science Department, University of Oviedo, Campus de Viesques s/n 33204 Gijón, Spain
{villarjose,mrsuarez}@uniovi.es
[2] Electromechanic Engineering Department, University of Burgos, Spain
jsedano@ubu.es
[3] Electric, Electronic, Computers and Systems Engineering Department, University of Oviedo, Campus de Viesques s/n 33204 Gijón, Spain
felipe@isa.uniovi.es

Abstract. Developing models and methods to manage data vagueness is a current effervescent research field. Some work has been done with supervised problems but unsupervised problems and uncertainty have still not been studied. In this work, an extension of the Fuzzy Mutual Information Feature Selection algorithm for unsupervised problems is outlined. This proposal is a two stage procedure. Firstly, it makes use of the fuzzy mutual information measure and Battiti's feature selection algorithm and of a genetic algorithm to analyze the relationships between feature subspaces in a high dimensional space. The second stage uses a simple ad hoc heuristic with the aim to extract the most relevant relationships. It is concluded, given the results from the experiments carried out in this preliminary work, that it is possible to apply frequent pattern mining or similar methods in the second stage to reduce the dimensionality of the data set.

Keywords: Unsupervised feature selection, genetic algorithms, data uncertainty, frequent pattern mining.

1 Introduction

Many real world applications include a high dimensional feature space. Moreover, it is well known that the data gathered from a real world process could contain uncertainty [13], that is, there could be missing data, the measures could be interval values, etc. Typically the uncertainty in the data has been nullified by means of crisp techniques, i.e. the different techniques to eliminate missing data. What we are really doing is losing information about the process, and this information could be relevant in decision processes or in association rule discovering, especially in unsupervised problems, which represent an effervescent research topic due to its scarcity in the reported techniques [4].

On the other hand, high dimensional feature space represents a big challenge as a reduced data set is needed in order to reduce the over fitting of the models to be

---

* Corresponding author.

E. Corchado et al. (Eds.): HAIS 2009, LNAI 5572, pp. 565–572, 2009.

obtained. Also, high dimensional feature spaces increase the computational time needed in modeling such problems. Several different techniques have been employed to reduce the dimension of the data sets; they are known as feature reduction techniques and are divided into two main types: the feature extraction and feature selection techniques [10, 4]. Feature extraction includes the techniques that involve transforming the feature space into a smaller one. The transformation comprises any linear or nonlinear combination of a feature subset. An example of this kind of techniques is feature extraction by means of Principal Component Analysis [15].

Feature selection includes any method that proposes a feature subset from the original data set without any kind of transformation. The reduced feature space is supposed to include most relevant features according to a certain measure.

In this work, a feature selection technique able to deal with the data uncertainty is detailed. It is based on the Fuzzy Extension of the Mutual Information measure presented in [13], and it is designed for unsupervised problems. A two stages algorithm overcomes the problem of the dimensionality of the original data set. The new algorithm –called Fuzzy Unsupervised Mutual Information Feature Selection, from now on referred to as FUMIFS– has been found valid compared with previous approaches and some conclusions to improve the second stage have been extracted.

This work is organized as follows. A brief review of the feature selection methods and data uncertainty is outlined in the following section. In Section 3 the FUMIFS method is detailed. Section 4 deals with the experiments run and commented results. Finally, conclusions and future work is presented.

2 Uncertainty and Feature Selection in Unsupervised Problems

There are several feature selection techniques available in the literature. According to how the method must be used, feature selection methods are classified as *filters* or as *wrappers* [10, 17]. A feature selection method is referred to as a filter method if it is designed as a prepossess method before the modeling algorithm, i.e. [13]. When the feature selection is ran within the modeling algorithm then it is referred to as a wrapper method, i.e. the SSGA method [3]. The former methods are usually faster than the latter, with lower computation costs. In general, the performance of the wrapper methods is better than that of the filter methods, especially if the model obtained will be used to model the problem. Therefore, the wrappers are essentially designed for supervised problems.

According to how the method searches the domain, there are three possibilities: the *Complete Search* methods, the *Heuristic Search* methods and the *Random Search* methods. Also, the search is known as *Sequential Forward Search* -from now on, SFS- or *Sequential Backward Search* -from now on, SBS-. A heuristic search is called SFS if initially the feature subset is empty, and in each step it is incremented in one feature, i.e. the Battiti method [2]. On the other hand, it is an SBS if at the beginning the feature subset is equal to the feature domain, and in each step the feature subset is reduced in one feature, i.e. the Fisher algorithm [12].

Although there are quite a lot of feature selection contributions reported in the literature, they are mainly designed for supervised problems [3, 17]. Moreover, the uncertainty included in the data is avoided in all of them, only crisp data is considered. Some unsupervised feature selection methods are also reported in the literature.

In [5] the threshold that maximizes the mutual information is used in an SFS, choosing the features with higher mutual information values. Mitra et al proposed clustering feature subsets with the so-called maximum information compression index and choosing the most compact feature from each cluster [11]. Despite the speed and performance of the algorithm, the method is only designed for crisp data. Li et al proposed a hybrid method including a filter stage –using the fuzzy feature evaluation index– and a wrapper stage –using feature clustering. Finally, an unsupervised feature ranking is detailed in [7], where a ranking of the features is calculated based on clustering feature subsets, which they refer to as multiple view. For generating the feature subsets they proposed the random subspace method [6].

Imprecision and vagueness in data have been included in feature selection for modeling problems. Fuzzy logic has been employed for such task in the Fuzzy extension of the Mutual Information measure, which has been used in [13] to extend Battiti's algorithm for data uncertainty. Perhaps the rough set theory is the most widely used technique [9], all of them for supervised problems. A review of the rough set theory and the dimensionality reduction can be obtained in [16]. An SFS feature selection method for unsupervised problems using the neighborhood rough set is detailed in [8], where a neighborhood matrix is used to choose the features that maximize the neighborhood dependency in an SFS like algorithm. This feature selection method is specially defined to accomplish with heterogeneous data sets, that is, data sets that include both real valued and discrete valued features.

Some drawbacks should be commented. The majority of the feature selection methods in the literature do not consider uncertainty in the data and are mainly prepared for supervised problems. To our knowledge, only the last mentioned work deals with unsupervised problems using rough set theory. In general, it has been found that the performance of the SFS methods gets worse with the dimension of the domain space. Particularly, as the FMIFS is an SFS method for supervised problems that uses fuzzy theory, it is also concerned with this drawback. Finally, an increase in the computational cost has to be considered to manage uncertainty in the data.

3   The Fuzzy Unsupervised Mutual Information Feature Selection Algorithm

In previous work an extension of Battiti's mutual information based feature selection method was proposed [13]. This extension, called Fuzzy Mutual Information Feature Selection –for short, FMIFS–, makes use of the fuzzy mutual information measure in order to deal with the uncertainty in the data. The robustness of the FMIFS performance against data uncertainty as missing data or interval-valued features within the dataset was shown. Unfortunately, the FMIFS also shows the above-mentioned drawbacks [14].

The FUMIFS is proposed to overcome these disadvantages. In the FMIFS, the fuzzy mutual information measure is used to establish the information relationship between each variable and the class feature. In each step the feature with the highest value of residual mutual information of the feature class was included in the feature subset. This last step is dependant of a real value parameter called β, which represents the way the residual information of the feature class is calculated according to Eq. 1, where S is the set of features already chosen –that is, the best valued features subset– f is a feature in the domain that has not been chosen and C is the class feature.

$$RMF(f,C) = MF(f,C) - \beta \oplus_{sf \in S} MF(f,sf) \qquad (1)$$

It has been found that the value of β is critical and problem dependant [14]. Moreover, when the number of features increases the residual mutual information is more influenced by the noisy variables. In such cases, the feature subset would include random variables, which are not related with the features in S. Nevertheless, if the number of features is relatively low the FMIFS behaves properly and it is a relatively fast method. Finally, the FMIFS is designed for supervised problems as reflected in Eq. 1 with the class feature C. Hence, the FUMIFS should exploit the behavior of the FMIFS when faced with relatively low dimension feature sets and must try to eliminate the influence of the random variables in high dimension feature sets. Also, as its main goal, the FUMIFS should manage unsupervised problems.

3.1 The Unsupervised Algorithm

The FUMIFS is based on some different approaches found in the literature. Firstly, the random sub-space method [6], which was also employed in [7], is used to choose a feature subset of lower dimension where the FUMIFS is intended to behave properly. The random sub-space method is applicable provided that there is no possibility of repeating the feature subset evaluation, which is to say that it should avoid evaluating a feature subset if it has already been evaluated. Secondly, a genetic algorithm is responsible for generating new feature subsets and evaluating them considering the restriction of the random sub-space method. So the individual is ranked according to how different it is compared to all the previously examined random subspaces, which in fact is the genetic fitness function. If an individual is found repeated then it is eliminated and a new one will be proposed.

Let N be the size of the random subspace. For each individual the FMIFS is run N times; in each run the feature from the random subspace used as objective feature is changed. Therefore, the K most relevant features according to the FIMFS are found for each feature in the feature subset. The value K represents the dimension of the feature subset proposed by the FMIFS in each run.

The individuals in the population are sorted according to their fitness. The genetic selection is carried out choosing individuals from the population with a probability that decreases with the position in the sorted population. The crossover follows a two points crossover schema: according to the crossover probability the vector of included features of both parents are swapped to generate the two offsprings provided no repeated feature is included. In this case, the offspring is completed with a random chosen feature. The mutation goes through the vector of features of the individual to be mutated, and randomly changes each feature with the mutation probability. The vectors of features are always sorted according to their position in the original dataset.

When a population is completed then the certainty table is updated. The *Certainty Table* –for short, *CT*– accumulates the certainty that a feature depends on another. Each run of the FMIFS for an individual has an objective feature –for short, *of*– and proposes a K dimensional vector –for short, *vf*– of the most relevant features according to their mutual information measure. Then the certainty table is updated by means of Eq. 2, Eq. 3 and Eq. 4. Each value in CT is an interval value initialized to the crisp value of 0.

$$a_i = \min(CT(vf[i], of).\min(), \frac{1}{i}) \quad \forall i = 1 \cdots K \tag{2}$$

$$b_i = \max(CT(vf[i], of).\max(), \frac{1}{i}) \quad \forall i = 1 \cdots K \tag{3}$$

$$CT(vf[i], of) = Interval(a_i, b_i) \quad \forall i = 1 \cdots K \tag{4}$$

Finally, some relationships are extracted from the CT given the following rules of thumb. Let {LOW, MEDIUM, HIGH} be the linguistic terms of a fuzzy variable, and Let be f_i and f_j a pair of features for which relationships are to be found. The linguistic rules used to extract the relationships are: "*if CT(f_i, f_j) is HIGH and CT(f_i, f_j) is HIGH then there exists an Equivalence between f_i and f_j*" and "*if CT(f_i, f_j) is HIGH and CT(f_i,f_j) is LOW then there exists a DEPENDENCE of f_j in f_i*". These rules are used to prove that the algorithm is valid; it could be easily improved using the frequent pattern matching or any other algorithm that outperforms these simple rules.

Both N and K are parameters given to the FUMIFS. If N is set to the dimension of the original data set then the FUMIFS behaves like the FMIFS. K is typically set to less than half the value of N. The value of β should also be given as a parameter so FMIFS could be executed. The number of iterations (nIter), the population size (pop-Size) and the crossover and mutation probabilities must also be given. Care must be taken in setting the FUMIFS parameters to avoid infinite loops in the genetic algorithm. As the random subspace method is used there should not be a repeated individual. To prevent such an occurrence the number of iterations and the population size are bounded to not search more than the possible combinations of feature subspaces.

4 Experiments and Results

The FUMIFS is to be compared with the FMIFS in order to test its goodness. So the same experimentation carried out in [13] is to be repeated for the FUMIFS. The datasets are available in the KEEL Project website [1]. To provide unsupervised datasets the class feature has been considered as an input feature. The datasets have been modified to introduce vagueness, and both versions, the crisp and the imprecise ones, have been tested. After a FUMIFS run two data sets are generated: the first with all the features for which a relationship has been found with the class feature and the second data set with only those features for which dependency of the class feature was found. Then FMIFS has been run to choose the same number of features as FUMIFS. The values in all output files from each run are crisp according to [13], using the central point to convert an interval into a crisp value. Due to the length of this work, neither the crisp data sets results nor the boxplot graphics have been included. However, the results are commented.

The same thirteen different fuzzy rule-learning algorithms have been considered, both heuristic and genetic algorithms-based. In all cases, the number of linguistic terms in each partition is set beforehand, and not optimized by the learning algorithm. The experiments have been repeated ten times for different permutations of the datasets (10cv experimental setup). The heuristic classifiers use weighted fuzzy rules: always 1 (H1), the same weight as the confidence (H2), differences between the confidences (H3, H4, H5), weights tuned by reward-punishment (RE) and analytical learning (AL). The genetic fuzzy classifiers are the Genetic selection of rules taken from HEU3 (GE), Michigan learning (MI) –with population size 25 and 1000 generations–, Pittsburgh learning (PI) –with population size 50, 25 rules each individual and 50 generations–, the Hybrid learning (HY) –same parameters as PI, macromutation with probability 0.8–, the Fuzzy Ababoost (AD) –less than 25 rules with a single consequent, fuzzy inference by sum of votes– and Fuzzy Logitboost (LO) –less than 10 rules with multiple consequents, fuzzy inference by sum of votes–.

In Table 1 the classification mean errors for the thirteen methods are shown with the imprecise data sets. For each data set and method four results are given: the FUMIFS with only the dependence relationships found for the class feature (fumifs_d), the FUMIFS with all the relationships found for the class feature (fmifs_r), and the FMIFS results with the same number of features (fmifs_d and fmifs_r, respectively). Comparing the results of the FMIFS and the FUMIFS, it can be seen that the classification mean error is quite similar in both cases: with the dependence relationships and with any relationship between input features and the class output. Also the experiments run with the crisp data sets produced analogous results. Although they could not be included due to space limitations, the statistics boxplot graphics showed that results are totally comparable and it can not be concluded which method is better.

Table 1. FMIFS and FUMIFS classification mean error with the German Credit, the Ionosphere and the Pima Indian dyabetes data sets when vagueness is introduced in the data

	germandn0				ionosphere				pima			
	fmifs_r	fumifs_r	fmifs_d	fumifs_d	fmifs_r	fumifs_r	fmifs_d	fumifs_d	fmifs_r	fumifs_r	fmifs_d	fumifs_d
H1	**0.259**	0.300	0.317	0.300	**0.126**	0.140	0.204	0.129	0.314	0.140	0.357	**0.129**
H2	**0.286**	0.300	0.288	0.300	**0.177**	0.209	0.240	0.194	0.338	0.209	0.361	**0.194**
H3	**0.285**	0.300	0.302	0.300	0.183	0.203	0.241	**0.197**	0.345	0.203	0.351	**0.197**
H4	**0.285**	0.300	0.302	0.300	0.183	0.203	0.241	**0.197**	0.345	0.203	0.351	**0.197**
H5	**0.285**	0.300	0.302	0.300	0.183	0.203	0.206	**0.197**	0.345	0.203	0.351	**0.197**
RE	**0.272**	0.300	0.282	0.300	**0.129**	0.143	0.205	0.154	0.305	**0.143**	0.312	0.154
AL	**0.274**	0.300	0.288	0.300	0.137	0.140	0.208	**0.137**	0.328	0.140	0.295	**0.137**
GE	**0.279**	0.300	0.293	0.300	0.251	**0.126**	0.299	0.157	0.325	**0.126**	0.355	0.157
MI	**0.300**	**0.300**	**0.300**	**0.300**	**0.114**	0.311	0.230	0.309	0.350	0.311	0.350	**0.309**
PI	**0.286**	0.300	0.300	0.300	**0.089**	0.183	0.219	0.203	0.333	**0.183**	0.350	0.203
H	**0.288**	0.300	0.300	0.300	**0.143**	0.143	0.219	0.203	0.330	**0.143**	0.350	0.203
AD	**0.265**	0.292	0.277	0.294	**0.149**	0.149	0.218	0.200	0.230	**0.149**	0.288	0.200
LO	**0.271**	0.273	0.273	0.285	0.143	**0.134**	0.211	0.194	0.229	**0.134**	0.266	0.194

5   Conclusions and Future Works

In this work an unsupervised feature selection method has been described. It makes use of the Fuzzy Mutual Information measure and Battiti's algorithm which, combined with a genetic algorithm, generates a new data set that is to be post processed. It is proposed to use a frequent pattern matching method, but for this work only two rules of thumb were used. Results show that the FUMIFS behaves similarly to the previous work FMIFS. The FUMIFS is really influenced by the simple rules of thumb used. Also, both the aggregation method and the generation of the so-called certainty table have not been optimized. Nevertheless, this unsupervised feature selection method behaves properly, and the results encourage the authors to apply frequent pattern matching in order to improve the goodness of the relationships found.

Acknowledgments. This work was funded by the Spanish Min. of Science and Technology, grants TIN2005-08036-C05-05 and TIN2008-06681-C06-04.

References

[1] Alcala-Fdez, J., Sanchez, L., Garcia, S., Jesus, M.J.D., Ventura, S., Garrell, J.M., Otero, J., Romero, C., Bacardit, J., Rivas, V.M., Fernandez, J.C., Herrera, F.: KEEL: A Software Tool to Assess Evolutionary Algorithms to Data Mining Problems. Soft Computing 13(3), 307–318 (2009)

[2] Battiti, R.: Using mutual information for selecting features in supervised neural net learning. IEEE Transactions on Neural Networks 5(4), 537–550 (1994)

[3] Casillas, J., Cordon, O., Jesus, M.J.D., Herrera, F.: Genetic feature selection in a fuzzy rule-based classification system learning process for high-dimensional problems. Information Sciences 136, 135–157 (2001)

[4] Chow, T.W.S., Wang, P., Ma, E.W.M.: A New Feature Selection Scheme Using a Data Distribution Factor for Unsupervised Nominal Data. IEEE Transactions on Systems, Man and Cybernetics - PART B: Cybernetics 38(2), 499–509 (2008)

[5] Conaire, C.O., Connor, N.E.: Unsupervised feature selection for detection using mutual information thresholding. In: Ninth International Workshop on Image Analysis for Multimedia Interactive Services (2008)

[6] Ho, T.K.: The random subspace method for constructing decision forests. IEEE Transactions on Pattern Analysis and Machine Intelligence 20(8), 832–844 (1998)

[7] Hong, Y., Kwong, S., Chang, Y., Ren, Q.: Consensus unsupervised feature ranking from multiple views. Pattern Recognition Letters 29(5), 595–602 (2008)

[8] Hu, Q., Yu, D., Xie, Z., Liu, J.: Fuzzy Probabilistic Approximation Spaces and Their Information Measures. IEEE Transactions on Fuzzy Systems 14(2), 191–201 (2006)

[9] Jensen, R., Shen, Q.: Fuzzy-rough sets assisted attribute selection. IEEE Transactions on Fuzzy Systems 1(15), 73–89 (2007)

[10] Marcelloni, F.: Feature selection based on a modified fuzzy c-means algorithm with supervision. Information Sciences 151 (2003)

[11] Mitra, P., Murthy, C.A., Pal, S.K.: Unsupervised Feature Selection using Feature Similarity. IEEE Transactions on Pattern Analysis and Machine Intelligence 24(3), 301–312 (2002)

[12] Roubus, J.A., Setnes, M., Abonyi, J.: Learning fuzzy classification rules from labelled data. Information Sciences 150, 77–93 (2003)

[13] Sanchez, L., Suarez, M.R., Villar, J.R., Couso, I.: Mutual Information-based Feature Selection and Fuzzy Discretization of Vague Data. International Journal of Aproximate Reasoning (2008), http://dx.doi.org/10.1016/

[14] Sanchez, L., Villar, J.R., Couso, I.: Proceedings of the 12th International Conference on Information Processing and Management of Uncertainty in Knowledge-Based Systems. EUSFLAT, Genetic Feature Selection for Fuzzy Discretized Data (2008)

[15] Sedano, J., Villar, J.R., Corchado, E.S., Curiel, L., Bravo, P.M.: The application of a two-step AI model to an Automated Pneumatic Drilling Process. Accepted to be published in the International Journal of Computer Mathematics (2008)

[16] Thangavel, K., Pethalakshmi, A.: Dimensionality reduction based on rough set theory: A review. Applied Soft Computing 9(1), 1–12 (2009)

[17] Uncu, O., Turksen, I.: A novel feature selection approach: Combining feature wrappers and filters. Information Sciences 177, 449–466 (2007)

Non-dominated Multi-objective Evolutionary Algorithm Based on Fuzzy Rules Extraction for Subgroup Discovery

C.J. Carmona[1], P. González[1], M.J. del Jesus[1], and F. Herrera[2]

[1]Department of Computer Science of University of Jaen
{ccarmona,pglez,mjjesus}@ujaen.es
[2]Deparment of Computer Science and AI of University of Granada
herrera@decsai.ugr.es

Abstract. A new multi-objective evolutionary model for subgroup discovery with fuzzy rules is presented in this paper. The method resolves subgroup discovery problems based on the hybridization between fuzzy logic and genetic algorithms, with the aim of extracting interesting, novel and interpretable fuzzy rules. To do so, the algorithm includes different mechanisms for improving diversity in the population. This proposal focuses on the classification of individuals in fronts, based on non-dominated sort. A study can be seen for the proposal and other previous methods for different databases. In this study good results are obtained for subgroup discovery by this new evolutionary model in comparison with existing algorithms.

Keywords: Data mining, Subgroup Discovery, Multi-Objective Evolutionary Algorithms, Fuzzy Rules, Genetic Fuzzy Systems.

1 Introduction

Knowledge Discovery in Databases (KDD) is defined as the non-trivial process of identifying valid, novel, potentially useful, and ultimately understandable patterns in data [8]. Within the KDD process the data mining stage is responsible for high level automatic knowledge discovery using real data. In the KDD process two different tasks can be distinguished: predictive induction, whose objective is the discovery of knowledge for classification or prediction [16]; and descriptive induction, whose main objective is the extraction of interesting knowledge from the data. In descriptive induction, attention can be drawn to the discovery of association rules following an unsupervised learning model [1], subgroup discovery [14][19] and other approaches to non-classificated induction.

Subgroup discovery (SD) is a descriptive induction task [9] whose goal is the discovery of interesting individual patterns in relation to a specific property of interest for the user. The development of new models in this task are focusing in the use of soft computing techniques: genetic algorithm and fuzzy logic.

Genetic Algorithms (GAs) [10] are beginning to be used to solve SD problems [3][7][18] because they offer a set of advantages for knowledge extraction

E. Corchado et al. (Eds.): HAIS 2009, LNAI 5572, pp. 573–580, 2009.

and specifically for rule induction processes, although they were not specifically designed for learning.

In an SD algorithm, a fuzzy approach [20], which considers linguistic variables expressed in linguistic terms through descriptive fuzzy rules, allows us to obtain knowledge in a similar way to human reasoning. The use of fuzzy rules allows us to obtain more interpretable and actionable solutions in the field of SD, and in general in the analysis of data in order to establish relationships and identify patterns [12].

The hybridization between fuzzy logic and GAs, called genetic fuzzy systems (GFSs) [5], has attracted considerable attention in the computational intelligence community. GFSs provide an useful tools for pattern analysis and for the extraction of new types of useful information.

In [7] a mono-objective GFS within the iterative rule learning approach for SD is presented with proper results. In spite of that, the induction of rules describing subgroups can be considered as a multi-objective problem rather than a single objective one, since there are different quality measures which can be used for SD. The different measures used for evaluating a rule can be thought of as different objectives of the SD rule induction algorithm. In this sense, multi-objective evolutionary algorithms (MOEAs) are adapted to solve problems in which different objectives must be optimized [4]. In [3] a multi-objective GFS for SD based on SPEA II algorithm is proposed.

This paper describes a new proposal based on the NSGA-II algorithm for the induction of rules which describe subgroups, the Non-dominated Multi-objective Evolutionary algorithm based on Fuzzy rules extraction for Subgroup Discovery, NMEF-SD, which combines the approximated reasoning capacity of the fuzzy systems with the learning capacities of the MOEAs. This proposal tries to obtain a set of general and interesting fuzzy rules. The generality is obtained both with an operator which performs a biased initialization process and with biased genetic operators, while the diversity in the genetic population is increased with re-initialization based on coverage.

The paper is organized as follows: In Section 2, subgroup discovery is described. The new evolutionary approach to obtain fuzzy rules for SD is explained in Section 3. In Section 4 the results obtained are analyzed. Finally, the conclusions and further research are outlined.

2   Subgroup Discovery

The concept of SD was initially formulated by Klösgen [14] and Wrobel [19], and was defined as: *Given a population of individuals and a property of those individuals we are interested in, find population subgroups that are statistically "most interesting", e.g., are as large as possible and have the most unusual statistical characteristics with respect to the property of interest.*

Therefore, the objective in SD is to discover characteristics of the subgroups by constructing simple individual rules with high support and significance. These rules have the form: *Cond → Class.*

One of the most important aspects of any rule induction approach that describe subgroups is the selection of the quality measures to use. Although there is no consensus about which measures are more adapted for SD, the most common in the literature include: coverage [15], significance [14], unusualness [15], support [15] and confidence [7].

Some of the more interesting models which obtain description of subgroups represented in different forms are the classical deterministic algorithms like Apriori-SD [13] and CN2-SD [15] (available in KEEL[1] software tool [2]), and the evolutionary algorithms SDIGA [7] and MESDIF [3].

3   NMEF-SD: Non-dominated Multi-objective Evolutionary Algorithm Based on the Extraction of Fuzzy Rules for Subgroup Discovery

In this section a new evolutionary model, NMEF-SD, is described. This algorithm extracts descriptive fuzzy or crisp rules -depending on the nature of the features of the problem (continuous and/or nominal variables)- which describe subgroups.

The objective of this evolutionary process is to extract a variable number of different rules describing information of the examples belonging to the original set for each value of the target variable. As the objective is to obtain a set of rules which describe subgroups for all the values of the target variable, the algorithm must be executed as many times as the number of different values the target variable contains.

Each candidate solution is codified according to the "Chromosome = Rule" approach [5], representing only the antecedent part of the rule in the chromosome. The antecedent of a rule is composed of a conjunction of value-variable pairs. A special value is used to indicate that the variable is not considered for the rule. Fig. 1 shows a chromosome and the rule it codifies, for a problem with four features and three possible values for each one.

$$
\begin{array}{c}
Genotype \\
\begin{array}{|c|c|c|c|}
\hline
x_1 & x_2 & x_3 & x_4 \\
\hline
3 & 4 & 1 & 4 \\
\hline
\end{array}
\end{array}
\Rightarrow
\quad
\begin{array}{c}
Phenotype \\
\text{IF } (x_1 = 3) \text{ AND } (x_3 = 1) \text{ THEN } (x_{Obj} = FixedValue)
\end{array}
$$

Fig. 1. Representation of a rule in NMEF-SD

When the features are continuous, the model uses fuzzy rules, and the fuzzy sets corresponding to the linguistic labels are defined by means of the corresponding membership functions. These can be specified by the user or defined by means of a uniform partition if the expert knowledge is not available. In this paper, uniform partitions with triangular membership functions are used.

---

[1] http://www.keel.es

In this extraction process the objective is to obtain interpretable rules with high quality, precision and generality. To do so, two quality measures are selected as objectives:

Support [15]: Is defined as the frequency of correctly classified examples covered by the rule.

$$Sup_c N(R_i) = \frac{n(Class \cdot Cond_i)}{n(Class)} \tag{1}$$

where $n(Class \cdot Cond_i)$ is the number of examples which satisfy the conditions for the antecedent and $n(Class)$ is the number of examples for the target variable indicated in the consequent part of the rule.

Unusualness [15]: Measures the balance between the coverage of the rule and its accuracy gain.

$$WRAcc(R_i) = \frac{n(Cond_i)}{N} \left(\frac{n(Class \cdot Cond_i)}{n(Cond_i)} - \frac{n(Class)}{N} \right) \tag{2}$$

where $n(Cond_i)$ is the number of example which satisfy the antecedent part of the rule, N is the number of examples of the data set, and the weighted relative accuracy of a rule can be described for the coverage using the first part of the expression $\frac{n(Cond_i)}{N}$ and the accuracy gain using the second part $\frac{n(Class \cdot Cond_i)}{n(Cond_i)} - \frac{n(Class)}{N}$.

NMEF-SD is based on the NSGA-II approach [6], and its main purpose is to evolve the population based on the non-dominated sort of the solutions in fronts of dominance. The first front is composed of the non-dominated solutions of the population (the Pareto front), the second is composed of the solutions dominated by one solution, the third of solutions dominated by two, and so on.

The operating scheme of NMEF-SD, can be seen in Fig. 2.

NMEF-SD tries to obtain a set rules with high generality (one of the main objectives of SD) by introducing diversity in the population with different operators, since the diversity in the MOEAs is a handicap for these algorithms. The

```
BEGIN
     Create P_0 with biased initialization
     REPEAT
          Q_t ← ∅
          Tournament Selection (P_t)
          Q_tc ← Multi-point Crossover (P_t)
          Q_tm ← Biased Mutation (Q_tc)
          Q_t ← Q_tc + Q_tm
          Q_t ← Q_t + descendants
          R_t ← Join(P_t,Q_t)
          Fast-non-dominated-sort(R_t)
          IF F_1 evolves
               Introduce fronts in P_{t+1}
          ELSE
               Re-initialization based on coverage P_{t+1}
     WHILE (num-eval < Max-eval)
     RETURN F_1
END
```

Fig. 2. The NMEF-SD algorithm

generality is obtained both with an operator which performs a biased initialization process and with biased genetic operators, while the diversity is introduced with the crowding distance [6] and with re-initialization based on coverage.

In the following subsections the different parts of the algorithm are defined.

3.1 Initialization

The first step of the algorithm is to create an initial population (P_0) whose size is prefixed by an external parameter.

The purpose of this initialization is to generate part of the individuals of the population (75% of the total) using only a maximum percentage of the variables which form part of each rule (25% of the rule). The rest of the variables of the rule and the rest of the individuals of the population are randomly generated.

This operator allows the algorithm to obtain a set of rules with high generality because most of the generated individuals are rules with a low percentage of variables.

3.2 Genetic Operators

The model obtains the descendant population (Q_t), with the same size as the original one, by means of the Tournament Selection [17], Multi-point Crossover [11] and Biased Mutation operators.

Biased Mutation [7] is applied to the gene selected considering the mutation probability. This operator can be applied in two different ways: The first causes the elimination of the variable of the individual, in order to generate a more general rule; and the second randomly mutates the value of the variable. Either one of these two ways can be applied in each mutation, with the same probability.

3.3 Fast Non-dominated Sort

The algorithm joins the populations (P_t and Q_t) in a new population R_t, subsequently applying the non-dominated sort [6] to the new population R_t in order to obtain the classification in fronts of dominance.

This proposal achieves diversity in the population through the crowding distance [5][6] used in the sorting of the individuals belonging to the last front introduced in the main population (P_{t+1}) of the next generation.

3.4 Re-initialization Based on Coverage

The last step of the model is the obtaining of the population for the next generation (P_{t+1}). Before carrying out this step a check is needed on the pareto to see whether or not it evolves. We consider that the pareto evolves if it covers at least one example more than the pareto of the previous generation. If the pareto does not evolve during more than five percent of the evolutive process (quantified through number of evaluations) the re-initialization is performed.

Re-initialization based on coverage performs an elimination of those individuals repeated in the pareto which cover the same examples of the data set.

Table 1. Results of the experimentation

Database	Algorithm	Rul	Var	COV	SIGN	WRAcc	SUP$_c$N	FCNF
Australian	NMEF-SD	3,58	2,92	**0,454**	**23,178**	**0,171**	0,783	**0,930**
	MESDIF	10,00	3,52	0,311	7,594	0,060	0,577	0,807
	SDIGA	**2,68**	3,28	0,310	16,348	0,120	**0,803**	0,591
	CN2-SD	30,50	4,58	0,400	15,350	0,055	0,649	0,830
	AprioriSD	10,00	**2,02**	0,377	16,998	0,074	0,654	0,863
Balance	NMEF-SD	**2,30**	2,00	**0,362**	5,326	**0,070**	**0,530**	**0,698**
	MESDIF	28,10	3,08	0,163	3,516	0,022	0,318	0,557
	SDIGA	7,40	2,39	0,291	5,331	0,049	0,487	0,664
	CN2-SD	15,60	2,23	0,336	**8,397**	0,063	0,512	0,583
	AprioriSD	10,00	**1,20**	0,333	5,444	0,058	0,480	0,649
Echo	NMEF-SD	3,62	2,35	**0,428**	**1,293**	**0,043**	**0,628**	**0,757**
	MESDIF	19,74	3,30	0,164	0,877	0,017	0,355	0,591
	SDIGA	**2,32**	2,27	0,394	1,165	0,013	0,566	0,590
	CN2-SD	17,30	3,23	0,400	1,181	0,019	0,490	0,667
	AprioriSD	9,80	**1,70**	0,194	0,901	0,034	0,226	0,510
Vote	NMEF-SD	**1,10**	2,05	**0,577**	**21,974**	**0,217**	**0,946**	**0,979**
	MESDIF	7,86	3,44	0,429	19,937	0,187	0,827	0,957
	SDIGA	3,06	3,19	0,422	18,243	0,180	0,802	0,891
	CN2-SD	8,00	1,79	0,438	18,830	0,176	0,858	0,932
	AprioriSD	10,00	**1,44**	0,428	17,060	0,147	0,800	0,930

The rest of the individuals are copied in the population of the next generation (P_{t+1}). Repeated individuals which have been eliminated are replaced with new individuals generated through re-initialization based on coverage, introducing individuals which cover previously uncovered examples.

3.5 Stop Condition

The evolutionary process ends when the number of evaluations is reached. Then the algorithm returns the rules in the pareto which reach a predefined confidence [7] value threshold.

4 Experimentation

In order to analyze the behaviour of the proposed model an experimentation with different data sets available in UCI repository[2] has been carried out. The selected data sets have different numbers of features and classes, and different types of features (discrete and continuous). These data sets are: Australian, Balance, Echo and Vote.

As the NMEF-SD model is non-deterministic, it has been run five times, and the mean values of these runs are computed. In addition a 10 cross-validation is performed and the results are compared with those obtained by other SD algorithms: CN2-SD [15], Apriori-SD [13], SDIGA [7] and MESDIF [3].

In this experimentation, the parameters used in NMEF-SD are: a population size of 25 individuals, a maximum number of evaluations of 5000, crossover probability of 0.6 and mutation probability of 0.1.

Table 1 shows the average values obtained by the analyzed methods: number of rules (Rul), number of variables (Var), coverage (COV) [15], significance ($SIGN$) [14], unusualness ($WRAcc$) [15], support (SUP_cN) [15] and fuzzy confidence ($FCNF$) [7]. These measures have been chosen because in previous studies

---

[2] http://www.ics.uci.edu/~mlearn/MLRepository.html

they have been shown to be the most suitable measures for the SD task. The best results are shown in bold characters.

Results in table 1 show that the NMEF-SD model obtains the best results for the quality measures in almost all data sets with respect to the other algorithms. Considering the different characteristics of the data sets, NMEF-SD is able to obtain better results in generality and precision than CN2-SD, Apriori-SD and SDIGA. The results obtained by NMEF-SD are usually the best for the different measures for the databases used.

The subgroups obtained for NMEF-SD (rules and variables), are good, useful and representative. These characteristics, together with the previously mentioned one, make this new model a promising approach for the SD task.

5 Conclusions

In this paper a new multi-objective evolutionary model for the induction of fuzzy rules which describe subgroups is presented. NMEF-SD hybridizes soft-computing techniques like fuzzy logic and the GAs in a MOEA, which is able to obtain high quality results.

The model allows us to obtain small interpretable rule sets which may be fuzzy or crisp depending on the problem. These rules are obtained with a multi-objective model which considers only two quality measures used in SD, with quite good results. Furthermore, the results obtained are better than the results of the classical models considered in this paper.

The proposed model improves on the results obtained for every quality measures with other evolutionary models such as SDIGA and MESDIF. The combination of the NSGA-II approach makes NMEF-SD improves over other.

Moreover, NMEF-SD algorithm tries to obtain a set rules with high generality (one of the main objectives of SD) by introducing diversity in the population with different operators. The generality is obtained both with an operator which performs a biased initialization process and with biased genetic operators, while the diversity in the genetic population is increased with re-initialization based on coverage. All these characteristics make it a good proposal for SD regardless the quality measures the expert considers important for the process.

As future work we will study the use of other rule representation, such as DNF rules, in order to obtain more expressive and understandable results.

Acknowledgments

This work was supported by the Spanish Ministry of Education, Social Policy and Sports under projects TIN-2008-06681-C06-01 and TIN-2008-06681-C06-02, and by the Andalusian Research Plan under project TIC-3928.

References

1. Agrawal, R., Imieliski, T., Swami, A.: Mining association rules between sets of items in large databases. In: SIGMOD 1993, New York, NY, USA, pp. 207–216 (1993)

2. Alcalá-Fdez, J., Sánchez, L., García, S., del Jesus, M.J., Ventura, S., Garrell, J.M., Otero, J., Romero, C., Bacardit, J., Rivas, V.M., Fernández, J.C., Herrera, F.: KEEL: A Software Tool to Assess Evolutionary Algorithms for Data Mining Problems Soft Computing 13(3), 307–318 (2009)
3. Berlanga, F., del Jesus, M.J., González, P., Herrera, F., Mesonero, M.: Multiobjective Evolutionary Induction of Subgroup Discovery Fuzzy Rules: A Case Study in Marketing. In: Perner, P. (ed.) ICDM 2006. LNCS, vol. 4065, pp. 337–349. Springer, Heidelberg (2006)
4. Coello, C.A., Van Veldzhuizen, D.A., Lamont, G.B.: Evolutionary Algorithms for Solving Multi-Objective Problems, 2nd edn. Kluwer Academic Publishers, Dordrecht (2007)
5. Cordón, O., Herrera, F., Hoffmann, F., Magdalena, L.: Genetic Fuzzy Systems: Evolutionary Tuning and Learning of Fuzzy Knowledge Bases (2001)
6. Deb, K., Pratap, A., Agrawal, S., Meyarivan, T.: A fast and elitist multiobjective genetic algorithm: NSGA-II. IEEE Transactions Evolutionary Computation 6(2), 182–197 (2002)
7. del Jesus, M.J., González, P., Herrera, F., Mesonero, M.: Evolutionary Fuzzy Rule Induction Process for Subgroup Discovery: A case study in marketing. IEEE Transactions on Fuzzy Systems 15(4), 578–592 (2007)
8. Fayyad, U.M., Piatetsky-Shapiro, G., Smyth, P.: From data mining to knowledge discovery: an overview. Advances in knowledge discovery and data mining, pp.1–34 (1996)
9. Gamberger, D., Lavraĉ, N.: Expert-Guided Subgroup Discovery: Methodology and Application. Journal Artificial Intelligence Research 17, 501–527 (2002)
10. Golberg, D.E.: Genetic Algorithms in search, optimization and machine learning. Addison-Wesley, Reading (1989)
11. Holland, J.H.: Adaptation in natural and artificial systems. University of Michigan Press (1975)
12. Hüllermeier, E.: Fuzzy methods in machine learning and data mining: Status and prospects. Fuzzy Sets and Systems 156(3), 387–406 (2005)
13. Kavŝek, B., Lavraĉ, N.: APRIORI-SD: Adapting association rule learning to subgroup discovery. Applied Artificial Intelligence 20, 543–583 (2006)
14. Klösgen, W.: Explora: A Multipattern and Multistrategy Discovery Assistant. In: Fayyad, U., et al. (eds.) Advances in Knowledge Discovery and Data Mining, pp. 249–271 (1996)
15. Lavraĉ, N., Kavŝek, B., Flach, P.A., Todorovski, L.: Subgroup Discovery with CN2-SD. Journal of Machine Learning Research 5, 153–188 (2004)
16. Michie, D., Spiegelhalter, D.J., Tayloy, C.C.: Machine Learning. Ellis Horwood (1994)
17. Miller, B.L., Goldberg, D.E.: Genetic Algorithms, Tournament Selection, and the Effects of Noise. Complex System 9, 193–212 (1995)
18. Romero, C., González, P., Ventura, S., del Jesus, M.J., Herrera, F.: Evolutionary algorithm for subgroup discovery in e-learning: A practical application using Moodle data. Expert Systems with Applications 36, 1632–1644 (2009)
19. Wröbel, S.: An Algorithm for Multi-relational Discovery of Subgroups. In: Komorowski, J., Żytkow, J.M. (eds.) PKDD 1997. LNCS, vol. 1263, pp. 78–87. Springer, Heidelberg (1997)
20. Zadeh, L.A.: The concept of a linguistic variable and its applications to approximate reasoning, Parts I, II, III. Information Science 8-9, 199–249, 301–357,43–80 (1975)

A First Study on the Use of Interval-Valued Fuzzy Sets with Genetic Tuning for Classification with Imbalanced Data-Sets

J. Sanz[1,*], A. Fernández[2], H. Bustince[1], and F. Herrera[2]

[1] Dept. of Automatic and Computation, Public University of Navarra, Spain
Tel.:+34-948169839; Fax:+34-948168924
{joseantonio.sanz,bustince}@unavarra.es
[2] Dept. of Computer Science and A.I., University of Granada
{alberto,herrera}@decsai.ugr.es

Abstract. Classification with imbalanced data-sets is one of the recent challenging problems in Data Mining. In this framework, the class distribution is not uniform and the separability between the classes is often difficult. From the available techniques in the Machine Learning field, we focus on the use of Fuzzy Rule Based Classification Systems, as they provide an interpretable model for the end user by means of linguistic variables.

The aim of this work is to increase the performance of fuzzy modeling by adding a higher degree of knowledge by means of the use of Interval-valued Fuzzy Sets. Furthermore, we will contextualize the Interval-valued Fuzzy Sets with a post-processing genetic tuning of the amplitude of their upper bounds in order to enhance the global behaviour of this methodology.

Keywords: Fuzzy Rule-Based Classification Systems, Interval-valued Fuzzy Sets, Tuning, Genetic Algorithms, Imbalanced Data-Sets.

1 Introduction

When facing a classification problem, the user can choose among many techniques to solve it. One of them, known as Fuzzy Rule-Based Classification Systems (FRBCS)[1], is mostly employed because of its interpretability and the possibility of mixing different kinds of information as the one given by experts and the one that comes from mathematical models or empiric measures.

In this work, we will deal with one of the emergent challenging problems in Data Mining [2], the classification with imbalanced data-sets [3]. Specifically, we will focus in the two-class imbalanced problem which appears when one class (known as positive class) is represented by only a few examples, whereas the other (negative class) is described by many instances. Furthermore, it is common that the positive class is the most interesting one from the point of view of the learning task. We can find some recent works in the literature that study the effect of imbalance between the classes in the framework of FRBCSs [4].

---

* Corresponding author.

E. Corchado et al. (Eds.): HAIS 2009, LNAI 5572, pp. 581–588, 2009.

Standard classifier algorithms tend to be biased towards the negative class, since the rules that predict the highest number of examples are rewarded by the accuracy metric. Our aim here is to improve the performance of FRBCSs using the model of Interval-valued Fuzzy Sets (IVFSs) [5]. Specifically, we consider that the success of the use of fuzzy set theory depends on the choice of the membership function (MF) but, when experts do not have precise knowledge of the function to be taken, or it is defined ad-hoc, it can be appropriate to represent the membership degree of each element by means of an interval. Hence, not only vagueness (lack of sharp class boundaries) but also a feature of uncertainty (lack of information) can be addressed intuitively.

We will apply a post-processing step for tuning the amplitude of the upper bounds in the IVFSs, contextualizing the fuzzy partitions for the problem to solve. This is necessary because the data distribution is not necessary uniform and the amplitude of each label may be different.

To build the initial Knowledge Base (KB) we will employ the Chi et al.'s method [6] and we will compare the IVFS methodology (with and without tuning) against the results obtained with this initial KB. Furthermore, we will include the C4.5 decision tree in our experimental study, since it is an algorithm of reference in the field of imbalanced data-sets [7,8]. To do so, we will employ forty four data-sets from UCI repository [9], where multi-class data sets are modified to obtain two-class non-balanced problems, defining the joint of one or more classes as positive and the joint of one or more classes as negative. To evaluate our results we have applied the Area Under the Curve (AUC) metric [10] carrying out some non-parametric tests [11,12] to show the significance in the performance improvements obtained.

This work is organized as follows: in Section 2 we describe the problem of imbalanced data-sets. In Section 3 we define the IVFS model. Section 4 introduces our experimentation framework and shows the experimental study. In Section 5 we summarize the study carried out.

2   Imbalanced Data-Sets in Classification

The problem of imbalanced data-sets in classification [3] occurs when the class distribution is not uniform. In this situation, the number of examples that represents one class of the data-set (usually the concept of interest) is much lower than that of the other class. This situation has been recently identified as one important problem in data mining, since it is implicit in most real applications including telecommunications, finances, biology or medicine.

This scenario may suppose an added difficulty for the identification and discovery of rules covering the under-represented samples. In [4], the authors studied different configurations for FRBCSs in order to determine the most suitable model in this classification framework. Furthermore, it is shown the necessity to apply a re-sampling procedure; specifically, the "Synthetic Minority Oversampling Technique" (SMOTE) [13] obtains a very good behaviour.

As we stated before, most of proposals for automatic learning of classifiers use some kind of accuracy measure as the classification percentage over the example set. However, these measures can lead to erroneous conclusions over imbalanced data-sets since they don't take into account the proportion of examples for each class. Therefore, in this work we use the AUC metric [10], defined as

$$AUC = \frac{1 + TP_{rate} - FP_{rate}}{2},\qquad(1)$$

where TP_{rate} and FP_{rate} are the percentage of correctly and wrongly classified cases belonging to the positive class respectively.

3   Interval-Valued Fuzzy Sets and Amplitude Tuning

In this work we want to improve the performance of FRBCSs applying IVFS to represent the different fuzzy partitions. We will use the Chi et al.'s rule learning algorithm [6], where we represent fuzzy rules as:

Rule R_j : If x_1 is A_{j1} and ... and x_n is A_{jn} then Class $= C_j$ with RW_j,

$$(2)$$

where R_j is the label of the jth rule, $x = (x_1, \ldots, x_n)$ is an n-dimensional pattern vector, A_{ji} is an antecedent fuzzy set, C_j is a class label, and RW_j is the rule weight. We represent the MFs by triangular functions.

In the remaining of this section, we will first describe the IVFSs model and then we will present the genetic tuning of the amplitude for the fuzzy labels.

3.1   IVFSs Model

The IVFSs [5] are an extension of the theory of fuzzy sets which enables to manage additional knowledge in the fuzzy partitions. In the following we define this model with some detail:

We denote by $L([0,1])$ the set of all closed subintervals of the closed interval $[0,1]$; that is: $L([0,1]) = \{\mathbf{x} = [\underline{x}, \overline{x}] | (\underline{x}, \overline{x}) \in [0,1]^2 \text{ and } \underline{x} \leq \overline{x}\}$. $L([0,1])$ is a partially ordered set with respect to the relation $\leq_L$ defined in the following way; given $\mathbf{x}, \mathbf{y} \in L([0,1])$: $\mathbf{x} \leq_L \mathbf{y}$ if and only if $\underline{x} \leq \underline{y}$ and $\overline{x} \leq \overline{y}$. $(L([0,1]), \leq_L)$ is a complete lattice where the smallest element is $0_L = [0,0]$ and the largest is $1_L = [1,1]$.

Definition 1. *An Interval-valued fuzzy set (IVFS) A on the universe $U \neq \emptyset$ is a mapping $A : U \to L([0,1])$.*

Obviously, $A(u) = [\underline{A}(u), \overline{A}(u)] \in L([0,1])$ is the membership degree of $u \in U$.

We generate the initial KB following a simple rule learning algorithm (Chi el al.'s method in this case) and, from this KB, we include the IVFSs model by adding an upper bound for each fuzzy partition, centered in the maximum of the MF and with a higher amplitude. In our initial model, the amplitude of the upper bound will be 50% greater than that of the lower bound. We must remark that we note "upper" and "lower" bounds referring to the corresponding fuzzy labels.

Now, we are working with an interval when computing the matching degree between the antecedent of the rule and the example. In order to give a single output value, we obtain the mean between the lower and the upper matching degrees. Furthermore, the rule weight is composed by two numbers, associated to the lower and the upper bound respectively, and the same procedure will be employed in this case. Specifically, the rule weight is computed using the Penalized Certainty Factor defined in [14] as:

$$CF_{Lj} = \frac{\sum\limits_{x_p \in Class\,C_j} \underline{A_j}(x_p) - \sum\limits_{x_p \notin Class\,C_j} \underline{A_j}(x_p)}{\sum\limits_{p=1}^{m} \underline{A_j}(x_p)} \tag{3}$$

Note that for the rule weight computation of the upper bound, we may only replace $\underline{A_j}(x_p)$ with $\overline{A_j}(x_p)$.

3.2 Genetic Tuning of the Amplitude of Upper Bound of the IVFS

To improve the performance of the initial IVFSs model, we have to contextualize the fuzzy partitions for each problem. To do so, we propose a genetic tuning approach to perform slight changes of the original upper bound amplitude.

The modification of the amplitude is given by a number within the interval [0, 1], that is, from the overlapping of both bounds (value 0) to twice the amplitude of the upper with respect to the lower bound (value 1). The amplitude of the upper bound will be uniformly increased according to intermediate values.

In order to apply the genetic tuning, we will consider the use of CHC algorithm [15], which presents a good trade-off between diversity and convergence, being a good choice in complex problems. The components needed to design this process are explained below:

1. *Coding Scheme:* A real coding is considered, where each gene of the chromosome represents the amplitude modifier as defined above. Thus, there are as many genes as fuzzy partitions in the Data Base.
2. *Chromosome Evaluation:* The fitness function is the AUC metric.
3. *Initial Gene Pool:* The initial pool is obtained with the first individual having all genes with value '0.5' (the initial FRBCS). The second and the third individuals having all genes with values 0 and 1 respectively, whereas the remaining individuals are generated at random in [0, 1].
4. *Crossover Operator:* We consider the Parent Centric BLX (PCBLX) operator, which is based on the BLX-α. We consider the incest prevention mechanism, checking and modifying an initial threshold, in order to apply the PCBLX operator.
5. *Restarting approach:* When the threshold value is lower than zero, all the chromosomes are regenerated at random within the interval [0, 1]. Furthermore, the best global solution found is included in the population to increase the convergence of the algorithm.

4   Experimental Study

In this study, our intention is to show the improvement achieved in FRBCSs applying the IVFSs model. To do this we have to do a double analysis:

- We want to analyze whether the IVFSs model enhances the performance of a simple KB.
- We want to determine the significance of the tuning step in the IVFSs model.

In the remaining of this section, we will first present the experimental framework and all the parameters employed in this study and then we will show the empirical study for the IVFSs model in imbalanced data-sets.

4.1   Experimental Set-Up

To carry out the different experiments we consider a *5-folder cross-validation model*, i.e., 5 random partitions of data with a 20%, and the combination of 4 of them (80%) as training and the remaining one as test. For each data-set we consider the average results of the five partitions. Furthermore, a Wilcoxon's Signed-Ranks Test [16] is used for statistical comparison of our empirical results. In all cases the level of confidence (α) will be set at 0.05.

Table 1. Summary Description for Imbalanced Data-Sets

Data-set	#Ex.	#Atts.	Class (min., maj.)	%Class(min.; maj.)
Glass1	214	9	(build-win-non_float-proc; remainder)	(35.51, 64.49)
Ecoli0vs1	220	7	(im; cp)	(35.00, 65.00)
Wisconsin	683	9	(malignant; benign)	(35.00, 65.00)
Pima	768	8	(tested-positive; tested-negative)	(34.84, 66.16)
Iris0	150	4	(Iris-Setosa; remainder)	(33.33, 66.67)
Glass0	214	9	(build-win-float-proc; remainder)	(32.71, 67.29)
Yeast1	1484	8	(nuc; remainder)	(28.91, 71.09)
Vehicle1	846	18	(Saab; remainder)	(28.37, 71.63)
Vehicle2	846	18	(Bus; remainder)	(28.37, 71.63)
Vehicle3	846	18	(Opel; remainder)	(28.37, 71.63)
Haberman	306	3	(Die; Survive)	(27.42, 73.58)
Glass0123vs456	214	9	(non-window glass; remainder)	(23.83, 76.17)
Vehicle0	846	18	(Van; remainder)	(23.64, 76.36)
Ecoli1	336	7	(im; remainder)	(22.92, 77.08)
New-thyroid2	215	5	(hypo; remainder)	(16.89, 83.11)
New-thyroid1	215	5	(hyper; remainder)	(16.28, 83.72)
Ecoli2	336	7	(pp; remainder)	(15.48, 84.52)
Segment0	2308	19	(brickface; remainder)	(14.26, 85.74)
Glass6	214	9	(headlamps; remainder)	(13.55, 86.45)
Yeast3	1484	8	(me3; remainder)	(10.98, 89.02)
Ecoli3	336	7	(imU; remainder)	(10.88, 89.12)
Page-blocks0	5472	10	(remainder; text)	(10.23, 89.77)
Yeast2vs4	514	8	(cyt; me2)	(9.92, 90.08)
Yeast05679vs4	528	8	(me2; mit,me3,exc,vac,erl)	(9.66, 90.34)
Vowel0	988	13	(hid; remainder)	(9.01, 90.99)
Glass016vs2	192	9	(ve-win-float-proc; build-win-float-proc, build-win-non_float-proc,headlamps)	(8.89, 91.11)
Glass2	214	9	(Ve-win-float-proc; remainder)	(8.78, 91.22)
Ecoli4	336	7	(om; remainder)	(6.74, 93.26)
Yeast1vs7	459	8	(nuc; vac)	(6.72, 93.28)
Shuttle0vs4	1829	9	(Rad Flow; Bypass)	(6.72, 93.28)
Glass4	214	9	(containers; remainder)	(6.07, 93.93)
Page-blocks13vs2	472	10	(graphic; horiz.line,picture)	(5.93, 94.07)
Abalone9vs18	731	8	(18; 9)	(5.65, 94.25)
Glass016vs5	184	9	(tableware; build-win-float-proc, build-win-non_float-proc,headlamps)	(4.89, 95.11)
Shuttle2vs4	129	9	(Fpv Open; Bypass)	(4.65, 95.35)
Yeast1458vs7	693	8	(vac; nuc,me2,me3,pox)	(4.33, 95.67)
Glass5	214	9	(tableware; remainder)	(4.20, 95.80)
Yeast2vs8	482	8	(pox; cyt)	(4.15, 95.85)
Yeast4	1484	8	(me2; remainder)	(3.43, 96.57)
Yeast1289vs7	947	8	(vac; nuc,cyt,pox,erl)	(3.17, 96.83)
Yeast5	1484	8	(me1; remainder)	(2.96, 97.04)
Ecoli0137vs26	281	7	(pp,imL; cp,im,imU,imS)	(2.49, 97.51)
Yeast6	1484	8	(exc; remainder)	(2.49, 97.51)
Abalone19	4174	8	(19; remainder)	(0.77, 99.23)

Table 2. Results for FRBCSs and C4.5 in imbalanced data-sets. By columns we represent the Chi et al.'s algorithm with the lower bound (Chi_Low), Chi with the upper bound (Chi_Up), IVFSs model (Chi_IVFS), IVFS with tuning (Chi_IVFS_tun) and C4.5.

Data-set	Chi_Low		Chi_Up		Chi_IVFS		Chi_IVFS_tun		C4.5	
	AUC_{Tr}	AUC_{Tst}	AUC_{Tr}	AUC_{Tst}	AUC_{Tr}	AUC_{Tst}	AUC_{Tr}	AUC_{Tst}	AUC_{Tr}	AUC_{Tst}
Glass1	75.54	65.53	69.28	67.76	72.06	66.31	80.37	71.75	89.78	75.77
Ecoli0vs1	95.61	92.71	98.05	96.04	96.42	94.04	98.54	95.38	99.27	97.96
Wisconsin	98.07	89.19	97.08	96.03	97.32	96.03	98.24	96.43	98.32	95.45
Pima	72.64	67.66	67.23	65.93	70.03	67.69	76.04	70.80	84.11	71.45
Iris0	100.00	100.00	100.00	100.00	100.00	100.00	100.00	100.00	100.00	99.00
Glass0	71.68	69.74	72.18	72.18	72.01	71.83	73.92	71.81	94.33	78.56
Yeast1	70.12	69.44	67.57	67.81	68.55	68.71	72.96	70.97	80.49	71.09
Vehicle1	76.86	71.40	71.71	69.34	72.55	69.26	78.87	71.80	95.51	70.30
Vehicle2	88.13	85.55	83.30	81.45	83.83	82.29	93.56	88.28	98.95	94.92
Vehicle3	75.96	69.51	71.66	67.28	71.72	67.19	77.49	69.34	94.93	74.44
Haberman	66.98	60.60	65.79	55.06	67.13	59.89	70.75	61.06	74.26	63.09
Glass0123vs456	94.08	86.42	93.18	90.09	94.13	92.02	96.29	92.33	99.08	90.32
Vehicle0	88.75	86.96	80.82	80.07	82.27	81.62	90.98	87.13	98.97	91.18
Ecoli1	87.95	85.88	90.35	88.51	90.31	87.91	91.84	85.72	96.31	77.55
New-Thyroid2	94.80	90.60	96.73	96.31	97.15	90.87	99.51	98.02	99.57	96.59
New-Thyroid1	92.60	88.33	97.73	96.31	95.16	90.04	99.30	96.31	99.22	98.02
Ecoli2	89.68	88.26	89.09	87.68	89.14	88.34	91.54	87.44	95.17	91.62
Segment0	95.53	95.07	86.39	86.31	91.37	91.38	97.69	96.64	99.85	99.27
Glass6	95.07	84.69	92.14	84.74	93.99	85.01	97.33	84.42	99.59	84.50
Yeast3	91.43	90.22	85.35	84.55	88.21	86.69	93.15	91.21	95.65	88.76
Ecoli3	89.38	87.84	88.17	86.60	88.19	86.95	91.98	90.30	98.15	89.21
Page-Blocks0	81.89	81.40	80.34	80.32	81.22	80.95	84.35	83.68	98.46	94.85
Yeast2vs4	89.68	87.36	89.39	88.28	90.12	88.25	91.04	88.15	98.14	85.88
Yeast05679vs4	82.65	79.07	81.48	78.75	82.18	78.59	85.58	77.80	95.26	76.02
Vowel0	98.57	98.39	97.95	97.77	98.19	98.16	99.21	98.83	99.67	94.94
Glass016vs2	62.71	54.17	63.93	61.50	64.00	60.93	66.50	55.98	97.16	60.62
Glass2	66.54	55.30	67.50	68.28	67.43	67.76	70.16	60.38	95.71	54.24
Ecoli4	94.06	91.51	94.58	91.17	92.60	89.77	95.82	91.20	97.69	83.10
yeast1vs7	82.00	80.63	78.67	78.61	80.89	78.84	83.44	80.39	93.51	70.03
shuttle0vs4	100.00	99.12	100.00	99.57	100.00	99.57	100.00	99.57	99.99	99.97
Glass4	95.27	85.70	92.29	84.79	93.35	86.80	96.33	89.54	98.44	85.08
Page-Blocks13vs4	93.68	92.05	86.60	83.30	85.96	82.62	97.19	94.95	99.75	99.55
Abalone9-18	70.23	64.70	65.32	63.77	65.72	63.99	72.41	68.51	95.31	62.15
Glass016vs5	90.57	79.71	75.93	75.71	79.00	78.00	93.79	92.29	99.21	81.29
shuttle2vs4	95.00	90.78	100.00	98.78	98.88	95.58	98.98	96.38	99.90	99.17
Yeast1458vs7	71.25	64.65	68.25	62.76	68.63	66.10	74.80	60.78	91.58	53.67
Glass5	94.33	83.17	77.13	75.85	83.17	80.49	95.30	94.15	99.76	88.29
Yeast2vs8	78.61	77.28	77.39	77.39	77.39	77.39	79.24	77.39	91.25	80.66
Yeast4	83.58	83.15	85.42	84.34	85.65	83.53	87.15	82.18	91.01	70.04
Yeast1289vs7	74.70	77.12	76.18	75.75	76.71	76.79	80.27	78.24	94.65	68.32
Yeast5	94.68	93.58	96.48	96.49	96.70	96.63	97.46	96.01	97.77	92.33
Ecoli0137vs26	93.96	81.90	86.80	83.18	91.81	82.63	96.60	82.27	96.78	81.36
Yeast6	88.48	88.09	87.84	87.63	89.28	89.43	90.58	87.41	92.42	82.80
Abalone19	71.44	63.94	66.58	65.29	68.35	65.73	73.59	61.45	85.44	52.02
Global	85.56	81.33	83.18	81.35	84.06	81.65	88.18	**83.51**	95.46	82.17

We have selected forty-four data-sets from UCI repository [9]. The data are summarized in Table 1, showing the number of examples (#Ex.), and attributes (#Atts.), class name (minority and majority) and class attribute distribution.

In order to reduce the effect of imbalance, we will employ the SMOTE preprocessing method [13] for all our experiments balancing both classes to the 50% distribution.

We will employ the following configuration for the FRBCS: 3 labels per fuzzy partition, product T-norm as conjunction operator, together with the Penalized Certainty Factor approach for the rule weight and Fuzzy Reasoning Method of the winning rule. We have selected this fuzzy model as it achieved a good performance in previous studies for FRBCS on imbalanced data-sets [4].

The specific parameters for the genetic tuning of the amplitude are listed below:

- Number of evaluations: 5000 · number of variables.
- Population Size: 50 individuals.
- Number of Bits per Gene (for the gray codification): 30 bits.

Table 3. Wilcoxon's test to compare the IVFSs model (R^+), with and without tuning, against the Chi et al. method and C4.5 (R^-) in imbalanced data-sets

Comparison	R^+	R^-	Hypothesis ($\alpha = 0.05$)	p-value
Base IVFS				
Chi_IVFS vs. Chi_Low	506	440	Not Rejected	0.690
Chi_IVFS vs. Chi_Up	511	309	Not Rejected	0.175
Chi_IVFS vs. C4.5	435	555	Not Rejected	0.484
IVFS with Genetic Tuning of the Amplitude				
Chi_IVFS_tun vs. Chi_Low	798	148	Rejected for Chi_IVFS_tun	0.000
Chi_IVFS_tun vs. Chi_Up	596	224	Rejected for Chi_IVFS_tun	0.012
Chi_IVFS_tun vs. Chi_IVFS	642	149	Rejected for Chi_IVFS_tun	0.006
Chi_IVFS_tun vs. C4.5	611	379	Not Rejected	0.176

4.2 Analysis of the IVFSs Performance on Imbalanced Data-Sets

In the first part of our study, our aim is to analyze whether the use of the IVFSs improves the FRBCS performance by means of the comparison with the results obtained by the Chi et al.'s method, considering two amplitude values in the Data Base: using the standard MF ("lower bound" in IVFS) and a fuzzy label with a higher amplitude ("upper bound"). These results are shown in Table 2.

We observe the good behaviour of the IVFSs model, since it obtains very good results in most data-sets of the study. In order to check for significant differences between this approach and the basic FRBCSs, we carry out a Wilcoxon test (shown in Table 3) in which we observe that the rankings are very similar in all cases, concluding that the different methods have a similar performance.

When we apply the genetic tuning step, the results are enhanced considerably, obtaining the best mean result among all the algorithms of this study. The statistical analysis (also shown in Table 3) shows the goodness of this approach, since it have better behavior than the basic FRBCSs and the initial IVFSs model. Regarding C4.5, we achieve a higher ranking in this case, which implies that the IVFS with genetic tuning is a suitable methodology in order to deal with imbalanced data-sets with fuzzy models.

5 Conclusions

In this work we have analyzed the behavior of the IVFSs in the context of imbalanced data-sets. We start from an initial KB generated by a simple fuzzy rule learning method and we add a new level of fuzzy partitions in order to manage a higher knowledge for the problem.

Our experimental results have determined the goodness of this model, achieving better results than the base FRBCS. Furthermore, we have applied a post-processing step to adapt the amplitude of the upper bounds in order to contextualize this knowledge for each specific data-set by means of a genetic tuning. We have determined empirically that this methodology enhances our initial IVFSs model, outperforming the initial FRBCS and being highly competitive with the well-known C4.5 decision tree.

Acknowledgment

This work has been supported by the Spanish Ministry of Science and Technology under projects TIN2008-06681-C06-01 and TIN2007-65981.

References

1. Ishibuchi, H., Nakashima, T., Nii, M.: Classification and modeling with linguistic information granules: Advanced approaches to linguistic Data Mining. Springer, Heidelberg (2004)
2. Yang, Q., Wu, X.: 10 challenging problems in data mining research. International Journal of Information Technology and Decision Making 5(4), 597–604 (2006)
3. Chawla, N.V., Japkowicz, N., Kolcz, A.: Editorial: special issue on learning from imbalanced data sets. SIGKDD Explorations 6(1), 1–6 (2004)
4. Fernández, A., García, S., del Jesus, M.J., Herrera, F.: A study of the behaviour of linguistic fuzzy rule based classification systems in the framework of imbalanced data-sets. Fuzzy Sets and Systems 159(18), 2378–2398 (2008)
5. Bustince, H., Montero, J., Barrenechea, E., Gomez, D.: A survey of Interval-Valued Fuzzy Sets. In: Handbook of Granular Computing. Addison-Wesley, Reading (2008)
6. Chi, Z., Yan, H., Pham, T.: Fuzzy algorithms with applications to image processing and pattern recognition. World Scientific, Singapore (1996)
7. Su, C.T., Hsiao, Y.H.: An evaluation of the robustness of MTS for imbalanced data. IEEE Transactions on Knowledge Data Engineering 19(10), 1321–1332 (2007)
8. Batista, G.E.A.P.A., Prati, R.C., Monard, M.C.: A study of the behaviour of several methods for balancing machine learning training data. SIGKDD Explorations 6(1), 20–29 (2004)
9. Asuncion, A., Newman, D.: UCI machine learning repository, University of California, Irvine, School of Information and Computer Sciences (2007), http://www.ics.uci.edu/~mlearn/MLRepository.html
10. Huang, J., Ling, C.X.: Using AUC and accuracy in evaluating learning algorithms. IEEE Transactions on Knowledge and Data Engineering 17(3), 299–310 (2005)
11. Demšar, J.: Statistical comparisons of classifiers over multiple data sets. Journal of Machine Learning Research 7, 1–30 (2006)
12. García, S., Herrera, F.: An Extension on Statistical Comparisons of Classifiers over Multiple Data Sets for all Pairwise Comparisons. Journal of Machine Learning Research 9, 2677–2694 (2008)
13. Chawla, N.V., Bowyer, K.W., Hall, L.O., Kegelmeyer, W.P.: Smote: Synthetic minority over-sampling technique. Journal of Artificial Intelligent Research 16, 321–357 (2002)
14. Ishibuchi, H., Yamamoto, T.: Rule weight specification in fuzzy rule-based classification systems. IEEE Transactions on Fuzzy Systems 13, 428–435 (2005)
15. Eshelman, L.J.: The CHC adaptive search algorithm: How to have safe search when engaging in nontraditional genetic recombination. In: Foundations of Genetic Algorithms, pp. 265–283. Morgan Kaufmann, San Francisco (1991)
16. García, S., Fernández, A., Luengo, J., Herrera, F.: A study of statistical techniques and performance measures for genetics-based machine learning: Accuracy and interpretability. Soft Computing 13(10), 959–977 (2009)

Feature Construction and Feature Selection in Presence of Attribute Interactions*

Leila S. Shafti and Eduardo Pérez

Universidad Autónoma de Madrid, 28049 Madrid, Spain

Abstract. When used for data reduction, feature selection may success-fully identify and discard irrelevant attributes, and yet fail to improve learning accuracy because regularities in the concept are still opaque to the learner. In that case, it is necessary to highlight regularities by constructing new characteristics that abstract the relations among at-tributes. This paper highlights the importance of feature construction when attribute interaction is the main source of learning difficulty and the underlying target concept is hard to discover by a learner using only primitive attributes. An empirical study centered on predictive accu-racy shows that feature construction significantly outperforms feature selection because, even when done perfectly, detection of interacting at-tributes does not sufficiently facilitates discovering the target concept.

Keywords: feature construction, data reduction, genetic algorithms.

1 Introduction

An important preprocessing step in data mining and machine learning is data size reduction. A data reduction method aims to transform the original data set into a smaller one to achieve two major goals: saving running time in further data mining and machine learning steps, and improving the quality of data to achieve more accurate results in next steps. Instance replacement is one of the techniques used for data reduction when the data representation is primitive and some attributes are redundant or irrelevant [1]. It transforms the original representation of data into a new one with a smaller data dimension. This can be achieved by replacing original attributes with more abstract ones and/or eliminating redundant or irrelevant attributes.

Feature Selection (FS) and Feature Construction (FC) methods have been used for instance replacement. FS aims to reduce the data dimension by remov-ing redundant and irrelevant attributes. FC maps the original representation of data into a new one by constructing new attributes or *features* from original at-tributes. When FC is used for data reduction, each constructed feature replaces several attributes or an attribute with a larger amount of attribute values.

Recently, more attention has been devoted to FS [2]. FS is an important pre-processing step when irrelevant attributes are introduced; however, it is not al-ways enough for improving learning. This paper highlights the importance of FC

---

* Supported by the Spanish Ministry of Science and Technology, TIN2008-02081.

E. Corchado et al. (Eds.): HAIS 2009, LNAI 5572, pp. 589–596, 2009.

when primitive data representation makes relevant information difficult to discover, and reports an empirical comparison of the FC performed by MFE3/GA^{DR} and a perfect FS. Section 2 explains the significance of FC in presence of attribute interactions. Section 3 reviews FC methods, MFE3/GA and MFE3/GA^{DR}. Previous work [3] showed the advantages of MFE3/GA when applied to primitive data with interactions. The use of genetic algorithm (GA) as a global search helps MFE3/GA to deal with the problem of interactions better than other FCs. The constructed features successfully capture and encapsulate relations among attributes when the only available knowledge about the concept is primitive training data. MFE3/GA^{DR} is a data reduction FC that inherits several aspects from MFE3/GA which makes it distinguishable [4]. The empirical comparison of FS and FC is reported in Sect. 4. Conclusions are summarized in Sect. 5.

2   FS and FC in Presence of Attribute Interaction

Most learners assume attribute independence and consider attributes one by one. Consequently, these algorithms achieve high accuracy when available domain knowledge provides a data representation based on highly informative attributes, as in many of the UCI Databases used to benchmark machine learning algorithms [5,6]. Otherwise, their performance degrades. Since most real-world data are not particularly prepared for machine learning purposes, their representations are often primitive and not appropriate [7,8]. The primitive representation facilitates the existence of attribute interactions whose complexity makes the relevant information opaque to most learners. Interaction exists among attributes when the relation between one attribute and the target concept is not constant for all values of the other attributes [9,10]. Interaction becomes complex when changing the value of one attribute does not only change the relation between another attribute and the target concept, but also yields an opposite relation. Hence, there appears to be no relation between each attribute and the concept. Since an attribute by itself gives no information about the target concept, each attribute individually does not help to uncover the underlying complex patterns that define the target concept. The values of all attributes need to be considered simultaneously to predict the concept. Thus, the interaction complexity augments with an increasing number of interacting attributes.

Interaction becomes a stronger hindrance for learners when data contains irrelevant attributes, since interacting attributes can be easily mistaken as irrelevant attributes [10]. In this case, FS is crucial for highlighting the importance of the interacting attributes to the learner. Distinguishing interacting attributes from irrelevant ones has recently received attention, and FS methods have been designed to tackle attribute interaction problem [11]. These methods consider several attributes together and apply heuristics to distinguish between *subsets* of interacting and irrelevant attributes.

However, when interactions are complex, identifying relevant attributes may not be sufficient for improving learning accuracy. Interactions need to be highlighted in order to discover underlying complex patterns that define the target concept. When complex interaction exists among attributes, concept is dispersed;

i.e., each class label is scattered through many small regions of the instance space, each covering relatively few instances of the same class. Thus, regularities are less prominent. So, even if interacting attributes are identified, the underlying structure of the concept is still difficult to detect. This problem worsens when few training data are available. Moreover, when primitive attributes are provided for representing data, the concept description that is to be generated using such attributes tends to be large and complex [12]. So, it is likely that the learner makes mistakes in constructing such description. Thus, its accuracy will be low.

FC methods have been used to facilitate learning from primitive data representation. FC aims to generate more predictive attributes derived from the primitive attributes to improve a particular learner's performance [13]. An FC either replaces original attributes with constructed features or adds features as new attributes to the original set. If original attributes are replaced by more abstract features, FC reduces the data size and improves learning performance.

When FC is applied to concepts with complex interactions, it aims to discover opaque information about the relations between subsets of interacting attributes and the target concept. The discovered information is abstracted into a constructed feature. The new feature groups data samples of the same class, which could be scattered in the original data space. If FC finds the appropriate features, such a change of representation makes the instance space less dispersed and highly regular; thus, the concept is easy to learn.

3   Data Reduction Using MFE3/GA

MFE3/GA is an FC preprocessing method that highlights the interactions among attributes by constructing new features [3]. It receives as input the training data and the original attribute set; then, it searches through the space of attribute subsets to find subsets of interacting attributes and a function defined over each of the subsets found. The more promising functions are then added as new features to the original attribute set; and the new representation of data is given to a standard learner such as C4.5 [14] to proceed with the data mining process. The current version assumes that the class labels are binary and continuous attributes are transformed to nominal ones before running the system.

The search space for finding relevant attributes and constructing functions is large and with high variation when complex interactions exist in the concept. The importance of applying a global search to such a space for constructing features has been shown [15]. MFE3/GA uses GA as a global search to successfully find the optimal solution. Each individual in MFE3/GA is designed to represent a set of k attribute subsets. Each subset is represented by a bit-string of length N, where each bit shows the presence or absence (in the subset) of one of the N original attributes. Thus, each individual is a bit-string of length $k \cdot N$ ($k > 0$). Since each individual has different number of subsets, the length of individuals is variable. To avoid unnecessary growth of individuals, the number of subsets in each individual is limited by a user defined parameter K (by default $K = 5$).

Each attribute subset in an individual is associated with a function defined over the attributes in the subset and induced from the data. Such functions are expressed by a non-algebraic (operator-free) representation; that is, no algebraic operator is used for representing functions. Non-algebraic representation has been used successfully for expressing constructed functions by other FC methods [16,17,18]. This form of representation permits extracting part of the function's description from data and inducing the rest. The function f_i for any given subset $S_i = \{x_{i_1}, \ldots, x_{i_m}\}$ is defined by assigning binary class labels (as outcomes of the function) to all the tuples in the Cartesian product of attributes $x_{i_1} \times \ldots \times x_{i_m}$ (as inputs of the function). The class label assigned to each tuple t_j depends on the class labels of the training samples that match the tuple. A sample matches a tuple t_j if its values for attributes in S_i are equal to the corresponding values in the jth combination of attribute values in the Cartesian product. The class labels are assigned as discussed case by case next (see Fig. 1 for an example):

> *Case 1: Unknown tuple.* If there are no training samples matching t_j, a class label is assigned to $f_i(t_j)$ stochastically, according to the class distribution in the training data.
> *Case 2: Pure tuple.* If all training samples matching t_j belong to the same class, this is the class assigned to $f_i(t_j)$.
> *Case 3: Mixed tuple.* If there is a mixture of classes in the samples matching t_j, the class assigned to $f_i(t_j)$ depends on the numbers of tuples labeled by Case 2 as positive (class label '1') and negative (class label '0'), p_2 and n_2 respectively. If $p_2 > n_2$, the negative class is assigned; and otherwise, the positive class is assigned to $f_i(t_j)$

Note that the label of mixed tuples depends on the definitive labels in the function under construction, which are the labels of pure tuples. The opposite label to the most frequent label among pure tuples is selected. When all tuples' labels are defined, the function f_i is represented by a vector of values that shows the outcome of the function for each combination of attribute values in S_i.

GA aims to converge the population members toward the set of attribute subsets and their corresponding functions that best represent attribute interactions. Genetic operators are performed over attribute subsets. However, each attribute subset determines a function; thus, changing a subset in an individual by operators implies changing the associated function. The fitness measure evaluates the inconsistency and complexity of constructed functions based on

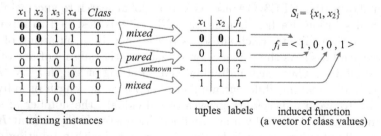

Fig. 1. Extracting function f_i, defined over $S_i = \{x_1, x_2\}$, from the training data

MDL principle [19]. The fitness of each individual $Ind = \langle S_1, \ldots, S_k \rangle$ is determined by evaluating the set of corresponding functions $\{f_1, \ldots, f_k\}$ as a theory and measuring two factors: the inconsistency of the set with the training data (exceptions produced representing data with the theory) and its complexity (the code needed to represent the theory). For details about the MFE3/GA's genetic operators and fitness evaluation, the reader is referred to [3,20]. When GA finishes the constructed features are *added* to the original attribute set. The output of MFE3/GA is a new data set represented using the new set of attributes which is then given to a standard learner such as C4.5 for learning.

Theoretical and empirical analysis in [3] proved that MFE3/GA successfully captures the relations among attributes and encapsulates them into a set of new features. Therefore, a learner can exploit the new representation to easily learn concepts that it could not learn accurately by using the original representation. Although, this approach increases the dimension of data by adding new features, MFE3/GA can be effective in data reduction too, as shown in [4]. We modified the output of MFE3/GA and called the new system MFE3/GA^{DR}. When GA finishes, MFE3/GA^{DR} *replaces* the original attribute set with the set of constructed features. Then, data are redescribed using the constructed feature set only; and, the new representation is given to C4.5 to proceed learning. Note that MFE3/GA constructs a set of maximum K boolean features where K is a user defined parameter. Replacing the original N nominal attributes, each of n_i possible values, with K boolean features by MFE3/GA^{DR} reduces the representation complexity from $\prod_{i=1}^{N} n_i$ to 2^K, or less since the actual number of constructed features is bound by K. Since the features constructed by MFE3/GA encapsulate the relations among original attributes and are highly informative, MFE3/GA^{DR} increases the accuracy after data reduction, as shown in [4].

4 Empirical Study

This section describes experiments performed to empirically compare FC and FS when complex interactions exist among attributes and few training data instances are provided. Synthetic concepts with complex interactions are used for experiments to allow analyzing results deeply. MFE3/GA^{DR} is applied as a preprocessing FC method to transform the representation of these concepts to a smaller one where the structure of the concept is more apparent. After data reduction by FC, C4.5 [14] is used as a standard learner to evaluate the predictive accuracy. To compare the results with those obtained after a perfect preprocessing FS method, for each concept, relevant attributes are selected manually. Then, C4.5 is forced to only use selected attributes for learning. Thus, the predictive accuracy is measured using the data representation obtained by a *perfect FS*.

Concepts are defined over 12 boolean attributes (see Appendix for concept definitions). These concepts represent sources of difficulty that appear in real-world domains when primitive attributes are used for representing data. Columns 1 to 4 of Table 1 give a summary of concepts. The numbers of relevant and irrelevant attributes are shown in columns 2 and 3, respectively. All relevant attributes participate in interactions. The 4th column shows the majority class percentage

Table 1. Average predictive accuracies (with standard deviations in parentheses)

Concept	♯ Rel Atts	♯ Irrel Atts	Maj %	C4.5	Perfect FS +C4.5	FC +C4.5
p_4	4	8	50	*58.7(10.9)*	100(0.0)	100(0.0)
p_6	6	6	50	*48.5(0.3)*	84.5(3.1)	**98.1(1.5)**
p_8	8	4	50	48.4(0.2)	49.2(0.5)	**76.7(2.7)**
$cp_{4,9}$	6	6	75	*73.1(2.7)*	91.3(4.1)	100(0.0)
$cp_{3,10}$	8	4	75	**73.1(2.6)**	71.0(2.7)	**100(0.0)**
$cp_{2,11}$	10	2	75	**72.6(4.1)**	64.0(3.1)	**97.3(6.1)**
$cdp_{3,11}$	6	6	62	*81.3(8.2)*	95.0(4.0)	**99.7(0.7)**
$cdp_{2,10}$	9	3	62	66.0(9.8)	67.9(5.4)	**90.3(5.7)**
$P_{3,6} \wedge (2)$	10	2	88	**87.9(1.6)**	83.1(2.1)	**92.8(6.2)**
$P_{3,6} \wedge (3)$	10	2	84	**84.4(0.1)**	77.6(2.3)	**95.4(3.9)**
$P_{3,6} \wedge (3or2)$	10	2	73	**68.7(2.0)**	66.8(1.7)	**93.2(4.2)**
$P_{3,6} \vee (2)$	10	2	62	59.4(5.0)	60.0(3.8)	**93.1(2.0)**
$P_{3,6} \vee (3)$	10	2	66	60.9(2.9)	60.9(3.7)	**93.7(1.8)**
$P_{3,6} \vee (3or2)$	10	2	77	**75.5(2.2)**	71.9(2.0)	**95.7(1.7)**
$\wedge(P_{1,4},P_{3,6})$	6	6	75	*72.5(3.2)*	90.7(3.1)	**99.8(0.5)**
$\wedge(P_{1,6},P_{3,8})$	8	4	75	**73.4(2.7)**	68.3(3.0)	**94.1(2.8)**
$\wedge(P_{1,3},P_{3,5},P_{4,6})$	6	6	88	*87.6(1.2)*	94.6(2.8)	**99.8(0.7)**
$\wedge(P_{1,4},P_{2,5},P_{3,6})$	6	6	88	*87.5(0.3)*	93.8(2.6)	**99.6(0.7)**
$\wedge(P_{1,4},P_{3,6},P_{5,8})$	8	4	88	**87.5(0.1)**	83.4(1.7)	**98.6(1.7)**
$\wedge(P_{1,6},P_{2,7},P_{3,8})$	8	4	88	86.6(1.8)	85.4(1.8)	**93.8(2.4)**
$\wedge(WL3_{1,5},WL3_{3,7})$	7	5	64	89.4(2.9)	91.2(1.8)	93.1(5.9)
$\wedge(WL3_{1,5},WL3_{4,8})$	8	4	68	*85.1(1.8)*	87.5(1.8)	89.9(9.6)
$\wedge(WL3_{1,5},WL3_{5,9})$	9	3	71	*83.3(1.9)*	85.8(2.0)	**93.5(7.0)**
$\wedge(WL3_{1,5},WL3_{6,10})$	10	2	75	*80.1(1.4)*	81.9(1.4)	**88.1(8.4)**
$\wedge(WL3_{1,4},WL3_{3,6},WL3_{5,8})$	8	4	58	*85.6(2.6)*	88.5(2.6)	**97.5(2.2)**
$\wedge(W23_{1,5},C_{4,7},WL3_{6,10})$	10	2	77	77.1(2.6)	78.3(2.7)	**88.7(8.9)**
Average	8	4	72	75.2	79.7	94.7

in the concept. This value is used as a trivial baseline performance to compare accuracies with. These concepts are hard because of complex interactions among attributes and existence of irrelevant attributes. Relevant attributes are easily confounded with irrelevant attributes. Note that most concepts are composed of several complex interactions. Thus, the underlying regularities in these concepts are complex when represented by primitive relevant attributes.

C4.5 is used with default parameter's values. Each experiment is run 20 times over 20 sets of shuffled data and the average accuracy is calculated. For each trial, 5% of all 2^{12} data are used for training and the rest (95%) are kept unseen as test data for evaluating predictive accuracy. Column 5 of Table 1 shows the average accuracy obtained by C4.5 for each concept using the original data set. The accuracy of C4.5 after the perfect FS is reported in the 6th column. Finally, column 7 reports the average accuracy of C4.5 after reducing data by FC using MFE3/GA^{DR}. For all experiments results for C4.5 are obtained after tree pruning since for all concepts the average accuracy with pruning was better than those without pruning. The accuracy of C4.5 with FS (column 6) is significantly better than those in *italic* and worse than those in **bold** (*t*-test, $\alpha = 0.02$).

Comparing the results of columns five and six shows that for some concepts, the FS, in addition to reducing the data size, improves learning accuracy. However, for few concepts (11 out of 26) this improvement is significant. When number of relevant attributes increases to 8 or more, after FS, C4.5 fails to construct a proper concept description, therefore, achieves a predictive accuracy less than or equal to the majority class percentage. Thus, if data are classified using the majority class, a better result is obtained, which means the learner does not learn the concept. This shows the complexity of these concepts for a standard learner even after FS. Identifying relevant attributes by an FS method is not enough for learning this kind of concepts because the number of interacting attributes is high and few training data are provided.

Column seven shows that FC with MFE3/GA^{DR} reduces the problem of learning concepts with complex interactions for all concepts. After data reduction using the constructed features, the learning accuracy is improved significantly comparing to those obtained by C4.5 using original attributes (column five) or selected attributes (column six). MFE3/GA^{DR} successfully recognizes interacting attributes from irrelevant attributes and constructs functions over them to encapsulate the interactions. The constructed features highlight the underlying structure of the concept to C4.5. Thus, the learner easily learns the concept and achieves high accuracy on test data.

5 Conclusion

The problem of primitive data representation is highlighted in this paper. It was explained that FS and FC have been used for data reduction to improve the data representation and achieve better accuracy; however, when the primitive representation of real-world data produces attribute interactions, then FS may reduce the data size but is not enough for improving learning accuracy.

When complex interactions exist among attributes and the only information provided about the concept is few primitive training data, regularities are opaque to the learner. Then an FC method is needed to construct features that abstract and encapsulate such occulted information into new features in order to highlight it. Each constructed feature works as an intermediate concept, which forms part of the theory that highlights interactions in primitive data representation.

Empirical evaluation over synthetic concepts with complex attribute interactions showed that even a perfect FS cannot improve learning these concepts significantly. However, the data reduction FC method, MFE3/GA^{DR}, with the help of GA successfully detects interacting attributes and constructs highly informative features that abstracts interactions. Thus, it significantly improves learning while it reduces the data size.

References

1. Liu, H., Motoda, H. (eds.): Feature Extraction, Construction and Selection: A Data Mining Perspective, vol. 453. Kluwer Academic Publishers, MA (1998)
2. Liu, H., Motoda, H. (eds.): Computational Methods of Feature Selection. Data Mining and Knowledge Discovery Series. Chapman & Hall/CRC, Boca Raton (2007)
3. Shafti, L.S., Pérez, E.: Fitness function comparison for GA-based feature construction. In: Borrajo, D., Castillo, L., Corchado, J.M. (eds.) CAEPIA 2007. LNCS, vol. 4788, pp. 249–258. Springer, Heidelberg (2007)

4. Shafti, L.S., Pérez, E.: Data reduction by genetic algorithms and non-algebraic feature construction: A case study. In: Proceedings of HIS (2008)
5. Blake, C., Merz, C.: UCI repository of machine learning databases (1998)
6. Holte, R.C.: Very simple classification rules perform well on most commonly used datasets. Machine Learning 11, 63–91 (1993)
7. Jakulin, A., Bratko, I., et al.: Attribute interactions in medical data analysis. In: Dojat, M., Keravnou, E.T., Barahona, P. (eds.) AIME 2003. LNCS, vol. 2780, pp. 229–238. Springer, Heidelberg (2003)
8. Danyluk, A.P., Provost, F.J.: Small disjuncts in action: Learning to diagnose errors in the local loop of the telephone network. In: Proceedings of ICML (1993)
9. Rendell, L.A., Seshu, R.: Learning hard concepts through constructive induction: Framework and rationale. Computational Intelligence 6, 247–270 (1990)
10. Freitas, A.A.: Understanding the crucial role of attribute interaction in data mining. Artificial Intelligence Review 16(3), 177–199 (2001)
11. Zhao, Z., Liu, H.: Searching for interacting features. In: Veloso, M.M. (ed.) Proceedings of IJCAI, Hyderabad, India, pp. 1156–1161 (January 2007)
12. Bloedorn, E., Michalski, R.S.: Data-driven constructive induction: Methodology and applications. In: [1], pp. 51–68
13. Michalski, R.S.: Pattern recognition as knowledge-guided computer induction. Technical Report 927, Dept. of Computer Science, University of Illinois (1978)
14. Quinlan, J.R.: C4.5: Programs for Machine Learning. Morgan Kaufmann, San Francisco (1993)
15. Vafaie, H., DeJong, K.: Feature space transformation using genetic algorithms. IEEE Intelligent Systems 13(2), 57–65 (1998)
16. Pérez, E., Rendell, L.A.: Using multidimensional projection to find relations. In: Proceedings of the Twelfth ICML, pp. 447–455. Morgan Kaufmann, San Francisco (1995)
17. Pazzani, M.: Constructive induction of cartesian product attributes. In: [1], pp. 341–354
18. Zupan, B., Bratko, I., et al.: Function decomposition in machine learning. Machine Learning and Its Applications, Advanced Lectures, 71–101 (2001)
19. Grunwald, P.D.: The Minimum Description Length Principle. MIT Press, Cambridge (2007)
20. Shafti, L.S.: Multi-feature construction based on genetic algorithms and non-algebraic feature representation to facilitate learning concepts with complex interactions. Ph.D thesis, EPS, Universidad Autonoma de Madrid (2008)

Appendix: Concept Definitions

Concepts in Sect. 4 are defined over boolean attributes x_1 to x_{12}. Let $w(x_{i..j})$ be the weight of attributes x_i to x_j, i.e., the number of ones in $\{x_i, \ldots, x_j\}$, then:

$$P_{i,j} \stackrel{\text{def}}{=} w(x_{i..j}) \text{ is an odd number, i.e., } parity(x_i, \ldots, x_j),$$

$$cp_{i,j} \stackrel{\text{def}}{=} P_{i,6} \wedge P_{7,j}, \qquad\qquad P_{i,j} \vee (l) \stackrel{\text{def}}{=} P_{i,j} \vee w(x_{7..12}) = l,$$

$$cdp_{i,j} \stackrel{\text{def}}{=} P_{i,4} \wedge (P_{\frac{i+j}{2},8} \vee P_{j,12}), \qquad P_{i,j} \wedge (l) \stackrel{\text{def}}{=} P_{i,j} \wedge w(x_{7..12}) = l.$$

The rest of the concepts are defined as conjunctions $\wedge(f_1, \ldots, f_n)$ where f_m is one of the followings:

$$WL3_{i,j} \stackrel{\text{def}}{=} w(x_{i..j}) < 3, \qquad\qquad W23_{i,j} \stackrel{\text{def}}{=} w(x_{i..j}) \in \{2,3\},$$

$$C_{i,i+3} \stackrel{\text{def}}{=} parity(x_i, x_{i+2}) \vee parity(x_{i+1}, x_{i+3}).$$

The reader is referred to [16] and [20] for more details about above concepts.

Multiobjective Evolutionary Clustering Approach to Security Vulnerability Assesments

G. Corral, A. Garcia-Piquer, A. Orriols-Puig, A. Fornells, and E. Golobardes

Grup de Recerca en Sistemes Intel·ligents
La Salle - Universitat Ramon Llull
c/ Quatre Camins 2, 08022 Barcelona, Spain
{guiomar,alvarog,aorriols,afornells,elisabet}@salle.url.edu

Abstract. Network vulnerability assessments collect large amounts of data to be further analyzed by security experts. Data mining and, particularly, unsupervised learning can help experts analyze these data and extract several conclusions. This paper presents a contribution to mine data in this security domain. We have implemented an evolutionary multiobjective approach to cluster data of security assessments. Clusters hold groups of tested devices with similar vulnerabilities to detect hidden patterns. Two different metrics have been selected as objectives to guide the discovery process. The results of this contribution are compared with other single-objective clustering approaches to confirm the value of the obtained clustering structures.

Keywords: Multiobjective Optimization, Evolutionary Algorithm, Unsupervised Learning, Clustering, Network Security, AI applications.

1 Introduction

Information Technology and the communication networks that support it have gradually changed into critical resources for organizations. The combination of computer and communication technologies offers many benefits, but introduces weaknesses. Consequently periodic audits and vulnerability assessments are needed. Vulnerability assessment is the process of identifying and quantifying vulnerabilities in systems or networks [16]. As time and cost may restrict its depth, the automation of the involved processes is essential, specially those related to the data analysis. In addition, a comprehensive network security analysis must coordinate diverse sources of information to support large scale visualization and intelligent response [7]. So security applications require some intelligence to detect malicious data, unauthorized traffic or vulnerabilities [8].

Artificial intelligence can be applied to vulnerability assessment results. The use of clustering for discovering hidden patterns through the identification of device groups with similar vulnerabilities has been demonstrated [3]. Different validity techniques to select the best clustering solution have been analyzed [4]. These contributions have been included in *Analia*, a computer-aided system to automate network security tests [3]. *Analia* helps security analysts, but it has a drawback. Two independent processes are needed: select (1) the clustering

E. Corchado et al. (Eds.): HAIS 2009, LNAI 5572, pp. 597–604, 2009.

approach and (2) the validity index. The best clustering solution depends on the selected validity index, as each index may pursue different goals. Moreover, the goals of clustering and the index may not be aligned. Analysts also ask for a process where configuration parameters not related to their domain are provided.

This paper presents an improvement of *Analia* based on including an evolutionary multiobjective (EMO) clustering algorithm [12] to group network devices with similar vulnerabilities after a vulnerability assessment. The optimization of the different validity indices will be used as the goal to cluster tested devices in groups with similar vulnerabilities. This new approach will allow security analysts to obtain the best clustering solution considering different criteria simultaneously. In addition, this selection will become a transparent process to analysts, due to the fact that this technique includes the optimization of the selected criteria in the clustering process itself. Then analysts will not need to care about the difference between validity indices and will be able to focus only on the obtained clustering results, which is their actual concern.

The remainder of this paper is organized as follows. Section 2 describes related work on machine learning in the security domain. Section 3 details our clustering multiobjective evolutionary approach. Section 4 describes *Analia* with single and multiple optimization clustering. Section 5 summarizes the experimentation and results. Conclusions and further work are given in Section 6.

2   Related Work

The large volume of data generated by vulnerability assessments has unleashed the need of using enhanced techniques to recognize malicious behavior patterns or unauthorized changes in data networks [8]. These domains are usually defined by sets of unlabeled examples, and experts aim at extracting novel and useful information about the network behavior that helps them detect vulnerabilities, among others. In this context, clustering appears as an appealing approach that permits grouping network devices with similar security vulnerabilities, thence, identifying potential threats to the network.

Several clustering techniques have been applied to the network security domain thus far. For example, K-means [13] has been used to group similar alarm records [2] and to detect network intrusions [15]. SOM [14] has been employed to detect computer attacks [8], network intrusions [9], and anomalous traffic [17]. Despite the success of these applications, all these clustering techniques guide the discovery process with a single criterion. For example, K-means minimizes the total within-cluster variance and tends to find spherical clusters [13]. Nevertheless, we are interested in obtaining clusterings that satisfy different criteria. For this purpose, several authors have proposed to run different clustering techniques to obtain different structures, and then, involve the network expert into the process in order to manually select the best structure according to certain predetermined validation methods.

In this paper, we propose to automatize this process by guiding the clustering process with different objectives. To achieve this, we employ a multiobjective clustering approach [12]. Among the different techniques for multiobjective optimization such as simulated annealing or ant colony optimization, we base on evolutionary algorithms since they (1) employ a population based-search, evolving a set of optimal trade-offs among objectives, (2) use a flexible knowledge representation that can be easily adapted to the type of data of our domain, and (3) are able to optimize different objectives without assuming any underlying structure of the objective functions. In addition, EMO clustering has been successfully applied to important real-world problems such as intrusion detection [1], formation of cluster-based sensing networks in wireless sensor networks [19], and creation of security profiles [11].

3   Evolutionary Multiobjective Clustering Approach

This section explains the design of the EMO approach employed to evolve data clusterings that optimize several objectives. Our approach is based on the MOCK system [12] which uses the PESA-II algorithm. PESA-II evolves a set of solutions, where each one defines a possible clustering configuration. In what follows, the knowledge representation, the process organization to obtain the Pareto set of solutions, and the method to recover the best solution among the ones in the Pareto set are briefly explained.

Representation. The system evolves a population of individuals of size N, where each individual represents a cluster structure for the problem. The individual is represented in a vector of n integers: $x_1, x_2, \ldots, x_n$. Then, x_i indicates that instance i is connected to instance x_i, that is, that they belong to the same cluster.

Process organization. The result of the algorithm is a Pareto set of solutions, that is, a set of individuals for which it does not exist any other individual in the population that dominates them[1]. The population is evolved as follows. We first initialize the population with an individual that represents the minimum spanning tree (MST) constructed from the undirected, fully connected labeled graph that represents the Euclidean distance between each pair of examples. In addition, we also create $N - 1$ individuals that progressively remove the links with highest distance from the original MST. Then, the population iteratively goes through a process of selection, crossover, and mutation as described in [12].

Selection of the best solution. After evolving a set of non-dominated solutions, we use the following criterion to recover the best solution. The system returns the solutions between 6 and 9 clusters because, according to the experts, the devices included in the dataset can be broadly categorized in those groups. Furthermore, these solutions are the best ones that optimize the couple connectivity/deviation.

4   Analia

This section explains the architecture of *Analia* and the inclusion of evolutionary multiobjective clustering to improve data analysis in this security domain.

---

[1] In multiobjective algorithms, a solution x dominates another solution y if all the objectives of x are better than the corresponding objectives of y.

4.1   Single-Objective Clustering in *Analia*

Analia is the data analysis module of *Consensus* [5]. Whereas *Consensus* gathers
security data, *Analia* includes AI to help analysts after a vulnerability assessment. *Analia* finds resemblances within tested devices and clustering aids analysts in the extraction of conclusions. Afterwards, the best results are selected
by applying cluster validity indices [4]. Then explanations of clustering results
are included to give a more comprehensive response [3]. Figure 1 depicts the
architecture of *Analia* and its interaction with *Consensus*.

Previous work has validated the incorporation of K-means, X-means and
SOM in *Analia* [4]. A drawback of this unsupervised domain is that no previous
knowledge of the possible existing classes is known. So several executions are
run to select the best one by using cluster validation techniques. The most used
indices found in the literature have been included in *Analia*: Dunn [10], Davies-
Boudin [6] and Silhouette [18]. Besides, two indices have been designed ad hoc
for this security domain: *Intracohesion* and *Intercohesion* factors [4]. The process
to obtain the best partition in *Analia* is summarized in the following steps:

1. Select the clustering approach and execute it on *Consensus* dataset
2. Select the executions to analyze
3. Calculate validation indices for each execution
4. Select the validation index as decision criterion and obtain the best execution

When having run a clustering algorithm several times varying parameters or
different clustering algorithms, several clustering solutions are obtained. Then,
any of the mentioned validity indices can be selected to obtain the best solution
over a set of executions. High values for *Dunn* and *Silhouette*, whereas low values for *DB* are preferred. Regarding *Cohesion factors*, the best solution should
consider the highest *Intracohesion factor* and the lowest *Intercohesion factor*.

Security analysts do not usually care about the selected index criteria, but
about the best clustering solution. Thus an automated mechanism has been
designed to combine the calculated validity factors based on a weighted voting
scheme. It ranks the list of options based on the number of votes each option
earns, considering that some votes carry more weight than others. We give more
importance to *Cohesion factors* by assigning a higher weight to them. Then,
analysts can easily get the best partition without any previous knowledge about
validity indices, knowledge not usually related with their study area.

The main drawback of this option is focused on the different decisions not directly related to their domain that security analysts must consider. This two-step
process of selecting a clustering techique and, afterwards, applying validation indices may slow down the whole process. Next section presents a contribution to
improve this statement. If validation indices are considered as initial goals of the
clustering approach, the obtained clustering solutions will optimize the selected
indices, thus reducing the process into a single step.

4.2   Multiobjective Clustering in *Analia*

We have integrated the designed EMO clustering approach into *Analia*, optimizing two complementary objectives: the overall deviation and the connectivity.

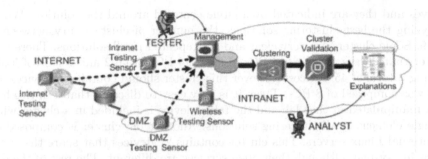

Fig. 1. Architecture of *Consensus* system and *Analia* data analysis module

The former is related to cluster compactness and the latter is based on connect-edness of clusters. These goals are aligned to *Cohesion* validation indices. The *Intracohesion factor* evaluates the cohesion between the elements of a cluster in terms of the vulnerabilities common to the members of that cluster [4], con-cept equivalent to cluster compactness. The *Intercohesion factor* evaluates the cohesion between clusters, considering the vulnerabilities common to elements of different clusters [4], concept equivalent to connectivity.

When including this EMO clustering in *Analia*, the process to obtain the best partition is summarized in a single step: execute the EMO clustering approach. All the other aforementioned steps are included in that single phase. The EMO clustering approach obtains a set of non-dominated solutions that optimize both deviation and connectivity in a single run. So the algorithm needs to be run only once. Note that, as the validation indices have been included in the search pro-cess, there is no need to calculate those indices afterwards. The best executions with the best number of clusters will be automatically obtained.

5   Experiments

The EMO clustering approach presented in this paper has run on the *Consensus* dataset, which contains information regarding port scanning, operating system fingerprinting and vulnerability testing of a data network. This dataset has been extracted from real security tests performed at La Salle (Universitat Ramon Llull) network. These assessments have been executed over 90 network devices, including public and internal servers, alumni laboratories and staff computers.

Alumni lab computers are the most restricted devices. The IT department installs a unique image on them, so any other software is forbidden. Thus lab devices should be grouped in the same cluster. If new software has been illegally installed or its configuration has been modified, it will be easy to identify as this rogue device should be separated in a new cluster. On the other hand, staff computers are administered by their owners and hence different patterns will be found. Several servers with different Internet services and different operating systems have been audited, so their classification may also vary.

A solution for an EMO clustering run on *Consensus* dataset is shown in Figure 2. The best executions should find good tradeoffs between the two objectives and they are indicated by a circle centered around the solution. When analyzing the best clustering solutions, the number of clusters vary between 6 and 9. Some clusters are very clear and get repeated in all solutions. There is a cluster that always contains 14 PCs of lab1, 24 PCs of lab2 and 7 PCs of lab3, making a total of 45 devices. However this cluster should contain 46 devices, as lab3 was composed of 8 PCs. Then, it is very easy to discover that a device has been manipulated in that lab and the faked device is included in a cluster with a single element in all clustering solutions. Another big cluster is composed of all internal Linux servers. This cluster contains 27 devices that share the same operating system, although their open services are different. The rest of the devices are grouped in small clusters, depending on their operating system and the offered services. It is remarkable that 3 devices are separated always in 3 single clusters, showing their dissimilarity in comparison with the rest of the elements in the dataset. They correspond to the fake device of a lab, to the wireless access control server which is a Linux device but with specific peculiarities and, finally, to a Sun Solaris server.

Cohesion factors have been calculated to evaluate the correctness of the different solutions and the alignment between these indices and the EMO clustering objectives, overall deviation and connectivity. Results have shown that the best executions of EMO clustering also obtain the best values of *Cohesion factors*, achieving the best values of Intracohesion = 0.896 and Intercohesion = 0.38. The range of these indices is [0..1], preferring high values for *Intracohesion* and low values for *Intercohesion* indices.

Single-objective clustering algorithms have also been run on the same dataset to compare their solutions. K-means has run for a range of different numbers of clusters $k \in 3..10$ and different seeds. Considering *Cohesion* factors, the best solutions also obtain a number of clusters between 6 and 9. However, the calculated *Cohesion factors* are lower than EMO results. The runs of X-means conclude in 7 for the best value of K. Again, the *Cohesion factors* are lower than EMO clustering results. Both partitioning algorithms show clustering structures without one-element clusters, or at least, only one unique cluster with one element.

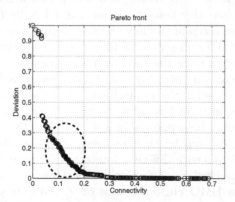

Fig. 2. EMO clustering output on *Consensus* with the deviation and the connectivity

Table 1. Summary of the *Cohesion* factors for K-means, X-means and EMO clustering for different number of clusters

Clusters	K-means		X-means		EMO Clustering	
	Intrach.	Interch.	Intrach.	Interch	Intrach.	Interch.
6	0.662	0.597	0.559	0.546	0.858	0.451
7	0.626	0.526	0.628	0.538	0.879	0.419
8	0.672	0.591	0.621	0.531	0.895	0.380
9	0.662	0.538	0.563	0.517	0.866	0.371

A summary of the obtained *Cohesion* factors of the different clustering approaches are shown in Table 1. The best values for all possible configurations of number of clusters are obtained with the EMO approach. This is the approach that tries to optimize the two objectives more directly related to *Cohesion* factors. On the other hand, partition methods minimize only overall deviation and thus *Cohesion* factor results are not as good as the EMO clustering approach.

Network security experts have also analyzed the results obtained after clustering the dataset. Regarding to EMO clustering solution, the high number of clusters with a single element allows the location of outlier devices. But this approach returns a higher number of clusters for the same dataset, compared to the partitioning methods.

6 Conclusions

This paper has presented the incorporation of an evolutive multiobjective clustering algorithm based on PESA-II to analyze data from vulnerability assessments in a network security domain. This approach benefits from the use of multiple objectives. The achieved clustering solutions overcome the results obtained with different single-objective clustering algorithms, like K-means or X-means. Besides, the use of this EMO approach permits reducing the efforts spent by security analysts in the clustering phase. Security analysts do not need to have previous knowledge of cluster validation indices in order to obtain the best solution of a set of clustering executions. When using this EMO clustering approach, the pursued goals that guide the search for the best solutions are aligned with the validity indices. Therefore, the obtained clustering solutions comply with validity requirements. Then, a subsequent phase where validity indices are applied is not necessary. Once the clustering results are shown to security analysts, their task starts analyzing the characteristics of the obtained clusters. Clusters will group devices with similar operating systems, open ports, and vulnerabilities.

Further work will focus on the inclusion of different metrics as objectives of the EMO clustering approach. *Cohesion* factors and other validity indices will be incorporated as input goals to be optimized.

Acknowledgements. This work has been supported by the MCYT-FEDER projects TIN2006-15140-C03-03, TIN2008-06681-C06-05 and by the Generalitat de Catalunya (2005SGR-302). We want to thank La Salle - URL for the support to our research group.

References

1. Anchor, K., Zydallis, J., Gunsch, G.: Extending the computer defense immune system: Network intrusion detection with a multiobjective evolutionary programming approach. In: 1st Conf. on Artificial Immune Systems, pp. 12–21 (2002)
2. Bloedorn, E., Talbot, L., DeBarr, D.: Data Mining Applied to Intrusion Detection: MITRE Experiences. In: Maloof, M.A. (ed.). Springer, Heidelberg (2005)
3. Corral, G., Armengol, E., Fornells, A., Golobardes, E.: Data security analysis using unsupervised learning and explanations. In: Innovations in Hybrid Intelligent Systems. Advances in Soft Computing, vol. 44, pp. 112–119. Springer, Heidelberg (2008)
4. Corral, G., Fornells, A., Golobardes, E., Abella, J.: Cohesion factors: improving the clustering capabilities of consensus. In: Corchado, E., Yin, H., Botti, V., Fyfe, C. (eds.) IDEAL 2006. LNCS, vol. 4224, pp. 488–495. Springer, Heidelberg (2006)
5. Corral, G., Zaballos, A., Cadenas, X., Grane, A.: A distributed vulnerability detection system for an intranet. In: Proceedings of the 39th IEEE International Carnahan Conference on Security Technology (ICCST 2005), pp. 291–295 (2005)
6. Davies, D.L., Bouldin, D.W.: A cluster separation measure. IEEE Transactions on Pattern Analysis and Machine Learning 4, 224–227 (1979)
7. Dawkins, J., Dale, J.: A systematic approach to multi-stage network attack analysis. In: 2nd. IEEE Int. Information Assurance Workshop (IWIA 2004) (2004)
8. DeLooze, L.: Classification of computer attacks using a self-organizing map. In: Proc. of the 2004 IEEE Workshop on Information Assurance, pp. 365–369 (2004)
9. Depren, M.O., Topallar, M., Anarim, E., Ciliz, K.: Network-based anomaly intrusion detection system using soms. In: Proc. of the IEEE 12th Signal Processing and Communications Applications Conference, pp. 76–79 (2004)
10. Dunn, J.C.: Well separated clusters and optimal fuzzy partitions. Journal of Cybernetics 4, 95–104 (1974)
11. Gupta, M., Rees, J., Chaturvedi, A., Chi, J.: Matching information security vulnerabilities to organizational security profiles: a genetic algorithm approach. Decision Support Systems 41(3), 592–603 (2006)
12. Handl, J., Knowles, J.: An evolutionary approach to multiobjective clustering. IEEE Transactions on Evolutionary Computation 11(1), 56–76 (2007)
13. Hartigan, J.A.: Clustering Algorithms. John Wiley and Sons, New York (1975)
14. Kohonen, T.: Self-Organizing Maps, 3rd edn. Springer, Heidelberg (2000)
15. Leung, K., Leckie, C.: Unsupervised anomaly detection in network intrusion detection using clusters. In: Proc. 28th Australasian CS Conf., vol. 38 (2005)
16. Peltier, T.R., Peltier, J., Blackley, J.: Managing a Network Vulnerability Assessment. Auerbach Publishers Inc. (2003)
17. Ramadas, M., Ostermann, S., Tjaden, B.C.: Detecting anomalous network traffic with self-organizing maps. In: Vigna, G., Krügel, C., Jonsson, E. (eds.) RAID 2003. LNCS, vol. 2820, pp. 36–54. Springer, Heidelberg (2003)
18. Rousseeuw, P.: Silhouettes: a graphical aid to the interpretation and validation of cluster analysis. J. of Comp. Applic. in Math 20, 53–65 (1987)
19. Yang, E., Erdogan, A., Arslan, T., Barton, N.: Multi-objective evolutionary optimizations of a space-based reconfigurable sensor network under hard constraints. In: Symp. on Bioinspired, Learning, and Int. Syst. for Security, pp. 72–75 (2007)

Beyond Homemade Artificial Data Sets

Núria Macià, Albert Orriols-Puig, and Ester Bernadó-Mansilla

Grup de Recerca en Sistemes Intel·ligents
La Salle - Universitat Ramon Llull
C/ Quatre Camins 2, 08022 Barcelona, Spain
{nmacia,aorriols,esterb}@salle.url.edu

Abstract. One of the most important challenges in supervised learning is how to evaluate the quality of the models evolved by different machine learning techniques. Up to now, we have relied on measures obtained by running the methods on a wide test bed composed of real-world problems. Nevertheless, the unknown inherent characteristics of these problems and the bias of learners may lead to inconclusive results. This paper discusses the need to work under a controlled scenario and bets on artificial data set generation. A list of ingredients and some ideas about how to guide such generation are provided, and promising results of an evolutionary multi-objective approach which incorporates the use of data complexity estimates are presented.

Keywords: Data complexity, artificial data sets, machine learning.

1 Introduction

Machine learning techniques have a practical application on a large variety of real-world problems. The diversity of domains—medicine, industry, learning—provides extremely disparate data sets regarding the type of features, volume of instances, and data distribution, among others. All of these characteristics have led to the implementation of different strategies to tackle each problem properly, since learner performance depends partly on the algorithm design. At present, the development of techniques has reached an advanced state of maturity offering thousands of methods, all of them very competitive, providing accurate models from data which are generalized from a sample of the problem at hand. Despite the headway progress in data classification, many questions remain unanswered such as how the intrinsic characteristics of the data sets affect learners. This, coupled with the little leeway for improvement and the uncertainty of the ability of techniques to fully capture the underlying knowledge of data, leads us to look toward other elements involved in the learning process. At this point, data steals the limelight from learners.

Some authors have started giving importance to the study of data complexity in supervised learning, and recent studies, compiled in [2], have shown that learner performance also depends on data complexity. This dependence has unleashed a new research line which is focused on the analysis of the nature of problems and whose aim is to characterize data and relate them to learner properties. This link,

E. Corchado et al. (Eds.): HAIS 2009, LNAI 5572, pp. 605–612, 2009.

based on complexity metrics, could help assess the ability of learners and recommend which learner should be applied to solve a specific problem. Nonetheless, again doubts arise about the reliability of these complexity estimates, since the characterization is made up of real-world problems and validated by machine learning techniques whose relationship is close but not yet well-defined [3]. Thereby, it highlights the need to create data sets of bounded difficulty in order to analyze complexity metrics and learners under a controlled scenario.

The purpose of this paper is to present a new technique to generate artificial data sets (ADS) diverse enough to provide a solid experimental framework where the analysis of learner behavior and leaner performance can be carried out. There have been the first attempts to build ADS based on complexity estimates by using heuristic searches [10] and genetic algorithms [9]. Despite of the fresh aroma of these proposals, the approaches just optimize one complexity dimension. In this work, we explode this idea and use a multi-objective evolutionary algorithm to make data sets that meet different types and levels of complexity.

The remainder of the paper is organized as follows. Section 2 briefly reviews data complexity analysis. Then, Section 3 discusses the why, what kind, and how to generate ADS. The mix of some points dealt with in the previous section results in a *"nouvel* generator of ADS" whose design and results are presented in Sections 4 and 5 respectively. Finally, Section 6 concludes the work with some future directions.

2   Data Complexity

In order to define the relationship between data and learners, some studies investigated problem characterization by means of different estimates based on difficulty factors. Ho and Basu first identified the sources of problem difficulty [7] and proposed the following classification: (1) class ambiguity, (2) boundary complexity, and (3) sample sparsity and feature space dimensionality. *Ambiguity* refers to the situation when there are examples whose features cannot permit distinguishing their classes. Usually, this ambiguity is due to the problem formulation in which the concepts are intrinsically inseparable or the set of attributes is no adequate or sufficient to describe the concepts. *Boundary complexity* is related to the description of the class boundary. Class separability and problem linearity are based on the geometrical complexity of data structure. Finally, *sample sparsity and feature space dimensionality* are concerned with the difficulty layer that an incomplete or sparse sample add to the problem, enkindling the importance of the sample representativity.

Among these sources of problem difficulty, investigations carried out by Ho and Basu focused on boundary complexity because of the difficulty to determine the class ambiguity and the real sparsity of a training set. Thus, they designed a set of measures to estimate the class boundary [7]. These measures evaluate different aspects such as (1) overlaps in features values from different classes, (2) separability of classes, and (3) geometry, topology, and density of manifolds. In later studies, the data characterization built upon this set of metrics, which provides a space of data complexity, permitted determining some relationships between certain estimates and certain learners from different paradigms [3].

Although preliminary results showed some correlations between these complexity estimates and learner accuracy, the link between data characteristics and learner properties is not mature enough. There are still too many relationships among data, complexity metrics, and learners, highly dependent and out of control [6]. Therefore, in order to avoid again getting partial conclusions through pieces of problems and apparent estimations, we have to resort to ADS. The generation of ADS is the procedure to establish the framework to study estimates of data complexity and learner performance.

3   The Why, What Kind, and How to Generate ADS

This section discusses ADS generation. We present a general picture of this incipient topic by answering why we need ADS, what kind of characteristics they should have, and how to provide ADS with these desirable requirements.

3.1   Why?

Over the last few decades, the machine learning community has designed and developed techniques to solve real-world problems and to extract knowledge from their data. To validate the efficiency of these techniques, the most usual methodology adopted by the community consists in testing new techniques on a collection of real-world problems and comparing the obtained accuracy with other learners. Nevertheless, this procedure may lead to inaccurate conclusions due to (1) real-world problems constraints and (2) data dependence of the learner.

Usually, learners are tested using real-world problems from public repositories. Even though sharing these problems benefits the obtaining of a common test bed for the experiments and facilitates the comparison between the own and the community results, these data sets may result in misleading conclusions. On the one hand, the current sets are composed of few problems whose independence is unknown, i.e, we ignore whether this set of problems is representative enough to cover the whole problem space. We cannot guarantee that these problems are diverse enough to test the learner limitations in an exhaustive way, since there are no studies that indicate what problems, regardless of the domain to which they belong, are structurally similar. On the other hand, the high cost of experiments, the difficulty of conducting them, or data privacy policies hinder data collection, resulting in complex data sets characterized by few instances, missing values, and imprecise data. The combination of these deficiencies in the data sample goes beyond our control, blurring our knowledge of to what extent the influence of these constraints negatively affects leaner performance.

Empirical results show that there exist learning paradigms more suitable to solve a type of problems than others. Nevertheless, the learner dependence on opaque data is responsible for our lack of knowledge of the relationship between data characteristics and learner properties, limiting us in the progress of learning techniques. To overcome this, we need to work under a controlled scenario, with a certain kind of data, where complexity is known.

3.2 What Kind?

Being aware of the need of artificial data sets to test learner performance, we have to define what kind of data set should be generated. To this end, we focus on the concepts for classification learning taking into consideration (1) data structure and (2) complexity factors.

Data structure present in real-world problems is significant in data analysis since these structures contain the underlying knowledge. Thus, we should force data sets to resemble real-world problems and attain such real structures in data. It means that data not only have to follow uniform or gaussian distributions but also have to include physic processes. Moreover, for classification problems, class labeling has to conform with clustering rules.

Complexity factors are related to the aforementioned aspects that are measured by the complexity metrics, such as the discriminative power of attributes, class separability, and geometry. Firstly, we have to generate well-defined problems with a known underlying concept and whose definition is complete and without ambiguity. After defining which characteristics have to describe data, constraints have to be introduced by varying their degree of difficulty to test different learner abilities. This implies relating difficulty factors to the type of performance that we want to assess, such as robustness, scalability, and predictive accuracy. For instance, noise, missing values, or ambiguity are suitable characteristics to test the learner robustness. Learner scalability would be tested by varying the number of features and the number of instances. By adding irrelevant or redundant attributes, a relevance analysis can be performed. Determining the number of classes of the problem adds another layer of difficulty, since some of the complexity factors have to be interpreted differently.

Generators of artificial data sets have to allow us to tune all these characteristics to test learner efficiency in particular cases and comprehend learner behavior in front of specific constraints.

3.3 How?

We pursue the generation of data sets which meet different complexity levels for different difficulty factors and whose structure resemble real-world problems.

The first requirement involves addressing the problem as an optimization problem in which each objective to minimize or maximize becomes a complexity metric. In this regard, multi-objective evolutionary algorithms (EMO) [4] are a natural support to conduct optimization problems with several objectives. In the problem definition, we consider a set of n unlabeled examples $\{e_1, e_2, ..., e_n\}$, and the system proceeds to search for the combination of class labels $\{c_1, c_2, ..., c_n\}$ that satisfies m predefined objectives, which correspond to m complexity metrics.

The second requirement, concerning the structure of the problems, could be achieved by generating the initial data set according to fixed distributions. Correlations among features can be set by users guided taking into account statistics extracted from real data [8]. The use of existing samples of real-world problems or of learning techniques such as instance selection and feature selection are alternatives to dynamically manage distributions. Regarding the class labeling, some instance classes could be previously fixed or grouped following real-world problems labeling.

4 EMO-Made Artificial Data Sets

In this section, we propose a data set generator based on a multi-objective evolutionary algorithm. In particular, we used the *non-dominated sorting genetic algorithm* (NSGA-II) [5] to tackle the optimization of data set complexities. However, the implementation or election of this method can change since the interest lie on the multi-objective concept.

In what follows, we first describe the meta-information required by the algorithm and the knowledge representation. Then, we detail the process organization and the genetic operators employed in our approach. And finally, we present empirical results to illustrate the outcome of the system.

4.1 Meta-information and Knowledge Representation

In our implementation, we aim at obtaining the class labeling for a data set that meets different degrees of values from the specified set of complexity metrics. For this purpose, we have to define (1) the meta-information that needs the system, (2) the genetic representation of the solution of the problem, and (3) the fitness function to evaluate each candidate solution.

Meta-information refers to data structure and data itself. The first step is to load a data set containing unlabeled examples whose dimensionality in terms of number of instances and number of attributes is predefined by the user. Each instance is defined by m continuous- or nominal-valued attributes, and samples can be randomly generated following any kind of distribution or using a real-world distribution directly.

As for the *genetic representation*, the EMO system evolves a population of N individuals. Each candidate solution represents a class labeling of the data set which is encoded by a k-ary array (k is a configuration parameter that indicates the maximum number of classes of the data set) where the position i corresponds to the class label of the ith instance. Note that the individual size is constant and is determined by the number of instances contained in the data set.

The EMO technique searches the best combination of labels that satisfies the required complexity for the specified metrics. The *fitness function* for each objective corresponds to the computation of different complexity metrics.

4.2 Process Organization and Genetic Operators

In the following, we briefly describe the process organization and how the genetic operators are combined.

The NSGA-II algorithm evolves a population P_t of N individuals which are initialized at random and evaluated. To avoid dealing with special cases in the first iteration of the algorithm, we also create an auxiliary population Q_t of N individuals whose individuals are initialized randomly and evaluated as well. Then, the procedure iteratively applies the following steps. First, populations P_t and Q_t are joined into population R_t, which contains $2N$ individuals. R_t is ranked according to the fast non-dominated sorting approach, which divides solutions up into different fronts. Then, starting from the first front, all the solutions of each front i are introduced into the new population P_{t+1} provided that there is enough room to allocate all the solutions of the given front. Otherwise, the solutions with the highest crowding distance of the front i are introduced into P_{t+1} until filling all the population; thence, no more solutions of higher fronts are added to the population.

Then, the offspring population Q_{t+1} is created using classical GA operators—selection, crossover, and mutation. The individuals are chosen from the parent population by means of s-wise tournament selection. Pairs of these parents are selected without replacement, and they undergo crossover and mutation with probabilities χ and μ respectively. If neither of both operators is applied, the parents are directly copied in the new population Q_{t+1}. Then, both populations constitute the population R_{t+1} whose individuals are reevaluated. This process iterates until the stop criterion is met, i.e., the number of generations is reached.

The EMO approach includes two sorting concepts: (1) the fast non-dominated sorting and (2) the crowding distance assignment. Thanks to these mechanisms the system can optimize different objectives in a single simulation run. The former organizes the population into different fronts, and the latter estimates the density of the solutions surrounding a particular solution in the population. Concerning the classical genetic operators, we used: (1) s-wise tournament selection, where tournaments of s randomly chosen parents are held, and the best parent, according to the crowded-comparison operator, is selected for recombination; (2) two-point crossover, which, provided two parents, randomly generates two cut points and uses them to shuffle the information of both parents; and (3) bit-wise mutation, which flips the value of the bit selected for mutation.

5   Experimental Results

The purpose of the experiments is to show how the system is able to provide a diverse test bed composed of data sets with different complexity levels across the required difficulty factors.

To this end, we fed the system with two types of meta-information: (1) a data set that follows a uniform distribution and (2) the iris problem [1] with a real-world problem structure. For all the runs, the system was initialized with a population of 400 individuals which was evolved during 50 generations. The probabilities of crossover and mutation were 0.85 and $1/n$ respectively, where n is the individual size, i.e., the number of instances of the data set.

Figure 1 plots the data sets characterized by different complexity metrics. The x-axis and y-axis represent the demanded objectives. In particular, we focus on optimizing three complexity metrics of the aforementioned set proposed by Ho and Basu: (1) the fraction of points on the class boundary (N1), (2) the ratio of average intra/inter nearest neighbor (NN) distance (N2), and (3) the ratio of the maximum Fisher's discriminant (F1).

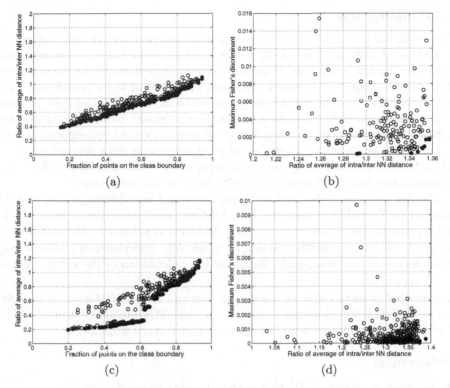

Fig. 1. Artificial data sets characterized by different complexity metrics. The upper plots refer to data sets whose structure follows a uniform distribution and, the lower plots correspond to the iris problem.

The solutions in the Pareto front are depicted with a black circle. We observe that, regardless of the distributions, the system finds solutions across the complexity space drawing the pareto-optimal. Figures 1(a) and 1(c) draw the optimization of two class separability measures, the maximization of N1 and the minimization of N2. In this case, we obtained data sets whose complexity estimated by N1 and N2 was ranged in [0.11,0.96] and [0.32,1.17] respectively. The greater the value of these measures is, the higher the complexity is. Thus, this variability permits analyzing the learner ability according to the density of the class boundary. In Figures 1(b) and 1(d), one measure of class separability N2 was maximized and one measure of feature overlap was minimized, in particular F1. Thanks to these data sets, we could test the learner robustness to the class separability for data sets in which all the attributes are relevant, since the lower the value of F1 is, the lower the discriminative power of the attributes is.

The results show that our proposal can build data sets with different characteristics defined under a set of difficulty factors to test specific learner properties. However, we have to enhance the system and provide it with mechanisms to control the class balance and to enable dynamic distributions.

6 Conclusions

This work has presented a new approach of an old practice. Design and implementation of ADS are familiar to any practitioner that has once created data sets to test specific properties of learning techniques or to highlight the discovery of a concrete behavior. However, the main drawback of these homemade data sets is that they are *ad hoc* to the given problem, and often because of their no formal design, they could be in doubt. Our proposal, based on evolutionary learning, attempts to go one step further by providing data set generation with a theoretical basis, which allows us to produce generic data sets whose characteristics are customized by means of complexity estimates. Hence, data sets generated can satisfy different complexity levels at the same time.

The discussion and the proposed approach could be a turning point in ADS generation. Nevertheless, there is still a long way until achieving a set of benchmark problems since data characteristic definition is subject to the maturity of the study of data complexity. Attaining a complete experimental platform may imply including new complexity factors or the redefinition of the existing ones.

Acknowledgments. The authors would like to thank the *Ministerio de Educación y Ciencia* for its support under the project TIN2008-06681-C06-05. They also acknowledge *Fundació Crèdit Andorrà* and *Govern d'Andorra*.

References

1. Asuncion, A., Newman, D.: UCI machine learning repository (2007)
2. Basu, M., Ho, T.K.: Data Complexity in Pattern Recognition. Springer, Heidelberg (2006)
3. Bernadó-Mansilla, E., Ho, T.K., Orriols-Puig, A.: Data complexity and evolutionary learning. In: Data Complexity in Pattern Recognition, pp. 115–134. Springer, Heidelberg (2006)
4. Coello, C.A., Lamont, G.B., Veldhuizen, D.A.V.: Evolutionary Algorithms for Solving Multi-Objective Problems, 2nd edn. Springer, New York (2007)
5. Deb, K., Pratap, A., Agarwal, S., Meyarivan, T.: A fast and elitist multiobjective genetic algorithm: NSGA-II. IEEE TEC 6, 182–197 (2002)
6. Ho, T.K.: Data complexity analysis: Linkage between context and solution in classification. In: Proceedings of the Joint IAPR International Workshops on Structural and Syntactic Pattern Recognition (SSPR 2008) and Statistical Techniques in Pattern Recognition, SPR 2008 (2008)
7. Ho, T.K., Basu, M.: Complexity measures of supervised classification problems. IEEE Transactions on PAMI 24(3), 289–300 (2002)
8. Jeske, D.R., Samadi, B., Lin, P.J., Ye, L.: Generation of synthetic data sets for evaluating the accuracy of knowledge discovery systems. In: 11th International Conference on Knowledge Discovery in Data mining, pp. 756–762 (2005)
9. Macià, N., Bernadó-Mansilla, E., Orriols-Puig, A.: Preliminary approach on synthetic datasets generation for classification. In: 2008 International Conference on Pattern Recognition. LNCS, vol. 5342, pp. 986–995. Springer, Heidelberg (2008)
10. Macià, N., Orriols-Puig, A., Bernadó-Mansilla, E.: Genetic-based synthetic data sets for the analysis of classifiers' behavior. In: Proceedings of the 2008 Hybrid Intelligent Systems Conference, pp. 507–512 (2008)

A Three-Objective Evolutionary Approach to Generate Mamdani Fuzzy Rule-Based Systems

Michela Antonelli, Pietro Ducange, Beatrice Lazzerini,
and Francesco Marcelloni

Dipartimento di Ingegneria dell'Informazione, Elettronica, Informatica, Telecomunicazioni
University of Pisa
Via Diotisalvi 2, 56122 Pisa
{michela.antonelli,pietro.ducange,beatrice.lazzerini,
francesco.marcelloni}@iet.unipi.it

Abstract. In the last years, several papers have proposed to adopt multi-objective evolutionary algorithms (MOEAs) to generate Mamdani fuzzy rule-based systems with different trade-offs between interpretability and accuracy. Since interpretability is difficult to quantify because of its qualitative nature, several measures have been introduced, but there is no general agreement on any of them. In this paper, we propose an MOEA to learn concurrently rule base and membership function parameters by optimizing accuracy and inter-pretability, which is measured in terms of number of conditions in the antecc-dents of rules and partition integrity. Partition integrity is evaluated by using a purposely-defined index based on the piecewise linear transformation exploited to learn membership function parameters. Results on a real-world regression problem are shown and discussed.

Keywords: Accuracy-Interpretability Trade-off, Interpretability Index, Multi-objective Evolutionary Algorithms, Piecewise Linear Transformation.

1 Introduction

The issue of balancing interpretability and accuracy of Mamdani fuzzy rule-based systems (MFRBSs) has arisen a growing interest in the fuzzy community [1]. In the literature, interpretability of an MFRBS has been defined in different ways. A common approach is to distinguish between interpretability of rule base (RB), also known as complexity, and interpretability of fuzzy partitions, also known as integrity of the data base (DB) [2]. Complexity is usually defined in terms of simple measures, such as number of rules in the RB and number of linguistic terms in the antecedent of rules [3]-[6]. On the other hand, integrity depends on some properties of the fuzzy partition, such as coverage, distinguishability and normality, which may be difficult to measure [2]. Anyway, as pointed out in [7][8], there does not exist a formal universally agreed definition of interpretability of MFRBs because of its very nature of being a qualitative concept. Nevertheless, during the last years, some interpretability indices have been proposed in the specialized literature. For example in [7] a set of heuristics for assessing the interpretability of MFRBs are implemented in a fuzzy rule-based system, while in [8] a partition integrity index for context adaptation applications is proposed.

E. Corchado et al. (Eds.): HAIS 2009, LNAI 5572, pp. 613–620, 2009.
© Springer-Verlag Berlin Heidelberg 2009

The issue of finding a good trade-off between accuracy and interpretability has been often tackled by using Multi-Objective Evolutionary Algorithms (MOEAs) [9]. Several approaches have been proposed to learn the RB, using either a predefined DB [3][4] or learning the membership function (MF) parameters [10], to perform concurrently rule selection and tuning of the DB [5][6] and to adapt the DB to a specific context [8]. None of these approaches takes both complexity and integrity into account.

In this paper, we propose an MOEA to generate MFRBSs with different trade-offs between complexity, accuracy and integrity. The RB, and the MF parameters of each fuzzy set are learnt concurrently during the evolutionary process. The learning of the MF parameters is performed by using a piecewise linear transformation [11], which allows us to obtain a high modeling capability with a limited number of parameters. Based on this transformation, we introduce a simple partition integrity index which measures how much the current partition is far from the uniform partition, considered as the most interpretable partition.

Preliminary results are shown on a real word dataset: our approach has provided Pareto fronts with solutions characterized by different trade-offs between accuracy, complexity and partition integrity. Further, we have shown that these Pareto fronts dominate in the accuracy-complexity plane the Pareto fronts generated by learning only the rule base with a fixed DB as described in [4]. Section 2 briefly introduces MFRBSs. In Section 3 we describe the MF parameters learning method. Section 4 and Section 5 discuss some MFRBS interpretability issues and the multi-objective approach, respectively. Section 6 shows the experimental results and Section 7 draws final conclusions.

2 Mamdani Fuzzy Rule-Based Systems

Let $\mathbf{X} = \{X_1, ..., X_f, ..., X_F\}$ be the set of input variables and X_{F+1} be the output variable. Let U_f, with $f = 1, ..., F+1$, be the universe of the fth variable. Let $P_f = \{A_{f,1}, ..., A_{f,T_f}\}$ be a fuzzy partition of T_f fuzzy sets on variable X_f. An MFRBS is composed of M rules expressed as:

$$R_m: \text{IF } X_1 \text{ is } A_{1, j_{m,1}} \text{ AND } ... \text{ AND } X_F \text{ is } A_{F, j_{m,F}} \text{ THEN } X_{F+1} \text{ is } A_{F+1, j_{m,F+1}} \tag{1}$$

where $j_{m,f} \in [1, T_f]$ identifies the index of the fuzzy set (among the T_f fuzzy sets of partition P_f), which has been selected for X_f in rule R_m.

To take the "don't care" condition into account, a new fuzzy set $A_{f,0}$ $(f = 1, ..., F)$ is added to all the F input partitions P_f. This fuzzy set is characterized by a membership function equal to 1 on the overall universe.

The terms $A_{f,0}$ allow generating rules which contain only a subset of the input variables. It follows that $j_{m,f} \in [0, T_f]$, $f = 1, ..., F$, and $j_{m,F+1} \in [1, T_{F+1}]$. Thus, an MFRBS can be completely described by a matrix $J \in \mathbf{N}^{M \times (F+1)}$ [4], where the generic element (m, f) indicates that fuzzy set $A_{f, j_{m,f}}$ has been selected for variable X_f in rule R_m. We adopt the product and the weighted average method as AND logical operator and defuzzification method, respectively.

Given a set of N input observations $\mathbf{x}_n = [x_{n,1}, ..., x_{n,F}]$, with $x_{n,f} \in \Re$, and the set of the corresponding outputs $x_{n,F+1} \in \Re$, $n = 1, ..., N$, we apply an MOEA which produces a set of MFRBSs with different trade-offs among accuracy, complexity and integrity by learning simultaneously the RB and the MF parameters.

3 MF Parameter Learning

We approach the problem of learning the MF parameters by using a piecewise linear transformation [11]. The transformation is described in Fig. 1 for a generic variable X_f. We adopt triangular fuzzy sets $A_{f,j}$ defined by the tuple ($a_{f,j}, b_{f,j}, c_{f,j}$), where $a_{f,j}$ and $c_{f,j}$ correspond to the left and right extremes of the support of $A_{f,j}$, and $b_{f,j}$ to the core. In the following, we assume that the interval ranges of the original and transformed variables are identical. Further, we consider each variable normalised in [0,1]. Finally, given a generic partition $P_f = \left\{ A_{f,1}, ..., A_{f,T_f} \right\}$, we assume that $a_{f,1} = b_{f,1}$, $b_{f,T_f} = c_{f,T_f}$, and for $j = 2 ... T_f - 1$, $b_{f,j} = c_{f,j-1}$ and $b_{f,j} = a_{f,j+1}$.

Before handing the input value x_f over to the fuzzy system, we apply the transformation $t(x_f)$. Thus, we have that $A_{f,j}(x_f) = \tilde{A}_{f,j}(t(x_f)) = \tilde{A}_{f,j}(\tilde{x}_f)$, where $\tilde{A}_{f,j}$ and $A_{f,j}$ are two generic fuzzy sets from the uniform and non-uniform fuzzy partitions, respectively. In those regions where t has a high value of the derivative

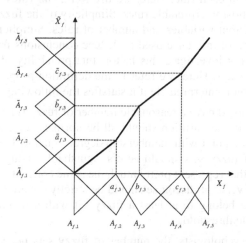

Fig. 1. An example of piecewise linear transformation

(high slope of the lines), the fuzzy sets $A_{f,j}$ are narrower; otherwise, the fuzzy sets $A_{f,j}$ are wider. To preserve the shape of the MFs, we force the change of slopes in t to coincide with the cores of the fuzzy sets in the partitions.

Let $b_{f,1},...,b_{f,T_f}$ and $\tilde{b}_{f,1},...,\tilde{b}_{f,T_f}$ be the cores of $A_{f,1},...,A_{f,T_f}$ and $\tilde{A}_{f,1},...,\tilde{A}_{f,T_f}$, respectively. Transformation $t()$ can be defined as:

$$t(x_f) = \frac{\tilde{b}_{f,j} - \tilde{b}_{f,j-1}}{b_{f,j} - b_{f,j-1}}(x_f - b_{f,j-1}) + \tilde{b}_{f,j-1}, \qquad b_{f,j-1} \le x_f < b_{f,j}, \qquad j = 2...T_f. \qquad (2)$$

The cores $\tilde{b}_{f,1},...,\tilde{b}_{f,T_f}$ are fixed and therefore known. Further, $b_{f,1}$ and b_{f,T_f} coincide with the extremes of the universe U_f of X_f. Thus, $t(x_f)$ depends on $T_f - 2$ parameters, that is, $t(x_f; b_{f,2},...,b_{f,T_f-1})$. Once fixed $b_{f,2},...,b_{f,T_f-1}$, the partition $P_f = \{A_{f,1},...,A_{f,T_f}\}$ can be obtained simply by transforming the three points $(\tilde{a}_{f,j}, \tilde{b}_{f,j}, \tilde{c}_{f,j})$, which describe the generic fuzzy set $\tilde{A}_{f,j}$ into $(a_{f,j}, b_{f,j}, c_{f,j})$ applying $t^{-1}(\tilde{x}_f)$.

4 Interpretability

As discussed in [12], the interpretability of fuzzy rule-based systems depends on four factors: (i) simplicity of fuzzy rules, (ii) simplicity of the fuzzy RB, (iii) simplicity of fuzzy reasoning and (iv) integrity of the fuzzy partitions.

Simplicity of the fuzzy rules is related to the type of fuzzy rules and to the number of inputs involved in each rule. Here, we use Mamdani rules which are universally recognised as the most interpretable rules. Simplicity of the fuzzy RB mainly depends on the number of input variables and number of rules. Simplicity of fuzzy reasoning depends on the type of inference used to deduce conclusions from facts and rules. In this paper, we do not investigate this factor. Interpretability of the partitions can be defined in several ways. Here, we refer to the definition proposed by [2], who states that a fuzzy partition is interpretable if it satisfies the following properties:

1. The partition should have a reasonable number of fuzzy sets;
2. The fuzzy sets in the partition should all be normal, i.e., for each fuzzy set there exists at least one point with membership degree equal to 1;
3. Each couple of fuzzy sets should be distinguishable enough, so that there are no two fuzzy sets that represent pretty much the same concept;
4. The overall universe of discourse should be strictly covered, i.e., each point of the universe should belong to at least a fuzzy set with a membership degree over a given reasonable threshold.

According to psychologists, the number of fuzzy sets per variable should not be higher than 9 due to a limit of human information processing capability. Actually, the number of linguistic terms should be in the range 7 ± 2 [7]. To increase interpretability, we chose the minimum value. Thus, we evaluate the performance of our approach setting $T_f = T = 5$. Further, we start from uniform partitions composed of normal triangular fuzzy sets and the piecewise linear transformation preserves the normality. Finally, items 3 and 4 are fully satisfied when partitions are uniform. On the other hand, the piecewise linear transformation tends to increase accuracy by adapting the MFs to the specific application context. Often, the MF adaptation process generates partitions which are quite far from being uniform, thus loosing in interpretability: the more the partition is different from a uniform partition, the less the partition is interpretable. To take these considerations into account, we introduce the following dissimilarity measure $D = \dfrac{1}{F+1} \sum\limits_{f}^{F+1} \sum\limits_{j=2}^{T-1} \left| b_{f,j} - \tilde{b}_{f,j} \right|$. D expresses how much on average the partitions generated by the MF parameter learning are different from the uniform partitions. From simple mathematical considerations, we derive that $0 \le D \le \dfrac{1}{2}(T-2)$. To evaluate the partition interpretability we define the integrity index $I = 1 - \dfrac{2D}{(T-2)}$. If the transformed partitions remain uniform, I is equal to 1. The higher the difference between the uniform partitions and the transformed partitions, the lower the value of I (at the minimum, $I = 0$). We observe that the covering of the universe is always guaranteed by the type of transformation.

To increase the partition integrity and the accuracy, and to reduce the complexity of MFRBSs, that is, to restrict the number of rules, input variables and conditions in the antecedent of rules, are often conflicting objectives. Thus, we approach the problem by using a three-objective evolutionary algorithm, where the three objectives are the half of MSE, the complexity and the I index, respectively.

5   The Three-Objective Evolutionary Approach

We adopt the (2+2)M-PAES proposed in [4]. Each solution is codified by a chromosome C composed of two parts (C_1, C_2), which define the RB and the piecewise linear transformations of all the variables, respectively. C_1 codifies matrix J and is composed of $M \cdot (F+1)$ natural numbers where M is the number of rules currently present in the RB. C_2 is a vector containing $F+1$ vectors of $T-2$ real numbers: the f th vector contains the $\left[b_{f,2}, ..., b_{f,T_f-1} \right]$ points which define the piecewise linear transformation for the linguistic variable X_f.

In order to generate the offspring populations, we exploit both crossover and mutation. We apply the one-point crossover defined in [4] to C_1 and the BLX-α crossover, with α = 0.5, to C_2. Possibly, we reorder the cores so as to preserve the label ordering. To constrain the search space, we fix the possible minimum and maximum numbers of rules to 5 and 30, respectively. The crossover is applied with probability 0.5.

As regards mutation, we apply for C_1 the two mutation operators described in [4]. If the crossover is not applied, the mutation is always applied to C_1; otherwise the mutation is applied with probability 0.2. The two mutations operators are applied with probabilities 0.55 and 0.45, respectively. The mutation applied to C_2 first chooses randomly a variable $f \in [1, F+1]$, then extracts a random value $j \in [2, T_f - 1]$ and changes the value of $b_{f,j}$ to a random value in the interval $\left[b_{f,j-1}, b_{f,j+1}\right]$. The probability of applying the mutation to C_2 is 0.2.

6 Experimental Results

We tested our approach on the real word regression problem described in [13] that consists of estimating the maintenance costs of medium voltage lines in a town. The data set contains 1059 patterns (4 input and 1 output variables). In order to asses the reliability of our approach, we performed a five-fold cross-validation, using each fold six times with different seeds for the random function generator (thirty trials in total). We fixed the archive size and the maximum number of iterations to 128 and 300,000, respectively. Figure 2 shows an example of the Pareto fronts achieved by the algorithm on the training and test sets, respectively. As expected, we can observe that, when the accuracy increases, the complexity increases and the integrity decreases.

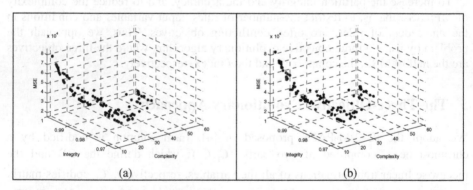

Fig. 2. An example of Pareto fronts obtained in the training (a) and test (b) sets, respectively

To assess the advantages of learning concurrently RB and MF parameters in the three-objective framework, we compared the results achieved by our approach with the ones obtained by applying the (2+2)M-PAES to learn only rules, that is, using only the first part of the chromosome. We denote these two approaches as PAES-3ob and PAES-RB, respectively. To perform the comparison statistically and not on a single trial, we introduced the idea of average Pareto fronts. These fronts are obtained as follows. First, the solutions in the Pareto front approximations produced on the training set in each of the thirty trials are ordered for increasing MSE values. Then, the corresponding solutions, that is the solutions with the same index, are averaged on the thirty Pareto front approximations. We plot for PAES-RB the twenty solutions with the lowest MSEs (the choice of considering only the twenty solutions with the lowest MSEs was motivated by the observation that the other solutions are in general characterized by quite high MSEs which make these solutions impractical). For PAES-3ob we plot on the MSE-complexity plane the average solutions in the same range of complexity of the twenty solutions generated by PAES-RB. We show the interpretability by using different grades of grey: black and white correspond to the highest and lowest values of the interpretability index, respectively. We can observe that the solutions of the average Pareto front generated by PAES-3ob completely dominate the solutions of the average Pareto front generated by PAES-RB, thus confirming the validity of our approach.

Figures 4a and 4b show examples of DB for, respectively, the most accurate MFRBS and one MFRBS with a medium value of the three objectives, generated by PAES-3ob on one fold. We represent the uniform and the transformed partitions with dashed and continuous lines, respectively. In Fig. 4a, we can observe that the partitions of two variables are quite different from the uniform partition, while in Fig. 4b the obtained partitions are very close to the uniform partition for all the variables.

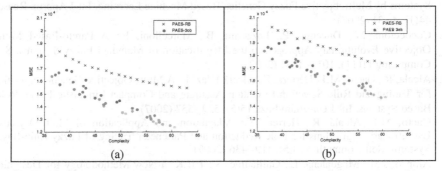

Fig. 3. Comparison between PAES-RB and PAES-3ob in the training (a) and test (b) sets

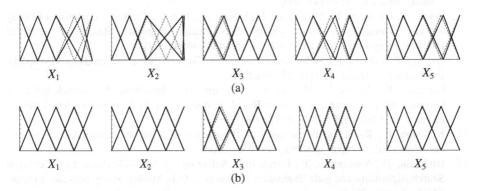

Fig. 4. An example of DB for one of the most accurate solutions (a) and for a solution with a good trade-off among the three objectives (b)

7 Conclusions

In this paper we have proposed a three-objective evolutionary algorithm to generate a
set of Mamdani rule-based fuzzy systems with different trade-offs among accuracy,
complexity and partition interpretability. We have exploited a modified version of the
(2+2)PAES and a chromosome consisting of two parts which codify the RB, and, for
each variable, the parameters of a piecewise linear transformation of the membership
functions. This approach has proved to be very efficient and effective, allowing both a
good exploitation of the solutions and an accurate exploration of the search space.
The algorithm has been tested on a real word regression problem and has provided
Pareto fronts that completely dominate in the accuracy-complexity plane the Pareto
fronts generated by learning only the rule base with a fixed DB.

References

1. Casillas, J., Cordón, O., Herrera, F., Magdalena, L.: Interpretability Issues in Fuzzy
   Modeling. Springer (2003)
2. de Oliveira, J.V., Semantic Constraints for Membership Function Optimization. IEEE
   Trans Syst Man Cybern Part A 29(1), 128--138 (1999)
3. Ishibuchi, H., Nojima, Y.: Analysis of Interpretability-Accuracy Tradeoff of Fuzzy
   Systems by Multiobjective Fuzzy Genetics-Based Machine Learning. Int J Approx Reason
   44(1), 4--31 (2007)
4. Cococcioni, M., Ducange, P., Lazzerini, B., Marcelloni, F.: A Pareto-Based Multi-
   Objective Evolutionary Approach to the Identification of Mamdani Fuzzy Systems. Soft
   Computing 11(11), 1013--1031 (2007)
5. Alcalá, R., Gacto, M.J., Herrera, F., Alcalá-Fdez, J.: A Multi-Objective Genetic Algorithm
   for Tuning and Rule Selection to Obtain Accurate and Compact Linguistic Fuzzy Rule-
   Based Systems. Int J Uncertain Fuzz 15(5), 539--557 (2007)
6. Gacto, M.J., Alcalá, R., Herrera, F.: Adaptation and Application of Multi-Objective
   Evolutionary Algorithms for Rule Reduction and Parameter Tuning of Fuzzy Rule-Based
   Systems. Soft Computing 13(5), 419--436 (2009)
7. Alonso, J.M., Magdalena, L., Guillaume, S.: HILK: a New Methodology for Designing
   Highly Interpretable Linguistic Knowledge Bases Using the Fuzzy Logic Formalism. Int.
   J. Intell. Syst. 23, 761--794 (2008)
8. Botta, A., Lazzerini, B., Marcelloni, F., Stefanescu, Dan C.: Context Adaptation of Fuzzy
   Systems Through a Multi-Objective Evolutionary Approach Based on a Novel
   Interpretability Index. Soft Computing 13(5), 437--449 (2009)
9. Herrera, F.: Genetic Fuzzy Systems: Taxonomy, Current Research Trends and Prospects.
   Evolutionary Intelligence 1(1), 27--46 (2008)
10. Ducange, P., Alcalá, R., Herrera, F., Lazzerini, B., Marcelloni, F.: Knowledge Base
    Learning of Linguistic Fuzzy Rule-Based Systems in a Multi-objective Evolutionary
    Framework. LNCS 5271, 747--754 (2008)
11. Klawonn, F.: Reducing the Number of Parameters of a Fuzzy System Using Scaling
    Functions. Soft Computing 10(9), 749--756 (2006)
12. Ishibuchi, H., Yamamoto, T.: Fuzzy Rule Selection by Multi-Objective Genetic Local
    Search Algorithms and Rule Evaluation Measures in Data Mining. Fuzzy Sets and Systems
    141(1), 59--88 (2004)
13. Cordón, O., Herrera, F., Sanchez, L.: Solving Electrical Distribution Problems using
    Hybrid Evolutionary Data Analysis Techniques. Applied Intelligence 10, 5--24 (1999)

A New Component Selection Algorithm Based on Metrics and Fuzzy Clustering Analysis

Camelia Şerban, Andreea Vescan, and Horia F. Pop

Babeş-Bolyai University, Department of Computer Science,
1 Kogalniceanu St., 400084, Cluj-Napoca, Romania
{camelia,avescan,hfpop}@cs.ubbcluj.ro
http://www.cs.ubbcluj.ro/~{camelia,avescan,hfpop}

Abstract. Component-Based Software Engineering is concerned with the assembly of preexisting software components that lead to software systems responding to client specific requirements. This paper presents a new algorithm for constructing a software system by assembling components. The process of selecting a component from a given set takes into account some quality attributes. Metrics are defined in order to quantify the considered attributes. Using these metrics values, a fuzzy clustering approach groups similar components in order to select the best candidate. We comparatively evaluate our results with a case study.

Keywords: Component Selection Problem, Metrics, Fuzzy Analysis.

1 Introduction

The main objective of Component-Based Software Engineering [1] is that of obtaining a better and more efficient system while having a shorter development time and using existing components rather than developing new ones. In this paper we address the problem of component selection. Informally, our problem is to select a subset of components satisfying the system requirements. The difficulty resides in the fact that each component had a related set of components that share similar functionalities and because of this, an algorithm for the decision process is needed. Fuzzy clustering analysis is used to classify the components based on the values of metrics that measure different attributes of the components. The choice of the best component is based on the obtained classifications.

We discuss the proposed approach as follows. Section 2 presents the theoretical background regarding the problem that we address. Section 3 presents our proposed algorithm for selecting a set of components that satisfies all the requirements. The approach uses fuzzy clustering analysis to help us decide which component should be selected. Section 4 presents a case study. We have compared the obtained solution by our algorithm with solutions obtained by other approaches. Finally, Section 5 summarizes the contributions of this work and outlines directions for further research.

E. Corchado et al. (Eds.): HAIS 2009, LNAI 5572, pp. 621–628, 2009.

2   Theoretical Background

Problem statement. Component Selection Problem (CSP) consists of choosing a number of components from a set of components such that their composition satisfies a set of objectives. The used notation for formally defining our problem is described in what follows.

Denote by SR the set of final system requirements $SR = \{r_1, r_2, ..., r_n\}$, and by SC the set of components available for selection $SC = \{c_1, c_2, ..., c_m\}$. Each component c_i may satisfy a subset of requirements from SR, $SR_{c_i} = \{r_{i_1}, r_{i_2}, ..., r_{i_k}\}$, with additional condition that $SR \cap SR_{c_i}$ is not empty. We also denote by $CR = (SR_{c_1}, SR_{c_2}, ..., SR_{c_m})$ the vector containing the requirements of all components.

In order to specify the component dependencies we use a dependency matrix D [5]. The dependencies specification table contains dependencies between each requirement in the set of all components requirements.

The goal is to find a set of components Sol in such a way that every requirement r_j from the set SR may have assigned a component c_i from Sol where r_j is in SR_{c_i}.

Different components may exist to satisfy the same needed requirement and our aim is to select the best available component.

Component classification based on metrics and fuzzy analysis. As it has already been mentioned above, we need to evaluate the available set of components in order select the best candidate. This evaluation is based on some criteria (quality attributes) that are important for the final system.

The main objective of CBSE is that of obtaining a more efficient system, with shorter development time and better quality products. These attributes are quantified using some of the metrics stated in the following. The *cost (C)* of a component metric is defined as the overall cost of acquisition and adaptation of that component. Regarding the reusability criterion, the selected metrics are *Provided Services Utilization (PSU)* and *Required Services Utilization (RSU)* [2]. The last metric considered in this study is *Functionality* metric (F) defined as the ratio between the number of required services of the system that are provided by the component and the number of required services of the system. The influence of these metrics values over the quality attributes previously mentioned has been discussed in [8].

In the following, each component c_i from SC is described by a 4-dimensional vector, $c_i = (C, PSU, RSU, F)$.

Next, our goal is to group similar components regarding the defined attributes. To obtain this, we use a clustering approach [4]. The objects to be clustered are components from our repository and the characteristics of these objects are the corresponding metrics values. The next problem is the selection of one component out of a set of possibilities.

3   Proposed Algorithm Description

In this section we propose a new algorithm for the component selection problem defined in section 2. Our algorithm is based on selected metrics and fuzzy clustering approach described before. The objects to be clustered are components, each component being identified by a vector of metrics values. Our focus is to group similar objects in order to select the best candidate. The fuzzy clustering algorithm (Fuzzy n-means algorithm) used to determine the fuzzy partition of the set of components is described in [3]. Taking these into account, we next give the proposed algorithm.

Two alternative approaches are possible: one that uses only one initial partition and the second one that recomputes metrics based on the update of SR and reclassifies the candidate components at each step of a component selection (from a set of candidates). The described situations are emphasized in the algorithm by the use of the *changePartition* input variable.

The Pseudocode of the approach (algorithms Metrics and Fuzzy-based Component Selection with Same Partition **MFbCSwSPA** and Metrics and Fuzzy-based Component Selection with Changed Partition **MFbCSwCPA**) is described in Algorithm 1.

Algorithm 1. Metrics and Fuzzy-based Component Selection Algorithm (MFbCSwSPA/MFbCSwCPA)

Require: SR,n; {SR-set of requirements,n-no. of requirements}
  SC, m; {SC-set of components and their metrics values, m-no. of components}
  CR, D; {CR- components requirements vector, D-dependency matrix}
  changePartition. {boolean value deciding if the algorithm recomputes the metrics and then recomputes the Fuzzy partitions using the remained components.}
Ensure: Sol. {obtained solution }

1: FuzzyPartitionDet(SC, m, A, B);
  {A, B fuzzy sets represented as vectors containing the membership degree of the components of the two clusters;}
2: startCompSet=StartComponentsSet(CR, m, D);
3: selectedComp=SelectComp(A,B,m,startCompSet);
4: AddToSol(Sol,selectedComp); { adds the component selectedComp to Sol;}
5: UpdateReqSet(SR, n, selectedComp, CR, m);
6: **while** (SR is not empty) **do**
7:    **if** (changePartition) **then**
8:       CurrentCompSet(SC,m,SR,n); { provides the components that may offer functionalities from SR;}
9:       ReComputeMetrics(SC, m);
10:      FuzzyPartitionDet(SC, m, A, B);
11:   **end if**
12:   posibleCompSet=PosibleComponentsSet(SR, n, CR, m);
13:   selectedComp=SelectComp(A,B,m,posibleCompSet)
14:   AddToSol(Sol,selectedComp); { adds the component to the solution;}
15:   UpdateReqSet(SR, n, selectedComp, CR, m);
16: **end while**

The $FuzzyPartitionDet(SC, m, A, B)$ subalgorithm computes a fuzzy partition of the set SC. The fuzzy sets obtained after the first splitting are A, B. The subalgorithm $StartComponentsSet$ determines the subset of start components

624   C. Şerban, A. Vescan, and H.F. Pop

from the set of given components set, those components that have no dependencies. The components with no dependencies are the "source" components, i.e. the components that read the input data of the algorithm.

The *SelectComp* subalgorithm selects the new component to be added to *Sol* from a set of possible components. Initially, the set of possible components to be added to *Sol* are the start components set. At the next step, those possible components are obtained by calling the subalgorithm *PosibleComponentsSet*.

The *SelectComp* subalgorithm is described in Subalgorithm 2. From this set of possible candidates we have to select one of them. There are two different cases that may appear: all these components belong to the same cluster, or some of them are in the first cluster and the others in the second one. For the first case, we select the component with the maximum membership degree for that class. Regarding the second case we proceed in the following way: the best candidate from each cluster is identified, and then some criteria are considered to choose one of them. For this reason, the component set is split in two clusters. In future research we will apply a divisive hierarchical algorithm and the initial partition will be further split.

Algorithm 2. SelectComp Subalgorithm

Require: A,B, m; {A, B fuzzy partition (vectors containing the membership degree of the components); m-no. of components;}
   compSet; {the set of possible components candidates.}
Ensure: comp. {the best candidate component.}

1: **if** (BelongsToTheSameCluster(compSet,A,B,m,firstCluster)) **then**
2:   **if** (firstCluster) **then**
3:     comp = MaxMembershipDegree(compSet,A,m);
4:   **else**
5:     comp = MaxMembershipDegree(compSet,B,m);
6:   **end if**
7: **else**
8:   compA = MaxMembershipDegree(compSet,A,m);
9:   compB = MaxMembershipDegree(compSet,B,m);
10:    comp = CriteriaBasedBestClusterCandidateSelection(compA,compB);
11: **end if**

The subalgorithm *CriteriaBasedBestClusterCandidateSelection* decides, based on some criteria which one of the two components is the best candidate. The criteria helps us decide which of the two clusters contains "good" components considering the metrics values. In this way the selected component is the best representative object from the "good" cluster.

4 Case Study

In order to validate our approach we have used the following case study. The set of requirements $SR = \{r_0, r_3, r_4, r_7, r_9, r_{12}\}$ and the set of components $SC = \{c_0, c_1, c_2, c_3, c_4, c_5, c_6, c_7, c_8, c_9\}$ are given.

Table 1 contains for each component the provided services (in term of requirements for the final system). Table 2 contains the dependencies between each requirement from the set of requirements. Table 3 contains the values of the metrics.

Table 1. Requirements

Comp.	Requirements
c_0	r_0, r_1, r_7
c_1	r_4, r_5, r_6, r_{12}
c_2	r_0
c_3	r_0, r_2, r_8, r_{10}
c_4	r_3, r_{11}
c_5	r_4, r_5, r_6, r_9
c_6	r_7, r_9, r_{12}
c_7	r_1, r_2, r_9, r_{12}
c_8	r_3, r_4, r_{10}, r_{11}
c_9	$r_0, r_5, r_6, r_8, r_9, r_{12}$

Table 2. Specification of the Requirements Dependencies

Depend.	r_0	r_1	r_2	r_3	r_4	r_5	r_6	r_7	r_8	r_9	r_{10}	r_{11}	r_{12}
r_0				√									
r_1										√	√		
r_2		√			√								
r_3													
r_4	√									√			
r_5											√		
r_6									√				
r_7				√									
r_8				√									
r_9	√												
r_{10}												√	
r_{11}		√											
r_{12}				√									

Table 3. Initial Metrics Values

Comp.	c_0	c_1	c_2	c_3	c_4	c_5	c_6	c_7	c_8	c_9
PSU	0.66	0.50	1.00	0.25	0.50	0.50	1.00	0.5	0.5	0.5
RSU	0.5	0.60	1.00	0.20	0.0	0.60	1.00	0.33	0.50	0.5
F	0.33	0.33	0.16	0.16	0.16	0.33	0.50	0.33	0.33	0.50
C	0.08	0.07	0.06	0.09	0.06	0.14	0.15	0.14	0.07	0.14

4.1   Solution Obtained by the Proposed Algorithm

The criteria that help us decide which of the two components should be chosen to be part of the final solution are based on three metrics values that quantify: functionality, reusability and cost. When the functionality criterion has the same value for both components, the reusability criterion is considered. If again, the reusability criteria has the same value, the cost criterion is used. If all the criteria are equal one of the components is randomly selected.

Solution obtained with the same partition and metrics values. Applying the [3] algorithm we obtained the results in Table 4.

In the first step of the algorithm the only two components having no dependencies are c_4 and c_8. Thus the start set of components (responsible for reading the input data and do no computation) contains only these two. Based on the partition, and because both of them are in the same cluster (A), the component with the highest membership degree is chosen, c_8.

Table 4. The final partition for the set of 10 components

Class	c_0	c_1	c_2	c_3	c_4	c_5	c_6	c_7	c_8	c_9	Representative components (R.C.)
A	0.82	0.86	0.59	0.72	0.77	0.11	0.27	0.26	0.91	0.15	$c_0, c_1, c_2, c_3, c_4, c_8$
B	0.18	0.14	0.41	0.28	0.23	0.89	0.73	0.74	0.09	0.85	c_5, c_6, c_7, c_9

The remaining set of requirements is: $\{r_0, r_7, r_9, r_{12}\}$. The components that offer these requirements are split in the following clusters: A contains $\{c_0, c_1, c_2, c_3\}$ with the most representative being c_1 and the cluster B contains $\{c_5, c_6, c_7, c_9\}$ with the most representative being c_5. To decide which of c_1 and c_5 components should be chosen, we consider the metrics-based criteria selection mention above. In this case, the c_1 component is chosen due to the cost criterion.

The new set of remaining requirements is $\{r_0, r_7, r_9\}$. The components that offer these requirements are again grouped in two clusters. Based on the same rules, between the c_0 and c_5 components, the component c_0 was chosen.

After choosing the c_0 component, the only requirement needs to be satisfied is r_9. All the components that offer this functionality are grouped in the same cluster, thus the chosen component is c_5.

The obtained solution contains the components: $\{c_0, c_1, c_5, c_8\}$ and has the cost 32. The reusability of this system solution is 5. We define reusability as the number of requirements of the components in the system, that are not in the current set of requirements.

Solution obtained with the changed partition and metrics values. Applying the [3] algorithm, we obtained the results in Table 4. The first step is the same as in the first version, component c_8 being chosen. At the second step the set of remaining requirements is $\{r_0, r_7, r_9, r_{12}\}$ and the set of the candidate components and the metrics values are stated in Table 5. The corresponding fuzzy partition is presented in Table 7.

Table 5. Second Step Metrics Values for candidate components

Comp.	c_0	c_1	c_2	c_3	c_5	c_6	c_7	c_9
PSU	0.66	0.25	1.00	0.25	0.25	1.00	0.5	0.5
RSU	0.5	0.2	1.00	0.2	0.2	1.00	0.33	0.5
F	0.33	0.16	0.16	0.16	0.2	0.5	0.33	0.5
C	0.08	0.07	0.06	0.09	0.14	0.15	0.14	0.14

Table 6. Third Step of Metrics Values for the candidate components

Comp.	c_0	c_2	c_3	c_9
PSU	0.33	1.00	0.25	0.16
RSU	0.25	1.00	0.20	0.16
F	0.16	0.16	0.16	0.16
C	0.08	0.06	0.09	0.14

Based on the new metrics values and the new partition of the candidate components, the c_6 component is chosen between c_3 and c_6. Updating the remaining set of requirements, the only requirement needed to be satisfied is r_0, requirement provided by the components $\{c_0, c_2, c_3, c_9\}$. In Table 6 the new values of the

Table 7. The second step partition for the set of candidate components

Class.	c_0	c_1	c_2	c_3	c_5	c_6	c_7	c_9	R.C.
A	0.47	0.92	0.40	0.96	0.84	0.12	0.54	0.23	c_1, c_3
									c_5, c_7
B	0.53	0.08	0.60	0.04	0.16	0.88	0.46	0.77	c_0, c_2
									c_6, c_9

Table 8. The third step partition for the set of candidate components

Class.	c_0	c_2	c_3	c_9	R.C.
A		0.93	0.01	0.98	0.91	c_0
						c_3
						c_9
B		0.07	0.99	0.02	0.09	c_2

metrics are given. The corresponding partition is presented in Table 8 and the criterion that differentiate the c_2 and c_3 is *PSU*. The c_2 component is chosen.

Thus, the obtained solution contains the components: $\{c_2, c_6, c_8\}$ and has the cost 28 and the reusability only 2.

4.2 Comparative Analysis of the Obtained Solutions by Other Approaches

The problem of selecting components from a set of available components was also discussed in several papers using various approaches. A Greedy algorithm [6] was used taking into consideration also the dependencies between the components. A Branch and Bound approach using the cost criterion and the number of remained requirements was proposed in [7].

An approach for the same selection problem that uses principles of evolutionary computation and multiobjective optimization was proposed in [6]. The problem is formulated as a multiple objective optimization problem having two objectives: the total cost of the components used and the number of components used. Both objectives are to be minimized. Besides these, the dependencies are also treated, but are considered as constraints.

Another evolutionary algorithm that uses the same metrics values from this paper was proposed in [8]. The problem was formulated as multiobjective, considering the metrics values. Two types of experiments were performed: first considering only two objectives and then considering all objective but with different population sizes and different number of generations.

In order to compare our approach with those mention before, we describe in Table 9 the obtained solutions with all approaches.

Table 9. Obtained solutions using different approaches

Algorithm	Solution	Cost	Reusability
Greedy	c_4, c_0, c_7, c_1	35	5
Branch and Bound	c_4, c_2, c_6, c_1	34	3
Genetic algorithm (only cost)	c_2, c_6, c_8	28	2
Genetic algorithm (only PSU and RSU)	c_0, c_7, c_8	29	4
MFbCSwSPA	c_0, c_1, c_5, c_8	32	5
MFbCSwCPA	c_2, c_6, c_8	28	2

The solutions obtained by our proposed algorithm are comparable with the ones provided by the others approaches, the main advantages being: the search space dimension is drastically reduced, the obtained partition suggesting the component that should be selected at a given step; the execution time needed for selecting the best component is reduced due to the reduced search space; the selection criteria of the components are based on several characteristics of components (several metrics may be defined).

5   Conclusions and Future Work

A new algorithm based on metrics and fuzzy clustering analysis that addresses the problem of component selection was proposed in this paper. We evaluate our approach using a case study. We will focus our future work on three main fronts: to apply this approach for more case studies; to apply other fuzzy clustering algorithms in order to obtain the needed classification, and to select more relevant metrics.

Acknowledgement

This material is based upon work supported by the Romanian National University Research Council under award PN-II no. ID-550/2007.

References

1. Crnkovic, I., Larsson, M.: Building Reliable Component-Based Software Systems. Artech House publisher (2002)
2. Hoek, A.v.d., Dincel, E., Medvidovic, N.: Using Service Utilization Metrics to Assess and Improve Product Line Architectures. In: 9th IEEE International Software Metrics Symposium (Metrics 2003), Sydney, Australia (2003)
3. Bezdek, J.: Pattern Recognition with Fuzzy Objective Function Algorithms. Plenum Press, New York (1981)
4. Zadeh, L.A.: Fuzzy sets. Information and Control 8, 338–353 (1965)
5. Vescan, A.: Dependencies in the Component Selection Problem. In: Proceedings of the 6th ICAM - International Conference on Applied Mathematics (2008) (accepted)
6. Vescan, A.: An evolutionary multiobjective approach for the Component Selection Problem. In: Proceedings of the First IEEE International Conference on the Applications of Digital Information and Web Technologies, pp. 252–257. IEEE Press, Los Alamitos (2008)
7. Vescan, A., Pop, H.F.: The Component Selection Problem as a Constraint Optimization Problem. In: Hruska, T., Madeyski, L., Ochodek, M. (eds.) Software Engineering Techniques in Progress, Wroclaw University of Technology, Wroclaw, Poland, pp. 203–211. IEEE Press, Los Alamitos (2008)
8. Vescan, A.: A Metrics-based Evolutionary Approach for the Component Selection Problem. In: Proceedings of the 11th International Conference on Computer Modelling and Simulation. IEEE Press, Los Alamitos (2009) (accepted)

Multi-label Classification with
Gene Expression Programming

J.L. Ávila, E.L. Gibaja, and S. Ventura

Department of Computer Science and Numerical Analysis, University of Córdoba

Abstract. In this paper, we introduce a Gene Expression Programming algorithm for multi label classification. This algorithm encodes each individual into a discriminant function that shows whether a pattern belongs to a given class or not. The algorithm also applies a niching technique to guarantee that the population includes functions for each existing class. In order to evaluate the quality of our algorithm, its performance is compared to that of four recently published algorithms. The results show that our proposal is the best in terms of accuracy, precision and recall.

1   Introduction

Classification is one of the most studied tasks in the machine learning and data mining fields. This task basically consists of finding a function which is able to identify the set of an object's attributes (predictive variables) with a label or class identification (categorical variable). In the simplest case, each learning example has only one associated label, l_i, of a set of labels, L, which has been previously defined. This label l_i defines a set of patterns that do not share any element with subsets which are defined by the other labels l_j ($\forall j \neq i$). Nevertheless, this is not the only possible hypothesis, because numerous problems can be found where a given pattern can be simultaneously mapped to more than one class label. Typical examples are semantic scene classification [1], text and sound categorization [2,3], protein and gene classification [4] or medical diagnosis [5]. All these problems, which involve assigning all possible proper labels to a given example from a set of prediction variables, are multi-label classification problems [6].

The problem of multi-label classification has been tackled from numerous points of view. On the one hand, some papers describe a pre-processing of the data set which transforms a multi-label problem into one with a single-label which admits the application of supervised learning algorithms [1,7]. On the other hand, a specifically designed approach for multi-label data can be carried out [8,9]. With regard to the techniques that have been used, it is worth highlighting decision trees [8,10], Bayesian classifiers [11], artificial neural networks [12,13] and support vector machines [1,3,4]. Techniques of lazy learning have also been used, particularly a multi-label version of the well-known K-nearest neighbor algorithm called ML-KNN [9], and associative classification methods [14].

E. Corchado et al. (Eds.): HAIS 2009, LNAI 5572, pp. 629–637, 2009.

Finally, it is worthwhile mentioning the emerging interest in applying ensemble methods to multi-label classification in order to improve predictions [7,15,16].

As has been shown, there is a great variety of approaches to solve this kind of problems. Nevertheless, it seems that multi-label evolutionary approaches have not been applied, despite that the fact they have solved successfully numerous problems in traditional classification. Therefore, the goal of this paper has been the application and the analysis of this type of algorithms in multi-label problems. We have focused specifically on the Gene Expression Programming [17], a paradigm that has been successfully applied to other classification problems [18]. The algorithm developed, called GEP-MLC, encodes a discriminant function in each individual and uses a niching algorithm to guarantee diversity in the solutions. As we will see later, its results are quite satisfactory in terms of accuracy, precision and recall, improving results obtained by other recent algorithms.

The paper is organized as follows. The next section introduces the proposed algorithm. Then, the experiments carried out will be described, as will the results of the experiments along a set of conclusions and proposals for future research.

2   Algorithm Description

In this section we specify different aspects which have been taken into account in the design of the GEP-MLC algorithm, such as individual representation, genetic operators, fitness function and evolutionary process.

2.1   Individual Representation

As mentioned above, the GEP-MLC learns discriminant functions. A discriminant function is a function which is applied to the input features of a pattern (predictive variables) and produces a numerical value associated with the class that the pattern belongs to. To establish this correspondence, a set of thresholds are defined, and intervals of values in the output space are mapped to classification labels. The simplest example is that of the binary classifier, where only one threshold is defined (usually zero). Values to the right of this threshold are associated with patterns belonging to the class, while values to the left will be associated with non-membership in the class.

$$if(f(\mathbf{X}) > 0) \; then \; \mathbf{X} \in class \; else \; \mathbf{X} \notin class \qquad (1)$$

In the case of multi-class problems (the number of classes $N > 2$), there are two approaches to tackle the problem. On the one hand, $N - 1$ thresholds and N intervals can be defined. On the other hand, $N - 1$ functions with only one threshold can be used and deal with the membership of an individual class as a binary classification problem. This last approach is the one that has been used in this study. So, each individual codes in its genotype the mathematical expression corresponding to a discriminant function (binary classifier), and the threshold value of zero has been assigned for all cases. As will be shown later, the class

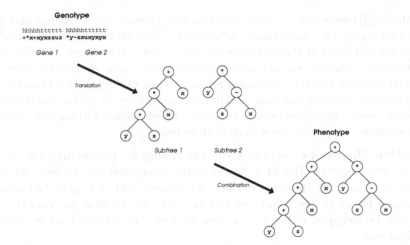

Fig. 1. Conversion from genes to subtrees

associated to each discriminant function is assigned during the evaluation of the individual.

Regarding individual representation, it must be said that in GEP-MLC, as in GEP algorithms, individuals have a dual encoding, that is, they have both genotype and phenotype. Genotype is a lineal string that consists of several genes, whose number and length is fixed and predetermined for each problem. Each gene is divided into two sections, *head* and *tail*. The first one contains terminal and non-terminal elements, whereas the second one can just contain terminal elements[1]. Head length is selected *a priori*, but tail size is calculated as

$$t = h(n-1) + 1 \qquad (2)$$

where t is tail size, h is head length and n the maximum arity (number of arguments) in the non-terminal set.

Phenotype is an expression tree obtained by a mapping process in which (a) each gene is converted into an expression subtree and (b) subtrees are combined by means of a connection function (in our case, the summation operator -see Figure 1 for an example).

2.2 Genetic Operators

GEP-MLC uses all the genetic operators defined in the standard GEP algorithm, that is, crossover, transposition and mutation. In the following their main features will be explained briefly.

Crossover Operators. choose two random parents and swap parts of their genotypes to generate two valid new offspring. GEP has three kinds of crossover

---

[1] In this context, terminal elements are functions without arguments, and non-terminal ones are functions with one or more arguments.

operators: (a) *One-point recombination* takes two parent chromosomes and splits them at exactly the same point. Afterwards the material downstream of the recombination point is exchanged between the two chromosomes. (b) *Two-point recombination* takes two points at random, and split chromosomes by them. The material between the recombination points is then exchanged between the two chromosomes, forming two new children chromosomes. (c) *Gene recombination* exchanges entire genes between two parent chromosomes, forming two children chromosomes containing genes from both parents.

Mutation Operators. perform a random change in a randomly selected element in the genotype. However, the operator must preserve the internal structure of the genotype and consequently, an element from the gene tail must be exchanged for an element from the terminal set, but an element from the gene head can be swapped either for an element from the terminal set or the nonterminal one.

Transposition Operators. are special mutation operators that select a substring from the genotype and transfer it to another position. GEP paradigm defines three transposition operators: (a) *IS Transposition* randomly selects one substring also with random length, this is inserted in any position in the gene head, other than the first element(the tree root). (b) *RIS Transposition* randomly chooses one fragment in the same way as IS transposition, but the element chosen should start with a function. This fragment is inserted into the first element of the gene head (the tree root). (c) *Gene Transposition* changes the position of two genes in individuals with more than one gene. This transposition randomly chooses two genes that swap their position in the genotype.

2.3   Individual Evaluation

During the evaluation phase, the discriminant function is mapped by a given individual, and its quality as a classifier is calculated for each class defined in the problem. The fitness function used is the F score, that is, the harmonic mean of precision and recall [19]:

$$fitness = \frac{2 \times precision \times recall}{precision + recall} \tag{3}$$

In contrast with single-label algorithms, that assign the label that produces the highest fitness value, the GEP-MLC algorithm calculates a fitness value for each label, storing N raw fitness values (one per label). As will be shown, the fitness value used in the selection phase is obtained by transforming these values during the token competition phase and taking the highest one.

2.4   Evolutionary Algorithm

Listing 1 shows the pseudocode of the GEP-MLC algorithm. As can be seen, its structure is similar to that of the standard GEP algorithm, but also takes the n

Algorithm 1. GEP-MLC pseudocode

Generate initial population $P(0)$
$g_{count} \leftarrow 0$
while $g_{count} < g_{max}$ **do**
    for all $l_i \in L$ (labels set) **do**
        Evaluate all individuals in $P(g_{count})$
    end for
    for all $l_i \in L$ **do**
        $total\_tokens$ = number of patterns belonging to l_i
        for all Individual Ind_j in $P(g_{count})$ **do**
            $tokens\_won_{Ind_j} = 0$
        end for
        for all positive pattern (token) **do**
            Select Ind_j with highest fitness that correctly classifies the pattern
            $tokens\_won_{Ind_j} + +$
        end for
        for all Individual Ind_j in $P(g_{count})$ **do**
            Update the fitness by using formula 4
        end for
    end for
    Do parents selection
    Apply genetic operators
    Update population
    $g_{count} + +$
end while
Generate classifier

fitness calculation into consideration. In addition, the algorithm also implements the *Token Competition* technique to correct individual fitness after its evaluation.

Token competition [20] is used in classification algorithms to emulate the niching effect, as shown in natural ecosystems; when a species finds a convenient place to live, it does not usually evolve abruptly, but tries to adapt to a mild environment and does not allow any other species to settle there. Token Competition tries to achieve a set of specialized individuals to classify patterns sub-sets. So, for each positive pattern, a *token* is a stake which is won by the individual with the highest fitness correctly classifying the pattern. When all the tokens are distributed, the algorithm corrects the fitness of each individual using the expression:

$$new\_fitness = \frac{original\_fitness \times tokens\_won}{total\_tokens} \tag{4}$$

Token Competition penalizes individuals that, despite their average fitness, do not contribute to the classifier. On the other hand, they help both the individuals with good fitness that correctly classify many patterns, and individuals specialized in classifying strange patterns, which are not usually correctly classified as the best individuals. In the proposed algorithm, there will be as many token competitions as labels in the training set, only using patterns with labels. Each token is played by every individual with the fitness associated to its label.

When the algorithm finishes, it is easy to find which individuals must be in the learned classifier. Only individuals that win any token are relevant for the classifier.

3   Experimental Section

The goal of the experiments is to establish if the proposed GEP-MLC algorithm is effective in dealing with multi-label classification problems, and its performance is compared to other multi-label classification algorithms. This section explains several details related with these experiments such as data sets and algorithmic details.

3.1   Data Sets

For the experimentation, the algorithm proposed has been tested with three multi-label data sets, *scene*, *emotions* and *yeast*. Scene data set contains a series of patterns about kinds of landscapes (beach, sunset, mountain, field, urban and fall foliage). Emotions data set is concerned with the classification of songs according to the emotions they evoke. Finally, Yeast includes information about protein function. Table 1 shows the main features (number of labels, attributes and patterns) of each data set. Label cardinality (the average number of labels per example) and label density (the same number divided by the total number of labels) are used to explain how much multi-label is a data set [6,7].

All data sets have been randomly split into 10 partitions in order to carry out a 10-fold cross validation. For each test, 5 different runs have been executed and an average value has been calculated in order to measure the performance of the evolutionary algorithm as independently as possible from its randomness.

3.2   Implementation

GEP-MPC implementation was made using the JCLEC library [21]. JCLEC is a software system that provides a framework to develop evolutionary algorithms. Our algorithm is based on the GEP module available at the library. The class responsible for fitness calculation and the evolutionary algorithm had to be implemented (the standard GEP algorithm did not implement the niching scheme).

Table 1. Data sets Features

Data set	#patterns	#train	#test	#attributes	#labels	density	cardinality
Scene	2407	1211	1196	294	6	1.061	0.176
Emotions	593	1500	917	103	6	1.868	0.311
Yeast	2417	391	202	78	14	4.228	0.302

Table 2. GEP-MLC algorithm parameters

Parameter	Value
Number of genes	6
Head size	35
Population size	1000
Max of generations	60
Tournament size	2
Mutation probability	0.2
Crossover probability	0.7
Transposition probability	0.4

A set of tests was made to find the optimal parameters of the algorithm (see Table 2). These parameters have been the ones used in the main experiments.

The rest of the algorithms used in the tests were available in the MULAN library. This is a Java package which contains several problem transformation and algorithm adaptation methods for multi-label classification, an evaluation framework that computes several evaluation measures and a class providing data set statistics. MULAN is built on top of the WEKA data mining tool [22] and is freely available at `http://mlkd.csd.auth.gr/multilabel.html`.

4 Results and Discussion

The performance of the proposed algorithm has been compared to four other methods for multi-label classification, namely, Binary Relevance (BR), Label Powerset (LP), RAKEL [7] and the ML-KNN method [9]. The measures of accuracy, precision and recall have been used to compare these methods. To extend these measures from single-label to multi-label we have used the macro-averaged approach proposed in [23], where precision and recall are first evaluated locally for each category, and then globally by averaging over the results of the different categories.

Table 3 shows the accuracy (acc), precision (prec) and recall (rec) results. As can be observed, the proposed algorithm obtains better results than the other ones for the three measures and data sets, being the difference more significant with the data sets which contain more multi-label patterns.

It can also be observed that the differences between the results obtained with the GEP based algorithm and the rest of the algorithms increase with the higher numbers of multi-label features (more density and cardinality). Thus, in the case of the scene data set, where density is very close to one (nearly a single label problem), the GEP algorithm obtains the best results in accuracy and recall, but these results can be compared to those obtained with RAKEL and ML-KNN. Nevertheless, when the results for emotion are analyzed, GEP is found to obtain the best scores for all measures, and the scores are much more higher than the scores of the algorithms studied. Furthermore, the same result can be observed with the yeast data set, the one with the highest number of labels and values of density and cardinality (the most multi-label data set).

Table 3. Experimental results

Algorithm	Scene			Emotions			Yeast		
	acc	prec	rec	acc	prec	rec	acc	prec	rec
Binary Relevance	0.538	0.630	0.623	0.203	0.280	0.253	0.141	0.192	0.129
Label Powerset	0.587	0.594	0.597	0.290	0.276	0.285	0.131	0.193	0.192
RAKEL	0.619	0.773	0.651	0.253	0.110	0.258	0.119	0.212	0.119
ML-Knn	0.647	0.799	0.675	0.126	0.321	0.029	0.113	0.114	0.113
GEP-MLC	0.709	0.746	0.744	0.903	0.724	0.695	0.738	0.715	0.649

5   Conclusions and Future Work

This study presents the GEP-MLC algorithm, an evolutionary algorithm for
multi-label classification. This algorithm, based on GEP, codifies discriminant
functions that indicate that a pattern belongs to a certain class in such a way
that the final classifier is obtained by combining several individuals from the
population. It uses a niching technique (token competition) to ensure that the
population will present functions representing all the classes present in a given
problem. Studies have been carried out to check the performance of our algorithm
and compare it with those of other available algorithms, to verify that GEP-MLC
renders the best performance of them all in terms of exactness, precision and
recall, and is at the same time much less insensitive to the degree of overlapping
in its classes, which is a very positive characteristic in this type of problem.
Regarding to future research, the algorithm is being tested in other domains
and, besides, the efficiency is being studied in order to be optimized.

Acknowledgment. This work has been financed in part by the TIN2008-06681-
C06-03 project of the Spanish Inter-Ministerial Commission of Science and Technology (CICYT) and FEDER funds.

References

1. Boutell, M.R., Luo, J., Shen, X., Brown, C.M.: Learning multi-label scene classification. Pattern Recognition 37(9), 1757–1771 (2004)
2. Li, T., Zhang, C., Zhu, S.: Empirical studies on multi-label classification. In: IC-
   TAI 2006: Proceedings of the 18th IEEE International Conference on Tools with
   Artificial Intelligence, Washington, DC, USA, pp. 86–92. IEEE Computer Society,
   Los Alamitos (2006)
3. Li, T., Ogihara, M.: Detecting emotion in music. In: Proceedings of the 14th intern.
   conference on music information retrieval (ISMIR 2003), Baltimore, USA (2003)
4. Elisseeff, A., Weston, J.: A kernel method for multi-labelled classification. Advances
   in Neural Information Processing Systems 14, 681–687 (2001)
5. Rak, R., Kurgan, L.A., Reformat, M.: Multilabel associative classification categorization of medline articles into mesh keywords. IEEE Engineering in Medicine and
   Biology Magazine 26(2), 47–55 (2007)

6. Tsoumakas, G., Katakis, I.: Multi label classification: An overview. International Journal of Data Warehousing and Mining 3(3), 1–13 (2007)
7. Tsoumakas, G., Vlahavas, I.: Random k-labelsets: An ensemble method for multi-label classification. In: Kok, J.N., Koronacki, J., Lopez de Mantaras, R., Matwin, S., Mladenič, D., Skowron, A. (eds.) ECML 2007. LNCS, vol. 4701, pp. 406–417. Springer, Heidelberg (2007)
8. Clare, A., King, R.D.: Knowledge discovery in multi-label phenotype data. In: Siebes, A., De Raedt, L. (eds.) PKDD 2001. LNCS, vol. 2168, p. 42. Springer, Heidelberg (2001)
9. Zhang, M.L., Zhou, Z.H.: A k-nearest neighbor based algorithm for multi-label classification., vol. 2, pp. 718–721. The IEEE Computational Intelligence Society, Los Alamitos (2005)
10. Noh, H.G., Song, M.S., Park, S.H.: An unbiased method for constructing multilabel classification trees. Computational Statistics & Data Analysis 47(1), 149–164 (2004)
11. Ghamrawi, N., Mccallum, A.: Collective multi-label classification. In: CIKM 2005: Proceedings of the 14th ACM international conference on Information and knowledge management, New York, USA, pp. 195–200. ACM Press, New York (2005)
12. Crammer, K., Singer, Y.: A family of additive online algorithms for category ranking. The Journal of Machine Learning Research 3, 1025–1058 (2003)
13. Zhang, M.L., Zhou, X.H.: Multilabel neural networks with applications to functional genomics and text categorization. IEEE Transactions on Knowledge and Data Engineering 18(10), 1338–1351 (2006)
14. Rak, R., Kurgan, L., Reformat, M.: A tree-projection-based algorithm for multi-label recurrent-item associative-classification rule generation. Data & Knowledge Engineering 64(1), 171–197 (2008)
15. Schapire, R.E., Singer, Y.: Boostexter: A boosting-based system for text categorization. Machine Learning 39(2/3), 135–168 (2000)
16. Johnson, M., Cipolla, R.: Improved image annotation and labelling through multi label boosting. In: Proceedings of the British Machine Vision Conference (16th BMVC), British Machine Vision Association (BMVA), Oxford, U.K (2005)
17. Ferreira, C.: Gene expression programming:a new adaptative algorithm for solving problems. Complex Systems 13(2), 87–129 (2001)
18. Zhou, C., Xiao, W., Tirpak, T.M., Nelson, P.C.: Evolving accurate and compact classification rules with gene expression programming. IEEE Transactions on Evolutionary Computation 7(6), 519–531 (2003)
19. Han, J., Kamber, M.: Data Mining: Methods and Techniques, 2nd edn. Morgan Kaufmann, San Francisco (2006)
20. Wong, M.L., Leung, K.S.: Data Mining Using Grammar-Based Genetic Programming and Applications. Genetic Programming Series. Kluwer Academic Publishers, Dordrecht (2002)
21. Ventura, S., Romero, C., Zafra, A., Delgado, J.A., Hervás, C.: JCLEC: A Java framework for evolutionary computation. Soft Computing 12(4), 381–392 (2008)
22. Witten, I.H., Frank, E.: Data Mining: Practical machine learning tools and techniques, 2nd edn. Morgan Kaufmann, San Francisco (2005)
23. Sebastiani, F.: Machine learning in automated text categorization. ACM Comput. Surv. 34(1), 1–47 (2002)

An Evolutionary Ensemble-Based Method for Rule Extraction with Distributed Data*

Diego M. Escalante, Miguel Angel Rodriguez, and Antonio Peregrin

Dept. of Information Technologies
University of Huelva
{diego.escalante,miguel.rodriguez,peregrin}@dti.uhu.es

Abstract. This paper presents a methodology for knowledge discovery from inherently distributed data without moving it from its original location, completely or partially, to other locations for legal or competition issues. It is based on a novel technique that performs in two stages: first, discovering the knowledge locally and second, merging the distributed knowledge acquired in every location in a common privacy aware maximizing the global accuracy by using evolutionary models. The knowledge obtained in this way improves the one achieved in the local stores, thus it is of interest for the concerned organizations.

1 Introduction

Information technologies peak has produced a huge amount of data, and mining them is one of the most successful areas of research in computer science.

Every now and then, data may be geographically distributed, and habitual data mining techniques need to centralize it, or to get benefits from the distributed computing, using distributed algorithms that move knowledge and training data [13] and [8].

Nevertheless, it is not always possible to move the data in a distributed system because competition or legal issues. For example, banking entities may be interested in global knowledge benefits to avoid credit card fraud, but they have to safeguard their clients data. Another example concerns the medical field, where global knowledge for diagnosis or research studies is desired considering that some pathologies may be different depending on geographical information, but the privacy of patients data must be guaranteed due to legal reasons. Also, in other cases it is not possible to merge all the data in a single system due to computational resources limitations.

On the other hand, model combination is the core idea behind classical machine learning methods such as Bagging [4], Boosting [5] or Stacking [6]. Models can be seen as experts and classificacion may be better if several experts opinions are combined. As far as we know, all these methods were designed to work in non

---

* This work has been supported by the Spanish Ministry of Innovation and Science under grant No. TIN2008-06681-C06-06, and the Andalusian government under grant No. P07-TIC-03179.

E. Corchado et al. (Eds.): HAIS 2009, LNAI 5572, pp. 638–645, 2009.

distributed environments. They have got the whole dataset at the beginning of the classification process, so they are not directly applicable to solve inherently distributed data problems.

Therefore, merging distributed knowledge without moving the data is an interesting research area. This paper proposes a novel method to learn from inherently distributed data without sending any of them from one place to another. It is based on making the classifiers locally, and then ensembling them in a single final one using an evolutionary methodology where the population of candidate classifiers is concurrently evaluated in a distributed way in each local node.

This document is organized as follows. Section 2 provides the concepts behind the solution presented. Section 3 focuses the proposed method describing all its components. Section 4 shows the experimental study developed and finally, we present some concluding remarks in Section 5.

2   Preliminaries

This section describes the theoretical concepts in which the proposed method is based on. First, an introduction to metalearning techniques is shown and finally genetic algorithms (GA) are introduced as a learning tool.

2.1   Metalearning

Metalearning [1] is a strategy that makes easier independent models combination and supports the data mining applications big scalability, so in some cases we can avoid the data movement issue by combining models instead of raw data. Two main policies are related in the literature [3] to perform model combination:

- Multiple Communication Round: Methods of this kind require a significant synchronization amount. They usually use a voting system, so sending examples throught the system is often necessary.
- Centralized Ensemble-based: This kind of algorithms can work generating the local classifiers first and combining them at a central site later. This is the one we use in our proposal.

2.2   Genetic Learning

GAs have achieved reputation of robustness in rule induction in common problems associated to real world mining (noise, outliers, incomplete data, etc). Initially, GA were not designed as machine learning algorithms but they can be easily dedicated to this task [9]. Typically the search space is seen as the entire possible hypothesis rule base that covers the data. The goodness can be related to a coverage function over a number of learning examples [8][13].

Regarding the representation of the solutions, the proposals in the specialized literature usually use two approaches in order to encode rules within a population of individuals:

- The "Chromosome = Set of rules", also called the Pittsburgh approach, in which each individual represents a rule set [10]. In this case, a chromosome evolves a complete rule set and they compete among them along the evolutionary process.
- The "Chromosome = Rule" approach, in which each individual codifies a single rule, and the whole rule set is provided by combining several individuals in a population (rule cooperation) or via different evolutionary runs (rule competition). In turn, within the "Chromosome = Rule" approach, there are three generic proposals:
  - The Michigan approach, in which each individual encodes a single rule. These kinds of systems are usually called learning classifier systems [11]. They are rule-based, message-passing systems that employ reinforcement learning and a GA to learn rules that guide their performance in a given environment. The GA is used for detecting new rules that replace the bad ones via the competition between the chromosomes in the evolutionary process.
  - The IRL (Iterative Rule Learning) approach, in which each chromosome represents a rule. Chromosomes compete in every GA run, choosing the best rule per run. The global solution is formed by the best rules obtained when the algorithm is run multiple times. SIA [12] is a proposal that follows this approach.
  - The GCCL (Genetic Cooperative-Competitive Learning) approach, in which the complete population or a subset of it encodes the rule base, In this model the chromosomes compete and cooperate simultaneously, [14] is an example of this approach.

3   Proposed Method

In this work we present an Evolutionary eNsemble-based method for Rule Extraction with Distributed Data (ENREDD). The ensemble-based process shares only the local models being a reasonable solution to privacy constraints. Also, it uses low bandwidth due to the low amount of data transmited (classifiers are sent instead of raw data).

Centralized ensemble-based metalearning processes are usually divided in two stages:

- Creating local classifiers from the distributed datasets.
- Aggregate local knowledge in a central node.

The algorithm resolves the first stage using a GA based on a GCCL approach. When local models are generated, they are sent to a central node where the second stage starts. Subsection 3.1 details the local learning system.

The central or master node uses an evolutionary algorithm to merge the rules from the local models. Because data stays in each local node, the algorithm must complete the task without moving the data, so it sends the candidate classifiers to the distributed nodes to evaluate their quality. Each local node sends the

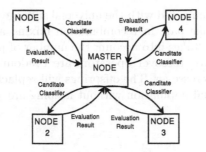

Fig. 1. Communications for candidate classifiers evaluation

accuracy obtained with its data. Once the master node has got the results, it averages the values to get a global measure of classifier quality (Fig. 1).

Master node uses a Pittsburgh approach to merge the rules from the local classifiers. A detailed description of this stage will be shown in Subsection 3.2.

3.1   Local Nodes

As was commented before, local nodes must build a classifier from the data they have got. The idea behind this method is that an acceptable rule in a local set is a candidate to compose the global classifier [2], so all the rules are at the central node. This method reduces the amount of communication and lets build an independent model generation.

ENREDD local learning process generates an initial population with a heuristic function based on local data. The chromosome has a binary representation and each gene represents a possible value for a given attribute that will be active if the value contributes to the rule.

In the example chromosome of Figure 2, the binary coding represents the rule *if c_1 in (v_1, v_3) and c_3 in (v_6) then class is v_{10}.*

c_1			c_2		c_3			Class	
V_1	v_2	v_3	v_4	v_5	v_6	v_7	v_8	v_9	v_{10}
1	0	1	0	0	1	0	0	0	1

Fig. 2. Local node chromosome representation

The evaluation function balances simplicity and quality with Equation 1

$$f(r) = \left(1 + \frac{zeros(r)}{length(r)}\right)^{-1*Cases-} \tag{1}$$

where Cases- is the number of covered examples predicted as false positives, Zeros(r) is the number of zeroes in the bit string representation of the rule r and length is the chromosome length expressed in bits. In order to force the winning rules to be as accurate as possible, the fitness is exponentially measured and just when Cases- is really low the length is taken into account.

The selection is implemented with the universal suffrage operator[13] selecting the individuals that take part in recombination mixing a vote process among the train examples that takes into account the number of positive covered cases. After the selection, two crossover operators are randomly applied, two point crossover and uniform crossover. The offsprings will replace a randomly selected individual in the original population without keeping any elite population.

3.2   Master Node

The master node collects all the local classifiers received, and then, it starts the genetic optimization process that includes two phases that are described next.

Rule merging. It solves the distributed rules aggregation. The GA task is to sort the rule list to get the best order possible.

The local classifiers rules are inserted together in a table without repetition, and an integer index is asigned to each rule. Thus an integer representation has been chosen for the GA chromosome. The order inside the chromosome will determine the rule application order.

The algorithm chosen is a CHC [7] based model. The initial population is generated randomly. The Hamming distance has been considered using the number of differing integer genes in the chromosome, so once it is calculated, the half differing genes can be swapped. The parents are only crossed when Hamming distance exceeds a threshold d.

For each chromosome, the evaluation function applies the classifier in each node to evaluate it with all the available data. Next, each node sends to the master node the accuracy percent obtained, and it averages the global quality of each chromosome.

Rule reduction. It deletes the rules that never get fired with the distributed data.

It sends the final classifier to the nodes and all the rules activated with the examples they have got are marked. Then, the classifier is returned to the master node and it deletes all the rules that have not been marked.

4   Experimental Study

This section describes the experimental study developed to test the proposed method and analyzes the results obtained.

In order to compare the behaviour of the presented method we propose to modify the well known Bagging [4]. It works selecting different samples of a single dataset named bags. A classifier for each bag is created and the final classifier output is the most voted in the samples classifiers. In order to apply Bagging to distributed data, we consider bags as distributed nodes, so we use T/N samples for each node, being T the dataset size and N the number of nodes. We name this modified version MBAG.

The aim of this preliminary study is to validate the proposal without real world complexities due to data distribution like unbalanced data, heterogeneous domain, local discretization, etc. To achieve this target the main dataset has been discretized using 10 fixed frequency values gaps with a 10 fold cross validation in a 70/30 training/test proportion.

Simulations with 5, 10 and 20 nodes have been performed. These values have been selected because higher values result in a few training examples and lower ones are no representative of a distributed configuration. For local nodes, the GA uses 250 individuals and 200 generations. The master node uses 50 individuals and 600 generations.

To compare ENREDD with MBAG we have used the Wilcoxon Signed-Ranks Test(WSRT) [15]. It is a non-parametric alternative to the paired t-test, which ranks the differences in performances of two classifiers for each dataset, ignoring the signs, and compares the ranks for the positive and the negative differences.

WSRT can reject the null hypothesys[16] (equal accuracy for compared algorithms in our study) with $\alpha = 0.05$ when parameter z is smaller than -1.96.

4.1 Results Analysis

Table 1 shows the training and test sets accuracy means for both methods. The BN columns are the trainining and test accuracy percentages for MBAG simulations with N nodes and the EN ones are the results for ENREDD.

Table 1. Average test accuracy

	TRAINING						TEST					
	B5	E5	B10	E10	B20	E20	B5	E5	B10	E10	B20	E20
Car	80.57	**97.22**	76.19	**94.49**	71.45	**92.78**	78.29	**93.97**	74.59	**90.91**	70.83	**89.89**
Cleveland	58.60	**82.83**	56.09	**79.35**	53.77	**77.92**	**52.78**	52.61	**54.00**	53.37	**54.44**	53.43
Credit	86.19	**91.72**	85.42	**94.67**	85.90	89.94	**85.60**	83.91	85.65	**86.38**	**86.28**	86.09
Ecoli	70.55	**89.72**	58.51	**85.25**	45.11	**79.66**	64.36	**66.84**	54.36	**63.05**	43.27	**60.36**
Glass	59.26	**85.43**	49.53	**79.85**	38.39	**87.70**	47.69	**50.54**	39.85	**44.51**	31.23	**38.23**
Haberman	74.58	**85.84**	73.93	**83.56**	73.64	**82.87**	**71.85**	68.52	**72.93**	70.35	**73.26**	69.22
House-votes	64.53	**99.19**	63.40	**98.92**	62.04	**98.30**	62.68	**96.80**	62.72	**97.12**	62.30	**96.00**
Iris	68.46	**96.11**	39.62	**98.06**	0	0	63.91	**76.09**	36.96	**68.09**	0	0
Krvskp	97.02	**98.74**	96.00	**98.50**	94.74	**98.29**	**96.94**	97.76	95.95	**97.72**	94.93	**97.57**
Monk	54.93	**84.72**	52.55	**82.07**	51.82	**75.92**	46.92	**65.50**	47.62	**63.98**	45.23	**64.27**
Mushroom	99.90	**100.0**	99.82	**99.99**	99.54	**100.0**	99.90	**100.0**	99.82	**99.95**	99.47	**99.99**
New-thyroid	88.60	**98.59**	77.13	**97.89**	70.20	**97.97**	85.69	**90.68**	74.92	**88.54**	68.77	**79.58**
Nursery	94.20	**97.99**	92.19	**98.08**	90.80	**96.96**	93.47	**96.12**	92.09	**96.53**	90.72	**96.40**
Pima	74.99	**86.59**	74.56	**83.39**	70.32	**81.28**	72.42	**79.87**	**72.03**	70.53	68.83	**71.41**
Segment	90.36	**96.23**	88.43	**93.13**	**85.67**	85.31	88.02	**92.19**	86.72	**89.04**	**83.8**	79.92
Soybean	91.66	**96.91**	88.25	**92.01**	77.31	**85.77**	91.15	**92.83**	**87.15**	87.10	75.05	**80.12**
Splice	94.22	**99.24**	93.84	**98.24**	92.75	**96.77**	93.32	**96.25**	93.25	**94.54**	91.79	**95.16**
Tic-tac-toe	88.88	**97.69**	74.13	**92.61**	70.42	**89.61**	85.35	**94.46**	70.94	**87.37**	69.03	**84.76**
Vehicle	66.44	**78.16**	63.61	**72.24**	60.57	**67.13**	55.28	**57.21**	**54.53**	53.61	**54.57**	50.25
Vote	95.59	**98.34**	95.63	**97.93**	95.39	**97.65**	**95.73**	95.16	**95.50**	94.93	**95.50**	94.00
Waveform	**79.08**	78.74	**80.65**	78.63	**80.76**	76.68	**75.88**	66.45	**78.18**	72.04	**78.93**	72.37
Wine	72.66	**96.56**	64.68	**96.35**	45.56	**100.0**	**61.85**	61.35	**57.59**	52.97	**39.63**	34.62
Wisconsin	93.26	**99.01**	89.41	**98.46**	84.96	**98.22**	91.02	**95.12**	88.39	**95.54**	84.83	**94.67**
Zoo	93.26	**99.85**	89.41	**99.68**	84.96	**99.08**	**91.02**	81.40	**88.39**	74.78	**84.83**	72.15

Table 2. Wilcoxon Signed Ranks Test

	E5 > B5	E10 > B10	E20 > B20
Negative Ranks	7	9	7
Positive Ranks	17	15	16
Ties	0	0	1
z	-2.743	-2.371	-2.403
p-value	0.006	0.018	0.016

The accuracy of both methods is lower than using standard classifing methods due to the impact of splitting datasets. Sometimes the distributed behaviour may be affected by the number of local datasets due to the fact that there is not enough representation of each class in a local node. Maybe some datasets show poor accuracy with both methods due to this fact. In the other datasets the distribution does not show any tendency regarding accuracy in ENREDD and shows a better accuracy than MBAG.

Table 2 shows the WSRT statistical test results obtained from the Table 1 data. It is shown that z is lower than -1.96 for all the cases, so we can reject the null hypothesis with $\alpha = 0.05$. For example, for a $E5 > B5$ accuracy, null hypothesis is rejected with a 99.4% confidence, because *p-value* is 0.006.

5   Conclusions and Future Work

A methodology for knowledge discovery from distributed data without moving it from its original location for legal or competition issues have been proposed. It generates local distributed classifiers using an evolutionary model, and after, it merges them using an additional evolutionary algorithm evaluating the candidate solutions with the distributed data sets in a distributed parallelized way. The knowledge discovered with this method may be of significance for some organizations interested in to collaborate to get common knowledge for some areas like security, frauds and so on.

As future work, we plan to get better the rule reduction mechanism in order to improve the interpretability of the models obtained and also, we will create synthesized data sets specifically created to simulate the geographically distributed data without the drawbacks of the generic datasets used in this work.

References

1. Brazdil, P., Giraud-Carrier, C., Soares, C., Vilalta, R.: Metalearning. In: Applications to Data Mining. Springer, Heidelberg (2009)
2. Provost, F., Hennessy, D.: Scaling up: Distributed machine learning with cooperation. In: Proceedings of the Thirteenth National Conference on Artificial Intelligence, pp. 74–79. AAAI Press, Menlo Park (1996)
3. Da Silva, J., Giannella, C., Bhargava, R., Kargupta, H., Klush, M.: Distributed Data Mining and Agents. Engineering Applications of Artificial Intelligence 18, 791–807 (2005)

4. Breiman, L.: Bagging predictors. Machine Learning 24(2), 123–140 (1996)
5. Freund, Y., Schapire, R.E.: Experiments with a new boosting algorithm. In: Proceedings of the Thirteenth International Conference on Machine Learning, pp. 148–156. Morgan Kaufmann, San Francisco (1996)
6. Wolper, D.H.: Stacked generalization. Neural Networks 5, 241–259 (1992)
7. Eshelman, L.J.: The CHC Adaptative Search Algorithm: how to have safe search when engaging in nontraditional genetic recombination Foundations of Genetic Algorithms I, pp. 265–283. Morgan Kaufmann Publishers, San Mateo (1991)
8. Peregrin, A., Rodriguez, M.A.: Efficient Distributed Genetic Algorithm for Rule Extraction Eighth International Conference on Hybrid Intelligent Systems, pp. 531–536 (2008)
9. Goldberg, D.E.: Genetic algorithms in search, optimization and machine learning. Addison-Wesley, New York (1989)
10. Smith, S.: A learning system based on genetic algorithms. PhD Thesis, University of Pittsburgh (1980)
11. Holland, J.H., Reitman, J.S.: Cognition Systems Based on Adaptive Algorithms. In: Waterman, D.A., Hayes-Roth, F. (eds.) Pattern-Directed Inference Systems, Academic Press, New York (1978)
12. Venturini, G.: SIA: a supervised inductive algorithm with genetic search for learning attribute based concepts. In: Proceedings of European conference on machine learning, Vienna, pp. 280–296 (1993)
13. Giordana, A., Neri, F.: Search-intensive concept induction. Evolutionary Computation, 375–416 (1995)
14. Greene, D.P., Smith, S.F.: Competition-based induction of decision models from examples. Machine Learning 12(23), 229–257 (1993)
15. Wilcoxon, F.: Individual comparisons by ranking methods. Biometrics 1, 80–83 (1945)
16. Zar, J.H.: Biostatistical Analysis. Prentice Hall, Englewood Cliffs (1999)
17. Merz, C.J., Murphy, P.M.: UCI repository of machine learning databases. University of Carolina Irvine, Department of Information and Computer Science (1996), http://kdd.ics.uci.edu

Evolutionary Extraction of Association Rules: A Preliminary Study on their Effectiveness

Nicolò Flugy Papè[1], Jesús Alcalá-Fdez[2], Andrea Bonarini[1], and Francisco Herrera[2]

[1] Dipartimento di Elettronica e Informazione, Politecnico di Milano, 20133 Milano, Italy
nicolo.flugy@mail.polimi.it, bonarini@elet.polimi.it
[2] Department of Computer Science and Artificial Intelligence, University of Granada,
18071 Granada, Spain
jalcala@decsai.ugr.es, herrera@decsai.ugr.es

Abstract. Data Mining is most commonly used in attempts to induce association rules from transaction data. Most previous studies focused on binary-valued transactions, however the data in real-world applications usually consists of quantitative values. In the last few years, many researchers have proposed Evolutionary Algorithms for mining interesting association rules from quantitative data. In this paper, we present a preliminary study on the evolutionary extraction of quantitative association rules. Experimental results on a real-world dataset show the effectiveness of this approach.

Keywords: Association Rules, Data Mining, Evolutionary Algorithms, Genetic Algorithms.

1 Introduction

Data Mining (DM) is the process for the automatic discovery of high level knowledge from real-world, large and complex datasets. Association Rules Mining (ARM) is one of the several DM techniques described in the literature [1].

Association rules are used to represent and identify dependencies between items in a database [2]. These are implications of the form $X \longrightarrow Y$, where X and Y are sets of items and $X \cap Y = \varnothing$. It means that if all the items in X exist in a transaction then all the items in Y are also in the transaction with a high probability, and X and Y should not have common items [3]. Many previous studies focused on databases with binary values, however the data in real-world applications usually consist of quantitative values. Designing DM algorithms, able to deal with various types of data, presents a challenge to workers in this research field.

Lately, many researchers have proposed Evolutionary Algorithms (EAs) [4] for mining association rules from quantitative data [5], [6], [7], [8]. EAs, particularly Genetic Algorithms (GAs) [9], [10], are considered as one of the most successful search techniques for complex problems and have proved to be an important technique for learning and knowledge extraction. The main motivation for applying EAs to knowledge extraction task is that they are robust and adaptive search methods that perform a global search in place of candidate solutions. Moreover, EAs let to obtain feasible solutions in a limited amount of time. Hence, they have been a growing interest in DM.

E. Corchado et al. (Eds.): HAIS 2009, LNAI 5572, pp. 646–653, 2009.

In this paper, we present a preliminary study on the evolutionary extraction of quantitative association rules. We perform an experimental study to show the behaviour of three GAs (*EARMGA*, *GAR*, and *GENAR*) along with two classical algorithms (a *Trie*-based implementation of *Apriori*, and *Eclat*) for the extraction of association rules on a real-world dataset. Moreover, several experiments have been carried out to analyse the scalability of these methods.

This paper is arranged as follows. The next section provides brief preliminaries on the genetic extraction of association rules. Section 3 describes the five analysed methods. Section 4 shows the results of the experimental study. Finally, Section 5 points out some conclusions.

2 Preliminaries: Genetic Extraction of Association Rules

Although GAs were not specifically designed for learning, but rather as global search algorithms, they offer a set of advantages for machine learning. Many methodologies for machine learning are based on the search of a good model among all possible models within this space. In this sense, they are very flexible because the same GA can be used with different representations. When considering a rule based system and focusing on learning rules, genetic learning methods follow two approaches in order to encode rules within a population of individuals [11]:

- The "Chromosome = Set of rules", also called the *Pittsburgh* approach, in which each individual represents a rules set [12]. In this case, a chromosome evolves a complete Rule Base (RB) and they compete among themselves along the evolutionary process. GABIL is a proposal that follows this approach [13].
- The "Chromosome = Rule" approach, in which each individual encodes a single rule, and the whole rule set is provided by combining several individuals in a population (rule cooperation) or via different evolutionary runs (rule competition). Within the "Chromosome = Rule" approach, there are three generic proposals:
  - *Michigan* approach, in which each individual stands for an association rule. These kinds of systems are usually called learning classifier systems [14]. They are rule-based, message-passing systems that employ reinforcement learning and a GA to learn rules that guide their performance in a given environment. The GA is used for detecting new rules that replace the bad ones via a competition among chromosomes throughout the evolutionary process.
  - *IRL (Iterative Rule Learning)* approach, in which each chromosome represents a rule. Chromosomes compete in every GA run, choosing the best rule per run. The global solution is formed by the best rules obtained when the algorithm has been executed multiple times. SIA [15] is a proposal that follows this approach.
  - *GCCL (genetic cooperative-competitive learning)* approach, in which the population, or one of its subsets, encodes the RB. Moreover, the chromosomes compete and cooperate simultaneously. COGIN [16] is an example of this approach.

In the literature we can find interesting works based on the Pittsburgh approach [17], the Michigan approach [5], and the IRL approach [6], [7] for mining association rules from quantitative data.

3 Association Rules Mining: Algorithms for the Analysis

In this section we introduce the five methods used for our experimental study. We can make out two types of algorithms:

- Classical algorithms: we analysed *Apriori* [18], [19] and *Eclat* [20].
- GAs for ARM: we analysed *EARMGA* [5], *GAR* [6], and *GENAR* [7].

We present the aforementioned methods in more details in the next subsections.

3.1 Association Rules Mining through Classical Algorithms: Apriori and Eclat

Among classical algorithms, it is worthwhile to mention *Apriori* because is the first successful algorithm used for mining association rules. Several implementations based on this method can be found in the literature, basically, with the aim of speeding up the support counting [19], [21], [22]. In this paper, we have used a fast implementation of Apriori which uses *Trie* [19]. Moreover, we have chosen to analyse *Eclat* [20] because it exploits a different strategy to search for frequent itemsets.

The main aim of *Apriori* is to explore the search space by means of the *downward closure property*. The latter states that any subset of a frequent itemset must also be frequent. As a consequence, it generates candidates for the current iteration by means of frequent itemsets collected from the previous iteration. Then, it enumerates all the subsets for each transaction and increments the support of candidates which match them. Finally, those having a user-specified minimum support are marked as frequent for the next iteration. This process is repeated until all frequent itemsets have been found. Thus, Apriori follows a *breadth-first* strategy to generate candidates.

On the other hand, *Eclat* employs a *depth-first* strategy. It generates candidates by extending prefixes of an itemset until an infrequent one is found. In that case, it simply backtracks on the next prefix and then recursively applies the above procedure. Unlike Apriori, the support counting is achieved by adopting a vertical layout. That is, for each item in the dataset, it first constructs the list of all transaction identifiers (*tid-list*) containing that item. Then, it counts the support by merely intersecting two or more tid-lists to check whether they have items in common. In that case, the support count is equal to the size of this resulting set.

Most of the classical algorithms usually identify relationships among transactions in datasets with binary values. In order to apply these methods on datasets containing real values, a pre-processing step must be accomplished.

3.2 Association Rules Mining through Evolutionary Algorithms: EARMGA, GAR, and GENAR

We have used the following GAs in the literature to perform the ARM task:

- *EARMGA* [5]. It is based on the discovery of quantitative association rules.
- *GAR* [6]. It searches for frequent itemsets by dealing with numerical domains.
- *GENAR* [7]. It mines directly association rules by handling numerical domains.

A chromosome in EARMGA encodes a generalized *k*-rule, where *k* indicates the desired length. Since we may handle association rules with more than one item in the consequent, the first gene stores an index representing the end of the antecedent part. In order to encode uniquely a rule into a chromosome, both antecedent and consequent attributes are sorted two-segmentally in an ascending order. On the contrary, the remaining k genes encode items. Each item is represented by a pair of values, where the first value is an attribute's index ranged from 1 to the number of attributes in the dataset, while the second one is a *gapped interval*. The authors have defined a gapped interval as the union of a finite number of *base intervals* obtained once a uniform discretization process has been accomplished on all attributes. Notice that we do not need to partition the domains of categorical attributes because here lower bound and upper bound basically coincide. Nevertheless, a base interval is always represented by an integer, apart from the kind of attributes we deal with. As a consequence, a gapped interval is a set of these values. Now we give some details of the genetic operators applied on each chromosome:

- *Selection*: a chromosome is selected only if the product of its fitness value with a random number is less than a given probability of selection (ps).
- *Crossover*: all the selected chromosomes have the chance to reproduce offsprings at a probability of crossover (pc). This operation simply consists in exchanging a segment of genes from the first chromosome to the second one and vice-versa, depending on two randomly generated crossover-points.
- *Mutation*: by considering both a probability of mutation (pm) and the fitness value, a chromosome is altered by changing the boundary between antecedent attributes and consequent attributes within the same rule. In addition, the operator randomly chooses a gene and modifies the attribute's index along with its gapped interval. Notice that the new gapped interval is always a union of base intervals which now form a sub-domain of the new attribute.

Finally, the ARM problem has been restated by the authors of this work because the algorithm searches for k-association rules by evaluating fitness values that represent measures of interest known as *positive confidence* [5].

On the contrary, *GAR* follows different strategies. First, a chromosome is composed of a variable number of genes, between 2 and n, where n is the maximum number of attributes. However, as we find frequent itemsets with this method, it is afterwards necessary to run another procedure for generating association rules. Moreover, we do not have to discrete a priori the domain of the attributes since each gene is represented by an upper bound and a lower bound along with an identifier for the attribute. We briefly recall the genetic operators for this method as follows:

- *Selection*: an elitist strategy is used for selecting a percentage (ps) of chromosomes which have the best fitness values from the current population. These are the first individuals that later form the new population.
- *Crossover*: the new population is completed by reproducing offsprings until reaching a desired size. To do that, the parents are randomly chosen at a probability of pc. Then, we obtain two different offsprings whenever their parents contain genes having the same attribute. In that case, their intervals could simply be exchanged by considering all the possible combinations between them, but, at last, always two chromosomes should be generated. Finally, only the best of them will be added to the population.

• *Mutation*: at a probability of *pm*, it alters one gene in the way that each limit could randomly decrease or increase its current value.

The fitness function tends to reward frequent itemsets that have a high support as well as a high number of attributes. In addition, it punishes frequent itemsets which have already covered a record in the dataset and which have intervals too large.

GENAR was the first attempt by the same authors of GAR in handling continuous domains. Here, a chromosome represents an association rules that always contain intervals. Nevertheless, the length of rules is fixed to the number of attributes and only the last attribute acts as consequent. The crossover operator employs a one-point strategy to reproduce offspring chromosomes, whereas the remaining operators are similar to those defined in GAR. For this reason, we do not give further details [7]. On the contrary, its fitness function only considers the support count of rules and punishes rules which have already covered the same records in the dataset.

4 Experimental Results

To evaluate the usefulness of the genetic extraction of association rules several experiments have been carried out on a real-world dataset named *House_16* (*HH*). It concerns a study to predict the median price of the house in the region by considering both the demographic composition and the state of housing market. This dataset contains 22,784 records with 17 quantitative attributes[1].

The parameters used for running the algorithms are:

• *Apriori* and *Eclat*: *minimum Support* = 0.1 and *minimum Confidence* = 0.8.
• *EARMGA*: *maxloop* = 100, *popsize* = 100, *k* = 4, *ps* = 0.75, *pc* = 0.7, *pm* = 0.1, and α = 0.01.
• *GAR*: *nItemset* = 100, *nGen* = 100, *popsize* = 100, *ps* = 0.25, *pc* = 0.7, *pm* = 0.1, *af* = 2.0, ω = 0.4, ψ = 0.7, and μ = 0.5.
• *GENAR*: *nRules* = 100, *nGen* = 100, *popsize* = 100, *ps* = 0.25, *pc* = 0.7, *pm* = 0.1, *af* = 2.0, and *pf* = 0.7.

Moreover, a uniform discretization in 4 intervals was accomplished on the continuous attributes in the dataset only if needed by one of these algorithms.

The results returned by all the methods are presented in Table 1, where *Itemsets* stands for the number of the discovered frequent itemsets, *Rules* for the number of the generated association rules, *Avg_Sup* and *Avg_Conf*, respectively, for the average support and the average confidence of the mined rules, *Avg_Amp* for the average length of the rules antecedents, and *%Records* for the percentage of records covered by these rules on the total records in the dataset. We remark that our results always refer to association rules which have minimum confidence greater or equal than 0.8.

---

[1] This database was designed on the basis of data provided by US Census Bureau [http://www.census.gov] (under Lookup Access [http://www.census.gov/cdrom/lookup]: Summary Tape File 1).

Table 1. Results for the dataset *HH*

Algorithm	Itemsets	Rules	Avg_Sup	Avg_Conf	Avg_Amp	%Records
Apriori	305229	1982211	0.22	0.96	7	**100.00**
Eclat	305229	1982211	0.22	0.96	7	**100.00**
EARMGA	-	**100**	0.24	**1.00**	2	**100.00**
GAR	100	167	**0.73**	0.94	2	**100.00**
GENAR	-	**100**	0.46	0.99	16	88.60

Analysing the results presented in Table 1, we highlight the following issues:

- The classical methods returned a large set of association rules (approximately 2 million) with the minimum support and confidence. On the contrary, the GAs let us to obtain a reduced set of high quality rules, thus denoting more interesting patterns. For instance, EARMGA mined 100 rules having the maximum average confidence and GAR mined 167 association rules with very high average support.
- EARMGA and GAR discovered association rules involving only few attributes in the antecedents, giving the advantage of a better understanding from the user's perspective. Notice that, GENAR considers rules of the maximal length due to the fact that it always involves all attributes in the dataset.
- The rules returned by the GAs achieved a good covering of the records although the number of rules is restricted by the population size. As an example, EARMGA and GAR covered 100% of the records in the dataset.

Several experiments have been also carried out to analyse the scalability of the algorithms. All the experiments were performed by using a Pentium Corel 2 Quad, 2.5GHz CPU with 4GB of memory and by running Linux. Fig. 1 and Fig. 2 show the relationship between the runtime and, respectively, the number of records and the number of attributes in the dataset for each algorithm.

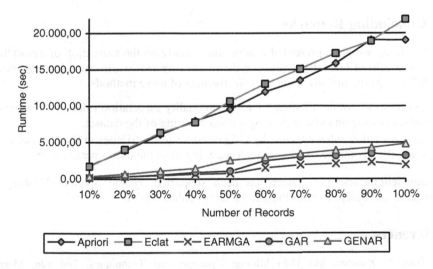

Fig. 1. Relationship between runtimes and number of records with all the attributes

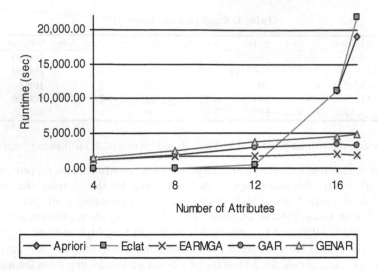

Fig. 2. Relationship between runtimes and number of attributes with 100% of the records

It can be easily seen from Fig. 1 that the runtime of the classical methods increase with respect to the runtime of the GAs since we increase the size of the problem. In Fig. 2 we notice that the classical algorithms expended a large amount of time when the number of attributes is also increased. On the contrary, from both the figures we can see how the GAs scaled quite linearly when performing the ARM task on the used dataset.

Finally, it is worthwhile to remark that the GAs expended a reasonable amount of time in mining a reduced set of high quality association rules. Nevertheless, increasing the population size of the GAs would increase the runtimes of these methods which, eventually, could be higher than those of Apriori and Eclat.

5   Concluding Remarks

In this paper, we have presented a preliminary study on the extraction of association rules by means of GAs. By evaluating the results on a real-world dataset, we point out the following conclusions about the effectiveness of these methods:

- GAs let us to obtain a reduced set of high quality association rules in a reasonable amount of time and also achieving a good covering of the dataset.
- The association rules returned by GAs consider few attributes in their antecedents, giving the advantage of a better understanding from the user's perspective.

Acknowledgments. This paper has been supported by the Spanish Ministry of Education and Science under Project TIN2008-06681-C06-01.

References

1. Han, J., Kamber, M.: Data Mining: Concepts and Techniques, 2nd edn. Morgan Kaufmann, San Francisco (2006)

2. Zhang, C., Zhang, S.: Association Rule Mining: Models and Algorithms. LNCS(LNAI), vol. 2307. Springer, Heidelberg (2002)
3. Agrawal, R., Imielinski, T., Swami, A.: Mining Association Rules between Sets of Items in Large Databases. In: ACM SIGMOD ICMD, pp. 207–216. ACM Press, Washington (1993)
4. Eiben, A.E., Smith, J.E.: Introduction to Evolutionary Computing, 1st edn. Natural Computing Series. Springer, Heidelberg (2003)
5. Yan, X., Zhang, C., Zhang, S.: Genetic algorithm-based strategy for identifying association rules without specifying actual minimum support. Expert Systems with Applications 36(2), 3066–3076 (2009)
6. Mata, J., Alvarez, J.L., Riquelme, J.C.: An Evolutionary Algorithm to Discover Numeric Association Rules. In: Proc. of ACM SAC 2002, Madrid, Spain, pp. 590–594 (2002)
7. Mata, J., Alvarez, J.L., Riquelme, J.C.: Mining Numeric Association Rules with Genetic Algorithms. In: 5th International Conference on Artificial Neural Networks and Genetic Algorithms, Prague, pp. 264–267 (2001)
8. Mata, J., Alvarez, J.L., Riquelme, J.C.: Discovering Numeric Association Rules via Evolutionary Algorithm. In: Chen, M.-S., Yu, P.S., Liu, B. (eds.) PAKDD 2002. LNCS, vol. 2336, pp. 40–51. Springer, Heidelberg (2002)
9. Goldberg, D.E.: Genetic Algorithms in Search, Optimization and Machine Learning. Addison-Wesley, New York (1998)
10. Holland, J.: Adaptation in Natural and Artificial Systems. The University of Michigan Press, London (1975)
11. Herrera, F.: Genetic Fuzzy Systems: Taxonomy, Current Research Trends and Prospects. Evolutionary Intelligence 1, 27–46 (2008)
12. Smith, S.: A learning system based on genetic algorithms. Ph.D. thesis. University of Pittsburgh (1980)
13. De Jong, K., Spears, W., Gordon, D.: Using genetic algorithms for concept learning. Machine Learning 13(2-3), 161–188 (1993)
14. Holland, J., Reitman, J.: Cognitive systems based on adaptive algorithms. In: Waterman, D.A., Hayes-Roth, F. (eds.) Patter-directed inference systems, pp. 1148–1158. Academic Press, London (1978)
15. Venturini, G.: SIA: a supervised inductive algorithm with genetic search for learning attrib-ute based concepts. In: ECML, Vienna, Austria, pp. 280–296 (1993)
16. Greene, D.P., Smith, S.F.: Competition-based induction of decision models from examples. Machine Learning 13(2-3), 229–257 (1993)
17. Pei, M., Goodman, E., Punch, W.: Pattern Discovery from Data using Genetic Algorithm. In: Proc. of PAKDD 1997, Singapore, pp. 264–276 (1997)
18. Srikant, R., Agrawal, R.: Mining Quantitative Association Rules in Large Relational Tables. In: Proc. of ACM SIGMOD ICMD 1996, pp. 1–12. ACM Press, Montreal (1996)
19. Borgelt, C.: Efficient Implementations of Apriori and Eclat. In: Workshop on Frequent Itemset Mining Implementations. CEUR Workshop Proc. 90, Florida, USA (2003)
20. Zaki, M.J., Parthasarathy, S., Ogihara, M., Li, W.: New Algorithms for Fast Discovery of Association Rules. Technical Report 651, University of Rochester (1997)
21. Bodon, F.: A trie-based APRIORI implementation for mining frequent item sequences. In: 1st International Workshop on Open Source Data Mining: Frequent Pattern Mining Implementations, Chicago, Illinois, USA, pp. 56–65. ACM Press, New York (2005)
22. Han, J., Pei, H., Yin, Y.: Mining Frequent Patterns without Candidate Generation. In: Proc. of ACM SIGMOD ICMD 2000, Dallas, TX. ACM Press, New York (2000)

A Minimum-Risk Genetic Fuzzy Classifier Based on Low Quality Data

Ana M. Palacios[1], Luciano Sánchez[1], and Inés Couso[2,*]

[1] Dpto. Informática, Univ. Oviedo, Gijón, Asturias, Spain
[2] Dpto. de Estadística e I.O. y D.M, Univ. Oviedo, Gijón, Asturias, Spain

Abstract. Minimum risk classification problems use a matrix of weights for defining the cost of misclassifying an object. In this paper we extend a simple genetic fuzzy system (GFS) to this case. In addition, our method is able to learn minimum risk fuzzy rules from low quality data. We include a comprehensive description of the new algorithm and discuss some issues about its fuzzy-valued fitness function. A synthetic problem, plus two real-world datasets, are used to evaluate our proposal.

1 Introduction

In the context of GFSs, fuzzy statistics and fuzzy logic have complementary functions. On the one hand, GFSs depend on fuzzy rule-based systems (FRBS), that deal with fuzzy logic and "IF-THEN" rules [3]. These FRBSs use fuzzy sets to describe subjective knowledge about a classifier or a regression model, which otherwise accept crisp inputs and produce crisp outputs. On the other hand, fuzzy statistics use fuzzy sets to represent imprecise knowledge about the data [1]. In a broad sense, we could say that FRBSs process crisp data with fuzzy algorithms, while fuzzy statistics process fuzzy data with crisp algorithms.

In order to process fuzzy data with an algorithm which is also described in fuzzy logic terms, we need to combine fuzzy logic and fuzzy statistics [14]. Up to date, there have been few bridges between GFSs and fuzzy statistics [10]. In this paper, we continue our prior work about generalizing Genetic Fuzzy Classifiers to low quality data [8], and introduce the minimum risk problem, where the cost of misclassifying an object depends on a matrix of weights, and the classifier system does not optimize the training error but a loss function that depends on the mentioned matrix [5].

The use of GFSs for obtaining minimum risk classifiers is scarce. Some works have dealt with the concept of "false positives" [9,13] or taken into account the confusion matrix in the fitness function [15]. There are also some works in fuzzy ordered classifiers [12], where an ordering of the class labels defines, in a certain sense, a risk function different than the training error. Minimum-risk classification is also related to the use of costs in classification of imbalanced data [4]. However, up to our knowledge, the

---

* This work was supported by the Spanish Ministry of Education and Science, under grants TIN2008-06681-C06-04, TIN2007-67418-C03-03, and by Principado de Asturias, PCTI 2006-2009.

E. Corchado et al. (Eds.): HAIS 2009, LNAI 5572, pp. 654–661, 2009.

inclusion of the matrix of misclassification costs in the fitness function has been part of GFSs neither in crisp nor low quality datasets.

The structure of this paper is as follows: in the next section we generalize a crisp GFS, defined in [7], to the minimum risk problem. In Section 3 we extend this last algorithm to imprecise data. In Section 4 we evaluate the generalized algorithm in both crisp and imprecise datasets. The paper finishes with the concluding remarks, in Section 5.

2   Minimum Risk Genetic Fuzzy Classifiers with Crisp Data

The GFS in Figure 1 is a simple Cooperative-Competitive algorithm [6], defined in [7]. Each chromosome codifies the antecedent of a single rule; the consequents are not part of the genetic individuals. Instead, the function "assignConsequent" (line 6) determines the class label that matches an antecedent for a maximum confidence. A second function "assignFitness," in turn, determines the winner rule for each object in the training set and increments the fitness of the corresponding individual if its consequent matches the class of the object. These last functions are described in Figures 2 and 3. The sum of the fitness of all the individuals is the number of correctly classified training examples.

```
function GFS
 1    Initialize population
 2    for iter in {1, . . . , Iterations}
 3        for sub in {1, . . . , subPop}
 4            Select parents
 5            Crossover and mutation
 6            assignConsequent(offspring)
 7        end for sub
 8        Replace the worst subPop individuals
 9        assignFitness(population,dataset)
10    end for iter
11    Purge unused rules
    return population
```

Fig. 1. Outline of the GFS that will be generalized [7]. Each chromosome codifies one rule. The fitness of the classifier is distributed among the rules at each generation.

Conversely, the minimum risk problem is based on a matrix of misclassification costs. We will assume that this matrix contains values between 0 and 1, and its diagonal is zero (because the cost of correctly classifying an object is null). The cost matrix needs not to be symmetric. Given a training set $\{(x_i, c_i)\}_{i=1...N}$, the objective of the classifier is to minimize the risk

$$\sum_{i=1}^{N} \text{loss}_i = \sum_{i=1}^{N} \text{cost}[\text{class}(x_i)][c_i] \tag{1}$$

where "class(x_i)" is the output of the FRBS in the i-th example. When all the costs but the diagonal are 1, this loss coincides with the training error.

Two changes are required in the GFS for coping with the cost matrix: (a) The assignment of the best consequent not only depends on the compatibility of the rule with the

```
function assignConsequent(rule)
1    for example in {1, ..., N}
2        m = membership(Antecedent,example)
3        weight[class(example)] =
            weight[class(example)] + m
4    end for example
5    mostFrequent = 0
6    for c in {1, ..., N_c}
7        if (weight[c]>weight[mostFrequent])
8            then mostFrequent = c
9        end if
10   end for c
11   Consequent = mostFrequent
return rule
```

```
function assignConsequentCost(rule)
1    for c in {1, ..., N_c}
2        weight[c] = 0
3        for example in {1, ..., N}
4            m = membership(Antecedent,example)
5            weight[c] = weight[c] +
                m ∧ (1-cost[c,class(example)])
6        end for example
7    end for c
8    lowestLoss = 0
9    for c in {1, ..., N_c}
10       if (weight[c]>weight[lowestLoss])
11           then lowestLoss = c
12       end if
13   end for c
14   Consequent = lowestLoss
return rule
```

Fig. 2. Left part: Training error-based algorithm. The consequent of a rule is not codified in the GA, but it is assigned the most frequent class label, between those compatible with the antecedent of the rule [7]. Right part: Cost-based algorithm. The compatibility between antecedent and consequent is affected by the cost matrix (line 5); the consequent that produces the lowest loss in the example is chosen.

```
function assignFitness(population,dataset)
1    for example in {1, ..., N}
2        wRule = winner rule
3        if (consequent(wRule)==class(example)) then
4            fitness[wRule] = fitness[wrRule] + 1
5        end if
6    end for example
return fitness
```

```
function assignFitnessCost(population,dataset)
1    for example in {1, ..., N}
2        wRule = winner rule
3        fitness[wRule] = fitness[wRule] +
            1 - cost[consequent[wRule],class(example)]
4    end for example
return fitness
```

Fig. 3. Left: Training error-based: The fitness of an individual is the number of examples that it classifies correctly. Single-winner inference is used, thus at most one rule changes its fitness when the rule base is evaluated in an example [7]. Right: Risk-based: The fitness of an individual is the number of examples minus the loss in eq. (1), according to the cost table. In line 3 we can see how the fitness of the rule includes a value of the cost table, depending on the output of the example and the consequent of the rule.

example, but also on the value "cost[c,class(example)]," which is the cost of assigning the class "c" to the training data number "example" (see Figure 2, right part, line 5) and (b) the fitness of the winner rule is increased an amount that depends on the cost of assigning the label in the consequent to the example (see Figure 3, right part, line 3).

3   Minimum Risk Genetic Fuzzy Classifiers with Low Quality Data

In this section we will extend the algorithm defined in the preceding section to low quality data, codified by intervals or fuzzy sets. Let us consider the fuzzy case first, which we will introduce with the help of an example. Consider the FRBS that follows:

$$\text{if TEMPERATURE is COLD then CLASS is A} \\ \text{if TEMPERATURE is WARM then CLASS is B} \tag{2}$$

where the membership functions of "COLD" and "WARM" are, respectively, $\text{COLD}(t) = 1 - t/100$ and $\text{WARM}(t) = 1 - \text{COLD}(t)$, for $t \in [0, 100]$. Let the actual temperature be $t = 48$, thus the degrees of truth of the two rules are 0.52 and 0.48, respectively, and the class of the object is the fuzzy set $0.52/A + 0.48/B$. Applying the winner-takes-all approach, $0.52 > 0.48$ and the output of the FRBS is "A."

Now suppose that we do not have a precise knowledge about the temperature: we perceive the triangular fuzzy set $\widetilde{T} = [45; 48; 51]$. Our information about the class of the object is limited to the fuzzy set

$$\widetilde{\text{class}}(\widetilde{T})(c) = \sup\{\widetilde{T}(u) \mid \text{class}(u) = c\}, \quad c \in \{1, \ldots, N_c\}, \tag{3}$$

thus $\widetilde{\text{class}}([45; 48; 51]) = 0.533/A + 0.464/B$. In words, the set $\widetilde{\text{class}}([45; 48; 51])$ contains the output of the FRBS for all the values compatibles with the fuzzy set $[45; 48; 51]$ (see [10] for a deeper discussion on the subject). It is immediate to see that, in this case, the loss in the classification of an example is also a fuzzy number. Let $\{(\widetilde{T}_i, c_i)\}_{i=1\ldots N}$ be an imprecise training set. The membership of the risk of the FRBS in the i-th example of this set is

$$\widetilde{\text{loss}}_i(x) = \sup\{\widetilde{\text{class}}(\widetilde{T}_i)(c) \mid x = \text{cost}[c][c_i]\} \quad x \in [0, 1]. \tag{4}$$

Observe that, if the training data is represented with intervals, then the loss is either a crisp number or a pair of values, as we will see in the implementation that is explained later in this section.

3.1 Computer Algorithm of the Generalized GFS

The fuzzy risk function defined in eq. (4) imposes certain changes in the procedures `assignConsequent` and `assignFitness`, that are detailed in Figure 4.

On the one hand, the crisp assigment of a consequent consisted in adding the terms $\widetilde{A}(x_i) \wedge (1 - \text{cost}[c][c_i])$, for every example (x_i, c_i) in the training set and all the values of "c". After that, the alternative with maximum weight was selected. The same approach can be used in the imprecise case, with the help of the fuzzy arithmetic. However, for determining the most compatible rule, we need to define an order in the set of weights of the rules. This order is expressed in the pseudocode by means of the operation "dominates" used in line 11. Generally speaking, we have to select one of the values in the set of nondominated confidences and use its corresponding consequent, thus we want an operator that induces a total order in the fuzzy subsets of $[0, 1]$ [2].

On the other hand, there are three ambiguities that must be resolved for computing the fitness of a rule (Figure 4, right part): (a) some different crisp values compatible the same example might correspond to different winner rules –line 5— (see also [8]), (b) these rules might have different consequents, thus we have to combine their losses –lines 7 to 14– and (c) we must assign credit to just one of these rules –lines 15 and 16.–

There are two other parts in the original algorithm that must be altered in order to use an imprecise fitness function: (a) the selection of the individuals in [7] is based on a

```
function assignImpConsequentCost(rule)
1    for c in {1, ..., N_c}
2       weight[c] = 0
3       for example in {1, ..., N}
4          m̃ = fuzmship(Antecedent,example)
5          weight[c] = weight[c] ⊕
              m̃ ∧ { cost[c][class(example)] }
6       end for example
7    end for c
8    mostFrequent = {1, ..., N_c}
9    for c in {1, ..., N_c}
10      for c_1 in {c+1, ..., N_c}
11         if (weight[c] dominates weight[c_1]) then
12            mostFrequent = mostFrequent - { c_1 }
13         end if
14      end for c_1
15   end for c
16   Consequent = select(mostFrequent)
return rule
```

```
function assignImpFitnessCost(population,dataset)
1    for ex in {1, ..., N}
2       for r in {1, ..., M}
3          rule.m̃ = fuzmship(Antecedent[r],ex)
4       end for r
5       setWinRule = set of indices of
              nondominated elements of rule.m̃
6       setOfCons= set of consequents of setWinRule
7       deltaFit= 0
8       if ({class(ex)} == setOfCons and
              size(setOfCons)==1) then
9          deltaFit = {1}
10      else
11         if ({class(ex)}∩ setOfCons ≠ ∅) then
12            deltaFit = deltaFit
                 ∪{ 1 - cost[c][class(ex)] | c ∈ setOfCons }
13         end if
14      end if
15      Select winRule ∈ setWinRule
16      fitness[winRule] = fitness[winRule] ⊕ deltaFit
17   end for ex
return fitness
```

Fig. 4. Left part: assignment of a consequent with costs and imprecise data. We induce a total order in the fuzzy subsets of $[0, 1]$ for selecting the best weight. The function fuzmship in line 4 computes the intersection of the antecedent and the fuzzy input. Right part: Generalization of the function "assignFitness" to imprecise data. The set of winner rules is determined in line 5 (see [8] for further details). The set of costs of these rules is found in lines 7-14 and one rule receives all the credit in lines 15-16.

tournament, that depends on a total order on the set of fitness values. And (b) the same happens with the removal of the worst individuals. In both cases, we have used a fuzzy ranking that defines a total order [2]. We leave for future works the application of a multicriteria genetic algorithm similar to those used in our previous works in regression modeling [11].

4 Numerical Results

First, we have evaluated the crisp cost-based algorithm in a synthetic dataset, with known Bayesian risk. Second, the GFS for low quality data has been compared with the results in [8], using the two real-world interval-valued datasets proposed in that reference. All the experiments have been run with a population size 100, probabilities of crossover and mutation of 0.9 and 0.1, respectively, and limited to 200 generations. The fuzzy partitions of the labels are uniform and their size is 3, except when mentioned otherwise.

4.1 Synthetic Dataset

The decision surfaces obtained by the training error-based GFS and the crisp version of the GFS in this paper, applied to the Gauss dataset described in [8], are plotted in Figure 5.

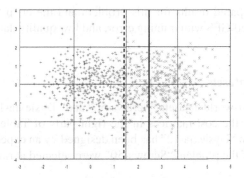

Fig. 5. Decision surfaces of the training error-based GFS (dashed) and the cost-based GFS (continuous). The cost based GFS has a small number of false positives.

Table 1. Results of the algorithm in Section 2 GFS in the synthetic dataset "Gauss"

	Train Error	Test Error	Train Risk	Test Risk	COST	
cost-based	0.441	0.437	0.013	**0.013**	0	0.03
training error-based	0.084	**0.086**	0.038	0.041	0.9	0

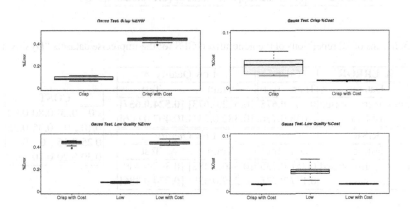

Fig. 6. Boxplots illustrating the dispersion of 10 repetitions of crisp GFS with costs, low quality data-based GFS with training error, and low quality data GFS with costs in the problem "Gauss", with 4 labels/partition. Upper row: Crisp GFS. Lower row: Low Quality-based GFS. Left parts: test error. Right parts: expected risk.

It can be observed that the cost-based classifier system (the cost matrix is defined in the right part of Table 1) produces a small number of false positives, but a much higher misclasification rate. The obtained decision surface is compatible with the theoretical surface derived from the probability density of the sample.

The first row of numbers in Table 1 contains the results of the cost-based GFS. The results of the training error-based GFS are in the second row. We have also included boxplots of the compared results for this problem and the crisp GFS with costs, the low quality data-based GFS with training error, and low quality data GFS with costs, in Figure 6.

4.2 Real World Datasets

We have used the real-world datasets "Screws-50" and "Dyslexia-12", described in [8], in combination with the cost matrices in the right parts of Tables 2 and 3. The cost matrix in the problem "Dyslexia-12" has been designed by an expert in the diagnosis of dyslexia; the costs of the problem "Screws-50" were selected at random.

Table 2. Means of 10 repetitions of the generalized GFS for the imprecise datasets "Screws-50"

% ERROR	Crisp		Low Quality	
Dataset	Train	Test	Train	Test
training error-based	0.077	**0.379**	[0.086,0.105]	**[0.350,0.379]**
cost-based	0.290	0.466	[0.189,0.207]	[0.406,0.425]
RISK	Crisp		Low Quality	
Dataset	Train	Test	Train	Test
training error-based	0.091	0.302	[0.057,0.073]	[0.214,0.236]
cost-based	0.097	**0.246**	[0.051,0.052]	**[0.190,0.207]**

COST		
0	0.90	0.90
0.15	0	0.85
0.40	0.20	0

Table 3. Means of 10 repetitions of the generalized GFS for the imprecise datasets "Dyslexia-12"

% ERROR	Crisp		Low Quality	
Dataset	Train	Test	Train	Test
training error-based	0.664	**0.675**	[0.512,0.673]	**[0.524,0.657]**
cost-based	0.648	0.726	[0.382,0.641]	[0.407,0.666]
RISK	Crisp		Low Quality	
Dataset	Train	Test	Train	Test
training error-based	0.251	0.235	[0.346,0.482]	[0.367,0.489]
cost-based	0.177	**0.225**	[0.259,0.438]	**[0.274,0.486]**

COST			
0	0.20	0.80	0.90
0.10	0	0.35	0.35
0.25	0.15	0	0.45
0.30	0.20	0.10	0

The real-world datasets show mixed results, that are mostly coherent with the cost matrix, albeit less conclusive than in the synthetic case. The average risk of the cost-based classifiers is better than the risk of the error-based GFSs. However, that difference is not significant.

5 Concluding Remarks and Future Work

The minimum risk problem posed in this paper has an immediate application to certain problems of medical diagnosis. We have briefly introduced the problem of diagnosis of dyslexia, where the professional suggested us a cost matrix. However, we plan to study more flexible cases, where the cost matrix is defined in linguistic terms.

On a different subject, our algorithm performed as expected in synthetic datasets, but the results were not conclusive in real-world datasets, and this suggests us that more work is needed to extend more modern GFSs than the simple algorithm used in this paper.

References

1. Bertoluzza, C., Gil, M.A., Ralescu, D.A. (eds.): Statistical Modeling, Analysis and Management of Fuzzy Data. Springer, Heidelberg (2003)
2. Bortolan, G., Degani, R.: A review of some methods for ranking fuzzy subsets. Fuzzy Sets and Systems 15, 1–19 (1985)
3. Cordón, O., Herrera, F., Hoffmann, F., Magdalena, L.: Genetic fuzzy systems. In: Evolutionary tuning and learning of fuzzy knowledge bases. World Scientific, Singapore (2001)
4. Fernandez, A., Garcia, S., del Jesus, M.J., Herrera, F.: A Study of the Behaviour of Linguistic Fuzzy Rule Based Classification Systems in the Framework of Imbalanced Data Sets. Fuzzy Sets and Systems 159(18), 2378–2398 (2008)
5. Hand, D.J.: Discrimination and Classification. Wiley, Chichester (1981)
6. Herrera, F.: Genetic Fuzzy Systems: Taxonomy, Current Research Trends and Prospects. Evolutionary Intelligence 1, 27–46 (2008)
7. Ishibuchi, H., Nakashima, T., Murata, T.: A fuzzy classifier system that generates fuzzy if-then rules for pattern classification problems. In: Proc. of 2nd IEEE CEC, pp. 759–764 (1995)
8. Palacios, A., Sánchez, L., Couso, I.: A baseline genetic fuzzy classifier based on low quality data. IFSA-EUSFLAT (2009) (Submitted)
9. Pulkkinen, P., Hytönen, J., Koivisto, H.: Developing a bioaerosol detector using hybrid genetic fuzzy systems. Engineering Applications of Artificial Intelligence 21(8), 1330–1346 (2008)
10. Sánchez, L., Couso, I.: Advocating the use of imprecisely observed data in genetic fuzzy systems. IEEE Transactions on Fuzzy Systems 15(4), 551–562 (2007)
11. Sánchez, L., Otero, J., Couso, I.: Obtaining linguistic fuzzy rule-based regression models from imprecise data with multiobjective genetic algorithms. Soft Computing 13(5), 467–479 (2008)
12. Van Broekhoven, E., Adriaenssens, V., De Baets, B.: Interpretability-preserving genetic optimization of linguistic terms in fuzzy models for fuzzy ordered classification: An ecological case study. International Journal of Approximate Reasoning 44(1), 65–90 (2007)
13. Verschae, R., Del Solar, J.R., Köppen, M., Garcia, R.V.: Improvement of a face detection system by evolutionary multi-objective optimization. In: Proc. HIS 2005, pp. 361–366 (2005)
14. Wu, B., Sun, C.: Interval-valued statistics, fuzzy logic, and their use in computational semantics. Journal of Intelligent and Fuzzy Systems 1–2(11), 1–7 (2001)
15. Teredesai, A., Govindaraju, V.: GP-based secondary classifiers. Pattern Recognition. 38(4), 505–512 (2005)

Performance Analysis of the Neighboring-Ant Search Algorithm through Design of Experiment[*]

Claudia Gómez Santillán[1,2], Laura Cruz Reyes[1], Eustorgio Meza Conde[2],
Claudia Amaro Martinez[1], Marco Antonio Aguirre Lam[1],
and Carlos Alberto Ochoa Ortíz Zezzatti[3]

[1] Centro de Investigación en Ciencia Aplicada y Tecnología Avanzada (CICATA), Carretera
Tampico-Puerto Industrial Alt., Km.14.5, Altamira,Tamps., México, Phone: 018332600124
[2] Instituto Tecnológico de Ciudad Madero (ITCM), 1ro. de Mayo y Sor Juana I. de la Cruz s/n
CP. 89440, Tamaulipas, México, Phone: (52) 833 3574820 Ext. 3024
[3] Instituto de Ingeniería y Tecnología, Universidad Autónoma de Ciudad Juárez
Henry Dunant 4016, Zona Pronaf Cd. Juárez, Chihuahua, México C.P. 32310
cggs71@hotmail.com, lcruzreyes@prodigy.net.mx, emezac@ipn.mx,
amarito@hotmail.com, marco@marcoaguirre.com.mx

Abstract. In many science fields such as physics, chemistry and engineering, the theory and experimentation complement and challenge each other. Algorithms are the most common form of problem solving in many science fields. All algorithms include parameters that need to be tuned with the objective of optimizing its processes. The NAS (Neighboring-Ant Search) algorithm was developed to route queries through the Internet. NAS is based on the ACS (Ant Colony System) metaheuristic and SemAnt algorithm, hybridized with local strategies such as: learning, characterization, and exploration. This work applies techniques of *Design of Experiments* for the analysis of NAS algorithm. The objective is to find out significant parameters for the algorithm performance and relations among them. Our results show that the probability distribution of the network topology has a huge significance in the performance of the NAS algorithm. Besides, the probability distributions of queries invocation and repositories localization have a combined influence in the performance.

Keywords: Semantic Query Routing, Metaheuristic Algorithm, Parameter Setting, Design of Experiment, Factorial Design.

1 Introduction

With the advent of Internet and WWW, the amount of information available on the Web grows daily due to the opportunities to access, publish and manage resources. It has become common for Internet users to form online communities to share resources without a centralized management. These systems are known as unstructured peer-to-peer networks (P2P), and together with the underlying Internet are considered complex networks for its size and interconnectivity that constantly evolve. In these circumstances, global information recollection is not a feasible approach to perform

---

[*] This research was supported in part by CONACYT, DGEST and IPN.

E. Corchado et al. (Eds.): HAIS 2009, LNAI 5572, pp. 662–669, 2009.
© Springer-Verlag Berlin Heidelberg 2009

queries on shared resources. A new approach to build query algorithms is to determine their behavior locally without using a global control mechanism. Whit this approach, we developed the NAS algorithm to perform text queries. Recently this type of query has been named Semantic Query Routing Problem (SQRP) [8].

NAS and many other algorithms have many parameters. Some researchers confirm [1][2][13], that the cost of not setting parameters with optimization algorithms, consumes a great amount of time and resources. It is well-known that parameter setting mainly depends on the problem characteristics. However, in adaptive complex systems, this is an open research [1][8][13]. The *Design of Experiment* (DOE) is an approach that can be adapted successfully to address the parameter setting problem for algorithms that find approximate solutions to optimization problems [1][2]. The objective in this work is to identify the SQRP main characteristics that affect the NAS algorithm performance.

2   Parameter Setting

The parameter setting can be classified into: parameter tuning and parameter control. *Parameter tuning* is the setting done before the algorithm runs that provides a global initial configuration. It evaluates the general performance of the algorithm but, it does not assure that the values for the parameters will be the best in each instant of the run of the algorithm. *Parameter Control* is the setting done during the run of the algorithm. This type of parameter setting supervises the local environmental changes and the current state of the algorithm to adapt locally the configuration to the local conditions [3][4].

The parameter tuning can be applied through three techniques: a) *by hand*, doing a sequence of experiments with different values of the parameters, and choosing the configuration with the best performance, b) *Meta-Evolution*, using an auxiliary metaheuristic algorithm to improve the perfomance of the main metaheuristic algorithm and c) *Design of Experiment* (DOE), this technique provides a great variety of statistics tests to make useful decisions [3][4].

2.1   Design of Experiments (DOE)

DOE is a statistics tool set, useful on making plans, running and interpreting an experiment, while searching for valid and impartial deductions. *An experiment* can be defined, as a planned test which introduces checked changes in the process or system variables, with the aim of analyzing changes that could happen over the system outputs. To apply the statistical approach in designing and analyzing an experiment, it is necessary to follow a general scheme of procedures. The items that Montgomery´s work [5] suggested are: identifying and enunciating the problem, selecting the factors and levels, selecting response variables, choosing the experimental design, doing the experiment, analyzing statistical data, conclusions and recommendations.

2.2   Experiment Strategy: Factorial Design

In statistics, a *factorial experiment* is an experiment whose design consists of two or more factors, each with discrete possible values or "levels", and whose experimental units take on all possible combinations of these levels across all such factors. A *factorial design* may also be called a *fully-crossed design*. Such an experiment allows studying the effect of each factor on the response variable, as well as the effects of interactions between factors on the response variable. The effect of the factor is defined as the change in the response variable, produced by a variation in the level of the factor. Its also known as *main effect* because it does reference to the primary factors in the experiment. Some experiments show that the difference in response between factors is not the same for all the factors and levels. When it happens, an interaction between them exists [5][6].

3   Ant Algorithms for Semantic Query Routing

Ant algorithms were inspired by the ants behavior, while searching for food. Because when they perform the search, each ant drops a chemical called pheromone which provides an indirect communication among the ants.

The basis of the ant algorithms can be found in a metaheuristic called ACO (Ant Colony Optimization) [7]. The ACO algorithm and its variants (e.g. ACS) were proposed to solve problems modeled as graphs. The ACS was designed for circuit-switched and packet-switched networks [8]. Each component of the network is represented by a *node* (or a vertex of the graph) and the interactions among themselves represent the *connections* (the edges of the graph). ACS needs to know information about all nodes in the network to select the destination node. However, in the SQRP the goal is to find one or more destination nodes for a query without having information from the complete network, requiring operating with local information.

3.1   Algorithm Neighboring-Ant Search

NAS is a metaheuristic algorithm, where a set of independent agents called ants cooperate indirectly and sporadically to achieve a common goal. The algorithm has two objectives: maximizes the found resources and minimizes the given steps. NAS guides the queries towards nodes that have better connectivity using the local structural metric *Degree Dispersion Coefficient* (DDC) [9]. The DDC measures the differences between the degree of a vertex and the degrees of its neighbors in order to minimize the hop count. Therefore, the more frequent is a query towards a resource, the better path is selected; that is to say, the degree of optimization of a query depends directly on its popularity. The NAS algorithm consists of two main phases [10].

Phase 1: the evaluation of results (lines 03-06 in pseudo code on Table 1). This phase implements the classic Lookahead technique. This is, ant k, located in a node s, queries to the neighboring nodes for the requested resource. If the resource is found, the result is retrieved. In case the evaluation phase fails, phase 2 is carried out.

Phase 2: the state transition (lines 07-17 in pseudo code on Table 1). This Phase selects through q, a neighbor node s. In the case that there is no node towards which to move (that is to say, the node is a leaf or all neighbor nodes had been visited) a hop backward on the path is carried out, otherwise the ant adds the node s to its path, and reduces TTL_k by one hop. The query process ends when the expected result has been satisfied or TTL_k is equal to cero, then the ant is killed indicating the end of the query.

Table 1. NAS algorithm pseudo code

```
01    for each query
02       for each ant
03          if Hits < maxResults and TTL > 0              // Phase 1
04             if the neighbor from edge s_k has results   // Lookahead strategy
05                append s_k to Path_k and TTL_k = TTL_k -1
06                Pheromone globalUpdate
07             else                                        // Phase 2
08                s_k = apply the transition rule with the DDC
09                if path does not exist or node was visited,
10                   remove the last node from Path_k
11                else,
12                   append s_k to Path_k and TTL_k = TTL_k -1
13                   Pheromone localUpdate
14                endif
15             endif
16          else
17             Kill ant
18          endif
19       endfor
20    endfor
```

4 Case Study: Determining SQRP Main Characteristics That Affect NAS Performance

4.1 Characteristics of SQRP

The problem of locating textual information in a network over the Internet is known as *semantic query routing problem* (SQRP). The goal of SQRP is to determine the shortest paths from a node that issues a query to nodes that can appropriately respond to the query [8]. The query traverses the network moving from the source node to a neighboring node and then, to a neighbor of a neighbor and so forth until locating the requested resource, or giving up in it absence. The challenge lies in the design of algorithms that navigate the internet in an intelligent and autonomous manner. In order to reach this goal, the NAS algorithm selects the next node to visit using information of the near-by nodes of the current node, for example, information on the local topology of the current node [10].

We considered three main characteristics of the SQRP and its probability distribution to study their impact in the performance of the NAS algorithm. First is the probability distribution of the node connections, which defines the topology of the network. Next, the probability distribution of the repository texts, which is the frequency of the distribution of the information contained in the nodes. Finally, the probability distribution of the queries, which defines how many times the same query, is repeated into the different nodes on the network. In order to simplify the experiment description, the following short names will be use respectively: topology, repository and query distributions. The most common distributions on Internet are: power law [11] and random [12].

4.2 Research Question and Hypothesis

The research question can be stated as: Does a set of characteristics of the SQRP affect the performance of the NAS algorithm? This question can be defined in the following research hypothesis.

Null Hypothesis H_0: The distributions of the topology, repositories and queries of the SQRP, have no significant effect on the average performance of the NAS algorithm.

Alternative Hypothesis H_1: The distributions of the topology, repositories and queries of the SQRP, have a significant effect on the average performance of the NAS algorithm.

4.3 Experimental Design

Hypothesis above require studying all combinations between the problem characteristic and probability distribution types. When it is needed to study the combined effect of two factors, the statistic literature recommends carrying out a full factorial 2^k experiment. In this kind of experiment, it is necessary to identify the factors with its levels, the variable responses and replicates. The impact of the factor levels is measured through the variable response in each replicate.

Two *factors* were studied: the probability distribution type and the problem characteristics. The *levels* of the first factor are: Scale free (SF) and Random (RM) and for the second factor the levels are: Topology, Repositories and Queries.

The *response1* variable is the average amount of discovered text (*hits*) and *response2* variable is the average on the number of visited nodes (*hops*). In the experiment the **replicates** are the different runs of the NAS metaheuristic.

4.4 Performing the Experiment

The NAS algorithm was implemented to solve SQRP instances. Each instance is formed by three files that correspond with the problem characteristics: Topology, Repositories and Queries. The generation of these files is explained below. In this experiment 2^3 instances were generated. These instances are the combinations of probability distributions of the characteristics of the problem. Each instance was replicated 7 times with NAS. The Topology and the Repositories were simulated in a static way, while the set of Queries was dynamic.

For the Topology, each no-uniform network with a *scale-free distribution* was generated using the model of Barabási et al. [11]. Similarly, the networks with r*andom distribution were generated* using the model of Erdös-Rényi [12]. In the *scale-free or power law* distribution, a reduced set of nodes has a very high degree connection and the rest of the nodes have a small degree. In the *random* distribution, all nodes have a degree very closed to the average degree. All networks were generated with 1,024 nodes and bi-directional links. This number of nodes was obtained from the recommendations of the literature [8][7].

For the repository texts *R*, the application is a distributed search machine. Each peer manages a local repository of resources, and offers its resources to other peers. The "topics" of the resources were obtained from ACM Computing Classification

System (ACMCCS) in which the database contains 910 topics. The R file was created with the two distributions before mentioned, but now the distributions are applied in how many times the topic was repeated. When the distribution was random the topics were duplicated 30 times in average. When the distribution was scale free, in each node was assigned a reduced set of topics with many copies, and the rest of the topics were assigned with few copies.

For the queries Q, each node has a list of possible resources to search. This list is limited by the total of topics of the ACM classification. During each algorithm step, each node has a probability of 0.1 to launch a query. The topics were selected randomly for each query. The probability distribution of Q determines how many times will be repeated a query in the network. When the distribution is random, the query is duplicated 30 times in average and, when the distribution is scale free, a reduced set of queries is much repeated and the rest of the queries is a few repeated.

Each simulation was run 10,000 steps. The average performance was calculated taking measures of the performance in progress each 100 steps. The performance was measure through two response variables. The *response*1 variable measures the average number of hops; in each query the maximal number of *hops* was settled in 25. The *response*2 variable measures the average of hits; the maximum number of hits was settled in 5.

The statistical experiment was performed with the package MINITAB. We created a full factorial 2^k design with 2 factors, 3 levels (k), 7 replicates, the significance level $\alpha = 0.05$, and the values of the two response variables for each instance. This factorial design was analyzed in order to determine the most important interactions of the factors, in other words, which combinations of the factors have the biggest effects on the performance. With this information, we ran the General Linear Model (GLM), which is based mainly on the ANOVA test. GLM was used to confirm, if the selected combinations of the factors are *statistically significant*. These results are described in the next section with plots of main effects and interactions.

5 Analyzing Statistics Results

The plots of the Figure 1 revel that the type of probability distribution of the main factors has *relevant effect* on the average performance of NAS. The Topology is the factor that has the biggest effect on the average of the quality and efficiency of NAS. The other two factors (Repositories and Queries) show *irrelevant effect* on the average of the quality and efficiency of NAS. The size and direction of the graphs confirm this affirmation.

The plots of the Figure 2 revel that the type of probability distribution of the factor interactions has *relevant effect* on the average performance of NAS. The interaction between *Repositories* and *Queries* has the biggest effect on the average of the quality and efficiency of NAS. The other two interactions show *irrelevant effect* on the average of the quality and efficiency of NAS. We demonstrated the statistical significance of the *relevant effect* with the GLM methodology.

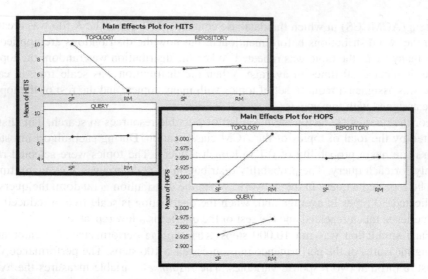

Fig. 1. Result of the Main effects: *response*1 (HITS) and *response*2 (HOPS)

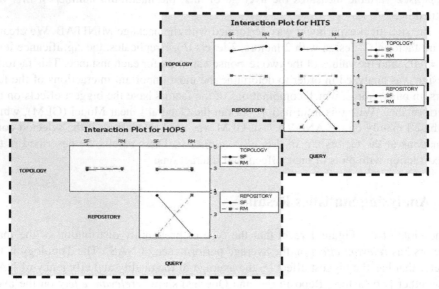

Fig. 2. Result of the characteristic interactions: *response*1 (HITS) and *response*2 (HOPS)

6   Conclusions and Future Works

A methodology based on Design of Experiment (DOE) was presented to answer the research question. The results obtained in the experimentation show that, the variability in the probability distributions of the SQRP characteristics (topology, query and repository) affects the performance of the NAS algorithm.

The experiment results showed that the topology distribution had a huge influence for the two response variables: number of resources founded for the queries and number of steps taken. We found out that the distributions of queries and repositories have strong relation ship in both response variables.

In addition, with the analysis of the results we identified which type of probability distribution is the most significant to the algorithm performance: a) The Scale-free, for the topological distributions, b) The random, for the query distributions and c) The scale-free for the repository distributions.

We are planning to use DOE, in order to analyze the control parameter of NAS algorithm. The objective of this future work is to find the best configuration to ensure the maximum NAS performance.

References

1. Birattari, M., Stutzle, T.: A Racing algorithm for Configuring Metaheuristics. Artificial life, 11–18 (2002)
2. Barr, R., Golden, J., Kelly, M.: Designing and Reporting Computational Experiments with Heuristics Methods. Journal of Heuristics 1, 9–32 (1995)
3. Michalewicz, Z.: How to solve it: Modern Heuristics. Springer, NY (2000)
4. Angeline, P.: Adaptative and Self-Adaptative Evolutionary Computations. IEEE, Computational Intelligence, 152–163 (1995)
5. Montgomery, D.C.: Design and Analysis of Experiments. John Wiley & Sons, New York (2001)
6. Manson, R.L., Gunst, R.F., James, L.H.: Statistical Design and Analysis of Experiments with Applications to Engineering and Science, 2nd edn. Wiley – Interscience, Chichester (2003)
7. Dorigo, M., Stützle, T.: Ant Colony Optimization. MIT Press, Cambridge (2004)
8. Michlmayr, E.: Ant Algorithms for Self-Organization in Social Networks. Doctoral Thesis, Women's Postgraduate College for Internet Technologies (WIT), Institute of Software Technology and Interactive Systems, Vienna University of Technology (2007)
9. Ortega, R., et al.: Impact of Dynamic Growing on the Internet Degree Distribution. Polish Journal of Environmental Studies 16, 117–120 (2007)
10. Cruz-Reyes, L., et al.: NAS Algorithm for Semantic Query Routing System for Complex Network. In: Advances in Soft Computing, vol. 50, pp. 284–292. Springer, Heidelberg (2008)
11. Barabási, A.L., Albert, R., Jeong, H.: Mean-Field theory for scale-free random networks. Physic A 272, 173–189 (1999)
12. Erdős, P., Rényi, A.: On random graphs. I. Publ. Math. Debrecen 6, 290–297 (1959)
13. Adenso-Díaz, B., Laguna, M.: Fine-Tuning of Algorithms Using Fractional Experimental Designs and Local Search. Operation Research, 99–114 (2004)

A New Approach to Improve the Ant Colony System Performance: Learning Levels

Laura Cruz R., Juan J. Gonzalez B., José F. Delgado Orta, Barbara A. Arrañaga C.,
and Hector J. Fraire H.

Instituto Tecnológico de Ciudad Madero
1°. de mayo s/n Col. Los Mangos, CP. 89100. Ciudad Madero Tamaulipas, México
lcruzreyes@prodigy.net.mx, jjgonzalezbarbosa@hotmail.com,
francisco.delgado.orta@gmail.com, aralia38@hotmail.com,
hfraire@prodigy.net.mx

Abstract. In this paper a hybrid ant colony system algorithm is presented. A new approach to update the pheromone trails, denominated learning levels, is incorporated. Learning levels is based on the distributed Q-learning algorithm, a variant of reinforcement learning, which is incorporated to the basic ant colony algorithm. The hybrid algorithm is used to solve the Vehicle Routing Problem with Time Windows. Experimental results with the Solomon's dataset of instances reveal that learning levels improve execution time and quality, respect to the basic ant colony system algorithm, 0.15% for traveled distance and 0.6% in vehicles used. Now we are applying the hybrid ant colony system in other domains.

Keywords: Ant Colony System (ACS), Distribued Q-learning (DQL), vehicle routing problem (VRP).

1 Introduction

Routing and scheduling of vehicles in delivery processes are problems in which, several customers need to be satisfied for their suppliers in efficient way, such as transportation represents from 5 to 20 percent of total costs of the prices of the products [1]. Scientists have modeled transportation problems in a subset of tasks, which have been modeled through the Vehicle Routing Problem, a known NP-Hard problem which has been object of study for many researchers.

The most known works has been approached the solution of VRP using heuristic algorithms. Bent approached VRP through an algorithm of hybrid local search [2], Braysy developed a deterministic algorithm of neighborhood variable search in [3], Pisinger developed a search algorithm in variable environments [4], A multi-objective ant colony system [5]. Hybrid and parallel genetic algorithm were developed by Berger and Homberger in [6,7], Rochat, Taillard and Cordeau developed heuristic algorithms based on tabu search [8,9,10] and a hybrid heuristic developed in [11]. They report the best results for the Solomon's dataset of test, a set of instances for the Vehicle Routing Problem with Time Windows, the most known variant of VRP. So, this work presents a study of ant colony system algorithm (ACS) [12] that uses a new

E. Corchado et al. (Eds.): HAIS 2009, LNAI 5572, pp. 670–677, 2009.

approach to update the pheromone trails denominated learning levels. This hybrid algorithm is used to solve the Vehicle Routing Problem with Time Windows. Learning levels is based on the distributed Q-learning algorithm. It includes a technique of reinforcement learning, which is incorporated to the basic ant colony algorithm. This hybrid approach is used to improve the solutions of the basic ACS. Section 2 defines the vehicle routing problem and its time windows variants. Section 3 describes the state of art related to ant algorithms; section 4 shows the distributed Q-Learning technique (DQL) while learning levels is presented in Section 5. Experimentation with VRPTW instances of the Solomon's dataset is presented in section 6. The conclusions of this work are presented in section 7.

2   Vehicle Routing Problem (VRP)

VRP, defined by Dantzig in [13], is a classic problem of combinatorial optimization, which consist of one or various depots, a fleet of m available vehicles and a set of n customers to be visited, joined through a graph $G(V,E)$, where $V=\{v_0, v_1, v_2, ...,v_n\}$ is the set of vertex v_i, such that v_0 the depot and the rest of the vertex represent the customers. Each customer has a demand q_i of products to be satisfied by the depot. $E=\{(v_i, v_j) \mid v_i, v_j \varepsilon V, i \neq j\}$ is the set of edges. Each edge has an associated value c_{ij} that represents the transportation cost from v_i to v_j. The VRP problem consists of obtaining a set R of routes with a total minimum cost such that: each route starts and ends at the depot, each vertex is visited only once by either route, or the length of each route must be less than or equal to a distance threshold named L.

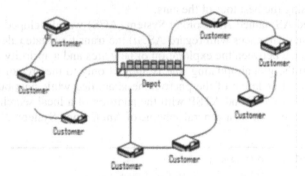

Fig. 1. The Vehicle Routing Problem

According to [14,15,16] the description of VRPTW is similar to the basic VRP described. The main difference consists of the addition of time Windows at the customer and depot facilities (schedules in which vehicles can be attended at the facilities). It is defined as a multi-objective problem which is solved hierarchically in two steps: 1) the number of used vehicles is minimized and 2) the total cost of travel is minimized to supply all demands at the customer facilities. VRPMTW is defined in [14] as a generalization of VRPTW. It adds various time Windows by customer. In real situations, this definition is interpreted as different shifts of the customers.

3   Ant Colony System Optimization: State of Art

Ant Colony Optimization (ACO) is a technique inspired by the behavior of the ants, which search the shortest path between the food sources and their anthill. Ants use two main characteristics: the heuristic information η_{rs}, used to measure the predilection to travel between a pair of vertex (r,s); and the trails of artificial pheromone τ_{rs}, used to compute the learned reference of traveling in a determined arc (r,s).

ACO was introduced initially in [17] through the creation of Ant System (AS). It is formed by three algorithms: Ant-density, Ant-quantity and Ant-Cycle. Ant-density and Ant-quantity uses the update of pheromone trails in every step of the ants, while Ant-Cycle makes updates after a complete cycle of the ant. These algorithms were tested with the Traveling Salesman Problem (TSP), with good results and the assurance that this method can be applied in several types of problems. A study of the correct configuration of AS for solving TSP was done in [18]. It concludes that the main parameter is β, which and establishes that the optimal number of ants is equivalent to the number of nodes of the problem. The properties of the Ant-cycle algorithm was proposed in [19] based on uniform and randomly distributions of the ants at the nodes whole results shown a few differences; however, randomly distribution obtains the best results.

Other ant algorithm is Ant-Q, which is based on Q-learning, is applied to AS [20]. It was applied for solving the TSP and Asymmetric TSP (ATSP) through the use of a table of values Q, equivalent to the table of values of Q-learning, which is used to indicate how good a determined movement towards a node s is since a node r. It applies a rule to choose the next node to be visited and reinforcement learning applied to the Q values using the best tour of the ants.

An improved AS named Ant Colony System (ACS) was developed in [21]. It presents three main differences with regard AS: a) the transition-state rule is modified to establish a balance between the exploration of new arcs and a apriority exploitation of the problem; b) the global updating rule is applied only to the arcs of the tour of the best ant; c) a local updating of the pheromone is applied while ants build a solution. ACS was applied to TSP and ATSP with the addition of a local search based on a 3-opt scheme. Figure 2 shows a general scheme of Ant Colony System.

```
Ant_Colony_System ( )
Initialize Data Structures
Do
     For each ant initialize its solution
   Do
        For   each   ant:   pseudo-randomly   rule   is
         applied to build a solution
        Local update (τ_rs )
   Until all ants have completed their solutions
   Global update (τ_rs )
Until stop criteria is reached
```

Fig 2. The Ant Colony System algorithm

Heuristic information of expression (1) is used to choose the most appropriated vehicle. It combines capacity, velocity and remaining service time of each vehicle v.

$$\eta_v = \left(nv_v \cdot (\overline{TM}_v + \overline{TR}_v) \cdot \frac{tr_v}{tt_v} \right)^{-1} \tag{1}$$

Where η_v is the value of the heuristic information or closeness for each vehicle v, nv_v is a bound of the quantity of travels required for the vehicle v to satisfy all the demands of the available nodes $N_k(r)$, $\overline{TM}_v$ is the average of the service time in $N_k(r)$, $\overline{TR}_v$ is the time trip average of the vehicle to $N_k(r)$, tr_v is the available for the vehicle v, tt_v is the time of attention of the vehicle v; tr_v/tt_v is a factor time of use/availability.

The transition-state rule establishes that given an ant k, located in a node r with q_0 a balancing value between the constructive focuses and q a random value; if $q \leq q_0$, exploitation of the learned knowledge is applied through a nearest neighborhood heuristic to choose a node s. Otherwise, a controlled exploration is applied to the ant with the best solution using expression (2).

$$p_{rs}^x = \begin{cases} \dfrac{\tau_{rs}\eta_{rs}^\beta}{\displaystyle\sum_{s \in N_k(r)} \tau_{rs}\eta_{rs}^\beta} & \text{if } s \in N_k(r) \\[4mm] 0 & \text{otherwise} \end{cases} \tag{2}$$

ACS uses evaporation trails of pheromone to reduce the possibilities of nodes to be chosen through global and local updates. Both update strategies use an evaporation rate ρ. Local update is done to generate different solutions the already obtained. The initial trail of pheromone τ_0, used in the local update, is computed like the inverse of the product of the length of the shortest global solution generated and the number of visited nodes. Local update is applied through expression (3).

$$\tau_{rs} \leftarrow (1 - \rho)\tau_{rs} + \rho\tau_0 \tag{3}$$

Then, when all ants have completed their solution, a global update process $\Delta\tau_{rs}$ is computed like the inverse of the length of the shortest global solution generated by the ants. It is used to memorize "preferred routes", which are used in each iteration to improve the global solution. This updating process is computed using expression (4).

$$\tau_{rs} \leftarrow (1 - \rho)\tau_{rs} + \rho\Delta\tau_{rs} \tag{4}$$

Max-Min Ant System algorithm (MMAS), developed in [22], uses the elements of Ant System. However, it modifies the original rules of AS, presenting three main differences: a) the updating rule was modified to choose the best tour of the ants and the best found solution during the execution of the algorithm, increasing with this the exploration; b) a upper limit for the pheromone trails was established, which permits that a subset of nodes not being chosen in recurrent form; and c) the pheromone trails were initialized with the permitted upper bound to choose only the best arcs, increasing their pheromone trails and ants rarely choose bad arcs to build a solution. Other variant of AS, named ASrank, was developed in [23]. It consist of a sorting of the ants (once that ants has built their solution) following the length of travel and to weight the contribution of the ant to the trail-pheromone updating level according a rate μ; it permits that only the best ω ants are considered, omitting the pheromone trails of some ants that uses suboptimal roads. This algorithm was tested with TSP instances.

This work approaches on a new technique applied to the Ant Colony System (ACS) for solving the vehicle routing problem; a generalized TSP with transportation constrains that defines the Vehicle Routing Problem with Time Windows (VRPTW), a known NP-hard problem object of study for many researchers. The developed technique is based on the Distributed Q-Leaning algorithm (DQL). Therefore, its application in the basic ACS is used to improve the performance for the basic ACS.

4   Distributed Q-Learning: A Learning-Based Approach

Distributed Q-Learning (DQL), presented in [24], is similar to the Ant-Q algorithm [20], but establishes three main differences: *a*) DQL does not use heuristics dependent of domain, reason why it does not require additional parameters, *b*) DQL updates the Q values only once with the best solution obtained for all the agents and *c*) DQL permits more exploration and a best exploitation.

DQL establishes that all the agents have an access to a temporal copy Qc of the evaluation functions (Q values) for each pair state-action. Each time that an agent has to choose an action, it observes the copy of the Q values to execute the next action, once that the next action is chosen the table Qc is updated. This process is similar to Ant-Q; however, all the agents of DQL update and share their common values Qc. When all the agents have found a solution, this copy is deleted and the original Q values are rewarded, using the best obtained solution by the agents. The DQL algorithm is shown in Fig. 3.

```
DQL_Algorithm ( )
Initialize Q(s,a)arbitratily
Repeat (for n episodes)
    Initialize s, Qc(s,a)← Q(s,a)
    Repeat (For each step of the episode)
        Repeat (for m agents)
            Execute action a, observe r, s'
            Qc(s,a)←Qc(s,a)+ α[γmaxₐ' Qc(s',a')-Qc(s,a)]
            s ← s';
    Until s is terminal
    Evaluarte the m proposed solutions
    Assign a reward to the best found solutión
        and update the Q-values:
        Q(s,a) ← Q(s,a)+ α[r + γ maxₐ' Q(s',a')-Q(s,a)]
```

Fig. 3. The DQL algorithm

The Qc values are used as a guide, which permits that the agents can observe more promissory states. So, SQL permits a bigger exploration and a better exploitation since only the best actions are rewarded. DQL was compared in [24] with Ant-Q and Q-learning solving TSP instances and it obtains a faster convergence with regard the other algorithms without needing the configuration of extra parameters. These characteristics are essential to define learning levels.

5 Learning Levels (LL)

The main similarity of DQL with ACS is the fact that both use a table of values in which the learned experience is storaged: in DQL is the table of Q-values while in ACS is the pheromone table τ_{rs} .This relation permits the implementation of DQL over the ACS algorithm; it permits the creation of a new technique, named learning levels. The scheme of ACS with learning levels is shown in Fig. 4.

```
ACS_Learning_Levels( )
Initialize Data Structures
Do
    For each ant: initialize its solution
    τrs copy ← τrs
    Do
        For each ant
        Apply the pseudo-randomly rule to
            build a solution
        local pheromone update ( τrs copy )
    until ∀ ants have complete their solution
    Global Pheromone Update ( τrs )
Until Stop criteria is reached
```

Fig 4. ACS with learning levels

Learning levels defines two levels of knowledge: the first level is equal to the values of the original pheromone table, which only contains the information of the best obtained solution for the ants and it is modified only in the global updating process; the second level equals to the copy of the pheromone table, which contains the local values of the pheromone and it is used for the ants as a guide in the search of better solutions. This level is updated locally for each ant in the local update process.

6 Experimentation and Results

The performance of the basic ACS, described in section 3 and proposed in [21] was compared versus ACS plus Learning levels (ACS+LL). All the algorithms were applied to solve the Vehicle Routing Problem with Time Windows of Solomon's dataset, which is formed of 56 instances grouped in six sets named: C1, C2, R1, R2, RC1, RC2; everyone with 9,8,12,11,8 and 8 cases with $n=100$ nodes respectively.

The group of instances is named according the characteristics of the cases that form them: the group C is formed by customers in one area. Instances of type R have sets of customers uniformly distributed over a squared area; the group RC has a combination of customers uniformly distributed and grouped in areas. The instances of type 1 have reduced time Windows and small capacity for the vehicles while instances of type 2 have wider time windows and bigger capacity for the vehicles.

The configuration of ACS was: $q_0 = 0.65, \beta = 6, \rho = 0.1$, with 15 generations per colony and 100 ants per generation. The number of ants was defined as the number of nodes of the instance according to [18]. The ACS developed was coded in C# and it was executed during 1800 seconds (30 minutes). Results are shown in table 1.

Table 1. Comparative of ACS and ACS plus Learning Levels (ACS-LL)

Instances	Traveled Distance		Used Vehicles	
	ACS	ACS+LL	ACS	ACS+LL
C1	883.30	887.17	10.00	10.00
C2	634.48	633.67	3.00	3.00
R1	1455.35	1437.93	13.58	13.42
R2	1221.16	1245.66	3.09	3.00
RC1	1627.36	1599.03	13.00	13.00
RC2	1452.36	1456.16	3.38	3.38
Accumulated	68559.44	68452.79	442	439

Table 1 reveals that ACS+LL improves 0.15% and 0.6% the solutions of the ACS according to the accumulated account with regard to the traveled distance and used vehicles respectively. This is used to prove the advantage of learning techniques to improve a basic heuristic as ant colony system.

7 Conclusions and Future Works

As a result of this work, it is concluded that the use of learning levels permits improve the results of ACS as a consequence of the bigger covering of the solution space of VRP and the exploitation of the knowledge obtained by the ants in the past. The use of this new approach (learning levels) allowed a faster convergence of the ACS. An implementation of this new approach in other optimization problems is proposed to study its behavior, as well as the application with other search techniques to develop a new hybrid algorithm. This could be able to improve solutions for VRPTW reported in this paper.

References

1. Toth, P., Vigo, D.: The Vehicle Routing Problem. In: Proc. SIAM Monographs on Discrete Mathematics and Applications (2002)
2. Bent, R., Van Hentenryck, P.: A Two-Stage Hybrid Local Search for the Vehicle Routing Problem with Time Windows. Transportation Science (2001)
3. Bräysy, O.: A Reactive Variable Neighborhood Search Algorithm for the Vehicle Routing Problem with Time Windows. Tech. report, SINTEF Applied Mathematics, Department of Optimization (2001)
4. Pisinger, D., Ropke, S.: A General Heuristic for Vehicle Routing Problems. Tech. report, Dept. of Computer Science, Univ. Copenhagen (2005)
5. Gambardella, L., Taillard, E., Agazzi, G.: MACS-VRPTW: A Múltiple Ant Colony System for Vehicle Routing Problems with Time Windows. Tech. report IDSIA-06-99, IDSIA (1999)
6. Homberger, J., Gehring, H.: Two Evolutionary Metaheuristics for the Vehicle Routing Problems with Time Windows. INFOR 37, 297–318 (1999)

7. Berger, J., Barkaoui, M.: A Memetic Algorithm for the Vehicle Routing Problem with Time Windows. In: The 7th International Command and Control Research and Technology Symposium (2002)
8. Rochat, Y., Taillard, E.: Probabilistic Diversification and Intensification in Local Search for Vehicle Routing. Journal of Heuristics 1, 147–167 (1995)
9. Taillard, E., et al.: A Tabu Search Heuristic for the Vehicle Routing Problem with Soft Time Windows. Transportation Science 31, 170–186 (1997)
10. Cordeau, F., et al.: The VRP with time windows. Technical Report Cahiers du GERAD G-99-13, Ecole des Hautes 'Etudes Commerciales de Montreal (1999)
11. Potvin, J.Y., Rousseau, J.M.: An Exchange Heuristic for Routeing Problems with Time Windows. Journal of the Operational Research Society 46, 1433–1446 (1995)
12. Dorigo, M., Gambardella, L.M.: Ant Colony System: A Cooperative Learning Approach to the Traveling Salesman Problem. Technical Report TR/IRIDIA/1996-5, IRIDIA, Université Libre de Bruxelles (1996)
13. Dantzig, G.B., Ramser, J.H.: The Truck Dispatching Problem. Management Science 6(1), 80–91 (1959)
14. Jong, C., Kant, G., Vliet, A.V.: On Finding Minimal Route Duration in the Vehicle Routing Problem with Multiple Time Windows. Tech. report. Dept. of Computer Science, Utrecht Univ. (1996)
15. Shaw, P.: Using Constraint Programming and Local Search Methods to Solve Vehicle Routing Problems. In: Maher, M.J., Puget, J.-F. (eds.) CP 1998. LNCS, vol. 1520, pp. 417–431. Springer, Heidelberg (1998)
16. Dorronsoro, B.: The VRP Web (2006),
    http://neo.lcc.uma.es/radi-aeb/WebVRP/
17. Dorigo, M.: Positive Feedback as a Search Strategy. Technical Report. No. 91-016. Politecnico Di Milano, Italy (1991)
18. Colorni, A., Dorigo, M., Matienzo, V.: Distributed Optimization by Ant Colonies. In: Varela, F.J., Bourgine, P. (eds.) Proc. First European Conference on Artificial Life, pp. 134–142. MIT Press, Cambridge (1992)
19. Colorni, A., Dorigo, M., Maniezzo, V.: An Investigation of Some Properties of an Ant Algorithm. In: Manner, R., Manderick, B. (eds.) Proceedings of PPSN-II, Second International Conference on Parallel Problem Solving from Nature, pp. 509–520. Elsevier, Amsterdam (1992)
20. Gambardella, L.M., Dorigo, M.: Ant-Q: A Reinforcement Learning Approach to the Traveling Salesman Problem. In: Prieditis, A., Russell, S. (eds.) Proceedings of ML1995, Twelfth International Conference on Machine Learning, Tahoe City, CA, pp. 252–260. Morgan Kaufmann, San Francisco (1995)
21. Dorigo, M., Gambardella, L.M.: Ant Colony System: A Cooperative Learning Approach to the Traveling Salesman Problem. Technical Report TR/IRIDIA/1996-5, IRIDIA, Université Libre de Bruxelles (1996)
22. Stützle, T., Hoos, H.H.: Improving the Ant System: A detailed report on the MAXMIN Ant System. Technical report AIDA-96-12, FG Intellektik, FB Informatik, TU Darmstadt (1996)
23. Bullnheimer, B., Hartl, R.F., Strauss, C.: A New Rank Based Version of the Ant System: A Computational Study, Technical report, Institute of Management Science, University of Vienna, Austria (1997)
24. Mariano, C., Morales, E.F.: DQL: A New Updating Strategy for Reinforcement Learning Based on Q-Learning. In: Flach, P.A., De Raedt, L. (eds.) ECML 2001. LNCS, vol. 2167, pp. 324–335. Springer, Heidelberg (2001)

Hybrid Algorithm to Data Clustering

Miguel Gil[1], Alberto Ochoa[2], Antonio Zamarrón[1], and Juan Carpio[1]

[1] Instituto Tecnológico de León, Av. Tecnológico s/n, c.p.37290 León, Guanajuato, México
[2] Instituto de Ingeniería y Tecnología, Universidad Autónoma de Ciudad Juárez

Abstract. In this research an N-Dimentional clustering algorithm based on ACE algorithm for large datasets is described. Each part of the algorithm will be explained and experimental results obtained from apply this algorithm are discussed. The research is focused on the fast and accurate clustering using real databases as workspace instead of directly loaded data into memory since this is very limited and insufficient when large data amount are used. This algorithm can be applied to a great variety and types of information i.e. geospatial data, medical data, biological data and others. The number of computations required by the algorithm is ~$O(N)$.

Keywords: classification, clustering, large datasets.

1 Introduction

Clustering takes an important part into the KDD (Knowledge Discovery into Databases) [1]. Today, databases size are very long and the number of records in a dataset can vary from some thousands to thousands of millions [2]. In [3] a new unsupervised and fast algorithm named ACE was proposed. In the paper, the authors describe the algorithm but, it is explained and exemplified only in one dimension and the tests made to the algorithm only include data loaded into memory and not using real databases. Based in the paper and algorithm above mentioned, a N-Dimensional Clustering algorithm called ACEND was developed. The motivations to research this job is because more existing clustering algorithm requires multiple iterations (scans) over de datasets previous to achieve a convergence [4], and many of them are sensitive to initial conditions, by example, the K-Means and ISO-Data algorithms [5, 6].

2 The ACEND Algorithm

In this section the ACEND algorithm is described, which is based on clustering data by a particle-mesh heuristic and how each data is represented as a point in a two-dimension space and how after is associated with one point of the mesh, and finally the procedure to conform the clusters is described.

2.1 Grid Creation and Weighting

The first step is to create a mesh by each dimension to be clustered. Each dimension represents one attribute or field from the dataset to be clustered. Next, we will call x_p

E. Corchado et al. (Eds.): HAIS 2009, LNAI 5572, pp. 678–685, 2009.
© Springer-Verlag Berlin Heidelberg 2009

to one grid-point from one dimension and x_i to one attribute from one data point. Then a data density by each grid-point (x_p) must be calculated. In the algorithm only a "order-zero" weighting function is considered, even though, other highest weighting orders as are mentioned in [2, 3, 4] and it can be used. In a zero-order weighting form, we must determinate the closest grid-point (x_p) to each data-point (x_i) measuring the distance between xi and xp and then between x_i and x_{p+1} using the equation (1):

$$d = |x_p - x_i| \tag{1}$$

$$\text{if } d(x_p - x_i) = 1 \text{ then } p(x_p) = p(x_p) + 1/h \text{ else } p(x_p + 1) = p(x_p + 1) + 1/h \tag{2}$$

Fig. 1. Data weighting to nearest grid points on a one-dimensional mesh. In nearest grid point weighting, the data point at xi is assigned to the nearest grid point at xp. In linear weighting, the data point at xi is shared between x_p and x_{p+1} according to linear interpolation.

2.2 Creating Preliminary Clusters

When the grid whit their corresponding weights and associated data points are calculated, it conform a set of preliminary clusters too. Then, the next step is executing a preliminary depuration process. This is a difference from the original ACE Algorithm. In ACE, authors proposal a set of rule-based agents where each agent search for the most weight grid point. At the end of search, each grid point targeted previously by each agent, comes to be a preliminary cluster centroid. However, preliminary cluster centroids are selected by the agents, the question "How many agents are required in order to perform the search?" is important and non explicit answer is described in the ACE paper. Instead of the use of agents, a SQL-Select instruction was used in order to retrieve the grid points ordered by their weight. In ACE the next step is combine the centroids to conform more dense clusters and it suggest that two "close" grid points can be combined into one cluster, but in this point, the original paper don't mention a standard measure to make this, by example a threshold grid points distance. Instead of the ACE combine, in ACEND the combination centroid procedure is described as follows and a set of intermediate clusters will be created by the combination of the preliminary centroids:

```
for i = 1 to last_gridpoint - 1 {
    for j = i + 1 to last_gridpoint {
        dist = distance between gridPoint[i] and gridPoint[j]
        if distance <= threshold {
            if weight of gridPoint[i] > weight of gridPoint[j]{
                associate the data points in gridpoints[j]
                with gridpoints[i];
                remove gridPoint[j];
                recalculate centroid[i];
            }
            else if weight of gridPoint[j] > weight of gridPoint[i] {
                associate the data points in gridpoints[j]
                with gridpoints[i];
                remove gridPoint[j];
                recalculate centroid[i];
            }
        }//Close if-brace open.
    } //End of sub-main for-loop
}//End of main for-loop
```

The distance can be calculated with the Euclidean formula:

$$d(x_1, x_2) = \sqrt{(x_2 - x_1)^2 + (y_2 - y_1)^2} \qquad (3)$$

The *threshold* value is a value arbitrary given by the user and it involves some domain knowledge. This value threshold is mentioned too in the original ACE algorithm. While the threshold value is higher, the number of final clusters will be decreased.

Finally, consider that if two grid-points have the same weigth, then both grid points remains because them are not combined.

2.3   Creating Intermediate and Final Clusters

This part is very different than the ACE original algorithm, because in ACE simply don't appear. The objective in this step conform a new group of "depurated" clusters based on the preliminary clusters built in 2.2. In this step we will begin with the resultant centroids after the algorithm in 2.2. Then, in this step we compare each data point with each grid point and finally we associate the data point with the closest grid point. At the end of this step, we have a new group of clusters, in other words, a group of grid points with their closest data points associated.

Finally, the last iteration of the algorithm is made repeating the steps in the paragraph above.

3   Implementation Using a Database Relational Model

In this part, the implementation of ACEND algorithm using a database as data storage system and too as workspace is described.

3.1   The Data Table

All experiments was realized using the MySQL engine. First, a table containing a auto-increment integer field as primary key and the fields required to make the clustering process must be created. We will call to this table t1.

Fig. 2. MySQL data table example

idTabla1	valorX	valorY
1	6.63	-0.71
2	-9.33	0.24
3	4.82	-2.64

Fig. 3. Fields used in the table that store the data to be clustered

idGridPoint	valueX	valueY	weigth
7	-5.27	2.25	18.45
85	5.18	4.58	27.9
89	4.54	-4.15	10.35

Fig. 4. Fields used in the table that store the grid points and their x and y values and the weight field

3.2 Auxiliary Tables

In order to keep the amount of grid points for read-write operations other table $t2$ must be created. In this table each grid point with their correspondent id grid point, value in x, value in y and weight fields must be stored. The values x and y corresponds at the position of the grid point in each dimension. When each grid point is weighting (as was explained in 2.1) the appropriate gridpoint record stored in this table must be updated. Finally, a third table $t3$ must be created in order to keep a relation between each grid point and their correspondent data points. This table contains the fields grid_point_id, data_point_id, and distance. This table is the relational table between data points table ($t1$) and grid points table ($t2$). The distance field is the distance between these data point and their correspondent grid point. The id field is only an auto-increment integer field that keeps one identifier to each record. Using this work schema in the algorithm, it gives the possibility to manage long amounts of records minimizing the risk of an insufficient main memory space problem since the main work area resides in the hard disk of the computer in form of a database.

4 Small Dataset Example

In order to make comprehensive and verifiable the implementation and modification to the original ACE algorithm the example proposed in the original ACE paper was replicated, and it is as follows: Consider a small dataset of 160 points P(x, y). The data consisted of one hundred of points randomly distributed in the interval between $-10 < x < 10$ and $-10 < y < 10$. In addition, as shown in Fig. 5, three artificial clusters of points (20 points in each cluster) were produced that were randomly distributed around the positions (4, 4), (-4, 4), and (4, -4). One more difference with the original ACE example dataset is that in our case, we use a MySQL database to store the data and measure the time using a real DBMS instead of data loaded into main memory.

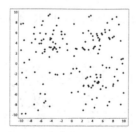

Fig. 5. Data points distribution with three visible clusters closest to the coordinates (-4, 4), (4, 4) and (4, -4)

4.1 Results with ACEND

We probe the ACEND algorithm with the data show above and we compare it with the classic K-Means Algorithm and the results are showed in next table:

Table 1. ACEND Statistics versus classic K-Means algorithm and their average error measure. The average error is the average of the distance between the data points and their centroid. Lower values in the avg. error is better.

Number of Points: 160	Centroids Expected in Positions: (-4, 4); (4, 4); (4, -4)			
Algorithm	Run Time (secs)	Number of Clusters	Centroids Positions	Avg. Error
ACEND	0.203	3	(-5.27, 2.25); (5.18, 4.58); (4.54, -4.15)	1.34
K-Means	0.344	3	(-5.17, 4.13); (5.44, 4.23); (2.08, 4.73)	1.56

With the results showed above, we can see that ACEND algorithm was more efficient than the K-Means algorithm.

4.2 Average Error

The average error is the average of the difference between each centroid position and their expected position. By example, for the first centroid finded by ACEND, the position was (-5.27, 2.25) and the expected position was ~(-4, 4). Successively he difference between each centroid position and their expected position was calculated and finally these differences was divided by three. The same procedure was made it with the K-Means results.

In this case, ACEND have a lowest average error than the K-Means algorithm. This means that clusters encountered by ACEND Algorithm are more "closed" in their internal data points and more separated between them.

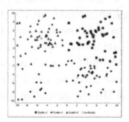

Fig. 6. Clusters formed by the K-Means algorithm

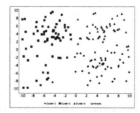

Fig. 7. Clusters formed by the ACEND algorithm

5 Large Dataset Test

To test the algorithm with a large dataset, a large dataset made up of 132, 264 records, was created and stored into a MySQL table.

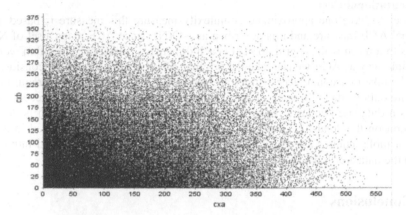

Fig. 8. Large dataset example. Consist of ~105 datapoints

The value's range is between 0 and 550 for the x-axis and between 0 and 375 for the y-axis. For the additional parameters the next values was setted: 85 for the x-axis cell distance, 75 for the y-axis cell distance and 270 as threshold value. Since the cell distances and threshold value are very important in order to determinate the final number of clusters, we try different values and finally select the values described above in order to find only six representative clusters on the entire dataset. The value's range is between 0 and 550 for the x-axis and between 0 and 375 for the y-axis. For the additional parameters we set the next values: 85 for the x-axis cell distance, 75 for the y-axis cell distance and 270 as threshold value. Since the cell distances and threshold value are very important in order to determinate the final number of clusters, we try different values and finally select the values described above in order to find only six representative clusters on the entire dataset. Using these values we create a grid with 7 gridpoints for the x-axis and 6 gridpoints for the y-axis. In total, we create a mesh consistent of 42 gridpoints.

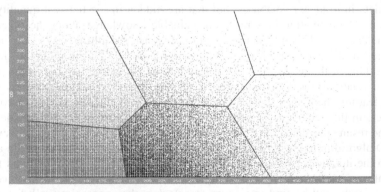

Fig. 9. Resultant clusters finded by ACEND algorithm

When the algorithm finished, only six grid points was leave as centroids for the representative groups (clusters) in the dataset. The total time took by ACEND in order to cluster the entire dataset was only one minute and six seconds and was able to find six clusters as show in the figure 4.2.

Computational Cost

In order to have one approximate complexity measure this measure is based in the original ACE measure and it is described as: $\sim O(N)$ since ACE maps a set of N data points to a mesh with N_g grid points in each dimension. Since our algorithm works in the same way as ACE we can take the original ACE equation. The next consideration must be take in account: access times are very different when a program requires read or write data loaded into main memory and when the program requires access to data across a database manager system or service. Finally, the second option is the most common on the "real-life" problems. Generally, working with data stored in a database or table requires a major access time in order to read or write it and minor time when the data are stored into main memory.

6 Conclusions

The methodology of ACEND is focused in the fast and accurate clustering using real databases as workspace instead of directly loaded data into memory since this is very limited and insufficient when we want work with large data amount. The efficiency of the ACE algorithm was verified. Future uses of this algorithm can be focused in use it with real databases, by example, in many Medical Data bases where the use of decision tree learning algorithms degrade their learning performance due to irrelevant, unreliable or uncertain data are introduced; or some focus on unvaried only, without taking the interdependent relationship among others into consideration; while some are limited in handling the attributes with discrete values. All these cases may be caused by improper pre-processing methods, where feature selection (FS) and continuous feature discrimination (CFD) are treated as the dominant issues. Even if a learning algorithm is able to deal with various cases, it is still better to carry out the pre-processing prior the learning algorithm, so as to minimize the information lost and increase the classification accuracy accordingly [7]. Other example in the Medical or Biological Data mining is the search of global patterns and the existent relationships among data in immense databases, but that are hidden inside the vast quantity of information. These relationships represent valuable knowledge about the objects that are in the database. This information is not necessarily a faithful copy of the information stored in the databases. Rather, is the information that one can deduce from the database. One of the main problems in data mining is that the number of the possible extracted relationships is exponential. Therefore, there are a great variety of machine's learning heuristics that have been proposed for the knowledge discovery in databases, in this technique is very usually used tree decision for obtain knowledge [10]. The results presented here have demonstrated the efficiency and accuracy of the ACEND algorithm to cluster long amount of data in a relative short time. In a future research the iterations over the entire dataset will tried to be decreased and focused principally in a fully-non-supervised algorithm. In other words, the main objective in future research will be find some way to calculate the appropriate cell distances by each dimension and the appropriate threshold values in an automatic way. Finally, since entropy is a complicated problem related with missing data, repeated data and inconsistent data, this affect to the performance of clustering methods, the veracity of the dataset to be clustered must be considered. This is because ACEND don't have some mechanism in order to detect wrong or invalid values.

References

1. Berry, M.J.A., Linoff, G.: Data Mining Techniques – for Marketing, Sales and Customer Support. John Wiley & Sons, New York (1997)
2. Birdsall, C.K., Landon, A.B.: Plasma Physics via Computer Simulation. Adam Hilger, Bristol (1991)
3. William, P., John, C., Clare, G.: New Unsupervised Clustering Algorithm for Large Data-sets. In: SIGKDD 2003, Washington, DC, USA, August 24-24, 2003. ACM, New York (2003)
4. Hockney, R.W., Eastwood, J.W.: Computer Simulation using Particles. Adam Hilger, Bristol (1988)
5. Bradley, P.S., Fayyad, U., Reina, C.: Scaling clustering algorithms to large databases. In: Proceedings of the Fourth International Conference on Knowledge Discovery and Data Mining, KDD 1998, New York, August 1998, pp. 9–15. AAAI Press, Menlo Park (1998)
6. Hartigan, J.A., Wong, M.A.: A k-means clustering algorithm. Applied Statistics 28, 100–108 (1979)
7. Chao, S., et al.: Discovery of the Latent Supportive Relevance in Medical Data Mining. In: Data Mining in Medical and Biological Research, I-Tech, Vienna, Austria, December 2008, p. 320 (2008) ISBN 978-953-7619-30-5

Financial Forecasting of Invoicing and Cash Inflow Processes for Fair Exhibitions

Dragan Simić[1,2], Ilija Tanackov[2], Vladeta Gajić[2], and Svetlana Simić[3]

[1] Novi Sad Fair, Hajduk Veljkova 11, 21000 Novi Sad, Serbia
dsimic@nsfair.co.yu
[2] University of Novi Sad, Faculty of Technical Sciences, Trg Dositeja Obradovića 6, 21000 Novi Sad, Serbia
ilijat@uns.na.ac.yu
[3] University of Novi Sad, Faculty of Medicine, Hajduk Veljkova 1-9, 21000 Novi Sad, Serbia
dsimic@eunet.rs

Abstract. The concept of case-based reasoning (CBR) system for financial forecasting, the dynamics of issuing invoices and the corresponding cash inflow for a fair exhibition is presented in this paper. The aim of the research is the development of a hybrid intelligent tool, based on the heuristic interpolating CBR *adaptation* phase and the data gravitation classification method in the *revise* phase. The previous experience with new forecasting is taken into account. Simulations performed are based on the already known behaviour of the adapted logistic curves followed during the last eight years of fair exhibitions. Methodological aspects have been practically tested as a part of the management information system development project of "Novi Sad Fair".

Keywords: Financial forecasting, cash inflow, case-based reasoning, data gravitation classification, heuristic interpolating, fair exhibition.

1 Introduction

The financial crisis that affected the world in the middle of the previous year also affects, more or less, every part of the world economy system. Risk and uncertainty are central to forecasting. In practice, business forecasting for manufacturing and service companies is used on a daily basis. Case-based reasoning (CBR), an artificial intelligence technique, is a suitable candidate for the development of a system forecasting such as events, due to their nonlinear structure and generalisation quality.

The concept of CBR system for financial forecasting, the dynamics of issuing invoices and corresponding cash inflow regarding fair exhibitions that is based on previous exhibition experience, is presented in this paper. The adapted logistic curve heuristic interpolation and data gravitation classification (DGC) method were used in the case *adaptation* and *revise* phases as part of a suggested forecasting system. Single hybrid architecture is formed by combining a heuristic method and artificial intelligence techniques. The deviation in financial forecasting for the maximum invoicing value is less than 1 %, while the deviation at financial forecasting for the maximum cash inflow value is less than 1 % after 400 days. So far, financial forecasting has not been discussed in business making in this manner.

E. Corchado et al. (Eds.): HAIS 2009, LNAI 5572, pp. 686–693, 2009.

The rest of the paper has a following structure: Section 2 overviews the CBR technique; Section 3 shows a short part of previous research. Section 4 describes the logistic function and the adapted logistic curve. Section 5 presents results and measurements, while Section 6 describes available analogous implementation. Section 7 concludes the paper and provides notes on future work.

2 Case Based Reasoning

Case-based reasoning is used for solving problems in domains where experience plays an important role. Generally speaking, CBR aims to solve new problems by adapting solutions that were successfully applied to similar problems in the past. The main supposition here is that problems that are similar have similar solutions.

The main phases of the case-based reasoning activities are described in the *CBR-cycle* [1]. In the *retrieve* phase the case that is most similar to the problem case is retrieved from the case memory, while in the *reuse* phase some modifications to the retrieved case are done in order to provide a better solution to the problem (case adaptation). As the case-based reasoning only suggests solutions, there may be a need for a confirmation or an external validation. That is the task of the *revise* phase. In the *retain* phase the knowledge and the information acquired are, integrated in the system. The main advantage and the strongest argument in favour of this technique is that it can be applied to almost any domain, where rules and connections between parameters are not known.

3 Our Previous Research

The presented research represents the continuation of a previous project (2003-2007) [2] [3]. In our previous research, we have presented the invoicing and cash inflow of the most important fair manifestation, taking place in May. The entire 400 days long process starts in December of the previous year, and is performed throughout the whole of the following year. The measurement of the invoicing and cash inflow values was performed every 4 days from the beginning of the invoice process; therefore every curve consists of approximately 100 points. The points were connected with the *cubic "spline"* as smooth curves. The invoice and cash inflow curves had been put at the same starting point, which simplified the *similarity* among the curves, but it did not succeed completely.

The previous researches contain discussions about the problem of the invoice curve and the cash inflow curve at the moment of the exhibition. Then, the invoice curve reaches its saturation point, while the cash inflow curve is still far away from it/the saturation point. The solution to this problem is the saturation point for the cash inflow curve.

4 Logistic Function and Adapted Logistic Curve

A logistic function or logistic curve is the most common sigmoid curve.

$$f(t) = \frac{a}{1+be^{-ct}} \tag{1}$$

688   D. Simić et al.

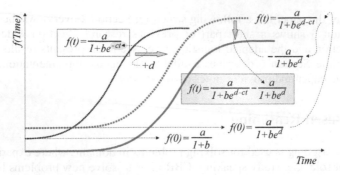

Fig. 2. The formation of the function for describing invoicing and cash inflow for the time function

The proposed function is easily adaptable to the upper horizontal asymptotic invoicing and cash inflow values via parameter a. Abscissa invoicing and cash inflow values for growth time period can be obtained by the increasing parameter b or the decreasing parameter c. Also, if the theoretical function of the asymptotic values is well matched, there is a possibility for the theoretical function of abscissa values to be matched to a smaller extent. That is why it is important to introduce the fourth parameter d (fig. 2).

The function defined in such a way gives a possibility to shift the whole function on abscissa. According to the latter, it should be noticed that the proposed functions (1) and (2) are always positive and non-declining. With the starting growth data being zero, it is necessary to subtract the starting value f(0) from the function's total value (2). The previously defined parameters a, b, c and d led to the modification of a general logistic curve through the observed processes.

$$f(t) = \frac{a}{1+be^{d-ct}} \qquad (2)$$

The adapted logistic function is:

$$f(t) = \frac{a}{1+be^{d-ct}} - \frac{a}{1+be^{d}} \qquad (3)$$

The adapted logistic curve is considered to be appropriate for explaining invoicing and cash inflow process of the fair manifestation.

5 Results and Measurements

The "International Agriculture Fair" (IAF) represents a fair manifestation suitable for financial analysis and forecasting because it comprises about 1800 clients and 7000 invoices. There is also a large number of cash inflows because the account holders pay for their services in instalments, using the approved time discount. A great number of the observed account holders, invoices and cash inflow ensures highly significant representative and analytical regularity. This fair manifestation brings the largest financial income to the company, and increases the seriousness of forecasting.

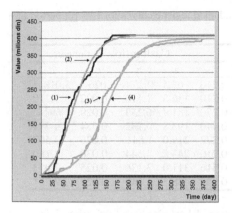
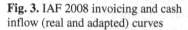
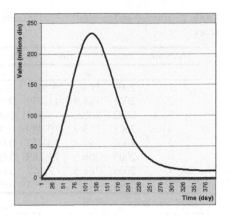

Fig. 3. IAF 2008 invoicing and cash inflow (real and adapted) curves

Fig. 4. Unsettled invoices curve for IAF 2008

The goal of this research is forecasting the dynamics of issuing invoices and receiving the actual cash inflow for the IAF 2008 exhibition, which is based on the observed behaviour of the adapted logistic curves in the last eight exhibitions, that is, from year 2000 to year 2007, which represents a *retrieve* phase from the *CBR-Cycle*.

An average cash inflow delay from the day of invoicing is 12 days, while deciding the first day of invoicing. The percentage of settled invoice process at the beginning of the exhibition is from 39 to 65%, which is 88-99% of the total invoice process in 400 days from the beginning of the invoicing process, representing very good financial results.

The measurement of the invoicing and cash inflow values was performed daily from the beginning of the invoice process and lasted for 400 days. The invoice and cash inflow curves start at different times representing a realistic business-making manner. The following curves are presented for the IAF 2008 exhibition: (1) the real invoicing; (2) the adapted logistic for invoicing; (3) the real cash inflow; (4) the adapted logistic for cash inflow (fig. 3). Figure 4 shows an unsettled invoices curve.

It is very important that a correlation coefficient is closer to 1, and that the coefficient before x should be ≈ 1 in the linear regression equation. Since both conditions have been met, it can be stated that there is high correlation between the real curves and the adapted logistics invoicing and cash inflow curves (table 1).

Table 1. "International Agriculture Fair" (2000 – 2007) invoicing and cash inflow

Linear regression equation for invoicing curve	Correlation coefficient invoicing curve	Linear regression equation for cash inflow curve	Correlation coefficient cash in flow curve
$y = 0.996 \, x + 2 \cdot 10^6$	$R^2 = 0.9902$	$y = 1.0181 \, x - 6 \cdot 10^6$	$R^2 = 0.9944$

Table 2. Invoice and cash inflow parameters for adapted logistic curves for IAF for period of 8 years

Number	Year	Invoice parameter				Cash inflow parameter			
		a	b	c	d	a	b	c	d
1	2000	74.1	13.567	0.044	0.210	62.5	0.239	0.026	5.018
2	2001	144.9	11.589	0.041	0.210	136.5	1.116	0.022	3.210
3	2002	208.0	3.388	0.046	0.270	181.7	6.253	0.029	1.691
4	2003	193.9	19.989	0.043	0.440	193.8	0.782	0.034	4.221
5	2004	230.3	13.457	0.040	0.272	219.0	1.541	0.026	3.507
6	2005	266.5	21.289	0.036	0.210	243.5	4.544	0.030	3.180
7	2006	361.5	7.277	0.034	0.516	311.0	3.136	0.025	2.933
8	2007	414.3	8.404	0.034	0.045	358.5	3.749	0.028	2.751

Suitable parameters (a, b, c, d) for the adapted logistics curves for invoicing and cash inflow for every year from year 2000 to year 2007 are shown in table 2. It can be observed that parameters b and d for the cash inflow in years 2000, 2002 and 2003 are quite different from the other values, and it is the consequence of turbulent social and economical occurrences in the environment, and internal company organisation.

According to the invoice and cash inflow parameters (a, b, c, d) for the adapted logistic curves for the period of last 8 years, suitable regression equations are formed and their correlation is presented in table 3.

Table 3. Invoice and cash inflow parameters for the adapted logistic curves for the IAF for period of 8 years

Para-meter	Linear regression equa-tion for invoicing curve	Invoicing coefficient correlation	Linear regression equa-tion for cash inflow curve	Cash inflow coefficient correlation
	Result (millions)		Result (millions)	
a	y = 43.772 x + 39,715	R^2 =0.9402	y = 37.558 x + 44.305	R^2 =0.9621
b	y = - 0.1253 x + 12.934	R^2 =0.0025	y = 0.3607 x + 1.0467	R^2 =0.1769
c	y = - 0.0016 x + 0.0472	R^2 =0.7964	y = 0.0003 x + 0.0485	R^2 =0.0485
d	y = 0.0003 x + 0.2702	R^2 =3 • 10^{-5}	y = - 0.1657 x + 4.0371	R^2 =0.1576

Based on the regression equations for the parameters in the last eight years, extrapolating parameters were obtained for year 2008 as *reuse* phases (*adaptation*). Equation for the adapted logistic curves was formed through the usage of the acquired parameters, and financial forecasting for the invoicing (4) and cash inflow (5) for IAF 2008 was conducted as well. After 400 days the deviation in financial forecasting for maximum invoicing value is less than 1 %, and the deviation in financial forecasting for maximum cash flow value is less than 4%.

$$i(t) = \frac{433.7}{1 + 11.822\, e^{0.279 - 0.032t}} - \frac{433.7}{1 + 11.822\, e^{0.279}} \tag{4}$$

$$p(t) = \frac{382.3}{1 + 4.29\, e^{2.680 - 0.028t}} - \frac{382.3}{1 + 4.29\, e^{2.680}} \tag{5}$$

To improve the financial forecasting for cash inflow, the hybrid solution of the case-based reasoning architecture based on the heuristic interpolation in the

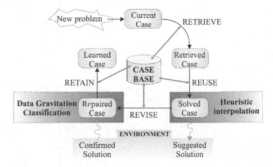

Fig. 5. The hybrid Case-Based Reasoning architecture - the heuristic interpolation in the *adaptation* phase and Data Gravitation Classification in the *revise* phase

adaptation phase and Data Gravitation Classification (AI) technique in the *revise* phase is presented in fig. 5.

By applying the bases of Data Gravitation Base Classification method (DGC) [5], according to which the correlation of data is based on the basic gravitational force calculation, the extrapolation value of the parameter *a* of the adapted logistic curve in 2008 was conducted (fig. 5), as a *revise* phase from the CBR-*Cycle*.

The basic mass, to which the value of mass '1' is added, represents the collection of time series of the parameters *a* of the logistic curve (from m: 2000 to m: 2007). The question is, which mass at which coordinate Y: 2007/2008, caused the mass movement m: 2000 with the position of Y: 2000 in 2001 within the position m: 2001 at the position of Y: 2001/2002. The force that caused this movement has the known direction D: 2000/2001 that is collinear with mass positions m: 2000 and m: 2001. The mass is positioned in the section of this direction and coordinate Y: 2007/2008.

The mass value M: 2001 is obtained from the known force value F: 2000/2001, whose intensity is equal to the vector intensity causing the mass M: 2000 to move to the position of the mass M: 2001. This intensity is obtained by standard calculation of the vector intensity, where the distance is equal to the length D: 2000/2001. The mass value M: 2001 is equal to:

$$M:2001 = \frac{F:2000/2001 \cdot |D:2000/2001|}{m:2000} \tag{6}$$

This procedure of the calculation is separately conducted for each (D:), (F:), (m:) of the appropriate time series. The coordinate calculation Y: 2007/2008 represents the mass centre DGC of the time series parameters *a*, with given coordinates and, according to Steiner's theorem it equals to:

$$Y:2007/2008 = \frac{\sum_{i=2000/2001}^{2006/2007} M_i Y_i}{\sum_{i=2000/2001}^{2006/2007} M_i} \tag{7}$$

The extrapolation of the parameter a Y: 2007/2008 is obtained by this method, which is represented by the mass collection of the mass parameter *a* and the series of force impulses moved from 2000 to 2007. The basic values used in these calculations are shown in fig. 6.

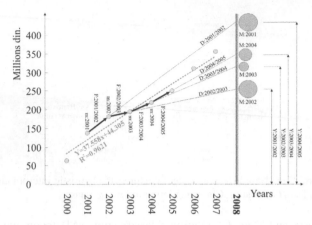

Fig. 6. The Data Gravitation Based Method for describing cash inflow parameter a

$$f_{forecast}^{2008}(a,b,c,d) = f_{Hybrid}^{2008}(\frac{a_{CBR}^{2008} + a_{DGC}^{2008}}{2}, \frac{b_{CBR}^{2008} + b_{DGC}^{2008}}{2}, \frac{c_{CBR}^{2008} + c_{DGC}^{2008}}{2}, \frac{d_{CBR}^{2008} + d_{DGC}^{2008}}{2})$$ (8)

The hybrid function and real cash inflow values are equal to $R^2=0.9549$. The regression of the declined linear correlation is $Y=1.0678x - 3 \cdot 10^7$

Table 4. Forecasting and real values parameters adapted logistic curves for invoice and cash inflow, heuristic interpolation CBR and DGC adapted, for the IAF 2008

	Year	Cash inflow parameter			
		a	b	c	d
Forecast CBR	2008	382.3	4.293	0.0288	2.680
Forecast DGC	2008	428.6	9.490	0.0299	3.948
Hybrid (CBR & DGC)		405.5	6,891	0,0293	3.314
Real	2008	404.0	4.257	0.0295	2.983

The extrapolation values of the parameter of the adapted logistic curve of cash inflow for 2008 can be described by the following hybrid solution: the *adaptation* and *revise* phases of CBR methodology, heuristic interpolation and DGC (AI) technique. And the function now is:

$$p(t)_{2008}^{CBR,DGC} = \frac{405.5}{1+6.89e^{3.314-0.0293t}} - \frac{405.5}{1+6.89e^{3.314}}$$ (9)

The deviation is less then 1 % after 400 days for maximum invoicing and maximum cash inflow values in the assessment collection for 2008.

6   Available Analogue Implementation

There are numerous systems based on different CBR techniques application, which are designed for various financial predictions: (1) Predicting the upcoming broiler market price using the adapted CBR approach in Taiwan [5]; (2) Gaussian CBR system for business failure prediction with empirical data in China [6]; (3) predicting a financial activity rate with a hybrid decision support model using the CBR augmented with genetic algorithms and the fuzzy k nearest neighbour methods [7]. There is also, one other AI technique, which uses clustering method on financial ratios on a balance sheet, solvency, profitability, and the return on the investment [8].

7 Conclusion and Notes on Future Work

The concept of the case-based reasoning system for financial forecasting the dynamics of issuing invoices and receiving actual cash inflow for the "International Agriculture Fair" 2008 exhibition has been presented in this paper. The performed simulations are based on the observed previous behaviour of adapted logistic curves based on the heuristic interpolating CBR *adaptation* phase and the data gravitation classification method in the *revise* phase in previous eight exhibitions, from year 2000 to year 2007. According to the parameters of those curves, the parameters for the new exhibition in 2008 have been adapted.

The deviation at financial forecasting for a maximum invoicing and for maximum cash inflow values is less than 1 %, in the hybrid CBR proposed architecture. Although the achieved results of the financial forecasting present a significantly positive outcome, the research on this project can be continued. Presented financial forecasting is not limited to this case-study but it can be applied to other financial values as well as expenses, profit or other business domains.

References

1. Aamodt, A., Plaza, E.: Case-Based Reasoning: Foundational Issues, Methodological Variation and System Approaches. In: AI Intelligence Commutations, pp. 39–58 (1994)
2. Simić, D., Kurbalija, V., Budimac, Z., Ivanović, M.: Case-Based Reasoning for Financial Prediction. In: Ali, M., Esposito, F. (eds.) IEA/AIE 2005. LNCS (LNAI), vol. 3533, pp. 839–841. Springer, Heidelberg (2005)
3. Simić, D., Simić, S.: An Approach to Efficient Business Intelligent System for Financial Prediction. Journal: Soft Computing 11(12), 1185–1192 (2007)
4. Peng, L., Yang, B.a., Chen, Y., Abraham, A.: Data gravitation based classification. Journal: Information Sciences 179, 809–819 (2009)
5. Huang, B.W., Shih, M.L., Nan-Hsing, C., Hu, W.Y., Chiu, C.: Price Information Evaluation and Prediction for Broiler Using Adapted Case-Based Reasoning Approach. Journal: Expert System with Applications 36, 1014–1019 (2009)
6. Li, H., Sun, L.: Gaussian Case-Based Reasoning for Business Failure Prediction with Empirical Data in China. Journal: Information Sciences 179, 89–108 (2009)
7. Li, S.T., Ho, H.F.: Predicting Financial Activity with Evolutionary Fuzzy Case-Based Reasoning. Journal: Expert System with Application 36, 411–422 (2009)
8. Wang, Y.J., Lee, H.S.: A clustering method to identify representative financial ratios. Journal: Information Sciences 178, 1087–1097 (2008)

A Hybrid Neural Network-Based Trading System

Nikos S. Thomaidis and Georgios D. Dounias

Management & Decision Analysis Laboratory, Dept. of Financial Engineering &
Management, University of the Aegean, 31 Fostini Str., GR-821 00, Chios, Greece
Tel.: +30-2271-0-35454 (35483); Fax: +30-2271-0-35499
{nthomaid,g.dounias}@fme.aegean.gr
http://fidelity.fme.aegean.gr/decision

Abstract. We present a hybrid intelligent trading system that combines artificial neural networks (ANN) and particle swarm optimisation (PSO) to generate optimal trading decisions. A PSO algorithm is used to train ANNs using objective functions that are directly linked to the performance of the trading strategy rather than statistical measures of forecast error (e.g. mean squared error). We experiment with several objective measures that quantify the return/risk associated with the trading system. First results from the application of this methodology to real data show that the out-of-sample performance of trading models is fairly consistent with respect to the objective function they derive from.

1   Introduction

Neural network and computational intelligent (CI) models have been widely appreciated in financial forecasting tasks, especially in the design of investment strategies and the implementation of trading systems. Although much has been written about the use of CI models in real-time trading, the majority of papers follow the standard route of first training the model using statistical measures of "goodness-of-fit" and, then, designing a trading strategy that exploits model's forecasts. However, in financial institutions practitioners are often interested in trading features, such as the cumulative return of the trading strategy, rather than the predictive accuracy of the model. Although it can be argued that the latter implies the former, this is not always the case. Many empirical studies show that low measures of model error do not necessarily imply better investment performance out-of-sample [4]. Therefore, an alternative route would be to train a computational intelligent model using a cost function that is directly linked to the economic performance of the associated trading strategy.

Much of the related work on training neural networks with trading-related objectives tends to concentrate on employing standard optimisation procedures to minimise a cost function related to the trading performance. Approaching this problem from a traditional optimisation point of view, somehow restricts the potentiality of applications as it requires that the cost function used to evaluate trading performance be differentiable and "well-behaved". Besides, quadratic

E. Corchado et al. (Eds.): HAIS 2009, LNAI 5572, pp. 694–701, 2009.

techniques are usually trapped into local optima and also show problems in incorporating complex constraints that often guarantee reasonable out-of-sample performance. What is thus of particular interest is to use a flexible optimisation technique that can effectively train models on any function of interest given arbitrary constraints on the parameters. This is where nature-inspired techniques, like genetic algorithms, ant colonies and particle swarm optimisation, may be proven very useful.

In this paper, we present a hybrid intelligent combination of PSO and ANNs for trading purposes. Particle swarm optimisation is employed for fine-tuning neural network architectures based on objective functions that are directly linked to the trading performance of the forecasting model. Because of its easy implementation and inexpensive computation, the PSO has been proven very effective in this optimisation task, without requiring too much user intervention. We experiment with different types of objective functions (Sortino Ratio, Sharpe Ratio, Cumulative Return, Maximum Drawdown) that incorporate various definitions of risk and return. We compare results with a neural network model trained using the mean squared error criterion.

The rest of this paper is organised as follows: section 2 discusses Particle Swarm Optimisation. Section 3 details the neural network model architecture employed in the implementation of trading strategies. In section 4 we present a number of alternative trading performance indicators that are used as objectives in the estimation of model parameters. Section 5 presents an empirical application of our methodology on a portfolio of stocks traded in the New York Stock Exchange. Finally, section 6 concludes the paper and proposes directions for future research.

2   Particle Swarm Optimisation

Particle Swarm Optimisation (PSO) is a popular computational intelligent methodology inspired by the social behaviour of organisms, such as birds and fishes. It was originally proposed by Kennedy and Eberhart [6,7] as a stochastic-search population-based algorithm able to consistently explore complex solution spaces.

Each particle represents a feasible solution or point in the D-dimensional solution space. Instead of applying genetic operators, like in other evolutionary computational algorithms, PSO flows particles in the search space with a velocity that is dynamically adjusted according to both their own and the swarm's experience. The dynamic equations that describe each particle's movement are:

$$\hat{\mathbf{x}}_i = \mathbf{x}_i + \hat{\mathbf{v}}_i \tag{1a}$$

$$\hat{\mathbf{v}}_i = w\mathbf{v}_i + c_1\mathbf{r}_1 \circ (\mathbf{x}_i^b - \mathbf{x}_i) + c_2\mathbf{r}_2 \circ (\mathbf{x}_g - \mathbf{x}_i) \tag{1b}$$

where $\hat{\mathbf{x}}_i$, $\mathbf{x}_i$ are the current and new position, $\hat{\mathbf{v}}_i, \mathbf{v}_i$ are the current and new velocity, $\mathbf{x}_i^b$ is the best solution found by the particle in its own history and $\mathbf{x}_g$ is the globally best solution found by the algorithm at each generation.

o is the Hadamard matrix multiplication operator, indicating element-by-element multiplication. w is the inertia parameter that controls the velocity change and r_1, r_2 are random vectors uniformly distributed in $[0, 1]$. The constants c_1, c_2 control how much the particle is on average attracted towards its personal best and global solution.

The inertia is often used as a parameter to control the trade-off between exploration and exploitation. Low values facilitate swarm's cohesion, while high values result in abrupt particle movements around target regions. In order to promote better exploitation of optimal regions, the inertia weight is often decreased with the number of iterations, using the formula:

$$w = w_{\max} - \frac{w_{\max} - w_{\min}}{iter_{max}} \times iter$$

where $w_{\max}$, $w_{\min}$ are the maximum (initial) and minimum (final) value of w, $iter$ is an iterations index and $iter_{max}$ is the maximum number of iterations.

The basic PSO and its variants have been successfully applied to a variety of financial problems, including portfolio optimisation [2,5], credit scoring [3], time-series forecasting and asset trading [8,9,10]. Of these studies, possibly the most relevant to our work are [8,9,10] who employ a particle swarm optimiser for the training of neural network models that generate one-step-ahead investment decisions. However, our approach is different from theirs in many aspects: First, in our methodology, trading decisions are made based not on "point" recommendations but on confidence intervals on the future value of an asset's price, which also takes into account the risk of trading positions. Moreover, we optimise neural networks with respect to different measures of trading performance, which enables the future user of our methodology to adjust the trading system towards his own preference towards risk. In this aspect, our approach is more closely related to [4].

3   Forecasting Models

Particle swarm optimisation can be used in a number of ways in neural network design, including input variable selection, specification of network architecture and parameter estimation. In our study, we treat PSO as a parameter estimation algorithm, while the network structure is being decided a priori.

The neural network models employed in this study relate the future value of a financial asset with its own history presented at different lags, thus take the general autoregressive form:

$$\hat{y}_{t+h} = f(y_t, y_{t-1}, ..., y_{t-L}) \tag{2}$$

where h is the forecasting horizon, y_t is the value of the asset at time t and L is the maximum lag length. One can produce h-step-ahead confidence bounds with this model by assuming that the future value follows a normal distribution with mean $\hat{y}_{t+h}$ and variance equal to the variance of model errors in the training sample.

4   Trading-Related Objective Functions

When training a neural network for trading purposes, the decision as to which objective to use is often unclear. This is because different trading measures concentrate on different aspects of the equity curve of the trading strategy[1]. Some measures, such as the total return at the end of the investment period, are profit-oriented. Others, like the Sharpe or the Shortino ratio, provide a trade-off between return and risk. Finally, there exist safety-first measures, such as the Value-at-Risk or the maximum drawdown, that exclusively focus on downside risk. Generally, there is no widely acceptable trading indicator and each measure has its own advantages and disadvantages. In this paper, we employ four different objective functions to train neural networks: the Sharpe Ratio (SR), the Sortino Ratio (SoR), the Cumulative Return (CR) and the Maximum Drawdown (MD). These are discussed below.

The *Sharpe Ratio* (SR) is a measure of the excess return (i.e. the return over the risk-free benchmark) gained on the trading strategy over the standard deviation of returns, i.e

$$SR = \frac{m - r_f}{s}$$

where r_f is the risk-free rate of return (i.e the return on a saving account) and m, s is the mean and standard deviation of the trading strategy's daily returns. Although SR is probably the most commonly applied measure of trading performance, it suffers from several disadvantages. Note that SR equally penalises both positive and negative returns, as the standard deviation does not distinguish between upside and downside volatility. The *Sortino Ratio* (SoR) is a modification of the SR that focuses on returns falling below a user-specified target. In our case, we employ the following version of the Sortino Ratio

$$SoR = \frac{m - r_f}{s^-}$$

in which only the standard deviation of negative returns (s^-) is penalised.

Another measure of trading performance that we employed in this study is the *Cumulative Return* (CR). This is simply the (annual equivalent percentage) compound return earned by the trading strategy at the end of the investment period

$$CR = 100 \times \left(\prod_{t=1}^{T}(1 + r_t) - 1 \right)$$

where r_t is the rate of return on each period and T is the sample size. CR is different in many aspects from the measures presented above, mainly in that it focuses on the net outcome of the trading strategy and does not take into account the risk associated with trading positions.

---

[1] The equity curve is the value of a trading account (as measured by the compound return) graphed over the investment period.

Another indicator of trading performance that belongs to the family of *"safety-first"* measures [1] is the Maximum Drawdown (MD). A drawdown is a decline from a historical peak in the equity curve since the inception of the trading strategy. It measures the percentage loss experienced by a strategy before it starts making profit again and driving the investment balance back up. The maximum drawdown is simply the largest drawdown experienced by a strategy during the period of time under study. Conceptually, the calculation looks at all subperiods of the equity curve and finds the largest distance between a historical "peak" and a subsequent "valley".

5   Empirical Study

Our study uses daily closing prices from 08-Nov-2002 until 22-Aug-2005 of four stocks traded in the New York Stock Exchange, covering a period of almost 700 trading days[2]. Initially, the entire data set was divided into a training and a test set of 500 and 200 observations, respectively. Based only on in-sample observations, we formed a portfolio of these stocks whose weights were determined using cointegration analysis, so that its value shows as high mean-reversion as possible. Portfolio weights were then kept constant in the test sample, resulting in a fairly stationary time-series.

Neural network autoregressive models (2) were applied to forecast the "turning-points" of the portfolio value. The parameters of the trading models were estimated based on their in-sample trading performance and then applied out-of-sample. All networks had 1 neuron in the hidden layer. The last 10 consecutive portfolio values were used as inputs to the network to predict the value of the portfolio 5 days ahead (i.e. $h = 5$). We trained four networks using each of the objective functions presented in section 4: the Sharpe Ratio (SR), the Sortino Ratio (SoR), the Cumulative Return (CR) and the Maximum Drawdown (MD). During the training phase we imposed several constraints on model parameters to guarantee stable performance of the network as well as a constraint on the cumulative return of the trading strategy. Specifically, networks were asked to maximise SR, SoR or minimise MD, while guaranteeing a 10% (annualised) return on the initial investment by the end of the training sample. In order to base the calculation of the equity curve and various risk measures on realistic market assumptions, we assumed a trading cost of 3 cents per stock and a constant annual risk-free rate of 3%.

To determine the optimal trading settings, we run the PSO algorithm 20 times using a different initial population. The best solution found in all runs was then applied out-of-sample. The parameter values of the PSO algorithm were chosen as follows: number of particles 10, number of generations 200, $c_1 = c_2 = 2$, $w_{max} = 0.2, w_{min} = 10^{-6}$ and $V_{max} = 0.5$. Those were mainly the result of trial-and-error and prior knowledge on reasonable values for the dynamic range of particle movements. Note that as the algorithm reaches the maximum number of iterations, the inertia value is significantly decreased ($w_{min} = 10^{-6}$). This

---

[2] Data were downloaded from finance.yahoo. Closing prices are adjusted for splits and dividends.

gradually makes the step size smaller so that the algorithm consistency explore the optimal region. The latter is also achieved by imposing a constraint on the maximum velocity of each particle ($V_{max} = 0.5$). Although we have not put much effort in deriving optimal settings values that would boost algorithm's performance, we paid special attention to creating an initial swarm of particles that would provide a broad coverage of the feasible parameter space. This has been found to be critical as to the successful convergence of the algorithm to the optimal parameter values.

Table 1 shows the in-sample performance of trading models. Each column of the table corresponds to the model specified using one of the objective functions discussed in section 4. For comparison purposes, we also report trading measures for the neural network model estimated using a mean squared error criterion (MSE). The confidence bounds of this model were set to the level where the in-sample compound return of the trading strategy slightly exceeds 10% per annum, the constraint also adopted in the models estimated using trading-related objective functions. Of these models that have a great percentage of the sample time period active trades in the market, the MSE model delivers the worst performance. Apart from the lowest cumulative return, it also suffers from comparatively low Sharpe and Sortino ratios. The SoR and SR objectives deliver models that are largely comparable, as seen from the trading figures. Among all models, the CR places the most and the MD the fewest trades. This result is somewhat expected if one takes into account the nature of these objectives (see also the discussion in section 4). CR can be characterised as a risky profit-seeking measure which focuses on the final outcome of the trading strategy. Although the model places much more winning than losing trades, which seems to be a necessary condition to achieve profitability, it does not take into account the efficiency of trades in terms of risk and return. This justifies the relatively low values for the Sharpe and Sortino ratio. The MD objective, on the other hand, is an entirely risk-averse measure that attempts to reduce the downside risk. This reduction is often at expense of the total return. As table 2 shows, the out-of-sample performance of most models is fairly consistent with the in-sample results

Table 1. The in-sample performance of trading models

Trading measures	SoR	SR	CR	MD	MSE
Cumulative return* (%)	39.02	32.55	47.24	11.72	10.70
Mean Return	0.001	0.001	0.001	0.000	0.000
Sharpe ratio*	2.52	2.71	2.17	0.33	0.66
Sortino ratio*	4.73	4.02	3.60	0.58	0.47
Number of trades	30	29	38	2	8
Number of winning trades	28	25	34	2	7
Number of losing trades	2	4	4	0	1
Maximum drawdown	-0.018	-0.016	-0.016	-0.003	-0.020
Percentage of time in the market	91.60	51.04	82.41	0.20	9.82

Notes:
* Annualised figures.

Table 2. The out-sample performance of trading models

Trading measures	SoR	SR	CR	MD	MSE
Cumulative return* (%)	10.7	8.20	5.57	1.046	8.16
Mean Return	0.001	0.001	0.000	0.000	0.000
Sharpe ratio*	0.85	0.61	0.38	-0.64	0.09
Sortino ratio*	1.22	0.90	0.60	-0.22	0.06
Number of trades	11	6	5	1	2
Number of winning trades	10	5	4	1	2
Number of losing trades	1	1	1	0	0
Maximum drawdown	-0.016	-0.007	-0.013	-0.004	-0.007
Percentage of time in the market	87.44	70.35	68.34	3.02	5.53

Notes:
* Annualised figures.

and the associated maximising objective. The only exception is the CR model, whose performance significantly deteriorates out-of-sample. What is interesting is that although this model continues to be very accurate in predicting changes in the portfolio value time-series (as seen by the high percentage of winning trades), it fails to fulfil its objective out-of-sample. The CR trading strategy delivers 5.57% at the end of the investment period, which is even less than the cumulative return of the MSE trading strategy, attained with only two trades. Both SoR and SR place more trades on the average and attain a better performance in terms of total return and risk-adjusted measures. As expected, the MD model is the most conservative among all.

6 Discussion-Further Research

In this paper, we show how a particle swarm (PS) optimiser can be used in the training of neural network models based on objective functions that are directly linked to the trading performance of models' forecasts. The introduction of trading measures as objectives for the neural network generally leads to "hard" optimisation problems, in which traditional quadratic techniques are quite unlikely to succeed. As such, the use of an intelligent optimisation algorithm, i.e. PSO, in the fine-tuning of the forecasting model is very much recommended. PSO-like algorithms, in particular, can be proven a very effective tool in this task, mainly because of their easy software implementation, the consistency of their performance and the limited requirement for user intervention (compared to other heuristic optimisation methodologies). We experiment with several objective functions (Sortino Ratio, Sharpe Ratio, Cumulative Return and Maximum Drawdown) that focus on different aspects of the equity curve. First results from the application of this methodology to real data seem very promising compared to the classical approach of trading based on models that minimise a statistical criterion (e.g. MSE).

Future work plans to extend the PS optimiser to other aspects of the neural network design, including selection of input variables and specification of the optimal network architecture. In this case, however, the optimisation task turns into a mixed integer-nonlinear programming problem, which becomes rather computationally demanding given a complex neural network architecture (many neurons in the hidden layer or many input variables). Of much interest would also be to apply other optimisation techniques (such as genetic algorithms, ant colonies, etc) and compare their performance with the particle swarm component proposed in this paper. This comparison should be made on the basis of multiple criteria, involving the probability of finding a near-optimum solution, convergence speed, etc.

References

1. Elton, E.J., Gruber, M.J., Brown, S.J., Goetzmann, W.N.: Modern Portfolio Theory and Investment Analysis, 6th edn. John Wiley & Sons, Chichester (2003)
2. Fischer, T., Roehrl, A.: Optimization of performance measures based on expected shortfall. Working paper (2005)
3. Gao, L., Zhou, C., Gao, H.B., Shi, Y.R.: Credit scoring module based on neural network with particle swarm optimization. Advances in Natural Computation 14, 76–79 (2006)
4. Harland, Z.: Using nonlinear neurogenetic models with profit-related objective functions to trade the US T-bond future. In: Abu-Mostafa, Y., LeBaron, B., Lo, A., Weigend, A. (eds.) Proceedings of the 6th International Conference Computational Finance 1999, pp. 327–342 (2000)
5. Kendall, G., Su, Y.: A particle swarm optimisation approach in the construction of optimal risky portfolios. In: Proceedings of the 23rd IASTED International Multi-Conference on Artificial Intelligence and Applications, pp. 140–145 (2005)
6. Kennedy, J.: Small worlds and mega-minds: Effects of neighborhood topology on particle swarm performance. In: Proceedings of the 1999 IEEE Congress on Evolutionary Computation, pp. 22–31. IEEE Service Center, Piscataway (1999)
7. Kennedy, J., Eberhart, R.C.: Particle swarm optimization. In: Proceeding of the IEEE International Conference on Neural Networks, Perth, Australia, pp. 12–13. IEEE Service Center, Los Alamitos (1995)
8. Nenortaite, J.: A particle swarm optimization approach in the construction of decision-making model. Information Technology and Control 1A 36, 158–163 (2007)
9. Nenortaite, J., Simutis, R.: Stocks' trading system based on the particle swarm optimization algorithm. In: Bubak, M., van Albada, G.D., Sloot, P.M.A., Dongarra, J. (eds.) ICCS 2004. LNCS, vol. 3039, pp. 843–850. Springer, Heidelberg (2004)
10. Nenortaite, J., Simutis, R.: Adapting particle swarm optimization to stock markets. In: Proceedings of the 2005 5th International Conference on Intelligent Systems Design and Applications (ISDA 2005), pp. 520–525 (2005)

Active Portfolio Management under a Downside Risk Framework: Comparison of a Hybrid Nature – Inspired Scheme

Vassilios Vassiliadis, Nikolaos Thomaidis, and George Dounias

Management and Decision Engineering Laboratory,
Department of Financial & Management Engineering, School of Business Studies,
University of the Aegean, 31 Fostini Str. GR-821 00, Greece
{v.vassiliadis,nthomaid}@fme.aegean.gr, g.dounias@aegean.gr

Abstract. Hybrid intelligent systems are becoming more and more popular in solving nondeterministic polynomial-time – hard optimization problems. Lately, the focus is on nature – inspired intelligent algorithms, whose main advantage is the exploitation of unique features of natural systems. One type of complex optimization problems is the active portfolio management, where the incorporation of complex, realistic constraints makes it difficult for traditional numerical methods to deal with it. In this paper we perform a computational study of a hybrid Ant Colony Optimization algorithm. The application is a specific formulation of the problem. Our main aim in this paper is to introduce a new framework of study in the field of active portfolio management, where the main interest lies in minimizing the risk of the portfolio return falling below the benchmark. Secondary, we provide some preliminary results regarding the use of a new hybrid nature – inspired scheme in solving this type of problem.

Keywords: Hybrid Ant Colony Optimization, Active Portfolio Management, tracking error, downside probability.

1 Introduction

Hybrid intelligent algorithms are methods that combine specific features from individual methodologies. They are used to enhance their capabilities for solving complex real – life problems. Each individual technique combines its unique characteristics for confronting a problem. For example, genetic algorithms have two characteristic processes in order to produce a new, maybe better, solution, the crossover and the mutation operator. These operators combine already existing solutions of the population so as to produce new ones in the next generation. Also, another paradigm from real life is the particle swarm optimization algorithm, which exploits the ability of birds to locate the best direction in order to reach their target place. However, when we combine two or more of these techniques, we could take advantage of each intelligent technique's characteristics in the best possible way, regarding the problem at hand, of course. Particularly,

E. Corchado et al. (Eds.): HAIS 2009, LNAI 5572, pp. 702–712, 2009.

nature – inspired algorithms are a category of intelligent techniques. Their main characteristics, their abilities to evolve and work for example, stem from real life systems. Nature – inspired intelligent algorithms have proven efficient in searching the solution space, when trying to solve nondeterministic polynomial-time – hard optimization problems [1].

One such type of problem is the portfolio management, where the potential investor selects an appropriate portfolio, i.e. asset selection and calculation of weights for these assets, in a way that optimizes his/her preference towards risk / return. In the classical portfolio optimization problem the objective of the investor is either to maximize the expected return or to minimize the variance of the portfolio, under certain constraints. Nowadays, active portfolio strategies are typical setups for the optimization problem. In the case of active strategies, the focus is on exploiting inequilibria or any mispricing so as to form portfolios which beat a benchmark index. Specifically, when dealing with stock indexes, the active portfolio is constructed using assets which are included in the benchmark index, so it tries to replicate it. Apart from imitating the index, the formed portfolio aims at outperforming it. This strategy is widely accepted, because investors are content to follow the average performance of the market. However, they are not willing to accept the risks entailed in buying individual assets. What is more, investing in all components of the index is quite complex, and not to mention incurs higher transaction costs. As a result, an answer to this question is to seek for a combination of the index's assets which imitates the index and in the same time has better performance.

In our study, we use a hybrid scheme, which combines a nature – inspired intelligent algorithm, namely the ant colony optimization algorithm, and a local search algorithm[1], which originates from the field of non – linear programming, in order to solve a different formulation of the active portfolio optimization problem. Specifically, we focus on minimizing the probability that the tracking error, i.e. the difference between the portfolio's returns and the index returns, takes negative values. By introducing a probability which measures the downside risk of the tracking error, we would like to propose a new framework of study in the field of active portfolio management, which focuses on the distribution of tracking error, and particularly the region around the tails of the distribution (extreme values). So far, little work has been done in this area, as we will show in the literature review. What is more, the problem becomes more complex when we introduce real – life constraints such as a limit on the maximum number of assets included in the portfolio, as well as upper and lower bounds on the assets weights. From a computational point of view, the problem becomes a challenge for traditional numerical optimization methods. So, the introduction of a hybrid intelligent scheme could tackle with this difficulty. Finally, we compare the results with other two methods, namely a simple ant colony optimization algorithm and a Monte Carlo simulation. The aim of this paper is to

---

[1] The local search algorithm that we use is based on the Levenberg – Marquardt method which combines the Gauss – Newton and the steepest descent method.

provide a preliminary study regarding the use of a distributional concept as the objective of the optimization problem at hand. Furthermore, the incorporation of cardinality constraints has two effects. Firstly, the optimization problem becomes even more realistic, because the aim of an investor may be to form sub - groups of stocks, out of an index, containing a specific number of stocks. Secondly, traditional numerical methods fail to satisfy this kind of constraint. So, a hybrid intelligent scheme is presented in contrast with other simplified methods in order to solve this real life problem. Results and further conclusions regarding the use of this hybrid scheme can only be considered as preliminary in the scope of this paper. Generally, the portfolio optimization problem consists of dealing with the optimization of two sub–problems: one is finding the assets of the portfolio (i.e. optimization in a discrete solution space) and the other is calculating optimal weights for these assets (i.e. optimization in a continuous solution space). In [10], a hybrid solver is focused on the discrete variables leaving the determination of the continuous one to a quadratic programming solver. In our case, the hybrid scheme aims at tackling the problem in the discrete space using an ACO algorithm, and the weights (continuous space) are found using a nonlinear programming algorithm. On the other hand, the use of a single intelligent metaheuristic for jointly solving the portfolio problem (i.e. asset selection and weight calculation) relies on empirical techniques which aim at solving one aspect of the problem (i.e. finding the weights of assets) and at the same time selects which assets should be included in the portfolio (e.g. if the asset's weight is too small, then remove this asset from the portfolio). To sum up, out research aims at dealing with the two optimization problems separately, and not jointly. As we have mentioned, this is a preliminary study, and in the future we are going to consider more aspects of this approach (e.g. apply other hybrid schemes, study the solution space etc.).

The paper is organized as follows. In section 2, we briefly present some literature review about the active portfolio management problem and the use of downside risk measures concept. Also, we present a few works regarding the use of hybrid intelligent schemes for this type of problem. In section 3, we show some methodological issues. In section 4, we present the results from our experiments, and any useful conclusion which can be drawn from them. Finally, in section 5 we conclude our study by summarizing the main results of our experiments, and giving some useful insights for future research opportunities.

2 Literature Review

A table consisting of some research studies on active portfolio optimization, the incorporation of the downside risk measures notion in this framework, as well as works related with the use of nature – inspired algorithms and hybrid schemes for this kind of application is presented below.

Table 1. Literature review

Papers	Brief description
Active Portfolio Management	
[2]	Active portfolio optimization with constraints in tracking error volatility
[3]	Particle swarm optimization is used to deal with active portfolio management problem:
	- maximize excess return
	- maximize excess return with constraints on portfolio's risk
	- maximize Sharpe ratio with a constraint in tracking error volatility
[4]	Active portfolio optimization with a value-at-risk constraint
[5]	Dynamic active portfolio optimization problem
[15]	Active portfolio management in continuous – time
Downside risk measures	
[6]	Application of conditional value-at-risk in portfolio optimization
[7]	An introduction in conditional drawdown
Hybrid artificial intelligent schemes for the portfolio optimization problem	
[8]	Simulated annealing combined with evolutionary strategies for the classical portfolio optimization problem
[9]	Memetic algorithm for the following portfolio optimization problem: maximize expected utility of the investor with a value-at-risk constraint.
[1]	Threshold accepting for the following portfolio optimization problem: maximize expected utility of the investor with constraints in value-at-risk and expected shortfall
[10]	Local search combined with a quadratic programming procedure for the classical portfolio optimization problem.
[11]	Genetic algorithm combined with simulated annealing for the classical portfolio optimization problem.
[12]	Hybrid genetic algorithm for passive portfolio management

All in all, we could state the following basic points. Firstly, the approach of active portfolio management is used by a number of financial managers, due to the fact that their objective is to outperform a benchmark index. Secondly, research has been done in the topic of using alternative measures of risk, apart from variance. These alternative measures focus on the tails of the distribution. In most of the studies, these metrics have been used as constraints in the classical portfolio optimization problem, and in the case of active management. The results are quite promising. However, little research has been done in the direction of dealing with these metrics from the point of view of the objective function for the active portfolio optimization problem. What is more, the incorporation of real life constraints such as cardinality constraints is very important for the optimization problem at hand, because constraints on the maximum number of assets may reflect a target objective of an investor manager. Also, traditional numerical methods fail to satisfy these restrictions. Finally, hybrid schemes have proved their efficiency in solving different formulations of the portfolio optimization problem. Their performance has been compared with traditional quadratic programming techniques and other numerical methods. However, the studies referring to hybrid nature – inspired intelligent methods are very few, and they concentrate on genetic algorithms along with other intelligent heuristics such as simulated annealing. So, the use of a hybrid ant colony optimization for a different formulation of the active portfolio management could be considered as a new approach in dealing with this complex optimization problem.

3 Methodological Issues

In this section we are going to present the framework of the application domain, in brief. Moreover, we will show the main functions of the proposed hybrid scheme.

3.1 Active Portfolio Management under a Downside Risk Framework

As we have mentioned above, the main objective of the active portfolio optimization problem is to outperform a benchmark index. We define *tracking error (TE)* as the difference between the portfolio's and index's returns: $TE_{(t)} = R_{p(t)} - R_{B(t)}$, where $R_{p(t)}$ refers to portfolios returns and $R_{B(t)}$ refers to benchmark's returns at time t. In a classical formulation of this problem the objective function would be the following: *maximize mean (TE)*, which is referred as excess return. However, using this framework, without imposing any constraints in the distribution of TE's, has a main drawback. It does not deal with extreme values of TE, as long as the average TE is maximized. In the case where a constraint on the overall volatility of TE's is imposed, we consider the whole distribution of TE's. In this study, we propose an objective function which deals with extreme values of the left side (tail) of TE's distribution. This can be explained as follows. Fund managers aim at constructing portfolios which systematically outperform the benchmark index. If the constructed portfolios fall below the benchmark in most cases, then this is an undesirable characteristic for the manager. The optimization problem can be formulated as follows:

$$\text{Minimize Probability (TE < 0)}$$

$$\text{s.t.}$$

$$\sum w=1, \text{ for any } w$$

$$w_l < w < w_u, \text{ for any } w$$

$$\text{Maximum number of assets} = k$$

where,
w, is the asset's weight
w_l and w_u, are lower and upper bounds for each weight
k, is the maximum number of assets included in a portfolio, also referred as the cardinality of the portfolio.

So, the main interest is to prevent portfolio's returns from dropping below index's returns, as long as it is possible. We try to form portfolios, whose tracking error has a little probability of getting negative values.

3.2 Hybrid Ant Colony Optimization Algorithm with a Local Search Technique

The incorporation of real - life constraints, as well as the use of a complex objective function, makes it hard for traditional methodologies to solve this problem reliably. More sophisticated techniques are required, which have the ability of searching efficiently the solution space and finally find a near optimal solution in a reasonable amount of time.

Ant Colony Optimization algorithm is based on the way real ant colonies behave and evolve, mainly when they search for food. The main characteristic of ant colonies is that they cooperate in order to find a better food source via a information exchange chemical process, called *pheromone deposit*. Specifically, each ant starts with searching randomly the solution space. When it founds a food source, it carries some of it back to the nest. In the way back, after the ant has evaluated the quality and quantity of the food, it leaves a pheromone trail which informs other ants from the colony about the particular food source. Eventually, more ants would tend to the best food source [13]. A similar concept is applied in the artificial ant colony, as well.

We propose a hybrid approach consisting of an ant colony optimization algorithm and a local search technique, namely the Levenberg – Marquardt method. The reason for doing this is that portfolio optimization deals with two kinds of problem.

The first one is the selection of a good combination of assets. The solution space is discrete and comprises the universe of stock, i.e. in our case all the stocks of the index. The ant colony algorithm aims at constructing portfolios having a specific number of stocks, i.e. as dictated by the cardinality constraint.

The second problem is to determine the weights of each asset already selected by the ant colony algorithm. In this case, the solution space is continuous and has lower and upper value limits, defined by the floor and ceiling constraints. For a given combination of assets, the aim of the Levenberg – Marquardt method is to find a vector of weights which minimizes the given objective function under certain constraints. In our case, the objective function refers to the probability the returns of the portfolio do not fall below the returns of the benchmark, which is found by the cumulative density function of the distribution of tracking error. The constraints imposed on the calculation of weights are the floor and ceiling constraints.

The hybrid scheme works as follows. In the first iteration, the ants select combinations of assets randomly, due to the fact that no information regarding the environment (i.e. which assets are better than others etc.) is available at this point. For each portfolio, proper weights are found in order to minimize the value of the objective function. Here, we have to notice that a complete solution is a constructed portfolio consisting of a vector of assets and their corresponding weights. At the next step, solutions are ranked based on values of the objective function and the best one is assigned to be the elitist solution of this iteration. Then, the pheromone matrix is updated. Pheromone matrix contains information about the performance (i.e. quality of solution found) of the population. At first, all assets are assigned the same value of pheromone. The pheromone update process is performed in two stages:

a. In stage one, pheromone values in all assets are reduced by a certain amount. This is called evaporation.

b. In stage two, a percentage of the best solutions in each iteration (e.g. 20% of best ants are chosen for reinforcement) is considered for reinforcement. Their pheromone value is increased by a certain amount depending on their objective function value. By doing so, solutions with better fitness values get more reinforcement, thus having larger probability to be chosen in next generations.

The above process is repeated for all iterations. It is obvious that assets with larger pheromone values are preferred in the construction of portfolios. In order to avoid deteriorating the searching ability of the algorithm, we incorporated a roulette wheel process in the stage where assets are selected. So, assets with smaller pheromone values may be selected. In this case, the ant colony optimization component searches a larger area of the solution space. What is more, the selection of the same portfolio twice by the algorithm is excluded.

```
Program Hybrid_ACO
Initialization of pheromone matrix
Repeat
 Repeat
    Select a combination of assets based on pheromone matrix
   Check whether this solution exists
   Calculate assets' weights
Until Ants = maximum_number_of_Ants

    Find best solution in iteration
  Update the pheromone matrix
Until Iterations = maximum_number_of_iterations
```

Fig. 1. Pseudo code of Hybrid Ant Colony Optimization algorithm

4 Computational Study

We applied the proposed active portfolio optimization problem to a data set of stocks comprising the FTSE 100 index. Data were downloaded from *finance.yahoo* for the time period 01/01/2005 – 31/12/2005. Also, stock prices were in daily basis and adjusted for splits and dividends. We constructed portfolios of various cardinalities using stocks from this benchmark index. Here, we have to note that due to missing values the universe of assets was reduced to 94 assets. Due to the fact that the ant colony optimization metaheuristic contains a stochastic part, namely the probability transition matrix which is used to select an asset, each run of the algorithm from a different initial population would probably result to different sub – optimal solutions. So, we conducted 100 independent runs of the algorithm for various portfolios cardinalities: 3, 10, 20, 30 and 50. Also, the lower and upper limits for weights were adjusted -0,6 and 0,6, respectively. These constraints are typical in the portfolio optimization problem. Specifically, it would be preferable to check for various cardinalities, from very small values to very large. However, it would be imperative to run more simulations for other values of these constraints in order to have a better view of the results. In the end we obtained the distribution of solutions over the independent runs, and performed a statistical analysis on the results. Prior to the presentation of configuration settings of the algorithm and the application, we have to say that the complexity of the problem rises as we increase the cardinality constraint (until a certain point), because the number of possible combinations increases, as well. To provide a sense of the complexity, in an index composed of $N = 94$ stocks, there exist $N! / [K!*(N-K)!] = 134044$ combinations of assets when cardinality is set to 3, and this value rises to $1.3511*10^{12}$ if the cardinality constraint is 20 assets. Finally, we compared the results of the hybrid scheme, with a simple ant colony

Table 2. Configuration settings

Configuration settings	
Hybrid Scheme	
Generations	100
Ants	50
Evaporation rate	0.2
Iterations for greedy algorithm	100
Monte Carlo	
Iterations	5000
Ant Colony Algorithm	
Generations	100
Ants	50

optimization algorithm and a Monte Carlo simulation. The simple ant colony optimization metaheuristic used an empirical scheme for jointly selecting assets and weights [14], [3]. The Monte Carlo simulation was used for selecting a combination of assets, whereas the Levenberg – Marquardt method, as mentioned above, was applied in finding the assets' weights.

In the following table, we present numerical results regarding the distribution of the objective function (Probability (TE < 0)). Apart form some statistical measures such as mean, standard deviation, minimum and maximum of the distribution of the objective function, we provide the percentiles of the distribution. The notion of percentiles can be described as follows. If percentile of X is a in 0.05 confidence level, then there is a probability of 5% that X will get values less than a.

If we look closer to the above results, we could draw some useful insights regarding the performance of each methodology.

Based on the mean of objective value alone, we could observe some basic things. Firstly, the simple ant colony optimization algorithm seems to perform better among the other methods, i.e. lowest mean value. An exception is the case of low cardinality (k=3), where all methods show similar performance. However, as the cardinality of the tracking portfolio increases, the gap between these methods increases, with the Hybrid Scheme and simple ant colony optimization algorithm being perhaps the most favorable one. This is explained by the fact that the number of possible asset combinations, and thus the complexity of the problem, grows exponentially with the cardinality of the portfolio. In all cases, the worst results were achieved by the Monte Carlo method.

A simple look at higher moments and percentiles of the empirical distributions shows that the mean value is not at all an indicative measure of performance. Using the standard deviation, we can extract some useful conclusions, regarding the dispersion around the mean. We can observe that in all cases, apart from the case of low cardinality (k=3), the simple ant colony optimization has a great value, meaning that the solutions are more dispersed from the mean. This is not a desirable result. We would prefer a distribution, where the solutions do not deviate a lot from the mean value as far as the right tail is concerned, i.e. there is a small possibility of getting extreme values in the right part of the distribution. On the other hand, because we deal with a minimization problem and we would like our solutions to be as small as possible, we would encourage distributions where values of the objective function are spread out more to the left of the mean. So, at first sight, the hybrid scheme achieves the lowest standard deviation, which is quite desirable.

Table 3. Simulation results

	Hybrid Scheme	Monte Carlo	ACO
k=3			
Mean	0.3792	0.3975	0.3848
Standard deviation	0.0031	0.0057	0.0050
Minimum (best)	0.3668	0.3769	0.3760
Maximum (worst)	0.3865	0.4067	0.3955
| best − worst |	0.0197	0.0298	0.0195
Percentiles (0.05, 0.50, 0.95)	[0.3738,0.3794,0.3839]	[0.3868,0.3985,0.4051]	[0.3785,0.3847,0.3930]
| $\text{perc}_{0.05}$ − $\text{perc}_{0.95}$ |	0.0101	0.0183	0.0145
k=10			
Mean	0.3611	0.3885	0.3412
Standard deviation	0.0064	0.0075	0.0095
Minimum (best)	0.3340	0.3699	0.3179
Maximum (worst)	0.3731	0.4026	0.3638
| best − worst |	0.0391	0.0327	0.0459
Percentiles (0.05, 0.50, 0.95)	[0.3502,0.3624,0.3684]	[0.3744,0.3894,0.3994]	[0.3265,0.3412,0.3561]
| $\text{perc}_{0.05}$ − $\text{perc}_{0.95}$ |	0.0182	0.0250	0.0296
k=20			
Mean	0.3766	0.3943	0.3373
Standard deviation	0.0052	0.0069	0.0105
Minimum (best)	0.3641	0.3733	0.3130
Maximum (worst)	0.3872	0.4075	0.3640
| best − worst |	0.0231	0.0343	0.0510
Percentiles (0.05, 0.50, 0.95)	[0.3668,0.3765,0.3847]	[0.3831,0.3944,0.4045]	[0.3193,0.3377,0.3542]
| $\text{perc}_{0.05}$ − $\text{perc}_{0.95}$ |	0.0179	0.0214	0.0349
k=30			
Mean	0.3843	0.4001	0.3350
Standard deviation	0.0058	0.0075	0.0117
Minimum (best)	0.3670	0.3799	0.3023
Maximum (worst)	0.3957	0.4141	0.3702
| best − worst |	0.0287	0.0342	0.0680
Percentiles (0.05, 0.50, 0.95)	[0.3732,0.3849,0.3928]	[0.3854,0.3999,0.4124]	[0.3173,0.3344,0.3512]
| $\text{perc}_{0.05}$ − $\text{perc}_{0.95}$ |	0.0196	0.0270	0.0339
k=50			
Mean	0.3951	0.4051	0.3270
Standard deviation	0.0069	0.0087	0.0097
Minimum (best)	0.3760	0.3769	0.3053
Maximum (worst)	0.4046	0.4202	0.3463
| best − worst |	0.0286	0.0433	0.0410
Percentiles (0.05, 0.50, 0.95)	[0.3793,0.3948,0.4022]	[0.3907,0.4060,0.4182]	[0.3103,0.3268,0.3432]
| $\text{perc}_{0.05}$ − $\text{perc}_{0.95}$ |	0.0229	0.0275	0.0329

What is more, we computed the distance between the best and worst solution, in absolute terms. This is an empirical measure of how wide is the distribution. We observe that in almost all cases the hybrid algorithm yielded a distribution of solutions which is narrow. On the other hand, distributions of the simple ant colony optimization are wider. The Monte Carlo method achieved moderate results.

Finally, we present indicative distribution percentiles for each method. In our case, because of the minimization problem, it is desirable that values of the objective function would gather on the left side of the distribution, and as few as possible values would lie on the right tail of the empirical distribution. In terms of percentile, this is interpreted as follows. For each confidence level it is preferable that the percentile gets very low values. Observing the percentiles and the difference between the percentile at 0.05 and 0.95, we can say that the simple ant colony optimization metaheuristic yields a distribution of solutions which is more spread to the left, and in almost all cases it has reached a global minimum point compared to other two techniques. Although hybrid scheme's distribution is more condensed, percentiles indicate that objective function values lean more to the right, thus resulting in inefficient solutions compared to simple ant colony optimization. Monte Carlo achieved the worst out of the three methodologies.

5 Conclusions and Further Research

Up to this point in this work, we have studied a particular formulation of the active portfolio selection problem. The focus is on the minimization of the downside probability, i.e the probability that the portfolio will underperform the benchmark. The interesting thing about the particular formulation of the active portfolio optimization problem is that it deviates from the classical mean-variance framework. Also, the behavior of a hybrid scheme comprising a nature – inspired intelligent algorithm, namely ant colony optimization, and a local search technique was tested. The performance of the hybrid algorithm was compared with a simple ant colony optimization metaheuristic and a Monte Carlo technique. This study gave us some insights into both the methodology and the problem domain.

Firstly, from a financial point of view, active portfolio management has concerned a lot of academia and a vast research in different formulations of the problem has been conducted. However, little effort has been done to tackle with this optimization problem from a distributional point of view. In our study, our aim was to minimize the probability that the portfolio's returns would fall below the benchmark's returns. This is very important for a financial manager who seeks methods to outperform a benchmark index. In the case where the manager achieves to systematically outperform the benchmark index, then his/her strategy can be considered as a success. However, this is not an easy task. In our study, we managed to drop this probability as low as 0.32. What is more, the incorporation of real life constraints, especially regarding the portfolio's cardinality, is very important, because the objective of a fund manager is to find combinations of assets, where the number of assets is specified and is definitely less than the total number of assets in the benchmark index. This is the point where traditional numerical methods yield inferior results.

As far as the methodology is concerned, the results concerning the performance of the hybrid algorithm can be considered as preliminary. The empirical distribution of objective values indicated that a simple ant colony optimization approach can achieve better results. However, more experiments are needed before clearer conclusions can be drawn.

Regarding the hybrid scheme, future research is focused on two topic areas. The first one is the adjustment of the parameters of the technique that explores the solution space of weights, as well as the ant colony optimization component of the hybrid. The settings we used in our study were more the result of a trial-and-error process, and hence are not optimal. The second one is the development of hybrid scheme which combines two or more nature – inspired algorithms, such as an ant colony optimization metaheuristic for asset selection hybridized with an ant colony optimization metaheuristic, too, for finding assets' weights. We have some first results which indicate good performance of this scheme. Thirdly, as far as the financial implications are concerned, this distributional approach of the active portfolio management is in its early stages. More aspects of this problem should be studied, in order to get better insights from a financial point of view.

References

1. Gilli, M., Kellezi, E.: A Global Optimization Heuristic for Portfolio Choise with VaR and Expected Shortfall. In: Computational Methods in Decision-making, Economics and Finance. Applied Optimization Series, pp. 167–183 (2001)
2. Jorion, P.: Portfolio Optimization with Tracking-Error Constraints. Financial Analysts Journal 59(5), 70–82 (2003)
3. Thomaidis, N.S., Angelidis, T., Vassiliadis, V., Dounias, G.: Active Portfolio Management with Cardinality Constraints: An Application of Particle Swarm Optimization. New Mathematics and Natural Computation, Working Paper (2008)
4. Gordon, J.A., Baptista, A.M.: Active Portfolio Management with benchmarking: Adding a value-at-risk constraint. Journal of Economic Dynamics & Control 32, 779–820 (2008)
5. Browne, S.: Risk Constrained Dynamic Active Portfolio Management. Management Science 46(9), 1188–1199 (2000)
6. Rockafellar, R.T., Uryasev, S.: Conditional Value-at-Risk for General Loss Distributions. Journal of Banikng and Finance 26(7), 1443–1471 (2002)
7. Chekhlov, A., Urasyev, S., Zabarankin, M.: Drawdown Measure in Portfolio Optimization. International Journal of Theoretical and Applied Finance 8(1), 13–58 (2005)
8. Maringer, D., Kelleler, H.: Optimization of Cardinality Constrained Portfolios with a Hybrid Local Search Algorithm. OR Spectrum 25, 481–495 (2003)
9. Maringer, D.: Distribution Assumptions and Risk Constraints in Portfolio Optimization. Computational Management Science 2, 139–152 (2005)
10. Gaspero, L., Tollo, G., Roli, A., Schaerf A.: A Hybrid Solver for Constrained Portfolio Selection Problems: preliminary report. In: Proceedings of Learning and Intelligent Optimization (LION 2007) (2007)
11. Gomez, M.A., Flores, C.X., Osorio, M.A.: Hybrid search for cardinality constrained portfolio optimization. In: 8th Annual Conference on Genetic and Evolutionary Computation, pp. 1865–1866 (2006)
12. Jeurissen, R., Berg, J.: Optimized index tracking using a hybrid genetic algorithm. In: IEEE Congress on Evolutionary Computation, pp. 2327–2334 (2008)
13. Blum, C.: Ant Colony Optimization: Introduction and recent trends. Physics of Life Reviews 2, 353–373 (2005)
14. Maringer, D., Oyewumi, O.: Index Tracking With Constraints Portfolios. Intelligent Systems in Accounting, Finance and Management 15, 57–71 (2007)
15. Brown, S.: Beating a moving target: Optimal portfolio strategies for outperforming a stochastic benchmark. Finance and Stochastics 3, 275–294 (1999)

Author Index